BIOMATERIALS SCIENCE

BIOMATERIALS SCIENCE

An Introduction to
Materials in Medicine

Edited by

Buddy D. Ratner and Allan S. Hoffman

Center for Bioengineering and
Department of Chemical Engineering
University of Washington
Seattle, Washington

Frederick J. Schoen

Department of Pathology
Brigham and Women's Hospital
and Harvard Medical School
Boston, Massachusetts

Jack E. Lemons

Departments of Biomaterials and Surgery
School of Dentistry and Medicine
University of Alabama at Birmingham
Birmingham, Alabama

Academic Press

San Diego   London   Boston   New York   Sydney   Tokyo   Toronto

Academic Press
An imprint of Elsevier Science
525 B Street, Suite 1900, San Diego, California 92101-4495, USA
http://www.academicpress.com

Academic Press
84 Theobalds Road, London WC1X 8RR, UK
http://www.academicpress.com

Library of Congress Cataloging-in-Publication Data

Biomaterials science : an introduction to materials in medicine /
    edited by Buddy D. Ratner .... [et al.].
            p.   cm.
    Includes index.
ISBN: 0-12-582460-2 (case: alk. paper)
ISBN: 0-12-582461-0 (paperback: alk. paper)
1. Biomedical materials I. Ratner, B. D. (Buddy D.). Date
R857.M3B5735   1996
610'.28—dc20   96-19088
                                       CIP

PRINTED IN THE UNITED STATES OF AMERICA
02   03   04   EB   9   8   7   6

CONTENTS

CONTRIBUTORS

Numbers in parentheses indicate the pages on which the authors' contributions begin.

Harold Alexander (94) Department of Bioengineering, Hospital for Joint Diseases Orthopaedic Institute, New York, New York 10003

James M. Anderson (165, 451) Institute of Pathology, Case Western Reserve University, Cleveland, Ohio 44106

Brian Bevacqua (420) Case Western Reserve University School of Medicine and Anesthesia Service, Veterans Affairs Medical Center, Cleveland, Ohio 44106

Stanley A. Brown (457) Food and Drug Administration, Center for Devices and Radiological Health, Office of Science and Technology, Division of Mechanics and Materials Science, Rockville, Maryland 20852

John B. Brunski (37) Department of Biomedical Engineering, Rensselaer Polytechnic Institute, Troy, New York 12180

John F. Burke (360) Massachusetts General Hospital, Harvard Medical School, Boston, Massachusetts 02114

Francis W. Cooke (11) Orthopaedic Research Institute, Wichita, Kansas 67214

Stuart L. Cooper (50) College of Engineering, University of Delaware, Newark, Delaware 19716

Arthur J. Coury (243) Focal, Inc., Lexington, Massachusetts 02173

A. Norman Cranin (426) Department of Dental and Oral Surgery, Brookdale University Hospital and Medical Center, The Dental Implant Group, Brooklyn, New York 11212

Paul Didisheim (283) Bioengineering Research Group, Division of Heart and Vascular Diseases, National Heart, Lung, and Blood Institute, Bethesda, Maryland 20892

Dennis Goupil (356) Datascope Corporation, Fairfield, New Jersey 07004

Linda M. Graham (420) Division of Vascular Surgery, Case Western Reserve University School of Medicine and Vascular Surgery Section, Veterans Affairs Medical Center, Cleveland, Ohio 44106

Anthony G. Gristina (205) Medical Sciences Research Institute, an affiliate of George Mason University, Herndon, Virginia 22070

Stephen R. Hanson (193, 228) Division of Hematology–Oncology, Emory University School of Medicine, Atlanta, Georgia 30322

Laurence A. Harker (193) Division of Hematology–Oncology, Emory University School of Medicine, Atlanta, Georgia 30322

Jorge Heller (346) Advanced Polymer Systems Research Institute, Redwood City, California 94063

Larry L. Hench (73) Department of Materials Science and Engineering, University of Florida, Gainesville, Florida 32611

Robert W. Hergenrother (50) Target Therapeutics, Inc., Freemont, California 94537

Allan S. Hoffman (37, 105, 124, 443) Center for Bioengineering and Department of Chemical Engineering, University of Washington, Seattle, Washington 98195

Thomas A. Horbett (133) Center for Bioengineering and Department of Chemical Engineering, University of Washington, Seattle, Washington 98195

Richard J. Johnson (173) Baxter Healthcare Corporation, Round Lake, Illinois 60073

Jeffrey B. Kane (360) Massachusetts General Hospital, Harvard Medical School, Boston, Massachusetts 02114

J. Lawrence Katz (335) Department of Biomedical Engineering, Case Western Reserve University, Cleveland, Ohio 44106

Michael Klein (426) Department of Dental and Oral Surgery, Brookdale University Hospital and Medical Center, The Dental Implant Group, Brooklyn, New York 11212

Joachim Kohn (64) Department of Chemistry, Wright and Rieman Laboratories, Rutgers University, Piscataway, New Jersey 08855

John B. Kowalski (415) Johnson & Johnson, Sterilization Science & Technology, New Brunswick, New Jersey 08906

Peggy A. Lalor (220) Howmedica, Inc., Rutherford, New Jersey 07070

Robert Langer (64) Department of Chemical Engineering, Massachusetts Institute of Technology, Cambridge, Massachusetts 02139

Jack E. Lemons (11, 283, 308, 457) Departments of Biomaterials and Surgery, School of Dentistry and Medicine, University of Alabama at Birmingham, Birmingham, Alabama 35294

Robert J. Levy (272) Department of Pediatrics, and Communicable Diseases and Pharmaceutics, University of Michigan, Ann Arbor, Michigan 48109

Paul S. Malchesky (400) International Center for Artificial Organs and Transplantation, Cleveland, Ohio 44106

Nancy B. Mateo (461) Statprobe, Inc., Ann Arbor, Michigan 48108

Carl R. McMillin (267) AcroMed Corporation, Cleveland, Ohio 44115

Katharine Merritt (188) Food and Drug Administration, Center for Devices and Radiological Health, Office of Science and Technology, Division of Life Sciences, Gaithersberg, Maryland 20878

Robert F. Morrissey (415) Johnson & Johnson, Sterilization Science & Technology, New Brunswick, New Jersey 08906

Kevin D. Murray (389) Department of Cardiothoracic Surgery, Washington University, St. Louis, Missouri 63110

Paul T. Naylor (205) Tennessee Orthopedic Clinic, Inc., Knoxville, Tennessee 37922

Steven M. Niemi (238) Genzyme Transgenics Corporation, Framingham, Massachusetts 01701

Sharon J. Northup (215) Baxter Healthcare Corporation, Baxter Technology Park, Round Lake, Illinois 60073

Stephen A. Obstbaum (435) New York University School of Medicine, and Department of Ophthalmology, Lenox Hill Hospital, New York, New York 10016

Don B. Olsen (389) Artificial Heart Research Laboratory and Institute for Biomedical Engineering and Division of Artificial Organs, University of Utah, Salt Lake City, Utah 84103

Yashwant Pathak (272) South Dakota State University, College of Pharmacy, Brookings, South Dakota 57007

Nikolaos A. Peppas (60) Biomaterials and Drug Delivery Laboratories, School of Chemical Engineering, Purdue University, West Lafayette, Indiana 47907

Buddy D. Ratner (1, 21, 105, 133, 215, 228, 243, 445, 465) Center for Bioengineering and Department of Chemical Engineering, University of Washington, Seattle, Washington 98195

Miguel F. Refojo (328) Schepens Eye Research Institute and Department of Ophthalmology, Harvard Medical School, Boston, Massachusetts 02114

Lois S. Robblee (371) EIC Laboratories, Inc., Norwood, Massachusetts 02062

Jeff M. Schakenraad (141) Department of Biomaterials and Biocompatibility, University of Groningen, 97 KZ Groningen, The Netherlands

Frederick J. Schoen (147, 165, 200, 272, 389, 415) Department of Pathology, Brigham and Women's Hospital and Harvard Medical School, Boston, Massachusetts 02115

Shalaby W. Shalaby (118) Poly-Med, Inc., Anderson, South Carolina 29625

Aram Sirakian (426) Department of Dental and Oral Surgery, Brookdale University Hospital and Medical Center, The Dental Implant Group, Brooklyn, New York 11212

Steven M. Slack (469) Department of Biomedical Engineering, The University of Memphis, Memphis, Tennessee 38152

Dennis C. Smith (319) Centre for Biomaterials, University of Toronto, Toronto, Ontario, Canada M5G 2P6

Myron Spector (220) Department of Orthopedic Surgery, Brigham and Women's Hospital, Harvard Medical School, Boston, Massachusetts 02115

James D. Sweeney (371) The Bioengineering Program, Department of Chemical, Biological, and Materials Engineering, Arizona State University, Tempe, Arizona 85287

Ronald G. Tompkins (360) Trauma and Burn Services, Massachusetts General Hospital, Harvard Medical School, Boston, Massachusetts 02114

Brad H. Vale (238) Johnson & Johnson Development Corporation, New Brunswick, New Jersey 08933

Susan A. Visser (50) Eastman Kodak Company, Rochester, New York 14650

Sung Wan Kim (297) Department of Pharmaceutics and Pharmaceutical Chemistry, Center for Controlled Chemical Delivery, University of Utah, Salt Lake City, Utah 84132

John T. Watson (283) Bioengineering Research Group, Division of Heart and Vascular Diseases, National Heart, Lung, and Blood Institute, Bethesda, Maryland 20892

Diana Whittlesey (420) Division of Cardiothoracic Surgery, Case Western Reserve University School of Medicine and Cardiothoracic Surgery Section, Veterans Affairs Medical Center, Cleveland, Ohio 44106

David F. Williams (260) Department of Clinical Engineering, University of Liverpool, Liverpool L7 8XP, United Kingdom

Rachel L. Williams (260) Department of Clinical Engineering, University of Liverpool, Liverpool L7 8XP, United Kingdom

John E. Willson (238) Corporate Office of Science and Technology, Johnson & Johnson Development Corporation, New Brunswick, New Jersey 08901

Paul Yager (375) Center for Bioengineering, University of Washington, Seattle, Washington 98195

Ioannis V. Yannas (84) Department of Mechanical Engineering, Massachusetts Institute of Technology, Cambridge, Massachusetts 02139

Martin L. Yarmush (360) Center for Engineering in Medicine, Massachusetts General Hospital, Boston, Massachusetts 02114

PREFACE

Biomaterials science, the study of the application of materials to problems in biology and medicine, is a field characterized by medical needs, basic research, advanced technological development, ethical considerations, industrial involvement, and federal regulation. It encompasses an unusually broad spectrum of ideas, sciences, and technologies, which contributes to its intellectual excitement. At the same time, it serves important humanitarian needs. This book is intended to provide an overview of the theory and practice of biomaterials science.

The need for a balanced book on the subject of biomaterials science has been escalating for some time. Those of us who teach biomaterials science have been forced to offer our classes compromises for background written material: textbooks by single authors that too strongly emphasize their areas of expertise and pay too little attention to other important subjects; articles from the literature that are difficult to weave into a cohesive curriculum; our own handout materials, often graphically crude, and again, slanted to the specific interests of each professor. In *Biomaterials Science: An Introduction to Materials in Medicine,* by combining the experience of many leaders in the biomaterials field, we endeavor to present a balanced perspective on an evolving field.

Over 50 biomaterials professionals from academia, industry, and government have contributed to this work. Certainly, such a distinguished group of authors provides the needed balance and perspective. However, including many authors can also lead to particular complexities in a project of this type. Do the various writing styles clash? Does the presentation of material, particularly controversial material, result in one chapter contradicting another? Even with so many authors, all subjects relevant to biomaterials cannot be addressed—which subjects should be included and which left out? How should such a project be refereed to ensure scientific quality, pedagogical effectiveness, and the balance we strive for? These are some of the problems the editors grappled with over the years from conception to publication. Compromises have often been reached, and the end product is different from that originally envisioned. Still, a unique volume has evolved from this process that the editors feel can make a special contribution to the development of the biomaterials field. An educational tool directed toward starting those new to biomaterials on a path to appreciating the scope, complexity, basic principles, and importance of this enterprise has been synthesized.

A few acknowledgments and thanks are in order. First, let me address the Society For Biomaterials that served as sponsor and inspiration for this book. The Society For Biomaterials is a model of "scientific cultural diversity" with engineers, physicians, scientists, veterinarians, industrialists, inventors, regulators, attorneys, educators, and ethicists all participating in an endeavor that is intellectually exciting, humanitarian, and profitable. So, too, this book attempts to bring together in one tutorial volume these many influences, stances, and ideas. Royalties from this volume are being returned to the Society For Biomaterials to further education and professional advancement related to biomaterials. For further information on the Society For Biomaterials, call the US telephone number 612-927-8108.

Next, a special thanks to those who invested time and effort in the compilation of the material that became this book. The many authors who contributed their expertise and perspectives are clearly the backbone of this work and they deserve the lion's share of the commendation. My fellow editors, Allan Hoffman, Jack Lemons, and Fred Schoen also deserve tremendous credit and accolades for their hard work, insights, advice, and leadership.

Finally, a number of individuals at the University of Washington have contributed to the assembly and production aspects of this work. I offer my special thanks to Thomas Menduni, Mady Lund, and Nancy Mateo for their assistance and special effort on this important component of the project.

The biomaterials field has always been wide open with opportunities, stimulation, compassion, and intellectual ideas. I, and my fellow editors and authors, hope the overview you now approach will stimulate in you as much excitement and satisfaction as it has in us.

Buddy D. Ratner

Biomaterials Science:
An Interdisciplinary Endeavor

BUDDY D. RATNER

A VERY SHORT HISTORY OF BIOMATERIALS

The modern field we call biomaterials is too new for a formal history to have been compiled. However, a few comments are appropriate to place both ancient history and rapidly moving contemporary history in perspective. The Romans, Chinese, and Aztec used gold in dentistry more than 2000 years ago. Through much of recorded history, glass eyes and wooden teeth have been in common use. At the turn of this century, synthetic plastics became available. Their ease of fabrication led to many implantation experiments, most of them, in light of our contemporary understanding of biomaterials toxicology, doomed to failure. Poly(methyl methacrylate) (PMMA) was introduced in dentistry in 1937. During World War II, shards of PMMA from shattered gunnery turrets, unintentionally implanted in the eyes of aviators, suggested that some materials might evoke only a mild foreign body reaction. Just after World War II, Voorhees experimented with parachute cloth (Vinyon N) as a vascular prosthesis. In 1958, in a cardiovascular surgery textbook by Rob, the suggestion was offered that surgeons might visit their local draper's shop and purchase Dacron fabric that could be cut with pinking shears to fabricate an arterial prosthesis. In the early 1960s Charnley used PMMA, ultrahigh-molecular-weight polyethylene, and stainless steel for total hip replacement. While these applications for synthetic materials in medicine spanned much of written history, the term "biomaterial" was not invoked.

It is difficult to pinpoint the precise origins of the term "biomaterial." However, it is probable that the field we recognize today was solidified through the early Clemson University biomaterials symposia in the late 1960s and early 1970s. The scientific success of these symposia led to the formation of the Society For Biomaterials in 1975. The individual physician-visionaries who implanted miscellaneous materials to find a solution to pressing, often life-threatening, medical problems

were, with these Clemson symposia, no longer the dominant force. We had researchers and engineers designing materials to meet specific criteria, and scientists exploring the nature of biocompatibility. Around this term "biomaterial" a unique scientific discipline evolved. The evolution of this field and the Society For Biomaterials were intimately connected. From biomaterials ideas, many of which originated at society meetings, other fields evolved. Drug delivery, biosensors, and bioseparations owe much to biomaterials. Now we have academic departments of biomaterials, many biomaterials programs, and research institutes devoted to education and exploration in biomaterials science and engineering (Society For Biomaterials Educational Directory, 1992). Paralleling the research and educational effort, hundreds of companies that incorporate biomaterials into devices have developed. This textbook looks at a now well-established biomaterials field, circa the 1990s.

BIOMATERIALS SCIENCE

Although biomaterials are primarily used for medical applications, which will be the focus of this text, they are also used to grow cells in culture, in apparatus for handling proteins in the laboratory, in devices to regulate fertility in cattle, in the aquaculture of oysters, and possibly in the near future they will be used in a cell-silicon "biochip" that would be integrated into computers. How do we reconcile these diverse uses of materials into one field? The common thread is the interaction between biological systems and synthetic (or modified natural) materials.

In medical applications, biomaterials are rarely used as simple materials and are more commonly integrated into devices. Although this is a text on materials, it will quickly become apparent that the subject cannot be explored without also considering biomedical devices. In fact, a biomaterial must always be considered in the context of its final fabricated, sterilized form. For example, when a polyurethane elastomer is cast from a solvent onto a mold to form a heart assist device, it can elicit different blood–material interactions than when injection molding is used to form the same device. A hemodialysis system serving as an artificial kidney requires materials that must function in contact with a patient's blood and exhibit appropriate membrane permeability and mass transport characteristics. It also must employ mechanical and electronic systems to pump blood and control flow rates.

Unfortunately, many aspects of the design of devices are beyond the scope of this book. Consider the example of the hemodialysis system. The focus here is on membrane materials and their biocompatibility; there is less information on mass transport through membranes, and little information on flow systems and monitoring electronics.

A few definitions and descriptions are in order and will be expanded upon in this and subsequent chapters.

Many definitions have been proposed for the term "biomaterial." One definition, endorsed by a consensus of experts in the field, is:

> A biomaterial is a nonviable material used in a medical device, intended to interact with biological systems. (Williams, 1987)

If the word "medical" is removed, this definition becomes broader and can encompass the wide range of applications suggested above.

A complementary definition essential for understanding the goal of biomaterials science, is that of "biocompatibility."

> Biocompatibility is the ability of a material to perform with an appropriate host response in a specific application. (Williams, 1987)

Thus, we are introduced to considerations that set a biomaterial apart from most materials explored in materials science. Table 1 lists a few applications for synthetic materials in the body. It includes many materials that are often classified as "biomaterials." Note that metals, ceramics, polymers, glasses, carbons, and composite materials are listed. Table 2 presents estimates of the numbers of medical devices containing biomaterials that are implanted in humans each year and the size of the commercial market for biomaterials and medical devices.

Four examples of applications of biomaterials are given here to illustrate important ideas. The specific devices discussed were chosen because they are widely used in humans, largely with good success. However, key problems with these biomaterial devices are also highlighted. Each of these examples is discussed in detail in later chapters.

EXAMPLES OF BIOMATERIALS APPLICATIONS

Substitute Heart Valves

Degeneration and other diseases of heart valves often make surgical repair or replacement necessary. Heart valve prosthe-ses are fabricated from carbons, metals, elastomers, fabrics, and natural (e.g., pig) valves and other tissues chemically pre-treated to reduce their immunologic reactivity and to enhance durability. More than 45,000 replacement valves are implanted each year in the United States because of acquired damage to the natural valve and congenital heart anomalies. Figure 1 shows a bileaflet tilting disk heart valve, the most widely used design. Generally, almost as soon as the valve is implanted, cardiac function is restored to near normal levels and the patient shows rapid improvement. In spite of the good overall success seen with replacement heart valves, there are problems with different types of valves; they include degeneration of tissue, mechanical failure, postoperative infection, and induction of blood clots.

Artificial Hip Joints

The human hip joint is subjected to high mechanical stresses and undergoes considerable abuse. It is not surprising that because of 50 years or more of cyclic mechanical stress, or because of degenerative or rheumatological disease, the natural joint wears out, leading to considerable loss of mobility and, often, confinement to a wheelchair. Hip joints are fabricated from titanium, specific high-strength alloys, ceramics, composites, and ultrahigh molecular weight polyethylene. Replacement hip joints (Fig. 2) are implanted in more than 90,000 humans each year in the United States alone. With some types of replacement hip joints and surgical procedures, ambulatory function is restored within days after surgery. For other types, a healing-in period is required for attachment between bone and the implant before the joint can bear the full weight of the body. In most cases, good function is restored, and even athletic activities are possible, although they are generally not advised. After 10–15 years, the implant may loosen, necessitating another operation.

Dental Implants

The widespread introduction of titanium implants (Fig. 3) has revolutionized dental implantology. These devices, which form an artificial tooth root on which a crown is affixed, are implanted in approximately 275,000 people each year, with some individuals receiving more than 12 implants. A special requirement of a material in this application is the ability to form a tight seal against bacterial invasion where the implant traverses the gingiva (gum). One of the primary advantages originally cited for the titanium implant was bonding with the bone of the jaw. In recent years, however, this attachment has been more accurately described as a tight apposition or mechanical fit and not true bonding. Wear, corrosion, and the mechanical properties of titanium have also been of concern.

Intraocular Lenses

Intraocular lenses (IOLs) made of poly(methyl methacrylate), silicone elastomer, or other materials are used to replace

TABLE 1 Some Applications of Synthetic Materials and
Modified Natural Materials in Medicine

Application	Types of materials
Skeletal system	
Joint replacements (hip, knee)	Titanium, Ti–Al–V alloy, stainless steel, polyethylene
Bone plate for fracture fixation	Stainless steel, cobalt–chromium alloy
Bone cement	Poly(methyl methacrylate)
Bony defect repair	Hydroxylapatite
Artificial tendon and ligament	Teflon, Dacron
Dental implant for tooth fixation	Titanium, alumina, calcium phosphate
Cardiovascular system	
Blood vessel prosthesis	Dacron, Teflon, polyurethane
Heart valve	Reprocessed tissue, stainless steel, carbon
Catheter	Silicone rubber, Teflon, polyurethane
Organs	
Artificial heart	Polyurethane
Skin repair template	Silicone–collagen composite
Artificial kidney (hemodialyzer)	Cellulose, polyacrylonitrile
Heart–Lung machine	Silicone rubber
Senses	
Cochlear replacement	Platinum electrodes
Intraocular lens	Poly(methyl methacrylate), silicone rubber, hydrogel
Contact lens	Silicone–acrylate, hydrogel
Corneal bandage	Collagen, hydrogel

a natural lens when it becomes cloudy and cataractous (Fig. 4). By the age of 75, more than 50% of the population suffers from cataracts severe enough to warrant IOL implantation. This translates to over 1.4 million implantations in the United States alone each year, and double that number worldwide. Good vision is generally restored almost immediately after the lens is inserted and the success rate with this device is high. IOL surgical procedures are well developed and implantation is often performed on an outpatient basis. Recent observations of implanted lenses using a biomicroscope show that inflammatory cells migrate to the surface of the lenses after periods of implantation. Thus, the conventional healing pathway is seen with these devices, as is observed with materials implanted in other sites in the body.

Many themes are illustrated by these four vignettes. Widespread application with good success is generally noted. A broad range of synthetic materials varying in chemical, physical, and mechanical properties are used in the body. Many anatomical sites are involved. The mechanisms by which the body responds to foreign bodies and heals wounds are observed in each case. Problems, concerns, or unexplained observations are noted for each device. Companies are manufacturing each of the devices and making a profit. Regulatory agencies are carefully looking at device performance and making policy intended to control the industry and protect the patient. Are there ethical or social issues that should be addressed? To set the stage for the formal introduction of biomaterials science, we will return to the four examples just discussed to examine the issues implicit to each case.

CHARACTERISTICS OF BIOMATERIALS SCIENCE

Interdisciplinary

More than any other field of contemporary technology, biomaterials science brings together researchers with diverse academic backgrounds who must communicate clearly. Figure 5 lists some of the disciplines that are encountered in the progression from identifying the need for a biomaterial or device to the manufacture, sale, and implantation of it.

Many Materials

The biomaterials scientist will have an appreciation of materials science. This may range from an impressive command of the theory and practice of the field demonstrated by the materials scientist, to a general understanding of the properties of materials that might be demonstrated by the physician biomaterials scientist.

A wide range of materials is routinely used (Table 1) and no one researcher will be comfortable synthesizing and designing with all these materials. Thus, specialization is the rule. However, a broad appreciation of the properties and applications of these materials, the palette from which the biomaterials scientist chooses, is a hallmark of professionals in the field.

There is a tendency to group the materials (and the researchers) into the "hard tissue replacement biomaterials" camp (e.g., metals, ceramics), typically repesented by those involved in orthopedic and dental materials, and the "soft tissue replace-

TABLE 2   The Biomaterials and Healthcare Market—Facts and Figures (per year)

Total U.S. health care expenditures (1990)	$666,200,000,000
Total U.S. health research and development (1990)	$22, 600,000,000
Number of employees in the medical device industry (1988)	194,250
Registered U.S. medical device manufacturers (1991)	19,300
Total medical device sales:	
Surgical appliances	$8,414,000,000
Surgical instruments	$6,444,000,000
Electromedical devices	$5,564,000,000
U.S. market for biomaterials (1992)	$402,000,000
Individual medical device sales:	
Catheters, U.S. market (1991)	$1,400,000,000
Angioplasty catheters (market by mid 1990s)	$1,000,000,000
Orthopedic, U.S. market (1990)	$2,200,000,000
Wound care products (1988 estimate)	$4,000,000,000
Biomedical sensor market (1991)	$365,000,000
Artificial pancreas (if one existed, and was used by 10% of the U.S. insulin-dependent diabetics; 1985 estimate)	$2,300,000,000
Numbers of devices:	
Intraocular lenses	1,400,000[a]
Contact lenses:	
Extended wear soft lens users	4,000,000[a]
Daily wear soft lens users	9,000,000[a]
Rigid gas-permeable users	2,600,000[a]
Vascular grafts	250,000[b]
Heart valves	45,000[a]
Pacemakers	460,000[a]
Blood bags	30,000,000[b]
Breast prostheses	544,000[a]
Catheters	200,000,000[b]
Oxygenators	500,000[b]
Renal dialyzers	16,000,000[b]
Orthopedic (knee, hip)	500,000[b]
Knee	816,000[a]
Hip	521,000[a]

[a]1990 estimate for United States.
[b]1981 estimate for western countries and Japan.

ment biomaterials" camp (e.g., polymers), which is often associated with cardiovascular and general plastic surgery materials. In practice, this division does not hold up well—a heart valve may be fabricated from polymers, metals, and carbons, while a hip joint will also be composed of metals and polymers and will be interfaced to the body via a polymeric bone cement. There is a need for a general understanding of all classes of materials, and this book will provide this background.

Development of Biomaterials Devices

Figure 5 illustrates interdisciplinary interactions in biomaterials and shows the usual progression in the devlepoment of a biomaterial or device. It provides a perspective on how different disciplines work together, starting from the identification of a need for a biomaterial through development, manufacture, implantation, and removal from the patient.

Magnitude of the Field

Magnitude expresses both a *magnitude of need* and *magnitude of a commercial market*. Needless to say, a conflict of interest can arise with pressures from both the commercial quarter and from ethical considerations. Consider three commonly used biomaterial devices: a contact lens, a hip joint, and a heart valve. All fill a medical need. The contact lens offers improved vision and in some cases a cosmetic enhancement. The hip joint offers mobility to the patient who would otherwise be confined to a bed or wheelchair. The heart valve offers life. The contact lens may sell for $100, and the hip joint and heart valve may sell for up to $3000 each. There will be 20 million contact lenses purchased each year, but only perhaps 100,000 heart valves (worldwide) and 500,000 total artificial hip prostheses. Here are the issues for consideration: a large number of devices, differing magnitudes of need, and differing (but large) commercial potential. There is no simple

FIG. 1. A replacement heart valve. (Photograph courtesy of St. Jude Medical, Inc.)

FIG. 3. A titanium dental implant. (Photograph courtesy of Dr. A. Norman Cranin, Brookdale Hospital Medical Center, Brooklyn, NY.)

answer to how these components are integrated in this field we call "biomaterials science." As you work your way through this volume, view each of the ideas and devices presented in the context of these considerations.

Along with these characteristics of biomaterials science—the interdisciplinary flavor, the magnitude of the need, and the sophisticated materials science—there are certain, often unique, subjects that occupy particularly prominent positions in our field. Let us review a few of these.

SUBJECTS INTEGRAL TO BIOMATERIALS SCIENCE

Toxicology

A biomaterial should not be toxic, unless it is specifically engineered for such requirements (e.g., a "smart bomb" drug release system that seeks out cancer cells and destroys them). Since the nontoxic requirement is the norm, toxicology for biomaterials has evolved into a sophisticated science. It deals with the substances that migrate out of biomaterials. For exam-

ple, for polymers, many low-molecular-weight "leachables" exhibit some level of physiologic activity and cell toxicity. It is reasonable to say that a biomaterial should not give off anything from its mass unless it is specifically designed to do so. Toxicology also deals with methods to evaluate how well

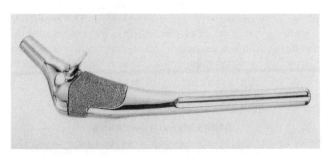

FIG. 2. A synthetic hip joint. (Photograph courtesy of Zimmer, Inc.)

FIG. 4. An intraocular lens. (Photograph courtesy of Alcon Laboratories, Inc.)

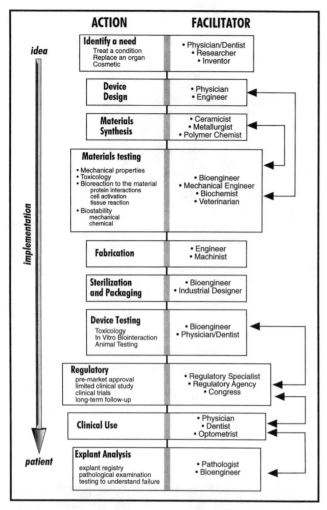

FIG. 5. Disciplines involved in biomaterials science and the path from a need to a manufactured medical device.

this design criterion is met when a new biomaterial is under development. Chapter 5.2 provides an overview of methods in biomaterials toxicology. The implications of toxicity are addressed in Chapters 4.2 and 4.4.

Biocompatibility

The understanding and measurement of biocompatibility is unique to biomaterials science. Unfortunately, we do not have precise definitions or accurate measurements of biocompatibility. More often than not, it is defined in terms of performance or success at a specific task. Thus, for a patient who is alive and doing well, with a vascular prosthesis that is unoccluded, few would argue that this prosthesis is, in this case, not "biocompatible." However, this operational definition offers us little to use in designing new or improved vascular prostheses. It is probable that biocompatibility may have to be specifically defined for applications in soft tissue, hard tissue, and the cardiovascular system (blood compatibility). In fact, biocom-

patibility may have to be uniquely defined for each application. The problems and meanings of biocompatibility will be explored and expanded upon throughout this textbook, in particular, see Chapters 4 and 5.

Healing

Special processes are invoked when a material or device heals in the body. Injury to tissue will stimulate the well-defined inflammatory reaction sequence that leads to healing. Where a foreign body (e.g., an implant) is involved, the reaction sequence is referred to as the "foreign body reaction" (Chapter 4.2). The normal response of the body will be modulated because of the solid implant. Furthermore, this reaction will differ in intensity and duration depending upon the anatomical site involved. An understanding of how a foreign object alters the normal inflammatory reaction sequence is an important concern for the biomaterials scientist.

Unique Anatomical Sites

Consideration of the anatomical site of an implant is essential. An intraocular lens may go into the lens capsule or the anterior chamber. A hip joint will be implanted in bone across an articulating joint space. A heart valve will be sutured into cardiac muscle. A catheter may be placed in a vein. Each of these sites challenges the biomedical device designer with special requirements for geometry, size, mechanical properties, and bioreaction. Chapter 3.4 introduces these ideas.

Mechanical and Performance Requirements

Each biomaterial and device has imposed upon it mechanical and performance requirements that originate from the physical (bulk) properties of the material. These requirements can be divided into three categories: mechanical performance, mechanical durability, and physical properties. First, consider mechanical performance. A hip prosthesis must be strong and rigid. A tendon material must be strong and flexible. A heart valve leaflet must be flexible and tough. A dialysis membrane must be strong and flexible, but not elastomeric. An articular cartilage substitute must be soft and elastomeric. Then, we must address mechanical durability. A catheter may only have to perform for 3 days. A bone plate may fulfill its function in 6 months or longer. A leaflet in a heart valve must flex 60 times per minute without tearing for the lifetime of the patient (it is hoped, for 10 or more years). A hip joint must not fail under heavy loads for more than 10 years. Finally, the bulk physical properties will address performance. The dialysis membrane has a specified permeability, the articular cup of the hip joint has a lubricity, and the intraocular lens has a clarity and refraction requirement. To meet these requirements, design principles are borrowed from mechanical engineering, chemical engineering, and materials science.

Industrial Involvement

At the same time as a significant basic research effort is under way to understand how biomaterials function and how to optimize them, companies are producing millions of implants for use in humans and earning billions of dollars on the sale of medical devices. Thus, although we are now only learning about the fundamentals of biointeraction, we manufacture and implant materials and devices. How is this dichotomy explained? Basically, as a result of considerable experience, trial and error, inspired guesses, and just plain luck, we now have a set of materials that performs satisfactorily in the body. The medical practitioner can use them with reasonable confidence, and the performance in the patient is largely acceptable. In essence, the complications of the devices are less than the complicatons of the original diseases. Companies make impressive profits on these devices. Yet, in some respects, the patient is trading one disease for another, and there is much evidence that better materials and devices can be made through basic science and engineering exploration. So, in the field of biomaterials, we always see two sides of the coin—a basic science and engineering effort, and a commercial sector.

The balance between the desire to alleviate loss of life and suffering, and the corporate imperative to turn a profit forces us to look further afield for guidance. Obviously, ethical concerns enter into the picture. Companies have large investments in the manufacture, quality control, clinical testing, regulatory clearance, and distribution of medical devices. How much of an advantage will be realized in introducing an improved device? The improved device may indeed work better for the patient. However, the company will incur a large expense that will, in the short term, be perceived by the stockholders as a cut in the profits. Moreover, product liability issues are a major concern of manufacturers. When looking at the industrial side of the biomaterials field, questions are asked about the ethics of withholding an improved device from people who need it, the market share advantages of having a better product, and the gargantuan costs (possibly nonrecoverable) of introducing a new product into the medical marketplace. If companies did not have the profit incentive, would there be medical devices, let alone improved ones, available for clinical application?

When the industrial segment of the biomaterials field is examined, we see other contributions to our field. Industry deals well with technological developments such as packaging, sterilization, and quality control and analysis. These subjects require a strong technological base, and have generated stimulating research questions. Also, many companies support in-house basic research laboratories and contribute in important ways to the fundamental study of biomaterials science.

Ethics

There are a wide range of other ethical considerations in biomaterials science. Some key ethical questions in biomaterials science are summarized in Table 3. Like most ethical questions, an absolute answer may be difficult to come by. Some articles have addressed ethical questions in biomaterials and debated the important points (Saha and Saha, 1987; Schiedermayer and Shapiro, 1989).

Regulation

The consumer (the patient) demands safe medical devices. To prevent inadequately tested devices and materials from

TABLE 3 Some Ethical Concerns Relevant to Biomaterials Science

Is the use of animal models justified? Specifically, is the experiment well designed and important so that the data obtained will justify the suffering and sacrifice of the life of a living creature?

How should research using humans be conducted to minimize risk to the patient and offer a reasonable risk-to-benefit ratio? How can we best ensure informed consent?

Companies fund much biomaterials research and own proprietary biomaterials. How can the needs of the patient be best balanced with the financial goals of a company? Consider that someone must manufacture devices—these would not be available if a company did not choose to manufacture them.

Since researchers often stand to benefit financially from a successful biomedical device and sometimes even have devices named after them, how can investigator bias be minimized in biomaterials research?

For life-sustaining devices, what is the tradeoff between sustaining life and the quality of life with the device for the patient? Should the patient be permitted to "pull the plug" if the quality of life is not satisfactory?

With so many unanswered questions about the basic science of biomaterials, do government regulatory agencies have sufficient information to define adequate tests for materials and devices and to properly regulate biomaterials?

coming on the market, and to screen out individuals clearly unqualified to produce biomaterials, a complex national regulatory system has been erected by the United States government through the Food and Drug Administration (FDA). Through the International Standards Organization (ISO), international regulatory standards have been developed for the world community. Obviously, a substantial base of biomaterials knowledge went into these standards. The costs to meet the standards and to demonstrate compliance with material, biological, and clinical testing are enormous. Introducing a new biomedical device to the market requires a regulatory investment of many millions of dollars. Are the regulations and standards truly addressing the safety issues? Is the cost of regulation inflating the cost of health care and preventing improved devices from reaching those who need them? Under this regulation topic, we see the intersection of all the players in the biomaterials community: government, industry, ethics, and basic science. The answers are not simple, but the problems are addressed every day. Chapters 10.2 and 10.3 expand on standards and regulatory concerns.

BIOMATERIALS LITERATURE

Over the past 40 years, the field of biomaterials has developed from individual medical researchers "trying things out," to the defined discipline we have today. Concurrent with the evolution of the discipline, a literature has also developed. A bibliography is provided at the end of this introduction to highlight key reference works and technical journals in the biomaterials field.

SUMMARY

This chapter provides a broad overview of the biomaterials field. It is intended to provide a vantage point from which the reader can begin to place all the subthemes (chapters) within the perspective of the larger whole.

To reiterate a key point, biomaterials science may be the most interdisciplinary of all the sciences. Consequently, biomaterials scientists must master material from many fields of science, technology, engineering, and medicine in order to be competent in this profession. The reward for mastering this volume of material is involvement in an intellectually stimulating endeavor that advances our understanding of basic sciences and also contributes to reducing human suffering.

Bibliography

References

Saha, S., and Saha, P. (1987). Bioethics and applied biomaterials. *J. Biomed. Mater. Res: Appl. Biomat.* **21**: 181–190.

Schiedermayer, D. L., and Shapiro, R. S. (1989). The artificial heart as a bridge to transplant: Ethical and legal issues at the bedside. *J. Heart Transplant* **8**: 471–473.

Society For Biomaterials Educational Directory (1992). Society For Biomaterials, Minneapolis, MN.

Williams, D. F., (1987). *Definitions in Biomaterials. Proceedings of a Consensus Conference of the European Society for Biomaterials,* Chester, England, March 3-5 1986, Vol. 4, Elsevier, New York.

Biomaterials Journals

Advanced Drug Delivery Reviews (Elsevier)
American Society of Artificial Internal Organs Transactions
Annals of Biomedical Engineering (Blackwell—Official Publication of the Biomedical Engineering Society)
Artificial Organs (Raven Press)
Artificial Organs Today (T. Agishi, ed., VSP Publishers)
Biofouling (Harwood Academic Publishers)
Biomaterial–Living System Interactions (Sevastianov, ed., BioMir)
Biomaterials (including *Clinical Materials*) (Elsevier)
Biomaterials, Artificial Cells and Artificial Organs (T. M. S. Chang, ed.)
Biomaterials Forum (Society For Biomaterials)
Biomaterials: Processing, Testing and Manufacturing Technology (Butterworth)
Biomedical Materials (Elsevier)
Biomedical Materials and Engineering (T. Yokobori, ed., Pergamon Press)
Biosensors and Bioelectronics (Elsevier)
Cell Transplantation (Pergamon)
Cells and Materials (Scanning Microscopy International)
Colloids and Surfaces B: Biointerfaces (Elsevier)
Drug Targeting and Delivery (Academic Press)
Frontiers of Medical and Biological Engineering (Y. Sakurai, ed., VSP Publishers)
International Journal of Artificial Organs (Wichtig Editore)
Journal of Applied Biomaterials (Wiley)*
Journal of Bioactive and Compatible Polymers (Technomics)
Journal of Biomaterials Applications (Technomics)
Journal of Biomaterials Science: Polymer Edition (VSP Publishers)
Journal of Biomedical Materials Research (Wiley—Official Publication of the Society For Biomaterials)
Journal of Controlled Release (Elsevier)
Journal of Drug Targeting (Harwood Academic Publishers)
Journal of Long Term Effects of Medical Implants (CRC Press)
Materials in Medicine (Chapman and Hall—Official Publication of the European Society for Biomaterials)
Medical Device and Diagnostics Industry (Canon Publications)
Medical Device Research Report (AAMI)
Medical Device Technology (Astor Publishing Corporation)
Medical Plastics and Biomaterials (Canon Communications, Inc.)
Nanobiology (Carfax Publishing Co.)
Nanotechnology (an Institute of Physics Journal)
Tissue Engineering (Mary Ann Liebert, Inc.)

Some Biomaterials Books

J. Black, *Biological Performance of Materials: Fundamentals of Biocompatibility,* 2nd ed., Marcel Dekker, New York, 1992.

J. W. Boretos, and M. Eden (eds.), *Contemporary Biomaterials— Material and Host Response, Clinical Applications, New Technology and Legal Aspects.* Noyes Publ., Park Ridge, NJ, 1984.

A. I. Glasgold, and F. H. Silver, *Applications of Biomaterials in Facial Plastic Surgery,* CRC Press, Boca Raton, FL, 1991.

G. Heimke, *Osseo-Integrated Implants.* CRC Press, Boca Raton, FL, 1990.

L. L. Hench, and E. C. Ethridge, *Biomaterials: An Interfacial Approach,* Academic Press, New York, 1982.

J. B. Park, *Biomaterials: An Introduction,* Plenum Publ., New York, 1979.

J. B. Park (ed.), *Biomaterials Science and Engineering.* Plenum Publ., New York, 1984.

F. J. Schoen, *Interventional and Surgical Cardiovascular Pathology: Clinical Correlations and Basic Principles,* W. B. Saunders, Philadelphia, 1989.

F. H. Silver and C. Doillon, *Biocompatibility: Interactions of Biological and Implanted Materials,* Vol. 1 - Polymers, VCH Publ., New York, 1989.

A. F. Von Recum, (ed.), *Handbook of Biomaterials Evaluation,* 1st ed., Macmillan, New York, 1986.

D. Williams (ed.), *Concise Encyclopedia of Medical and Dental Materials,* 1st ed., Pergamon Press, Oxford, UK, 1990.

T. Yamamuro, L. L. Hench, and J. Wilson, *CRC Handbook of Bioactive Ceramics.* CRC Press, Boca Raton, FL, 1990.

---

*Now a subsection of *Journal of Biomedical Materials Research.*

I

Materials Science and Engineering

1

Properties of Materials

FRANCIS W. COOKE, JACK E. LEMONS, AND BUDDY D. RATNER

1.1 INTRODUCTION

Jack E. Lemons

The bulk and surface properties of biomaterials that have been utilized for implants have been shown to directly influence, and in some cases, control the tissue interface dynamics from the time of initial *in vivo* placement until final disposition. Compatibility is recognized to be a two-way process between the biomaterials that have been fabricated into devices and the host environment.

It is critical to recognize that synthetic materials have specific bulk and surface characteristics that are property dependent. These characteristics must be known prior to any medical application, but also must be known in terms of changes that may take place over time *in vivo*. That is, changes with time must be anticipated at the outset and accounted for through selection of biomaterials and/or design of the device.

Information related to basic properties is available through national and international standards, plus handbooks and professional journals of various types. However, this information must be evaluated within the context of the intended biomedical use, since applications and host tissue responses are quite specific within areas, e.g., cardiovascular (flowing blood contact), orthopedic (functional load bearing), and dental (percutaneous).

The following two chapters provide basic information on the bulk and surface properties of biomaterials based on metallic, polymeric, and ceramic substrates. Also included are details about how some of these characteristics have been determined. The content of these chapters is intended to be relatively basic and more in-depth information is provided in later chapters and in the references.

1.2 BULK PROPERTIES OF MATERIALS

Francis W. Cooke

INTRODUCTION: THE SOLID STATE

Solids are distinguished from the other states of matter (liquids and gases) by the fact that their constituent atoms are held together by strong interatomic forces (Pauling, 1960). The electronic and atomic structures, and almost all the physical properties, of solids depend on the nature and strength of the interatomic bonds. Three different types of strong or primary interatomic bonds are recognized: ionic, covalent, and metallic.

Ionic Bonding

In the ionic bond, electron donor (metallic) atoms transfer one or more electrons to an electron acceptor (nonmetallic) atom. The two atoms then become a cation (e.g., metal) and an anion (e.g., nonmetal), which are strongly attracted by the electrostatic effect. This attraction of cations and anions constitutes the ionic bond (John, 1983).

In ionic solids composed of many ions, the ions are arranged so that each cation is surrounded by as many anions as possible to reduce the strong mutual repulsion of cations. This packing further reduces the overall energy of the assembly and leads to a highly ordered arrangement called a crystal structure. The loosely bound electrons of the atoms are now tightly held in the locality of the ionic bond. Thus, the electron structure of the atom is changed by the creation of the ionic bond. In addition, the bound electrons are not available to serve as charge carriers and ionic solids are poor electrical conductors. Finally, the low overall energy state of these substances endows them with relatively low chemical reactivity. Sodium fluoride (NaF) and magnesium chloride ($MgCl_2$) are examples of ionic solids.

Covalent Bonding

Elements that fall along the boundary between metals and nonmetals, such as carbon and silicon, have atoms with four valence electrons and about equal tendencies to donate and accept electrons. For this reason, they do not form strong ionic bonds. Rather, stable electron structures are achieved by sharing valence electrons. For example, two carbon atoms can each contribute an electron to a shared pair. This shared pair of electrons constitutes the covalent bond (Morrison *et al.*, 1983).

If a central carbon atom participates in four of these covalent bonds (two electrons per bond), it has achieved a stable outer shell of eight valence electrons. More carbon atoms can be added to the growing aggregate so that every atom has four nearest neighbors with which it shares one bond each. Thus, in a large grouping, every atom has a stable electron structure and four nearest neighbors. These neighbors often form a tetrahedron, and the tetrahedra in turn are assembled in an orderly repeating pattern (i.e., a crystal). This is the structure of both diamond and silicon. Diamond is the hardest of all materials, which shows that covalent bonds can be very strong. Once again, the bonding process results in a particular electronic structure (all electrons in pairs localized at the covalent bonds) and a particular atomic arrangement or crystal structure. As with ionic solids, localization of the valence electrons in the covalent bond renders these materials poor electrical conductors.

Metallic Bonding

The third and least understood of the strong bonds is the metallic bond. Metal atoms, being strong electron donors, do not bond by either ionic or covalent processes. Nevertheless, many metals are very strong (e.g., cobalt) and have high melting points (e.g., tungsten), suggesting that very strong interatomic bonds are at work here, too. The model that accounts for this bonding envisions the atoms arranged in an orderly, repeating, three-dimensional pattern, with the valence electrons migrating between the atoms like a gas.

It is helpful to imagine a metal crystal composed of positive ion cores, atoms without their valence electrons, about which the negative electrons circulate. On the average, all the electrical charges are neutralized throughout the crystal and bonding arises because the negative electrons act like a glue between the positive ion cores. This construct is called the free electron model of metallic bonding. Obviously, the bond strength increases as the ion cores and electron "gas" become more tightly packed (until the inner electron orbits of the ions begin to overlap). This leads to a condition of lowest energy when the ion cores are as close together as possible.

Once again, the bonding leads to a closely packed (atomic) crystal structure and a unique electronic configuration. In particular, the nonlocalized bonds within metal crystals permit plastic deformation (which strictly speaking does not occur in any nonmetals), and the electron gas accounts for the chemical reactivity and high electrical and thermal conductivity of metallic systems (John, 1983).

Weak Bonding

In addition to the three strong bonds, there are several weak secondary bonds that significantly influence the properties of some solids, especially polymers. The most important of these are van der Waals bonding and hydrogen bonding, which have strengths 3 to 10% that of the primary C—C covalent bond.

Atomic Structure

The three-dimensional arrangement of atoms or ions in a solid is one of the most important structural features that derives from the nature of the solid-state bond. In the majority of solids, this arrangement constitutes a crystal. A crystal is a solid whose atoms or ions are arranged in an orderly repeating pattern in three dimensions. These patterns allow the atoms to be closely packed [i.e., have the maximum possible number of near (contacting) neighbors] so that the number of primary bonds is maximized and the energy of the aggregate is minimized.

Crystal structures are often represented by repeating elements or subdivisions of the crystal called unit cells. Unit cells have all the geometric properties of the whole crystal (Fig. 1). A model of the whole crystal can be generated by simply stacking up unit cells like blocks or hexagonal tiles. Note that the representations of the unit cells in Fig. 1 are idealized in that atoms are shown as small circles located at the atomic centers. This is done so that the background of the structure can be understood. In fact, all nearest neighbors are in contact, as shown in Fig. 1B (John, 1983).

MATERIALS

The technical materials used to build most structures are divided into three classes, metals, ceramics (including glasses), and polymers. These classes may be identified only roughly with the three types of interatomic bonding.

Metals

Materials that exhibit metallic bonding in the solid state are metals. Mixtures or solutions of different metals are alloys.

About 85% of all metals have one of the crystal structures shown in Fig. 1. In both face-centered cubic (FCC) and hexagonal close-packed (HCP) structures, every atom or ion is surrounded by twelve touching neighbors, which is the closest packing possible for spheres of uniform size. In any enclosure filled with close-packed spheres, 74% of the volume will be occupied by the spheres. In the body-centered cubic (BCC) structure, each atom or ion has eight touching neighbors or eightfold coordination. Surprisingly, the density of packing is only reduced to 68% so that the BCC structure is nearly as densely packed as the FCC and HCP structures (John, 1983).

Ceramics

Ceramic materials are usually solid inorganic compounds with various combinations of ionic or covalent bonding. They also have tightly packed structures, but with special requirements for bonding such as fourfold coordination for covalent solids and charge neutrality for ionic solids (i.e., each unit cell must be electrically neutral). As might be expected, these additional requirements lead to more open and complex crystal structures.

Carbon is often included with ceramics because of its many ceramiclike properties, even though it is not a compound and conducts electrons in its graphitic form. Carbon is an interesting material since it occurs with two different crystal structures.

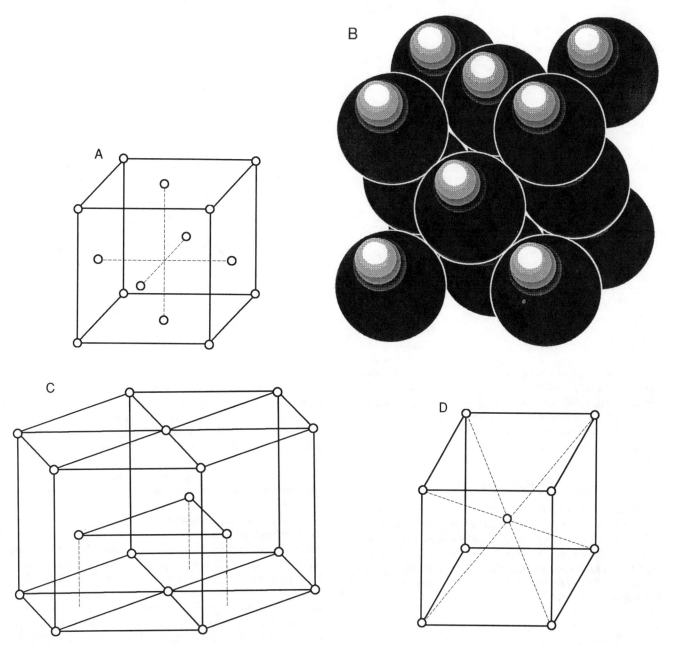

FIG. 1. Typical metal crystal structures (unit cells). (A) Face-centered cubic (FCC). (B) Full size atoms in FCC. (C) Hexagonal close packed (HCP). (D) Body-centered cubic (BCC).

In the diamond form, the four valance electrons of carbon lead to four nearest neighbors in tetrahedral coordination. This gives rise to the diamond cubic structure (Fig. 2A). An interesting variant on this structure occurs when the tetrahedral arrangement is distorted into a nearly flat sheet. The carbon atoms in the sheet have a hexagonal arrangement, and stacking of the sheets (Fig. 2B) gives rise to the graphite form of carbon. The (covalent) bonding within the sheets is much stronger than the bonding between sheets.

The existence of an element with two different crystal struc-

tures provides a striking opportunity to see how physical properties depend on atomic and electronic structure (Table 1) (Reed-Hill, 1992).

Inorganic Glasses

Some ceramic materials can be melted and upon cooling do not develop a crystal structure. The individual atoms have nearly the ideal number of nearest neighbors, but an orderly

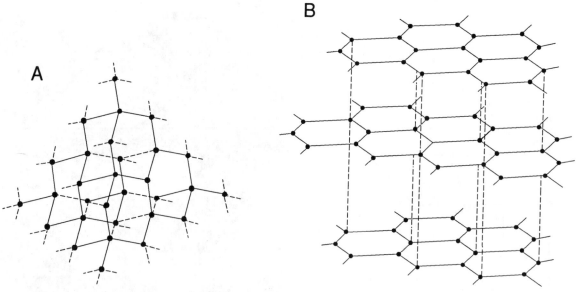

FIG. 2. Crystal structures of carbon. (A) Diamond (cubic). (B) Graphite (hexagonal).

repeating arrangement is not maintained over long distances throughout the three-dimensional aggregates of atoms. Such noncrystals are called glasses or, more accurately, inorganic glasses and are said to be in the amorphous state. Silicates and phosphates, the two most common glass formers, have random three-dimensional network structures.

Polymers

The third category of solid materials includes all the polymers. The constituent atoms of classic polymers are usually carbon and are joined in a linear chainlike structure by covalent bonds. The bonds within the chain require two of the valance electrons of each atom, leaving the other two bonds available for adding a great variety of atoms (e.g., hydrogen), molecules, functional groups, etc.

TABLE 1   Relative Physical Properties of Diamond and Graphite[a]

Property	Diamond	Graphite
Hardness	Highest known	Very low
Color	Colorless	Black
Electrical conductivity	Low	High
Density (g/cm³)	3.51	2.25
Specific heat (cal/gm atm/deg.C)	1.44	1.98

[a]Adapted from D. L. Cocke and A. Clearfield, eds., *Design of New Materials,* Plenum Publ., New York, 1987, with permission.

Based on the organization of these chains, there are two classes of polymers. In the first, the basic chains have little or no branching. Such "straight" chain polymers can be melted and remelted without a basic change in structure (an advantage in fabrication) and are called thermoplastic polymers. If side chains are present and actually form (covalent) links between chains, a three-dimensional network structure is formed. Such structures are often strong, but once formed by heating will not melt uniformly on reheating. These are thermosetting polymers.

Usually both thermoplastic and thermosetting polymers have intertwined chains so that the resulting structures are quite random and are also said to be amorphous like glass, although only the thermoset polymers have sufficient cross linking to form a three-dimensional network with covalent bonds. In amorphous thermoplastic polymers, many atoms in a chain are in close proximity to the atoms of adjacent chains, and van der Waals and hydrogen bonding holds the chains together. It is these interchain bonds that are responsible for binding the substance together as a solid. Since these bonds are relatively weak, the resulting solid is relatively weak. Thermoplastic polymers generally have lower strengths and melting points than thermosetting polymers (John, 1983; Budinski, 1983).

Microstructure

Structure in solids occurs in a hierarchy of sizes. The internal or electronic structures of atoms occur at the finest scale, less than 10^{-4} μm (which is beyond the resolving power of the most powerful direct observational techniques) and are responsible for the interatomic bonds. At the next higher size level, around 10^{-4} μm (which is detectable by X-ray diffraction, field ion microscopy, scanning tunneling microscopy, etc.) the long-range, three-dimensional arrangement of atoms in crystals and glasses can be observed.

At even larger sizes, 10^{-3} to 10^2 μm (detectable by light and electron microscopy), another important type of structural organization exists. When the atoms of a molten sample are incorporated into crystals during freezing, many small crystals are formed initially and then grow until they impinge on each other and all the liquid is used up. At that point the sample is completely solid. Thus, most crystalline solids (metal and ceramics) are composed of many small crystals or crystallites called grains that are tightly packed and firmly bound together. This is the microstructure of the material that is observed at magnifications where the resolution is between 1 and 100 μm.

In pure elemental materials, all the crystals have the same structure and differ from each other only by virtue of their different orientations. In general, these crystallites or grains are too small to be seen except with a light microscope. Most solids are opaque, however, so the common transmission (biological) microscope cannot be used. Instead, a metallographic or ceramographic reflecting microscope is used. Incident light is reflected from the polished metal or ceramic surface. The grain structure is revealed by etching the surface with a mildly corrosive medium that preferentially attacks the grain boundaries. When this surface is viewed through the reflecting microscope the size and shape of the grains, i.e., the microstructure, is revealed.

Grain size is one of the most important features that can be evaluated by this technique because fine-grained samples are generally stronger than coarse-grained specimens of a given material. Another important feature that can be identified is the coexistence of two or more phases in some solid materials. The grains of a given phase will all have the same chemical composition and crystal structure, but the grains of a second phase will be different in both these respects. This never occurs in samples of pure elements, but does occur in mixtures of different elements or compounds where the atoms or molecules can be dissolved in each other in the solid state just as they are in a liquid or gas solution.

For example, some chromium atoms can substitute for iron atoms in the FCC crystal lattice of iron to produce stainless steel, a solid solution alloy. Like liquid solutions, solid solutions exhibit solubility limits; when this limit is exceeded, a second phase precipitates. For example, if more Cr atoms are added to stainless steel than the FCC lattice of the iron can accommodate, a second phase that is chromium rich precipitates. Many important biological and implant materials are multiphase (Reed-Hill, 1992). These include the cobalt-based and titanium-based orthopedic implant alloys and the mercury-based dental restorative alloys, i.e., amalgams.

MECHANICAL PROPERTIES OF MATERIALS

Solid materials possess many kinds of properties (e.g., mechanical, chemical, thermal, acoustical, optical, electrical, magnetic). For most (but not all) biomedical applications, the two properties of greatest importance are strength (mechanical) and reactivity (chemical). The chemical reactivity of biomateri-

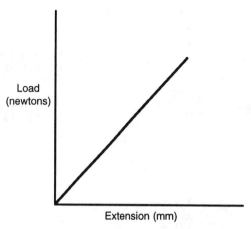

FIG. 3. Extension is proportional to load according to Hooke's law.

als will be discussed in Chapters 1.3 and 6. The remainder of this section will, therefore, be devoted to mechanical properties, their measurement, and their dependence on structure. It is well to note that the dependence of mechanical properties on microstructure is so great that it is one of the fundamental objectives of materials science to control mechanical properties by modifying microstructure.

Elastic Behavior

The basic experiment for determining mechanical properties is the tensile test. In 1678, Robert Hooke showed that a solid material subjected to a tensile (distraction) force would extend in the direction of traction by an amount that was proportional to the load (Fig. 3). This is known as Hooke's law and simply expresses the fact that most solids behave in an elastic manner (like a spring) if the loads are not too great.

Stress and Strain

The extension for a given load varies with the geometry of the specimen as well as its composition. It is, therefore, difficult to compare the relative stiffness of different materials or to predict the load-carrying capacity of structures with complex shapes. To resolve this confusion, the load and deformation can be normalized. To do this, the load is divided by the cross-sectional area available to support the load, and the extension is divided by the original length of the specimen. The load can then be reported as load per unit of cross-sectional area, and the deformation can be reported as the elongation per unit of the original length over which the elongation occurred. In this way, the effects of specimen geometry can be normalized.

The normalized load (force/area) is stress (σ) and the normalized deformation (change in length/original length) is strain (ε) (Fig. 4).

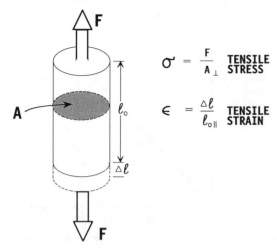

FIG. 4.   Tensile stress and tensile strain.

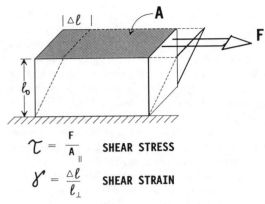

FIG. 5.   Shear stress and shear strain.

Tension and Compression

In tension and compression the area supporting the load is perpendicular to the loading direction (tensile stress), and the change in length is parallel to the original length (tensile strain).

If weights are used to provide the applied load, the stress is calculated by adding up the total number of pounds-force (lb) or newtons (N) used and dividing by the perpendicular cross-sectional area. For regular specimen geometries such as cylindrical rods or rectangular bars, a measuring instrument, such as a micrometer, is used to determine the dimensions. The units of stress are pounds per inch squared (psi) or newtons per meter squared (N/m^2). The N/m^2 unit is also known as the pascal (Pa).

The measurement of strain is achieved, in the simplest case, by applying reference marks to the specimen and measuring the distance between with calipers. This is the original length, l_o. A load is then applied, and the distance between marks is measured again to determine the final length, l_n. The strain, ε, is then calculated by:

$$\varepsilon = \frac{l_n - l_o}{l_o} = \frac{\Delta l}{l_o}. \tag{1}$$

This is essentially the technique used for flexible materials like rubbers, polymers, and soft tissues. For stiff materials like metals, ceramics, and bone, the deflections are so small that a more sensitive method is needed (i.e., the electrical resistance strain gage).

Shear

For cases of shear, the applied load is parallel to the area supporting it (shear stress, τ), and the dimensional change is perpendicular to the reference dimension (shear strain, γ) (Fig. 5).

Elastic Constants

By using these definitions of stress and strain, Hooke's law can be expressed in quantitative terms:

$$\sigma = E\,\varepsilon, \text{ tension or compression,} \tag{2a}$$
$$\tau = G\,\gamma, \text{ shear.} \tag{2b}$$

E and G are proportionality constants that may be likened to spring constants. The tensile constant, E, is the tensile (or Young's) modulus and G is the shear modulus. These moduli are also the slopes of the elastic portion of the stress versus strain curve (Fig. 6). Since all geometric influences have been removed, E and G represent inherent properties of the material. These two moduli are direct macroscopic manifestations of the strengths of the interatomic bonds. Elastic strain is achieved by actually increasing the interatomic distances in the crystal (i.e., stretching the bonds). For materials with strong bonds (e.g., diamond, Al_2O_3, tungsten), the moduli are high and a given stress produces only a small strain. For materials with weaker bonds (e.g., polymers and gold), the moduli are lower (John, 1983). The tensile elastic moduli for some important biomaterials are presented in Table 2.

Isotropy

The two constants, E and G, are all that are needed to fully characterize the stiffness of an isotropic material, (i.e., a material whose properties are the same in all directions).

Single crystals are anisotropic (not isotropic) because the stiffness varies as the orientation of applied force changes relative to the interatomic bond directions in the crystal. In polycrystalline materials (e.g., most metallic and ceramic specimens), a great multitude of grains (crystallites) are aggregated with multiply distributed orientations. On the average, these aggregates exhibit isotropic behavior at the macroscopic level, and values of E and G are highly reproducible for all specimens of a given metal, alloy, or ceramic.

On the other hand, many polymeric materials and most tissue samples are anisotropic (not the same in all directions) even at the macroscopic level. Bone, ligament, and sutures are

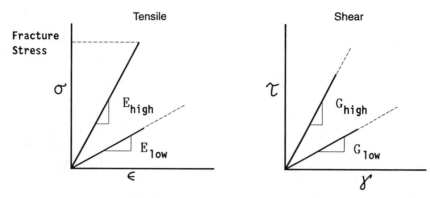

FIG. 6. Stress versus strain for elastic solids.

all stronger and stiffer in the fiber (longitudinal) direction than they are in the transverse direction. For such materials, more than two elastic constants are required to relate stress and strain properties.

MECHANICAL TESTING

To conduct controlled load-deflection (stress-strain) tests, a load frame is used that is much stiffer and stronger than the specimen to be tested (Fig. 7). One cross-bar or cross-head is moved up and down by a screw or a hydraulic piston. Jaws that provide attachment to the specimen are connected to the frame and to the movable cross-head. In addition, a load cell to monitor the force being applied is placed in series with the specimen. The load cell functions like a stiff spring scale to measure the applied loads.

Tensile specimens usually have a reduced gage section over which strains are measured. For a valid determination of fracture properties, failure must also occur in this reduced section and not in the grips. For compression testing, the direction of cross-head movement is reversed and cylindrical or prismatic specimens are simply squeezed between flat anvils. Standard-

TABLE 2 Mechanical Properties of Some Implant Materials and Tissues

	Elastic modulus (GPa)	Yield strength (MPa)	Tensile strength (MPa)	Elongation to failure (%)
Al$_2$O$_3$	350	—	1,000 to 10,000	0
CoCr Alloy[a]	225	525	735	10
316 S.S.[b]	210	240 (800)[c]	600 (1000)[c]	55 (20)[c]
Ti 6Al–4V	120	830	900	18
Bone (cortical)	15 to 30	30 to 70	70 to 150	0–8
PMMA	3.0	—	35 to 50	0.5
Polyethylene	0.4	—	30	15–100
Cartilage	[d]	—	7 to 15	20

[a]28% Cr, 2% Ni, 7% Mo, 0.3% C (max.), Co balance.
[b]Stainless steel, 18% Cr, 14% Ni, 2 to 4% Mo, 0.03 C (max), Fe balance.
[c]Values in parenthesis are for the cold-worked state.
[d]Strongly viscoelastic.

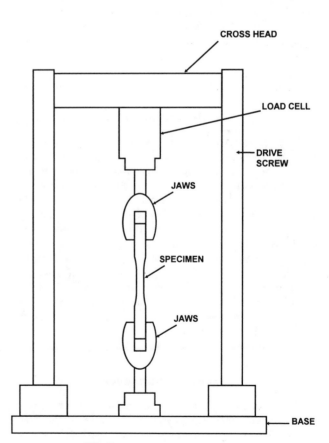

FIG. 7. Mechanical testing machine.

ized specimens should be used for all mechanical testing to ensure reproducibility of results (see the publications of the American Society for Testing and Materials, 100 Barr Harbor Dr., West Conshohocken, PA 19428-2959).

Another useful test that can be conducted in a mechanical testing machine is the bend test. In bend testing, the outside of the bowed specimen is in tension and the inside in compression. The outer fiber stresses can be calculated from the load and the specimen geometry (see any standard text on strength of materials; Levinson, 1971). Bend tests are useful because no special specimen shapes are required and no special grips are necessary. Strain gages can also be used to determine the outer fiber strains. The available formulas for the calculation of stress states are only valid for elastic behavior. Therefore, they cannot be used to describe any nonelastic strain behavior.

Some mechanical testing machines are also equipped to apply torsional (rotational) loads, in which case torque versus angular deflection can be determined and used to calculate the torsional properties of materials. This is usually an important consideration when dealing with biological materials, especially under shear loading conditions (John, 1983).

Elasticity

The tensile elastic modulus, E (for an isotropic material) can be determined by the use of strain gages, an accurate load cell, and cyclic testing in a standard mechanical testing machine. To do so, Hooke's law is rearranged as follows:

$$E = \frac{\sigma}{\varepsilon}. \qquad (3)$$

Brittle Fracture

In real materials, elastic behavior does not persist indefinitely. If nothing else intervenes, microscopic defects, which are present in all real materials, will eventually begin to grow rapidly under the influence of the applied tensile or shear stress, and the specimen will fail suddenly by brittle fracture. Until this brittle failure occurs, the stress-strain diagram does not deviate from a straight line, and the stress at which failure occurs is called the fracture stress (Fig. 6). This behavior is typical of many materials, including glass, ceramics, graphite, very hard alloys (scalpel blades), and some polymers like polymethymethacrylate (bone cement) and unmodified polyvinyl chloride (PVC). The number and size of defects, particularly pores, is the microstructural feature that most affects the strength of brittle materials.

Plastic Deformation

For some materials, notably metals and alloys, the process of plastic deformation sets in after a certain stress level is reached but before fracture occurs. During a tensile test, the stress at which 0.2% plastic strain occurs, is called the 0.2%

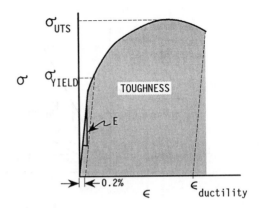

FIG. 8.   Stress versus strain for a ductile material.

offset yield strength. Once plastic deformation starts, the strains produced are very much greater than those during elastic deformation (Fig. 8), they are no longer proportional to the stress and they are not recovered when the stress is removed. This happens because whole arrays of atoms under the influence of an applied stress are forced to move, irreversibly, to new locations in the crystal structure. This is the microstructural basis of plastic deformation. During elastic straining, on the other hand, the atoms are displaced only slightly by reversible stretching of the interatomic bonds.

Large scale displacement of atoms without complete rupture of the material, i.e., plastic deformation, is only possible in the presence of the metallic bond so only metals and alloys exhibit true plastic deformation. Since long-distance rearrangement of atoms under the influence of an applied stress cannot occur in ionic or convolutely bonded materials, ceramics and many polymers exhibit only brittle behavior.

Plastic deformation is very useful for shaping metals and alloys and is called ductility or malleability. The total permanent (i.e., plastic) strain exhibited up to fracture by a material is a quantitative measure of its ductility (Fig. 8). The strength,

TABLE 3   Mechanical Properties Derivable from a Tensile Test

Property	Units		
		International	English
1. Elastic modulus (E)	F/A^a	N/m² (Pa)	lbf/in.² (psi)
2. Yield strength (YS)	F/A	N/m² (Pa)	lbf/in.² (psi)
3. Ultimate tensile strength (UTS)	F/A	N/m² (Pa)	lbf/in.² (psi)
4. Ductility	%	%	%
5. Toughness (work to fracture per unit volume)	$F \times l/V$	J/m³	in lbf/in.³

[a] lbf, pounds force; F, force; A, area; l, length; and V, volume.

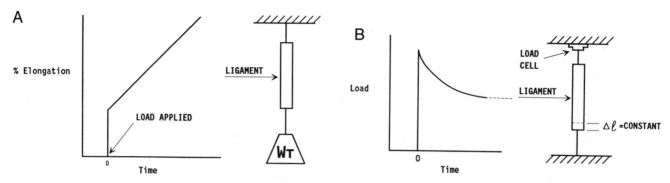

FIG. 9.   (A) Elongation versus time at constant load (creep) of ligament. (B) Load versus time at constant elongation (stress relaxation) for ligament.

particularly the 0.2% offset yield strength, can be increased significantly by reducing the grain size as well as by prior plastic deformation or cold work. The introduction of alloying elements and multiphase microstructures are also potent strengthening mechanisms.

Other properties can be derived from the tensile stress-strain curve. The tensile strength or the ultimate tensile stress (UTS) is the stress that is calculated from the maximum load experienced during the tensile test (Fig. 8).

The area under the tensile curve is proportional to the work required to deform a specimen until it fails. The area under the entire curve is proportional to the product of stress and strain, and has the units of energy (work) per unit volume of specimen. The work to fracture is a measure of toughness and reflects a material's resistance to crack propagation (Fig. 8). The important mechanical properties derived from a tensile test are listed in Table 3.

Creep and Viscous Flow

For all the mechanical behaviors considered to this point, it has been tacitly assumed that when a stress is applied, the strain response is instantaneous. For many important biomaterials, including polymers and tissues, this is not a valid assumption. If a weight is suspended from a specimen of ligament, the ligament continues to elongate for a considerable time even though the load is constant (Fig. 9A). This continuous, time-dependent extension under load is called creep.

Similarly, if the ligament is extended in a tensile machine to a fixed elongation and the load is monitored, the load drops

continuously with time (Fig. 9B). The continuous drop in load at constant extension is called stress relaxation. Both these responses are the result of viscous flow in the material. The mechanical analog of viscous flow is a dashpot or cylinder and piston (Fig. 10A). Any small force is enough to keep the piston moving. If the load is increased, the rate of displacement will increase.

Despite this liquid-like behavior, these materials are functionally solids. To produce such a combined effect, they act as though they are composed of a spring (elastic element) in series with a dashpot (viscous element) (Fig. 10B). Thus, in the creep test, instantaneous strain is produced when the weight is first applied (Fig. 9A). This is the equivalent of stretching the spring to its equilibrium length (for that load). Thereafter, the additional time-dependent strain is modeled by the movement of the dashpot. Complex arrangements of springs and dashpots are often needed to adequately model actual behavior.

Materials that behave approximately like a spring and dashpot system are viscoelastic. One consequence of viscoelastic behavior can be seen in tensile testing where the load is applied at some finite rate. During the course of load application, there is time for some viscous flow to occur along with the elastic strain. Thus, the total strain will be greater than that due to the elastic response alone. If this total strain is used to estimate the Young's modulus of the material ($E = \sigma/\varepsilon$), the estimate will be low. If the test is conducted at a more rapid rate, there will be less time for viscous flow during the test and the apparent modulus will increase. If a series of such tests is conducted at ever higher loading rates, eventually a rate can be reached where no detectable viscous flow occurs and the

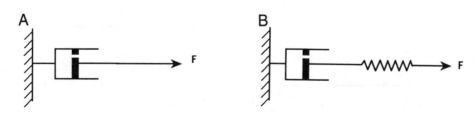

FIG. 10.   (A) Dash pot or cylinder and piston model of viscous flow. (B) Dash pot and spring model of a viscoelastic material.

modulus determined at this critical rate will be the true elastic modulus, i.e., the spring constant of the elastic component. Tests at even higher rates will produce no further increase in modulus. For all viscoelastic materials, moduli determined at rates less than the critical rate are "apparent" moduli and must be identified with the strain rate used. Failure to do this is one reason why values of tissue moduli reported in the literature may vary over wide ranges.

Finally, it should be noted that it may be difficult to distinguish between creep and plastic deformation in ordinary tensile tests of highly viscoelastic materials (e.g., tissues). For this reason, the total nonelastic deformation of tissues or polymers may at times be loosely referred to as plastic deformation even though some viscous flow may be involved.

OTHER IMPORTANT PROPERTIES OF MATERIALS

Fatigue

It is not uncommon for materials, including tough and ductile ones like 316L stainless steel, to fracture even though the service stresses imposed are well below the yield stress. This occurs when the loads are applied and removed for a great number of cycles, as happens to prosthetic heart valves and prosthetic joints. Such repetitive loading can produce microscopic cracks that then propagate by small steps at each load cycle.

The stresses at the tip of a crack or even a sharp corner are locally enhanced by the stress-raising effect. Under repetitive loading, these local high stresses (or strains) actually exceed the strength of the material over a small region. This phenomenon is responsible for the stepwise propagation of the cracks. Eventually, the load-bearing cross-section becomes so small that the part finally fails completely.

Fatigue, then, is a process by which structures fail as a result of cyclic stresses that may be much less than the ultimate tensile stress. Fatigue failure plagues many dynamically loaded structures, from aircraft to bones (march- or stress-fractures) to cardiac pacemaker leads.

The susceptibility of specific materials to fatigue is deter-mined by testing a group of identical specimens in cyclic tension or bending (Fig. 11A) at different maximum stresses. The number of cycles to failure is then plotted against the maximum applied stress (Fig. 11B). Since the number of cycles to failure is quite variable for a given stress level, the prediction of fatigue life is a matter of probabilities. For design purposes, the stress that will provide a low probability of failure after 10^6 or 10^7 cycles is often adopted as the fatigue strength or endurance limit of the material. This may be as little as one third or one fourth of the single-cycle yield strength. The fatigue strength is sensitive to environment, temperature, corrosion, deterioration (of tissue specimens), and cycle rate (especially for viscoelastic materials). Careful attention to these details is required if laboratory fatigue results are to be successfully transferred to biomedical applications (John, 1983).

Toughness

The ability of a material to plastically deform under the influence of the complex stress field that exists at the tip of a crack is a measure of its toughness. If plastic deformation does occur, it serves to blunt the crack and lower the locally enhanced stresses, thus hindering crack propagation. To design "failsafe" structures with brittle materials, it has become necessary to develop an entirely new system for evaluating service worthiness. This system is fracture toughness testing and requires the testing of specimens with sharp notches. The resulting fracture toughness parameter is a function of the apparent crack propagation stress and the crack depth and shape. It is called the critical stress intensity factor (K_{Ic}) and has units of $Pa\sqrt{m}$ or $N \cdot m^{3/2}$. Since fracture toughness depends on both the strength of the material and its ductility (ability to blunt cracks), there is an empirical correspondence between K_{Ic} and the area under the stress-strain curve for some materials and conditions. The energy absorbed in impact fracture is also a measure of toughness, but at higher loading rates (Brick *et al.*, 1977).

Effect of Fabrication on Strength

A general concept to keep in mind in considering the strength of materials is that the process by which a material

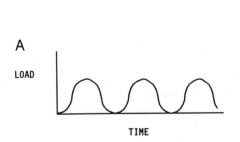

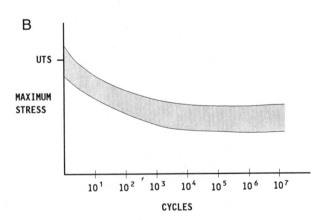

FIG. 11.   (A) Stress versus time in a fatigue test. (B) Fatigue curve: fatigue stress versus cycles to failure.

is produced has a major effect on its structure and hence its properties (John, 1983). For example, plastic deformation of most metals at room temperature flattens the grains and produces strengthening while reducing ductility. Subsequent high-temperature treatment (annealing) can reverse this effect. Polymers drawn into fibers are much stronger in the drawing direction than are undrawn samples of the same material.

Because strength properties depend on fabrication history, it is important to realize that there is no unique set of strength properties for each generic material (e.g., 316L stainless steel, PET, Al_2O_3). Rather, there is a range of properties that depends on the fabrication history and the microstructures produced.

CONCLUSION

The determination of mechanical properties is not only an exercise in basic materials science but is indispensable to the practical design and understanding of load-bearing structures. Designers must determine the service stresses in all structural members and be sure that at every point these stresses are safely below the yield strength of the material. If cyclic loads are involved (e.g., lower-limb prostheses, teeth, heart valves), the service stresses must be kept below the fatigue strength.

In subsequent chapters where the properties and behavior of materials are discussed in detail, it is well to keep in mind that this information is indispensable to understanding the mechanical performance (i.e., function) of both biological and manmade structures.

Bibliography

Budinski, K. (1995). *Engineering Materials, Properties and Selection,* 5th Ed. Prentice–Hall, Old Tappan, NJ.

John, V. B. (1983). *Introduction to Engineering Materials,* 2nd Ed. MacMillan Co., New York.

Levinson, I. J. (1971). *Statics and Strength of Materials.* Prentice–Hall, Englewood Cliffs, NJ.

Morrison, R. T., and Boyde, R. N. (1992). *Organic Chemistry,* 6th Ed. Simon & Schuster, New York.

Pauling, L. (1960). *The Nature of the Chemical Bond and the Structure of Molecules and Crystals.* Cornell Univ. Press, Ithaca, NY.

Reed-Hill, R. E. (1992). *Physical Metallurgy Principles.* Van Nostrand, New York.

1.3 SURFACE PROPERTIES OF MATERIALS
Buddy D. Ratner

CHARACTERIZATION OF BIOMATERIAL SURFACES

In developing biomedical implant devices and materials, we are concerned with function, durability, and biocompatibility. Understanding function (e.g., mechanical strength, permeability, elasticity) is relatively straightforward—the tools of engineers and materials scientists are appropriate to address this concern. Durability, particularly in a biological environment, is less well understood. Still, the tests we need to evaluate

durability are clear (see Chapters 1.2, 6.2, and 6.3). Biocompatibility represents a frontier of knowledge in this field, and its study is often assigned to the biochemist, biologist, and physician. However, the important question in biocompatibility is how the device or material "transduces" its structural makeup to direct or influence the response of proteins, cells, and the organism to it. For devices and materials that do not leach undesirable substances in sufficient quantities to influence cells and tissues (i.e., that have passed routine toxicological evaluation; see Chapter 5.2), this transduction occurs through the surface structure—the body "reads" the surface structure and responds. For this reason we must understand the surface structure of biomaterials. Chapter 9.7 elaborates on the biological implications of this idea.

A few general points about surfaces are useful at this point. First, the surface region of a material is known to be uniquely reactive (Fig. 1). Catalysis and microelectronics both capitalize on surface reactivity, and it would be naive to expect the biology not to respond to it. Second, the surface of a material is inevitably different from the bulk. Thus, the traditional techniques used to analyze the bulk structure of materials are not suitable for surface determination. Third, surfaces readily contaminate. Under ultrahigh vacuum conditions we can retard this contamination. However, in view of the atmospheric pressure conditions under which all biomedical devices are used, we must learn to live with some contamination. The key questions here are whether we can make devices with constant, controlled levels of contamination and avoid undesirable contaminants. This is critical so that a laboratory experiment on a biomaterial generates the same results when repeated after 1 day, 1 week, or 1 year, and so that the biomedical device performs for the physician in a constant manner over a reasonable shelf life. Finally, the surface structure of a material is often mobile. The movement of atoms and molecules near the surface in response to the outside environment is often highly significant. In response to a hydrophobic environment, (e.g., air) more hydrophobic (lower energy) components may migrate to the surface of a material. In response to an aqueous environment, the surface may reverse its structure and point polar groups outward to interact with the polar water molecules. An example of this is schematically illustrated in Fig. 2.

The nature of surfaces is a complex subject in its own right and the subject of much independent investigation. The reader is referred to one of many excellent monographs on this important subject for a complete and rigorous introduction (see Somorjai, 1981; Adamson, 1990; Andrade, 1985).

Parameters to Be Measured

Many parameters describe a surface, as shown in Fig. 3. The more of these parameters we measure, the more we can piece together a complete description of the surface. A complete characterization requires the use of many techniques to compile all the information needed. Unfortunately, we cannot yet specify which parameters are most important for understanding biological responses to surfaces. Studies have been published on the importance of roughness, wettability, surface mobility, chemical composition, crystallinity, and heterogeneity to bio-

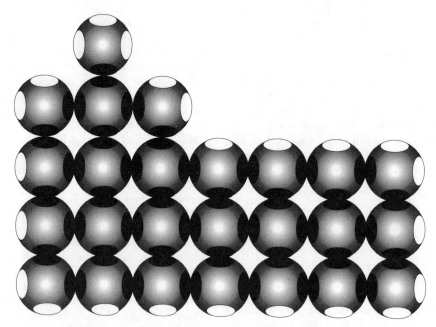

FIG. 1. A two-dimensional crystal lattice illustrating bonding orbitals (gray or black ovals). For atoms in the center (bulk) of the crystal (black ovals), all binding sites are associated. At planar exterior surfaces, one of the bonding sites is unfulfilled (gray ovals). At corners, two bonding sites are unfulfilled. The single atom on top of the crystal has three unfulfilled valencies. Energy is minimized where more of these unfulfilled valencies can interact.

logical reaction. Since we cannot be certain which surface factors are predominant in each situation, the controlling variable or variables must be independently ascertained.

MEASUREMENT TECHNIQUES

General Principles

A number of general ideas can be applied to all surface analysis. They can be divided into the categories of sample preparation and analysis described in the following paragraphs.

Sample Preparation

In sample preparation the sample should resemble, as closely as possible, the material or device being subjected to biological testing or implantation. Needless to say, fingerprints on the surface of the sample will cover up everything that might be of interest. If the sample is placed in a package for shipping or storage prior to surface analysis, it is critical to know whether the packaging material can induce surface contamination. Plain paper in contact with most specimens will transfer material (often metal ions) to the surface of the material. Many plastics are processed with silicone oils or other additives that can be transferred to the specimen. The packaging material used should be examined by surface analysis methods to ascertain its purity. Samples can be surface analyzed prior to and after storage or shipping in containers to ensure that the surface composition measured is not due to the container. As a general rule, the polyethylene press-close bags used in electron microscopy and cell culture plasticware are clean storage containers.

However, abrasive contact must be avoided and each brand must be evaluated so that a meticulous specimen preparation is not ruined by contamination.

Sample Analysis

Two general principles guide sample analysis. First, all methods used to analyze surfaces also have the potential to alter the surface. It is essential that the analyst be aware of the damage potential of the method used. Second, because of the potential for artifacts and the need for many pieces of information to construct a complete picture of the surface (Fig. 3), more than one method should be used whenever possible. The data derived from two or more methods should always be corroborative. When data are contradictory, be suspicious and question why. A third or fourth method may then be necessary to draw confident conclusions about the nature of a surface.

These general principles are applicable to all materials. There are properties (only a few of which will be presented here) that are specific to specific classes of materials. Compared with metals, ceramics, glasses, and carbons, organic and polymeric materials are more easily damaged by surface analysis methods. Polymeric systems also exhibit greater surface molecular mobility than inorganic systems. The surfaces of inorganic materials are contaminated more rapidly than polymeric materials because of their higher surface energy. Electrically conductive metals and carbons will often be easier to characterize than insulators using the electron, X-ray, and ion interaction methods. Insulators accumulate a surface electrical charge that requires special methods (e.g., a low energy electron beam) to neutralize. To learn about other concerns in surface analysis that are specific to specific classes

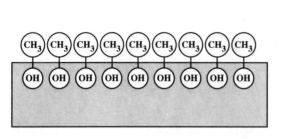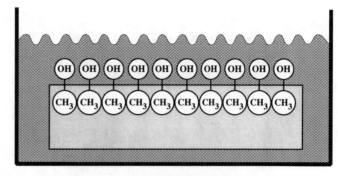

In air

Under water

FIG. 2. Many materials can undergo a reversal of surface structure when transferred from air into a water environment. In this schematic illustration, a hydroxylated polymer exhibits a surface rich in methyl groups (from the polymer chain backbone) in air, and a surface rich in hydroxyl groups under water. This has been observed experimentally (see Ratner *et al., J. Appl. Polym. Sci.,* **22**: 643, 1978).

of materials, published papers become a valuable resource for understanding the pitfalls that can lead to artifact or inaccurate results.

Table 1 summarizes the characteristics of many common surface analysis methods, including their depth of analysis and their spatial resolution (spot size analyzed). A few of the more frequently used techniques are described in the next section. However, space limitations prevent a fully developed discussion of these methods. The reader is referred to many comprehensive books on the general subject of surface analysis and on each of the primary methods (Andrade, 1985; Briggs and Seah, 1983; Feldman and Mayer, 1986).

Contact Angle Methods

The force balance between the liquid–vapor surface tension (γ_{lv}) of a liquid drop and the interfacial tension between a solid and the drop (γ_{sl}), manifested through the contact angle (θ) of the drop with the surface, can be used to characterize the energy of the surface (γ_{sv}). The basic relationship describing this force balance is:

$$\gamma_{sv} = \gamma_{sl} + \gamma_{lv} \cos \theta.$$

The energy of the surface, which is directly related to its wettability, is a useful parameter that has often correlated strongly with biological interaction. Unfortunately, γ_{sv} cannot be directly obtained since this equation contains two unknowns, γ_{sl} and γ_{sv}. Therefore, the γ_{sv} is usually approximated by the Zisman method for obtaining the critical surface tension (Fig. 4), or calculated by solving simultaneous equations with data from liquids of different surface tensions. Some critical surface tensions for common materials are listed in Table 2.

Experimentally, there are a number of ways to measure the contact angle and some of these are illustrated in Fig. 5. Contact angle methods are inexpensive, and, with some practice, easy to perform. They provide a "first line" characterization of materials and can be performed in any laboratory. Contact angle measurements provide unique insight into how the sur-

face will interact with the external world. However, in performing such measurements, a number of concerns must be addressed to obtain meaningful data (Table 3). There are a number of review articles on contact angle measurement for surface characterization (Andrade, 1985; Neumann and Good, 1979; Zisman, 1964; Ratner, 1985).

Electron Spectroscopy for Chemical Analysis

Electron spectroscopy for chemical analysis (ESCA) provides unique information about a surface that cannot be obtained by other means (Andrade, 1985; Ratner, 1988; Dilks, 1981; Ratner and McElroy, 1986). In contrast to the contact angle technique, ESCA is expensive and generally requires considerable training to perform the measurements. However, since ESCA is available from commercial laboratories, university analytical facilities, national centers, and specialized research laboratories, most biomaterials scientists can get access to it to have their samples analyzed. The data can be interpreted in a simple but still useful fashion, or more rigorously. They have shown unquestioned value in the development of biomedical implant materials and understanding the fundamentals of biointeraction.

The ESCA method (also called X-ray photoelectron spectroscopy, XPS) is based upon the photoelectric effect, properly described by Einstein in 1905. X-rays are focused upon a specimen. The interaction of the X-rays with the atoms in the specimen causes the emission of a core level (inner shell) electron. The energy of this electron is measured and its value provides information about the nature and environment of the atom from which it came. The basic energy balance describing this process is given by the simple relationship:

$$BE = h\nu - KE,$$

where *BE* is the energy binding the electron to an atom (the value desired), *KE* is the kinetic energy of the emitted electron (the value measured in the ESCA spectrometer), and *h*ν is the energy of the X-rays, a known value. A simple schematic

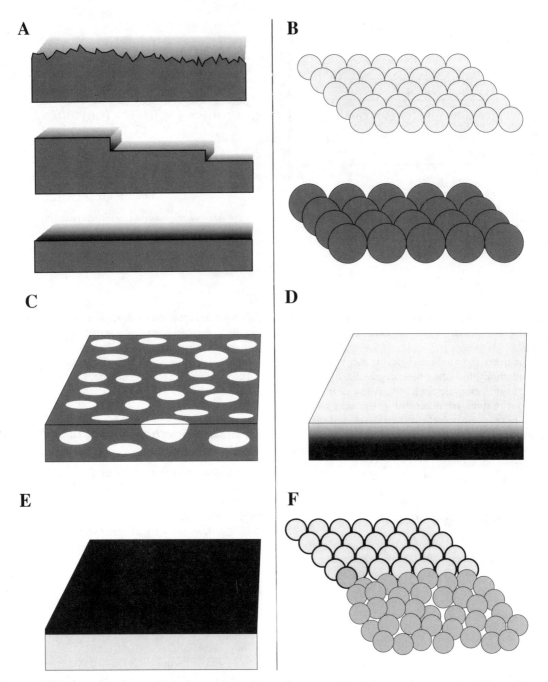

FIG. 3.   Some possibilities for surface structure. (A) Surfaces can be rough, smooth, or stepped. (B) Surfaces can be composed of different chemistries (atoms and molecules). (C) Surfaces may be structurally or compositionally inhomogeneous in the plane of the surface. (D) Surfaces may be inhomogeneous with depth into the specimen. (E) Surfaces may be covered by an overlayer. (F) Surfaces may be highly crystalline or disordered.

diagram illustrating an ESCA instrument is shown in Fig. 6. Table 4 lists some of the types of information about the nature of a surface that can be obtained by using ESCA. The origin of the surface sensitivity of ESCA is described in Fig. 7.

ESCA has many advantages, and a few disadvantages, for studying biomaterials. The advantages include the speed of analysis, the high information content, the low damage poten-

tial, and the ability to analyze samples with no specimen preparation. The latter advantage is particularly important since it means that many biomedical devices (or parts of devices) can be inserted, as fabricated and sterilized, directly in the analysis chamber for study. The disadvantages include the need for vacuum compatibility (i.e., no outgassing of volatile components), the possibility of sample damage if long analysis times

TABLE 1 Common Methods of Characterizing Biomaterial Surfaces

Method	Principle	Depth analyzed	Spatial resolution	Analytical sensitivity	Cost[a]
Contact angles	Liquid wetting of surfaces is used to estimate the surface energy	3–20 Å	1 mm	Low or high depending on the chemistry	$
ESCA	X-rays cause the emission of electrons of characteristic energy	10–250 Å	10–150 μm	0.1 Atom %	$$$
Auger electron spectroscopy[b]	A focused electron beam causes the emission of Auger electrons	50–100 Å	100 Å	0.1 Atom %	$$$
SIMS	Ion bombardment leads to the emission of surface secondary ions	10 Å–1μm[c]	100 Å	Very high	$$$
FTIR–ATR	IR radiation is adsorbed in exciting molecular vibrations	1–5 μm	10 μm	1 Mole %	$$
STM	Measurement of the quantum tunneling current between a metal tip and a conductive surface	5 Å	1 Å	Single atoms	$$
SEM	Secondary electron emission caused by a focused electron beam is measured and spatially imaged	5 Å	40 Å typically	High, but not quantitative	$$

[a]$, up to $5000; $$, $5000–$100,000; $$$, >$100,000.
[b]Auger electron spectroscopy is damaging to organic materials, and best used for inorganics.
[c]Static SIMS ≈ 10 Å, dynamic SIMS to 1 μm.

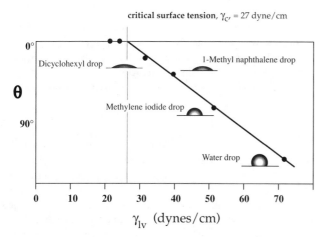

FIG. 4. The Zisman method permits a critical surface tension value, an approximation to the solid surface tension, to be measured. Drops of liquids of different surface tensions are placed on the solid, and the contact angles of the drops are measured. The plot of liquid surface tension versus angle is extrapolated to zero contact angle to give the critical surface tension value.

TABLE 2 Critical Surface Tension Values for Common Polymers[a]

Material	Critical surface tension (dynes/cm)
Polytetrafluoroethylene	19
Poly(dimethyl siloxane)	24
Poly(vinylidine fluoride)	25
Poly(vinyl fluoride)	28
Polyethylene	31
Polystyrene	33
Poly(hydroxyethyl methacrylate)	37
Poly(vinyl alcohol)	37
Poly(methyl methacrylate)	39
Poly(vinyl chloride)	39
Polycaproamide (nylon 6)	42
Poly(ethylene oxide)-diol	43
Poly(ethylene terephthalate)	43
Polyacrylonitrile	50

[a]Values from Table 5.5 in *Polymer Interface and Adhesion*, S. Wu, ed. Marcel Dekker, New York, Table 5.5, 1982.

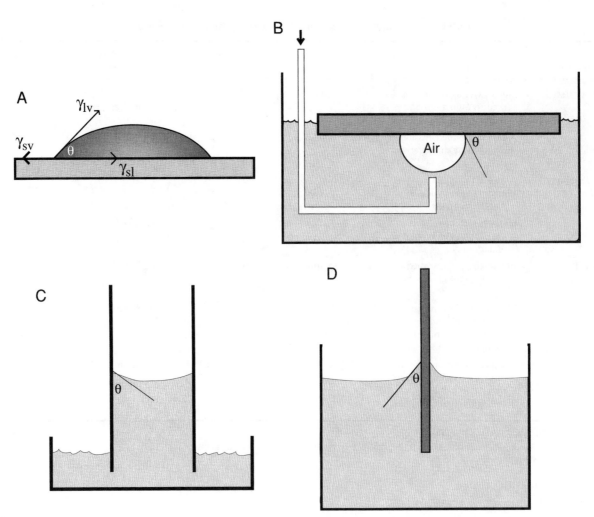

FIG. 5.   Four ways that the contact angle can be measured. (A) Sessile drop. (B) Captive air bubble method. (C) Capillary rise method. (D) Wilhelmy plate method.

TABLE 3   Concerns in Contact Angle Measurement

The measurement is operator dependent.

Surface roughness influences the results.

Surface heterogeneity influences the results.

The liquids used are easily contaminated (typically reducing their γ_{lv}).

The liquids used can reorient the surface structure.

The liquids used can absorb into the surface, leading to swelling.

The liquids used can dissolve the surface.

Few sample geometries can be used.

Information on surface structure must be inferred from the data obtained.

are used, the need for experienced operators, and the cost associated with the analysis. The vacuum compatibility limitation can be sidestepped by using an ESCA system with a cryogenic sample stage. At liquid nitrogen temperatures, samples with volatile components, or even wet, hydrated samples can be analyzed.

The use of ESCA is best illustrated with a brief example. A poly(methyl methacrylate) (PMMA) ophthalmologic device is to be examined. Taking care not to touch or damage the surface of interest, the device is inserted into the ESCA instrument introduction chamber. The introduction chamber is then pumped down to 10^{-6} torr pressure. A gate valve between the introduction chamber and the analytical chamber is opened and the specimen is moved into the analysis chamber. In the analysis chamber, at 10^{-9} torr pressure, the specimen is positioned (on contemporary instruments, using a microscope or TV camera) and the X-ray source is turned on. The ranges of electron energies to be observed are controlled (by computer) with the retardation lens on the spectrometer. First, a wide

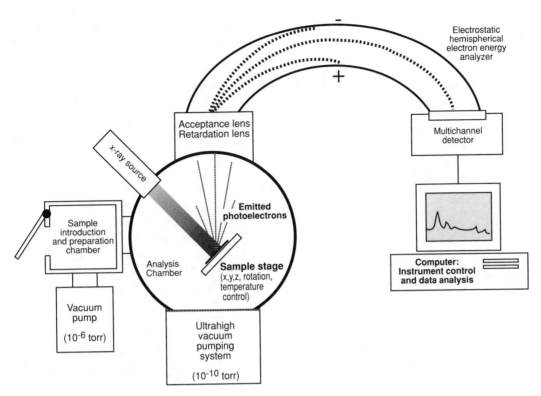

FIG. 6. A schematic diagram of a monochromatized ESCA instrument.

TABLE 4 Information Derived from an ESCA Experiment

In the outermost 100 Å of a surface, ESCA can provide:

Identification of all elements (except H and He) present at concentrations > 0.1 atomic %

Semiquantitative determination of the approximate elemental surface composition (±10%)

Information about the molecular environment (oxidation state, bonding atoms, etc.)

Information about aromatic or unsaturated structures from shake-up ($\pi^* \rightarrow \pi$) transitions

Identification of organic groups using derivatization reactions

Nondestructive elemental depth profiles 100 Å into the sample and surface heterogeneity assessment using angular-dependent ESCA studies and photoelectrons with differing escape depths

Destructive elemental depth profiles several thousand angstroms into the sample using argon etching (for inorganics)

Lateral variations in surface composition (spatial resolution 8–150 μm, depending upon the instrument)

"Fingerprinting" of materials using valence band spectra and identification of bonding orbitals

Studies on hydrated (frozen) surfaces

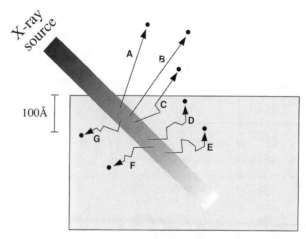

FIG. 7. ESCA is a surface-sensitive method. Although the X-ray beam can penetrate deeply into a specimen, electrons emitted deep in the specimen (D, E, F, G) will lose their energy in inelastic collisions and never emerge from the surface. Only those electrons emitted near the surface that lose no energy (A, B) will contribute to the ESCA signal used analytically. Electrons that lose some energy, but still have sufficient energy to emerge from the surface (C) contribute to the background signal.

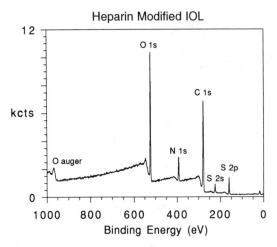

FIG. 8.   An ESCA wide scan of a surface-modified poly(methyl methacrylate) ophthalmologic device.

TABLE 5   Analytical Capabilities of SIMS

	Static SIMS	Dynamic SIMS
Identify hydrogen	✔	✔
Identify other elements (often must be inferred from the data)	✔	✔
Suggest molecular structures (inferred from the data)	✔	
Observe extremely high mass fragments (proteins, polymers)	✔	
Detection of extremely low concentrations	✔	✔
Depth profile to 1 μm into the sample		✔
Observe the outermost 1–2 atomic layers	✔	
High spatial resolution (features as small as approximately 500 Å)	✔	✔
Semiquantitative analysis (for limited sets of specimens)	✔	
Useful for polymers	✔	
Useful for inorganics (metals, ceramics, etc.)	✔	✔
Useful for powders, films, fibers, etc.	✔	✔

scan is made in which the energies of all emitted electrons are detected (Fig. 8). Then, narrow scans are made in which each of the elements detected in the wide scan is examined in higher resolution (Fig. 9).

From the wide scan, we learn that the specimen contains carbon, oxygen, nitrogen, and sulfur. The presence of sulfur and nitrogen is unexpected for PMMA. We can calculate elemental ratios from the wide scan. The sample surface contains 58.2% carbon, 27.7% oxygen, 9.5% nitrogen, and 4.5% sulfur. The narrow scan for the carbon region (C1s spectrum) suggests four species; hydrocarbon, carbons singly bonded to oxygen (the predominant species), carbons in amidelike environments, and carbons in acid or ester environments. This is

different from the spectrum expected for pure PMMA. An examination of the peak position in the narrow scan of the sulfur region (S2p spectrum) suggests sulfonate-type groups. The shape of the C1s spectrum, the position of the sulfur peak, and the presence of nitrogen all suggest that heparin was immobilized to the surface of the PMMA device. Since the stoichiometry of the lens surface does not match that for pure heparin, this suggests that we are seeing either some of the PMMA substrate through a >100 Å layer of heparin, or we are seeing some of the bonding links used to immobilize the heparin to the lens surface. Further ESCA analysis will permit the extraction of more detail about this surface-modified device, including an estimate of surface modification thickness, further confirmation that the coating is indeed heparin, and additional information about the nature of the immobilization chemistry.

Secondary Ion Mass Spectrometry

Secondary ion mass spectrometry (SIMS) is a recent addition to the armamentarium of tools that the surface analyst can bring to bear on a biomedical problem. SIMS produces a mass spectrum of the outermost 10 Å of a surface. Like ESCA, it requires complex instrumentation and an ultrahigh vacuum chamber for the analysis. However, it provides unique information that is complementary to ESCA and greatly aids in understanding surface composition. Some of the analytical capabilities of SIMS are summarized in Table 5. Review articles on SIMS are available (Ratner, 1985; Scheutzle et al., 1984;

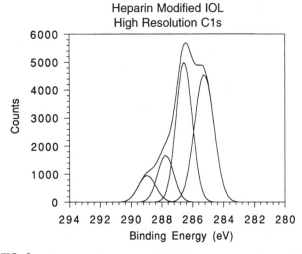

FIG. 9.   The carbon 1s narrow scan ESCA spectrum of a surface-modified poly(methyl methacrylate) ophthalmologic device. Narrow scan spectra can be generated for each element seen in low-energy resolution mode in Fig. 8.

Briggs, 1986; Davies and Lynn, 1990; Vickerman *et al.*, 1989; Benninghoven, 1983).

The SIMS method involves bombarding a surface with a beam of accelerated ions. The collision of these ions with the atoms and molecules in the surface zone can transfer enough energy to them so they sputter from the surface into the vacuum phase. The process is analogous to the way racked pool balls are ejected by the impact of the cue ball; the harder the cue ball hits the rack of balls, the more balls are emitted from the rack. In SIMS, the "cue balls" are accelerated ions (xenon, argon, cesium, and gallium ions are commonly used). The particles ejected from the surface are positive and negative ions (secondary ions), radicals, excited states, and neutrals. Only the secondary ions are measured in SIMS. In ESCA, the energy of emitted particles (electrons) is measured. SIMS measures the mass of emitted ions (more rigorously, the ratio of mass to charge, m/z) using a quadrupole mass analyzer or a time-of-flight (TOF) mass analyzer.

There are two kinds of SIMS, depending on the ion dose used: dynamic and static. Dynamic SIMS uses high ion doses. The primary ion beam sputters so much material from the surface that the surface erodes at an appreciable rate. We can capitalize on this to do a depth profile into a specimen. The intensity of the m/z peak of a species of interest (e.g., sodium ion, $m/z = 23$) might be followed as a function of time. If the ion beam is well controlled and the sputtering rate is constant, the amount of sodium detected at any time will be directly related to the erosion depth of the ion beam into the specimen. A depth profile can be constructed with information from the outermost atoms to a micron or more into the specimen. However, owing to the vigorous, damaging nature of the high-flux ion beam, only atomic fragments can be detected. Also, as the beam erodes deeper into the specimen, more artifacts are introduced in the data by knock-in and scrambling of atoms.

Static SIMS, by comparison, induces minimal surface destruction. The ion dose is adjusted so that during the period of the analysis less than one monolayer of surface atoms is sputtered. Since there are typically 10^{13}–10^{15} atoms in 1 cm^2 of surface, a total ion dose of less than 10^{13} ions/cm^2 during the analysis period is appropriate. Under these conditions, extensive degradation and rearrangement of the chemistry at the surface does not take place, and large, relatively intact molecular fragments can be ejected into the vacuum for measurement. Examples of large molecular fragments are shown in Fig. 10. This figure also introduces some of the ideas behind SIMS spectral interpretation. A more complete introduction to the concepts behind static SIMS spectral interpretation can be found in any of the standard texts on mass spectrometry.

Scanning Electron Microscopy

Scanning electron microscopy (SEM) images of surfaces have great resolution and depth of field, with a three-dimensional quality that offers a visual perspective familiar to most users. SEM images are widely used and much has been written about the technique. The comments here are primarily oriented toward SEM as a surface analysis tool.

SEM functions by focusing and rastering a relatively high-energy electron beam (typically, 5–100 keV) on a specimen. Low-energy secondary electrons are emitted from each spot where the focused electron beam impacts. The detectable

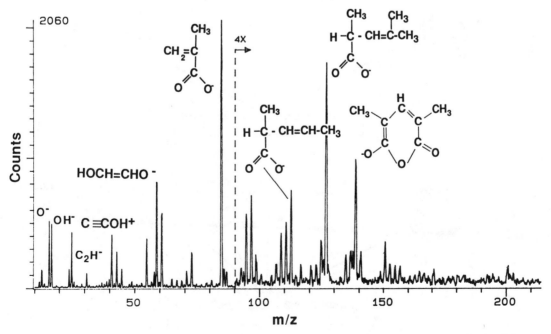

FIG. 10. A negative ion static SIMS spectrum of poly(2-hydroxyethyl methacrylate) with the primary peaks identified.

intensity of the secondary electron emission is a function of the atomic composition of the sample and the geometry of the features under observation. SEM images surfaces by spatially reconstructing on a phosphor screen the intensity of the secondary electron emission. Because of the shallow penetration depth of the low-energy secondary electrons produced by the primary electron beam, only secondary electrons generated near the surface can escape from the bulk and be detected (this is analogous to the surface sensitivity described in Fig. 7). Consequently, SEM is a surface analysis method.

Nonconductive materials observed in the SEM are typically coated with a thin, electrically grounded layer of metal to minimize negative charge accumulation from the electron beam. However, this metal layer is always so thick (>300 Å) that the electrons emitted from the sample beneath cannot penetrate. Therefore, in SEM analysis of nonconductors, the surface of the metal coating is, in effect, being monitored. If

the metal coat is truly conformal, a good representation of the surface geometry will be conveyed. However, all effects of the specimen surface chemistry on secondary electron emission are lost. Also, at very high magnifications, the texture of the metal coat and not the surface may be under observation.

SEM, in spite of these limitations in providing true surface information, is an important corroborative method to use in conjunction with other surface analysis methods. Surface roughness and texture can have a profound influence on data from ESCA, SIMS, and contact angle determinations. Therefore, SEM provides important information in the interpretation of data from these methods.

The recent development of low-voltage SEM offers a technique to truly study the surface chemistry and geometry of nonconductors. With the electron accelerating voltage lowered to approximately 1 keV, charge accumulation is not so critical and metallization is not required. Low-voltage SEM has been used to study platelets and phase separation in polymers. Also,

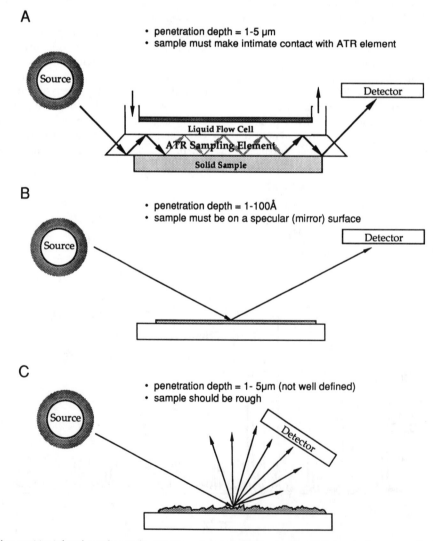

FIG. 11.   Three surface-sensitive infrared sampling modes. (A) Attenuated total reflectance mode. (B) External reflectance mode. (C) Diffuse reflectance mode.

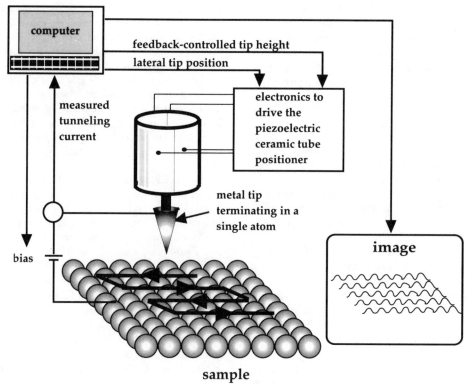

FIG. 12. A schematic diagram illustrating the principle of the scanning tunneling microscope.

the environmental SEM (ESEM) permits wet, uncoated specimens to be studied.

The primary electron beam also results in the emission of X-rays. These X-rays are used to identify elements with the technique called energy-dispersive X-ray analysis (EDXA). However, the high-energy primary electron beam penetrates deeply into a specimen (a micron or more). The X-rays produced from the interaction of these electrons with atoms deep in the bulk of the specimen can penetrate through the material and be detected. Therefore, EDXA is not a surface analysis method—it measures bulk atomic concentrations.

Infrared Spectroscopy

Infrared spectroscopy (IRS) provides information on the vibrations of atomic and molecular units. It is a standard analytical method that can reveal information on specific chemistries and the orientation of structures. By using a Fourier transform infrared (FTIR) spectrometer, great improvements in the signal-to-noise ratio (S/N) and spectral accuracy can be realized. However, even with this high S/N, the small absorption signal associated with the extremely small mass of material in a surface region can challenge the sensitivity of the spectrometer. Also, the problem of separating the massive bulk absorption signal from the surface signal must be dealt with.

Surface FTIR studies couple the infrared radiation to the sample surface to increase the intensity of the surface signal and reduce the bulk signal (Allara, 1982; Layden and Murthy, 1987; Nguyen, 1985). Some of these sampling modes, and the limitations associated with them, are illustrated in Fig. 11.

The attenuated total reflectance (ATR) mode of sampling has been used most often in biomaterials studies. The penetration depth into the sample is 1–5 μm. Therefore, ATR is not truly surface sensitive, but observes a broad region near the surface. However, it does offer the wealth of rich structural information common to infrared spectra. With extremely high S/N FTIR instruments, ATR studies of proteins and polymers under water have been performed. In these experiments, the water signal (which is typically 99% or more of the total signal) is subtracted from the spectrum to leave only the surface material (e.g., adsorbed protein) under observation.

Newer Methods

New methods, although still evolving, are showing the potential for making important contributions to biomaterials studies. Foremost among them are scanning tunneling microscopy (STM) and atomic force microscopy (AFM). General review articles (Binnig and Rohrer, 1986; Avouris, 1990; Albrecht et al., 1988) and articles oriented toward biological studies with these methods have been written (Hansma et al., 1988; Miles et al., 1990; Rugar and Hansma, 1990).

The STM uses quantum tunneling to generate an atomscale electron density image of a surface. A metal scanning tip terminating in a single atom is brought within 5–10 Å of an electrically conducting surface. At these distances, the electron cloud of the atom at the tip will significantly overlap the electron cloud of an atom on the surface. If a potential is applied between the tip and the surface, an electron tunneling current will be established whose magnitude, J, follows the proportionality:

$$J \alpha\, e^{(-Ak_0S)},$$

where A is a constant, k_0 is an average inverse decay length (related to the electron affinity of the metals), and S is the separation distance in angstrom units. For most metals, a 1-Å change in the distance of the tip to the surface results in an order of magnitude change in tunneling current. Even though this current is small, it can be measured with good accuracy.

To image a surface, this quantum tunneling current is used in one of two ways. In constant current mode, a piezoelectric driver moves a tip over a surface. When the tip approaches an atom protruding above the plane of the surface, the current rapidly increases, and a feedback circuit moves the tip up to keep the current constant. Then, a plot is made of the tip height required to maintain constant current versus distance along the plane. In constant height mode, the tip is moved over the surface and the change in current with distance traveled along the plane of the surface is directly recorded. A schematic diagram of an STM is presented in Fig. 12. The two scanning modes are illustrated in Fig. 13.

The AFM uses a similar piezo drive mechanism. However, instead of recording tunneling current, the deflection of a lever arm due to van der Waals (electron cloud) repulsion between an atom at the tip and an atom on the surface is measured. This measurement can be made by reflecting a laser beam off a mirror on the lever arm. A one-atom length deflection of the lever arm can easily be magnified by monitoring the position of the laser reflection on a spatially resolved photosensitive detector.

The STM measures electrical current and therefore is well suited for conductive and semiconductive surfaces. However, biomolecules (even proteins) on conductive substrates appear amenable to imaging. It must be remembered that STM does not "see" atoms, but monitors electron density. The conductive and imaging mechanism for proteins is not well understood. Still, Fig. 14 suggests that important images of biomolecules on surfaces can be obtained. Since the AFM measures force, it can be used with both conductive and nonconductive specimens. Since force must be applied to bend a lever, AFM is subject to artifacts caused by damage to fragile structures on the surface. Both methods can function well for specimens under water, in air, or under vacuum. For exploring biomolecules or mobile organic surfaces, the "pushing around" of

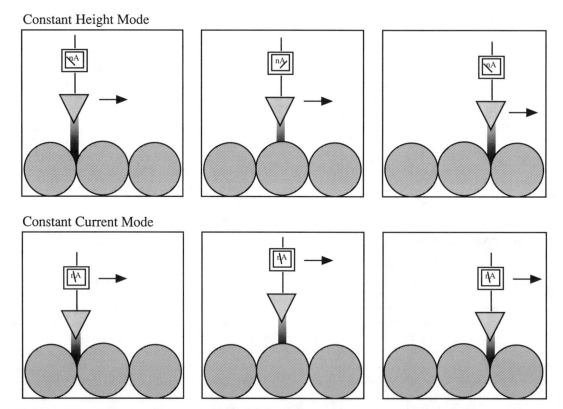

FIG. 13.   Scanning tunneling microscopy can be performed in two modes. In constant-height mode, the tip is scanned a constant distance from the surface (typically 5–10 Å) and the change in tunneling current is recorded. In constant current mode, the tip height is adjusted so that the tunneling current is always constant, and the tip distance from the surface is recorded as a function of distance traveled in the plane of the surface.

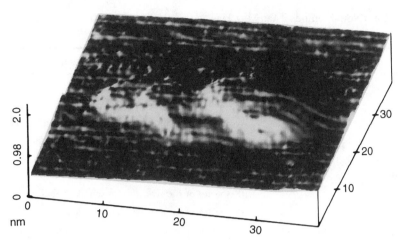

FIG. 14. A scanning tunneling micrograph image of a fibrinogen molecule on a gold surface, under buffer (image by Dr. K. Lewis).

structures by the tip is a significant concern. With both methods it is difficult to achieve good-quality, reproducible images of organic surfaces. However, some of the successes to date are exciting enough that the future of these methods in biomedical research is ensured.

There are many other surface characterization methods that may become important in future years. Some of these are listed in Table 6.

Studies with Surface Methods

Hundreds of studies have appeared in the literature in which surface methods have been used to enhance the understanding of biomaterial systems. Four studies are briefly described here. A symposium proceedings volume on this subject is also available (Ratner, 1988)

Platelet Consumption

Using a baboon arteriovenous shunt model of platelet interaction with surfaces, a first-order rate constant of reaction of platelets with a series of polyurethanes was measured. This rate constant, the platelet consumption by the material, correlated in an inverse linear fashion with the fraction of hydrocarbon-type groups in the ESCA C1s spectra of the polyurethanes (Hanson *et al.*, 1982). Thus, surface analysis revealed a chemical parameter about the surface that could be used to predict long-term biological reactivity of materials in a complex *ex vivo* environment.

Contact-Angle Correlations

The adhesion of a number of cell types, including bacteria, granulocytes, and erythrocytes, has been shown, under certain conditions, to correlate with solid-vapor surface tension as determined from contact-angle measurements. In addition, immunoglobulin G adsorption is related to γ_{sv} (Neumann *et al.*, 1983).

Contamination of Intraocular Lenses

Commercial intraocular lenses were examined by ESCA. The presence of sulfur, sodium, and excess hydrocarbon at their surfaces suggested contamination by sodium dodecyl sulfate (SDS) during the manufacture of the lenses (Ratner, 1983). A cleaning protocol was developed using ESCA to monitor results that produced a lens surface of clean PMMA.

Titanium

The discoloration sometimes observed on titanium implants after autoclaving was examined by ESCA and SIMS (Lausmaa *et al.*, 1985). The discoloration was found to be related to accelerated oxide growth, with oxide thicknesses to 650 Å. The oxide contained considerable fluorine, along with alkali metals and silicon. The source of the oxide was the cloth used to wrap the implant storage box during autoclaving. Since fluorine strongly affects oxide growth, and since the oxide layer has been associated with the biocompatibility of titanium implants, the authors advise avoiding fluorinated materials during sterilization of samples.

CONCLUSIONS

Contemporary methods of surface analysis can provide invaluable information about biomaterials and medical devices. The information obtained can be used to monitor contamination, ensure surface reproducibility, and explore fundamental aspects of the interaction of biological systems with living systems. Considering that biomedical experiments are typically expensive to perform, the costs for surface analysis are small for the assurance that the identical surface is being examined from experiment to experiment. Also, with routine surface analysis on medical devices, the physician can be confident that differences in the performance of a device are not related to changes in the surface structure.

TABLE 6 New Methods That May Have Applicability for Surface Characterization
of Biomaterials

Method	Information obtained
Second-harmonic generation (SHG)	Detect submolayer amounts of adsorbate at any light-accessible interface (air–liquid, solid–liquid, solid–gas)
Sum-frequency generation	Use the SHG method for spectroscopy using a tunable laser
Surface-enhanced raman spectroscopy (SERS)	High-sensitivity Raman at rough metal interfaces
Ion scattering spectroscopy (ISS)	Elastically reflected ions probe only the outermost atomic layer
Laser desorption mass spectrometry	Mass spectra of adsorbates at surfaces
IR diffuse reflectance	IR spectra of surfaces with no sample preparation
IR photoacoustic spectroscopy	IR spectra of surfaces with no sample preparation
High-resolution electron energy loss spectroscopy	Vibrational spectroscopy of a highly surface-localized region, under ultrahigh vacuum
X-Ray reflection	Structural information on order at surfaces and interfaces
Neutron reflection	Structural information on interfaces
Extended X-ray absorption fine structure (EXAFS)	Atomic-level chemical and nearest-neighbor (morphological) information

Acknowledgment

Support was received from NIH grant RR01296 during the preparation of this chapter and for some of the studies described herein.

Bibliography

Adamson, A. W. (1990). *Physical Chemistry of Surfaces*. Wiley-Interscience, New York.

Albrecht, T. R., Dovek, M. M., Lang, C. A., Grutter, P., Quate, C. F., Kuan, S. W. J., Frank, C. W., and Pease, R. F. W. (1988). Imaging and modification of polymers by scanning tunneling and atomic force microscopy. *J. Appl. Phys.* **64**: 1178–1184.

Allara, D. L. (1982). Analysis of surfaces and thin films by IR, Raman, and optical spectroscopy *ACS Symp. Ser.* **199**: 33–47.

Andrade, J. D. (1985). *Surface and Interfacial Aspects of Biomedical Polymers*, Vol. 1: *Surface Chemistry and Physics*, Plenum Publ., New York.

Avouris, P. (1990). Atom-resolved surface chemistry using the scanning tunneling microscope. *J. Phys. Chem.* **94**: 2246–2256.

Benninghoven, A. (1983). Secondary ion mass spectrometry of organic compounds (review) in *Springer Series of Chemical Physics: Ion Formation from Organic Solids*, A. Benninghoven, ed., Springer-Verlag, Berlin, Vol. 25, pp. 64–89.

Binnig, G. and Rohrer, H. (1986). Scanning tunneling microscopy. *IBM J. Res. Develop.* **30**: 355–369.

Briggs, D. (1986). SIMS for the study of polymer surfaces: a review *Surf. Interface Anal.* **9**: 391–404.

Briggs, D. and Seah, M. P. (1983). *Practical Surface Analysis*. Wiley, Chichester, England.

Dilks, A. (1981). X-ray photoelectron spectroscopy for the investigation of polymeric materials in *Electron Spectroscopy: Theory, Techniques, and Applications*, A. D. Baker and C. R. Brundle, eds., Academic Press, London, Vol. 4, pp. 277–359.

Davies, M. C. and Lynn, R. A. P. (1990). Static secondary ion mass spectrometry of polymeric biomaterials. *CRC Crit. Rev. Biocompat.* **5**: 297–341.

Feldman, L. C. and Mayer, J. W. (1986). *Fundamentals of Surface and Thin Film Analysis*. North-Holland, New York.

Hansma, P. K., Elings, V. B., Marti, O., and Bracker, C. E. (1988). Scanning tunneling microscopy and atomic force microscopy: application to biology and technology. *Science* **242**: 209–216.

Hanson, S. R., Harker, L. A., Ratner, B. D., and Hoffman, A. S. (1982). Evaluation of artificial surfaces using baboon arteriovenous shunt model in *Biomaterials 1980, Advances in Biomaterials*, G. D. Winter, D. F. Gibbons, and H. Plenk, eds., Wiley, Chichester, Vol. 3, pp. 519–530.

Lausmaa, J., Kasemo, B., and Hansson S. (1985). Accelerated oxide growth on titanium implants during autoclaving caused by fluorine contamination. *Biomaterials* **6**: 23–27.

Leyden, D. E. and Murthy, R. S. S. (1987). Surface-selective sampling techniques in Fourier transform infrared spectroscopy. *Spectroscopy* **2**: 28–36.

Miles, M. J., McMaster, T., Carr, H. J., Tatham, A. S., Shewry, P. R., Field, J. M., Belton, P. S., Jeenes, D., Hanley, B., Whittam, M., Cairns, P., Morris, V. J., and Lambert, N. (1990). Scanning tunneling microscopy of biomolecules. *J. Vac. Sci. Technol. A* **8**: 698–702.

Neumann, A. W. and Good, R. J. (1979). Techniques of measuring contact angles. in *Surface and Colloid Science—Experimental Methods*, R. J. Good and R. R. Stromberg, eds. Plenum Publ., New York, Vol. 11, pp. 31–61.

Neumann, A. W., Absolom, D. R., Francis, D. W., Omenyi, S. N., Spelt, J. K., Policova, Z., Thomson, C., Zingg, W., and Van Oss, C. J. (1983). Measurement of surface tensions of blood cells and proteins. *Ann. New York Acad. Sci.* **416**: 276–298.

Nguyen, T. (1985). Applications of Fourier transform infrared spectroscopy in surface and interface studies. *Prog. Org. Coat.* **13**: 1–34.

Ratner, B. D. (1983). Analysis of surface contaminants on intraocular lenses. *Arch. Ophthal.* **101**: 1434–1438.

Ratner, B. D. (1988). *Surface Characterization of Biomaterials*, Elsevier, Amsterdam.

Ratner, B. D. and McElroy, B. J. (1986). Electron spectroscopy for chemical analysis: applications in the biomedical sciences in *Spectroscopy in the Biomedical Sciences,* R. M. Gendreau, ed., CRC Press, Boca Raton, FL, pp. 107–140.

Rugar, D. and Hansma, P. (1990). Atomic force microscopy. *Physics Today* **43**: 23–30.

Scheutzle, D., Riley, T. L., deVries, J. E., and Prater, T. J. (1984). Applications of high-performance mass spectrometry to the surface analysis of materials. *Mass Spectror.* 3: 527–585.

Somorjai, G. A. (1981). *Chemistry in Two Dimensions—Surfaces,* Cornell Univ. Press, Ithaca, NY.

Vickerman, J. C., Brown, A., and Reed, N. M. (1989). *Secondary Ion Mass Spectrometry, Principles and Applications,* Clarendon Press, Oxford.

Zisman, W. A. (1964). Relation of the equilibrium contact angle to liquid and solid constitution. in *Contact Angle, Wettability and Adhesion,* ACS Advances in Chemistry Series, F. M. Fowkes, ed., American Chemical Society, Washington, DC, Vol. 43, pp. 1–51.

2

Classes of Materials Used in Medicine

Harold Alexander, John B. Brunski, Stuart L. Cooper, Larry L. Hench,
Robert W. Hergenrother, Allan S. Hoffman, Joachim Kohn, Robert Langer,
Nikolaos A. Peppas, Buddy D. Ratner, Shalaby W. Shalaby, Susan A. Visser,
and Ioannis V. Yannas

2.1 INTRODUCTION

Allan S. Hoffman

The wide diversity and sophistication of materials currently used in medicine and biotechnology is testimony to the significant technological advances which have occurred over the past 25 years. As little as 25 years ago, common, commercial polymers and metals were being used in implants and medical devices. There was relatively little stimulus or motivation for development of new materials. However, a relatively small group of "biomaterials scientists" with a strong interest in medicine, in collaboration with a like-minded group of physicians, evolved out of traditional fields such as chemistry, chemical engineering, metallurgy, materials science and engineering, physics and medicine. They recognized not only the need for new and improved materials, implants and devices, but also the challenges and opportunities involved. With the early support of the National Institutes of Health and a few enlightened companies, a wide range of new and exciting biomaterials began to emerge, and over the past 15–20 years, the field, its diversity, and the number of professionals working in the field have grown enormously. Materials and systems for biological use have been synthesized and fabricated in a wide variety of shapes and forms, including composites and coated systems. Some of the new materials and technologies have been developed especially for biological uses, while others have been borrowed from such unexpected areas as space technology. This section covers the background and most recent developments in the science and engineering of biomaterials.

2.2 METALS

John B. Brunski

Metallic implant materials have a significant economic and clinical impact on the biomaterials field. The total U.S. market for implants and instrumentation in orthopedics was about $2.098 million in 1991, according to recent estimates. This includes $1.379 million for joint prostheses made of metallic materials, plus a variety of trauma products ($340 million), instrumentation devices ($266 million), bone cement accessories ($66 million), and bone replacement materials ($29 million). Projections for 2002 indicate that the total global biomaterials market will be $6 billion. The clinical numbers are equally impressive. Of the 3.6 million orthopedic operations per year in the U.S., four of the ten most frequent involve metallic implants: open reduction of a fracture and internal fixation (first on the list), placement or removal of an internal fixation device without reduction of a fracture (sixth), arthroplasty of the knee or ankle (seventh), and total hip replacement or arthroplasty of the hip (eighth).

Besides orthopedics, there are other markets for metallic implants and devices, including oral and maxillofacial surgery (e.g., dental implants, craniofacial plates and screws) and cardiovascular surgery (e.g., parts of artificial hearts, pacemakers, balloon catheters, valve replacements, aneurysm clips). Interestingly, in 1988, about 11 million Americans (about 4.6% of the civilian population) had at least one implant (Moss *et al.*, 1990).

In view of this wide utilization of metallic implants, the objective of this chapter is to describe the composition, structure, and properties of current metallic implant alloys. A major emphasis is on the metallurgical principles underlying fabrication and structure-property relationships.

STEPS IN THE FABRICATION OF IMPLANTS

Understanding the structure and properties of metallic implant materials requires an appreciation of the metallurgical significance of the material's processing history (Fig. 1). Since each metallic device differs in the details of its manufacture, "generic" processing steps are presented in Fig. 1.

Biomaterials Science

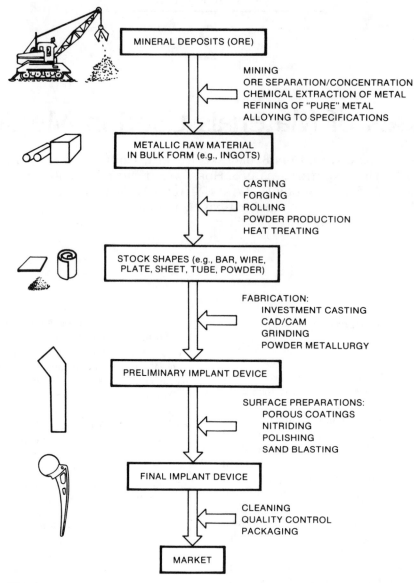

FIG. 1.   General processing history of a typical metallic implant device, in this case a hip implant.

Metal-Containing Ore to Raw Metal Product

With the exception of the noble metals (which do not represent a major fraction of implant metals), metals exist in the Earth's crust in mineral form and are chemically combined with other elements, as in the case of metal oxides. These mineral deposits (ore) must be located and mined, and then separated and enriched to provide ore suitable for further processing into pure metal.

With titanium, for example, certain mines in the southeastern United States yield sands containing primarily common quartz as well as mineral deposits of zircon, titanium, iron, and rare earths. The sandy mixture is concentrated by using water flow and gravity to separate out the metal-containing sands into titanium-containing compounds such as rutile

(TiO_2) and ilmenite ($FeTiO_3$). To obtain rutile, which is particularly good for making metallic titanium, further processing typically involves electrostatic separations. Then, to extract titanium metal from the rutile, one method involves treating the ore with chlorine to make titanium tetrachloride liquid, which in turn is treated with magnesium or sodium to produce chlorides of the latter metals and bulk titanium "sponge" according to the Kroll process. At this stage, the titanium sponge is not of controlled purity. Depending on the purity grade desired in the final titanium product, it is necessary to refine it further by using vacuum furnaces, remelting, and additional steps. All of this can be critical in producing titanium with the appropriate properties. The four grades of "commercially pure" (CP) titanium differ in oxygen content by only tenths of a percent, which in turn leads to significant differences in

TABLE 1 Chemical Compositions of Stainless Steels Used for Implants

Material	ASTM designation	Common/trade names	Composition (wt. %)	Notes
Stainless steel	F55 (bar, wire) F56 (sheet, strip) F138 (bar, wire) F139 (sheet, strip)	AISI 316LVM 316L 316L 316L	60–65 Fe 17.00–19.00 12.00–14.00 2.00–3.00 max 2.0 Mn max 0.5 Cu max 0.03 C max 0.1 N max 0.025 P max 0.75 Si max 0.01 S	F55, F56 specify 0.03 max for P, S. F138, F139 specify 0.025 max for P and 0.010 max for S. LVM = low vacuum melt.
Stainless steel	F745	Cast stainless steel cast 316L	60–69 Fe 17.00–19.00 11.00–14.00 2.00–3.00 max 0.06 C max 2.0 Mn max 0.045 P max 1.00 Si max 0.030 S	

mechanical properties. The resulting raw metal product eventually emerges in some type of bulk form such as ingots for supply to manufacturers.

In the case of multicomponent metallic implant alloys, the raw metal product will have to be processed further. Processing steps include remelting, the addition of alloying elements, and solidification to produce an alloy that meets certain chemical specifications. For example, to make ASTM (American Society for Testing and Materials) F138 316L stainless steel, iron is alloyed with specific amounts of carbon, silicon, nickel, and chromium. To make ASTM F75 or F90 alloy, cobalt is alloyed with specific amounts of chromium, molybdenum, carbon, nickel, and other elements. Table 1 lists the chemical compositions of some metallic alloys for surgical implants.

Raw Metal Product to Stock Metal Shapes

A manufacturer further processes the bulk raw metal product (metal or alloy) into "stock" shapes, such as bars, wire, sheet, rods, plates, tubes, and powders. These shapes are then sold to specialty companies (e.g., implant manufacturers) who need stock metal that is closer to the actual final form of the implant.

Bulk forms are turned into stock shapes by a variety of processes, including remelting and continuous casting, hot rolling, forging, and cold drawing through dies. Depending on the metal, there may also be heat-treating steps (heating and cooling cycles) designed to facilitate further working or shaping of the stock, relieve the effects of prior plastic deformation (e.g., annealing), or produce a specific microstructure and properties in the stock material. Because of the chemical reactivity of

some metals at elevated temperatures, high-temperature processes may require vacuum conditions or inert atmospheres to prevent unwanted uptake of oxygen by the metal. For instance, in the production of fine powders of ASTM F75 Co–Cr–Mo alloy, molten metal is ejected through a small nozzle to produce a fine spray of atomized droplets that solidify while cooling in an inert argon atmosphere.

For metallic implant materials in general, stock shapes are chemically and metallurgically tested to ensure that the chemical composition and microstructure of the metal meet industry standards for surgical implants (ASTM Standards), as discussed later in this chapter.

Stock Metal Shapes to Preliminary and Final Metal Devices

Typically, an implant manufacturer will buy stock material and then fabricate it into preliminary and final forms. Specific steps depend on a number of factors, including the final geometry of the implant, the forming and machining properties of the metal, the costs of alternative fabrication methods, and the company doing the fabrication.

Fabrication methods include investment casting (the "lost wax" process), conventional and computer-based machining (CAD/CAM), forging, powder metallurgical processes (hot isostatic pressing, or HIP), and a range of grinding and polishing steps. A variety of fabrication methods are required because not all implant alloys can be feasibly or economically made in the same way. For instance, cobalt-based alloys are extremely difficult to machine into the complicated shapes of some implants and are therefore frequently shaped into implant forms

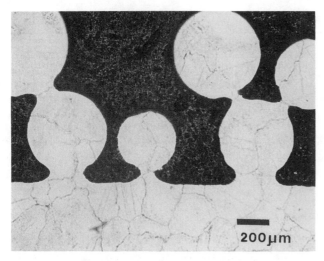

FIG. 2. Low-power view of the interface between a porous coating and solid substrate in the ASTM F75 Co–Cr–Mo alloy system. Note the nature of the necks joining the beads to one another and to the substrate. Metallographic cross section cut perpendicular to the interface; lightly etched to show the microstructure. (Photo courtesy of Smith & Nephew Richards, Inc. Memphis, TN.)

by investment casting or powder metallurgy. On the other hand, titanium is relatively difficult to cast and therefore is frequently machined even though it is not generally considered to be an easily machinable metal.

Another aspect of fabrication, which is actually an end-product surface treatment, involves the application of macro- or microporous coatings on implants. This has become popular in recent years as a means to facilitate fixation of implants in bone. The porous coatings can take various forms and require different fabrication technologies. In turn, this part of the processing history will contribute to metallurgical properties of the final implant device. In the case of alloy beads or "fiber metal" coatings, the manufacturer will apply the coating material over specific regions of the implant surface (e.g., on the proximal portion of the femoral stem), and then attach the coating to the substrate by a process such as sintering. Generally, sintering involves heating the construct to about one-half or more of the alloy's melting temperature to enable diffusive mechanisms to form necks that join the beads to one another and to the implant's surface (Fig. 2).

An alternative surface treatment to sintering is plasma or flame spraying a metal onto an implant's surface. A hot, high-velocity gas plasma is charged with a metallic powder and directed at appropriate regions of an implant surface. The powder particles fully or partially melt and then fall onto the substrate surface, where they solidify rapidly to form a rough coating (Fig. 3).

Other surface treatments are also available, including ion implantation (to produce better surface properties) and nitriding. In nitriding, a high-energy beam of nitrogen ions is directed at the implant under vacuum. Nitrogen atoms penetrate the surface and come to rest at sites in the substrate. Depending on the alloy, this process can produce enhanced properties. These treatments are commonly used to increase surface hardness and wear properties.

Finally, metallic implant devices usually undergo a set of finishing steps. These vary with the metal and manufacturer, but typically include chemical cleaning and passivation (i.e., rendering the metal inactive) in appropriate acid, or electrolytically controlled treatments to remove machining chips or impurities that may have become embedded in the implant's surface. As a rule, these steps are conducted according to good manufacturing practice (GMP) and ASTM specifications for cleaning and finishing implants. In addition, these steps can be extremely important to the overall biological performance of the implant.

MICROSTRUCTURES AND PROPERTIES OF IMPLANT METALS

In order to understand the properties of each alloy system in terms of microstructure and processing history, it is important to know (1) the chemical and crystallographic identities of the phases present in the microstructure; (2) their relative amounts, distribution, and orientation; and (3) their effects on properties. In this chapter, the mechanical properties of metals used in implant devices will be emphasized. Although other properties, such as surface properties, must also be considered, these are covered in Chapter 1.3. The following discussion of implant alloys is divided into the stainless steels, cobalt-based alloys, and titanium-based alloys.

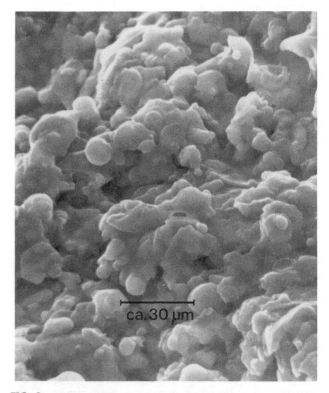

FIG. 3. Scanning electron micrograph of a titanium plasma spray coating on an oral implant. (Photo courtesy of A. Schroeder, E. Van der Zypen, H. Stich, and F. Sutter, *Int. J. Oral Maxillofacial Surgery* 9:15, 1981.)

Stainless Steels

Composition

While several types of stainless steels are available for implant use (Table 1), in practice the most common is 316L (ASTM F138, F139), grade 2. This steel has less than 0.030% (wt. %) carbon in order to reduce the possibility of *in vivo* corrosion. The "L" in the designation 316L denotes its low carbon content. The 316L alloy is predominantly iron (60–65%) alloyed with major amounts of chromium (17–19%) and nickel (12–14%), plus minor amounts of nitrogen, manganese, molybdenum, phosphorous, silicon, and sulfur.

The rationale for the alloying additions involves the metal's surface and bulk microstructure. The key function of chromium is to permit the development of a corrosion-resistant steel by forming a strongly adherent surface oxide (Cr_2O_3). However, the downside is that chromium tends to stabilize the ferritic (BCC, body-centered cubic) phase, which is weaker than the austenitic (FCC, face-centered cubic) phase. Molybdenum and silicon are also ferrite stabilizers. To counter this tendency to form ferrite, nickel is added to stabilize the austenitic phase.

The most important reason for the low carbon content relates to corrosion. If the carbon content of the steel significantly exceeds 0.03%, there is increased danger of formation of carbides such as $Cr_{23}C_6$. These tend to precipitate at grain boundaries when the carbon concentration and thermal history have been favorable to the kinetics of carbide growth. In turn, this carbide precipitation depletes the adjacent grain boundary regions of chromium, which has the effect of diminishing formation of the protective chromium-based oxide Cr_2O_3. Steels in which such carbides have formed are called "sensitized" and are prone to fail through corrosion-assisted fractures that originate at the sensitized (weakened) grain boundaries.

Microstructure and Mechanical Properties

Under ASTM specifications, the desirable form of 316L is single-phase austenite (FCC). There should be no free ferritic (BCC) or carbide phases in the microstructure. Also, the steel should be free of inclusions such as sulfide stringers. The latter can arise primarily from unclean steel-making practices and predispose the steel to pitting-type corrosion at the metal–inclusion interfaces.

The recommended grain size for 316L is ASTM #6 or finer. The ASTM grain size number n is defined by the formula:

$$N = 2^{n-1}, \qquad (1)$$

where N is the number of grains counted in 1 in.^{2} at $100\times$ magnification (0.0645 mm^2 actual area). $n = 6$ means a grain size of about 100 μm or less. Furthermore, the grain size should be relatively uniform throughout (Fig. 4a). The emphasis on a fine grain size is explained by a Hall–Petch-type[1] relationship between mechanical yield stress and grain diameter:

$$t_y = t_i + kd^{-m}, \qquad (2)$$

where t_y and t_i are the yield and friction stress, respectively; d is the grain diameter; k is a constant associated with propagation of deformation across grain boundaries; and m is approximately 0.5. From this equation it follows that higher yield stresses may be achieved by a metal with a smaller grain diameter d, all other things being equal. A key determinant of grain size is manufacturing history, including details on solidification conditions, cold-work, annealing cycles, and recrystallization.

Another notable microstructural feature of 316L is plastic deformation within grains (Fig. 4b). The metal is ordinarily used in a 30% cold-worked state because cold-worked metal has a markedly increased yield, ultimate tensile, and fatigue strength relative to the annealed state (Table 2). The trade-off is decreased ductility, but ordinarily this is not a major concern in implant products.

In specific orthopedic devices such as bone screws made of 316L, texture may also be apparent in the microstructure. Texture means a preferred orientation of deformed grains. For example, stainless steel bone screws show elongated grains in metallographic sections taken parallel to the long axis of the screws (Fig. 5). This finding is consistent with cold drawing or a similar cold-working operation in the manufacture of the bar stock from which screws are usually machined. In metallographic sections taken perpendicular to the screw's long axis, the grains appear more equiaxial. A summary of representative mechanical properties of 316L stainless is provided in Table 2.

Cobalt-Based Alloys

Composition

Cobalt-based alloys include Haynes-Stellite 21 and 25 (ASTM F75 and F90, respectively), forged Co–Cr–Mo alloy (ASTM F799), nd multiphase (MP) alloy MP35N (ASTM F562). The F75 and F799 alloys are virtually identical in composition (Table 3), each being about 58–69% Co and 26–30% Cr. The key difference is their processing history, which is discussed later. The other two alloys, F90 and F562, have slightly less Co and Cr, but more Ni in the case of F562, and more tungsten in the case of F90.

Microstructures and Properties

ASTM F75 The main attribute of this alloy is corrosion resistance in chloride environments, which is related to its bulk composition and the surface oxide (nominally Cr_2O_3). This alloy has a long history in the aerospace and biomedical implant industries.

When F75 is cast into shape by investment casting, the alloy is melted at 1350–1450°C and then poured or driven into ceramic molds of the desired shape (e.g., femoral stems for artificial hips, oral implants, dental partial bridgework). The molds are made by fabricating a wax pattern to near-final dimensions and then coating (or investing) the pattern with a special ceramic. A ceramic mold remains after the wax is burned. Then molten metal is poured into the mold. Once the metal has solidified into the shape of the mold, the ceramic

---

[1]Hall and Petch were the researchers who first observed this relationship in experiments on metallic systems. For more information, see E. O. Hall (1951). *Proc. Phys. Soc.* **64B**: 747; N. J. Petch (1953). *J. Iron Steel Inst.* **173**:25.

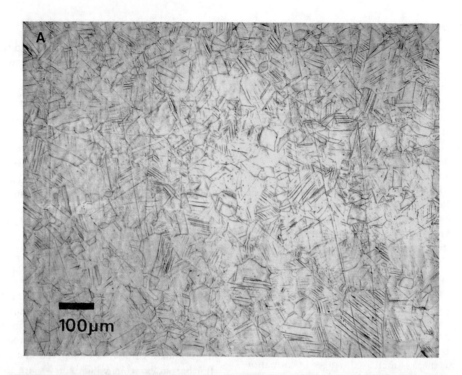

FIG. 4.   (A) Typical microstructure of cold-worked 316L stainless steel, ASTM F138, in a transverse section taken through a spinal distraction rod. (B) Detail of grains in cold-worked 316L stainless steel showing evidence of plastic deformation. (Photo courtesy of Zimmer USA, Warsaw, IN.)

TABLE 2 Typical Mechanical Properties of Implant Metals[a]

Material	ASTM designation	Condition	Young's modulus (GPa)	Yield strength (MPa)	Tensile strength (MPa)	Fatigue endurance limit (at 10^7 cycles, R = −1) (MPa)
Stainless steel	F745	Annealed	190	221	483	221–280
	F55, F56, F138, F139	Annealed	190	331	586	241–276
		30% Cold worked	190	792	930	310–448
		Cold forged	190	1213	1351	820
Co–Cr alloys	F75	As-cast/annealed	210	448–517	655–889	207–310
		P/M HIP[b]	253	841	1277	725–950
	F799	Hot forged	210	896–1200	1399–1586	600–896
	F90	Annealed	210	448–648	951–1220	Not available
		44% Cold worked	210	1606	1896	586
	F562	Hot forged	232	965–1000	1206	500
		Cold worked, aged	232	1500	1795	689–793 (axial tension R = 0.05, 30 Hz)
Ti alloys	F67	30% Cold-worked Grade 4	110	485	760	300
	F136	Forged annealed	116	896	965	620
		Forged, heat treated	116	1034	1103	620–689

[a]Data collected from references noted at the end of this chapter, especially Table 1 in Davidson and Georgette (1986).
[b]P/M HIP = Powder metallurgy product, hot-isostatically pressed.

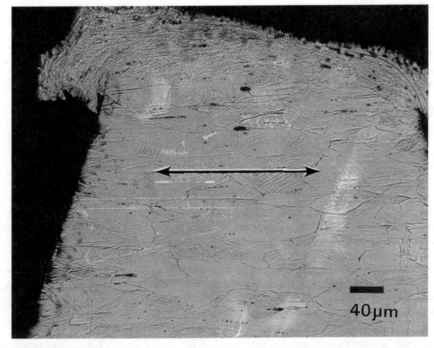

FIG. 5. Evidence of textured grain structure in 316L stainless steel ASTM F138, as seen in a longitudinal section through a cold-worked bone screw. The long axis of the screw is indicated by the arrow.

TABLE 3 Chemical Compositions of Co-Based Alloys for Implants

Material	ASTM designation	Common/trade names	Composition (wt %)	Notes
Co–Cr–Mo	F75	Vitallium Haynes-Stellite 21 Protasul-2 Micrograin-Zimaloy	58.9–69.5 Co 27.0–30.0 Cr 5.0–7.0 Mo max 1.0 Mn max 1.0 Si max 1.0 Ni max 0.75 Fe max 0.35 C	Vitallium is a trademark of Howmedica, Inc. Haynes-Stellite 21 (HS 21) is a trademark of Cabot Corp. Protasul-2 is a trademark of Sulzer AG, Switzerland. Zimaloy is a trademark of Zimmer USA.
Co–Cr–Mo	F799	Forged Co–Cr–Mo Thermomechanical Co–Cr–Mo FHS	58–59 Co 26.0–30.0 Cr 5.0–7.00 Mo max 1.00 Mn max 1.00 Si max 1.00 Ni max 1.5 Fe max 0.35 C max 0.25 N	FHS means "forged high strength" and is a trademark of Howmedica, Inc.
Co–Cr–W–Ni	F90	Haynes-Stellite 25 Wrought Co–Cr	45.5–56.2 Co 19.0–21.0 Cr 14.0–16.0 W 9.0–11.0 Ni max 3.00 Fe 1.00–2.00 Mn 0.05–0.15 C max 0.04 P max 0.40 Si max 0.03 S	Haynes-Stellite 25 (HS25) is a trademark of Cabot Corp.
Co–Ni–Cr–Mo–Ti	F562	MP 35 N Biophase Protasul-10	29–38.8 Co 33.0–37.0 Ni 19.0–21.0 Cr 9.0–10.5 Mo max 1.0 Ti max 0.15 Si max 0.010 S max 1.0 Fe max 0.15 Mn	MP35 N is a trademark of SPS Technologies, Inc. Biophase is a trademark of Richards Medical Co. Protasul-10 is a trademark of Sulzer AG, Switzerland

mold is cracked away and processing continues toward the final device. Depending on the exact casting details, this process can produce at least three microstructural features that can strongly influence implant properties.

First, as-cast F75 alloy (Figs. 6 and 7a) typically consists of a Co-rich matrix (alpha phase) plus interdendritic and grain-boundary carbides (primarily $M_{23}C_6$, where M represents Co, Cr, or Mo). There can also be interdendritic Co and Mo-rich sigma intermetallic, and Co-based gamma phases. Overall, the relative amounts of the alpha and carbide phases should be approximately 85% and 15%, respectively, but owing to non-equilibrium cooling, a "cored" microstructure can develop. In this situation, the interdendritic regions become solute (Cr, Mo, C) rich and contain carbides, while the dendrites become Cr depleted and richer in Co. This is an unfavorable electrochemical situation, with the Cr-depleted regions being anodic

with respect to the rest of the microstructure. (This is also an unfavorable situation if a porous coating will subsequently be applied by sintering.) Subsequent solution-anneal heat treatments in 1225°C for 1 hr can help alleviate this situation.

Second, the solidification during the casting process not only results in dendrite formation, but also in a relatively large grain size. This is generally undesirable because it decreases the yield strength via a Hall–Petch-type relationship between yield strength and grain diameter (see the section on stainless steel). The dendritic growth patterns and large grain diameter (about 4 mm) can be easily seen in Fig. 7a, which shows a hip stem manufactured by investment casting.

Third, casting defects may arise. Figure 7b shows an inclusion in the middle of a femoral hip stem. The inclusion was a particle of the ceramic mold (investment) material, which presumably broke off and became entrapped within the interior

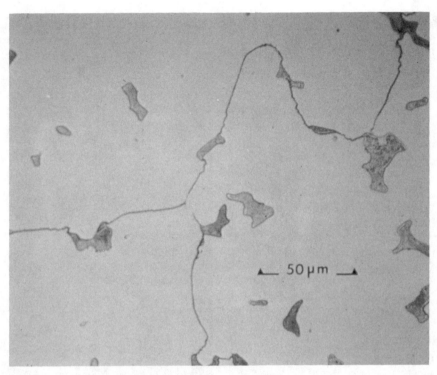

FIG. 6. Microstructure of as-cast Co–Cr–Mo ASTM F75 alloy, showing a large grain size plus grain boundary and matrix carbides. (Photo courtesy of Zimmer USA, Warsaw, IN.)

of the mold while the metal was solidifying. This contributed to a fatigue fracture of the implant device *in vivo*, most likely because of stress concentrations and fatigue crack sites associated with the ceramic inclusion. For similar reasons, it is also desirable to avoid macro- and microporosity arising from metal shrinkage upon solidification of castings.

To avoid these problems with cast F75, powder metallurgical methods have been used to improve the alloy's microstructure and mechanical properties. For example, in hot isostatic pressing, a fine powder of the F75 alloy is compacted and sintered together under appropriate pressure and temperature conditions (about 100 MPa at 1100°C for 1 hr) and then forged to final shape. The typical microstructure (Fig. 8) shows a much smaller grain size (about 8 μm) than the as-cast material. Again, according to a Hall–Petch-type relationship, this microstructure gives the alloy higher yield, and better ultimate and fatigue properties than the as-cast alloy (Table 2). Generally speaking, the improved properties of the HIP versus cast F75 result from both the finer grain size and a finer distribution of carbides, which has a hardening effect as well.

In porous-coated prosthetic devices based on F75 alloy, the microstructure will depend on both the prior manufacturing history of the beads and substrate metal, and the sintering process used to join the beads together and to the underlying bulk substrate. With Co–Cr–Mo alloys, for instance, sintering can be difficult, requiring temperatures near the melting point (1225°C). Unfortunately, these high temperatures can decrease the fatigue strength of the substrate alloy. For example, cast-solution-treated F75 has a fatigue strength of about 200–250

MPa, but it decreases to about 150 MPa after porous coating treatments. The reason for this probably relates to further phase changes in the nonequilibrium cored microstructure in the original cast F75 alloy. However, a modified sintering treatment can return the fatigue strength back up to about 200 MPa (Table 2). Beyond these metallurgical issues, a related concern with porous-coated devices is the potential for decreased fatigue performance as a result of stress concentrations where particles are joined with the substrate (Fig. 2).

ASTM F799 This is basically a modified F75 alloy that has been mechanically processed by hot forging (at about 800°C) after casting. It is sometimes known as thermomechanical Co–Cr–Mo alloy and has a composition slightly different from ASTM F75. The microstructure reveals a more worked grain structure than as-cast F75 and a hexagonal close-packed (HCP) phase that forms via a shear-induced transformation of FCC matrix to HCP platelets; this is not unlike that which occurs in MP35N (see ASTM F562). The fatigue, yield, and ultimate tensile strengths of this alloy are approximately twice those for as-cast F75 (Table 2).

ASTM F90 This alloy, also known as Haynes Stellite 25 (HS-25), is a Co–Cr–W–Ni alloy. Tungsten and nickel are added to improve machinability and fabrication properties. In the annealed state, its mechanical properties approximate those of F75 alloy, but when cold worked to 44%, the properties more than double (Table 2).

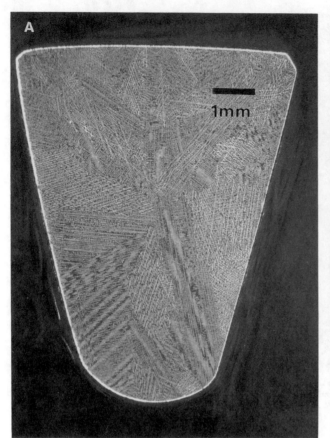

FIG. 7.   (A) Macrophoto of a metallographically polished and etched cross section of a cast Co–Cr–Mo ASTM F75 femoral hip stem, showing dendritic structure and large grain size. (B) Macrophoto of the fracture surface of the same Co–Cr–Mo ASTM F75 hip stem as in (A). Arrow indicates large inclusion within the central region of the cross section. Fracture of this hip stem occurred *in vivo*.

ASTM F562   Known as MP35N, this alloy is primarily Co (29–38.8%) and Ni (33–37%), with significant amounts of Cr and Mo. The "MP" in the name refers to the multiple phases in its microstructure. The alloy can be processed by thermal treatments and cold working to produce a controlled microstructure and a high-strength alloy, as explained later.

To start with, pure solid cobalt is (under equilibrium conditions) FCC above 419°C and HCP below 419°C. However, the solid-state transformation from FCC to HCP is sluggish and occurs by a martensitic-type shear reaction in which the HCP phase forms with its basal planes {0001} parallel to the close-packed {111} planes in FCC. The ease of this transformation is affected by the stability of the FCC phase, which in turn is affected by both plastic deformation and alloying additions.

When cobalt is alloyed to make MP35N, the processing includes 50% cold work, which increases the driving force for the transformation of retained FCC to the HCP phase. The HCP emerges as fine platelets within FCC grains. Because the FCC grains are small (0.01–0.1 μm, Fig. 9) and the HCP platelets further impede dislocation motion, the resulting structure is significantly strengthened (Table 2). It can be strengthened even further (as in the case of Richards Biophase) by an aging treatment at 430–650°C. This produces Co$_3$Mo precipitates on the HCP platelets. Hence, the alloy is multiphasic and derives strength from the combination of a cold-worked matrix phase, solid solution strengthening, and precipitation hardening. The resulting mechanical properties make the family of MP35N alloys among the strongest available for implant applications.

Titanium-Based Alloys

Composition

CP titanium (ASTM F67) and extra-low interstitial (ELI) Ti–6Al–4V alloy (ASTM F136) are the two most common titanium-based implant biomaterials. The F67 CP Ti is 98.9–99.6% titanium (Table 4). Oxygen content of CP Ti affects its yield and fatigue strength significantly. For example, at 0.18% oxygen (grade 1), the yield strength is about 170 MPa, while at 0.40% (grade 4), the yield strength increases to about 485 MPa. Similarly, at 0.085 wt.% oxygen (slightly purer than grade 1) the fatigue limit (10^7 cycles) is about 88.2

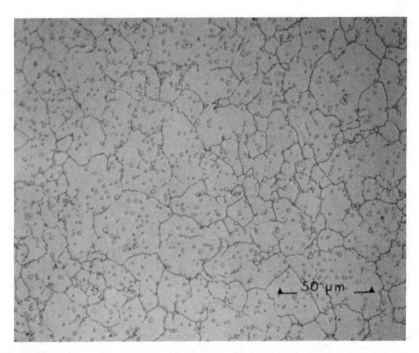

FIG. 8. Microstructure of the Co–Cr–Mo ASTM F75 alloy made via hot isostatic pressing (HIP), showing the much smaller grain size relative to that in Fig. 6. (Photo courtesy of Zimmer USA, Warsaw, IN.)

MPa, while at 0.27 wt.% oxygen (slightly purer than grade 2) the fatigue limit (10^7 cycles) is about 216 MPa.

With Ti–6Al–4V ELI alloy, the individual Ti–Al and Ti–V phase diagrams suggest the effects of the alloying additions in the ternary alloy. Al is an alpha (HCP) phase stabilizer while V is a beta (BCC) phase stabilizer. The 6Al–4V alloy used for implants is an alpha-beta alloy, the properties of which vary with prior treatments.

FIG. 9. Microstructure of Co-based MP35N, ASTM F562, Biophase. (Photo courtesy of Smith & Nephew Richards, Inc., Memphis, TN.)

Microstructure and Properties

ASTM F67 For relatively pure titanium implants, as exemplified by many current dental implants, typical microstructures are single-phase alpha (HCP), showing evidence of mild (30%) cold work and grain diameters in the range of 10–150 μm (Fig. 10), depending on manufacturing. The nominal mechanical properties are listed in Table 2. Interstitial elements (O, C, N) in titanium and the 6Al–4V alloy strengthen the metal through interstitial solid solution strengthening mechanisms, with nitrogen having approximately twice the hardening effect (per atom) of either carbon or oxygen.

There is increasing interest in the chemical and physical nature of the oxide on the surface of titanium and its 6Al–4V alloy. The nominal composition of the oxide is TiO_2. The oxide provides corrosion resistance and may also contribute to the biological performance of titanium at molecular and tissue levels, as suggested in the literature on osseointegrated oral and maxillofacial implants by Brånemark and co-workers in Sweden (Kasemo and Lausmaa, 1988).

ASTM F136 This is an alpha-beta alloy, the microstructure of which depends upon heat treating and mechanical working. If the alloy is heated into the beta phase field (e.g., above 1000°C, the region where only BCC beta is thermodynamically stable) and then cooled slowly to room temperature, a two-phase Widmanstatten structure is produced (Fig. 11). The HCP alpha phase (which is rich in Al and depleted in V) precipitates out as plates or needles having a specific crystallographic orientation within grains of the beta (BCC) matrix.

TABLE 4   Chemical Compositions of Ti-Based Alloys for Implants

Material	ASTM designation	Common/trade names	Composition (wt %)	Notes
Pure Ti	F67	CP Ti	Balance Ti max 0.10 C max 0.5 Fe max 0.0125–0.015 H max 0.05 N max 0.40 O	CP Ti comes in four grades according to oxygen content—only Grade 4 is listed.
Ti–6Al–4V	F136	Ti–6Al–4V	88.3–90.8 Ti 5.5–6.5 Al 3.5–4.5 V max 0.08 C 0.0125 H max 0.25 Fe max 0.05 N max 0.13 O	

Alternatively, if cooling from the beta phase field is very fast (as in oil quenching), a "basketweave" microstructure will develop, owing to martensitic or bainitic (nondiffusional, shear) solid-state transformations. Most commonly, the F136 alloy is heated and worked at temperatures near but not exceeding the beta transus, and then annealed to give a microstructure of fine-grained alpha with beta as isolated particles at grain boundaries (mill annealed, Fig. 12).

Interestingly, all three of the above-noted microstructures in Ti–6Al–4V alloy lead to about the same yield and ultimate tensile strengths, but the mill-annealed condition is superior in high-cycle fatigue (Table 2), which is a significant consideration.

Like the Co-based alloys, the above microstructural aspects for the Ti systems need to be considered when evaluating the structure-property relationships of porous-coated or plasma-sprayed implants. Again, there is the technical problem of successfully attaching some type of coating onto the metal

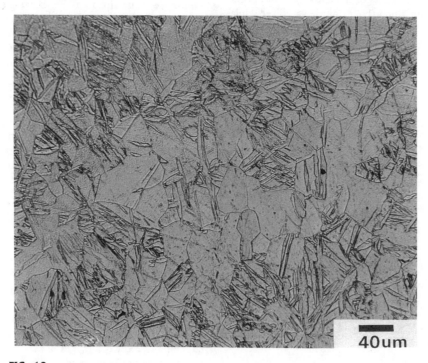

FIG. 10. Microstructure of moderately cold-worked commercial purity titanium, ASTM F67, used in an oral implant.

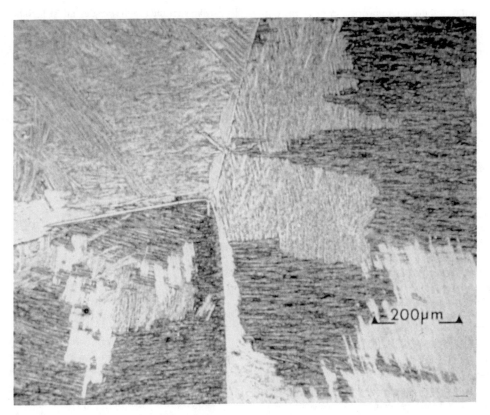

FIG. 11.   Widmanstatten structure in cast Ti–Al–4V, ASTM F136. Note prior beta grains (three large grains are shown in the photo) and platelet alpha structure within grains. (Photo courtesy of Zimmer USA, Warsaw, IN.)

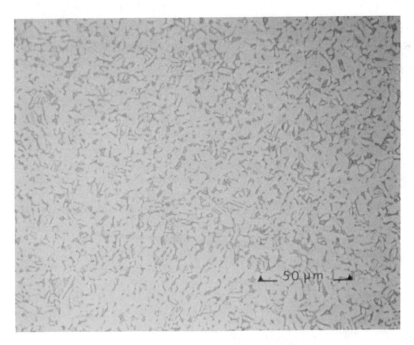

FIG. 12.   Microstructure of wrought and mill-annealed Ti–6Al–4V, showing small grains of alpha (light) and beta (dark). (Photo courtesy of Zimmer USA, Warsaw, IN.)

substrate while maintaining adequate properties of both coating and substrate. For example, optimizing the fatigue properties of Ti–6Al–4V porous-coated implants, becomes an interdisciplinary problem involving not only metallurgy but also surface properties and fracture mechanics.

CONCLUDING REMARKS

It should be evident that metallurgical principles guide our understanding of structure-property relationships in metallic implants, just as they would in the study of any metallic device. While this chapter's emphasis has been on mechanical properties (for the sake of specificity), other properties, in particular surface properties, are receiving increasing attention in relation to biological performance of implants.

Another point to remember is that the intrinsic material properties of metallic implants are not the sole determinant of implant performance and success. Existing implant metals and alloys have all been used in both successful and unsuccessful implant designs. The reasons for failures can include faulty or inappropriate use of the implant, surgical error, and inadequate mechanical design of the implant. Therefore, debates about which implant metal is "superior" often miss the point; implant design is a true multifaceted design problem in which the selection of materials is only a part—albeit an important part—of the total problem.

Bibliography

American Society for Testing and Materials (1978). *ASTM Standards for Medical and Surgical Materials and Devices.* Authorized Reprint from Annual Book of ASTM Standards, ASTM, Philadelphia, PA.

Beevers, C. J. and Robinson, J. L. (1969). Some observations on the influence of oxygen content on the fatigue behavior of α-titanium. *J. Less Common Metals* 17: 345–352.

Compte, P. (1984). Metallurgical observations of biomaterials. in *Contemporary Biomaterials,* J. W. Boretos and M. Eden, eds. Noyes Publ., Park Ridge, NJ, pp. 66–91.

Cox, D. O. (1977). The fatigue and fracture behavior of a low stacking fault energy cobalt-chromium-molybdenum-carbon casting alloy used for prosthetic devices. Ph.D. Dissertation, Engineering, University of California at Los Angeles.

Davidson, J. A., and Georgette, F. S. (1986). State-of-the-art materials for orthopaedic prosthetic devices. in *Implant Manufacturing and Material Technology.* Proc. Soc. of Manufacturing Engineering, Itasca, IL.

Hamman, G., and Bardos, D. I. (1980). Metallographic quality control of orthopaedic implants. in *Metallography as a Quality Control Tool,* J. L. McCall and P. M. French, eds. Plenum Publ., New York, pp. 221–245.

Honeycombe, R. W. K. (1968). *The Plastic Deformation of Metals.* St. Martin's Press, New York, p. 234.

Kasemo, B., and Lausmaa, J. (1988). Biomaterials from a surface science perspective. in *Surface Characterization of Biomaterials,* B. D. Ratner, ed. Elsevier, New York, Ch. 1, pp. 1–12.

Moss, A. J., Hamburger, S., Moore, R. M. *et al.* (1990). Use of selected medical device implants in the United States, 1988. Advance data from vital and health statistics. no. 191. National Center for Health Statistics, Hyattsville, MD.

Pilliar, R. M., and Weatherly, G. C. (1984). Developments in implant alloys. *CRC Critical Reviews in Biocompatibility* 1(4): 371–403.

Richards Medical Company (1985). *Medical Metals.* Richards Medical Company Publication No. 3922, Richards Medical Co., Memphis, TN. [Note: This company is now known as Smith & Nephew Richards, Inc.]

Zimmer USA (1984a). *Fatigue and Porous Coated Implants.* Zimmer Technical Monograph, Zimmer USA, Warsaw, IN.

Zimmer USA (1984b). *Metal Forming Techniques in Orthopaedics.* Zimmer Technical Monograph, Zimmer USA, Warsaw, IN.

Zimmer USA (1984c). *Physical and Mechanical Properties of Orthopaedic Alloys.* Zimmer Technical Monograph, Zimmer USA, Warsaw, IN.

Zimmer USA (1984d). *Physical Metallurgy of Titanium Alloy.* Zimmer Technical Monograph, Zimmer USA, Warsaw, IN.

2.3 POLYMERS

*Susan A. Visser, Robert W. Hergenrother,
and Stuart L. Cooper*

Polymers are long-chain molecules that consist of a number of small repeating units. The repeat units or "mers" differ from the small molecules which were used in the original synthesis procedures, the monomers, in the loss of unsaturation or the elimination of a small molecule such as water or HCl during polymerization. The exact difference between the monomer and the mer unit depends on the mode of polymerization, as discussed later.

The wide variety of polymers includes such natural materials as cellulose, starches, natural rubber, and deoxyribonucleic acid (DNA), the genetic material of all living creatures. While these polymers are undoubtedly interesting and have seen widespread use in numerous applications, they are sometimes eclipsed by the seemingly endless variety of synthetic polymers that are available today.

The task of the biomedical engineer is to select a biomaterial with properties that most closely match those required for a particular application. Because polymers are long-chain molecules, their properties tend to be more complex than their short-chain counterparts. Thus, in order to choose a polymer type for a particular application, the unusual properties of polymers must be understood.

This chapter introduces the concepts of polymer characterization and property testing as they are applied to the selection of biomaterials. Examples of polymeric biomaterials currently used by the medical community are cited and discussed with regard to their solid-state properties and uses.

MOLECULAR WEIGHT

In polymer synthesis, a polymer is usually produced with a distribution of molecular weights. To compare the molecular weights of two different batches of polymer, it is useful to define an average molecular weight. Two statistically useful definitions of molecular weight are the number average and weight average molecular weights. The number average molecular weight (M_n) is the first moment of the molecular weight distribution and is an average over the number of molecules. The weight average molecular weight (M_w) is the second moment of the molecular weight distribution and is an average over the weight of each polymer chain. Equations 1 and 2 define the two averages:

$$M_n = \frac{\sum N_i M_i}{\sum N_i} \tag{1}$$

$$M_w = \frac{\sum N_i M_i^2}{\sum N_i M_i}, \tag{2}$$

where N_i is the number of moles of species i, and M_i is the molecular weight of species i.

The ratio of M_w to M_n is known as the polydispersity index and is used as a measure of the breadth of the molecular weight distribution. Typical commercial polymers have polydispersity indices of 1.5–50, although polymers with polydispersity indices of less than 1.1 can be synthesized with special techniques. A molecular weight distribution for a typical polymer is shown in Fig. 1.

Linear polymers used for biomedical applications generally have M_n in the range of 25,000 to 100,000 and M_w from 50,000 to 300,000. Higher or lower molecular weights may be necessary, depending on the ability of the polymer chains to exhibit secondary interactions such as hydrogen bonding. The secondary interactions can give polymers additional strength. In general, increasing molecular weight corresponds to increasing physical properties; however, since melt viscosity also increases with molecular weight, processibility will decrease and an upper limit of useful molecular weights is usually reached.

SYNTHESIS

Methods of polymer preparation fall into two categories: addition polymerization (chain reaction) and condensation po-

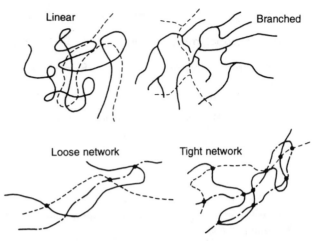

FIG. 2. Polymer arrangements. (From F. Rodriguez, *Principles of Polymer Systems,* Hemisphere Publ., 1982, p. 21, with permission.)

lymerization (stepwise growth). In addition polymerization, unsaturated monomers react through the stages of initiation, propagation, and termination to give the final polymer product. The initiators can be free radicals, cations, anions, or stereospecific catalysts. The initiator opens the double bond of the monomer, presenting another "initiation" site on the opposite side of the monomer bond for continuing growth. Rapid chain growth ensues during the propagation step until the reaction is terminated by reaction with another radical, a solvent molecule, another polymer, an initiator, or an added chain transfer agent.

Condensation polymerization is completely analogous to condensation reactions of low-molecular-weight molecules. Two monomers react to form a covalent bond, usually with elimination of a small molecule such as water, hydrochloric acid, methanol, or carbon dioxide. The reaction continues until almost all of one type of reactant is used up.

The choice of polymerization method strongly affects the polymer obtained. In free radical polymerization, a type of addition polymerization, the molecular weights of the polymer chains are difficult to control with precision. Added chain transfer agents are used to control the average molecular weights, but molecular weight distributions are usually broad. In addition, chain transfer reactions with other polymer molecules in the batch can produce undesirable branched products (Fig. 2) that affect the ultimate properties of the polymeric material. In contrast, molecular architecture can be controlled very precisely in anionic polymerization. Regular linear chains with polydispersity indices of close to unity can be obtained.

Polymers produced by addition polymerization can be homopolymers—polymers containing only one type of repeat unit—or copolymers of two or more types of repeat units. Depending on the reaction conditions and the reactivity of each monomer type, the copolymers can be random, alternating, or block copolymers, as illustrated in Fig. 3. Random copolymers exhibit properties that approximate the weighted average of the two types of monomer units, whereas block copolymers

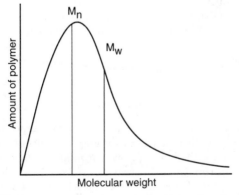

FIG. 1. Typical molecular weight distribution of a polymer.

Homopolymer   –A–A–A–A–A–A–A–

Random copolymer   –A–B–B–A–B–A–B–

Alternating copolymer –A–B–A–B–A–B–A–

Block copolymer   –A–A–A–A–B–B–B–

FIG. 3. Possible monomer arrangements in polymer materials.

tend to phase separate into a monomer-A-rich phase and a monomer-B-rich phase, displaying properties unique to each of the homopolymers.

Condensation polymerization can also result in copolymer formation. The properties of the condensation copolymer depend on three factors: the type of monomer units; the molecular weight of the polymer product, which can be controlled by the ratio of one reactant to another and by the time of polymerization; and the distribution of the molecular weight of the copolymer chains. The use of bifunctional monomers gives rise to linear polymers, while multifunctional monomers may be used to form covalently cross-linked networks.

Postpolymerization cross-linking of addition or condensation polymers is also possible. Natural rubber, for example, consists mostly of linear molecules that can be cross-linked to a loose network with 1–3% sulfur (vulcanization) or to a hard rubber with 40–50% sulfur (Fig. 2). In addition, physical, rather than chemical, cross-linking of polymers can be achieved in the presence of microscrystalline regions or through incorporation of ionic groups in the polymer (Fig. 4).

THE SOLID STATE

Tacticity

Polymers are long-chain molecules and, as such, are capable of assuming many conformations through rotation of valence bonds. The extended chain or planar zig-zag conformation of polypropylene is shown in Fig. 5. This figure illustrates the concept of tacticity. Tacticity refers to the arrangement of substituents (methyl groups in the case of polypropylene) around the extended polymer chain. Chains in which all substituents are located on the same side of the zigzag plane are isotactic, while syndiotactic chains have substituents alternating from side to side. In the atactic arrangement, the substituent groups appear at random on either side of the extended chain backbone.

Atactic polymers usually cannot crystallize, and an amorphous polymer results. Isotactic and syndiotactic polymers may crystallize if conditions are favorable. Crystalline polymers also possess a higher level of structure characterized by folded chain lamellar growth that results in the formation of spherulites. These structures can be visualized in a polarized light microscope.

Crystallinity

Polymers can be either amorphous or semicrystalline. They can never be completely crystalline owing to lattice defects that form disordered, amorphous regions. The tendency of a polymer to crystallize is enhanced by the small side groups and chain regularity. The presence of crystallites in the polymer usually leads to enhanced mechanical properties, unique thermal behavior, and increased fatigue strength. These properties make semicrystalline polymers (often referred to simply as crystalline polymers) desirable materials for biomedical applications.

Mechanical Properties

The tensile properties of polymers can be characterized by their deformation behavior (stress-strain response (Fig. 6). Amorphous, rubbery polymers are soft and reversibly extensible. The freedom of motion of the polymer chain is retained at a local level while a network structure resulting from chemical cross-links and chain entanglements prevents large-scale movement or flow. Thus, rubbery polymers tend to exhibit a lower modulus, or stiffness, and extensibilities of several hundred percent. Rubbery materials may also exhibit an increase of stress prior to breakage as a result of strain-induced crystallization assisted by molecular orientation in the direction of stress. Glassy and semicrystalline polymers have higher moduli and lower extensibilities.

The ultimate mechanical properties of polymers at large deformations are important in selecting particular polymers for biomedical applications. The ultimate strength of polymers is the stress at or near failure. For most materials, failure is catastrophic (complete breakage). However, for some semicrystalline materials, the failure point may be defined by the stress point where large inelastic deformation starts (yielding). The toughness of a polymer is related to the energy absorbed at failure and is proportional to the area under the stress-strain curve.

The fatigue behavior of polymers is also important in evaluating materials for applications where dynamic strain is applied. For example, polymers that are used in the artificial heart must be able to withstand many cycles of pulsating motion before failure. Samples that are subjected to repeated cycles of stress and release, as in a flexing test, fail (break) after a certain number of cycles. The number of cycles to failure decreases as the applied stress level is increased, as shown in Fig. 7 (see also Chapter 6.4). For some materials, a minimum stress exists below which failure does not occur in a measurable number of cycles.

Thermal Properties

In the liquid or melt state, a noncrystalline polymer possesses enough thermal energy for long segments of each polymer to move randomly (Brownian motion). As the melt is cooled, the temperature is eventually reached at which all long-range segmental motions cease. This is the glass transition temperature (T_g), and it varies from polymer to polymer. Poly-

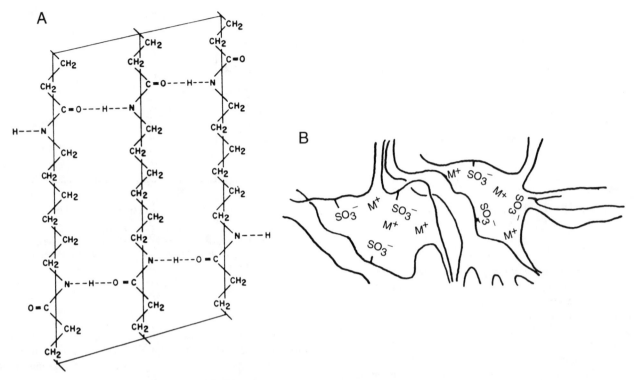

FIG. 4. (A) Hydrogen bonding in nylon 6,6 molecules in a triclinic unit cell: σ-form. (From L. Mandelkern, *An Introduction to Macromolecules*, Springer-Verlag, 1983, p. 43, with permission.) (B) Ionic aggregation giving rise to physical cross-links in ionomers.

mers used below their T_g tend to be hard and glassy, while polymers used above their T_g are rubbery. Polymers with any crystallinity will also exhibit a melting temperaure (T_m) owing to melting of the crystalline phase. Thermal transitions in polymers can be measured by differential scanning calorimetry (DSC), as discussed in the section on characterization techniques.

The viscoelastic responses of polymers can also be used to classify their thermal behavior. The modulus versus temperature curves shown in Fig. 8 illustrate behaviors typical of linear amorphous, cross-linked, and semicrystalline polymers. The response curves are characterized by a glassy modulus below T_g of approximately 3×10^9 Pa. For linear amorphous polymers, increasing temperature induces the onset of the glass transition region where, in a 5–10°C temperature span, the modulus drops by three orders of magnitude, and the polymer is transformed from a stiff glass to a leathery material. The relatively

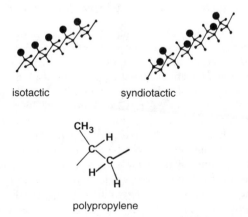

FIG. 5. Schematic of stereoisomers of polypropylene. (From F. Rodriguez *Principles of Polymer Systems*, Hemisphere Publ., 1982, p. 22, with permission.)

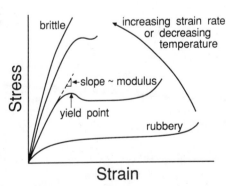

FIG. 6. Tensile properties of polymers.

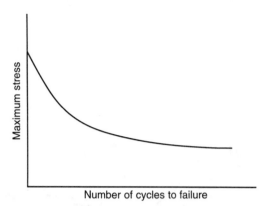

FIG. 7.   Fatigue properties of polymers.

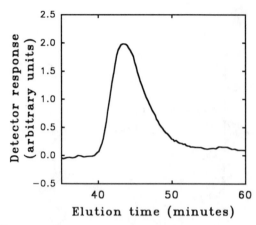

FIG. 9.   A typical trace from a gel permeation chromatography run for a poly(tetramethylene oxide)/toluene diisocyanate-based polyurethane. The response of the ultraviolet detector is directly proportional to the amount of polymer eluted at each time point.

constant modulus region above T_g is the rubbery plateau region where long-range segmental motion is occurring but thermal energy is insufficient to overcome entanglement interactions that inhibit flow. This is the target region for many biomedical applications. Finally, at high enough temperatures, the polymer begins to flow, and a sharp decrease in modulus is seen over a narrow temperature range.

Crystalline polymers exhibit the same general features in modulus versus temperature curves as amorphous polymers; however, crystalline polymers possess a higher plateau modulus owing to the reinforcing effect of the crystallites. Crystalline polymers tend to be tough, ductile plastics whose properties are sensitive to processing history. When heated above their flow point, they can be melt processed and will become rigid again upon cooling.

Chemically cross-linked polymers exhibit modulus versus temperature behavior analogous to that of linear amorphous polymers until the flow regime is approached. Unlike linear polymers, chemically cross-linked polymers do not display flow behavior; the cross links inhibit flow at all temperatures below the degradation temperature. Thus, chemically cross-linked

polymers cannot be melt processed. Instead, these materials are processed as reactive liquids or high-molecular-weight amorphous gums that are cross-linked during molding to give the desired product.

Copolymers

In contrast to the thermal behavior of homopolymers discussed earlier, copolymers can exhibit a number of additional thermal transitions. If the copolymer is random, it will exhibit a T_g that approximates the weighted average of the T_gs of the two homopolymers. Block copolymers of sufficient size and incompatible block types will exhibit T_gs characteristic of each homopolymer but slightly shifted owing to incomplete phase separation.

CHARACTERIZATION TECHNIQUES

Determination of Molecular Weight

Gel permeation chromatography (GPC), a type of size exclusion chromatography, involves passage of a dilute polymer solution over a column of porous beads. High-molecular-weight polymers are excluded from the beads and elute first whereas lower molecular weight molecules pass through the pores of the bead, increasing their elution time. By monitoring the effluent of the column as a function of time using an ultraviolet or refractive index detector, the amount of polymer eluted during each time interval can be determined. Comparison of the elution time of the samples with those of monodisperse samples of known molecular weight allows the entire molecular weight distribution to be determined. A typical GPC trace is shown in Fig. 9.

Osmotic pressure measurements can be used to measure M_n. The principle of membrane osmometry is illustrated in

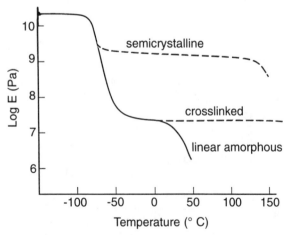

FIG. 8.   Dynamic mechanical behavior of polymers.

Fig. 10. A semipermeable membrane is placed between two chambers. Only solvent molecules flow freely through the membrane. Pure solvent is placed in one chamber, and a dilute polymer solution of known concentration is placed in the other chamber. The lowering of the activity of the solvent in solution with respect to that of the pure solvent is compensated by applying a pressure π on the solution. π is the osmotic pressure and is related to M_n by:

$$\frac{\pi}{c} = RT\left[\frac{1}{M_n} + A_2 c + A_3 c^2 + \cdots\right], \tag{3}$$

where c is the concentration of the polymer in solution, R is the gas constant, T is temperature, and A_2 and A_3 are virial coefficients relating to pairwise and triplet interactions of the molecules in solution. In general, a number of polymer solutions of decreasing concentration are prepared, and the osmotic pressure is extrapolated to zero:

$$\lim_{c \to 0} \frac{\pi}{c} = \frac{RT}{M_n}. \tag{4}$$

A plot of π/c versus c then gives as its intercept the number average molecular weight.

A number of other techniques, including vapor pressure osmometry, ebulliometry, cryoscopy, and end-group analysis can be used to determine the M_n of polymers up to molecular weights of about 40,000.

Light-scattering techniques are used to determine M_w. In dilute solution, the scattering of light is directly proportional to the number of molecules. The scattered intensity i_o observed at a distance r and an angle θ from the incident beam I_o is characterized by Rayleigh's ratio R_θ:

$$R_\theta = \frac{i_o r^2}{I_o}. \tag{5}$$

The Rayleigh ratio is related to M_w by:

$$\frac{K_c}{R_\theta} = \frac{1}{M_w} + 2 A_2 c + 3 A_2 c^2 + \cdots. \tag{6}$$

A number of solutions of varying concentrations are measured, and the data are extrapolated to zero concentration to determine M_w.

Determination of Structure

Infrared (IR) spectroscopy is often used to characterize the chemical structure of polymers. Infrared spectra are obtained by passing infrared radiation through the sample of interest and observing the wavelength of the absorption peaks. These peaks are caused by the absorption of the radiation and its conversion into specific motions, such as C–H stretching. The infrared spectrum of a polyurethane is shown in Fig. 11, with a few of the bands of interest marked.

Nuclear magnetic resonance (NMR), in which the magnetic spin energy levels of nuclei of spin 1/2 or greater are probed, may also be used to analyze chemical composition. NMR is also used in a number of more specialized applications relating to local motions of polymer molecules.

Wide-angle X-ray scattering (WAXS) techniques are useful for probing the local structure of a semicrystalline polymeric solid. Under appropriate conditions, crystalline materials diffract X-rays, giving rise to spots or rings. According to Bragg's law, these can be interpreted as interplanar spacings. The interplanar spacings can be used without further manipulation or the data can be fit to a model such as a disordered helix or an extended chain. The crystalline chain conformation and atomic placements can then be accurately inferred.

Small-angle X-ray scattering (SAXS) is used in determining the structure of many multiphase materials. This technique requires an electron density difference to be present between two components in the solid and has been widely applied to morphological studies of copolymers and ionomers. It can probe features of 10–1000 Å in size. With appropriate modeling of the data, SAXS can give detailed structural information unavailable with other techniques.

Electron microscopy of thin sections of a polymeric solid can also give direct morphological data on a polymer of interest, assuming that (1) the polymer possesses sufficient electron density contrast or can be appropriately stained without changing the morphology and (2) the structures of interest are sufficiently large.

Mechanical and Thermal Property Studies

In stress-strain or tensile testing, a dog bone-shaped polymer sample is subjected to a constant elongation, or strain, rate, and the force required to maintain the constant elongation rate is monitored. As discussed earlier, tensile testing gives information about modulus, yield point, and ultimate strength of the sample of interest.

Dynamic mechanical analysis (DMA) provides information about the small deformation behavior of polymers. Samples are subjected to cyclic deformation at a fixed frequency in the range of 1–1000 Hz. The stress response is measured while the

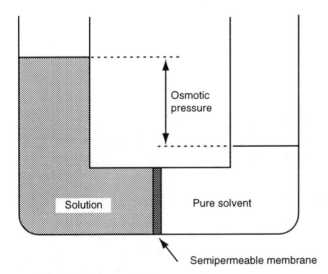

FIG. 10. The principle of operation of a membrane osmometer.

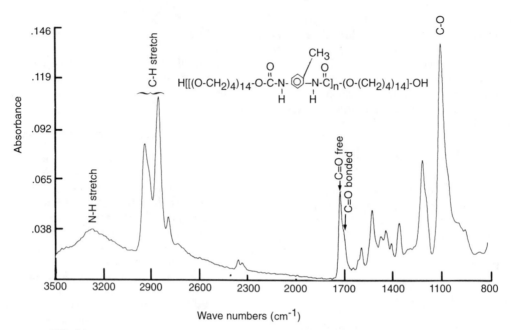

FIG. 11.   Infrared spectrum of a poly(tetramethylene oxide)/toluene diisocyanate-based polyurethane.

cyclic strain is applied and the temperature is slowly increased (typically at 2–3°/min). If the strain is a sinusoidal function of time given by:

$$\varepsilon(\omega) = \varepsilon_o \sin(\omega t), \qquad (7)$$

where ε is the time-dependent strain, ε_o is the strain amplitude, ω is the frequency of oscillation, and t is time, the resulting stress can be expressed by:

$$\sigma(\omega) = \sigma_o \sin(\omega t + \delta), \qquad (8)$$

where σ is the time-dependent stress, σ_o is the amplitude of stress response, and δ is the phase angle between stress and strain. For Hookean solids, the stress and strain are completely in phase ($\delta = 0$), while for purely viscous liquids, the stress response lags by 90°. Real materials demonstrate viscoelastic behavior where δ has a value between 0° and 90°.

A typical plot of tan δ versus temperature will display maxima at T_g and at lower temperatures where small-scale motions (secondary relaxations) can occur. Additional peaks above T_g, corresponding to motions in the crystalline phase and melting, are seen in semicrystalline materials. DMA is a sensitive tool for characterizing polymers of similar chemical composition or for detecting the presence of moderate quantities of additives.

Differential scanning calorimetry is another method for probing thermal transitions of polymers. A sample cell and a reference cell are supplied energy at varying rates so that the temperatures of the two cells remain equal. The temperature is increased, typically at a rate of 10–20°/min over the range of interest, and the energy input required to maintain equality of temperature in the two cells is recorded. Plots of energy supplied versus average temperature allow determination of

T_g, crystallization temperature (T_c), and T_m. T_g is taken as the temperature at which one half the change in heat capacity, ΔC_p, has occurred. The T_c and T_m are easily identified, as shown in Fig. 12. The areas under the peaks can be quantitatively related to enthalpic changes.

Surface Characterization

Surface characteristics of polymers for biomedical applications are critically important. The surface composition is inevi-

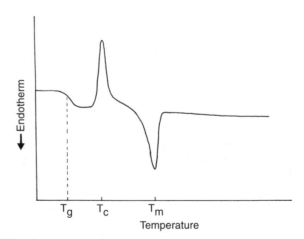

FIG. 12.   Differential scanning calorimetry thermogram of a semicrystalline polymer.

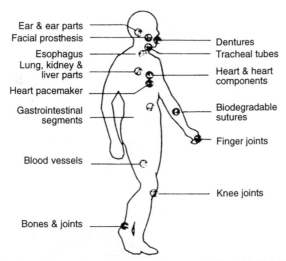

Ear & ear parts
Facial prosthesis

Esophagus
Lung, kidney &
liver parts

Heart pacemaker

Gastrointestinal
segments

Blood vessels

Bones & joints

Dentures
Tracheal tubes

Heart & heart
components

Biodegradable
sutures

Finger joints

Knee joints

Ear & ear parts: acrylic, polyethylene, silicone, poly(vinyl chloride) (PVC)
Dentures: acrylic, ultrahigh molecular weight polyethylene (UHMWPE), epoxy
Facial prosthesis: acrylic, PVC, polyurethane (PUR)
Tracheal tubes: acrylic, silicone, nylon
Heart & heart components: polyester, silicone, PVC
Heart pacemaker: polyethylene, acetal
Lung, kidney & liver parts: polyester, polyaldehyde, PVC
Esophagus segments: polyethylene, polypropylene (PP), PVC
Blood vessels: PVC, polyester
Biodegradable sutures: PUR
Gastrointestinal segments: silicones, PVC, nylon
Finger joints: silicone, UHMWPE
Bones & joints: acrylic, nylon, silicone, PUR, PP, UHMWPE
Knee joints: polyethylene

FIG. 13. Common clinical applications and types of polymers used in medicine. (From D. V. Rosato, in *Biocompatible Polymers, Metals, and Composites,* M. Szycher, ed., Technomic Publ., 1983, p. 1022, with permission.)

tably different from the bulk, and the surface of the material is generally all that is contacted by the body. The main surface characterization techniques for polymers are X-ray photoelectron spectroscopy (XPS), contact angle measurements, attenuated total reflectance Fourier transform infrared (ATR–FTIR) spectroscopy, and scanning electron microscopy (SEM). The techniques are discussed in detail in Chapter 1.3.

CLASSES OF POLYMERS USED IN MEDICINE

Many types of polymers are used for biomedical purposes. Figure 13 illustrates the variety of clinical applications for polymeric biomaterials. This section discusses some of the polymers used in medicine.

Homopolymers

Homopolymers are composed of a single type of monomer. Figure 14 shows the repeat units of many of the homopolymers used in medicine.

Poly(methyl methacrylate) (PMMA) is a hydrophobic, linear chain polymer that is glassy at room temperature and may be more easily recognized by such trade names as Lucite or Plexiglas. It has very good light transmittance, toughness, and stability, making it a good material for intraocular lenses and hard contact lenses.

Soft contact lenses are made from the same family of polymers, with the addition of a —CH$_2$OH group to the methyl methacrylate side group, resulting in 2-hydroxyethyl methacrylate (HEMA). The additional methylol group causes the polymer to be hydrophilic. For soft contact lenses, the poly(HEMA) is slightly cross-linked with ethylene glycol dimethyacrylate (EGDM) to prevent the polymer from dissolving when it is hydrated (Rodriguez, 1982). Fully hydrated, it is a swollen hydrogel. This class of polymers is discussed in more detail in Chapter 2.4.

Polyethylene (PE) is used in its high-density form in biomedical applications because low-density material cannot withstand sterilization temperatures. It is used in tubing for drains and catheters, and in very high-molecular-weight form as the acetabular component in artificial hips. The material has good toughness, resistance to fats and oils, and a relatively low cost.

Polypropylene (PP) is closely related to PE and has high rigidity, good chemical resistance, and good tensile strength.

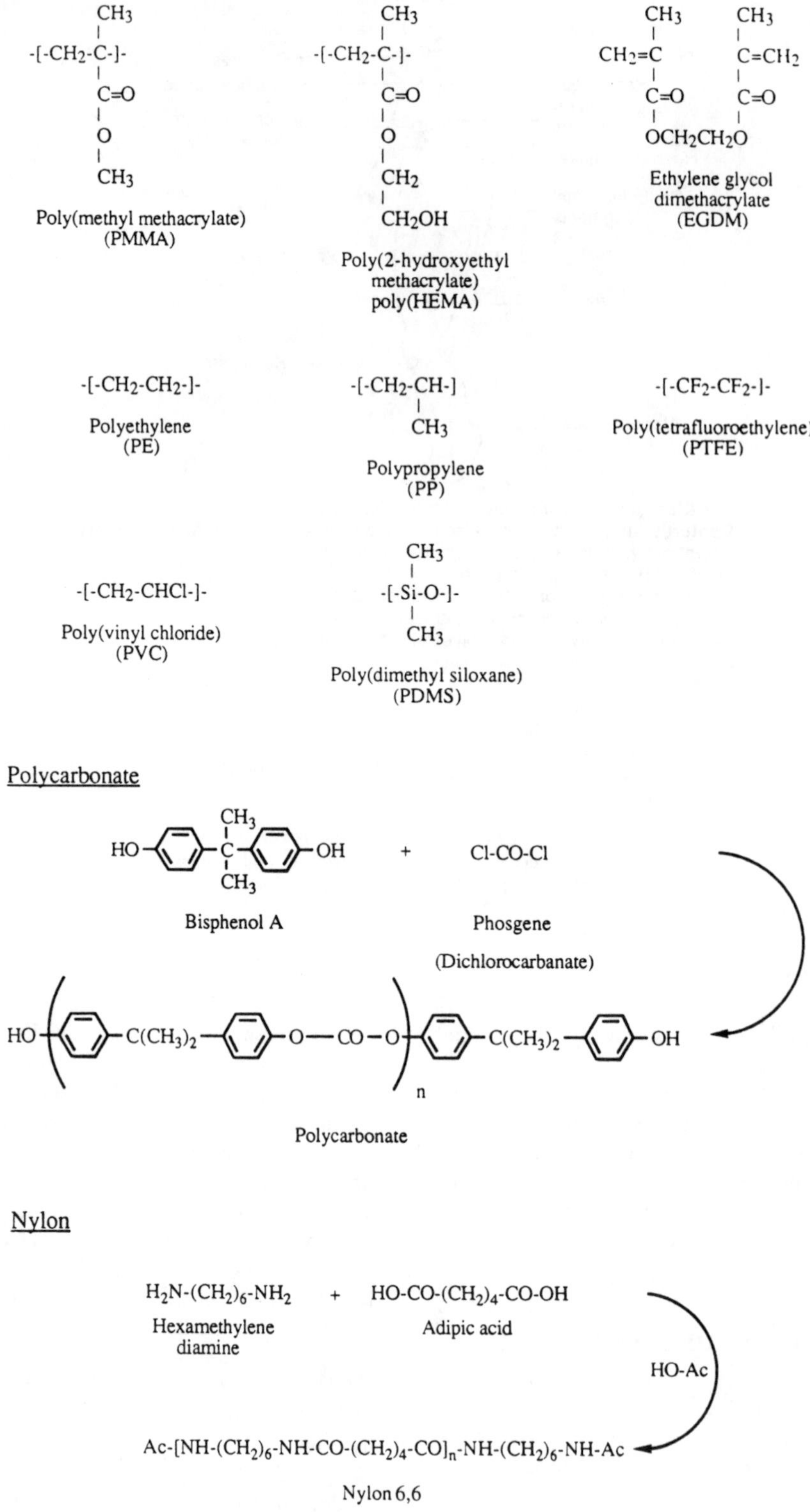

$$-[-CH_2-C-]-$$
with pendant groups, Poly(methyl methacrylate) (PMMA):

```
      CH3
       |
-[-CH2-C-]-
       |
      C=O
       |
      O
       |
      CH3
```
Poly(methyl methacrylate)
(PMMA)

```
      CH3
       |
-[-CH2-C-]-
       |
      C=O
       |
      O
       |
      CH2
       |
      CH2OH
```
Poly(2-hydroxyethyl
methacrylate)
poly(HEMA)

```
  CH3        CH3
   |          |
CH2=C       C=CH2
   |          |
  C=O        C=O
   |          |
   OCH2CH2O
```
Ethylene glycol
dimethacrylate
(EGDM)

```
-[-CH2-CH2-]-
```
Polyethylene
(PE)

```
-[-CH2-CH-]
        |
       CH3
```
Polypropylene
(PP)

```
-[-CF2-CF2-]-
```
Poly(tetrafluoroethylene)
(PTFE)

```
-[-CH2-CHCl-]-
```
Poly(vinyl chloride)
(PVC)

```
      CH3
       |
-[-Si-O-]-
       |
      CH3
```
Poly(dimethyl siloxane)
(PDMS)

Polycarbonate

HO—⬡—C(CH3)2—⬡—OH　　+　　Cl-CO-Cl

Bisphenol A　　　　　　　　　　　Phosgene

(Dichlorocarbanate)

HO—(⬡—C(CH3)2—⬡—O—CO—O)n—⬡—C(CH3)2—⬡—OH

Polycarbonate

Nylon

$H_2N-(CH_2)_6-NH_2$　　+　　$HO-CO-(CH_2)_4-CO-OH$

Hexamethylene　　　　　　　Adipic acid
diamine

HO-Ac

$Ac-[NH-(CH_2)_6-NH-CO-(CH_2)_4-CO]_n-NH-(CH_2)_6-NH-Ac$

Nylon 6,6

FIG. 14.　Homopolymers used in medicine.

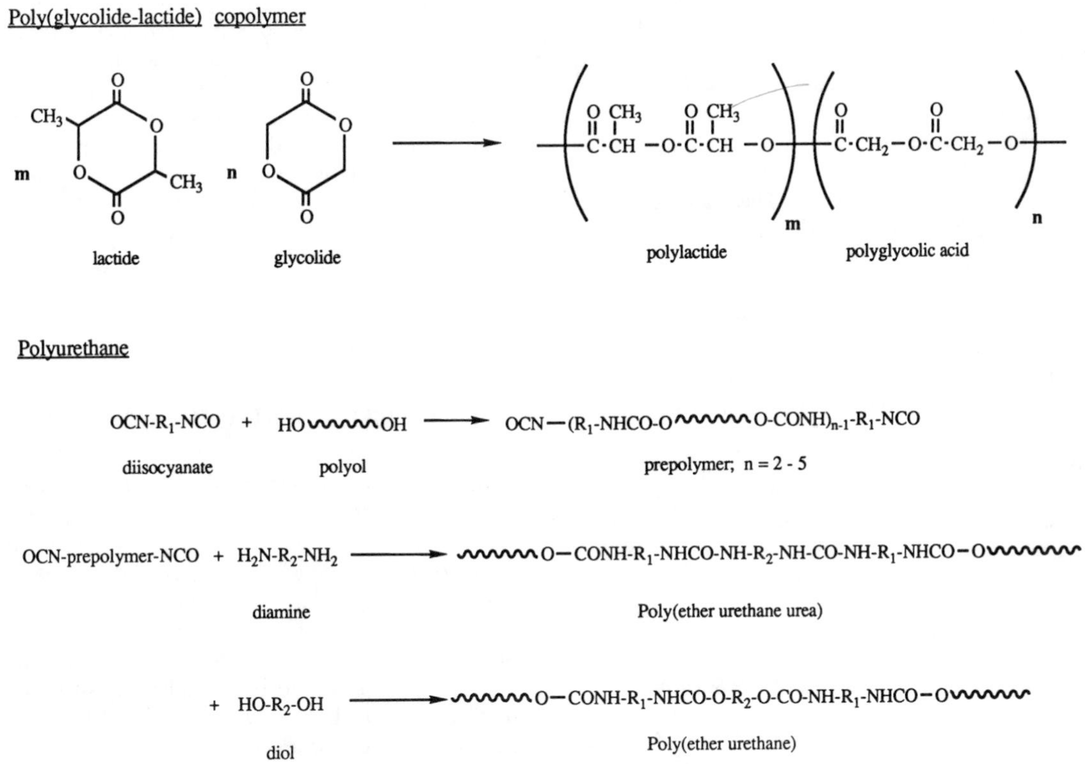

Poly(glycolide-lactide) copolymer

Polyurethane

FIG. 15. Copolymers and their base monomers used in medicine.

Its stress cracking resistance is superior to that of PE, and it is used for many of the same applications as PE.

Poly(tetrafluoroethylene) (PTFE), also known as Teflon, has the same structure as PE, except that the hydrogen in PE is replaced by fluorine. PTFE is a very stable polymer, both thermally and chemically, and as a result it is very difficult to process. It is very hydrophobic and has excellent lubricity. In microporous (Gore-Tex) form, it is used in vascular grafts.

Poly(vinyl chloride) (PVC) is used mainly in tubing in biomedical applications. Typical tubing uses include blood transfusion, feeding, and dialysis. Pure PVC is a hard, brittle material, but with the addition of plasticizers, it can be made flexible and soft. PVC can pose problems for long-term applications because the plasticizers can be extracted by the body. While these plasticizers have low toxicities, their loss makes the PVC less flexible.

Poly(dimethyl siloxane) (PDMS) is an extremely versatile polymer. It is unique in that it has a silicon-oxygen backbone instead of a carbon backbone. Its properties are less temperature sensitive than other rubbers because of its lower T_g. PDMS is used in catheter and drainage tubing, in insulation for pacemaker leads, and as a component in some vascular graft systems. It is used in membrane oxygenators because of its high oxygen permeability. Because of its excellent flexibility and stability, it is also used in a variety of prostheses such as finger joints, blood vessels, heart valves, breast implants, outer ears, and chin and nose implants (Rosato, 1983).

Polymerization of bisphenol A and phosgene produces poly-

carbonate, a clear, tough material. Its high impact strength dictates its use as lenses for eyeglasses and safety glasses, and housings for oxygenators and heart-lung bypass machine.

Nylon is the name given by Du Pont to a family of polyamides. Nylons are formed by the reaction of diamines with dibasic acids or by the ring opening polymerization of lactams. Nylons are used in surgical sutures.

Copolymers

Copolymers are another important class of biomedical materials. Fig. 15 shows two different copolymers used in medicine. Poly(glycolide lactide) (PGL) is a random copolymer used in resorbable surgical sutures. PGL polymerization occurs via a ring-opening reaction of a glycolide and a lactide, as illustrated in Fig. 15. The presence of ester linkages in the polymer backbone allows gradual hydrolytic degradation (resorption). In contrast to the natural resorbable suture material poly(glycolic acid), or catgut, a homopolymer, the PGL copolymer retains more of its strength over the first 14 days after implantation (Chu, 1983).

A copolymer of tetrafluoroethylene and hexafluoropropylene (FEP) is used in many applications similar to those of PTFE. FEP has a crystalline melting point near 265°C compared with 327°C for PTFE. This enhances the processibility of FEP compared with PTFE while maintaining the excellent chemical inertness and low friction characteristic of PTFE.

Polyurethanes are block copolymers containing "hard" and "soft" blocks. The "hard" blocks, having T_gs above room temperature and acting as glassy or semicrystalline reinforcing blocks, are composed of a diisocyanate and a chain extender. The diisocyanates most commonly used are 2,4-toluene diisocyanate (TDI) and methylene di(4-phenyl isocyanate) (MDI), with MDI being used in most biomaterials. The chain extenders are usually shorter aliphatic glycol or diamine materials with 2–6 carbon atoms. The "soft" blocks in polyurethanes are typically polyether or polyester polyols whose T_gs are much less than room temperature, allowing them to give a rubbery character to the materials. Polyether polyols are more commonly used for implantable devices because they are stable to hydrolysis. The polyol molecular weights tend to be on the order of 1000 to 2000.

Polyurethanes are tough elastomers with good fatigue and blood-containing properties. They are used in pacemaker lead insulation, vascular grafts, heart assist balloon pumps, and artificial heart bladders.

Bibliography

Billmeyer, F. W., Jr. (1984). *Textbook of Polymer Science,* 3rd ed. Wiley-Interscience, New York.

Bovey, F. A., and Winslow, F. H. (1979). *Macromolecules: An Introduction to Polymer Science.* Academic Press, Orlando, FL.

Chu, C. C. (1983). Survey of clinically important wound closure biomaterials, in *Biocompatible Polymers, Metals, and Composites,* M. Szycher, ed. Technomic Publ., Lancaster, PA, pp. 477–523.

Flory, P. J. (1953). *Principles of Polymer Chemistry.* Cornell University Press, London.

Lelah, M. D., and Cooper, S. L. (1986). *Polyurethanes in Medicine.* CRC Press, Boca Raton, FL.

Mandelkern, L. (1983). *An Introduction to Macromolecules.* Springer-Verlag, New York.

Rodriguez, F. (1982). *Principles of Polymer Systems,* 2nd ed. McGraw-Hill, New York.

Rosato, D. V. (1983). Polymers, processes and properties of medical plastics: including markets and applications, in *Biocompatible Polymers, Metals, and Composites,* M. Szycher, ed. Technomic Publ., Lancaster, PA, pp. 1019–1067.

Seymour, R. B., and Carraker, C. E. Jr. (1988). *Polymer Chemistry: An Introduction,* 2nd ed. Marcel Dekker, New York.

Sperling, L. H. (1986). *Introduction to Physical Polymer Science.* Wiley-Interscience, New York.

Stokes, K., and Chem. B. (1984). Environmental stress cracking in implanted polyether polyurethanes, in *Polyurethanes in Biomedical Engineering,* H. Planck, G. Engbers, and I. Syré, eds. Elsevier, Amsterdam.

2.4 HYDROGELS

Nikolaos A. Peppas

Hydrogels are water-swollen, cross-linked polymeric structures produced by the simple reaction of one or more monomers or by association bonds such as hydrogen bonds and strong van der Waals interactions between chains (Peppas, 1987). Hydrogels have received significant attention, especially in the past 30 years, because of their exceptional promise in biomedical applications. The classic book by Andrade (1976) offers some of the best work that was available prior to 1975. The more recent book by Peppas (1987) addresses the preparation, structure, and characterization of hydrogels. In this chapter, we concentrate on some features of the preparation of hydrogels, as well as characteristics of their structure and chemical and physical properties.

CLASSIFICATION AND BASIC STRUCTURE

Hydrogels may be classified in several ways, depending on their method of preparation, ionic charge, or physical structure features. Based on the method of preparation, they are (1) homopolymer hydrogels, (2) copolymer hydrogels, (3) multipolymer hydrogels, and (4) interpenetrating polymeric hydrogels. Homopolymer hydrogels are cross-linked networks of one type of hydrophilic monomer unit, whereas copolymer hydrogels are produced by cross-linking of two comonomer units, one of which must be hydrophilic. Multipolymer hydrogels are produced from three or more comonomers reacting together. Finally, interpenetrating polymeric hydrogels are produced by swelling a first network in a monomer and reacting the latter to form a second intermeshing network structure. Based on their ionic charges, hydrogels may be classified (Ratner and Hoffman, 1976) as (1) neutral hydrogels, (2) anionic hydrogels, (3) cationic hydrogels, and (4) ampholytic hydrogels. Based on physical structural features of the system, they can be classified as (1) amorphous hydrogels, (2) semicrystalline hydrogels, and (3) hydrogen-bonded structures. In amorphous hydrogels, the macromolecular chains are randomly arranged, whereas semicrystalline hydrogels are characterized by dense regions of ordered macromolecular chains (crystallites). Often, hydrogen bonds may be responsible for the three-dimensional structure formed.

Structural evaluation of hydrogels reveals that ideal networks are only rarely observed. Figure 1a shows an ideal macromolecular network (hydrogel) indicating tetrafunctional cross-links (junctions) produced by covalent bonds. However, the possibility exists of multifunctional junctions (Fig. 1b) or physical molecular entanglements (Fig. 1c) playing the role of semipermanent junctions. Hydrogels with molecular defects are always possible. Figures 1d and 1e indicate two such effects: unreacted functionalities with partial entanglements (Fig. 1d) and chain loops (Fig. 1e). Neither of these effects contributes to the mechanical or physical properties of a polymer network.

The terms "junction" and "cross-link" (an open circle symbol in Fig. 1d) indicate the connection points of several chains. This junction may be ideally a carbon atom, but it is usually a small chemical bridge [e.g., an acetal bridge in the case of poly(vinyl alcohol)] of molecular weight much smaller than that of the cross-linked polymer chains. In other situations, a junction may be an association of macromolecular chains caused by van der Waals forces, as in the case of the glycoproteinic network structure of natural mucus, or an aggregate

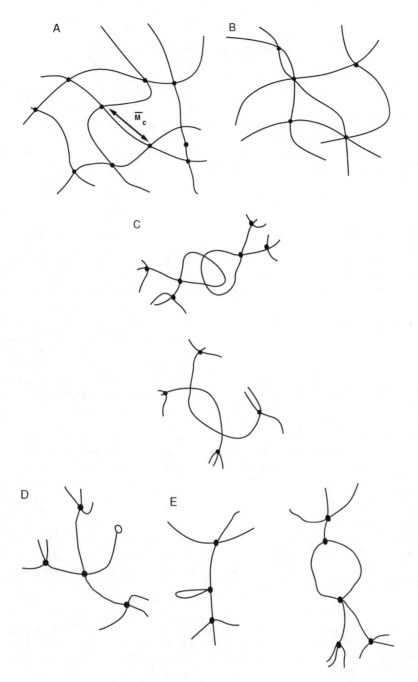

FIG. 1.   (A) Ideal macromolecular network of a hydrogel. (B) Network with multifunctional junctions. (C) Physical entanglements in a hydrogel. (D) Unreacted functionality in a hydrogel. (E) Chain loops in a hydrogel.

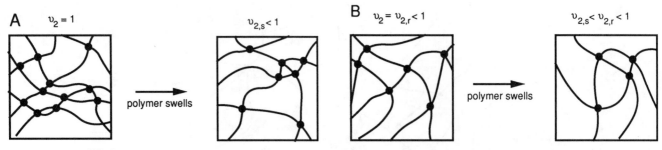

FIG. 2.   (A) Swelling of a network prepared by cross-linking in dry state. (B) Swelling of a network prepared by cross-linking in solution.

formed by hydrogen bonds, as in the case of aged microgels formed in polymer solutions.

Finally, the structure may include effective junctions that can be either simple, physical entanglements of permanent or semipermanent nature, or ordered chains forming crystallites. Thus, the junctions should never be considered as a "volumeless point," the usual depiction applied when developing structural models for analysis of the cross-linked structure of hydrogels (Flory, 1953).

PREPARATION

Hydrogels are prepared by swelling cross-linked structures in water or biological fluids containing large amounts of water. In many situations, the water may be present during the initial formation of the cross-linked structure. There are many methods of preparing cross-linked hydrogels, such as irradiative cross-linking and chemical reactions.

Radiation reactions (Chapiro, 1962) utilize electron beams, gamma-rays, X-rays, or ultraviolet light to excite a polymer and produce a cross-linked structure. Chemical cross-linking requires the use of at least one difunctional, small-molecular-weight, cross-linking agent. This agent usually links two longer molecular weight chains through its di- or multifunctional groups. The second method is a copolymerization–cross-linking reaction between one or more abundant monomers and one multifunctional monomer that is present in very small quantities. A third variation of these techniques involves using a combination of monomer and linear polymeric chains that are cross-linked by means of an interlinking agent, as in the production of polyurethanes.

SWELLING BEHAVIOR

An integral part of the physical behavior of hydrogels is their swelling behavior in water, since upon preparation they must be brought in contact with water to yield the final, solvated network structure. Figure 2 shows one of the two possible processes of swelling. A dry, hydrophilic cross-linked network is placed in water. Then, the macromolecular chains interact with the solvent molecules owing to the relatively good thermodynamic compatibility. Thus, the network expands to the solvated state. The Flory–Huggins theory can be used to calculate thermodynamic quantities related to that mixing process (Flory, 1953).

This thermodynamic swelling force is counterbalanced by the retractive force of the cross-linked structure. The latter is usually described by the Flory rubber elasticity theory and its variations (Flory, 1953). Equilibrium is attained in a particular solvent at a particular temperature when the two forces become equal. The volume degree of swelling, Q (i.e., the ratio of the actual volume of a sample in the swollen state divided by its volume in the dry state) can then be determined.

Several researchers working with hydrogels, especially for biomedical applications, prefer to use other parameters to define the equilibrium swelling behavior. For example, Yasuda *et al.* (1969) propagated the use of the so-called hydration ratio, H, which has been accepted by those researchers who use hydrogels for contact lenses. Another definition is that of the weight degree of swelling, q, which is the ratio of the weight of the swollen sample over that of the dry sample (Flory, 1953).

In general, highly swollen hydrogels are those of cellulose derivatives, poly(vinyl alcohol), poly(*N*-vinyl 2-pyrrolidone) (PNVP), and poly(ethylene glycol), among others. Moderately and poorly swollen hydrogels are those of poly(hydroxyethyl methacrylate) (PHEMA) and many of its derivatives. Of course, one may copolymerize a basic hydrophilic monomer with other more or less hydrophilic monomers to achieve desired swelling properties.

Such processes have led to a wide range of swellable hydrogels, as Gregonis *et al.* (1976), Peppas (1987), and others have pointed out. Knowledge of the swelling characteristics of a polymer is of utmost importance in biomedical and pharmaceutical applications since the equilibrium degree of swelling influences (1) the solute diffusion coefficient through these hydrogels, (2) the surface properties and surface mobility, (3) the optical properties, especially in relation to contact lens applications, and (4) the mechanical properties.

DETERMINATION OF STRUCTURAL CHARACTERISTICS

The parameter that describes the basic structure of the hydrogel is the molecular weight between cross-links, $\overline{M}_c$, as shown in Figure 1a. This parameter defines the average molecular size between two consecutive junctions regardless of the nature of those junctions. Additional parameters of importance in structural analysis of hydrogels are the cross-linking density, ρ_x, which is defined by Eq. 1, and the effective number of cross-links, ν_e, per original chain (Eq. 2).

$$\rho_x = \frac{1}{\overline{v}\,\overline{M}_c} \qquad (1)$$

$$\nu_e = \left(\frac{\overline{M}_n}{\overline{M}_c}\right) - 1 \qquad (2)$$

In these equations, $\overline{v}$ is the specific volume of the polymer (i.e., the reciprocal of the amorphous density of the polymer), and $\overline{M}_n$ is the initial molecular weight of the uncross-linked polymer.

PROPERTIES OF SOME BIOMEDICALLY AND PHARMACEUTICALLY IMPORTANT HYDROGELS

The multitude of hydrogels available leaves numerous choices for polymeric formulations. The best approach for developing a hydrogel with the desired characteristics is to correlate the macromolecular structures of the polymers available with the swelling and mechanical characteristics desired.

The most widely used hydrogel is water-swollen, cross-linked PHEMA, which was introduced as a biological material by Wichterle and Lim (1960). The PHEMA structure permits a water content similar to living tissue. The hydrogel is inert to normal biological processes, shows resistance to degradation, is permeable to metabolites, is not absorbed by the body, withstands heat sterilization without damage, and can be prepared in a variety of shapes and forms.

The swelling, mechanical, diffusional, and biomedical characteristics of PHEMA gels have been studied extensively. The properties of these hydrogels are dependent upon their method of preparation, polymer volume fraction, degree of cross-linking, temperature, and swelling agent.

Other hydrogels of biomedical interest include polyacrylamides. Tanaka (1979) has done extensive studies on the abrupt swelling and deswelling of partially hydrolyzed acrylamide gels with changes in swelling agent composition, curing time, degree of cross-linking, degree of hydrolysis, and temperature. These studies have shown that the ionic groups produced in an acrylamide gel upon hydrolysis give the gel a structure that shows a discrete transition in equilibrium swollen volume with environmental changes.

Discontinuous swelling in partially hydrolyzed polyacrylamide gels has been studied by Gehrke *et al.* (1986). They have utilized polyacrylamide gels in gel extraction processes as a method of concentrating dilute aqueous solutions. The solution to be concentrated is added to a small, unswollen gel particle. These gels then swell in water, often up to six times their original weight. The concentrated solution is then withdrawn from around the gel. Acid is added to shrink the gel and release the water; the gel particles are removed and treated with base; and the process is repeated. These gels may be used repeatedly for the same extraction process.

Besides HEMA and acrylamides, *N*-vinyl-2-pyrrolidone (NVP), methacrylic acid (MAA), methyl methacrylate (MMA), and maleic anhydride (MAH) have all been proven useful as monomers for hydrogels in biomedical applications. For instance, PNVP is used in soft contact lenses. Small amounts of MAA as a comonomer have been shown to dramatically increase the swelling of PHEMA polymers. Owing to the hydrophobic nature of MMA, copolymers of MMA and HEMA have a lower degree of swelling then pure PHEMA (Brannon-Peppas and Peppas, 1991). All of these materials have potential use in advanced technology applications, including biomedical separations, and biomedical and pharmaceutical devices.

APPLICATIONS

The physical properties of hydrogels make them attractive for a variety of biomedical and pharmaceutical applications. Their biocompatibility allows them to be considered for medical applications, whereas their hydrophilicity can impart desirable release characteristics to controlled and sustained release formulations.

Hydrogels exhibit properties that make them desirable candidates for biocompatible and blood-compatible biomaterials (Merrill *et al.*, 1987). Nonionic hydrogels for blood contact applications have been prepared from poly(vinyl alcohol), polyacrylamides, PNVP, PHEMA, and poly(ethylene oxide). Heparinized polymer hydrogels also show promise as materials for blood-compatible applications (Sefton, 1987).

One of the earliest biomedical applications of hydrogels was in contact lenses (Tighe 1976; Peppas and Yang, 1981) because of their relatively good mechanical stability, favorable refractive index, and high oxygen permeability.

Other applications of hydrogels include (Peppas, 1987) artificial tendon materials, wound-healing bioadhesives, artificial kidney membranes, articular cartilage, artificial skin, maxillofacial and sexual organ reconstruction materials, and vocal cord replacement materials.

Pharmaceutical hydrogel applications have become very popular in recent years. Pharmaceutical hydrogel systems can be classified into various types. The category of **equilibrium-swollen hydrogels** includes matrices that have a drug incorporated in them and are swollen to equilibrium. The category of **solvent-activated**, matrix-type, controlled-release devices comprises two important types of systems: swellable and swelling-controlled devices. In general, a system prepared by incorporating a drug into a hydrophilic, glassy polymer can be swollen when brought in contact with water or a simulant of biological fluids. This swelling process may or may not be the controlling mechanism for diffusional release, depending on the magnitude of the macromolecular relaxation of the polymer.

In **swelling-controlled release systems,** the bioactive agent is dispersed into the polymer to form nonporous films, disks, or spheres. Upon contact with an aqueous dissolution medium, a distinct front (interface) is observed that corresponds to the water penetration front into the polymer and separates the glassy from the rubbery (gel-like) state of the material. Under these conditions, the macromolecular relaxations of the polymer influence the diffusion mechanism of the drug through the rubbery state. This water uptake can lead to considerable swelling of the polymer with a thickness that depends on time. The swelling process proceeds toward equilibrium at a rate determined by the water activity in the system and the structure of the polymer. If the polymer is cross-linked or of sufficiently high molecular weight (so that chain entanglements can maintain structural integrity), the equilibrium state is a water-swollen gel. The equilibrium water content of such hydrogels can vary up to more than 90%. If the dry hydrogel contains a water-soluble drug, the drug is essentially immobile in the glassy matrix, but begins to diffuse out as the polymer swells with water. Drug release thus depends on two simultaneous rate processes: water migration into the device and drug diffusion outward through the swollen gel. Since some water uptake must occur before the drug can be released, the initial burst effect frequently observed in matrix devices is moderated, although it may still be present. The continued swelling of the matrix causes the drug to diffuse increasingly easily, ameliorating the slow tailing off of the release curve. The net effect of the swelling process is to prolong and linearize the release curve. Additional discussion of controlled release systems for drug delivery can be found in Chapter 7.8.

Details of these experimental techniques have been presented by Korsmeyer and Peppas (1981) for poly(vinyl alchol) systems, and by Peppas (1981) for PHEMA systems and their copolymers.

Bibliography

Andrade, J. D. (1976). *Hydrogels for Medical and Related Applications.* ACS Symposium Series, Vol. 31, American Chemical Society, Washington, DC.

Brannon-Peppas, L., and Peppas, N. A. (1991). Equilibrium swelling behavior of dilute ionic hydrogels in electrolytic solutions. *J. Controlled Release* **16**: 319–330.

Chapiro, A. (1962). *Radiation Chemistry of Polymeric Systems.* Interscience, New York.

Flory, P. J. (1953). *Principles of Polymer Chemistry.* Cornell University Press, Ithaca, NY.

Gehrke, S. H., Andrews, G. P., and Cussler, E. L. (1986). Chemical aspects of gel extraction. *Chem. Eng. Sci.* **41**: 2153–2160.

Gregonis, D. E., Chen, C. M., and Andrade, J. D. (1976). The chemistry of some selected methacrylate hydrogels, in *Hydrogels for Medical and Related Applications.* J. D. Andrade, ed. ACS Symposium Series, Vol. 31, pp. 88–104, American Chemical Society, Washington, DC.

Ilavsky, M. (1982). Phase transition in swollen gels. *Macromolecules* **15**: 782–788.

Korsmeyer, R. W., and Peppas, N. A. (1981). Effects of the morphology of hydrophilic polymeric matrices on the diffusion and release of water soluble drugs. *J. Membr. Sci* **9**: 211–227.

Merrill, E. W., Pekala, P. W., and Mahmud, N. A. (1987). Hydrogels for blood contact, in *Hydrogels in Medicine and Pharmacy,* N. A. Peppas, ed. CRC Press, Boca Raton, FL, Vol. 3, pp. 1–16.

Peppas, N. A. (1987). *Hydrogels in Medicine and Pharmacy.* CRC Press, Boca Raton, FL.

Peppas, N. A., and Yang, W. H. M. (1981). Properties-based optimization of the structure of polymers for contact lens applications. *Contact Intraocular Lens Med. J.* **7**: 300–321.

Ratner, B. D., and Hoffman, A. S. (1976). Synthetic hydrogels for biomedical applications. in *Hydrogels for Medical and Related Applications,* J. D. Andrade, ACS Symposium Series, American Chemical Society, Washington, DC, Vol. 31, pp. 1–36.

Sefton, M. V. (1987). Heparinized hydrogels. in *Hydrogels in Medicine and Pharmacy,* N. A. Peppas, ed. CRC Press, Boca Raton, FL, Vol. 3, pp. 17–52.

Tanaka, T. (1979). Phase transitions in gels and a single polymer. *Polymer* **20**: 1404–1412.

Tighe, B. J. (1976). The design of polymers for contact lens applications. *Brit. Polym. J.* **8**: 71–90.

Wichterle, O., and Lim, D. (1960). Hydrophilic gels for biological use. *Nature* **185**: 117–118.

Yasuda, H., Peterlin, A., Colton, C. K., Smith, K. A., and Merrill, E. W. (1969). Permeability of solutes through hydrated polymer membranes. III. Theoretical background for the selectivity of dialysis membranes. *Makromol. Chemie* **126**: 177–186.

Yoshio, N. Hirohito, N., and Matsuhiko, M. (1986). Properties of swelling and shrinking. *J. Chem. Eng. Japan* **19**: 274–280.

2.5 BIORESORBABLE AND BIOERODIBLE MATERIALS

Joachim Kohn and Robert Langer

TYPES OF IMPLANTS

Since a degradable polymeric implant does not have to be removed surgically once it is no longer needed, degradable polymers are of value in short-term applications that require only the temporary presence of a polymeric implant. An additional advantage is that the use of degradable implants can circumvent some of the problems related to the long-term safety of permanently implanted devices. Some typical short-term applications are listed in Table 1. From a practical perspective, it is convenient to distinguish among four main types of degrad-

TABLE 1 Some "Short-Term" Medical Applications of Degradable Polymeric Biomaterials

Application	Comments
Sutures	The earliest, successful application of synthetic, degradable polymers in human medicine.
Drug delivery devices	One of the most widely investigated medical applications for degradable polymers.
Orthopedic fixation devices	Requires polymers of exceptionally high mechanical strength and stiffness.
Adhesion prevention	Requires polymers that can form soft membranes or films.
Temporary vascular grafts and stents	Only investigational devices are presently available. Blood compatibility is a major concern.

able implants: the temporary scaffold, the temporary barrier, the drug delivery device, and the multifunctional implant.

The Temporary Scaffold

The use of a temporary scaffold can be envisioned in those circumstances where the natural tissue bed has been weakened by disease, injury, or surgery and requires some artificial support. A healing wound, a broken bone, or a damaged blood vessel are examples of such situations. Sutures, bone fixation devices (e.g., bone nails, screws, or plates), and vascular grafts would be examples of the corresponding support devices. In all of these instances, the degradable implant would provide temporary, mechanical support until the natural tissue healed and regained its strength. In order for a temporary scaffold to work properly, a gradual stress transfer should occur: as the natural tissue heals, the degradable implant should gradually weaken. The need to adjust the degradation rate of the temporary scaffold to the healing of the surrounding tissue represents one of the major challenges in the design of a temporary scaffold.

Currently, sutures represent the most successful example of a temporary scaffold-type implant in human medicine. The first synthetic, degradable sutures were made of poly(glycolic acid) (PGA) and became available under the trade name Dexon in 1970. This represented the first routine use of a degradable polymer in a major clinical application (Frazza and Schmitt, 1971). Later copolymers of PGA and poly(lactic acid) (PLA) were developed. The widely used Vicryl suture, for example, is a 90 : 10 copolymer of PGA/PLA, introduced into the market in 1974. Sutures made of polydioxanone (PDS) became available in the United States in 1981. In spite of extensive research efforts in many laboratories, no other degradable polymers are currently used to any significant extent in the formulation of degradable sutures.

The Temporary Barrier

The major medical application of a temporary barrier is in adhesion prevention. Surgical adhesions between two tissue sections are caused by clotting of blood in the extravascular tissue space, which is followed by inflammation and fibrosis. If this natural healing process occurs between surfaces that were not meant to bond together, the resulting adhesion can cause pain, functional impairment, and problems during subsequent surgery. Adhesions are a common problem after cardiac, spinal, and tendon surgery. A temporary barrier could take the form of a thin polymeric film or a meshlike device that would be placed between adhesion-prone tissues at the time of surgery. Artificial skin for the treatment of burns and other skin lesions is another widely investigated application for temporary barrier-type devices.

The Drug Delivery Device

Since implantable drug delivery devices are by necessity temporary devices, the development of implantable drug delivery systems is probably the most widely investigated application of degradable polymers. One can expect that the future acceptance of implantable drug delivery devices by physicians and patients alike will depend on the availability of degradable systems that do not have to be explanted surgically.

In an attempt to shorten the regulatory process, poly(lactic acid) and poly(glycolic acid) are often considered first, although a wide range of other polymers have been explored. Several implanted, controlled-release formulations based on degradable polymers are undergoing advanced clinical trials. Particularly noteworthy is an intracranial polyanhydride device used for administering a chemotherapeutic agent to patients suffering from glioblastoma multiformae, a usually lethal form of brain cancer.

Multifunctional Devices

Over the past few years, there has been a trend toward increasingly sophisticated applications for degradable biomaterials. Usually these applications envision the combination of several functions within the same device (hence the name "multifunctional devices") and require the design of custommade materials with a narrow range of predetermined properties. For example, the availability of biodegradable bone nails and bone screws made of ultrahigh-strength poly(lactic acid) opens the possibility of combining the mechanical support function with a site-specific drug delivery function: A biodegradable bone nail that holds the fractured bone in place can simultaneously stimulate the growth of new bone tissue at the fracture site by slowly releasing bone growth factors (e.g., bone morphogenic protein or transforming growth factor-β) throughout its degradation process.

Likewise, biodegradable stents for implantation into coronary arteries are being investigated. The stents are designed to mechanically prevent the collapse and restenosis (reblocking) of arteries that have been opened by balloon angioplasty. Ultimately, the stents could deliver an anti-inflammatory or antithrombogenic agent directly to the site of vascular injury. Again, it might be possible to combine a mechanical support function with site-specific drug delivery.

DEFINITIONS

Currently four different terms (biodegradation, bioerosion, bioabsorption, and bioresorption) are being used to indicate that a given material or device will eventually disappear after being introduced into a living organism. However, when reviewing the literature, no clear distinctions in the meaning of these four terms are evident. Likewise, the meaning of the prefix "bio" is not well established, leading to the often interchangeable use of the terms "degradation" and "biodegradation," or "erosion" and "bioerosion." Although efforts have been made to establish generally applicable and widely accepted definitions for all aspects of biomaterials research (Williams, 1987), there is still significant confusion even between experienced researchers in the field as to the correct terminology for various degradation processes.

In the context of this chapter, we follow the usage suggested by the Consensus Conference of the European Society for Biomaterials (Williams, 1987) and refer to "biodegradation" only when we wish to emphasize that a biological agent (enzyme or microbe) is a dominant component in the degradation process. Consequently, the degradation of poly(lactic acid) to lactic acid should not be described as "biodegradation" since this degradation process is caused by hydrolytic cleavage of the polymer backbone, with little or no evidence for the active participation of enzymes. In correspondence with Heller's suggestion (Heller, 1987), we define a "bioerodible polymer" as a water-insoluble polymer that is converted *under physiological conditions* into water-soluble material(s) without regard to the specific mechanism involved in the erosion process. "Bioerosion" includes therefore both physical processes (such as dissolution) and chemical processes (such as backbone cleavage). Here the prefix "bio" indicates that the erosion occurs under physiological conditons, as opposed to other erosion processes, caused, for example, by high temperature, strong acids or bases, UV light or weather conditions. The terms "bioresorption" and "bioabsorption" are used interchangeably and often imply that the polymer or its degradation products are removed by cellular activity (e.g., phagocytosis) in a biological environment. These terms are somewhat superfluous and have not been clearly defined.

CURRENTLY AVAILABLE DEGRADABLE POLYMERS

From the beginning of the material sciences, the development of highly stable materials has been a major research challenge. Today, many polymers are available that are virtually nondestructible in biological systems, e.g., Teflon, Kevlar, or poly(ether-ether ketone). On the other hand, the development of degradable biomaterials is a relatively new area of research. The variety of available, degradable biomaterials is still too limited to cover a wide enough range of diverse material properties. Thus, the design and synthesis of new, degradable biomaterials is an important research challenge.

Degradable materials must fulfill more stringent requirements in terms of their biocompatibility than nondegradable materials. In addition to the potential problem of toxic contaminants leaching from the implant (residual monomers, stabilizers, polymerizaton initiators, emulsifiers, etc.), one must also consider the potential toxicity of the degradation products and subsequent metabolites. The practical consequence of this consideration is that only a limited number of nontoxic, monomeric starting materials have been successfully used to prepare degradable biomaterials.

Over the past decade, dozens of hydrolytically unstable polymers have been suggested as degradable biomaterials; however, in most cases no attempts have been made to develop these new materials for specific medical applications. Thus, detailed toxicological studies *in vivo*, investigations of degradation rate and mechanism, and careful evaluations of the physicomechanical properties have so far been published for only

a very small fraction of those polymers. An even smaller number of synthetic, degradable polymers has so far been approved by the U.S. Food and Drug Administration (FDA) for use in clinical studies involving humans, and only three synthetic, degradable polymers (PLA, PGA, and PDS) are used routinely for a narrow range of applications in human medicine.

Recent research has led to a number of well-established investigational polymers that may find practical applications as degradable implants within the next decade. Representative examples of these polymers are described in the following paragraphs. In addition, structural formulas (Fig. 1) and important mechanical properties (Table 2) are provided. It is interesting that a large proportion of the currently investigated, degradable polymers are polyesters. It remains to be seen whether some of the alternative backbone structures such as polyanhydrides, polyphosphazenes, polyphosphonates, polyamides, or polyiminocarbonates will be able to challenge the dominant position of the polyesters in the future.

Polyhydroxybutyrate (PHB), Polyhydroxyvalerate (PHV), and Copolymers

These polymers are examples of bioerodible polyesters that are derived from microorganisms. PHB and its copolymers with up to 30% of 3-hydroxyvaleric acid are now commercially available under the trade name Biopol. PHB and PHV are intracellular storage polymers that provide a reserve of carbon and energy. The polymers can be degraded by soil bacteria but are relatively stable in ambient conditions. The rate of degradation can be controlled by varying the copolymer composition. *In vivo*, PHB degrades to D-3-hydroxybutyric acid, which is a normal constituent of human blood. The low toxicity of PHB may be at least partly due to this fact.

PHB homopolymer is very crystalline and brittle while the copolymers of PHB with hydroxyvaleric acid are less crystalline, more flexible and more readily processible. The polymers have been considered in several biomedical applications such as controlled drug release, sutures, and artificial skin, as well as industrial applications such as paramedical disposables.

Polycaprolactone

Polycaprolactone became available commercially following efforts at Union Carbide to identify synthetic polymers that could be degraded by microorganisms. It is a semicrystalline polymer. The high solubility of polycaprolactone, its low melting point (59–64°C) and exceptional ability to form blends has stimulated research on its application as a biomaterial. Polycaprolactone degrades at a slower pace than PLA and can therefore be used in drug delivery devices that remain active for over a year. The release characteristics of polycaprolactone have been investigated in detail by Pitt and his co-workers. The Capronor system, a 1-year implantable contraceptive device (Pitt, 1990), has undergone Phase I and Phase II clinical trials in the United States and may become commercially available in Europe in the near future. The toxicology of polycaprolactone has been extensively studied as part of the evaluation of Capronor. Based on a large number of tests, ε-caprolactone

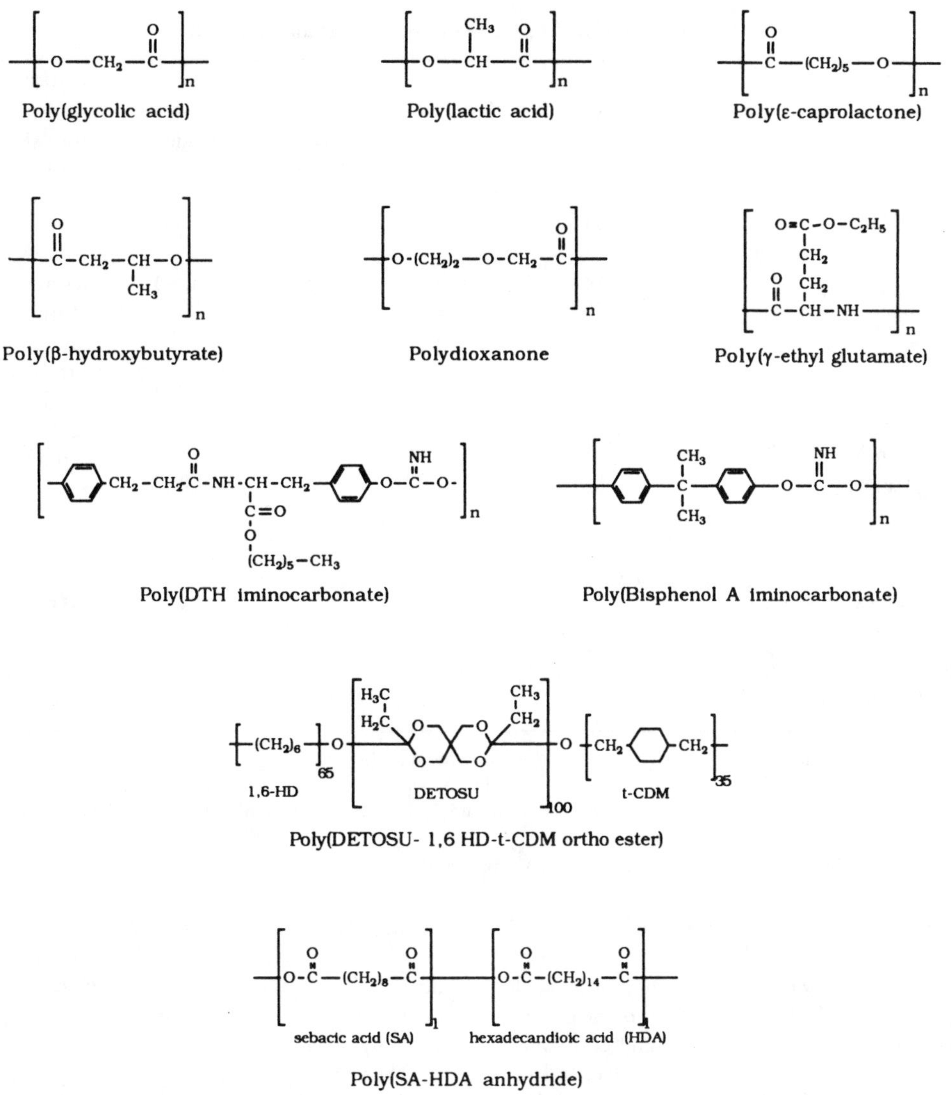

FIG. 1. Chemical structures of the degradable polymers listed in Table 2.

and polycaprolactone are currently regarded as nontoxic and tissue-compatible materials. In Europe, polycaprolactone is already in clinical use as a degradable staple (for wound closure), and it stands to reason that polycaprolactone, or blends and copolymers containing polycaprolactone, will find additional medical applications in the future.

Polyanhydrides

Polyanhydrides were explored as possible substitutes for polyesters in textile applications but ultimately failed owing to their pronounced hydrolytic instability (Conix, 1958). It was this property that prompted Langer and his co-workers to explore polyanhydrides as degradable implant materials (Domb *et al.*, 1988). Aliphatic polyanhydrides degrade within days whereas some aromatic polyanhydrides degrade over several years. Thus aliphatic-aromatic copolymers, having intermediate rates of degradation are usually employed.

Polyanhydrides are among the most reactive and hydrolytically unstable polymers currently used as biomaterials. The high chemical reactivity is both an advantage and a limitation of polyanhydrides. Because of their high rate of degradation, many polyanhydrides degrade by surface erosion without the need to incorporate various catalysts or excipients into the device formulation. On the other hand, polyanhydrides will react with drugs containing free amino groups or other nucleophilic functionalities, especially during high-temperature processing. The potential reactivity of the polymer matrix toward nucleophiles limits the types of drugs that can be successfully incorporated into a polyanhydride matrix by melt processing techniques.

A comprehensive evaluation of their toxicity showed that, in general, the polyanhydrides possess excellent *in vivo* biocom-

TABLE 2 Mechanical Properties of Some Degradable Polymers[a]

Polymer	Glass transition (°C)	Melting temperature (°C)	Tensile strength (MPa)	Tensile modulus (MPa)	Flexural modulus (MPa)	Elongation Yield (%)	Elongation Break (%)
Poly(glycolic acid) (MW: 50,000)	35	210	n/a	n/a	n/a	n/a	n/a
Poly(lactic acids)							
L-PLA (MW: 50,000)	54	170	28	1200	1400	3.7	6.0
L-PLA (MW: 100,000)	58	159	50	2700	3000	2.6	3.3
L-PLA (MW: 300,000)	59	178	48	3000	3250	1.8	2.0
D,L-PLA (MW: 20,000)	50	—	n/a	n/a	n/a	n/a	n/a
D,L-PLA (MW: 107,000)	51	—	29	1900	1950	4.0	6.0
D,L-PLA (MW: 550,000)	53	—	35	2400	2350	3.5	5.0
Poly(β-hydroxybutyrate) (MW: 422,000)	1	171	36	2500	2850	2.2	2.5
Poly(ε-caprolactone) (MW: 44,000)	−62	57	16	400	500	7.0	80
Polyanhydrides[b]							
Poly(SA-HDA anhydride) (MW: 142,000)	n/a	49	4	45	n/a	14	85
Poly(ortho esters)[c]							
DETOSU : t-CDM : 1,6-HD (MW: 99,700)	55	—	20	820	950	4.1	220
Polyiminocarbonates[d]							
Poly(BPA iminocarbonate) (MW: 105,000)	69	—	50	2150	2400	3.5	4.0
Poly(DTH iminocarbonate) (MW: 103,000)	55	—	40	1630	n/a	3.5	7.0

[a]Based on data published by Engelberg and Kohn (1991). n/a = not available, (—) = not applicable.
[b]A 1:1 copolymer of sebacic acid (SA) and hexadecanedioic acid (HDA) was selected as a specific example.
[c]A 100:35:65 copolymer of 3,9-bis(ethylidene 2,4,8,10-tetraoxaspiro[5,5] undecane) (DETOSU), *trans*-cyclohexane dimethanol (t-CDM) and 1,6-hexanediol (1,6-HD) was selected as a specific example.
[d]BPA: Bisphenol A; DTH: desaminotyrosyl-tyrosine hexyl ester. For detailed structures, see Fig. 1.

patibility (Laurencin *et al.*, 1990). The most immediate applications are in drug delivery. Drug-loaded devices are best prepared by compression molding or microencapsulation. A wide variety of drugs and proteins, including insulin, bovine growth factors, angiogenesis inhibitors (e.g., heparin and cortisone), and enzymes (e.g., alkaline phosphatase and β-galactosidase) have been incorporated into polyanhydride matrices and their *in vitro* and *in vivo* release characteristics have been evaluated (Chasin *et al.*, 1990). One particularly important application is in the delivery of bis-chloroethylnitrosourea (BCNU) to the brain for the treatment of glioblastoma multiformae, a universally fatal brain cancer. For this application, polyanhydrides derived from bis-*p*-(carboxyphenoxy propane) and sebacic acid have been approved by the FDA for Phase III clinical trials in a remarkably short time (Chasin *et al.*, 1990).

Poly(Ortho Esters)

This is a family of synthetic, degradable polymers that have been under development for a number of years (Heller *et al.*, 1990). Devices made of poly(ortho esters) can be formulated in such a way that the device undergoes surface erosion. Since surface eroding, slablike devices tend to release drugs embedded within the polymer at a constant rate, poly(ortho esters) appear to be particularly useful for controlled-release drug delivery applications. For example, poly(ortho esters) have been used for the controlled delivery of cyclobenzaprine and steroids and a significant number of publications describe the

use of poly(ortho esters) for various drug delivery applications. Since the ortho ester link is far more stable in base than in acid, Heller and his co-workers controlled the rate of polymer degradation by incorporating acidic or basic excipients into the polymer matrix.

There are two major types of poly(ortho esters): Initially, Choi and Heller prepared the polymers by the transesterification of 2,2'-dimethoxyfuran with a diol (Cho and Heller, 1978). The next generation of poly(ortho esters) was based on an acid-catalyzed addition reaction of diols with diketeneacetals. The properties of the polymers can be controlled to a large extent by the choice of the diols used in the synthesis. For example, the glass transition temperature of poly(ortho esters) containing *trans*-cyclohexanedimethanol can be reduced from about 100°C to below 20°C by replacing *trans*-cyclohexanedimethanol with 1,6-hexanediol.

Poly(Amino Acids) and "Pseudo"-Poly(Amino Acids)

Since proteins are composed of amino acids, it was an obvious idea to explore the possible use of poly(amino acids) in biomedical applications (Anderson *et al.*, 1985). Poly(amino acids) were regarded as promising candidates since the amino acid side chains offer sites for the attachment of drugs, cross-linking agents, or pendent groups that can be used to modify the physicomechanical properties of the polymer. In addition, poly(amino acids) usually show a low level of systemic toxicity, owing to their degradation to naturally occurring amino acids.

Poly(amino acids) have been investigated as suture materials, as artificial skin substitutes, and as drug delivery systems. Various drugs have been attached to the side chains of poly(amino acids) usually by a spacer unit that distances the drug from the backbone. Poly(amino acid)-drug combinations investigated include poly(L-lysine) with methotrexate and pepstatin (Campbell *et al.*, 1980), and poly(glutamic acid) with adriamycin and norethindrone (van Heeswijk *et al.*, 1985).

Despite their apparent potential as biomaterials, poly(amino acids) have actually found few practical applications. Most are highly insoluble and nonprocessible materials. Since poly(amino acids) have a pronounced tendency to swell in aqueous media, it can be difficult to predict drug release rates. Furthermore, the antigenicity of polymers containing three or more amino acids excludes their use in biomedical applications (Anderson *et al.*, 1985). Owing to these difficulties, only a few poly(amino acids), usually derivatives of poly(glutamic acid) carrying various pendent chains at the γ-carboxylic acid group, are being investigated as implant materials.

In an attempt to circumvent the problems associated with conventional poly(amino acids), backbone-modified "pseudo"-poly(amino acids) were introduced in 1984 (Kohn and Langer, 1984). The first "pseudo"-poly(amino acids) investigated were a polyester from *N*-protected *trans*-4-hydroxy-L-proline, and a poly(iminocarbonate) derived from tyrosine dipeptide (Kohn and Langer, 1987). Recent studies indicate that the backbone modification of poly(amino acids) may be a generally applicable approach for improving the physicomechanical properties of conventional poly(amino acids). For example, tyrosine-derived polycarbonates are high-strength materials that may be useful in the formulation of degradable orthopedic implants (Ertel and Kohn, 1994).

Polycyanoacrylates

These materials are used as bioadhesives and have also been intensively investigated as potential drug delivery matrices; however, the general tendency of polycyanoacrylates to induce a significant inflammatory response at the implantation site has discouraged their use as degradable implant materials.

Polyphosphazenes

This is a group of inorganic polymers whose backbone consists of nitrogen–phosphorus bonds. These polymers have unusual material properties and have found industrial applications. Their use for controlled drug delivery has been investigated (Allcock, 1990).

POLY(LACTIC ACID) AND POLY(GLYCOLIC ACID): EXAMPLES FOR WIDELY INVESTIGATED, BIOERODIBLE POLYMERS

Poly(glycolic acid) and poly(lactic acid) are currently the most widely investigated, and most commonly used synthetic,

bioerodible polymers. In view of their importance in the field of biomaterials, their properties and applications are described in more detail.

Poly(glycolic acid) (PGA) is the simplest linear, aliphatic polyester (Fig. 1). Since PGA is highly crystalline, it has a high melting point and low solubility in organic solvents. PGA was used in the development of the first totally synthetic, absorbable suture. PGA sutures have been commercially available under the trade name Dexon since 1970. A practical limitation of Dexon sutures is that they tend to lose their mechanical strength rapidly, typically over a period of 2 to 4 weeks after implantation. PGA was also used in the design of internal bone fixation devices (bone pins). These pins have become commercially available under the trade name Biofix.

In order to adapt the materials properties of PGA to a wider range of possible applications, copolymers of PGA with the more hydrophobic poly(lactic acid) (PLA) were intensively investigated (Gilding and Reed, 1981). The hydrophobicity of PLA limits the water uptake of thin films to about 2% and reduces the rate of backbone hydrolysis compared with PGA. Copolymers of glycolic acid and lactic acid have been developed as alternative sutures (trade names Vicryl and Polyglactin 910).

It is noteworthy that there is no linear relationship between the ratio of glycolic acid to lactic acid and the physicomechanical properties of the corresponding copolymers. Whereas PGA is highly crystalline, crystallinity is rapidly lost in copolymers of glycolic acid and lactic acid. These morphological changes lead to an increase in the rates of hydration and hydrolysis. Thus, 50:50 copolymers degrade more rapidly than either PGA or PLA.

Since lactic acid is a chiral molecule, it exists in two stereoisomeric forms which give rise to four morphologically distinct polymers: the two stereoregular polymers, D-PLA and L-PLA, and the racemic form D,L-PLA. A fourth morphological form, meso-PLA, can be obtained from D,L lactide but is rarely used in practice.

The polymers derived from the optically active D and L monomers are semicrystalline materials, while the optically inactive D,L-PLA is always amorphous. Generally, L-PLA is more frequently employed than D-PLA, since the hydrolysis of L-PLA yields L(+) lactic acid, which is the naturally occurring stereoisomer of lactic acid.

The differences in the crystallinity of D,L-PLA and L-PLA have important practical ramifications: Since D,L-PLA is an amorphous polymer, it is usually considered for such applications as drug delivery, where it is important to have a homogeneous dispersion of the active species within a monophasic matrix. On the other hand, the semicrystalline L-PLA is preferred in applications where high mechanical strength and toughness are required, such as sutures and orthopedic devices. Bone pins based on high-molecular-weight L-PLA are being developed by several companies and may become commercially available in the United States.

PHYSICAL MECHANISMS OF BIOEROSION

Within the context of this chapter, we limit our discussion to the case of a solid, polymeric implant. The transformation of such an implant into water-soluble material(s) is best described by the term "bioerosion." The bioerosion process of a solid, polymeric implant is associated with macroscopic changes in the appearance of the device; changes in the physicomechanical properties of the polymeric material; physical processes such as swelling, deformation, or structural disintegration; weight loss and the eventual loss of function.

All of these phenomena represent distinct and often independent aspects of the complex bioerosion behavior of a specific polymeric device. It is important to note that the bioerosion of a solid device is not necessarily due to the chemical cleavage of the polymer backbone, or the chemical cleavage of cross-links or side chains. Rather, simple solubilizaton of the intact polymer, for instance, as a result of changes in pH, may also lead to the erosion of a solid device.

Two distinct modes of bioerosion have been described in the literature (Heller, 1987). In bulk erosion, the rate of water penetration into the solid device exceeds the rate at which the polymer is transformed into water-soluble material(s). Consequently, the uptake of water is followed by an erosion process that occurs throughout the entire volume of the solid device. Owing to the rapid penetration of water into the matrix of hydrophilic polymers, most of the currently available polymers will give rise to bulk eroding devices. In a typical bulk erosion process, cracks and crevices will form throughout the device, which may rapidly crumble into pieces. A good illustration for a typical bulk erosion process is the disintegration of a sugar cube that has been placed into water. Depending on the specific application, the often uncontrollable tendency of bulk eroding devices to crumble into little pieces can be a disadvantage.

Alternatively, in surface erosion, the rate at which water penetrates into the polymeric device is slower than the rate of transformation of the polymer into water-soluble material(s). In this case, the transformation of the polymer into water-soluble material(s) is limited to the outer surface of the solid device. The device will therefore become thinner with time, while maintaining its structural integrity throughout much of the erosion process. In order to observe surface erosion, the polymer must be hydrophobic enough to impede the rapid imbibition of water into the interior of the device. In addition, the rate at which the polymer is transformed into water-soluble material(s) has to be reasonably fast. Under these conditions, scanning electron microscopic evaluation of surface eroding devices has sometimes shown a sharp border between the eroding surface layer and the intact polymer in the core of the device (Mathiowitz et al., 1990).

Surface eroding devices have so far been obtained only from a small number of polymers containing hydrolytically highly reactive bonds in the backbone. A possible exception to this general rule is enzymatic surface erosion. Reportedly, the inability of enzymes to penetrate into the interior of a solid, polymeric device may result in an enzyme-mediated surface erosion mechanism. Enzymatic surface erosion has so far been observed only in the case of a polymeric device made of cross-linked polycaprolactone (Pitt et al., 1984) Currently, polyanhydrides and poly(ortho esters) are the best-known examples of polymers that can be fabricated into surface eroding devices.

MECHANISMS OF CHEMICAL DEGRADATION

Although bioerosion can be caused by the solubilization of an intact polymer, chemical degradation of the polymer is usually the underlying cause for the bioerosion of a solid, polymeric device. Several distinct types of chemical degradation mechanisms have been identified (Fig. 2) (Rosen et al., 1988). Chemical reactions can lead to cleavage of cross-links between water-soluble polymer chains (mechanism I), to cleavage of polymer side chains resulting in the formation of polar or charged groups (mechanism II), or to the cleavage of the polymer backbone (mechanism III). Obviously, combinations of these mechanisms are possible: for instance, a cross-lined polymer may first be partially solubilized by the cleavage of cross-links (mechanism I), followed by the cleavage of the backbone itself (mechanism III).

Since the chemical cleavage reactions described here can be mediated by water or by biological agents such as enzymes and microorganisms, it is possible to distinguish between hydrolytic degradation and biodegradation, respectively. It has often been stated that the availability of water is virtually constant in all soft tissues and varies little from patient to patient. On the other hand, the levels of enzymatic activity may vary widely not only from patient to patient but also among different tissue sites in the same patient. Thus polymers that undergo hydrolytic cleavage tend to have more predictable *in vivo* erosion rates than polymers whose degradation is mediated predominantly by enzymes. The latter polymers tend to be generally less useful as degradable medical implants.

FACTORS THAT INFLUENCE THE RATE OF BIOEROSION

Although the solubilization of intact polymer as well as several distinct mechanisms of chemical degradation have been recognized as possible causes for the observed bioerosion of a solid, polymeric implant, virtually all currently available implant materials (Table 2) erode as a result of the hydrolytic cleavage of the polymer backbone (mechanism III in Fig. 2). We therefore limit the following discussion to solid devices that bioerode as a result of the hydrolytic cleavage of the polymer backbone.

In this case, the main factors that determine the overall rate of the erosion process are the chemical stability of the polymer backbone; the hydrophobicity of the monomer; the morphology of the polymer; the initial molecular weight of the polymer; the fabrication process; the presence of catalysts, additives, or plasticizers; and the geometry of the implanted device.

The susceptibility of the polymer backbone toward hydrolytic cleavage is probably the most fundamental parameter. Generally speaking, anhydride bonds tend to hydrolyze faster than ester bonds, which in turn hydrolyze faster than amide bonds. Thus, polyanhydrides will tend to degrade faster than polyesters, which in turn will have a higher tendency to bioerode than polyamides. Based on the known susceptibility of the polymer backbone structure toward hydrolysis, it is possi-

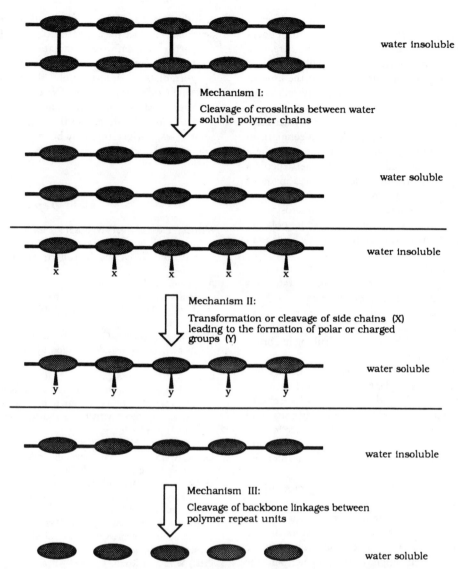

water insoluble

Mechanism I:
Cleavage of crosslinks between water soluble polymer chains

water soluble

water insoluble

Mechanism II:
Transformation or cleavage of side chains (X) leading to the formation of polar or charged groups (Y)

water soluble

water insoluble

Mechanism III:
Cleavage of backbone linkages between polymer repeat units

water soluble

FIG. 2. Mechanisms of chemical degradation. Mechanism I involves the cleavage of degradable cross-links between water-soluble polymer chains. Mechanism II involves the cleavage or chemical transformation of polymer side chains, resulting in the formation of charged or polar groups. The presence of charged or polar groups leads then to the solubilization of the intact polymer chain. Mechanism III involves the cleavage of unstable links in the polymer backbone, followed by solubilization of the low-molecular-weight fragments.

ble to predict the tendency of any given polymer to undergo bioerosion.

However, the actual erosion rate of a solid polymer cannot be predicted on the basis of the polymer backbone structure alone. The observed erosion rate is strongly dependent on the ability of water molecules to penetrate into the polymeric matrix. The hydrophobicity of the polymer, which is a function of the structure of the monomeric starting materials, can therefore have an overwhelming influence on the observed bioerosion rate. For instance, the erosion rate of polyanhydrides can be slowed by about three orders of magnitude when the hydrophilic sebacic acid is replaced by hydrophobic bis(carboxy phenoxy)propane as the monomeric starting material (Fig. 2). Like-

wise, devices made of poly(glycolic acid) erode faster than identical devices made of the more hydrophobic poly(lactic acid), although the ester bonds have about the same chemical reactivity toward water in both polymers.

The observed bioerosion rate is further influenced by the morphology of the polymer. Polymers can be classified as either semicrystalline or amorphous. At body temperature (37°C), amorphous polymers with T_g above 37°C will be in a glassy state, while polymers with T_g below 37°C will be in a rubbery state. In this discussion it is therefore necessary to consider three distinct morphological states: crystalline, amorphous-glassy, and amorphous-rubbery.

In the crystalline state, the polymer chains are most densely packed and offer the highest resistance to the penetration of water into the polymer matrix. Consequently, the rate of backbone hydrolysis tends to be higher in the amorphous regions of a semicrystalline polymer than in the crystalline regions. This phenomenon is of particular importance to the erosion of poly(glycolic acid) sutures whose degree of crystallinity is about 50%.

Another good illustration of the influence of polymer morphology on the rate of bioerosion is provided by a comparison of poly(L-lactic acid) and poly(D,L-lactic acid): although these two polymers have chemically identical backbone structures and an identical degree of hydrophobicity, devices made of poly(L-lactic acid) tend to degrade much slower than identical devices made of poly(D,L-lactic acid). The bioerosion of poly(L-lactic acid) is slower because this stereoregular polymer is semicrystalline, while the racemic poly(D,L-lactic acid) is an amorphous polymer.

Likewise, a polymer in its glassy state is less permeable to water than the same polymer when it is in its rubbery state. This observation could be important in cases where an amorphous polymer has a glass transition temperture that is not far above body temperature (37°C). In this situation, water sorption into the polymer could lower its T_g below 37°C, resulting in abrupt changes in the bioerosion rate.

The manufacturing process may also have a significant effect on the erosion profile. For example, Mathiowitz and co-workers (Mathiowitz *et al.*, 1990) showed that polyanhydride microspheres produced by melt encapsulation were very dense and eroded slowly, whereas when the same polymers were formed into microspheres by solvent evaporation, the microspheres were very porous (and therefore more water permeable) and eroded more rapidly.

These examples illustrate an important technological principle in the design of bioeroding devices: The bioerosion rate of a given polymer is not an unchangeable property, but depends to a very large degree on readily controllable factors such as the presence of plasticizers or additives, the manufacturing process, the initial molecular weight of the polymer, and the geometry of the device.

STORAGE STABILITY, STERILIZATION, AND PACKAGING

Finally, it is important to consider the need to minimize premature polymer degradation during fabrication and storage. Traces of moisture can seriously degrade even relatively stable polymers such as poly(bisphenol A carbonate) during injection molding or extrusion. Degradable polymers are particularly sensitive to hydrolytic degradation during high-temperature processing. The industrial production of degradable implants therefore often requires the construction of controlled atmosphere facilities in which the moisture content of the polymer and the ambient humidity can be strictly controlled.

After fabrication, γ-irradiation or exposure to ethylene oxide can be considered for the sterilization of degradable implants. Both methods have disadvantages and, as a general rule, the choice is between the lesser of two evils. γ-Irradiation

at a dose of 2 to 3 Mrad can result in significant backbone degradation. Since the aliphatic polyesters PLA, PGA, and PDS are particularly sensitive to radiation damage, these materials are usually sterilized by exposure to ethylene oxide and not by γ-irradiation. Unfortunately, the use of the highly toxic ethylene oxide gas is a serious safety hazard.

After sterilization, degradable implants are usually packaged in air-tight, aluminum-backed, plastic foil pouches. In some cases, refrigeration may be required to prevent degradation of the backbone during storage.

Bibliography

Allcock, H. R. (1990). Polyphosphazenes as new biomedical and bioactive materials. in *Biodegradable Polymers as Drug Delivery Systems,* M. Chasin and R. Langer, eds. Dekker, New York, pp. 163–193.

Anderson, J. M., Spilizewski, K. L., and Hiltner, A. (1985). Poly-α amino acids as biomedical polymers. in *Biocompatibility of Tissue Analogs,* D. F. Williams, ed. CRC Press, Boca Raton, FL, pp. 67–88.

Campbell, P., Glover, G. I., and Gunn, J. M. (1980). Inhibition of intracellular protein degradation by pepstatin, poly(L-lysine) and pepstatinyl-poly(L-lysine). *Arch Biochem. Biophys.* **203**: 676–680.

Chasin, M., Domb, A., Ron, E., Mathiowitz, E., Langer, R., Leong, K., Laurencin, C., Brem, H., and Grossman, S. (1990). Polyanhydrides as drug delivery systems. in *Biodegradable Polymers as Drug Delivery Systems,* M. Chasin and R. Langer, eds. Dekker, New York, pp. 43–70.

Cho, N. S., and Heller, J. (1978). US Patent 4,079,038.

Conix, A. (1958). Aromatic polyanhydrides, a new class of high melting fibre-forming polymers. *J. Polym. Sci.* **29**: 343–353.

Domb, A. J., Gallardo, C. F., and Langer, R. (1989). Poly(anhydrides). 3. Poly(anhydrides) based on aliphatic-aromatic diacids. *Macromolecules* **22**: 3200–3204.

Engelberg, I., and Kohn, J. (1991). Physicomechanical properties of degradable polymers used in medical applications: A comparative study. *Biomaterials* **12**: 292–304.

Ertel, S. I., and Kohn, J. (1994). Evaluation of a series of tyrosine-derived polycarbonates for biomedical applications. *J. Biomed. Mater. Res.* **28**: 919–930.

Frazza, E. J., and Schmitt, E. E. (1971). A new absorbable suture. *J. Biomed Mater. Res.* **1**: 43–58.

Gilding, D. K., and Reed, A. M. (1981). Biodegradable polymers for use in surgery—Poly(glycolic acid)/poly(lactic acid) homo- and copolymers. 2. In vitro degradation. *Polymer* **22**: 494–498.

Heller, J. (1987). Use of polymers in controlled release of active agents. in *Controlled Drug Delivery, Fundamentals and Applications,* 2nd ed., J. R. Robinson and V. H. L. Lee, eds. Dekker, New York, pp. 180–210.

Heller, J., Sparer, R. V., and Zentner, G. M. (1990). Poly(ortho esters). in *Biodegradable Polymers as Drug Delivery Systems,* M. Chasin and R. Langer, eds. Dekker, New York, pp. 121–162.

Kohn, J., and Langer, R. (1984). A new approach to the development of bioerodible polymers for controlled release applications employing naturally occurring amino acids. in *Proceedings of the ACS Division of Polymeric Materials, Science and Engineering,* Am. Chem. Soc., Washington, DC, pp. 119–121.

Kohn, J., and Langer, R. (1987). Polymerization reactions involving the side chains of α-L-amino acids. *J. Am. Chem. Soc.* **109**: 817–820.

Laurencin, C., Domb, A., Morris, C., Brown, V., Chasin, M., McConnel, R., Lange, N., and Langer, R. (1990). Poly(anhydride) administration in high doses *in vivo*: Studies of biocompatibility and toxicology. *J. Biomed. Mater. Res.* **24**: 1463–1481.

Mathiowitz, E., Kline, D., and Langer, R. (1990). Morphology of polyanhydride microsphere delivery systems. *J. Scanning Microsc.* **4**(2): 329–340.

Pitt, C. G. (1990). Poly-ε-caprolactone and its copolymers. in *Biodegradable Polymers as Drug Delivery Systems,* M. Chasin and R. Langer, eds. Dekker, New York, pp. 71–120.

Pitt, C. G., Hendren, R. W., and Schindler, A. (1984). The enzymatic surface erosion of aliphatic polyesters. *J. Control. Rel.* **1**: 3–14.

Rosen, H., Kohn, J., Leong, K., and Langer, R. (1988). Bioerodible polymers for controlled release systems. in *Controlled Release Systems: Fabrication Technology,* D. Hsieh, ed. CRC Press, Boca Raton, FL, pp. 83–110.

van Heeswijk, W. A. R., Hoes, C. J. T., Stoffer, T., Eenink, M. J. D., Potman, W., and Feijen, J. (1985). The synthesis and characterization of polypeptide–adriamycin conjugates and its complexes with adriamycin. Part 1. *J. Control. Rel.* **1**: 301–315.

Williams, D. F. (1987). *Definitions in Biomaterials—Proceedings of a Consensus Conference of the European Society for Biomaterials.* Elsevier, New York.

2.6 CERAMICS, GLASSES, AND GLASS-CERAMICS

Larry L. Hench

Ceramics, glasses, and glass-ceramics include a broad range of inorganic/nonmetallic compositions. In the medical industry, these materials have been essential for eyeglasses, diagnostic instruments, chemical ware, thermometers, tissue culture flasks, and fiber optics for endoscopy. Insoluble porous glasses have been used as carriers for enzymes, antibodies, and antigens, offering the advantages of resistance to microbial attack, pH changes, solvent conditions, temperature, and packing under high pressure required for rapid flow (Hench and Ethridge, 1982).

Ceramics are also widely used in dentistry as restorative materials such as in gold-porcelain crowns, glass-filled ionomer cements, and dentures. These dental ceramics are discussed by Phillips (1991).

This chapter focuses on ceramics, glasses, and glass-ceramics used as implants. Although dozens of compositions have been explored in the past, relatively few have achieved clinical success. This chapter examines differences in processing and structure, describes the chemical and microstructural basis for their differences in physical properties, and relates properties and tissue response to particular clinical applications. For a historical review of these biomaterials, see Hulbert *et al.* (1987).

TYPES OF BIOCERAMICS–TISSUE ATTACHMENT

It is essential to recognize that no one material is suitable for all biomaterial applications. As a class of biomaterials, ceramics, glasses, and glass-ceramics are generally used to repair or replace skeletal hard connective tissues. Their success depends upon achieving a stable attachment to connective tissue.

TABLE 1 Types of Implant–Tissue Response

If the material is toxic, the surrounding tissue dies.

If the material is nontoxic and biologically inactive (nearly inert), a fibrous tissue of variable thickness forms.

If the material is nontoxic and biologically active (bioactive), an interfacial bond forms.

If the material is nontoxic and dissolves, the surrounding tissue replaces it.

The mechanism of tissue attachment is directly related to the type of tissue response at the implant–tissue interface. No material implanted in living tissue is inert because all materials elicit a response from living tissues. There are four types of tissue response (Table 1) and four different means of attaching prostheses to the skeletal system (Table 2).

A comparison of the relative chemical activity of the different types of bioceramics, glasses, and glass-ceramics is shown in Fig. 1. The relative reactivity shown in Fig. 1,A correlates very closely with the rate of formation of an interfacial bond of ceramic, glass, or glass-ceramic implants with bone (Fig. 1,B). Figure 1,B is discussed in more detail in the section on bioactive glasses and glass-ceramics in this chapter.

The relative level of reactivity of an implant influences the thickness of the interfacial zone or layer between the material and tissue. Analyses of implant material failures during the past 20 years generally show failure originating at the biomaterial–tissue interface. When biomaterials are nearly inert (type 1 in Table 2 and Fig. 1) and the interface is not chemically or biologically bonded, there is relative movement and progressive development of a fibrous capsule in soft and hard tissues. The presence of movement at the biomaterial–tissue interface eventually leads to deterioration in function of the implant or the tissue at the interface, or both. The thickness of the nonadher-

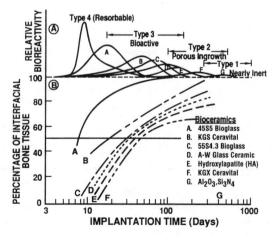

FIG. 1. Bioactivity spectra for various bioceramic implants: (A) relative rate of bioreactivity, (B) time dependence of formation of bone bonding at an implant interface.

TABLE 2   Types of Bioceramic–Tissue Attachment and Their Classification

Type of attachment	Example
1. Dense, nonporous, nearly inert ceramics attach by bone growth into surface irregularities by cementing the device into the tissues or by press-fitting into a defect (termed "morphological fixation").	Al_2O_3 (single crystal and polycrystalline)
2. For porous inert implants, bone ingrowth occurs that mechanically attaches the bone to the material (termed "biological fixation").	Al_2O_3 (polycrystalline) Hydroxyapatite-coated porous metals
3. Dense, nonporous surface-reactive ceramics, glasses, and glass-ceramics attach directly by chemical bonding with the bone (termed "bioactive fixation").	Bioactive glasses Bioactive glass-ceramics Hydroxyapatite
4. Dense, nonporous (or porous) resorbable ceramics are designed to be slowly replaced by bone.	Calcium sulfate (plaster of paris) Tricalcium phosphate Calcium-phosphate salts

ent capsule varies, depending upon both material (Fig. 2) and extent of relative motion.

The fibrous tissue at the interface of dense Al_2O_3 (alumina) implants is very thin. Consequently, as discussed later, if alumina devices are implanted with a very tight mechanical fit and are loaded primarily in compression, they are very successful. In contrast, if a type 1 nearly inert implant is loaded so that interfacial movement can occur, the fibrous capsule can become several hundred micrometers thick, and the implant can loosen very quickly.

The mechanism behind the use of nearly inert microporous materials (type 2 in Table 2 and Fig. 1) is the ingrowth of tissue into pores on the surface or throughout the implant. The increased interfacial area between the implant and the tissues results in an increased resistance to movement of the device in the tissue. The interface is established by the living tissue in the pores. Consequently, this method of attachment is often termed "biological fixation." It is capable of withstanding more complex stress states than type 1 implants with "morphological fixation." The limitation with type 2 porous implants, however, is that for the tissue to remain viable and healthy, it is necessary for the pores to be greater than 50 to 150 μm

(Fig. 2). The large interfacial area required for the porosity is due to the need to provide a blood supply to the ingrown connective tissue (vascular tissue does not appear in pore sizes less than 100 μm). Also, if micromovement occurs at the interface of a porous implant and tissue is damaged, the blood supply may be cut off, the tissues will die, inflammation will ensue, and the interfacial stability will be destroyed. When the material is a porous metal, the large increase in surface area can provide a focus for corrosion of the implant and loss of metal ions into the tissues. This can be mediated by using a bioactive ceramic material such as hydroxyapatite (HA) as a coating on the metal. The fraction of large porosity in any material also degrades the strength of the material proportional to the volume fraction of porosity. Consequently, this approach to solving interfacial stability works best when materials are used as coatings or as unloaded space fillers in tissues.

Resorbable biomaterials (type 4 in Table 2 and Fig. 1) are designed to degrade gradually over a period of time and be replaced by the natural host tissue. This leads to a very thin or nonexistent interfacial thickness (Fig. 2). This is the optimal biomaterial solution, if the requirements of strength and short-term performance can be met, since natural tissues can repair and replace themselves throughout life. Thus, resorbable biomaterials are based on biological principles of repair that have evolved over millions of years. Complications in the development of resorbable bioceramics are (1) maintenance of strength and the stability of the interface during the degradation period and replacement by the natural host tissue, and (2) matching resorption rates to the repair rates of body tissues (Fig. 1,A) (e.g., some materials dissolve too rapidly and some too slowly). Because large quantities of material may be replaced, it is also essential that a resorbable biomaterial consist only of metabolically acceptable substances. This criterion imposes considerable limitations on the compositional design of resorbable biomaterials. Successful examples of resorbable polymers include poly(lactic acid) and poly(glycolic acid) used for sutures, which are metabolized to CO_2 and H_2O and therefore are able to function for an appropriate time and then dissolve and disappear (see Chapters 2, 6, and 7 for other examples). Porous or particulate calcium phosphate ceramic materials such as tricalcium phosphate (TCP), have proved successful for resorb-

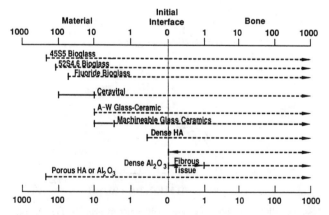

FIG. 2.   Comparison of interfacial thickness (μm) of reaction layer of bioactive implants of fibrous tissue of inactive bioceramics in bone.

TABLE 3 Bioceramic Material Characteristics and Properties

Composition

Microstructure
    Number of phases
    Percentage of phases
    Distribution of phases
    Size of phases
    Connectivity of phases

Phase State
    Crystal structure
    Defect structure
    Amorphous structure
    Pore structure

Surface
    Flatness
    Finish
    Composition
    Second phase
    Porosity

Shape

able hard tissue replacements when low loads are applied to the material.

Another approach to solving problems of interfacial attachment is the use of bioactive materials (type 3 in Table 2 and Fig. 1). Bioactive materials are intermediate between resorbable and bioinert. A bioactive material is one that elicits a specific biological response at the interface of the material, resulting in the formation of a bond between the tissues and the material. This concept has now been expanded to include a large number of bioactive materials with a wide range of rates of bonding and thicknesses of interfacial bonding layers (Figs. 1 and 2). They include bioactive glasses such as Bioglass; bioactive glass-ceramics such as Ceravital, A-W glass-ceramic, or machinable glass-ceramics; dense HA such as Durapatite or Calcitite; and bioactive composites such as HA-polyethylene, HA-Bioglass, Palavital, and stainless steel fiber-reinforced Bioglass. All of these materials form an interfacial bond with adjacent tissue. However, the time dependence of bonding, the strength of bond, the mechanism of bonding, and the thickness of the bonding zone differ for the various materials.

It is important to recognize that relatively small changes in the composition of a biomaterial can dramatically affect whether it is bioinert, resorbable, or bioactive. These compositional effects on surface reactions are discussed in the section on bioactive glasses and glass-ceramics.

CHARACTERISTICS AND PROCESSING OF BIOCERAMICS

The types of implants listed in Table 2 are made using different processing methods. The characteristics and properties of the materials, summarized in Table 3, differ greatly, depending upon the processing method used.

The primary methods of processing ceramics, glasses, and glass-ceramics are summarized in Fig. 3. These methods yield five categories of microstructures:

1. Glass
2. Cast or plasma-sprayed polycrystalline ceramic
3. Liquid-phase sintered (vitrified) ceramic
4. Solid-state sintered ceramic
5. Polycrystalline glass-ceramic

Differences in the microstructures of the five categories are primarily a result of the different thermal processing steps required to produce them. Alumina and calcium phosphate bioceramics are made by fabricating the product from fine-grained particulate solids. For example, a desired shape may be obtained by mixing the particulates with water and an organic binder, then pressing them in a mold. This is termed "forming." The formed piece is called green ware. Subsequently, the temperature is raised to evaporate the water (i.e., drying) and the binder is burned out, resulting in bisque ware. At a very much higher temperature, the part is densified during firing. After cooling to ambient temperature, one or more finishing steps may be applied, such as polishing. Porous ceramics are produced by adding a second phase that decomposes prior to densification, leaving behind holes or pores (Schors and Holmes, 1993), or transforming natural porous organisms, such as coral, to porous HA by hydrothermal processing (Roy and Linnehan, 1974).

The interrelation between microstructure and thermal processing of various bioceramics is shown in Fig. 3, which is a binary phase diagram consisting of a network-forming oxide such as SiO_2 (silica), and some arbitrary network modifier oxide (MO) such as CaO. When a powdered mixture of MO and SiO_2 is heated to the melting temperature T_m, the entire mass will become liquid (L). The liquid will become homogeneous when held at this temperature for a sufficient length of time. When the liquid is cast (paths 1B, 2, 5), forming the shape of the object during the casting, either a glass or a polycrystalline microstructure will result. Plasma spray coating follows path 1A. However, a network-forming oxide is not necessary to produce plasma-sprayed coatings such as hydroxyapatites, which are polycrystalline (Lacefield, 1993).

If the starting composition contains a sufficient quantity of network former (SiO_2), and the casting rate is sufficiently slow, a glass will result (path 1B). The viscosity of the melt increases greatly as it is cooled, until at approximately T_1, the glass transition point, the material is transformed into a solid.

If either of these conditions is not met, a polycrystalline microstructure will result. The crystals begin growing at T_L and complete growth at T_2. The final material consists of the equilibrium crystalline phases predicted by the phase diagram. This type of cast object is not often used commercially because the large shrinkage cavity and large grains produced during cooling make the material weak and subject to environmental attack.

If the MO and SiO_2 powders are first formed into the shape of the desired object and fired at a temperature T_3, a liquid-phase sintered structure will result (path 3). Before firing, the composition will contain approximately 10–40% porosity, depending upon the forming process used. A liquid will be formed first at grain boundaries at the eutectic temperature, T_2. The liquid will penetrate between the grains, filling the pores, and will draw the grains together by capillary attraction. These effects decrease the volume of the powdered compact. Since the mass remains unchanged and is only rearranged, an increased density results. Should the compact be heated for a sufficient length of time, the liquid content can be predicted from the phase diagram. However, in most ceramic processes, liquid formation does not usually proceed to equilibrium owing to the slowness of the reaction and the expense of long-term heat treatments.

The microstructure resulting from liquid-phase sintering, or vitrification as it is commonly called, will consist of small grains from the original powder compact surrounded by a liquid phase. As the compact is cooled from T_3 to T_2, the liquid phase will crystallize into a fine-grained matrix surrounding the original grains. If the liquid contains a sufficient concentration of network formers, it can be quenched into a glassy matrix surrounding the original grains.

A powder compact can be densified without the presence of a liquid phase by a process called solid-state sintering. This is the process usually used for manufacturing alumina and dense HA bioceramics. Under the driving force of surface energy gradients, atoms diffuse to areas of contact between particles. The material may be transported by either grain boundary diffusion, volume diffusion, creep, or any combination of these, depending upon the temperature or material involved. Because long-range migration of atoms is necessary, sintering temperatures are usually in excess of one-half of the melting point of the material: $T > T_L/2$ (path 4).

The atoms move so as to fill up the pores and open channels between the grains of the powder. As the pores and open channels are closed during the heat treatment, the crystals become tightly bonded together, and the density, strength, and fatigue resistance of the object improve greatly. The microstructure of a material that is prepared by sintering consists of crystals bonded together by ionic-covalent bonds with a very small amount of remaining porosity.

The relative rate of densification during solid-state sintering is slower than that of liquid-phase sintering because material transport is slower in a solid than in a liquid. However, it is possible to solid-state sinter individual component materials such as pure oxides since liquid development is not necessary. Consequently, when high purity and uniform fine-grained microstructures are required (e.g., for bioceramics) solid-state sintering is essential.

The fifth class of microstructures is called glass-ceramics because the object starts as a glass and ends up as a polycrystalline ceramic. This is accomplished by first quenching a melt to form the glass object. The glass is transformed into a glass-ceramic in two steps. First, the glass is heat treated at a temperature range of 500–700°C (path 5a) to produce a large concentration of nuclei from which crystals can grow. When sufficient nuclei are present to ensure that a fine-grained structure will be obtained, the temperature of the object is raised to a range of 600–900°C, which promotes crystal growth (path 5b). Crystals grow from the nuclei until they impinge and up to 100% crystallization is achieved. The resulting microstructure is nonporous and contains fine-grained, randomly oriented crystals that may or may not correspond to the equilibrium crystal phases predicted by the phase diagram. There may also be a residual glassy matrix, depending on the duration of the ceraming heat treatment. When phase separation occurs (composition B in Fig. 3), a nonporous, phase-separated, glass-in-glass microstructure can be produced. Crystallization of phase-separated glasses results in very complex microstructures. Glass-ceramics can also be made by pressing powders and a grain boundary glassy phase (Kokubo, 1993). For additional details on the processing of ceramics, see Reed (1988) or Onoda and Hench (1978), and for processing of glass-ceramics, see McMillan (1979).

NEARLY INERT CRYSTALLINE CERAMICS

High-density, high-purity (>99.5%) alumina is used in load-bearing hip prostheses and dental implants because of

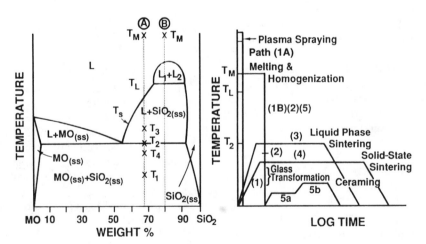

FIG. 3.   Relation of thermal processing schedules of various bioceramics to equilibrium phase diagram.

TABLE 4 Physical Characteristics of Al_2O_3 Bioceramics

	High alumina ceramics	ISO standard 6474
Alumina content (% by weight)	>99.8	≥99.50
Density (g/cm³)	>3.93	≥3.90
Average grain size (μm)	3–6	<7
Ra (μm)[a]	0.02	
Hardness (Vickers hardness number, VHN)	2300	>2000
Compressive strength (MPa)	4500	
Bending strength (MPa) (after testing in Ringer's solution)	550	400
Young's modulus (GPa)	380	
Fracture toughness (K_1C) (MPa¹²)	5–6	
Slow crack growth	10–52	

[a]Surface roughness value.

its excellent corrosion resistance, good biocompatibility, high wear resistance, and high strength (Christel *et al.*, 1988; Hulbert, 1993; Hulbert *et al.*, 1987; Miller *et al.*, 1996). Although some dental implants are single-crystal sapphires (McKinney and Lemons, 1985), most Al_2O_3 devices are very fine-grained polycrystalline α-Al_2O_3 produced by pressing and sintering at $T = 1600$–$1700°C$. A very small amount of MgO (<0.5%) is used to aid sintering and limit grain growth during sintering.

Strength, fatigue resistance, and fracture toughness of polycrystalline α-Al_2O_3 are a function of grain size and percentage of sintering aid (i.e., purity). Al_2O_3 with an average grain size of <4 μm and >99.7% purity exhibits good flexural strength and excellent compressive strength. These and other physical properties are summarized in Table 4, along with the International Standards Organization (ISO) requirements for alumina implants. Extensive testing has shown that alumina implants that meet or exceed ISO standards have excellent resistance to dynamic and impact fatigue and also resist subcritical crack growth (Dörre and Dawihl, 1980). An increase in average grain size to >7 μm can decrease mechanical properties by about 20%. High concentrations of sintering aids must be avoided because they remain in the grain boundaries and degrade fatigue resistance.

Methods exist for lifetime predictions and statistical design of proof tests for load-bearing ceramics. Applications of these techniques show that load limits for specific prostheses can be set for an Al_2O_3 device based upon the flexural strength of the material and its use environment (Ritter *et al.*, 1979). Load-bearing lifetimes of 30 years at 12,000 N loads have been predicted (Christel *et al.*, 1988). Results from aging and fatigue studies show that it is essential that Al_2O_3 implants be produced at the highest possible standards of quality assurance, especially

if they are to be used as orthopedic prostheses in younger patients.

Alumina has been used in orthopedic surgery for nearly 20 years (Miller *et al.*, 1996). Its use has been motivated largely by two factors: (1) its excellent type 1 biocompatibility and very thin capsule formation (Fig. 2), which permits cementless fixation of prostheses; and its exceptionally low coefficients of friction and wear rates

The superb tribiologic properties (friction and wear) of alumina occur only when the grains are very small (<4 μm) and have a very narrow size distribution. These conditions lead to very low surface roughness values ($R_a \leq 0.02$ μm, Table 4). If large grains are present, they can pull out and lead to very rapid wear of bearing surfaces owing to local dry friction.

Alumina on load-bearing, wearing surfaces, such as in hip prostheses, must have a very high degree of sphericity, which is produced by grinding and polishing the two mating surfaces together. For example, the alumina ball and socket in a hip prosthesis are polished together and used as a pair. The long-term coefficient of friction of an alumina–alumina joint decreases with time and approaches the values of a normal joint. This leads to wear on alumina-articulating surfaces being nearly 10 times lower than metal–polyethylene surfaces (Fig. 4).

Low wear rates have led to widespread use in Europe of alumina noncemented cups press-fitted into the acetabulum of the hip. The cups are stabilized by the growth of bone into grooves or around pegs. The mating femoral ball surface is also made of alumina, which is bonded to a metallic stem. Long-term results in general are good, especially for younger patients. However, Christel *et al.* (1988) caution that stress shielding, owing to the high elastic modulus of alumina, may be responsible for cancellous bone atrophy and loosening of the acetabular cup in old patients with senile osteoporosis or rheumatoid arthritis. Consequently, it is essential that the age of the patient, nature of the disease of the joint, and biomechanics of the repair be considered carefully before any prosthesis is used, including alumina ceramics.

Zirconia (ZrO_2) is also used as the articulating ball in total hip prostheses. The potential advantages of zirconia in load-bearing prostheses are its lower modulus of elasticity and higher strength (Hench and Wilson, 1993). There are insufficient data to determine whether these properties will result in higher clinical success rates over long times (>15 years).

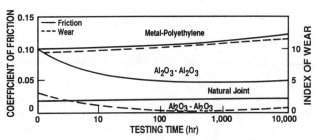

FIG. 4. Time dependence of coefficient of friction and wear of alumina–alumina vs. metal–polyethylene hip joint (*in vitro* testing).

Other clinical applications of alumina prostheses reviewed by Hulbert *et al.* (1987) include knee prostheses; bone screws; alveolar ridge and maxillofacial reconstruction; ossicular bone substitutes; keratoprostheses (corneal replacements); segmental bone replacements; and blade, screw, and post dental implants.

POROUS CERAMICS

The potential advantage offered by a porous ceramic implant (type 2, Table 2, Figs. 1 and 2) is its inertness combined with the mechanical stability of the highly convoluted interface that develops when bone grows into the pores of the ceramic. The mechanical requirements of prostheses, however, severely restrict the use of low-strength porous ceramics to nonload-bearing applications. Studies reviewed by Hench and Ethridge (1982), Hulbert *et al.* (1987), and Schors and Holmes (1993) have shown that when load-bearing is not a primary requirement, porous ceramics can provide a functional implant. When pore sizes exceed 100 μm, bone will grow within the interconnecting pore channels near the surface and maintain its vascularity and long-term viability. In this manner, the implant serves as a structural bridge or scaffold for bone formation.

The microstructures of certain corals make an almost ideal material for obtaining structures with highly controlled pore sizes. White and co-workers (White and Schors, 1986) developed the replamineform process to duplicate the porous microstructure of corals that have a high degree of uniform pore size and interconnection. The first step is to machine the coral with proper microstructure into the desired shape. The most promising coral genus, *Porites,* has pores with a size range of 140–160 μm, with all the pores interconnected. Another interesting coral genus, *Goniopora,* has a larger pore size, ranging from 200 to 1000 μm. The machined coral shape is fired to drive off CO_2 from the limestone ($CaCO_3$), forming CaO while maintaining the microstructure of the original coral. The CaO structure serves as an investment material for forming the porous material. After the desired material is cast into the pores, the CaO is easily removed by dissolving in dilute HCl. The primary advantages of the replamineform process are that the pore size and microstructure are uniform and controlled, and there is complete interconnection of the pores. Replamineform porous materials of $\alpha\text{-}Al_2O_3$, TiO_2, calcium phosphates, polyurethane, silicone rubber, poly(methyl methacrylate) (PMMA), and Co–Cr alloys have been used as bone implants, with the calcium phosphates being the most acceptable. Porous hydroxyapatite is also made from coral by using hydrothermal processing to transform $CaCO_3$ to HA (Schors and Holmes, 1993; Roy and Linnehan, 1974).

Porous materials are weaker than the equivalent bulk form in proportion to the percentage of porosity, so that as the porosity increases, the strength of the material decreases rapidly. Much surface area is also exposed, so that the effects of the environment on decreasing the strength become much more important than for dense, nonporous materials. The aging of porous ceramics, with their subsequent decrease in strength, requires bone ingrowth to stabilize the structure of the implant.

Clinical results for non-load-bearing implants are good (Schors and Holmes, 1993).

BIOACTIVE GLASSES AND GLASS-CERAMICS

Certain compositions of glasses, ceramics, glass-ceramics, and composites have been shown to bond to bone (Hench and Ethridge, 1982; Gross *et al.*, 1988; Yamamuro *et al.*, 1990; Hench, 1991; Hench and Wilson, 1993). These materials have become known as bioactive ceramics. Some even more specialized compositions of bioactive glasses will bond to soft tissues as well as bone (Wilson *et al.*, 1981). A common characteristic of bioactive glasses and bioactive ceramics is a time-dependent, kinetic modification of the surface that occurs upon implantation. The surface forms a biologically active carbonated HA layer that provides the bonding interface with tissues.

Materials that are bioactive develop an adherent interface with tissues that resist substantial mechanical forces. In many cases, the interfacial strength of adhesion is equivalent to or greater than the cohesive strength of the implant material or the tissue bonded to the bioactive implant.

Bonding to bone was first demonstrated for a compositional range of bioactive glasses that contained SiO_2, Na_2O, CaO, and P_2O_5 in specific proportions (Hench *et al.*, 1972) (Table 5). There are three key compositional features to these bioactive glasses that distinguish them from traditional soda–lime–silica glasses: (1) less than 60 mol% SiO_2, (2) high Na_2O and CaO content, and (3) a high CaO/P_2O_5 ratio. These features make the surface highly reactive when it is exposed to an aqueous medium.

Many bioactive silica glasses are based upon the formula called 45S5, signifying 45 wt. % SiO_2 (S = the network former) and 5 : 1 ratio of CaO to P_2O_5. Glasses with lower ratios of CaO to P_2O_5 do not bond to bone. However, substitutions in the 45S5 formula of 5–15 wt. % B_2O_3 for SiO_2 or 12.5 wt.% CaF_2 for CaO or heat treating the bioactive glass compositions to form glass-ceramics have no measurable effect on the ability of the material to form a bone bond. However, adding as little as 3 wt. % Al_2O_3 to the 45S5 formula prevents bonding to bone.

The compositional dependence of bone and soft tissue bonding on the Na_2O–CaO–P_2O_5–SiO_2 glasses is illustrated in Fig. 5. All the glasses in Fig. 5 contain a constant 6 wt. % of P_2O_5. Compositions in the middle of the diagram (region A) form a bond with bone. Consequently, region A is termed the bioactive bone-bonding boundary. Silicate glasses within region B (e.g., window or bottle glass, or microscope slides) behave as nearly inert materials and elicit a fibrous capsule at the implant–tissue interface. Glasses within region C are resorbable and disappear within 10 to 30 days of implantation. Glasses within region D are not technically practical and therefore have not been tested as implants.

The collagenous constituent of soft tissues can strongly adhere to the bioactive silicate glasses that lie within the dashed line region in Fig. 5. The interfacial thicknesses of the hard tissue–bioactive glasses are shown in Fig. 2 for several compositions. The thickness decreases as the bone-bonding boundary is approached.

TABLE 5 Composition of Bioactive Glasses and Glass-Ceramics (in Weight Percent)

	45S5 Bioglass	45S5F Bioglass	45S5.4F Bioglass	40S5B5 Bioglass	52S4.6 Bioglass	55S4.3 Bioglass	KGC Ceravital	KGS Ceravital	KGy213 Ceravital	A-W GC	MB GC
SiO$_2$	45	45	45	40	52	55	46.2	46	38	34.2	19–52
P$_2$O$_5$	6	6	6	6	6	6				16.3	4–24
CaO	24.5	12.25	14.7	24.5	21	19.5	20.2	33	31	44.9	9–3
Ca(PO$_3$)$_2$							25.5	16	13.5		
CaF$_2$		12.25	9.8							0.5	
MgO							2.9			4.6	5–15
MgF$_2$											
Na$_2$O	24.5	24.5	24.5	24.5	21	19.5	4.8	5	4		3–5
K$_2$O							0.4				3–5
Al$_2$O$_3$									7		12–33
B$_2$O$_3$				5							
Ta$_2$O$_5$/TiO$_2$									6.5		
Structure	Glass	Glass	Glass	Glass	Glass		Glass-Ceramic	Glass-Ceramic		Glass-Ceramic	Glass-Ceramic
Reference	Hench et al. (1972)	Hench et al. (1972)	Hench et al. (1972)	Hench et al. (1972)	Hench et al. (1972)	Hench et al. (1972)	Gross et al. (1988)	Gross et al. (1988)		Nakamura et al. (1985)	Höhland and Vogel (1993)

Gross *et al.* (1988) and Gross and Strunz (1985) have shown that a range of low-alkali (0 to 5 wt.%) bioactive silica glass-ceramics (Ceravital) also bond to bone. They found that small additions of Al$_2$O$_3$, Ta$_2$O$_5$, TiO$_2$, Sb$_2$O$_3$, or ZrO$_2$ inhibit bone bonding (Table 5, Fig. 1). A two-phase silica–phosphate glass-ceramic composed of apatite [Ca$_{10}$(PO$_4$)$_6$(OH$_1$F$_2$)] and wollastonite [CaO · SiO$_2$] crystals and a residual silica glassy matrix, termed A-W glass-ceramic (A-WGC) (Nakamura *et al.*, 1985; Yamamuro *et al.*, 1990; Kokubo, 1993), also bonds with bone. The addition of Al$_2$O$_3$ or TiO$_2$ to the A-W glass-ceramic also inhibits bone bonding, whereas incorporation of a second phosphate phase, B-whitlockite (3CaO · P$_2$O$_5$), does not.

Another multiphase bioactive phosphosilicate containing

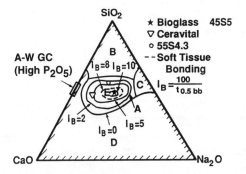

FIG. 5. Compositional dependence (in wt. %) of bone bonding and soft tissue bonding of bioactive glasses and glass-ceramics. All compositions in region A have a constant 6 wt. % of P$_2$O$_5$. A-W glass ceramic has higher P$_2$O$_5$ content (see Table 5 for details). I$_B$, Index of bioactivity.

phlogopite (Na,K) Mg$_3$[AlSi$_3$O$_{10}$]F$_2$ and apatite crystals bonds to bone even though Al is present in the composition (Hohland and Vogel, 1993). However, the Al^{3+} ions are incorporated within the crystal phase and do not alter the surface reaction kinetics of the material. The compositions of these various bioactive glasses and glass-ceramics are compared in Table 5.

The surface chemistry of bioactive glass and glass-ceramic implants is best understood in terms of six possible types of surface reactions (Hench and Clark, 1978). A high silica glass may react with its environment by developing only a surface hydration layer. This is called a type I response (Fig. 6). Vitreous silica (SiO$_2$) and some inert glasses at the apex of region B (Fig. 5) behave in this manner when exposed to a physiological environment.

When sufficient SiO$_2$ is present in the glass network, the surface layer that forms from alkali-proton exchange can repolymerize into a dense SiO$_2$-rich film that protects the glass from further attack. This type II surface (Fig. 6) is characteristic of most commercial silicate glasses, and their biological response of fibrous capsule formation is typical of many within region B in Fig. 5.

At the other extreme of the reactivity range, a silicate glass or crystal may undergo rapid, selective ion exchange of alkali ions, with protons or hydronium ions leaving a thick but highly porous and nonprotective SiO$_2$-rich film on the surface (a type IV surface) (Fig. 6). Under static or slow flow conditions, the local pH becomes sufficiently alkaline (pH >9) that the surface silica layer is dissolved at a high rate, leading to uniform bulk network or stoichiometric dissolution (a type V surface). Both type IV and V surfaces fall into region C of Fig. 5.

Type IIIA surfaces are characteristic of bioactive silicates (Fig. 6). A dual protective film rich in CaO and P$_2$O$_5$ forms

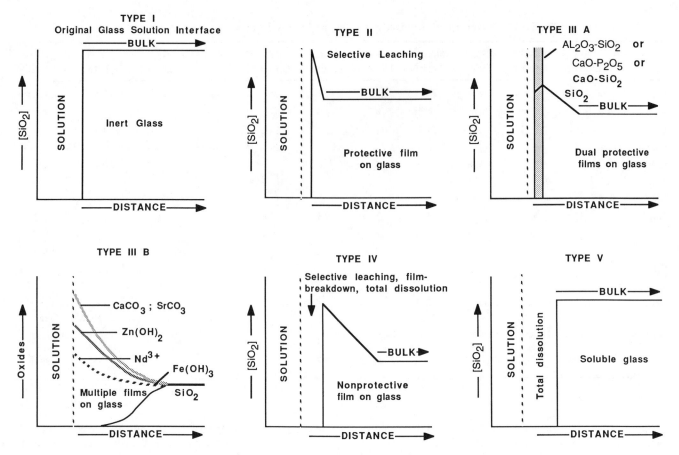

FIG. 6. Types of silicate glass interfaces with aqueous or physiological solutions.

on top of the alkali-depleted SiO_2-rich film. When multivalent cations such as Al^{3+}, Fe^{3+}, and Ti^{4+} are present in the glass or solution, multiple layers form on the glass as the saturation of each cationic complex is exceeded, resulting in a type IIIB surface (Fig. 6), which does not bond to tissue.

A general equation describes the overall rate of change of glass surfaces and gives rise to the interfacial reaction profiles shown in Fig. 6. The reaction rate (R) depends on at least four terms (for a single-phase glass). For glass-ceramics, which have several phases in their microstructures, each phase will have a characteristic reaction rate, R_i.

$$R = -k_1 t^{0.5} - k_2 t^{1.0} + k_3 t^{1.0} + k_4 t^y + k_n t^z \qquad (1)$$
$$\underset{\text{Stage 1}}{\quad} \underset{\text{Stage 2}}{\quad} \underset{\text{Stage 3}}{\quad} \underset{\text{Stage 4}}{\quad}$$

The first term describes the rate of alkali extraction from the glass and is called a stage 1 reaction. A type II nonbonding glass surface (region B in Fig. 6) is primarily undergoing stage 1 attack. Stage 1, the initial or primary stage of attack, is a process that involves an exchange between alkali ions from the glass and hydrogen ions from the solution, during which the remaining constituents of the glass are not altered. During stage 1, the rate of alkali extraction from the glass is parabolic ($t^{1/2}$) in character.

The second term describes the rate of interfacial network dissolution that is associated with a stage 2 reaction. A type IV surface is a resorbable glass (region C in Fig. 5) and is

experiencing a combination of stage 1 and stage 2 reactions. A type V surface is dominated by a stage 2 reaction. Stage 2, the second stage of attack, is a process by which the silica structure breaks down and the glass totally dissolves at the interface. Stage 2 kinetics are linear ($t^{1.0}$).

A glass surface with a dual protective film is designated type IIIA (Fig. 6). The thickness of the secondary films can vary considerably—from as little as 0.01 μm for Al_2O_3–SiO_2-rich layers on inactive glasses, to as much as 30 μm for CaO–P_2O_5-rich layers on bioactive glasses.

A type III surface forms as a result of the repolymerization of SiO_2 on the glass surface by the condensation of the silanols (Si—OH) formed from the stage 1 reaction. For example:

$$Si{-}OH + OH{-}Si \rightarrow Si{-}O{-}Si + H_2O \qquad (2)$$

Stage 3 protects the glass surface. The SiO_2 polymerization reaction contributes to the enrichment of surface SiO_2 that is characteristic of types II, III, and IV surface profiles (Fig. 6). It is described by the third term in Eq. 1. This reaction is interface controlled with a time dependence of $+k_3 t^{1.0}$. The interfacial thickness of the most reactive bioactive glasses shown in Fig. 2 is largely due to this reaction.

The fourth term in Eq. 1, $+k_4 t^y$ (stage 4), describes the precipitation reactions that result in the multiple films characteristic of type III glasses. When only one secondary film forms, the surface is type IIIA. When several additional films form, the surface is type IIIB.

In stage 4, an amorphous calcium phosphate film precipitates on the silica-rich layer and is followed by crystallization to form carbonated HA crystals. The calcium and phosphate ions in the glass or glass-ceramic provide the nucleation sites for crystallization. Carbonate anions (CO_3^{2-}) substitute for OH^- in the apatite crystal structure to form a carbonate hydroxyapatite similar to that found in living bone. Incorporation of CaF_2 in the glass results in incorporation of fluoride ions in the apatite crystal lattice. Crystallization of carbonate HA occurs around collagen fibrils present at the implant interface and results in interfacial bonding.

In order for the material to be bioactive and form an interfacial bond, the kinetics of reaction in Eq. 1, and especially the rate of stage 4, must match the rate of biomineralization that normally occurs *in vivo*. If the rates in Eq. 1 are too rapid, the implant is resorbable, and if the rates are too slow, the implant is not bioactive.

By changing the compositionally controlled reaction kinetics (Eq. 1), the rates of formation of hard tissue at an implant interface can be altered, as shown in Fig. 1. Thus, the level of bioactivity of a material can be related to the time for more than 50% of the interface to be bonded ($t_{0.5bb}$) [e.g., I_B index of bioactivity: $= (100/t_{0.5bb})$] (Hench, 1988). It is necessary to impose a 50% bonding criterion for an I_B since the interface between an implant and bone is irregular (Gross *et al.*, 1988). The initial concentration of macrophages, osteoblasts, chondroblasts, and fibroblasts varies as a function of the fit of the implant and the condition of the bony defect (see Chapters 3 and 4). Consequently, all bioactive implants require an incubation period before bone proliferates and bonds. The length of this incubation period varies widely, depending on the composition of the implant.

The compositional dependence of I_B indicates that there are isoI_B contours within the bioactivity boundary, as shown in Fig. 5 (Hench, 1988). The change of I_B with the $SiO_2/(Na_2O + CaO)$ ratio is very large as the bioactivity boundary is approached. The addition of multivalent ions to a bioactive glass or glass-ceramic shrinks the isoI_B contours, which will contract to zero as the percentage of Al_2O_3, Ta_2O_5, ZrO_2, or other multivalent cations increases in the material. Consequently, the isoI_B boundary shown in Fig. 5 indicates the contamination limit for bioactive glasses and glass-ceramics. If the composition of a starting implant is near the I_B boundary, it may take only a few parts per million of multivalent cations to shrink the I_B boundary to zero and eliminate bioactivity. Also, the sensitivity of fit of a bioactive implant and length of time of immobilization postoperatively depends on the I_B value and closeness to the $I_B = 0$ boundary. Implants near the I_B boundary require more precise surgical fit and longer fixation times before they can bear loads. In contrast, increasing the surface area of a bioactive implant by using them in particulate form for bone augmentation expands the bioactive boundary. Small (<200 μm) bioactive glass granules behave as a partially resorbable implant and stimulate new bone formation (Hench, 1994).

Bioactive implants with intermediate I_B values do not develop a stable soft tissue bond; instead, the fibrous interface progressively mineralizes to form bone. Consequently, there

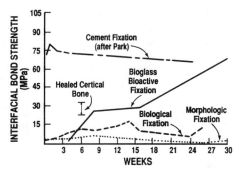

FIG. 7. Time dependence of interfacial bond strength of various fixation systems in bone. (After Hench, 1987.)

appears to be a critical isoI_B boundary beyond which bioactivity is restricted to stable bone bonding. Inside the critical isoI_B boundary, the bioactivity includes both stable bone and soft tissue bonding, depending on the progenitor stem cells in contact with the implant. This soft tissue–critical isoI_B limit is shown by the dashed contour in Fig. 5.

The thickness of the bonding zone between a bioactive implant and bone is proportional to its I_B (compare Fig. 1 with Fig. 2). The failure strength of a bioactively fixed bond appears to be inversely dependent on the thickness of the bonding zone. For example, 45S5 Bioglass with a very high I_B develops a gel bonding layer of 200 μm, which has a relatively low shear strength. In contrast, A-W glass-ceramic, with an intermediate I_B value, has a bonding interface in the range of 10–20 μm and a very high resistance to shear. Thus, the interfacial bonding strength appears to be optimum for I_B values of ~4. However, it is important to recognize that the interfacial area for bonding is time dependent, as shown in Fig. 1. Therefore, interfacial strength is time dependent and is a function of such morphological factors as the change in interfacial area with time, progressive mineralization of the interfacial tissues, and resulting increase of elastic modulus of the interfacial bond, as well as shear strength per unit of bonded area. A comparison of the increase in interfacial bond strength of bioactive fixation of implants bonded to bone with other types of fixation is given in Fig. 7 (Hench, 1987).

Clinical applications of bioactive glasses and glass-ceramics are reviewed by Gross *et al.* (1988), Yamamuro *et al.* (1990), Hench and Wilson (1993) (Table 6). The 8-year history of successful use of Ceravital glass-ceramics in middle ear surgery (Reck *et al.*, 1988) is especially encouraging, as is the 10 year use of A-W glass-ceramic in vertebral surgery (Yamamuro *et al.*, 1990) and 10 year use of 45S5 Bioglass in endosseous ridge maintenance (Stanley *et al.*, 1996) and middle ear replacement and 6 year success in repair of periodontal defects (Hench and Wilson, 1996; Wilson *et al.* 1994).

CALCIUM PHOSPHATE CERAMICS

Calcium phosphate-based bioceramics have been used in medicine and dentistry for nearly 20 years (Hulbert *et al.*,

TABLE 6 Present Uses of Bioceramics

Orthopedic load-bearing applications Al_2O_3	Coatings for tissue ingrowth (Cardiovascular, orthopedic, dental and maxillofacial prosthetics) Al_2O_3
Coatings for chemical bonding (Orthopedic, dental and maxillofacial prosthetics) HA Bioactive glasses Bioactive glass-ceramics	Temporary bone space fillers Tricalcium phosphate Calcium and phosphate salts
Dental implants Al_2O_3 HA Bioactive glasses	Periodontal pocket obliteration HA HA-PLA composite Trisodium phosphate Calcium and phosphate salts Bioactive glasses
Alveolar ridge augmentations Al_2O_3 HA HA-autogenous bone composite HA-PLA composite Bioactive glasses	Maxillofacial reconstruction Al_2O HA HA-PLA composite Bioactive glasses
Otolaryngological Al_2O_3 HA Bioactive glasses Bioactive glass-ceramics	Percutaneous access devices Bioactive glasses Bioactive composites
Artifical tendon and ligament PLA-carbon fiber composite	Orthopedic fixation devices PLA-carbon fibers PLA-calcium/phosphorous-base glass fibers

$$4\,Ca_3(PO_4)_2 \text{ (solid)} + 2H_2O \rightarrow Ca_{10}(PO_4)_6(OH) \text{ (surface)} + 2\,Ca^{2+} + 2\,HPO_4^{2-}. \tag{3}$$

Thus, the solubility of a TCP surface approaches the solubility of HA and decreases the pH of the solution, which further increases the solubility of TCP and enhances resorption. The presence of micropores in the sintered material can increase the solubility of these phases.

Sintering of calcium phosphate ceramics usually occurs in the range of 1000–1500°C following compaction of the powder into a desired shape. The phases formed at high temperature depend not only on temperature but also on the partial pressure of water in the sintering atmosphere. This is because with water present HA can be formed and is a stable phase up to 1360°C, as shown in the phase equilibrium diagram for CaO and P_2O_5 with 500 mm Hg partial pressure of water (Fig. 8). Without water, D_4P and C_3P are the stable phases.

The temperature range of stability of HA increases with the partial pressure of water, as does the rate of phase transitions of C_3P or C_4P to HA. Because of kinetics barriers that affect the rates of formation of the stable calcium phosphate phases, it is often difficult to predict the volume fraction of high-temperature phases that are formed during sintering and their relative stability when cooled to room temperature.

Starting powders can be made by mixing in an aqueous solution the appropriate molar ratios of calcium nitrate and ammonium phosphate, which yield a precipitate of stoichiometric HA. The Ca^{2+}, PO_4^{3-}, and OH^- ions can be replaced by other ions during processing or in physiological surroundings; for example: fluorapatite, $Ca_{10}(PO_4)_6(OH)_{2-x}$ with $O < x < 2$; and carbonate apatite, $Ca_{10}(PO_4)_6(OH)_{2-2x}(CO_3)_x$ or $Ca_{10-x+y}(PO_4)_{6-x}(OH)_{2-x-2y}$, where $O < x < 2$ and $O < y < \frac{1}{2}x$ can be formed. Fluorapatite is found in dental enamel, and carbonate hydroxyapatite is present in bone. For a discussion of the structure of these complex crystals, see Le Geros (1988).

The mechanical behavior of calcium phosphate ceramics strongly influences their application as implants. Tensile and

1987; de Groot, 1983, 1988; de Groot *et al.*, 1990; Jarcho, 1981; Le Geros, 1988; Le Geros and Le Geros, 1993). Applications include dental implants, periodontal treatment, alveolar ridge augmentation, orthopedics, maxillofacial surgery, and otolaryngology (Table 6). Different phases of calcium phosphate ceramics are used, depending upon whether a resorbable or bioactive material is desired.

The stable phases of calcium phosphate ceramics depend considerably upon temperature and the presence of water, either during processing or in the use environment. At body temperature, only two calcium phosphates are stable when in contact with aqueous media such as body fluids. At pH < 4.2, the stable phase is $CaHPO_4 \cdot 2H_2O$ (dicalcium phosphate or brushite, C_2P), while at pH ≥ 4.2 the stable phase is $Ca_{10}(PO_4)_6(OH)_2$ (HA). At higher temperatures, other phases such as $Ca_3(PO_4)_2$ (β-tricalcium phosphate, C_3P, or TCP) and $Ca_4P_2O_9$ (tetracalcium phosphate, C_4P) are present. The unhydrated high-temperature calcium phosphate phases interact with water or body fluids at 37°C to form HA. HA forms on exposed surfaces of TCP by the following reaction:

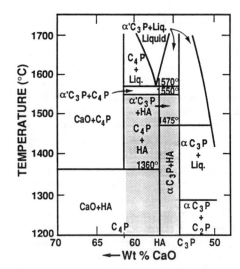

FIG. 8. Phase equilibrium diagram of calcium phosphates in a water atmosphere. Shaded area is processing range to yield HA-containing implants. (After K. de Groot, *Ann. New York Acad. Sci.* 523: 227, 1988.)

compressive strength and fatigue resistance depend on the total volume of porosity. Porosity can be in the form of micropores (<1 μm diameter, due to incomplete sintering) or macropores (>100 μm diameter, created to permit bone growth). The dependence of compressive strength (σ_c) and total pore volume (V_p) is described in de Groot *et al.* (1990) by:

$$\sigma_c = 700 \exp - 5V_p \text{ (in MPa)}, \qquad (4)$$

where V_p is in the range of 0–0.5.

Tensile strength depends greatly on the volume fraction of microporosity (V_m):

$$\sigma_t = 220 \exp - 20\,V_m \text{ (in MPa)}. \qquad (5)$$

The Weibull factor (n) of HA implants is low in physiological solutions ($n = 12$), which indicates low reliability under tensile loads. Consequently, in clinical practice, calcium phosphate bioceramics should be used as (1) powders; (2) small, unloaded implants such as in the middle ear; (3) with reinforcing metal posts, as in dental implants; (4) coatings (e.g., composites); or (5) low loaded porous implants where bone growth acts as a reinforcing phase.

The bonding mechanism of dense HA implants appears to be very different from that described above for bioactive glasses. The bonding process for HA implants is described by Jarcho (1981). A cellular bone matrix from differentiated osteoblasts appears at the surface, producing a narrow amorphous electron-dense band only 3 to 5 μm wide. Between this area and the cells, collagen bundles are seen. Bone mineral crystals have been identified in this amorphous area. As the site matures, the bonding zone shrinks to a depth of only 0.05 to 0.2 μm (Fig. 2). The result is normal bone attached through a thin epitaxial bonding layer to the bulk implant. TEM image analysis of dense HA bone interfaces show an almost perfect epitaxial alignment of some of the growing bone crystallites with the apatite crystals in the implant. A consequence of this ultrathin bonding zone is a very high gradient in elastic modulus at the bonding interface between HA and bone. This is one of the major differences between the bioactive apatites and the bioactive glasses and glass-ceramics. The implications of this difference for the implant interfacial response to Wolff's law is discussed in Hench and Ethridge (1982, Chap. 14).

RESORBABLE CALCIUM PHOSPHATES

Resorption or biodegradation of calcium phosphate ceramics is caused by three factors:

1. Physiochemical dissolution, which depends on the solubility product of the material and local pH of its environment. New surface phases may be formed, such as amorphous calcium phosphate, dicalcium phosphate dihydrate, octacalcium phosphate, and anionic-substituted HA.
2. Physical disintegration into small particles as a result of preferential chemical attack of grain boundaries.

3. Biological factors, such as phagocytosis, which causes a decrease in local pH concentrations (de Groot and Le Geros, 1988).

All calcium phosphate ceramics biodegrade to varying degrees in the following order:

$$\text{increasing rate} \rightarrow$$
$$\alpha\text{-TCP} > \beta\text{-TCP} \gg \text{HA}.$$

The rate of biodegradation increases as:

1. Surface area increases (powders > porous solid > dense solid)
2. Crystallinity decreases
3. Crystal perfection decreases
4. Crystal and grain size decrease
5. There are ionic substitutions of CO_3^{2-}, Mg^{2+}, and Sr^{2+} in HA

Factors that tend to decrease the rate of biodegradation include (1) F^- substitution in HA, (2) Mg^{2+} substitution in β-TCP, and (3) lower β-TCP/HA ratios in biphasic calcium phosphates.

Bibliography

Christel, P., Meunier, A., Dorlot, J. M., Crolet, J. M., Witvolet, J., Sedel, L., and Boritin, P. (1988). Biomechanical compatability and design of ceramic implants for orthopedic surgery, in *Bioceramics: Material Characteristics versus In-Vivo Behavior*, P. Ducheyne and J. Lemons, eds. Ann. New York Acad. Sci., Vol. 523, p. 234.

Dörre, E., and Dawihl, W. (1980). Ceramic hip endoprostheses, in *Mechanical Properties of Biomaterials*, G. W. Hastings and D. Williams, eds. Wiley, New York, pp. 113–127.

de Groot, K. (1983). *Bioceramics of Calcium-Phosphate*, CRC Press, Boca Raton, FL.

de Groot, K. (1988). Effect of porosity and physicochemical properties on the stability, resorption, and strength of calcium phosphate ceramics, in *Bioceramics: Material Characteristics versus In-Vivo Behavior*, Ann. New York Acad. Sci., Vol. 523, p. 227.

de Groot, K., and Le Geros, R. (1988). Position papers in *Bioceramics: Materials Characteristics versus In-Vivo Behavior*, P. Ducheyne and J. Lemons, eds. Ann. New York Acad. Sci., Vol. 523, pp. 227, 268, 272.

de Groot, K., Klein, C. P. A. T., Wolke, J. G. C., and de Blieck-Hogervorst, J. (1990). Chemistry of calcium phosphate bioceramics, in *Handbook on Bioactive Ceramics*, T. Yamamuro, L. L. Hench, and J. Wilson, eds. CRC Press, Boca Raton, FL, Vol. II, Ch. 1.

Gross, V., Kinne, R., Schmitz, H. J., and Strunz, V. (1988). The response of bone to surface active glass/glass-ceramics. *CRC Crit. Rev. Biocompatibility* 4: 2.

Gross, V., and Strunz, V. (1985). The interface of various glasses and glass-ceramics with a bony implantation bed. *J. Biomed. Mater. Res.* 19: 251.

Hench, L. L. (1987). Cementless fixation, in *Biomaterials and Clinical Applications*, A. Pizzoferrato, P. G. Marchetti, A. Ravaglioli, and A. J. C. Lee, eds. Elsevier, Amsterdam, p. 23.

Hench, L. L. (1988). Bioactive ceramics, in *Bioceramics: Materials Characteristics versus In-Vivo Behavior*, P. Ducheyne and J. Lemons, eds. Ann. New York Acad. Sci., Vol. 523, p. 54.

Hench, L. L. (1991). Bioceramics: From concept to clinic. *J. Am. Ceram. Soc.* 74: 1487–1510.

Hench, L. L. (1994). *Bioactive ceramics: Theory and clinical applications in Bioceramics-7,* O. H. Anderson and A. Yli-Urpo, eds. Butterworth–Heinemann, Oxford, England, pp. 3–14.

Hench, L. L., and Clark, D. E. (1978). Physical chemistry of glass surfaces. *J. Non-Cryst. Solids* 28(1):83–105.

Hench, L. L., and Ethridge, E. C. (1982). *Biomaterials: An Interfacial Approach.* Academic Press, New York.

Hench, L. L., and Wilson, J. W. (1993). *An Introduction to Bioceramics,* World Scientific, Singapore.

Hench, L. L., and Wilson, J. W. (1996). Clinical Performance of Skeletal Prostheses, Chapman and Hall, London.

Hench, L. L., Splinter, R. J., Allen, W. C., and Greenlec, Jr. T. K. (1972). Bonding mechanisms at the interface of ceramic prosthetic materials. *J. Biomed. Res. Symp.* No. 2. Interscience, New York, p. 117.

Holand, W., and Vogel, V. (1993). Machineable and phosphate glass-ceramics, in *An Introduction to Bioceramics,* L. L. Hench and J. Wilson, eds. World Scientific, Singapore, pp. 125–138.

Hulbert, S. (1993). The use of alumina and zirconia in surgical implants. in *An Introduction to Bioceramics,* L. L. Hench and J. Wilson, eds. World Scientific, Singapore, pp. 25–40.

Hulbert, S. F., Bokros, J. C., Hench, L. L., Wilson, J., and Heimke, G. (1987). Ceramics in clinical applications: Past, present, and future, in *High Tech Ceramics,* P. Vincenzini, ed. Elsevier, Amsterdam, pp. 189–213.

Jarcho, M. (1981). Calcium phosphate ceramics as hard tissue prosthetics. *Clin. Orthop. Relat. Res.* **157:** 259.

Kokubo, T. (1993). A/W glass-ceramics: Processing and properties. in *An Introduction to Bioceramics,* L. L. Hench and J. Wilson, eds. World Scientific, Singapore, pp. 75–88.

Lacefield, W. R. (1993). Hydroxylapatite coatings. in *An Introduction to Bioceramics,* L. L. Hench and J. Wilson, eds. World Scientific, Singapore, pp. 223–238.

Le Geros, R. Z. (1988). Calcium phosphate materials in restorative densitry: a review. *Adv. Dent. Res.* **2:** 164–180.

Le Geros, R. Z., and Le Geros, J. P. (1993). Dens hydroxyapatite. in *An Introduction to Bioceramics,* L. L. Hench and J. Wilson, eds. World Scientific, Singapore, pp. 139–180.

McKinney, Jr., R. V., and Lemons, J. (1985). *The Dental Implant,* PSG Publ., Littleton, MA.

McMillan, P. W. (1979). *Glass-Ceramics,* Academic Press, New York.

Miller, J. A., Talton, J. D., and Bhatia, S. (1996). in *Clinical Performance of Skeletal Prostheses,* L. L. Hench and J. Wilson, eds. Chapman and Hall, London, pp. 41–56.

Nakamura, T., Yamumuro, T., Higashi, S., Kokubo, T., and Itoo, S. (1985). A new glass-ceramic for bone replacement: Evaluation of its bonding to bone. *J. Biomed. Mater. Res.* **19:** 685.

Onoda, G., and Hench, L. L. (1978). *Ceramic Processing Before Firing.* Wiley, New York.

Phillips, R. W. (1991). *Skinners Science of Dental Materials,* 9th Ed., Ralph W. Phillips, ed. Saunders, Philadelphia.

Reck, R., Storkel, S., and Meyer, A. (1988). Bioactive glass-ceramics in middle ear surgery: an 8-year review. in *Bioceramics: Materials Characteristics versus In-Vivo Behavior,* P. Ducheyne and J. Lemons, eds. Ann. New York Acad. Sci., Vol. 523, p. 100.

Reed, J. S. (1988). *Introduction to Ceramic Processing.* Wiley, New York.

Ritter, J. E., Jr., Greenspan, D. C., Palmer, R. A., and Hench, L. L. (1979). Use of fracture mechanics theory in lifetime predictions for alumina and bioglass-coated alumina. *J. Biomed. Mater. Res.* **13:** 251–263.

Roy, D. M., and Linnehan, S. K. (1974). Hydroxyapatite formed from coral skeletal carbonate by hydrothermal exchange. *Nature* **247:** 220–222.

Schors, E. C., and Holmes, R. E. (1993). Porous hydroxyapatite. in *An Introduction to Bioceramics,* L. L. Hench and J. Wilson, eds. World Scientific, Singapore, pp. 181–198.

Stanley, H. R., Clark, A. E., and Hench, L. L. (1996). Alveolar ridge maintenance implants. in *Clinical Performance of Skeletal Prostheses,* Chapman and Hall, London, pp. 237–254.

White, E., and Schors, E. C. (1986). Biomaterials aspects of interpore-200 porous hydroxyapatite. *Dent. Clin. North Am.* **30:** 49–67.

Wilson, J. (1994). Clinical Applications of Bioglass Implants, in Bioceramics-7, O. H. Andersson, ed. Butterworth–Heinemann, Oxford, England.

Wilson, J., Pigott, G. H., Schoen, F. J., and Hench, L. L. (1981). Toxicology and biocompatibility of bioglass. *J. Biomed. Mater. Res.* **15:** 805.

Yamamuro, T., Hench, L. L., Wilson, J. (1990). *Handbook on Bioactive Ceramics,* Vol. I: Bioactive Glasses and Glass-Ceramics, Vol. II: Calcium-Phosphate Ceramics, CRC Press, Boca Raton, FL.

2.7 NATURAL MATERIALS

Ioannis V. Yannas

Natural polymers offer the advantage of being very similar, often identical, to macromolecular substances which the biological environment is prepared to recognize and to deal with metabolically (Table 1). The problems of toxicity and stimulation of a chronic inflammatory reaction, which are frequently provoked by many synthetic polymers, may thereby be suppressed. Furthermore, the similarity to naturally occurring substances introduces the interesting capability of designing biomaterials which function biologically at the molecular, rather than the macroscopic, level. On the other hand, natural polymers are frequently quite immunogenic. Furthermore, because they are structurally much more complex than most synthetic polymers, their technological manipulation is much more elaborate. On balance, these opposing factors have conspired to lead to a substantial number of biomaterials applications in which naturally occurring polymers, or their chemically modified versions, have provided unprecedented solutions.

An intriguing characteristic of natural polymers is their ability to be degraded by naturally occurring enzymes, a virtual guarantee that the implant will be eventually metabolized by physiological mechanisms. This property may, at first glance, appear as a disadvantage since it detracts from the durability of the implant. However, it has been used to advantage in biomaterials applications in which it is desired to deliver a specific function for a temporary period of time, following which the implant is expected to degrade completely and to be disposed of by largely normal metabolic processes. Since, furthermore, it is possible to control the degradation rate of the implanted polymer by chemical cross-linking or other chemical modifications, the designer is offered the opportunity to control the lifetime of the implant.

A disadvantage of proteins on biomaterials is their frequently significant immunogenicity, which, of course, derives precisely from their similarity to naturally occurring substances. The immunological reaction of the host to the implant is directed against selected sites (antigenic determinants) in the protein molecule. This reaction can be mediated by molecules in solution in body fluids (immunoglobulins). A single such molecule (antibody) binds to single or multiple determinants on an antigen. The immunological reaction can also be mediated by molecules which are held tightly to the surface of immune cells (lymphocytes). The implant is eventually degraded. The reaction can be virtually eliminated provided that the antigenic determinants have been previously modified chemically. The immunogenicity of polysaccharides is typically

TABLE 1 General Properties of Certain Natural Polymers

Polymer	Incidence	Physiological function
A. Proteins		
Silk	Synthesized by arthropods	Protective cocoon
Keratin	Hair	Thermal insulation
Collagen	Connective tissues (tendon, skin, etc.)	Mechanical support
Gelatin	Partly amorphous collagen	(Industrial product)
Fibrinogen	Blood	Blood clotting
Elastin	Neck ligament	Mechanical support
Actin	Muscle	Contraction, motility
Myosin	Muscle	Contraction, motility
B. Polysaccharides		
Cellulose (cotton)	Plants	Mechanical support
Amylose	Plants	Energy reservoir
Dextran	Synthesized by bacteria	Matrix for growth of organism
Chitin	Insects, crustaceans	Provides shape and form
Glycosaminoglycans	Connective tissues	Contributes to mechanical support
C. Polynucleotides		
Deoxyribonucleic acids (DNA)	Cell nucleus	Direct protein biosynthesis
Ribonucleic acids (RNA)	Cell nucleus	Direct protein biosynthesis

far lower than that of proteins. The collagens are generally weak immunogens relative to the majority of proteins.

Another disadvantage of proteins as biomaterials derives from the fact that these polymers typically decompose or undergo pyrolytic modification at temperatures below their melting point, thereby precluding the convenience of high-temperature thermoplastic processing methods, such as melt extrusion, during the manufacturing of the implant. However, processes for extruding these temperature-sensitive polymers at room temperature have been developed. Another serious disadvantage is the natural variability in structure of macromolecular substances which are derived from animal sources. Each of these polymers appears as a chemically distinct entity not only from one species to another (species specificity) but also from one tissue to the next (tissue specificity). This testimonial to the elegance of the naturally evolved design of the mammalian body becomes a problem for the manufacturer of implants, which are typically required to adhere to rigid specifications from one batch to the next. Consequently, relatively stringent control methods must be used for the raw material.

Most of the natural polymers in use as biomaterials today are constituents of the extracellular matrix (ECM) of connective tissues such as tendons, ligaments, skin, blood vessels, and bone. These tissues are deformable, fiber-reinforced composite materials of superior architectural sophistication whose main function in the adult animal appears to be the maintenance of organ shape as well as of the organism itself. In the relatively crude description of these tissues as if they were man-made composites, collagen and elastin fibers mechanically reinforce a "matrix" that primarily consists of protein–polysaccharides (proteoglycans) highly swollen in water. Extensive chemical bonding connects these macromolecules to each other, rendering these tissues insoluble and, therefore, impossible to characterize with dilute solution methods unless the tissue is chemi-

cally and physically degraded. In the latter case, the solubilized components are subsequently extracted and characterized by biochemical and physicochemical method. Of the various components of extracellular materials which have been used to fashion biomaterials, collagen is the one most frequently used.

Almost inevitably, the physicochemical processes used to extract the individual polymer from tissues, as well as subsequent deliberate modifications, alter the native structure, sometimes significantly. The following description emphasizes the features of the naturally occurring, or native, macromolecular structures. Certain modified forms of these polymers are also described.

STRUCTURE OF NATIVE COLLAGEN

Structural order in collagen, as in other proteins, occurs at several discrete levels of the structural hierarchy. The collagen in the tissues of a vertebrate occurs in at least ten different forms, each of these being dominant in a specific tissue. Structurally, these collagens share the characteristic triple helix, and variations among them are restricted to the length of the nonhelical fraction, the length of the helix itself, and the number and nature of carbohydrate attachments on the triple helix. The collagen in skin, tendon, and bone is mostly type I collagen. Type II collagen is predominant in cartilage, while type III collagen is a major constituent of the blood vessel wall as well as being a minor contaminant of type I collagen in skin. In contrast to these collagens, all of which form fibrils with the distinct collagen periodicity, type IV collagen, a constituent of the basement membrane which separates epithelial tissues from mesodermal tissues is largely nonhelical and does not form fibrils. We follow here the nomenclature which was proposed by W. Kauzmann (1959) to describe in a general way the

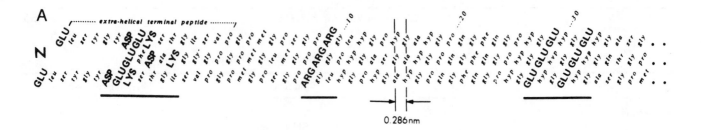

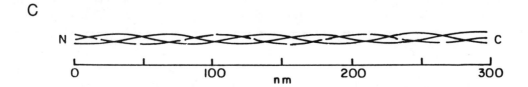

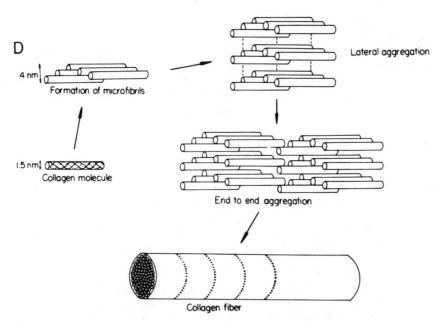

FIG. 1. Collagen, like other proteins, is distinguished by several levels of structural order. (A) Primary structure—the complete sequence of amino acids along each polypeptide chain. An example is the triple chain sequence of type I calf skin collagen at the N-end of the molecule. Roughly 5% of a complete molecule is shown. No attempt has been made to indicate the coiling of the chains. Amino acid residues participating in the triple helix are numbered, and the residue-to-residue spacing (0.286 nm) is shown as a constant within the triple helical domain, but not outside it. Bold capitals indicate charged residues which occur in groups (underlined) (Reprinted from J. A. Chapman and D. J. S. Hulmes, in *Ultrastructure of the Connective Tissue Matrix,* A Ruggeri and P. M. Motta, eds., Martinus Nijhoff, 1984, Chap. 1, Fig. 1, p. 2, with permission.)

structural order in proteins, and we specialize it to the case of type I collagen (Fig. 1).

The primary structure denotes the complete sequence of amino acids along each of three polypeptide chains as well as the location of interchain cross-links in relation to this sequence. Approximately one-third of the residues are glycine and another quarter or so are proline or hydroxyproline. The structure of the bifunctional interchain cross-link is the relatively complex condensation product of a reaction involving lysine and hydroxylysine residues; this reaction continues as the organism matures, thereby rendering the collagens of older animals more difficult to extract.

The secondary structure is the local configuration of a polypeptide chain that results from satisfaction of stereochemical angles and hydrogen-bonding potential of peptide residues. In collagen, the abundance of glycine residues (Gly) plays a key configurational role in the triplet Gly-X-Y, where X and Y are frequently proline or hydroxyproline, respectively, the two amino acids that direct the chain configuration locally by the rigidity of their ring structures. On the other hand, the absence of a side chain in glycine permits the close approach of polypeptide chains in the collagen triple helix. The tertiary structure refers to the global configuration of the polypeptide chains; it represents the pattern according to which the secondary structures are packed together within the complete macromolecule and it constitutes the structural unit that can exist as a physicochemically stable entity in solution, namely, the triple helical collagen molecule.

In type I collagen, two of the three polypeptide chains have identical amino acid composition, consist of 1056 residues, and are termed a1(I) chains, while the third has a different composition; it consists of 1038 residues and is termed a2(I). The three polypeptide chains fold to produce a left-handed helix while the three-chain supercoil is actually right-handed with an estimated pitch of about 100 nm (30–40 residues). The helical structure extends over 1014 of the residues in each of the three chains, leaving the remaining residues at the ends (telopeptides) in a nonhelical configuration. The residue spacing is 0.286 nm and the length of the helical portion of the molecule is, therefore, about 1014×0.286 or 290 nm.

The fourth-order or quaternary structure denotes the repeating supermolecular unit structure, comprising several molecules packed in a specific lattice, which constitutes the basic element of the solid state (microfibril). Collagen molecules are packed in a quasi-hexagonal lattice at an interchain distance of about 1.3 nm which shrinks considerably when the microfibril is dehydrated. Adjacent molecules in the microfibril are approximately parallel to the fibril axis; they all point in the same direction along the fibril and are staggered regularly, giving rise to the well-known D-period of collagen, about 64 nm, which is visible in the electron microscope. Higher levels of order, eventually leading to gross anatomical features which can be readily seen with the naked eye, have been proposed but there is no agreement on their definition.

PHYSICAL MODIFICATIONS OF THE NATIVE STRUCTURE OF COLLAGEN

Crystallinity in collagen can, according to Fig. 1, be detected at two discrete levels of structural order: the tertiary (triple helix) (Fig. 1C) and the quaternary (lattice of triple helices) (Fig. 1D). Each of these levels of order corresponds, interestingly enough, to a separate melting transformation. A solution of collagen triple helices is thus converted to the randomly coiled gelatin by heating above the helix-coil transition temperature, which is approximately 37°C for bovine collagen, or by exceeding a critical concentration of certain highly polarizable anions, e.g., bromide or thiocyanate, in the solution of collagen molecules. Infrared spectroscopic procedures, based on helical marker bands in the mid- and far infrared, have been developed to assay the gelatin content of collagen in the solid or semisolid states in which collagen is commonly used as an implant. Since implanted gelatin is much more rapidly degradable than collagen, these assays are essential tools for quality control of collagen-based biomaterials. Frequently such biomaterials have been processed under manufacturing conditions which may threaten the integrity of the triple helix.

Collagen fibers also exhibit a characteristic banding pattern

(B) Secondary structure—the local configuration of a polypeptide chain. The triplet sequence Gly-Pro-Hyp illustrates elements of collagen triple-helix stabilization. The numbers identify peptide backbone atoms. The conformation is determined by *trans* peptide bonds (3-4, 6-7, and 9-1); fixed rotation angle of bond in proline ring (4-5); limited rotation of proline past the C=O group (bond 5-6); interchain hydrogen bonds (dots) involving the NH hydrogen at position 1 and the C=O at position 6 in adjacent chains; and the hydroxy group of hydroxyproline, possibly through water-bridged hydrogen bonds. (Reprinted from K. A. Piez and A. H. Reddi, eds., *Extracellular Matrix Biochemistry*. Elsevier, 1984, Chap. 1, Fig. 1.6, p. 7, with permission.) (C) Tertiary structure—the global configuration of polypeptide chains, representing the pattern according to which the secondary structures are packed together within the unit substructure. A schematic view of the type I collagen molecule, a triple helix 300 nm long. (Reprinted from K. A. Piez and A. H. Reddi, eds., *Extracellular Matrix Biochemistry*, Elsevier, 1984, Chap. 1, Fig. 1.22, p. 29, with permission.) (D) Quaternary structure—the unit supermolecular structure. The most widely accepted unit is one involving five collagen molecules (microfibril). Several microfibrils aggregate end to end and also laterally to form a collagen fiber which exhibits a regular banding pattern in the electron microscope with a period of 65 nm. (Reprinted from M. E. Nimni, ed., *Collagen*, Vol. I, *Biochemistry*. CRC Press, Boca Raton, 1988, Chap. 1, Fig. 10, p. 14, with permission.)

with a period of 65 nm (quarternary structure). This pattern is lost reversibly when the pH of a suspension of collagen fibers in acetic acid is lowered below 4.25 ± 0.30. Transmission electron microscopy or small-angle X-ray diffraction can be used to determine the fraction of fibrils which possess banding as the pH of the system is altered. During this transformation, which appears to be a first-order thermodynamic transition, the triple helical structure remains unchanged. Changes in pH can, therefore, be used to selectively abolish the quarternary structure while maintaining the tertiary structure intact.

This experimental strategy has made it possible to show that the well-known phenomenon of blood platelet aggregation by collagen fibers (the reason for using collagen sponges as hemostatic devices) is a specific property of the quarternary rather than the tertiary structure. Thus collagen which is thromboresistant *in vitro* has been prepared by selectively "melting out" the packing order of helices while preserving the triple helices themselves. Figure 2 illustrates the banding pattern of such collagen fibers. Notice that short segments of banded fibrils persist even after very long treatment at low pH, occasionally interrupting long segments of nonbanded fibrils (Fig. 2, inset).

The porosity of collagenous implants normally makes an indispensable contribution to its performance. A porous structure provides an implant with two critical functions. First, pore channels are ports of entry for cells migrating from adjacent tissues into the bulk of the implant or for the capillary suction of blood from a hemorrhaging blood vessel nearby. Second, pores endow a solid with a frequently enormous specific surface which is made available either for specific interactions with invading cells (e.g., collagen-glycosaminoglycan (CG) copolymers which induce regeneration of skin in burned patients) or for interaction with coagulation factors in blood flowing into the device (e.g., hemostatic sponges).

Pores can be incorporated by first freezing a dilute suspension of collagen fibers and then inducing sublimation of the ice crystals by exposing the suspension to a low-temperature vacuum. The resulting pore structure is a negative replica of the network of ice crystals (primarily dendrites). It follows that control of the conditions for ice nucleation and growth can lead to a large variety of pore structures (Fig. 3).

In practice, the average pore diameter decreases with decreasing temperature of freezing while the orientation of pore channel axes depends on the magnitude of the heat flux vector during freezing. In experimental implants, the mean pore diameter has ranged between about 1 and 800 mm; pore volume fractions have ranged up to 0.995; the specific surface has been varied between about 0.01 and 100 m^2/g dry matrix; and the orientation of axes of pore channels has ranged from strongly uniaxial to highly radial. The ability of collagen-glycosaminoglycans to induce regeneration of tissues such as skin and nerve depends critically, among other factors, on the adjustment of the pore structure to desired levels, e.g., about 20–125 μm for skin regeneration and less than 10 μm for sciatic nerve regeneration. Determination of pore structure is based on principles of stereology, the discipline which allows the quantitative statistical properties of three-dimensional implant structures to be related to those of two-dimensional projections, e.g., sections used for histological analysis.

CHEMICAL MODIFICATION OF COLLAGEN

The primary structure of collagen is made up of long sequences of some 20 different amino acids. Since each amino acid has its own chemical identity, there are 20 types of pendant side groups, each with its own chemical reactivity, attached to the polypeptide chain backbone. As examples, there are carboxylic side groups (from glutamic acid and aspartic acid residues), primary amino groups (lysine, hydroxylysine, and arginine residues), and hydroxylic groups (tyrosine and hydroxylysine). The collagen molecule is therefore subject to modification by a large variety of chemical reagents. Such versatility comes with a price: Even though the choice of reagents is large, it is imporant to ascertain that use of a given reagent has led to modification of a given fraction of the residues of a certain amino acid in the molecule. This is equivalent to proof that a reaction has proceeded to a desired "yield." Furthermore, proof that a given reagent has attacked only a specific type of amino acid, rather than all amino acid residue types carrying the same functional group, also requires chemical analysis.

Historically, the chemical modification of collagen has been practiced in the leather industry (since about 50% of the protein content of cowhide is collagen) and in the photographic gelatin industry. Today, the increasing use of collagen in biomaterials applications has provided renewed incentive for novel chemical modification, primarily in two areas. First, implanted collagen is subject to degradative attack by collagenases, and chemical cross-linking is a well-known means of decelerating the degradation rate. Second, collagen extracted from an animal source elicits production of antibodies (immunogenicity). Although it is widely accepted that collagen elicits synthesis of a far smaller concentration of antibodies than other proteins (e.g., albumin), treatment with specific reagents, including enzymatic treatment, is occasionally used to reduce the immunogenicity of collagen.

Collagen-based implants are normally degraded by collagenases, naturally occurring enzymes which attack the triple helical molecule at a specific location. Two characteristic products result, namely, the N-terminal fragment which amounts to about two thirds of the molecule, and the one-quarter C-terminal fragment. Both of these fragments become spontaneously transformed (denatured) to gelatin at physiological temperatures via the helix-coil transition and the gelatinized fragments are then cleaved to oligopeptides by naturally occurring enzymes which degrade several other tissue proteins (nonspecific proteases).

Collagenases are naturally present in healing wounds and are credited with a major role in the degradation of collagen fibers at the site of trauma. At about the same time that degradation of collagen and of other ECM components proceeds in the wound bed, these components are being synthesized *de novo* by cells in the wound bed. Eventually, new architectural arrangements, such as scar tissue, are synthesized. While it is not a replica of the intact tissue, scar tissue forms a stable endpoint to the healing process, and forms a tissue barrier between adjacent organs which allows the healed organ to continue functioning at a nearly physiological level. The combined process of collagen degradation and scar synthesis is

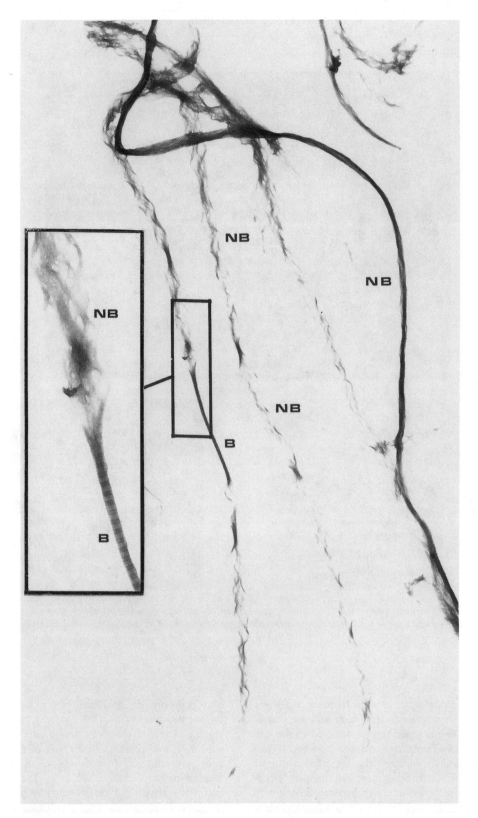

FIG. 2. Following exposure to pH below 4.25 ± 0.30, the banding pattern of type I bovine hide collagen practically disappears. Short lengths of banded collagen (B) do, however, persist next to very long lengths of nonbanded collagen (NB) which has tertiary but not quaternary structure. This preparation does not include platelet aggregation provided that the fibers are prevented from recrystallizing to form banded structures when the pH is adjusted to neutral in order to perform the platelet assay. Stained with 0.5 wt.% phosphotungstic acid. Banded collagen period, about 65 nm. ×12,750. Inset: ×63,750. (Reprinted from M. J. Forbes, M. S. dissertation, Massachusetts Institute of Technology, 1980, courtesy of MIT.)

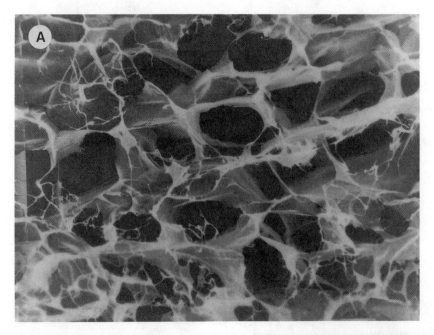

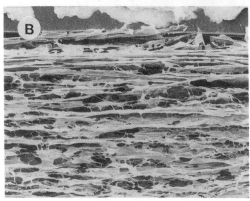

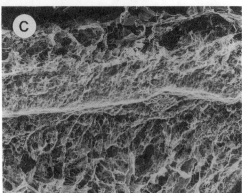

FIG. 3. Illustration of the variety of porous structures which can be obtained with collagen-GAG copolymers by adjusting the kinetics of crystallization of ice to the appropriate magnitude and direction. Pores form when the ice dendrites are eventually sublimed. Scanning electron microscopy. (Courtesy of MIT.)

often referred to as remodeling. One of the frequent challenges in the design of collagen implants is to modify collagen chemically in a way which either accelerates or slows down the rate of its degradation at the implantation site to a desired level.

An effective method for reducing the degradation rate of collagen by naturally occurring enzymes is chemical cross-linking. A very simple self-cross-linking procedure, dehydrative cross-linking, is based on the fact that removal of water below about 1 wt.% insolubilizes collagen as well as gelatin by inducing formation of interchain peptide bonds. The nature of the cross-links formed can be inferred from the results of studies using chemically modified gelatins. Gelatin which had been modified either by esterification of the carboxylic groups of aspartyl–glutamyl residues or by acetylation of the ε-amino groups of lysyl residues remained soluble in aqueous solvents

after exposure of the solid protein to high temperature, while unmodified gelatins lost their solubility. Insolubilization of collagen and gelatin following severe dehydration has been, accordingly, interpreted as the result of drastic removal of the aqueous product of a condensation reaction which led to the formation of interchain amide links. The proposed mechanism is consistent with results, obtained by titration, showing that the number of free carboxylic groups and free amino groups in collagen are both significantly decreased following high-temperature treatment.

Removal of water to the extent necessary to achieve a density of cross-links in excess of 10^{-5} moles of cross-links/g dry gelatin, which corresponds to an average molecular weight between cross-links, M_c, of about 70 kDa, can be achieved within hours by exposure to temperatures in excess of 105°C under atmospheric pressure. The possibility that the cross-linking achieved under these conditions is caused by a pyrolytic reaction has been ruled out. Furthermore, chromatographic data have shown that the amino acid composition of collagen remains intact after exposure to 105°C for several days. In fact, it has been observed that gelatin can be cross-linked by exposure to temperatures as low as 25°C provided that a sufficiently high vacuum is present to achieve the drastic moisture removal which appears to drive the cross-linking reaction.

Exposure of highly hydrated collagen to temperatures in excess of about 37°C is known to cause reversible melting of the triple helical structure, as described earlier. The melting point of the triple helix increases with the collagen-diluent ratio from 37°C, the helix-coil transition of the infinitely dilute solution, to about 120°C for collagen swollen with as little as 20% wt. diluent and up to about 210°C, the melting point of anhydrous collagen. Accordingly, it is possible to cross-link collagen using the drastic dehydration procedure described earlier without loss of the triple helical structure. It is sufficient to adjust the moisture content of collagen to a low enough level prior to exposure to the high temperature levels required for rapid dehydration.

Dialdehydes have been long known in the leather industry as effective tanning agents and in histological laboratories as useful fixatives. Both of these applications are based on the reaction between the dialdehyde and the ε-amino group of lysyl residues in the protein, which induces formation of interchain cross-links. Glutaraldehyde cross-linking is a relatively widely used procedure. The nature of the cross-link formed has been the subject of controversy, primarily because of the complex, apparently polymeric, character of this reagent. Considerable evidence supports the proposed anabilysine structure, which is derived from two lysine side chains and two molecules of glutaraldehyde:

modulus of films that have been treated to induce cross-linking and have subsequently been gelatinized by treatment in 1 M NaCl at 70°C. Under such conditions, only films which have been converted into a three-dimensional network support an equilibrium tensile force; by contrast, uncross-linked specimens dissolve readily in the hot medium.

The immunogenicity of the collagen used in implants is significant and has been studied assiduously using laboratory preparations. However, the clinical significance of such immunogenicity has been shown to be very low, and is often considered to be negligible. This immense simplification of the immunological problem of using collagen as a biomaterial was recognized a long time ago by manufacturers of collagen-based sutures. The apparent reason for the low antigenicity of type I collagen stems from the small species difference among type I collagens (e.g., cow vs. human). Such similarity is, in turn, probably understandable in terms of the inability of the triple helical configuration to incorporate the substantial amino acid substitutions which characterize species differences with other proteins. The relative constancy of the structure of the triple helix among the various species is, in fact, the reason why collagen is sometimes referred to as a "successful" protein in terms of its evolution or, rather, the lack of it.

In order to modify the immunogenicity of collagen, it is useful to consider the location of its antigenic determinants, i.e., the specific chemical groups which are recognized as foreign by the immunological system of the host animal. The configurational (or conformational) determinants of collagen depend on the presence of the intact triple helix and, consequently, are abolished when collagen is denatured into gelatin; the latter event (see earlier discussion) occurs spontaneously after the triple helix is cleaved by a collagenase. Gelatinization exposes the sequential determinant of collagen over the short period during which gelatin retains its macromolecule character, before it is cleared away following attack by one of several non-

$$HO_2C—CH(NH_2)—[CH_2]_4—N \qquad N^+—[CH_2]_4—CH(NH_2)—CO_2H$$

Evidence for other mechanisms has been presented. Compared with other aldehydes, glutaraldehyde has shown itself to be a particularly effective cross-linking agent, as judged, for example, by its ability to increase the cross-link density. The M_c values provide the experimenter with a series of collagens in which the enzymatic degradation rate can be studied over a wide range, thereby affording implants which effectively disappear from tissue between a few days and several weeks following implantation. Although the mechanism of the reaction between glutaraldehyde and collagen at neutral pH is understood in part, the reaction in acidic media has not been studied extensively. Evidence that covalent cross-linking is involved comes from measurements of the equilibrium tensile

specific proteases. Controlling the stability of the triple helix during processing of collagen, therefore, prevents the display of the sequential determinants.

Sequential determinants also exist in the nonhelical end (telopeptide region) of the collagen molecule and this region has been associated with most of the immunogenicity of collagen-based implants. Several enzymatic treatments have been devised to cleave the telopeptide region without destroying the triple helix. Treating collagen with glutaraldehyde not only reduces its degradation rate by collagenase but also appears to reduce its antigenicity as well. The mechanism of this effect is not well understood. Certain applications of collagen-based biomaterials are shown in Table 2.

TABLE 2 Certain Applications of Collagen-Based Biomaterials

Application	Physical state
Sutures	Extruded tape (Schmitt, 1985)
Hemostatic agents	Powder, sponge, fleece (Stengel et al., 1974; Chvapil, 1979)
Blood vessels	Extruded collagen tube, processed human or animal blood vessel (Nimni, 1988)
Heart valves	Processed porcine heart valve (Nimni, 1988)
Tendon, ligaments	Processed tendon (Piez, 1985)
Burn treatment (dermal regeneration)	Porous collagen-glycosaminoglycan (GAG) polymer[a] (Yannas et al., 1981, 1982, and 1989)
Peripheral nerve regeneration	Porous collagen-GAG copolymer (Chang and Yannas, 1992)
Meniscus regeneration	Porous collagen-GAG copolymers (Stone et al., 1989)
Intradermal augmentation	Injectable suspension of collagen particles (Piez, 1985)
Gynecological applications	Sponges (Chvapil, 1979)
Drug-delivery systems	Various forms (Stenzel et al., 1974, Chvapil, 1979)

[a]See also Chapter 7.10.

PROTEOGLYCANS AND GLYCOSAMINOGLYCANS

Glycosaminoglycans (GAGs) occur naturally as polysaccharide branches of a protein chain, or protein core, to which they are covalently attached via a specific oligosaccharide link. The entire branched macromolecule, which has been described as having a "bottle brush" configuration, is known as a proteoglycan and has a molecular weight of about 10^3 kDa.

The structure of GAGs can be generically described as that of an alternating copolymer, the repeat unit consisting of a hexosamine (glucosamine or galactosamine) and of another sugar (galactose, glucuronic acid or iduronic acid). Individual GAG chains are known to contain occasional substitutions of one uronic acid for another; however, the nature of the hexosamine component remains invariant along the chain. There are other deviations from the model of a flawless alternating copolymer, such as variations in sulfate content along the chain. It is, nevertheless, useful for the purpose of getting acquainted with the GAGs to show their typical (rather, typified) repeat unit structure, as in Fig. 4. The molecular weights of GAGs are in the range of 5–60 kDa, with the exception of hyaluronic acid, the only GAG which is not sulfated; it exhibits molecular weights in the range of 50–500 kDa. Sugar units along GAG chains are linked by α or β glycosidic bonds and are 1, 3, or 1, 4 (Fig. 4). There are several naturally occurring enzymes which degrade specific GAGs, the most well-known

being hyaluronidase. These enzymes are primarily responsible for the physiological turnover rate of GAGs, which is in the range of 2–14 days.

The nature of the oligosaccharide link appears to be identical for the GAGs, except for keratan sulfate, and is a galactosyl–galactosyl–xylose, with the latter glycosidically linked to the hydroxyl group of serine in the protein core.

The very high molecular weight of hyaluronic acid is the basis of most uses of this GAG as a biomaterial: almost all applications make use of the exceptionally high viscosity and the facility to form gels which characterize this polysaccharide. Hyaluronic acid gels have found considerable use in ophthalmology because they facilitate cataract surgery as well as retinal reattachment. Other uses of this GAG reported are the treatment of degenerative joint dysfunction in horses and experimental treatment of certain orthopedic dysfunctions in humans. On the other hand, sulfated GAGs are anionically charged and can induce precipitation of collagen at acidic pH levels, a process which yields collagen–GAG coprecipitates that can be subsequently freeze dried and covalently cross-linked to yield biomaterials which have been shown capable of inducing regeneration of skin (dermis), peripheral nerve, and the meniscus of joints (Table 2).

ELASTIN

Elastin is the least soluble protein in the body, consisting as it does of a three-dimensional cross-linked network. It can be extracted from tissues by dissolving and degrading all adjacent substances. It appears to be highly amorphous and thus eludes elucidation of its structure by crystallographic methods. Fortunately, it exhibits ideal rubber elasticity and it thus becomes possible to study certain features of the macromolecular network. For example, mechanical measurements have shown that the average number of amino acid units between crosslinks is 71–84. Insoluble elastic preparations can be degraded by the enzyme elastase but the soluble preparations made this way have not yet been used extensively as biomaterials.

GRAFT COPOLYMERS OF COLLAGEN AND GLYCOSAMINOGLYCANS

The preceding discussion has focused on the individual macromolecular components of ECM. By contrast, naturally occurring ECMs are insoluble networks comprising several macromolecular components. Several types of ECMs are known to play critical roles during organ development. During the past several years certain analogs of ECMs have been synthesized and have been studied as implants. This section summarizes the evidence for the unusual biological activity of a small number of ECM analogs.

In the 1970s it was discovered that a highly porous graft copolymer of type I collagen and chondroitin 6-sulfate was capable of modifying dramatically the kinetics and mechanism of healing of full-thickness skin wounds in rodents (Yannas et al., 1977). In the adult mammal, full-thickness skin wounds

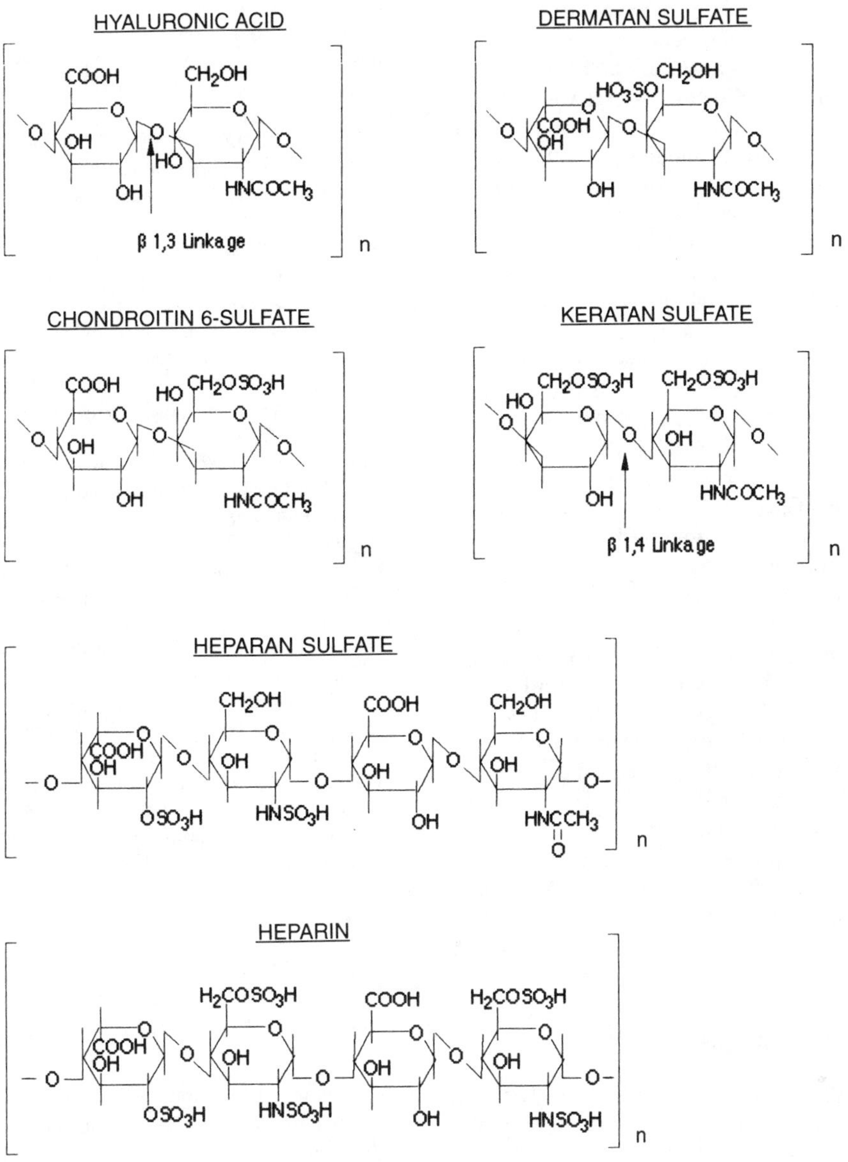

FIG. 4. Repeat units of glycosaminoglycans. (Reprinted for J. E. Siebert, 1987, with permission.)

represent areas that are demonstrably devoid of both epidermis and dermis, the outer and inner layers of skin, respectively. Such wounds normally close by contraction of wound edges and by synthesis of scar tissue. Previously, collagen and various glycosaminoglycans, each prepared in various forms such as powder and films, had been used to cover such deep wounds without observing a significant modification in the outcome of the wound healing process (compare the historical review of Schmitt, 1985).

Surprisingly, grafting of these wounds with the porous CG copolymer on guinea pig wounds blocked the onset of wound contraction by several days and led to synthesis of new connective tissue within about 3 weeks in the space occupied by the copolymer (Yannas et al., 1981, 1982). The copolymer underwent substantial degradation under the action of tissue collagenases during the 3-week period, at the end of which it had degraded completely at the wound site. Studies of the connective tissue synthesized in place of the degraded copolymer eventually showed that the new tissue was distinctly different from scar and was very similar, though not identical, to physiological dermis. In particular, new hair follicles and new sweat glands had not been synthesized. This marked the first instance where scar synthesis was blocked in a full-thickness skin wound of an adult mammal and, in its place, a physiological dermis had been synthesized. That this result was not confined to guinea pigs was confirmed by grafting the same copolymer on full-thickness skin wounds in other adult mammals, including pigs (Butler *et al.,* 1995) and, most importantly, human victims of massive burns (Burke *et al.,* 1981).

Although a large number of CG copolymers were synthesized and studied as grafts, it was observed that only one possessed the requisite activity to modify dramatically the wound healing process in skin. The structure of the CG copolymer required specification at two scales: at the nanoscale, the average molecular weight of the cross-linked network which was required to induce regeneration of the dermis was described by an average molecular weight between cross-links of 12,500 ± 5,000; at the microscale, the average pore diameter was between 20 and 120 μm. Relatively small deviations from these structural features, either at the nanoscale or the microscale, led to loss of activity (Yannas *et al.,* 1989). In view of the nature of its unique activity this biologically active macromolecular network has been referred to as skin regeneration template (SRT). (See also Chapter 7.10.)

The regeneration of dermis was followed by regeneration of a quite different organ, the peripheral nerve (Yannas *et al.,* 1987; Chang and Yannas, 1992). This was accomplished using a distinctly different ECM analog, termed nerve regeneration template (NRT). Although the chemical composition of the two templates was identical there were significant differences in other structural features. NRT degrades considerably slower than SRT (about 6 weeks for NRT compared to about 2 weeks for SRT) and is also characterized by a much smaller average pore diameter (about 5 μm compared to 20–120 μm for SRT). A third ECM analog was shown to induce regeneration of the knee meniscus in the dog (Stone *et al.,* 1990). The combined findings showed that the activity of individual ECM analogs was organ specific. It also suggested that other ECM analogs, still to be discovered, could induce regeneration of organs such as a kidney or the pancreas.

Bibliography

Burke, J. F., Yannas, I. V., Quinby, W. C., Jr., Bondoc, C. C., and Jung, W. K. (1981). Successful use of a physiologically acceptable artificial skin in the treatment of extensive burn injury. *Ann. Surg.* **194:** 413–428.

Butler, C. E., Compton, C. C., Yannas, I. V., and Orgill, D. P. (1995). The effect of keratinocyte seeding of collagen-glycosaminoglycan membranes on the regeneration of skin in a porcine model. 27th Annual Meeting of the American Burn Association, Albuquerque, NM, April 19–21.

Chang, A. S., and Yannas, I. V. (1992). Peripheral nerve regeneration. in *Encyclopedia of Neuroscience,* B. Smith and G. Adelman, eds., Birkhaüser, Boston. Suppl. 2, pp. 125–126.

Chvapil, M. (1979). Industrial uses of collagen. in *Fibrous Proteins: Scientific, Industrial and Medical Aspects,* D. A. D. Parry and L. K. Creamer, eds., Academic Press, London, vol. 1, pp. 247–269.

Davidson, J. M. (1987). Elastin, structure and biology. in *Connective Tissue Disease,* J. Uitto and A. J. Perejda, eds. Marcel Dekker, New York, Ch. 2 pp. 29–54.

Kauzmann, W. (1959). Some factors in the interpretation of protein denaturation, *Adv. Protein Chem.* **14:** 1–63.

Nimni, M. E., ed. (1988). *Collagen, Vol III, Biotechnology.* CRC Press, Boca Raton, FL.

Piez, K. A. (1985). Collagen. in *Encyclopedia of Polymer Science and Technology* **3:** 699–727.

Schmitt, F. O. (1985). Adventures in molecular biology. *Ann. Rev. Biophys. Biophys. Chem.* **14:** 1–22.

Silbert, J. E. (1987). Advances in the biochemistry of proteoglycans. in *Connective Tissue Disease,* J. Uitto and A. J. Perejda, eds. Marcel Dekker, New York, Ch. 4, pp. 83–98.

Stenzel, K. H., Miyata, T., and Rubin, A. L. (1974). Collagen as a biomaterial. in *Annual Review of Biophysics and Bioengineering,* L. J. Mullins, ed., Annual Reviews Inc., Palo Alto, CA, Vol. 3, pp. 231–252.

Stone, K. R., Rodkey, W. G., Webber, R. J., McKinney, L., and Steadman, J. R. (1990). Collagen-based prostheses for meniscal regeneration. *Clin. Orth.* **252:** 129–135.

Yannas, I. V. (1972). Collagen and gelatin in the solid state. *J. Macromol. Sci.-Revs. Macromol. Chem.,* C7(1), 49–104.

Yannas, I. V., Burke, J. F., Gordon, P. L., and Huang, C. (1977). Multilayer membrane useful as synthetic skin. U.S. Patent 4,060,081; Nov. 29.

Yannas, I. V., Burke, J. F., Warpehoski, M., Stasikelis, P., Skrabut, E. M., Orgill, D. P., and Giard, D. J. (1981). Prompt, long-term functional replacement of skin. *Trans. Am. Soc. Artif. Intern. Organs* **27:** 19–22.

Yannas, I. V., Burke, J. F., Orgill, D. P., and Skrabut, E. M. (1982). Wound tissue can utilize a polymeric template to synthesize a functional extension of skin. *Science* **215:** 174–176.

Yannas, I. V., Orgill, D. P., Silver, J., Norregaard, T. V., Zervas, N. T., and Schoene, W. C. (1987). Regeneration of sciatic nerve across 15 mm gap by use of a polymeric template. in *Advances in Biomedical Polymers,* C. G. Gebeleim, ed. Plenum, New York, pp. 1–9.

Yannas, I. V., Lee, E., Orgill, D. P., Skrabut, E. M., and Murphy, G. F. (1989). Synthesis and characterization of a model extracellular matrix that induces partial regeneration of adult mammalian skin. *Proc. Natl. Acad. Sci. U.S.A.* **86:** 933–937.

2.8 COMPOSITES

Harold Alexander

The word "composite" means "consisting of two or more distinct parts." At the atomic level, materials such as metal alloys and polymeric materials could be called composite materials in that they consist of different and distinct atomic groupings. At the microstructural level (about 10^{-4} to 10^{-2} cm), a metal alloy such as a plain-carbon steel containing ferrite and pearlite could be called a composite material since the ferrite and pearlite are distinctly visible constituents as observed in the optical microscope. At the molecular and microstructural level, tissues such as bone and tendon are certainly composites with a number of levels of hierarchy.

In engineering design, a composite material usually refers to a material consisting of constituents in the micro- to macrosize range, favoring the macrosize range. For the purpose of discussion in this chapter, composites can be considered to be materials consisting of two or more chemically distinct constituents, on a macroscale, having a distinct interface separating them. This definition encompasses the fiber and particulate composite materials of primary interest as biomaterials. Such composites consist of one or more discontinuous phases embedded within a continuous phase. The discontinuous phase is usually harder and stronger than the continuous phase and is called the rein-

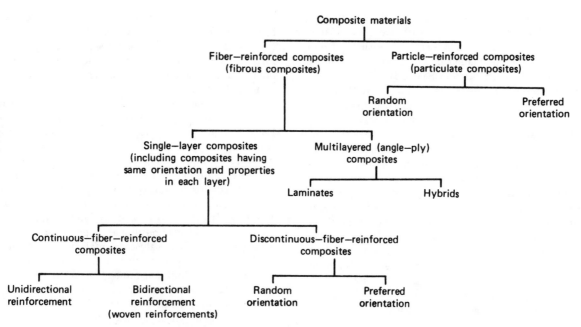

FIG. 1. Classification of composite materials. (Reprinted with permission from B. D. Agarwal and L. J. Broutman, *Analysis and Performance of Fiber Composites*, p. 4, Fig. 1.1, 1980, John Wiley & Sons, Inc.)

forcement or reinforcing material, whereas the continuous phase is termed the "matrix."

The properties of composites are strongly influenced by the properties of their constituent materials, their distribution, and the interaction among them. The properties of a composite may be the volume fraction sum of the properties of the constituents, or the constituents may interact in a synergistic way owing to geometric orientation to provide properties in the composite that are not accounted for by a simple volume fraction sum. Thus, in describing a composite material, besides specifying the constituent materials and their properties, one needs to specify the geometry of the reinforcement, its concentration, distribution, and orientation.

Most composite materials are fabricated to provide mechanical properties such as strength, stiffness, toughness, and fatigue resistance. Therefore, it is natural to study together the composites that have a common strengthening mechanism. This mechanism strongly depends upon the geometry of the reinforcement. It is quite convenient to classify composite materials on the basis of the geometry of a representative unit of reinforcement. Figure 1 shows a commonly accepted classification scheme.

With regard to this classification, the distinguishing characteristic of a particle is that it is nonfibrous in nature. It may be spherical, cubic, tetragonal, or other regular or irregular shapes, but it is approximately equiaxial. A fiber is characterized by its length being much greater than its cross-sectional dimensions. Particle-reinforced composites are sometimes referred to as particulate composites. Fiber-reinforced composites are, understandably, called fibrous composites.

In addition to the classic engineering concerns regarding composites, biomedical composites have other unique properties that affect their performance. As with all biomaterials, the question of biocompatibility (tissue response to the composite) is paramount among them. Being composed of two or more materials, composites carry an enhanced probability of causing adverse tissue reactions. Also, the fact that one constituent (the reinforcement) usually has dimensions on the cellular scale always leaves open the possibility of cellular ingestion of particulate debris that can result in either the production of tissue-lysing enzymes or transport into the lymph system. One class of composites, absorbable composites (to be discussed in some detail later in this chapter), overlay additional requirements for degradation rate kinetics, binding between matrix and reinforcement, and biocompatibility of the degradation products.

REINFORCING SYSTEMS

The main reinforcing materials that have been used in biomedical composites are carbon fibers, polymer fibers, ceramics, and glasses. Depending upon the application, the reinforcements have been either inert or absorbable.

Carbon Fiber

Carbon fiber for biomedical use is produced from polyacrylonitrile (PAN) precursor fiber in a three-step process: (1) stabilization, (2) carbonization, and (3) graphitization. In the stabilization stage, the PAN fibers are first stretched to align the fibrillar networks within each fiber parallel to the fiber axis, and then they are oxidized in air at about 200–220°C while held in tension. The second stage in the production of high-strength carbon fibers is carbonization. In this process, the stabilized PAN-based fibers are pyrolyzed (heated) in a controlled environment until they become transformed into carbon fibers by the elimination of O, H, and N from the precursor fiber. The carbonization heat treatment is usually carried out in an inert atmosphere in the 1000–1500°C range. During the graphitization process, turbostratic graphitelike fibrils or ribbons are formed within each fiber, greatly increasing the tensile strength of the material.

In recent years, carbon fiber has been recognized as a biocompatible material with many exciting applications in medicine. Several commercial products have used carbon fiber as a reinforcing material to enhance the mechanical properties of the polymeric resin systems in which it is included. It has been used to reinforce porous polytetrafluoroethylene for soft tissue augmentation and as a surface coating for the attachment of orthopedic implants. It has also been used to reinforce ultra-high-molecular-weight (UHMW) polyethylene used as a bearing surface in total joint prostheses, as a tendon and ligament repair material, to reinforce fracture fixation devices, and to reinforce total joint replacement components.

Polymer Fibers

For the majority of applications, polymer fibers are not strong or stiff enough to be used to reinforce other polymers. The possible exceptions are aramid fibers, UHMW polyethylene fibers, and certain fibers that have been used for their absorbability, not their mechanical superiority. Aramid is the generic name for aromatic polyamide fibers. Aramid fibers were introduced commercially under the trade name Kevlar. Kevlar composites are used commercially where high strength and stiffness, damage resistance, and resistance to fatigue and stress rupture are important. Potential biomedical applications are in hip prostheses stems, fracture fixation devices, and ligament and tendon prostheses. UHMW polyethylene fibers have recently become available. To date, these fibers have not seen extensive use in biomedical composites. Bulk UHMW polyethylene demonstrates excellent biocompatibility. However, there are preliminary data suggesting a less favorable response to UHMW polyethylene fibers (Shieh et al., 1990). Questions are always raised when bulk and fiber properties are equated. While in theory the basic materials should be the same, differences associated with surface characteristics and the details of manufacturing and processing can be significant.

Absorbable polymer fibers have been used to reinforce absorbable polymers in fabricating fully absorbable fracture fixation systems (Vert et al., 1986). Polylactic acid polymer and polyglycolic acid polymer have been used for this purpose. Both of these polymer fibers have been used for a number of years in absorbable sutures. More recently, they have also been utilized as scaffolds for cellular support in experimental hybrid implants that combine synthetics with living cells (Vacanti et al., 1991).

Ceramics

Several different ceramic materials have been used to reinforce biomedical composites. Since most biocompatible ceramics, when loaded in tension or shear, are relatively weak and brittle materials compared with metals, the preferred form for this reinforcement has usually been particulate. These reinforcements have included various calcium phosphates, aluminum- and zinc-based phosphates, glass and glass ceramics, and bone mineral. Minerals in bone are numerous. In the past, bone has been defatted, ground, and calcined or heated to yield a relatively pure mix of the naturally occurring bone minerals. It was recognized early that this mixture of natural bone mineral was poorly defined and extremely variable. Consequently, its use as an implant material was limited.

The calcium phosphate ceramic system has been the most intensely studied ceramic system. Of particular interest are the calcium phosphates having calcium to phosphorus ratios of 1.5–1.67. Tricalcium phosphate and hydroxyapatite form the boundaries of this compositional range. At present, these two materials are used clinically for dental and orthopedic applications. Tricalcium phosphate has a nominal composition of $Ca_3(PO_4)_2$. The common mineral name for this material is whitlockite. It exists in two crystallographic forms, α- and β-whitlockite. In general, it has been used in the β form.

The ceramic hydroxyapatite has received a great deal of attention. Hydroxyapatite is, of course, the major mineral component of bone. The nominal composition of this material is $Ca_{10}(PO_4)_6(OH)_2$. Some investigators have suggested that there may be a direct bonding of synthetic hydroxyapatite with host bone (Jarcho et al., 1977). Most researchers agree that both tricalcium phosphate and hydroxyapatite are extremely compatible materials. This is particularly true when the materials are in contact with bone. This, of course, is not surprising since both materials display an apatitic surface structure.

Glasses

Glass fibers are used to reinforce plastic matrices in forming structural composites and molding compounds. Commercial glass fiber–plastic composite materials have the following favorable characteristics: high strength-to-weight ratio; good dimensional stability; good resistance to heat, cold, moisture, and corrosion; good electrical insulation properties; ease of fabrication; and relatively low cost. Most of these reasons for using glass fiber reinforcement are not relevant to biomedical composites; consequently, conventional glass fiber reinforcement is not generally practiced. However, recently Zimmerman et al. (1991) and Lin (1986) introduced an absorbable polymer composite reinforced with an absorbable calcium phosphate glass fiber. This allowed the fabrication of a completely absorbable composite implant material. Commercial glass fiber produced from a lime–aluminum–borosilicate glass typically has a tensile strength of about 3 GPa and a modulus of elasticity of 72 GPa. Lin (1986) estimates the absorbable glass fiber to have a modulus of 48 GPa, which compares favorably with the commercial fiber. The tensile strength, however, was significantly lower—approximately 500 MPa.

MATRIX SYSTEMS

Biomedical composites have been fabricated with both absorbable and nonabsorbable matrices. The most common matrices are synthetic nonabsorbable polymers. However, natural polymers and calcium salts have all been used. By far the largest literature exists for the use of polysulfone, ultrahigh-molecular weight (UHMW) polyethylene, polytetrafluoroethylene, poly(ether ether ketone), and poly(methyl methacrylate). These matrices, reinforced with carbon fibers and ceramics, have been used as prosthetic hip stems, fracture fixation devices, artificial joint bearing surfaces, artificial tooth roots, and bone cements.

Absorbable composite implants can be produced from absorbable α-polyester materials such as poly lactic and poly glycolic polymers. Previous work has demonstrated that for most applications, it is necessary to reinforce these polymers to obtain adequate mechanical strength. Poly(glycolic acid) (PGA) was the first biodegradable polymer synthesized (Frazza and Schmitt, 1971). It was followed by poly(lactic acid) (PLA) and copolymers of the two (Gilding and Reed, 1979). These α-polyesters have been investigated for use as sutures and as implant materials for repairing a variety of osseous and soft tissues. Important biodegradable polymers developed in recent years include polyorthoesters (POEs), synthesized by Heller and co-workers (Heller *et al.*, 1980), and a new class of bioerodable dimethyl-trimethylene carbonates (DMTMCs) (Tang *et al.*, 1990). A good review of absorbable polymers by Barrows (1986) included PLA, PGA, poly(lactide-coglycolide), polydioxanone, poly(glycolide-cotrimethylene carbonate), poly(ethylene carbonate), poly(imino carbonates), polycaprolactone, polyhydroxybutyrate, poly(amino acids), poly(ester amides), poly(orthoesters), poly(anhydrides), and cyanoacrylates.

Absorbable polymers of natural origin have also been utilized in biomedical composites. Purified bovine collagen, because of its biocompatibility, resorbability, and availability in a well-characterized implant form, has been used as a composite matrix, mainly as a ceramic composite binder (Lemons *et al.*, 1984). A commercially available fibrin adhesive (Bochlogyros *et al.*, 1985) and calcium sulfate (Alexander *et al.*, 1987) have similarly been used for this purpose.

FABRICATION OF FIBER-REINFORCED COMPOSITES

Fiber-reinforced composites are produced commercially by one of two classes of fabrication techniques: open or closed molding. Most of the open-molding techniques are not appropriate to biomedical composites because of the character of the matrices used (mainly thermoplastics) and the need to produce materials that are resistant to water intrusion. Consequently, the simplest techniques, the hand lay-up and spray-up procedures, are seldom, if ever, used to produce biomedical composites. The two open-molding techniques that may find application in biomedical composites are the vacuum bag–autoclave process and the filament-winding process.

Vacuum Bag–Autoclave Process

This process is used to produce high-performance laminates, usually of fiber-reinforced epoxy. Composite materials produced by this method are currently used in aircraft and aero-

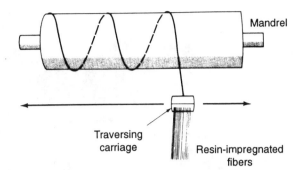

FIG. 2. Filament-winding process for producing fiber-reinforced composite materials. (Reprinted with permission from William F. Smith, *Principles of Materials Science and Engineering*, p. 721, Fig. 13.22, 1986, McGraw-Hill.)

space applications. The first step in this process, and indeed in many other processes, is the production of a "prepreg." This basic structure is a thin sheet of matrix imbedded with uniaxially oriented reinforcing fibers. When the matrix is epoxy, it is prepared in the partially cured state. Pieces of the prepreg sheet are cut out and placed on top of each other on a shaped tool to form a laminate. The layers, or plies, may be placed in different directions to produce the desired strength and stiffness.

After the laminate is constructed, the tooling and attached laminate are vacuum bagged, with a vacuum being applied to remove trapped air from the laminated part. Finally, the vacuum bag enclosing the laminate and the tooling is put into an autoclave for the final curing of the epoxy resin. The conditions for curing vary, depending upon the material, but the carbon fiber–epoxy composite material is usually heated at about 190°C at a pressure of about 700 kPa. After being removed from the autoclave, the composite part is stripped from its tooling and is ready for further finishing operations. This procedure is potentially useful for the production of fracture fixation devices and total hip stems.

Filament-Winding Process

Another important open-mold process to produce high-strength hollow cylinders is the filament-winding process. In this process, the fiber reinforcement is fed through a resin bath and then wound on a suitable mandrel (Fig. 2). When sufficient layers have been applied, the wound mandrel is cured. The molded part is then stripped from the mandrel. The high degree of fiber orientation and high fiber loading with this method produce extremely high tensile strengths. Biomedical applications for this process include intramedullary rods for fracture fixation and prosthetic hip stems.

Closed-Mold Processes

There are many closed-mold methods for producing fiber-reinforced plastic materials. The methods of most importance to biomedical composites are compression and injection mold-

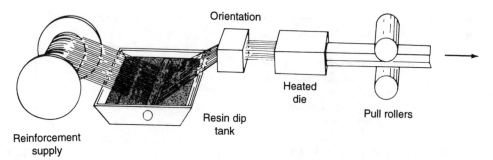

FIG. 3.   The pultrusion process for producing fiber-reinforced polymer composite materials. Fibers impregnated with polymer are fed into a heated die and then are slowly drawn out as a cured composite material with a constant cross-sectional shape. (Reprinted with permission from William F. Smith, *Principles of Materials Science and Engineering,* p. 723, Fig. 13.26, 1986, McGraw-Hill.)

ing and continuous pultrusion. In compression molding, the previously described prepregs are arranged in a two-piece mold that is then heated under pressure to produce the laminated part. This method is particularly useful for thermoplastic matrix applications.

In injection molding, the fiber–matrix mix is injected into a mold at elevated temperature and pressure. The finished part is removed after cooling. This is an extremely fast and inexpensive technique that has application to chopped fiber-reinforced thermoplastic composites. It offers the possibility of producing composite devices, such as bone plates and screws, at much lower costs than comparable metallic devices.

Continuous pultrusion is a process used to manufacture fiber-reinforced plastics of constant cross section such as structural shapes, beams, channels, pipe, and tubing. In this process, continuous-strand fibers are impregnated in a resin bath and then are drawn through a heated die, which determines the shape of the finished stock (Fig. 3). Highly oriented parts cut from this stock can then be used in other structures or they can be used alone in such applications as intramedullary rodding or pin fixation of bone fragments.

MECHANICAL AND PHYSICAL PROPERTIES OF COMPOSITES

Continuous Fiber Composites

Laminated, continuous fiber-reinforced composites are described from either a micro- or macromechanical point of view. Micromechanics is the study of composite material behavior in which the interaction of the constituent materials is examined on a local basis. Macromechanics is the study of composite material behavior in which the material is presumed homogeneous and the effects of the constituent materials are detected only as averaged apparent properties of the composite. Both the micromechanics and macromechanics of experimental laminated composites will be discussed.

Micromechanics

There are two basic approaches to the micromechanics of composite materials: the mechanics of materials and the elastic-

ity approach. The mechanics-of-materials approach embodies the concept of simplifying assumptions regarding the hypothesized behavior of the mechanical system. It is the simpler of the two and the traditional choice for micromechanical evaluation.

The most prominent assumption made in the mechanics-of-materials approach is that strains in the fiber direction of a unidirectional fibrous composite are the same in the fibers and the matrix. This assumption allows the planes to remain parallel to the fiber direction. It also allows the longitudinal normal strain to vary linearly throughout the member with the distance from the neutral axis. Accordingly, the stress will also have a linear distribution.

Some other important assumptions are as follows:

1. The lamina is macroscopically homogeneous, linearly elastic, orthotropic, and initially stress-free.
2. The fibers are homogeneous, linearly elastic, isotropic, regularly spaced, and perfectly aligned.
3. The matrix is homogeneous, linearly elastic, and isotropic.

In addition, no voids are modeled in the fibers and the matrix, or between them.

The mechanical properties of a lamina are determined by fiber orientation. The laminate coordinate system that is used the most often has the length of the laminate in the x direction and the width in the y direction. The principal fiber direction is the 1 direction, and the 2 direction is normal to that. The angle between the x and 1 directions is ϕ. A counterclockwise rotation of the 1–2 system yields a positive ϕ.

The mechanical properties of the lamina are dependent on the material properties and the volume content of the constituent materials. The equations for the mechanical properties of a lamina in the 1–2 directions are

$$E_1 = E_f V_f + E_m V_m \tag{1}$$

$$E_2 = \frac{E_f E_m}{V_m E_f} + V_f E_m \tag{2}$$

$$\nu_{12} - V_m \nu_m + V_f \nu_f \tag{3}$$

$$\nu_{21} E_1 = \nu_{12} E_2 \tag{4}$$

$$G_{12} = \frac{G_f G_m}{V_m G_\nu} + V_f G_\nu \qquad (5)$$

$$V_m = 1 - V_f, \qquad (6)$$

where E is the modulus, G is the shear modulus, V is the volume fraction, ν is Poisson's ratio, and subscripts f and m are the fiber and matrix properties, respectively. These equations are based on the law of mixtures for composite materials.

Macromechanics of a Lamina

The generalized Hooke's law relating stresses to strains is

$$\sigma_i = C_{ij}\varepsilon_j \quad ij = 1, 2, \ldots, 6, \qquad (7)$$

where σ_i are the stress components, C_{ij} is the stiffness matrix, and ε_j are the strain components. An alternative form of the stress-strain relationship is

$$\varepsilon_j = S_{ij}\sigma_j \quad ij = 1, 2, \ldots, 6, \qquad (8)$$

where S_{ij} is the compliance matrix.

Given that $C_{ij} = C_{ji}$, the stiffness matrix is symmetric, thus reducing its population of 36 elements to 21 independent constants. We can further reduce the matrix size by assuming the laminae are orthotropic. There are nine independent constants for orthotropic laminae. In order to reduce this three-dimensional situation to a two-dimensional situation for plane stress, we have

$$\tau_3 = 0 = \sigma_{23} = \sigma_{13}, \qquad (9)$$

thus reducing the stress-strain relationship to

$$\begin{vmatrix} \varepsilon_1 \\ \varepsilon_2 \\ \psi_{12} \end{vmatrix} = \begin{vmatrix} S_{11} & S_{12} & 0 \\ S_{21} & S_{22} & 0 \\ 0 & 0 & S_{66} \end{vmatrix} \cdot \begin{vmatrix} \sigma_1 \\ \sigma_2 \\ \tau_{12} \end{vmatrix}. \qquad (10)$$

This stress-strain relation can be inverted to obtain

$$\begin{vmatrix} \sigma_1 \\ \sigma_2 \\ \tau_{12} \end{vmatrix} = \begin{vmatrix} Q_{11} & Q_{12} & 0 \\ Q_{21} & Q_{22} & 0 \\ 0 & 0 & Q_{66} \end{vmatrix} \cdot \begin{vmatrix} \varepsilon_1 \\ \varepsilon_2 \\ \psi_{12} \end{vmatrix}, \qquad (11)$$

where Q_{ij} are the reduced stiffnesses. The equations for these stiffnesses are

$$Q_{11} = \frac{E_1}{1 - \nu_{21}\nu_{12}} \qquad (12)$$

$$Q_{12} = Q_{21} = \frac{\nu_{21}E_1}{1 - \nu_{21}\nu_{12}} \qquad (13)$$

$$Q_{22} = \frac{E_2}{1 - \nu_{21}\nu_{12}} \qquad (14)$$

$$Q_{66} = G_{12}. \qquad (15)$$

The material directions of the lamina may not coincide with

the body coordinates. The equations for the transformation of stresses in the 1–2 direction to the x–y direction are

$$\begin{vmatrix} \sigma_x \\ \sigma_y \\ \tau_{xy} \end{vmatrix} = [T^{-1}] \cdot \begin{vmatrix} \sigma_1 \\ \sigma_2 \\ \tau_{12} \end{vmatrix}, \qquad (16)$$

where $[T^{-1}]$ is

$$[T^{-1}] = \begin{vmatrix} \cos^2 \phi & \sin^2 \phi & -2\sin\phi\cos\phi \\ \sin^2 \phi & \cos^2 \phi & 2\sin\phi\cos\phi \\ \sin\phi\cos\phi & \sin\phi\cos\phi & \cos^2\phi - \sin^2\phi \end{vmatrix}. \qquad (17)$$

The x and 1 axes form angle ϕ. This matrix is also valid for the transformation of strains,

$$\begin{vmatrix} \varepsilon_x \\ \varepsilon_y \\ \psi_{xy} \end{vmatrix} = [T^{-1}] \cdot \begin{vmatrix} \varepsilon_1 \\ \varepsilon_2 \\ \psi_{12} \end{vmatrix}. \qquad (18)$$

Finally, it can be demonstrated that

$$\begin{vmatrix} \sigma_x \\ \sigma_y \\ \tau_{xy} \end{vmatrix} = [\overline{Q}_{ij}] \cdot \begin{vmatrix} \varepsilon_x \\ \varepsilon_y \\ \psi_{xy} \end{vmatrix}, \qquad (19)$$

where $[\overline{Q}_{ij}]$ is the transformed reduced stiffness. The transformed reduced-stiffness matrix is

$$[\overline{Q}_{ij}] = \begin{vmatrix} \overline{Q}_{11} & \overline{Q}_{12} & \overline{Q}_{16} \\ \overline{Q}_{21} & \overline{Q}_{22} & \overline{Q}_{26} \\ \overline{Q}_{16} & \overline{Q}_{26} & \overline{Q}_{66} \end{vmatrix}. \qquad (20)$$

$\overline{Q}_{16} = \overline{Q}_{26} = 0$ for a laminated symmetric composite, where

$$\overline{Q}_{11} = Q_{11}\cos^4\phi + 2(Q_{12} + 2Q_{66})\sin^2\phi\cos^2 f \\ + Q_{22}\sin^4\phi \qquad (21)$$

$$\overline{Q}_{12} = (Q_{11} + Q_{22} - 4Q_{66})\sin^2\phi\cos^2\phi \\ + Q_{12}(\sin^4\phi + \sin^4\phi) \qquad (22)$$

$$\overline{Q}_{22} = Q_{11}\sin^4\phi + 2(Q_{12} + 2Q_{66})\sin^2\phi\cos^2\phi \\ + Q_{22}\cos^4\phi \qquad (23)$$

$$\overline{Q}_{16} = (Q_{11} - Q_{12} - 2Q_{66})\sin\phi\cos^3\phi \\ + (Q_{12} - Q_{22} + 2Q_{66})\sin^3\phi\cos\phi \qquad (24)$$

$$\overline{Q}_{26} = (Q_{11} - Q_{12} - 2Q_{66})\sin^3\phi\cos\phi \\ + (Q_{12} - Q_{22} + 2Q_{66})\sin\phi\cos^3\phi \qquad (25)$$

$$\overline{Q}_{66} = (Q_{11} + Q_{22} - 2Q_{12} - 2Q_{66})\sin^2\phi \\ \times \cos^2\phi + Q_{66}(\sin^4\phi + \cos^4\phi). \qquad (26)$$

The transformation matrix $[T^{-1}]$ and the transformed reduced stiffness matrix $[\overline{Q}_{ij}]$ are very important matrices in the macromechanical analysis of both laminae and laminates. These matrices play a key role in determining the effective in-plane and bending properties and how a laminate will perform when subjected to different combinations of forces and moments.

Macromechanics of a Laminate

The development of the A, B, and D matrices for laminate analysis is important for evaluating the forces and moments to which the laminate will be exposed and in determining the stresses and strains of the laminae. As given in Eq. 19,

$$\langle \sigma_k \rangle = [Q_{ij}](\varepsilon_k), \tag{27}$$

where σ are normal stresses, ε are normal strains, and $[Q_{ij}]$ is the stiffness matrix. The A, B, and D matrices are equivalent to

$$[A_{ij}] = \sum_{k=1}^{N} (Q_{ij})k(h_k - h_{k-1}) \tag{28}$$

$$[B_{ij}] = \tfrac{1}{2} \sum_{k=1}^{N} (Q_{ij})^k(h_k^2 - h_{k-1}^2) \tag{29}$$

$$[D_{ij}] = \tfrac{1}{3} \sum_{k=1}^{N} (Q_{ij})^k(h_k^3 - h_{k-1}^3) \tag{30}$$

The $[A_{ij}]$ matrix expresses the extensional stiffnesses (in-plane properties), the $[B_{ij}]$ matrix the coupling stiffnesses, and the $[D_{ij}]$ matrix the bending stiffnesses. The letter k denotes the number of laminae in the laminate with a maximum number (N). The letter h represents the distances from the neutral axis to the edges of the respective laminae. A standard procedure for numbering laminae is used in which the 0 lamina is at the bottom of a plate and the Kth lamina is at the top.

The resultant laminate forces and moments are

$$\begin{vmatrix} N_x \\ N_y \\ N_{xy} \end{vmatrix} = [A_{ij}] \cdot \begin{vmatrix} \varepsilon_x \\ \varepsilon_y \\ \psi_{xy} \end{vmatrix} + [B_{ij}] + \begin{vmatrix} k_x \\ k_y \\ k_{xy} \end{vmatrix} \tag{31}$$

$$\begin{vmatrix} M_x \\ M_y \\ M_{xy} \end{vmatrix} = [B_{ij}] \cdot \begin{vmatrix} \varepsilon_x \\ \varepsilon_y \\ \psi_{xy} \end{vmatrix} + [D_{ij}] + \begin{vmatrix} k_x \\ k_y \\ k_{xy} \end{vmatrix}. \tag{32}$$

The k vector represents the respective curvatures of the various planes. The resultant forces and moments of a loaded composite can be analyzed given the ABD matrices. If the laminate is assumed symmetric, the force equation reduces to

$$\begin{vmatrix} N_x \\ N_y \\ N_{xy} \end{vmatrix} = [A_{ij}] \cdot \begin{vmatrix} \varepsilon_x \\ \varepsilon_y \\ \psi_{xy} \end{vmatrix}. \tag{33}$$

Once the laminate strains are determined, the stresses in the x–y direction for each lamina can be calculated. The most useful information gained from the ABD matrices involves the determination of generalized in-plane and bending properties of the laminate.

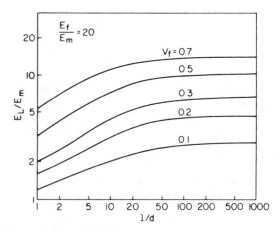

FIG. 4. Variations of longitudinal modulus of short-fiber composites against aspect ratio for different fiber volume fractions ($E_f/E_m = 20$). (Reprinted with permission from B. D. Agarwal and L. J. Broutman, *Analysis and Performance of Fiber Composites*, p. 80, Fig. 3.8, 1980, John Wiley & Sons, Inc.)

Short-Fiber Composites

A distinguishing feature of the unidirectional laminated composites discussed earlier is that they have higher strength and modulus in the fiber direction, and thus their properties are amenable to alteration to produce specialized laminates. However, in some applications, unidirectional multiple-ply laminates may not be required. It may be advantageous to have isotropic lamina. An effective way of producing an isotropic lamina is to use randomly oriented short fibers as the reinforcement. Of course, molding compounds consisting of short fibers that can be easily molded by injection or compression molding may be used to produce generally isotropic composites. The theory of stress transfer between fibers and matrix in short-fiber composites goes beyond this text; it is covered in detail by Agarwal and Broutman (1980). However, from the theory, the longitudinal and transverse moduli for an aligned short-fiber lamina can be found from

$$\frac{E_L}{E_m} = \frac{1 + (2l/d)\eta_L V_f}{1 - \eta_L V_f} \tag{34}$$

$$\frac{E_T}{E_m} = \frac{1 + 2\eta_T V_f}{1 - \eta_T V_f} \tag{35}$$

$$\eta_L = \frac{E_f/E_m - 1}{E_f/E_m + 2(l/d)} \tag{36}$$

$$\eta_T = \frac{E_f/E_m - 1}{E_f/E_m + 2}. \tag{37}$$

For a ratio of fiber to matrix modulus of 20, the variation of longitudinal modulus of an aligned short-fiber lamina as a function of fiber aspect ratio for different fiber volume fractions is shown in Fig. 4. It can be seen that approximately 85% of the modulus obtainable from a continuous fiber lamina is attainable with an aspect ratio of 20.

The problem of predicting properties of randomly oriented

short-fiber composites is more complex. The following empirical equation can be used to predict the modulus of composites containing fibers that are randomly oriented in a plane,

$$E_{random} = \tfrac{3}{8}E_L + \tfrac{5}{8}E_T, \qquad (38)$$

where E_L and E_T are, respectively, the longitudinal and transverse moduli of an aligned short-fiber composite having the same fiber aspect ratio and fiber volume fraction as the composite under consideration. Moduli E_L and E_T can either be determined experimentally or calculated using Eqs. 34 and 35.

ABSORBABLE MATRIX COMPOSITES

Absorbable matrix composites have been used in situations where absorption of the matrix is desired. Matrix absorption may be desired in order to expose surfaces to tissue or to release admixed materials such as antibiotics or growth factors (drug release) (Yasko *et al.*, 1992). However, the most common reasons for using this class of matrices for composites has been to accomplish time-varying mechanical properties and ensure complete dissolution of the implant, eliminating long-term biocompatibility concerns. A typical clinical example is fracture fixation (Daniels *et al.*, 1990).

Fracture Fixation

Rigid internal fixation of fractures has conventionally been accomplished with metallic plates, screws, and rods. During the early stages of fracture healing, rigid internal fixation maintains alignment and promotes primary osseous union by stabilization and compression. Unfortunately, as healing progresses, or after healing is complete, rigid fixation may cause bone to undergo stress protection atrophy. This can result in significant loss of bone mass and osteoporosis. In addition, there may be a basic mechanical incompatibility between the metal implants and bone. The elastic modulus of cortical bone ranges from 17 to 24 GPa, depending upon age and location of the specimen, while the commonly used alloys have moduli ranging from 110 GPa (titanium alloys) to 210 GPa (316L steel). This large difference in stiffness can result in disproportionate load sharing, relative motion between the implant and bone upon loading, as well as high stress concentrations at bone–implant junctions.

Another potential problem is that the alloys currently used corrode to some degree. Ions so released have been reported to cause adverse local tissue reactions as well as allogenic responses, which in turn raises questions of adverse effects on bone mineralization as well as adverse systemic responses such as local tumor formation (Bauer *et al.*, 1987). Consequently, it is usually recommended that a second operation be performed to remove hardware.

The advantages of absorbable devices are thus twofold. First, the devices degrade mechanically with time, reducing stress protection and the accompanying osteoporosis. Second, there is no need for secondary surgical procedures to remove adsorbable devices. The state of stress at the fracture site gradually returns to normal, allowing normal bone remodeling.

Absorbable fracture fixation devices have been produced from PLA polymer, PGA polymer, and polydioxinone. An excellent review of the mechanical properties of biodegradable polymers was prepared by Daniels and co-workers (Daniels *et al.*, 1990; see Figs. 5 and 6). Their review revealed that unreinforced biodegradable polymers are initially 36% as strong in tension as annealed stainless steel, and 54% in bending, but only 3% as stiff in either test mode. With fiber reinforcement, the highest initial strengths exceeded those of stainless steel. Stiffness reached 62% of stainless steel with nondegradable carbon fibers, 15% with degradable inorganic fibers, but only 5% with degradable polymeric fibers.

Most previous work on absorbable composite fracture fixation has been performed with PLA polymer. PLA possesses three major characteristics that make it a potentially attractive biomaterial:

1. It degrades in the body at a rate that can be controlled.
2. Its degradation products are nontoxic, biocompatible, easily excreted entities. PLA undergoes hydrolytic deesterification to lactic acid, which enters the lactic acid cycle of metabolites. Ultimately it is metabolized to carbon dioxide and water and is excreted.
3. Its rate of degradation can be controlled by mixing it with PGA polymer.

PLA polymer reinforced with randomly oriented chopped carbon fiber was used to produce partially degradable bone plates (Corcoran *et al.*, 1981). It was demonstrated that the plates, by virtue of the fiber reinforcement, exhibited mechanical properties superior to those of pure polymer plates. *In vivo*, the PLA matrix degraded and the plates lost rigidity, gradually transferring load to the healing bone. However, the mechanical properties of such chopped fiber plates were relatively low; consequently, the plates were only adequate for low-load situations. If a composite plate of these materials were to be successful in a high-load situation, it was determined, an improved design was necessary. Hence, a study was organized to investigate the possibility of using a long-fiber, angle-ply-laminated composite of carbon fiber and PLA. The results of that study were reported by Zimmerman *et al.* (1987). Composite theory was used to determine an optimum fiber layup for a composite bone plate. Composite analysis suggested the mechanical superiority of a 0°/±45° laminae layup. Although the 0°/±45° carbon/PLA composite possessed adequate initial mechanical properties, water absorption and subsequent delamination degraded the properties rapidly in an aqueous environment (Fig. 7). The fibers did not chemically bond to the PLA. Consequently, future work on biocompatible polymer–fiber coupling agents would be necessary.

In an attempt to develop a totally absorbable composite material, a calcium–phosphate-based glass fiber has been used to reinforce PLA. Experiments were pursued to determine the biocompatibility and *in vitro* degradation properties of the composite (Zimmerman *et al.*, 1991). These studies showed that the glass fiber–PLA composite was biocompatible, but its degradation rate was too high for use as an orthopedic implant.

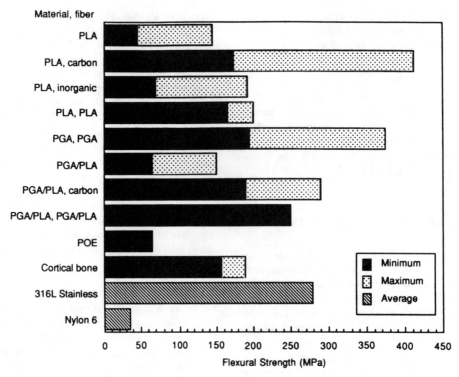

FIG. 5. Representative flexural strengths of absorbable polymer composites. [Reprinted with permission from A. U. Daniel, M. K. O. Chang, and K. P. Andriano, Mechanical properties of biodegradable polymers and composites proposed for internal fixation of bone. *J. Appl. Biomater.* 1(1)57–78, Fig. 26, 1990, John Wiley & Sons, Inc.]

More in-depth research is needed concerning the optimization of composite degradation rates.

NONABSORBABLE MATRIX COMPOSITES

Nonabsorbable matrix composites are generally used as biomaterials to provide specific mechanical properties unattainable with homogeneous materials. Particulate and chopped-fiber reinforcement has been used in bone cements and bearing surfaces to stiffen and strengthen these structures. To date, although they have been introduced into clinical practice, these materials have not been generally accepted and are not in widespread use.

For fracture fixation, reduced-stiffness carbon-fiber-reinforced epoxy bone plates to reduce stress protection osteoporosis have been made. These plates have also been used clinically, but were found to not be as reliable or biocompatible as stainless steel plates. Consequently, they have not generally been accepted in clinical use. By far, the most studied and potentially most valuable use of nonabsorbable composites has been in total joint replacement.

Total Joint Replacement

Bone resorption in the proximal femur that leads to aseptic loosening is an all-too-common occurrence associated with the implantation of metallic femoral hip replacement components. It has been suggested that proximal bone loss may be related to the state of stress and strain in the femoral cortex. It has long been recognized that bone adapts to functional stress by remodeling to reestablish a stable mechanical environment. When applied to the phenomenon of bone loss around implants, one can postulate that the relative stiffness of the metallic component is depriving bone of its accustomed load. Clinical and experimental results have shown the significant role that the elastic characteristics of implants play in allowing the femur to attain a physiologically acceptable stress state. Femoral stem stiffness has been indicated as an important determinant of cortical bone remodeling (Cheal *et al.*, 1992). Composite materials technology offers the ability to alter the elastic characteristics of an implant and provide a better mechanical match with the host bone, potentially leading to a more favorable bone remodeling response.

By using different polymer matrices reinforced with carbon fiber, a large range of mechanical properties is possible. St. John (1983) reported properties for ±15° laminated test specimens (Table 1) with moduli ranging from 18 to 76 GPa. However, the best-reported study involved a novel press-fit device constructed of carbon fiber–polysulfone composite (Magee *et al.*, 1988). The femoral component designed and used in this study utilized composite materials with documented biologic profiles. These materials demonstrated strength commensurate with a totally unsupported implant region and elastic properties commensurate with a fully bone-supported implant region. These

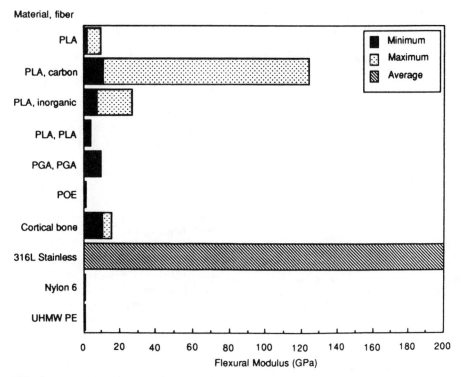

FIG. 6.   Representative flexural moduli of absorbable polymer composites. [Reprinted with permission from A. U. Daniel, M. K. O. Chang, and K. P. Andriano, Mechanical properties of biodegradable polymers and composites proposed for internal fixation of bone. *J. Appl. Biomater.* 1(1)57–78, Fig. 27, 1990, John Wiley & Sons, Inc.]

FIG. 7.   Scanning electron micrograph of laminae buckling and delamination (D) between lamina in a carbon fiber-reinforced PLA fracture fixation plate.

TABLE 1   Typical Mechanical Properties of Polymer-Carbon
             Composites (Three-Point Bending)

Polymer	Ultimate strength (MPa)	Modulus (GPa)
PMMA	772	55
Polysulfone	938	76
Epoxy		
Stycast	535	30
Hysol	207	24
Polyurethane	289	18

properties were designed to produce constructive bone remodeling. The component contained a core of unidirectional carbon–polysulfone composite enveloped with a bidirectional braided layer composed of carbon–polysulfone composite covering the core. These regions were encased in an outer coating of pure polysulfone (Fig. 8).

Finite-element stress analysis predicted that this construction would cause minimal disruption of the normal stresses in the intact cortical bone. Canine animal studies carried out to 4 years showed a favorable bone remodeling response. The authors suggested that implants fabricated from carbon–polysulfone composites should have the potential for use in load-bearing applications. An implant with appropriate elastic

properties provides an opportunity for a natural bone remodeling response to enhance implant stability.

SUMMARY

Biomedical composites have demanding properties that allow few, if any, "off-the-shelf" materials to be used. The designer must almost start from scratch. Consequently, few biomedical composites are yet in general clinical use. Those that have been developed to date have been fabricated from fairly primitive materials with simple designs. They are simple laminates or chopped-fiber- or particulate-reinforced systems with no attempts made to react or bond the phases together. Such bonding may be accomplished by altering the surface texture of the filler or by introducing coupling agents (i.e., molecules that can react with both filler and matrix). However, concerns about the biocompatibility of coupling agents and the high development costs of surface texture alteration procedures have curtailed major developments in this area. It is also possible to provide three-dimensional reinforcement with complex fiber weaving and impregnation procedures now regularly used in high-performance aerospace composites. Unfortunately, the high development costs associated with these techniques have restricted their application to biomedical composites.

Because of the high development costs and the small market available, to date, few materials have been designed specifically for biomedical use. Biomedical composites, because of their unique requirements, will probably be the first general class of materials developed exclusively for implantation purposes.

Bibliography

Agarwal, B. D., and Broutman, L. J. (1980). *Analysis and Performance of Fiber Composites.* Wiley-Interscience, New York.

Alexander, H., Parsons, J. R., Ricci, J. L., Bajpai, P. K. Weiss, A. B. (1987). Calcium-based ceramics and composites in bone reconstruction. *CRC Crit. Rev. Biocompat.* 4(1): 43–77.

Barrows, T. H. (1986). Degradable implant materials: a review of synthetic absorbable polymers and their applications. *Clin. Mater.* 1: 233.

Bauer, T. W., Manley, M. T., Stern, L. S., *et al.* (1987). Osteosarcoma at the site of total hip replacement. *Trans. Soc. Biomater.* 10: 36.

Bochlogyros, P. M., Hensher, R., Becker, R., and Zimmerman, E. (1985). A modified hydroxyapatite implant material. *J. Maxillofac. Surg.* 13(5): 213.

Cheal, E. J., Spector, M., and Hayes, W. C. (1992). Role of loads and prosthesis material properties on the mechanics of the proximal femur after total hip arthroplasty. *J. Orthop. Res.* 10: 405–422.

Corcoran, S., Koroluk, J., Parsons, J. R., Alexander, H., and Weiss, A. B. (1981). The development of a variable stiffness, absorbable composite bone plate. in *Current Concepts for Internal Fixation of Fractures,* H. K. Uhthoff, ed. Springer-Verlag, New York, p. 136.

Daniels, A. U., Melissa, K. O., and Andriano, K. P. (1990). Mechanical properties of biodegradable polymers and composites proposed for internal fixation of bone. *J. Appl. Biomater.* 1(1): 57–78.

Frazza, E. J., and Schmitt, E. E. (1971). A new absorbable suture. *J. Biomed. Mater. Res.* 1: 43.

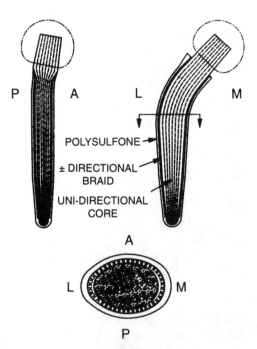

FIG. 8.   Construction details of a femoral stem of a composite total hip prosthesis. [Reprinted with permission from F. P. Magee, A. M. Weinstein, J. A. Longo, J. B. Koeneman, and R. A. Yapp, *Clin. Orthopaed. Related Res.* 235: 237–252, Fig. 1A, 1988, J. B. Lippincott.]

Gilding, D. K., and Reed, A. M. (1979). Biodegradable polymers for use in surgery: PGA/PLA homo- and copolymers, 1. *Polymer* **20**: 1459.

Heller, J., Penhale, D. W. H., and Helwing, R. F. (1980). Preparation of poly(ortho esters) by the reaction of diketene acetals and polyols. *J. Polym. Sci. Plym. Lett. Ed.* **18**: 619.

Jarcho, M., Kay, J. F., Gumaer, K. L., *et al.* (1977). Tissue cellular and subcellular events at a bone-ceramic hydroxyapatite interface. *J. Bioeng.* **1**: 79.

Lemons, J. E., Matukas, V. J., Nieman, K. M. W., Henson, P. G., and Harvey, W. K. (1984). Synthetic hydroxylapatite and collagen combinations for the surgical treatment of bone. in *Biomedical Engineering*. Pergamon, New York, vol. 3, p. 13.

Lin, T. C. (1986). Totally absorbable fiber reinforced composite from internal fracture fixation devices. *Trans. Soc. Biomater.* **9**: 166.

Magee, F. P., Weinstein, A. M., Longo, J. A., Koeneman, J. B., and Yapp, R. A. (1988). A canine composite femoral stem. *Clin. Orthop. Rel. Res.* **235**: 237.

St. John, K. R. (1983). Applications of advanced composites in orthopaedic implants. in *Biocompatible Polymers, Metals, and Composites,* M. Szycher, ed. Technomic Publ., Lancaster, PA, p. 861.

Shieh, S.-J., Zimmerman, M. C., and Parson, J. R. (1990). Preliminary characterization of bioresorbable and nonresorbable synthetic fibers for the repair of soft tissue injuries. *J. Biomed. Mater. Res.* **24**(7): 789–808.

Tang, R., Boyle, Jr., W. J., Mares, F., and Chiu, T.-H. (1990). Novel bioresorbable polymers and medical devices. *Trans. 16th Ann. Mtg. Soc. Biomater.* **13**: 191.

Vacanti, C. A., Langer, R., Schloo, B., and Vacanti, J. P. (1991). Synthetic polymers seeded with chondrocytes provide a template for new cartilage formation. *Plast. Reconstr. Surg.* **88**(5): 753–759.

Vert, M., Christel, P., Garreau, H. Audion, M., Chanavax, M., and Chabot, F. (1986). Totally bioresorbable composites systems for internal fixation of bone fractures. in *Polymers in Medicine,* C. Migliaresi and L. Nicolais, eds. Plenum Publ., New York, vol. 2, pp. 263–275.

Yasko, A., Fellinger, E., Waller, S., Tomin, A., Peterson, M., Wange, E., and Lane, J. (1992). Comparison of biological and synthetic carriers for recombinant human BMP induced bone formation. *Trans. Orth. Res. Soc.* **17**: 71.

Zimmerman, M. C., Parsons, J. R., and Alexander, H. (1987). The design and analysis of a laminated partially degradable composite bone plate for fracture fixation. *J. Biomed. Mater. Res. Appl. Biomater.* **21A**(3): 345.

Zimmerman, M. C., Alexander, H., Parsons, J. R., and Bajpai, P. K. (1991). The design and analysis of laminated degradable composite bone plates for fracture fixation. in *Hi-Tech Textiles,* T. Vigo, ed. American Chemical Society, Washington, DC.

2.9 THIN FILMS, GRAFTS, AND COATINGS

Buddy D. Ratner and Allan S. Hoffman

Much effort goes into the design, synthesis, and fabrication of biomaterials and devices to ensure that they have the appropriate mechanical properties, durability, and functionality. To cite a few examples, a hip joint should withstand high stresses, a hemodialyzer should have the requisite permeability characteristics, and the pumping bladder in an artificial heart should flex for millions of cycles without failure. The bulk structures of the materials govern these properties.

The biological responses to biomaterials and devices, on the other hand, are controlled largely by their surface chemistry and structure (see Chapters 1.3 and 9.7). The rationale for the surface modification of biomaterials is therefore straightforward: to retain the key physical properties of a biomaterial while modifying only the outermost surface to influence the biointeraction. If such surface modification is properly effected, the mechanical properties and functionality of the device will be unaffected, but the tissue-interface-related biocompatibility will be improved or changed.

Materials can be surface modified by using biological or physicochemical methods. Many biological surface modification schemes are covered in Chapter 2.11. Generalized examples of physicochemical surface modifications, the focus of this chapter, are illustrated schematically in Fig. 1. Surface modification with Langmuir–Blodgett (LB) films has elements of both biological modification and physicochemical modification. LB films will be discussed later in this chapter. Some applications for surface-modified biomaterials are listed in Table 1. Physical and chemical surface modification methods, and the types of materials to which they can be applied, are listed in Table 2. Methods to modify or create surface texture or roughness will not be explicitly covered here.

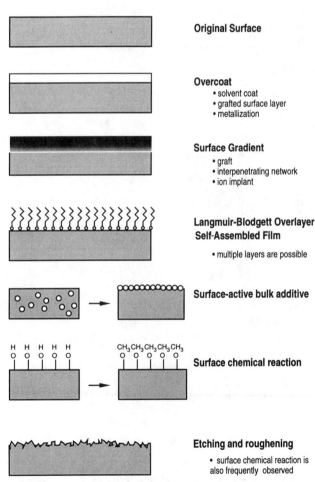

FIG. 1. Schematic representations of methods to modify surfaces.

TABLE 1 Examples of Surface-Modified Biomaterials

To Modify Blood Compatibility

Octadecyl group attachment to surfaces (albumin affinity)
Silicone-containing block copolymer additive
Plasma fluoropolymer deposition
Plasma siloxane polymer deposition
Radiation-grafted hydrogels
Chemically modified polystyrene for heparin-like activity

To Influence Cell Adhesion and Growth

Oxidized polystyrene surface
Ammonia plasma-treated surface
Plasma-deposited acetone or methanol film
Plasma fluoropolymer deposition (reduce endothelial adhesion to IOLs)

To Control Protein Adsorption

Surface with immobilized poly(ethylene glycol) (reduce adsorption)
Treated ELISA dish surface (enhance adsorption strength)
Affinity chromatography particulates (reduce adsorption or enhance specific adsorption)
Surface-cross-linked contact lens (reduce adsorption)

To Improve Lubricity

Plasma treatment
Radiation-grafted hydrogels
Interpenetrating polymeric networks

To Improve Wear Resistance and Corrosion Resistance

Ion implantation
Diamond deposition
Anodization

To Alter Transport Properties

Plasma depositions (methane, fluoropolymer, siloxane)

To Modify Electrical Characteristics

Plasma depositions (insulation layer)
Solvent coatings (insulator or conductor)
Parylene (insulation layer)

GENERAL PRINCIPLES

Surface modifications fall into two categories: (1) chemically or physically altering the atoms, compounds, or molecules in the existing surface (treatment, etching, chemical modification), or (2) overcoating the existing surface with a material having a different composition (coating, grafting, thin film deposition) (Fig. 1). A few general principles provide guidance when undertaking surface modification.

Thin Surface Modifications

Thin surface modifications are desirable. The modified zone at the surface of the material should be as thin as possible. Modified surface layers that are too thick can change the me-

chanical and functional properties of the material. Thick coatings are also more subject to delamination. How thin should a surface modification be? Ideally, alteration of only the outermost molecular layer (3–10 Å) should be sufficient. In practice, thicker films than this will be necessary since it is difficult to ensure that all of the original surface is uniformly covered when coatings and treatments are so thin. Also, extremely thin layers may be more subject to surface reversal (see later discussion) and mechanical erosion. Some coatings intrinsically have a specific thickness. For example, the thickness of LB films is related to the length of the surfactant molecules that comprise them (25–50 Å). Other coatings, such as poly(ethylene glycol) protein-resistant layers, may require a minimum thickness (i.e., a dimension related to the molecular weight of chains) to function. In general, surface modifications should be the minimum thickness needed for uniformity, durability, and functionality, but no thicker. This must be experimentally defined for each system.

Delamination Resistance

The surface-modified layer should be resistant to delamination. This is achieved by covalently bonding the modified region to the substrate, intermixing the components of the substrate and the surface film at an interfacial zone, incorporating a compatibilizing ("primer") layer at the interface, or incorporating appropriate functional groups for strong intermolecular adhesion between a substrate and an overlayer (Wu, 1982).

Surface Rearrangement

Surface rearrangement occurs readily. Surface chemistries and structures can change as a result of the diffusion or translation of surface atoms or molecules in response to the external environment (see Chapter 1.3 and Fig. 2 in that chapter). A newly formed surface chemistry can migrate from the surface into the bulk, or molecules from the bulk can diffuse to cover the surface. Such reversals occur in metallic and other inorganic systems, as well as in polymeric systems. Terms such as "reconstruction," "relaxation," and "surface segregation" are often used to describe mobility-related alterations in surface structure and chemistry (Ratner and Yoon, 1988; Garbassi *et al.*, 1989; Somorjai, 1990, 1991). The driving force for these surface changes is thermodynamic—to minimize the interfacial energy. However, sufficient atomic or molecular mobility must exist for the surface changes to occur in reasonable periods of time. For a modified surface to remain as it was designed, surface reversal must be prevented or inhibited. This can be done by cross-linking, sterically blocking the ability of surface structures to move, or by incorporating a rigid, impermeable layer between the substrate material and the surface modification.

Surface Analysis

Surface analysis is needed. The surface-modified region is usually thin and consists of only minute amounts of material. Undesirable contamination can be readily introduced during

TABLE 2 Physical and Chemical Surface Modification Methods

	Polymer	Metal	Ceramic	Glass
Noncovalent coatings				
Solvent coating	✔	✔	✔	✔
Langmuir–Blodgett film deposition	✔	✔	✔	✔
Surface-active additives	✔	✔	✔	✔
Vapor deposition of carbons and metals[a]	✔	✔	✔	✔
Vapor deposition of Parylene (*p*-xylylene)	✔	✔	✔	✔
Covalently attached coatings				
Radiation grafting (electron accelerator and gamma)	✔	—	—	—
Photografting (UV and visible sources)	✔	—	—	✔
Plasma (gas discharge) (RF, microwave, acoustic)	✔	✔	✔	✔
Gas phase deposition				
Ion beam sputtering	✔	✔	✔	✔
Chemical vapor deposition (CVD)	—	✔	✔	✔
Flame spray deposition	—	✔	✔	✔
Chemical grafting (e.g., ozonation + grafting)	✔	✔	✔	✔
Silanization	✔	✔	✔	✔
Biological modification (biomolecule immobilization)	✔	✔	✔	✔
Modifications of the original surface				
Ion beam etching (e.g., argon, xenon)	✔	✔	✔	✔
Ion beam implantation (e.g., nitrogen)	—	✔	✔	✔
Plasma etching (e.g., nitrogen, argon, oxygen, water vapor)	✔	✔	✔	✔
Corona discharge (in air)	✔	✔	✔	✔
Ion exchange	✔[b]	✔	✔	✔
UV irradiation	✔	✔	✔	✔
Chemical reaction				
Nonspecific oxidation (e.g., ozone)	✔	✔	✔	✔
Functional group modifications (oxidation, reduction)	✔	—	—	—
Addition reactions (e.g., acetylation, chlorination)	✔	—	—	—
Conversion coatings (phosphating, anodization)	—	✔	—	—

[a]Some covalent reaction may occur.

[b]For polymers with ionic groups.

modification reactions. The potential for surface reversal to occur during surface modification is also high. The reaction should be monitored to ensure that the intended surface is indeed being formed. Since conventional analytical methods are often not sensitive enough to detect surface modifications, special surface analytical tools are called for (Chapter 1.3).

Commercializability

The end products of biomaterials research are devices and materials that are mass produced for use in humans. A surface modification that is too complex will be difficult and expensive to commercialize. It is best to minimize the number of steps in a surface modification process and to design each step to be relatively insensitive to small changes in reaction conditions.

METHODS FOR MODIFYING THE SURFACES OF MATERIALS

General methods to modify the surfaces of materials are illustrated in Fig. 1, with many examples listed in Table 2. A few of the more widely used of these methods are briefly described here. Some of the conceptually simpler methods, such as solution coating a polymer on a substrate or metallization by sputtering or thermal evaporation, are not elaborated upon here.

Chemical Reaction

There are hundreds of chemical reactions that can be used to modify the chemistry of a surface. In the context of this chapter, chemical reactions are those reactions performed with reagents that react with atoms or molecules at the surface, but do not overcoat those atoms or molecules with a new layer. Chemical reactions can be classified as nonspecific and specific.

Nonspecific reactions leave a distribution of different functional groups at the surface. An example of a nonspecific surface chemical modification is the chromic acid oxidation of polyethylene surfaces. Other examples include the corona discharge modification of materials in air; radiofrequency glow discharge (RFGD) treatment of materials in oxygen, argon,

A

Alkylation of poly(chlorotrifluoroethylene)

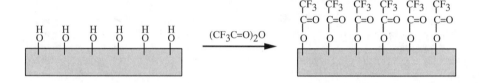

B

Trifluoroacetic anhydride reaction of a hydroxylated surface

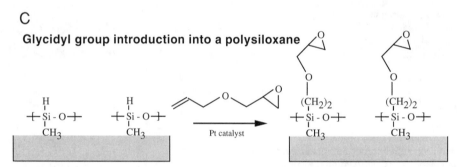

C

Glycidyl group introduction into a polysiloxane

FIG. 2.   Some specific chemical reactions to modify surfaces. (A) Dias and McCarthy, *Macromolecules* **18**, 1826, 1985. (B) Chilkoti *et al., Chem. Mater.* **3**, 51, 1991. (C) Coqueret *et al. Eur. Polymer J.* **24**, 1137, 1988.

nitrogen, carbon dioxide or water vapor environments; and the oxidation of metal surfaces to a mixture of suboxides.

Specific chemical surface reactions change only one functional group into another with a high yield and few side reactions. Examples of specific chemical surface modifications for polymers are presented in Fig. 2.

Radiation Grafting and Photografting

Radiation grafting and related methods have been widely used for the surface modification of biomaterials, and comprehensive review articles are available (Ratner, 1980; Hoffman *et al.,* 1983; Stannett, 1990). Within this category, three types of reactions can be distinguished: grafting using ionizing radiation sources (most commonly, a cobalt-60 gamma radiation source), grafting using UV radiation (photografting) (Matsuda and Inoue, 1990; Dunkirk *et al.,* 1991), and grafting using high-energy electron beams. In all cases, similar processes occur. The radiation breaks chemical bonds in the material to be grafted, forming free radicals, peroxides, or other reactive species. These reactive surface groups are then exposed to a

monomer. The monomer reacts with the free radicals at the surface and propagates as a free radical chain reaction, incorporating other monomers into a surface-grafted polymer.

Three distinct reaction modes can be described: (1) In the mutual irradiation method, the substrate material is immersed in a solution (monomer ± solvent) that is then exposed to the radiation source. (2) The substrate materials can also be exposed to the radiation under an inert atmosphere or at low temperatures. In this case, the materials are later contacted with a monomer solution to initiate the graft process. (3) Finally, the exposure to the radiation can take place in air or oxygen, leading to the formation of peroxide groups on the surface. Heating the material to be grafted in the presence of a monomer or the addition of a redox reactant (e.g., Fe^{2+}) will decompose the peroxide groups to form free radicals that can initiate the graft polymerization.

Graft layers formed by energetic irradiation of the substrate are often thick (>1 μm). However, they are well bonded to the substrate material. Since many polymerizable monomers are available, a wide range of surface chemistries can be created. Mixtures of monomers can form unique graft copolymers (Ratner and Hoffman, 1980). For example, the hydrophilic/

hydrophobic ratio of surfaces can be controlled by varying the ratio of a hydrophilic and a hydrophobic monomer in the grafting mixture (Ratner and Hoffman, 1980; Ratner et al., 1979).

Photoinitiated grafting (usually with visible or UV light) represents a unique subcategory of surface modifications for which there is growing interest. There are many approaches to effect this photoinitiated covalent coupling. For example, a phenyl azide group can be converted to a highly reactive nitrene upon UV exposure. This nitrene will quickly react with many organic groups. If a synthetic polymer is prepared with phenyl azide side groups and this polymer is exposed simultaneously to UV light and a substrate polymer or polymeric medical device, the polymer containing the phenyl azide side groups will be immobilized to the substrate (Matsuda and Inoue, 1990). Another method involves the coupling of a benzophenone molecule to a hydrophilic polymer (Dunkirk et al., 1991). In the presence of UV irradiation, the benzophenone is excited to a reactive triplet state that can covalently couple to many polymers.

Radiation, electron, and photografting have frequently been used to bond hydrogels to the surfaces of hydrophobic polymers (Matsuda and Inoue, 1990; Dunkirk et al., 1991) (see also Chapter 2.4). The protein interactions (Horbett and Hoffman, 1975), cell interactions (Ratner et al., 1975; Matsuda and Inoue, 1990), blood compatibility (Chapiro, 1983; Hoffman et al., 1983), and tissue reactions (Greer et al., 1979) of hydrogel graft surfaces have been investigated.

RFGD Plasma Deposition and Other Plasma Gas Processes

RFGD plasmas, as used for surface modification, are low-pressure ionized gas environments typically at ambient (or slightly above ambient) temperature. They are also referred to as glow discharge or gas discharge depositions or treatments. Plasmas can be used to modify existing surfaces by ablation or etching reactions or, in a deposition mode, to overcoat surfaces (Fig. 1). Good review articles on plasma deposition and its application to biomaterials are available (Yasuda and Gazicki, 1982; Hoffman, 1988; Ratner et al., 1990). Some biomedical applications of plasma-modified biomaterials are listed in Table 3. Since we believe that RFGD plasma surface modifications have special promise for the development of improved biomaterials, they will be emphasized in this chapter.

The specific advantages of plasma-deposited films (and to some extent, plasma-treated surfaces) for biomedical applications are:

1. They are conformal. Because of the penetrating nature of a low-pressure gaseous environment in which transport of mass is governed by both molecular (line-of-sight) diffusion and convective diffusion, complex geometric shapes can be treated.
2. They are free of voids and pinholes. This continuous barrier structure is suggested by transport and electrical property studies (Charlson et al., 1984).
3. Plasma-deposited polymeric films can be placed upon almost any solid substrate, including metals, ceramics,

TABLE 3 Biomedical Applications of Glow Discharge Plasma-Induced Surface Modification Processes

A. Plasma treatment (etching)
  1. Clean
  2. Sterilize
  3. Cross-link surface molecules

B. Plasma treatment (etching) and plasma deposition
  1. Form barrier films
    Protective coating
    Electrically insulating coating
    Reduce absorption of material from the environment
    Inhibit release of leachables
    Control drug delivery rate

  2. Modify cell and protein reactions
    Improve biocompatibility
    Promote selective protein adsorption
    Enhance cell adhesion
    Improve cell growth
    Form nonfouling surfaces
    Increase lubricity

  3. Provide reactive sites
    For grafting or polymerizing polymers
    For immobilizing biomolecules

and semiconductors. Other surface grafting or surface modification technologies are highly dependent upon the chemical nature of the substrate.
4. They exhibit good adhesion to the substrate. The energetic nature of the gas phase species in the plasma reaction environment can induce mixing, implantation, penetration, and reaction between the overlayer film and the substrate.
5. Unique film chemistries can be produced. The chemical structure of the polymeric overlayer films produced by the plasma deposition usually cannot be synthesized by conventional organic chemical methods.
6. They can serve as excellent barrier films because of their pinhole-free and dense, cross-linked nature.
7. Plasma-deposited layers generally show low levels of leachables. Owing to their highly cross-linked nature, plasma-deposited films contain negligible amounts of low-molecular-weight components that might lead to an adverse biological reaction and can also prevent leaching of low-molecular-weight material from the substrate.
8. These films are easily prepared. Once the apparatus is set up and optimized for a specific deposition, treatment of additional substrates is rapid and simple.
9. There is a mature technology for the production of these coatings. The microelectronics industry has made extensive use of inorganic plasma-deposited films (Sawin and Reif, 1983).
10. Although they are chemically complex, plasma surface modifications can be characterized by infrared (IR) (Inagaki et al., 1983; Haque and Ratner, 1988), nuclear

magnetic resonance (NMR) (Kaplan and Dilks, 1981), electron spectroscopy for chemical analysis (ESCA) (Chilkoti *et al.*, 1991a), chemical derivatization studies (Gombotz and Hoffman, 1988; Griesser and Chatelier, 1990; Chilkoti *et al.*, 1991a), and static secondary ion mass spectrometry (SIMS) (Chilkoti *et al.*, 1991b, 1992).

11. Plasma-treated surfaces are sterile when removed from the reactor, offering an additional advantage for cost-efficient production of medical devices.

It would be inappropriate to cite all these advantages without also discussing some of the disadvantages of plasma deposition and treatment for surface modification. First, the chemistry produced on a surface can be ill defined. For example, if tetrafluoroethylene gas is introduced into the reactor, polytetrafluoroethylene will not be deposited on the surface. Rather, a complex, branched fluorocarbon polymer will be produced. This scrambling of monomer structure has been addressed in studies dealing with retention of monomer structure in the final film (Lopez and Ratner, 1991, 1992). Second, the apparatus used to produce plasma depositions can be expensive. A good laboratory-scale reactor will cost $10,000–$30,000, and a production reactor can cost $100,000 or more. Third, a uniform reaction within long, narrow pores can be difficult to achieve. Finally, contamination can be a problem and care must be exercised to prevent extraneous gases and pump oils from entering the reaction zone. However, the advantages of plasma reactions outweigh these potential disadvantages for many types of modifications that cannot be accomplished by any other method.

THE NATURE OF THE PLASMA ENVIRONMENT

Plasmas are atomically and molecularly dissociated gaseous environments. A plasma environment contains positive ions, negative ions, free radicals, electrons, atoms, molecules, and photons. Typical conditions within the plasma include an electron energy of 1–10 eV, a gas temperature of 25–60°C, an electron density of 10^{-9} to $10^{-12}/cm^2$, and an operating pressure of 0.025–1.0 torr.

A number of processes can occur on the substrate surface that lead to surface modification or deposition. First, a competition takes place between deposition and etching by the high-energy gaseous species (ablation) (Yasuda, 1979). When ablation is more rapid than deposition, no deposition will be observed. Because of its energetic nature, the ablation or etching process can result in substantial chemical and morphological changes to the substrate.

A number of mechanisms have been postulated for the deposition process. A reactive gaseous environment may create free radical and other reactive species on the substrate surface that react with and polymerize molecules from the gas phase. Alternatively, reactive small molecules in the gas phase could combine to form higher molecular weight units or particulates that may settle or precipitate onto the surface. Most likely the depositions observed are formed by some combination of these two processes.

PRODUCTION OF PLASMA ENVIRONMENTS FOR DEPOSITION

Many experimental variables relating both to reaction conditions and to the substrate onto which the deposition is placed affect the final outcome of the plasma deposition process (Fig. 3). A diagram of a typical inductively coupled radio frequency plasma reactor is presented in Fig. 3. The major subsystems that comprise this apparatus are a gas introduction system (control of gas mixing, flow rate, and mass of gas entering the reactor), a vacuum system (measurement and control of reactor pressure and inhibition of backstreaming of components from the pumps), an energizing system to efficiently couple energy into the gas phase within the reactor, and a reactor zone in which the samples are treated. Radio frequency, acoustic, or microwave energy can be coupled to the gas phase. Devices for monitoring the molecular weight of the gas phase species (mass spectrometers), the optical emission from the glowing plasma (spectrometers), and the deposited film thickness (ellipsometers, vibrating quartz crystal microbalances) are also commonly found on plasma reactors.

RFGD PLASMAS FOR THE IMMOBILIZATION OF MOLECULES

Plasmas have often been used to introduce organic functional groups (e.g., amine, hydroxyl) on a surface that can be activated to attach biomolecules (see Chapter 2.11). Certain reactive gas environments can also be used to directly immobilize organic molecules such as surfactants. For example, a poly(ethylene glycol-propylene glycol) block copolymer surfactant will adsorb to polyethylene via the propylene glycol block. If the polyethylene surface with the adsorbed surfactant is briefly exposed to an argon plasma, the poly(propylene glycol) block will be cross-linked, thereby leading to the covalent attachment of pendant poly(ethylene glycol) chains (Sheu *et al.*, 1992).

HIGH-TEMPERATURE AND HIGH-ENERGY PLASMA TREATMENTS

The plasma environments described here are of relatively low energy and low temperature. Consequently, they can be used to deposit organic layers on polymeric or inorganic substrates. Under higher energy conditions, plasmas can effect unique and important inorganic surface modifications on inorganic substrates. For example, flame-spray deposition involves injecting a high-purity, relatively finely divided (~100 mesh) metal powder into a high-velocity plasma or flame. The melted or partially melted particles hit the surface and solidify rapidly (see Chapter 2.2 for additional information).

Silanization

The proposed chemistry of a typical silane surface modification reaction is illustrated in Fig. 4. Silane reactions can be

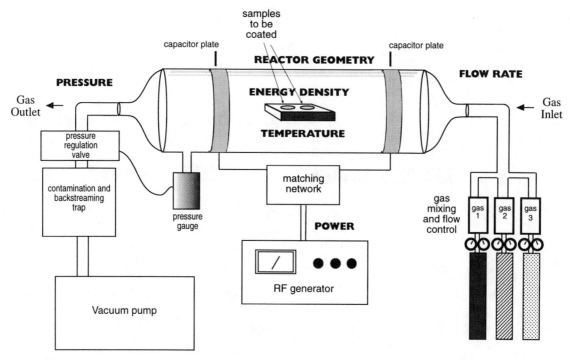

FIG. 3. A diagram of a typical inductively coupled RF plasma reactor. Important experimental variables are indicated in bold typeface.

used to modify hydroxylated or amine-rich surfaces. Since glass, silicon, germanium, alumina, and quartz surfaces, as well as many metal oxide surfaces, are all rich in hydroxyl groups, silanes are particularly useful for modifying these materials. Direct evidence for surface modification on these substrates is observed by an increase in contact angles, particularly where alkyl and fluoroalkyl silanes are used. A wide range of different silanes are available, permitting many different chemical functionalities to be incorporated on surfaces (Table 4). The advantages of silane reactions are their simplicity and stability, which are attributed to their covalent, cross-linked structure. However, the link between a silane and a hydroxyl group is also readily subject to basic hydrolysis, and film breakdown under some conditions must be considered (Wasserman *et al.*, 1989).

Silanes can form two types of surface film structures. If only surface reaction occurs (perhaps catalyzed by traces of adsorbed surface water), a structure similar to that shown in Fig. 4 can be formed. However, if more water is present, a thicker silane layer can be formed that consists of both Si–O groups bonded to the surface and silane units participating in a "bulk," three-dimensional, polymerized network. The initial stages in the formation of a thicker silane film are suggested by the reaction of the group at the right side of Fig. 4D. A new class of silane-modified surfaces based upon the former (monolayer) class of silane films and yielding self-assembled, highly ordered structures has been attracting considerable attention (Maoz *et al.*, 1988). These self-assembled monolayers are described in more detail later in this chapter. Many general reviews on surface silanization are available (Arkles, 1977; Plueddemann, 1980).

TABLE 4 Silanes for Surface Modification of Biomaterials

$$X-\underset{\underset{X}{|}}{\overset{\overset{X}{|}}{Si}}-R$$

X = leaving group	R = functional group
—Cl	—$(CH_2)_nCH_3$
—OCH_3	—$(CH_2)_3NH_2$
—OCH_2CH_3	—$(CH_2)_2(CF_2)_5CF_3$
	—$(CH_2)_3O-\underset{\underset{O}{\|}}{C}-\overset{\overset{CH_3}{|}}{C}=CH_2$
	—CH_2CH_2—◯

Ion Beam Implantation

The ion beam method injects accelerated ions with energies ranging from 10^1 to 10^6 eV (1 eV = 1.6×10^{-19} joules) into the surface zone of a material to alter its surface properties. It is largely, but not exclusively, used with metals and other inorganic systems. Ions formed from most of the atoms in the periodic table can be implanted, but not all provide useful modifications of the surface properties. Important potential

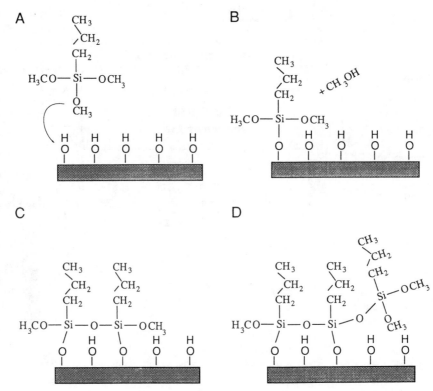

FIG. 4.   The chemistry of a typical silane surface modification reaction. (A) A hydroxylated surface is immersed in a solution containing *n*-propyl trimethoxysilane (nPTMS). (B) One of the methoxy groups of the nPTMS couples with a hydroxyl group, releasing methanol. (C) Two of the methoxy groups on another molecule of the nPTMS have reacted, one with a hydroxyl group and the other with a methoxy group from the first nPTMS molecule. (D) A third nPTMS molecule has reacted only with a methoxy group. This molecule is tied into the silane film network, but is not directly bound to the surface.

applications for biomaterials include modification of hardness (wear), lubricity, toughness, corrosion, conductivity, and bioreaction.

If an ion with an energy greater than a few electron volts hits a surface, the probability that it will enter the surface is high. High energy densities are also transferred to a localized surface zone over short periods. Some considerations for the ion implantation process are illustrated in Fig. 5. These surface changes must be understood quantitatively for precise engineering of new surface characteristics. Many review articles are available on ion implantation processes for tailoring surface properties (Picraux and Pope, 1984; Sioshansi, 1987).

Specific examples of biomaterials that have been surface altered by ion implantation processes are plentiful. Iridium was ion implanted in a Ti-6Al-4V alloy to improve corrosion resistance (Buchanan *et al.*, 1990). Implanting nitrogen into titanium greatly reduces wear (Sioshansi, 1987). The ion implantation of boron and carbon into type 316L stainless steel improves the high-cycle fatigue life of these alloys (Sioshansi, 1987).

Langmuir–Blodgett Deposition

The Langmuir–Blodgett (LB) deposition method covers a surface with a highly ordered layer. Each of the molecules

that assembles into this layer contains a polar head group and a nonpolar region. The deposition of an LB film using an LB trough is illustrated schematically in Fig. 6. By pulling the vertical plate through the air–water interface, and then pushing the plate down through the interface, keeping the surface film at the air–water interface compressed at all times (as illustrated in Fig. 6), multilayer structures can be created. Some compounds that form organized LB layers are shown in Fig. 7. The advantages of films deposited on surfaces by this method are their high degree of order and uniformity. Also, since a wide range of chemical structures can form LB films, there are many options for incorporating new chemistries at surfaces. The stability of LB films can be improved by cross-linking or polymerizing the molecules together after film formation, often through double bonds in each molecule (Meller *et al.*, 1989). A number of research groups have investigated LB films for biomedical applications (Hayward and Chapman, 1984; Bird *et al.*, 1989; Cho *et al.*, 1990). Many general reviews on these surface structures are available (Knobler, 1990; Ulman, 1991).

Self-Assembled Monolayers

Self-assembled monolayers (SAMs) are surface-coating films that spontaneously form as highly ordered structures (two-dimensional crystals) on specific substrates (Maoz *et al.*, 1988; Ulman, 1991; Whitesides *et al.*, 1991). In some ways SAMs resemble LB films, but there are important differences. Examples of SAM films include *n*-alkyl silanes on hydroxylated surfaces (silica, glass, alumina), alkane thiols [e.g., CH$_3$(CH$_2$)$_n$SH] and dithiols on some metals (gold, silver, copper), amines and alcohols on platinum, and carboxylic acids on aluminum oxide and silver. Most molecules that form SAMs have the general characteristics illustrated in Fig. 8.

Two processes are particularly important for the formation of SAMs (Ulman, 1991): a strong, exothermic adsorption of an anchoring chemical group to the surface (typically 30–100 kcal/mol), and van der Waals interaction of the alkyl chains. The strong bonding to the substrate (chemisorption) provides a driving force to fill every site on the surface and to displace contaminants from the surface. This process is analogous to the compression of the LB film by the movable barrier in the trough. Once every adsorption site is filled on the surface, the chains will be in sufficiently close proximity to each other so that the weaker van der Waals interactive forces between chains can exert their influence and lead to a crystallization of the alkyl groups. Molecular mobility is an important consideration in this coating formation process so that (1) the molecules have sufficient time to maneuver into position for a tight packing of the binding end groups at the surface and (2) the chains can enter the quasi-crystal. The advantages of SAMs are their ease of formation, their chemical stability (often considerably higher than comparable LB films), and the many options for changing the outermost

group that interfaces with the world. Although the discovery of SAMs is relatively recent, biomaterials applications have already been suggested (Lewandowska *et al.*, 1989; Prime and Whitesides, 1991).

Surface-Modifying Additives

Certain components can be added in low concentrations to a material during fabrication and will spontaneously rise to and dominate the surface (Ward, 1989). These surface-modifying additives (SMAs) are well known for both organic and inorganic systems. The driving force to concentrate the SMA at the surface after blending the SMA with a biomaterial to be surface modified (the bulk material) is energetic—the SMA should reduce the interfacial energy. To do this, two factors must be taken into consideration. First, the magnitude of the difference in interfacial energy between the system without the additive and the same system with the SMA at the surface will determine the strength of the driving force leading to a SMA-dominated surface. Second, the mobility of the bulk material and the SMA additive molecules within the bulk will determine the rate at which the SMA reaches the surface, or if it will get there at all. An additional concern is the durability and stability of the SMA at the surface.

A typical SMA designed to alter the surface properties of a polymeric material will be a relatively low-molecular-weight diblock copolymer (see Chapter 2.3). The "A" block will be soluble in, or compatible with, the bulk material into

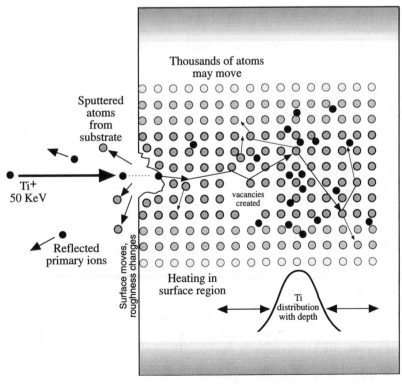

FIG. 5. Some considerations for the ion implantation process.

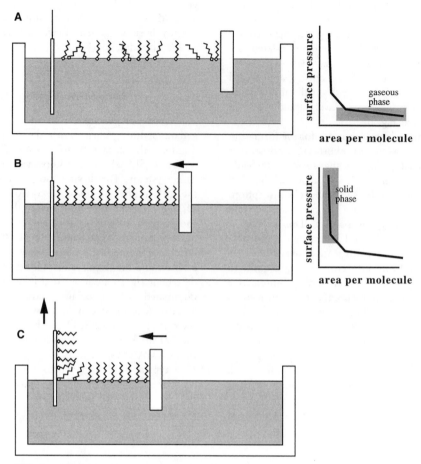

FIG. 6.   Deposition of a lipid film onto a glass slide by the Langmuir–Blodgett technique. (A) The lipid film is floated on the water layer. (B) The lipid film is compressed by a moveable barrier. (C) The vertical glass slide is withdrawn while pressure is maintained on the floating lipid film with the movable barrier.

which the SMA is being added. The "B" block will be incompatible with the bulk material and have lower surface energy. Thus, the A block will anchor the B block into the material to be modified at the interface. This is suggested schematically in Fig. 9. During initial fabrication, the SMA might be distributed uniformly throughout the bulk. After a period for curing or an annealing step, the SMA will migrate to the surface.

For example, an SMA for a polyurethane might have a low-molecular-weight polyurethane A block and a poly(dimethyl siloxane) (PDMS) B block. The A block will anchor the SMA in the polyurethane bulk (the polyurethane A block should be reasonably compatible with the bulk polyurethane), while the low-surface-energy, highly flexible, silicone B block will be exposed at the air surface to lower the interfacial energy (note that air is "hydrophobic"). The A block anchor should confer stability to this system. However, if the system is placed in an aqueous environment, a low-surface-energy (in air) polymer (the B block) is now in contact with water—a high interfacial energy situation. If the system,

after fabrication, still exhibits sufficient chain mobility, it might phase invert to bring the bulk polyurethane or the A block to the surface. Unless the system is specifically engineered to do such a surface phase reversal, this inversion is undesirable. Proper choice of the bulk polymer and the A block can impede surface phase inversion.

Many SMAs for inorganic systems are known. For example, very small quantities of nickel will completely alter the structure of a silicon (111) surface (Wilson and Chiang, 1987). Copper will accumulate at the surface of gold alloys (Tanaka *et al.*, 1988). Also, in stainless steels, chromium will concentrate (as the oxide) at the surface, imparting corrosion resistance.

There are a number of additives that spontaneously surface-concentrate, but are not necessarily designed as SMAs. A few examples for polymers include PDMS, some extrusion lubricants (Ratner, 1983), and some UV stabilizers (Tyler *et al.*, 1992). The presence of such additives at the surface of a polymer may be unexpected and they will not necessarily form stable, durable surface layers.

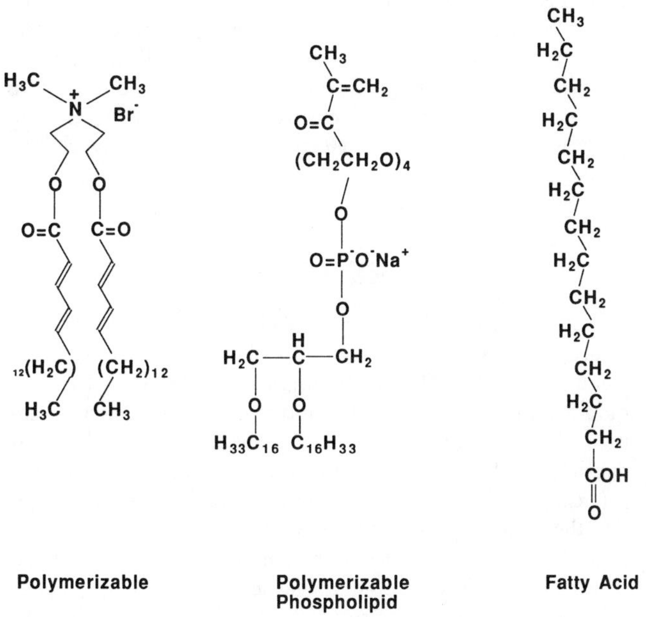

anodized in chromic, oxalic, or sulfuric acid electrolytes. Anodization may also be useful for surface-modifying titanium and Ti-Al alloys (Bardos, 1990; Kasemo and Lausmaa, 1985).

The conversion of metallic surfaces to "oxide-like," electrochemically passive states is a common practice for base–metal alloy systems used as biomaterials. Standard and recommended techniques have been published (e.g., ASTM F4-86) and are relevant for most musculoskeletal load-bearing surgical implant devices. The background literature supporting these types of surface passivation technologies has been summarized (von Recum, 1986).

Base–metal alloy systems, in general, are subject to electrochemical corrosion ($M \rightarrow M^+ + e^-$) within saline environments. The rate of this corrosion process is reduced 10^3–10^6 times by the presence of a minimally conductive, relatively inert oxide surface. For many metallic devices, exposure to a mineral acid (e.g., nitric acid in water) for up to 30 min will provide a passivated surface (i.e., protected by its own oxide).

The reason that many of these surface modifications are called oxide-like is that the structure is complex, including OH, H, and subgroups that may or may not be crystalline. Since most passive surfaces are thin films (50–5000×10^{-8} cm), and are transparent or metallic in color, the surface appears similar before and after passivation. Further details on surfaces of this type can be found in Chapters 1.2, 2.2, and 6.3.

FIG. 7. Three examples of molecules that form organized Langmuir–Blodgett films.

Polymerizable   Polymerizable Phospholipid   Fatty Acid

Conversion Coatings

Conversion coatings modify the surface of a metal into a dense oxide-rich layer that imparts corrosion protection, enhanced adhesivity, and sometimes lubricity to the metal. Steel is frequently phosphated (treated with phosphoric acid) or chromated (with chromic acid). Aluminum is electrochemically

Parylene Coating

Parylene (*para*-xylylene) coatings occupy a unique niche in the surface modification literature because of their frequent application and the good quality of the thin film coatings formed (Loeb *et al.*, 1977a; Nichols *et al.*, 1984). The deposition method is also unique and involves the simultaneous evap-

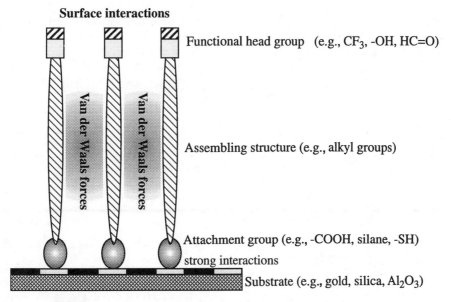

FIG. 8. General characteristics of molecules that form self-assembled monolayers.

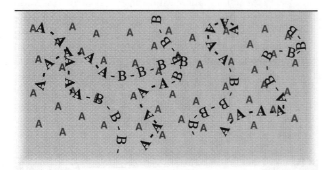

during Fabrication

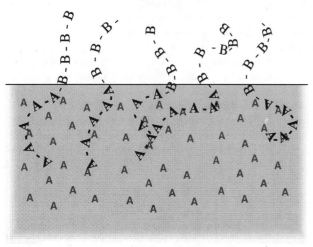

post-Fabrication

FIG. 9. A block copolymer surface-modifying additive within an A block and a B block is blended into a support polymer (the bulk) with a chemistry similar to A block. During fabrication, the block copolymer is randomly distributed throughout the support polymer. After curing or annealing, the A block anchors the surface-modifying additive into the support, while the low-energy B block migrates to the air–polymer interface.

oration, pyrolysis, deposition, and polymerization of the monomer, di-*para*-xylene (DPX), according to the following reaction:

CH₂ — ⬡ — CH₂ → CH₂ ═ ⬡ ═ CH₂

CH₂ — ⬡ — CH₂

Di-para-xylylene   **Para-xylylene**

1) Vaporize   2) Pyrolize

→ — [CH₂ — ⬡ — CH₂]ₙ

Poly (para-xylylene)

3) Deposition

The DPX monomer is vaporized at 175°C and 1 torr, pyrolized at 700°C and 0.5 torr, and finally deposited on a substrate at 25°C and 0.1 torr. The coating has excellent electrical insulation and moisture barrier properties, and has been used to protect implanted electrodes (Loeb *et al.*, 1977b; Nichols *et al.*, 1984) and electronic circuitry (Spivack and Ferrante, 1969).

Laser Methods

Lasers can rapidly and specifically induce surface changes in organic and inorganic materials (Picraux and Pope, 1984; Dekumbis, 1987). The advantages of using lasers for such modification are the precise control of the frequency of the light, the wide range of frequencies available, the high energy density, the ability to focus and raster the light, the possibilities for using both heat and specific excitation to effect change, and the ability to pulse the source and control reaction time. Lasers commonly used for surface modification include ruby, neodymium : yttrium aluminum garnet (Nd : YAG), argon, and CO_2. Treatments are pulsed (100 nsec to picoseconds pulse times) and continuous wave (CW), with interaction times often less than 1 msec. Laser-induced surface alterations include annealing, etching, deposition, and polymerization. The major considerations in designing a laser surface treatment include the absorption (coupling) between the laser energy and the material, the penetration depth of the laser energy into the material, the interfacial reflection and scattering, and heating induced by the laser.

CONCLUSIONS

Surface modifications are being widely explored to enhance the biocompatibility of biomedical devices and improve other aspects of performance. Since a given medical device may already have appropriate performance characteristics, physical properties, and clinical familiarity, surface modification provides a means to alter only the biocompatibility of the device without the need for redesign, retooling for manufacture, and retraining of medical personnel.

Acknowledgment

The suggestions and assistance of Professor J. Lemons have enhanced this chapter and are appreciated.

Bibliography

Arkles, B. (1977). Tailoring surfaces with silanes. *Chemtech* 7: 766–778.

Bardos, D. I. (1990). Titanium and titanium alloys. in *Concise Encyclopedia of Medical and Dental Materials,* 1st Ed. E. Williams, R. W. Cahn, and M. B. Bever, eds. Pergamon Press, Oxford, pp. 360–365.

Bird, R. R., Hill, B., Hobbs, K. E. F., and Chapman, D. (1989). New haemocompatible polymers assessed by thrombelastography. *J. Biomed. Eng.* **11**: 231–234.

Buchanan, R. A., Lee, I. S., and Williams, J. M. (1990). Surface modification of biomaterials through noble metal ion implantation. *J. Biomed. Mater. Res.* **24**: 309–318.

Chapiro, A. (1983). Radiation grafting of hydrogels to improve the thromboresistance of polymers. *Eur. Polym. J.* **19**: 859–861.

Charlson, E. J., Charlson, E. M., Sharma, A. K., and Yasuda, H. K. (1984). Electrical properties of glow-discharge polymers, parylenes, and composite films. *J. Appl. Polym. Sci. Appl. Polym. Symp.* **38**: 137–148.

Chilkoti, A., Ratner, B. D., and Briggs, D. (1991a). Plasma-deposited polymeric films prepared from carbonyl-containing volatile precursors: XPS chemical derivatization and static SIMS surface characterization. *Chem. Mater.* **3**: 51–61.

Chilkoti, A., Ratner, B. D., and Briggs, D. (1991b). A static secondary ion mass spectrometric investigation of the surface structure of organic plasma-deposited films prepared from stable isotope-labeled precursors. Part I. Carbonyl precursors. *Anal. Chem.* **63**: 1612–1620.

Chilkoti, A., Ratner, B. D., Briggs, D., and Reich, F. (1992). Static secondary ion mass spectrometry of organic plasma deposited films created from stable isotope-labeled precursors. Part II. Mixtures of acetone and oxygen. *J. Polym. Sci. Polym. Chem. Ed.* **30**: 1261–1278.

Cho, C. S., Takayama, T., Kunou, M., and Akaike, T. (1990). Platelet adhesion onto the Langmuir–Blodgett film of poly(gamma-benzyl-L-glutamate)-poly(ethylene oxide)-poly(gamma-benzyl-L-glutamate) block copolymer. *J. Biomed. Mater. Res.* **24**: 1369–1375.

Dekumbis, R. (1987). Surface treatment of materials by lasers. *Chem. Eng. Prog.* **83**: 23–29.

Dunkirk, S. G., Gregg, S. L., Duran, L. W., Monfils, J. D., Haapala, J. E., Marcy, J. A., Clapper, D. L., Amos, R. A., and Guire, P. E. (1991). Photochemical coatings for the prevention of bacterial colonization. *J. Biomater. Appl.* **6**: 131–156.

Garbassi, F., Morra, M., Occhiello, E., Barino, L., and Scordamaglia, R. (1989). Dynamics of macromolecules: A challenge for surface analysis. *Surf. Interface Anal.* **14**: 585–589.

Gombotz, W. R., and Hoffman, A. S. (1988). Functionalization of polymeric films by plasma polymerization of allyl alcohol and allylamine. *J. Appl. Polym. Sci. Appl. Polym. Symp.* **42**: 285–303.

Greer, R. T., Knoll, R. L., and Vale, B. H. (1979). Evaluation of tissue-response to hydrogel composite materials. *SEM* **2**: 871–878.

Griesser, H. J., and Chatelier, R. C. (1990). Surface characterization of plasma polymers from amine, amide and alcohol monomers. *J. Appl. Polym. Sci. Appl. Polym. Symp.* **46**: 361–384.

Haque, Y., and Ratner, B. D. (1988). Role of negative ions in the RF plasma deposition of fluoropolymer films from perfluoropropane. *J. Polym. Sci. Polym. Phys. Ed.* **26**: 1237–1249.

Hayward, J. A., and Chapman, D. (1984). Biomembrane surfaces as models for polymer design: The potential for haemocompatibility. *Biomaterials* **5**: 135–142.

Hoffman, A. S. (1988). Biomedical applications of plasma gas discharge processes. *J. Appl. Polym. Sci. Appl. Polym. Symp.* **42**: 251–267.

Hoffman, A. S., Cohn, D. C., Hanson, S. R., Harker, L. A., Horbett, T. A., Ratner, B. D., and Reynolds, L. O. (1983). Application of radiation-grafted hydrogels as blood-contacting biomaterials. *Radiat. Phys. Chem.* **22**: 267–283.

Horbett, T. A., and Hoffman, A. S. (1975). Bovine plasma protein adsorption on radiation-grafted hydrogels based on hydroxyethyl methacrylate and N-vinyl-pyrrolidone. in *Applied Chemistry at Protein Interfaces, Advances in Chemistry Series,* R. E. Baier, ed. Am. Chem. Soc., Washington, DC, pp. 230–254.

Inagaki, N., Nakanishi, T., and Katsuura, K. (1983). Glow discharge polymerizations of tetrafluoroethylene, perfluoromethylcyclohexane and perfluorotoluene investigated by infrared spectroscopy and ESCA. *Polym. Bull.* **9**: 502–506.

Kaplan, S., and Dilks, A. (1981). A solid state nuclear magnetic resonance investigation of plasma-polymerized hydrocarbons. *Thin Solid Films* **84**: 419–424.

Kasemo, B., and Lausmaa, J. (1985). Metal selection and surface characteristics. in *Tissue-Intergrated Prostheses,* P. I. Branemark, G. A. Zarb, and T. Albrektsson, eds. Quintessence, Chicago, pp. 99–116.

Knobler, C. M. (1990). Recent developments in the study of monolayers at the air–water interface. *Adv. Chem. Phys.* **77**: 397–449.

Lewandowska, K., Balachander, N., Sukenik, C. N., and Culp, L. A. (1989). Modulation of fibronectin adhesive functions for fibroblasts and neural cells by chemically derivatized substrata. *J. Cell. Physiol.* **141**: 334–345.

Loeb, G. E., Bak, M. J., Salcman, M., and Schmidt, E. M. (1977a). Parylene as a chronically stable, reproducible microelectrode insulator. *IEEE Trans. Biomed. Eng.* **BME-24**: 121–128.

Loeb, G. E., Walker, A. E., Uematsu, S., and Konigsmark, B. W. (1977b). Histological reaction to various conductive and dielectric films chronically implanted in the subdural space. *J. Biomed. Mater. Res.* **11**: 195–210.

Lopez, G. P., and Ratner, B. D. (1991). Substrate temperature effects of film chemistry in plasma deposition of organics. I. Nonpolymerizable precursors. *Langmuir* **7**: 766–773.

Lopez, G. P., and Ratner, B. D. (1992). Substrate temperature effects on film chemistry in plasma deposition of organics. II. Polymerizable precursors. *J. Polym. Sci. Polym. Chem. Ed.* **30**: 2415–2425.

Maoz, R., Netzer, L. Gun, J., and Sagiv, J. (1988). Self-assembling monolayers in the construction of planned supramolecular structures and as modifiers of surface properties. *J. Chim. Phys.* **85**: 1059–1064.

Matsuda, T., and Inoue, K. (1990). Novel photoreactive surface modification technology for fabricated devices. *Trans. Am. Soc. Artif. Int. Organs* **36**: M161–M164.

Meller, P., Peters, R., and Ringsdorf, H. (1989). Microstructure and lateral diffusion in monolayers of polymerizable amphiphiles. *Coll. Polym. Sci.* **267**: 97–107.

Nichols, M. F., Hahn, A. W., James, W. J., Sharma, A. K., and Yasuda, H. K. (1984). Evaluating the adhesion characteristics of glow-discharge plasma-polymerized films by a novel voltage cycling technique. *J. Appl. Polym. Sci. Appl. Polym. Symp.* **38**: 21–33.

Picraux, S. T., and Pope, L. E. (1984). Tailored surface modification by ion implantation and laser treatment. *Science* **226**: 615–622.

Pleuddemann, E. P. (1980). Chemistry of silane coupling agents. in *Silylated Surfaces,* D. E. Leyden, ed. Gordon & Breach, New York, pp. 31–53.

Prime, K. L., and Whitesides, G. M. (1991). Self-assembled organic monolayers: Model systems for studying adsorption of proteins at surfaces. *Science* **252**: 1164–1167.

Ratner, B. D. (1980). Characterization of graft polymers for biomedical applications. *J. Biomed. Mater. Res.* **14**: 665–687.

Ratner, B. D. (1983). ESCA studies of extracted polyurethanes and polyurethane extracts: Biomedical implications. in *Physicochemical Aspects of Polymer Surfaces,* K. L. Mittal, ed. Plenum, New York, pp. 969–983.

Ratner, B. D., and Hoffman, A. S. (1980). Surface grafted polymers for biomedical applications. in *Synthetic Biomedical Polymers.*

Concepts and Applications, M. Szycher and W. J. Robinson, eds. Technomic, Westport, CT, pp. 133–151.

Ratner, B. D., Hoffman, A. S., Hanson, S. R., Harker, L. A., and Whiffen, J. D. (1979). Blood compatibility–water content relationships for radiation grafted hydrogels. *J. Polym. Sci. Polym. Symp.* **66**: 363–375.

Ratner, B. D., and Yoon, S. C. (1988). Polyurethane surfaces: Solvent and temperature induced structural rearrangements. in *Polymer Surface Dynamics,* J. D. Andrade, ed. Plenum, New York, pp. 137–152.

Ratner, B. D., Horbett, T. A., Hoffman, A. S., and Hauschka, S. D. (1975). Cell adhesion to polymeric materials: Implications with respect to biocompatibility. *J. Biomed. Mater. Res.* **9**: 407–422.

Ratner, B. D., Chilkoti, A., and Lopez, G. P. (1990). Plasma deposition and treatment for biomaterial applications. in *Plasma Deposition, Treatment and Etching of Polymers,* R. D'Agostino, Academic Press, San Diego, pp. 463–516.

Sawin, H. H., and Reif, R. (1983). A course on plasma processing in integrated circuit fabrication. *Chem. Eng. Ed.* **17**: 148–152.

Sheu, M.-S., Hoffman, A. S., and Feijen, J. (1992). A glow discharge process to immobilize PEO/PPO surfactants for wettable and nonfouling biomaterials. *J. Adhes. Sci. Technol.* **6**: 995–1101.

Sioshansi, P. (1987). Surface modification of industrial components by ion implantation. *Mater. Sci. Eng.* **90**: 373–383.

Somorjai, G. A. (1990). Modern concepts in surface science and heterogeneous catalysis. *J. Phys. Chem.* **94**: 1013–1023.

Somorjai, G. A. (1991). The flexible surface. Correlation between reactivity and restructuring ability. *Langmuir* **7**: 3176–3182.

Spivack, M. A., and Ferrante, G. (1969). Determination of the water vapor permeability and continuity of ultrathin parylene membranes. *J. Electrochem. Soc.* **116**: 1592–1594.

Stannett, V. T. (1990). Radiation grafting—State-of-the-art. *Radiat. Phys. Chem.* **35**: 82–87.

Tanaka, T., Atsuta, M., Nakabayashi, N., and Masuhara, E. (1988). Surface treatment of gold alloys for adhesion. *J. Prosthet. Dent.* **60**: 271–279.

Tyler, B. J., Ratner, B. D., Castner, D. G., and Briggs, D. (1992). Variations between Biomer™ lots. 1. Significant differences in the surface chemistry of two lots of a commercial polyetherurethane. *J. Biomed. Mater. Res.* **26**: 273–289.

Ulman, A. (1991). *An Introduction to Ultrathin Organic Films.* Academic Press, Boston.

von Recum, A. F. (1986). *Handbook of Biomaterials Evaluation,* 1st Ed. Macmillan Co., New York.

Ward, R. S. (1989). Surface modifying additives for biomedical polymers. *IEEE Eng. Med. Bio.* **June:** 22–25.

Wasserman, S. R., Tao, Y.-T., and Whitesides, G. M. (1989). Structure and reactivity of alkylsiloxane monolayers formed by reaction of alkyltrichlorosilanes on silicon substrates. *Langmuir* **5**: 1074–1087.

Whitesides, G. M., Mathias, J. P., and Seto, C. T. (1991). Molecular self-assembly and nanochemistry: A chemical strategy for the synthesis of nanostructures. *Science* **254**: 1312–1319.

Wilson, R. J., and Chiang, S. (1987). Surface modifications induced by adsorbates at low coverage: A scanning-tunneling-microscopy study of the Ni/Si(111) $\sqrt{19}$ surface. *Phys. Rev. Lett.* **58**: 2575–2578.

Wu, S. (1982). *Polymer Interface and Adhesion.* Dekker, New York.

Yasuda, H. K. (1979). Competitive ablation and polymerization (CAP) mechanisms of glow discharge polymerization. in *ACS Symposium Series 108: Plasma Polymerization,* M. Shen, and A. T. Bell, eds. Am. Chem. Soc., Washington, DC, pp. 37–52.

Yasuda, H. K., and Gazicki, M. (1982). Biomedical applications of plasma polymerization and plasma treatment of polymer surfaces. *Biomaterials* **3**: 68–77.

2.10 FABRICS

Shalaby W. Shalaby

The use of fabrics and other fibrous forms as biomaterials dates back to the early Egyptians and Indians. Linen sutures and strips was used by the Egyptians with natural adhesives to draw the edges of wounds together to achieve proper healing and retention of original strength. The American Indians used horsehair, cotton, and thin leather strips (Shalaby, 1985) for a similar purpose. More recent use of fabrics as biomaterials was generally viewed as an extended application of the traditional woven and knitted forms of textiles. Prior to the development of the polyethylene terepthalate-based vascular grafts (Hoffman, 1977; Williams and Roaf, 1973), woven, nonwoven and knitted cellulosic fabrics represented the major types of fibrous materials used by the health care industry. Over the past few decades, with the development of sophisticated polymer and fiber processing technologies, nontraditional forms of fabrics, and fabriclike fibrous products have become available and used successfully as biomaterials in old and new applications (Boretos and Edeen, 1984). Hence, it is an objective of this chapter to survey the major traditional and nontraditional forms of fabric constructions and related products and provide brief descriptions of the constituent materials, their processing and properties. A list of physical and biological characterization and test methods is also provided.

TYPES OF FABRICS AND THEIR CONSTRUCTION

Textile fabrics of woven, nonwoven, and knitted types have been used in one or more biomedical applications. These fabrics are made from a wide range of natural and synthetic fibers, as described in fiber and textile science publications (Joseph, 1981, 1984; Labarthe, 1975; Moncrieff, 1975). Descriptions of these fibers and their parent polymers are given in Tables 1-A to 1-C. The processing and characterization of fabrics are addressed in detail in these publications. In a review of fibrous materials for biomedical applications by Shalaby (1985), major types of materials were highlighted. The formation and characterization of unconventional constructions (some of which are not assembled by fiber processing), such as expanded porous poly(tetrafluoroethylene) (Gore-Tex, W. L. Gore and Assoc., Inc.) and hollow fibers, are discussed in a few reviews (Collier, 1970; Hoffman, 1977; Shalaby, 1985; Shalaby *et al.*, 1984). The characterization and testing of fibrous devices and fabric surfaces have been reported in a few reviews (Cooper and Peppas, 1982; Hoffman, 1977; Hastings and Williams, 1980). Important aspects of these constructions are outlined in Table 2.

Cellulose fibers from cotton or wood pulp are the natural fibers most commonly used in the production of biomedical fabrics and related construction. Highly absorbent cellulose fibers, obtained in recent years by fermentation, may find use in certain sanitary products such as napkins. The small production of these fibers, however, may limit their application. Although cellulose acetate and viscose rayon are less commonly used as fibers than cellulose, interest in other regenerated natu-

TABLE 1-A Typical Examples of Modified Natural Polymers, Important Features, and Useful Forms

Polymer		Construction/useful forms	Sterilization method[a]	Comments/applications
Type	Chemical and physical aspects			
Chitosan	A biodegradable, partially deacetylated product of the natural polysaccharide chitin, based on free and acetylated glucosamine units (commonly 70 and 30%, respectively).	Regenerated fibers, used as braids or woven fabrics.	Ethylene oxide	Used experimentally as surgical sutures and meshes. (Skjak-Braek and Sanford, 1989).
Alginates	Family of copolymers based on β-D-manuronic and α-L-gluluronic acid residues, extracted from seaweed.	Regenerated fibers mostly in nonwoven form.	Ethylene oxide	Used as wound-dressing components. (Shalaby and Shah, 1991).

[a]The most common method of sterilization.

TABLE 1-B Typical Examples of Synthetic Nonabsorbable Polymers, Important Features, and Useful Forms

Polymer		Construction/useful forms	Sterilization method[a]	Comments/applications
Type	Chemical and physical aspects			
Carbonized polymers	Made usually by the graphitization of organic polymers; display exceptional tensile strength and modulus.	Produced mostly as continuous multifilament yarns or chopped fibers.	E.O., G, E.B.	Used experimentally as load-bearing composite prostheses.
Polyethylene (PE)	High-density PE (HDPE) melting temperature ($T_m = 125°$). Low-density PZ (LDPE) $T_m = 110°$, and linear low-density (LLDPE).	Melt-spun into continuous yarns for woven fabric and/or melt blown to nonwoven fabrics.	E.O., G.	The HDPE, LDPE and LLDP are used in a broad range of health care products.
	Ultrahigh high molecular weights PE (UHMWPE) ($T_m = 140–150°$, exceptional tensile strength and modulus.	Converted to very high tenacity yarn by gel spinning.		Used experimentally as reinforced fabrics in lightweight orthopedic casts, ligament prostheses, and load-bearing composites.
Polypropylene (PP)	Predominantly isotactic, $T_m = 165–175°$; higher fracture toughness than HDPE.	Melt spun to monofilaments and melt blown to nonwoven fabrics.	E.O.	Sutures, surgical drapes, and gown.
		Hollow fibers		Plasma filtration.
Poly(tetrafluoroethylene) (PTFE)	High melting ($T_m = 325°$) and high crystallinity polymer (50–75% for processed material).	Special procedures used (see Table 2) for production of expanded PTFE.	E.O., A.	Vascular grafts, periodontal inserts.
Nylon 6	$T_g = 45°$, $T_m = 220°$ Thermoplastic, hydrophilic	Monofilaments, braids	E.O., G.	Sutures
Nylon 66	$T_g = 50°$, $T_m = 265°$ Thermoplastic, hydrophilic	Monofilaments, braids	E.O., G.	Sutures
Aramids or Kelvar	A family of aromatic polyamides that form liquid-crystalline solution.	Multifilament yarns with ultrahigh strength	G.	Used experimentally in composites.
Poly(ethylene terephthalate) (PET)	Excellent fiber-forming properties, $T_m = 265°$, $T_g = 65–105°$	Multifilament yarn for weaving, knitting and braiding	E.O., G.	Sutures, meshes and vascular grafts.

[a]Most common methods of sterilization listed in a decreasing order of industrial acceptance: E.O. = Ethylene oxide, G. = Gamma rays from Co-60 source, A. = autoclaving.

TABLE 1-C Typical Examples of Synthetic, Absorbable Polymers, Important Features, and Useful Forms

Polymer		Construction/useful forms	Sterilization method[a]	Comments/applications
Type	Chemical and physical aspects			
Poly(glycolide) (PGA)	Thermoplastic crystalline polymer ($T_m = 225°$, $T_g = 40-45°$)	Multifilament yarns, for weaving, knitting and braiding, sterilized by ethylene oxide.	E.O.	Absorbable sutures and meshes (for defect repairs and periodontal inserts).
10/90 Poly(l-lactide coglycolide) (Polyglactin 910)	Thermoplastic crystalline copolymer, ($T_m = 205°$, $T_g = 43°$)	Multifilament yarns, for weaving, knitting and braiding, sterilized by ethylene oxide.	E.O.	Absorbable sutures and meshes.
Poly(p-dioxanone) (PDS)	Thermoplastic crystalline polymer ($T_m = 110-115°$, $T_g = 10°$)	Melt spun to monofilament yarn.	E.O.	Sutures, intramedullary pins and ligating clips.
Poly(alkylene oxalates)	A family of absorbable polymers with T_m between 64 and 104°	Can be spun to monofilament and multifilament yarns.	E.O., G.	Experimental sutures.
Isomorphic poly(hexamethylene co-trans-1,4-cyclohexane dimethylene oxalates)	A family of crystalline polymers with T_m between 64 and 225°	Can be spun to monofilament and multifilament yarns.	E.O., G.	Experimental sutures.

[a]Most common methods of sterilization listed in a decreasing order of industrial acceptance: E.O. = Ethylene oxide, G. = Gamma rays from Co-60 source.

ral polysaccharides and their derivatives as fibrous biomaterials is growing rapidly. Among these are the alginates, chitosans, and dextran, which are obtained from brown algae, crab shells, and bacterial fermentation, respectively (Shalaby and Shah, 1991). Absorbable surgical sutures and meshes have been made of chitosan (Skjak-Braek and Sanford, 1989). Chitosan and alginate fibers are formed by coagulation of streams of polymer solutions in typical solution spinning processes. Alginates, in fibrous forms, were described as useful components of wound dressings. Similarly, regenerated collagen, a major protein in most living tissue, has been used to produce absorbable sutures, foams, and meshes. Another natural fibrous protein, silk, which is produced by the silkworm, has been processed into braided sutures. In contrast to these natural polymers, which degrade or denature upon heating at 120°C or less, biosynthetic thermoplastic polyesters which can be melt processed, can be produced by fermentation. These include poly(β-hydroxybutyrate) (PHB) and its copolymers with β-hydroxyvalerate. Fibers made of PHB were described as biodegradable materials, with potential use in biomedical applications (see Shalaby, 1985, p. 89).

With the exception of synthetic absorbable polymers, which degrade to nontoxic by-products, no synthetic polymers were manufactured specifically for the production of textile biomaterials. Hence, commercial textile fibers are virtually the only available source for the production of nonabsorbable biomedical textiles. Therefore, these fabrics and fibers normally contain typical additives such as dyes, antistatic agents, delustrants, photostabilizers, and antioxidants, which are used to achieve specific textile characteristics that may not be necessary in biomedical uses. The major polymers used in the production of absorbable and nonabsorbable biomaterials are summarized in Tables 1-A to 1-C. A few experimental absorbable polymers and their projected applications are also included in those tables.

PROCESSING AND CHARACTERISTICS OF MAJOR TYPES OF CONSTRUCTIONS

Natural and synthetic fibers can be converted to different forms and fabric constructions, as shown in Table 3. Woven fabrics usually display low elongation and high breaking strength. They are more stable mechanically than the flexible, stretchable, knitted fabrics. The tensile or burst strength of knitted fabrics is usually inferior to that of woven fabrics of comparable densities. On the other hand, knitted structures have superior elastic recovery and good wrinkle and crush resistance and allow free circulation of gases and nonviscous fluids. Needle felts display poor mechanical properties and are used primarily as insulators or for liquid absorption. The relevance of the fabric construction to the performance of vascular grafts and new approaches to improving their functional performance are discussed in the section on biomedical applications.

Fiber bonding is a major technique which is used to produce large-volume fiber-bonded fabrics, otherwise known as nonwoven fabrics. These fabrics, made by different processes, are used in large-volume health-care products where modest or low fabric strength is required. Thus, they are used to produce diapers, sanitary napkins, gauze, and bandages. Nonwoven fabrics can also be (1) spun-bonded fabrics made directly from thermoplastic polymers such as poly(ethylene terephthalate) (PET), polypropylene (PP), polyethylene (PE), and nylon 6; (2) spun-laced, light fabrics; and (3) solvent-bonded fabrics, where a solvent is used to produce a fiber–fiber interface.

Other less common or experimental processes for producing specialty woven fabrics and/or intricate forms of nonwoven devices with microfiber constituents of micron dimension include (1) melt-blowing, in which a relatively low-viscosity polymer such as PE or PP is extruded at high speed while being injected with a high-pressure air jet, the resulting ultrafine "fibrillar" melt is collected onto a moving or rotating screen;

(2) solution-blowing, which is similar to melt-blowing except that a polymer solution is extruded into a coagulating bath; and (3) electrostatic spinning, in which fine streams of a polymer solution are further subdivided into finer ones in the presence of a strong elastostatic field before phase separation onto a rotating mandril to eventually form microporous devices. Using the latter process, a segmented polyurethane can be converted to microporous vascular grafts (Berry, 1987; Fisher, 1987; Hess *et al.*, 1991). These fabrics and their biodegradable counterparts have been evaluated as grafts for arterial reconstruction (van der Lei and Wilderuur, 1991; Hess *et al.*, 1991). Melt-blown, nonwoven fabrics are used in the production of surgical gowns and masks.

In addition to the fabric constructions discussed here, microporous or porous systems can be made to provide fabric-like properties. These include the expanded Teflon and micropo-

TABLE 2 Specialty Fabrics, Their Constructions and Applications

Composition	Construction, finishing and distinct properties	Present and potential applications	References
Carbonized polymers	Different types of carbon fibers used in the form of chopped fibers, continuous yarns, and occasionally woven fabrics	Components of many experimental implants and devices, mostly orthopedic prostheses	Bradley and Hastings (1980)
Collagen	Highly purified, lyophilized collagen fleece	Used for repair of mandibular defects	Joos, Vogel and Ries (1980)
Polyethylene (PE)	High-density PE (HDPE) yarn converted to woven tape (0.5 mm thickness, 3 mm width, and 200 m² pore size).	Evaluated as artificial tendons	Hudge and Wade (1980)
Poly(ethylene terephthate) (PET) Dacron	Random flock made of 300 × 17 mm fibers Knitted mesh coated with a poly(urethane) elastomer Knitted, woven, and velour-type (double, internal and external) conduits	Flocked blood pump bladder Composite for reconstructive surgery, particularly mandibular implants Principal constructions for arterial substitutes	Poirier (1980) Leake *et al.* (1980) Guidoin *et al.* (1980)
Polypropylene (PP)	Microfabrics made by extruding blends of PP, poly(ethylene-coacrylic acid) (ionomer, used as sodium salt) and glycerine. Sheets are extracted with water to product PP random microfiber mesh which yields nonwoven thin PP microfiber web (microfabric) by freeze-drying and tentering.	Cell seeding substrates in blood circulatory devices after coating with parylene (a vapor phase-deposited polymer of *p*-xylylene). Rate-controlling membrane in transdermal drug delivery.	Tittman and Beach (1980)
Poly(tetrafluoroethylene) (PTFE)	Microporous, expanded forms made by mixing resin with a solvent binder, cold extrusion of a billet, driving off the solvent, expanding and stretching followed by sintering. the microporosity varies with processing techniques.	Vascular prostheses and sutures	Snyder and Helmus (1988)

TABLE 3 Conversion of Commercial fibers[a]

Major process	Resulting constructions	Comments
Weaving	Plain (with rib and basket variations), twill, satin and doup (or gauze).	Using simple or shuttleless loom.
	Pile (filling or warp) weave	
	Double weaves	At least 2 sets of filling yarn and 2 sets warp yarns are interfaced.
	Triaxial fabrics	Require 3 sets of yarn to interlace at less than 90° angle, usually 2 sets of wrap yarn and one set of filling yarn.
Knitting	Plain single knit	Using filling (or weft) knitting on a flat-bed or circular machine.
	Double knit (by interlock stitch)	
Needle felting	Needle felts (or needle-punched fabrics)	This form of nonwoven fabric has intimate 3-dimensional fiber entanglement and is produced by the action of barbed needles (not the simple mechanical interlocking used in wool felting).
Braiding	Braids with or without lightly twisted cores	Mostly for surgical sutures.

[a]For details see Joseph (1981) and Labarthe (1975).

rous polypropylene noted in Table 2. Expanded Teflon has been explored extensively for use in vascular prosthesis (Kogel *et al.*, 1991).

CHARACTERIZATION, TESTING, AND EVALUATION

For biomedical systems, some of the characterization and testing techniques are similar to nonbiomedical ones. However, certain evaluation techniques can be unique to biomaterials (Shalaby, 1985).

General Characterization Methods

General characterization methods usually address the constituent polymer and fiber identity and include (1) microscopic examination, (2) color spot testing, (3) solubility and swelling behavior in synthetic media simulating biological fluids or components thereof, (4) spectroscopic analysis (infrared, nuclear magnetic resonance, UV-visible), (5) thermal analysis: thermal gravimetric analysis (TGA), differential scanning calorimetry (DSC), thermal mechanical analysis (TMA), and (6) diffraction techniques (e.g., X-ray). Molecular weights of the constituent polymers can be determined using gel permeation chromatography, osmometry, and viscometry. The methods used to determine chemical identity, molecular weight, and concentration and type of extractable impurities are critical to biomedical uses of fabrics.

Biological Evaluation and Simulated in Vitro Testings

Biological evaluation and simulated *in vitro* tests are usually designed to test prototypes or finished products. They vary with the type of the device. However, most of them pertain to

cytotoxicity, mechanical compatibility, blood compatibility, and degradation and retention of intended mechanical properties.

MAJOR BIOMEDICAL APPLICATIONS

From the perspective of an application site, the biomedical uses of fabrics can be dealt with as external and internal applications.

External Applications

For applications which involve direct or close contact with intact skin, the following types of products are being used:

Surgical gowns, made mostly from woven and nonwoven cellulose, polyethylene, and polypropylene fibers

Masks and shoe covers made of gauze and nonwoven fabrics, respectively

Sheets and packs based on laminates of plastic with nonwoven fabrics

Adhesive tapes consisting of an adhesive, elastomeric film, woven or knitted fabric strip, and nonwoven fabrics.

Internal Applications

Internal applications are primarily associated with (1) wound repair and reconstruction of soft tissues such as sutures, reinforcing meshes, and hemostatic devices; (2) cardiovascular prostheses; (3) orthopedic prostheses, such as tendons and ligaments; and (4) hollow fibers as in a typical dialysis unit. Other applications associated with limited exposure to injured tissues or mucous membranes include the use of dressings, woven sponges, blood filters, and fiber-optic instruments.

From a functional viewpoint, the biomedical applications

of fabrics can be divided into those dealing with (1) general surgical applications, (2) cardiovascular system applications, (3) musculoskeletal system applications, and (4) percutaneous and cutaneous applications. Hoffman (1982) has presented a detailed review of this functional classification. A summary of this treatment and an update based on more recent reviews (Shalaby *et al.*, 1984; Shalaby, 1985; von Recum, 1986) are given in the following paragraphs.

General Surgical Applications

General surgical applications include (1) sutures and threads used to close wounds, (2) ligating threads to tie off bleeding vessels, and (3) fiber- or fabric-reinforced implants for reconstructive and reparative surgery of soft tissues. Commercial sutures are available as monofilaments or braid constructions. They can be absorbable (or biodegradable), such as the polyglycolide, braided sutures, or nonabsorbable as in case of polypropylene monofilament sutures. Ligatures are similar to monofilament sutures, but carry no needles. Among the common implants for reconstructive and repairative surgery of soft tissue are those made of silastic (a polysiloxane) reinforced with a Dacron (polyethylene terephthelate) mesh or that has a Dacron felt backing.

Cardiovascular System Applications

Included in cardiovascular system applications are those dealing with vascular prostheses, prosthetic heart valves, and heart assist devices. The most important applications of fabrics in this category are those associated with vascular grafts (Kogel, 1991). These can be made of woven, knitted, or microporous constructions. Other forms of vascular grafts are made of microporous polyurethane and biodegradable polymers. For heart assist devices, Dacron fibers are often used as a component of a fiber-reinforced polymeric (e.g., a polysiloxane or polyurethane) diaphragm of a blood-pumping chamber. Dacron and to a lesser extent Teflon (TFE), nylon 66, and polypropylene have been used in prosthetic heart valves as "sewing rings" (to suture the valve to surrounding tissues).

Musculoskeletal System Applications

Nonwoven graphite–Teflon fibrous mats have been used as matrices for tissue ingrowth in stabilization of dental or orthopedic implants. Similarly, these matrices have found applications in reconstructive and maxillofacial surgeries. An important application of fiber constructions is their use in artificial tendons and ligaments. So far these applications are limited to nonabsorbable materials. However, the use of absorbable polymers is being explored.

Percutaneous and Cutaneous Applications

Among the most common percutaneous systems are the shunts which have been developed to provide access to the circulation for routine dialysis of the blood of kidney patients. In most of these shunts, a short Dacron felt or velour cuff is used in conjunction with a silastic tube as it exits from the blood vessel. A successful application of fabrics in the development of artificial skin, for use as a temporary burn dressing, consists of a nylon velour on a synthetic polypeptide backing.

Bibliography

Berry, J. P. (1987). Eur. Pat. Appl. (to Ethicon Inc.), 0, 223, 374.

Boretos, J. W., and Eden, M. (eds.) (1984). *Contemporary Biomaterials.* Noyes Publ., Park Ridge, NJ.

Bradley, J. S., and Hastings, G. W. (1980). Carbon fiber-reinforced plastics for orthopedic implants, in *Mechanical Properties of Biomaterials,* G. W. Hastings and D. F. Williams, eds. Wiley, New York, p. 379.

Collier, A. M. (1979). *A Handbook of Textiles.* Pergamon Press, New York.

Cooper, S. L., and Peppas, N. A. (1982). *Biomaterials: Interfacial Phenomena and Applications.* ACS Adv. Chem. Ser. American Chemical Society, Washington, DC, Vol. 199.

Fisher, A. C. (1987). U.S. Pat. (to Ethicon, Inc.) 4, 657, 793.

Guidoin, R., Gosselin, C., Roy, J., Gagnon, D., Marois, M., Noel, H. P., Roy, P., Martin, L., Awad, J., Bourassa, S. and Rauleau, C. Structural and mechanical properties of Dacron prostheses as arterial substitutes, in *Mechanical Properties of Biomaterials,* G. W. Hastings and D. T. Williams, eds. Wiley, New York, p. 547.

Hastings, G. W., and Williams, D. T. (eds.) (1980). *Mechanical Properties of Biomaterials.* Wiley, New York.

Hess, F., Jerusalem, C., Braun, B., Grande, P., Steeghs, S., and Reinders, O. (1991). Vascular prostheses made of polyurethanes, in *The Prosthetic Substitution of Blood Vessels,* H. C. Kogel, ed. Quintessen-Verlags, München, Germany, p. 87.

Hoffman, A. S. (1977). Medical application of polymeric fibers, *J. Appl. Polym. Sci., Appl. Polym. Symp.* 31, 313.

Hoffman, A. S. (1982). in *Macromolecules,* H. Benoit and P. Rempp, eds. Pergamon Press, New York.

Hudge, J. W., and Wade, C. W. R. (1980). in *Synthetic Biomedical Polymers: Concepts and Applications,* M. Szycher and W. J. Robinson, eds. Technomic Publ., Westport, CT, p. 201.

Joos, U., Vogel, D. and Ries, P. (1980). Collagen fleece as a biomaterial for mandibular defects, in *Mechanical Properties of Biomaterials,* G. W. Hastings and D. T. Williams, eds. Wiley, New York, p. 515.

Joseph, M. L. (1980). *Textile Science,* 4th ed. Holt, Rinehardt & Winston, New York.

Joseph, M. L. (1984). *Essentials of Textiles,* 3rd ed. Holt, Rinehart & Winston, New York.

Kogel, H. C. (ed.) (1991). *The Prosthetic Substitution of Blood Vessels.* Quintessenz-Verlags, München, Germany.

Kogel, H., Vollmar, J. F., Proschek, M. B., Scharf, G., and Büttel, H. M. (1991). Healing pattern of small caliber arterial prostheses and their endothelialization, in *Prosthetic Substitution of Blood Vessels,* H. Kogel, ed. Quintessen-Verlags, München, Germany, p. 143.

Labarthe, J. (1975). *Elements of Textiles.* Macmillan, New York.

Leake, D., Habal, M. B., Schwartz, H., Michieli, S. and Pizzoferrato, A. (1980). Urethane elastomer coatec cloth mesh, in *Mechanical Properties of Biomaterials,* G. W. Hastings and D. T. Williams, eds. Wiley, New York, p. 419.

Moncrieff, R. W. (1975). *Man-Made Fibers,* 6th ed., Wiley, New York.

Poirier, V., (1980). Fabrication and testing of flocked blood pump bladder, in *Synthetic Biomedical Polymers: Concept and Applications,* M. Szycher and W. J. Robinson, eds. Technomic Publ., Westport, CT, p. 73.

Snyder, R. W., and Helmus, M. N. (1988). Vascular prostheses, in *Encyclopedia of Medical Devices and Instrumentation,* J. G. Webster, ed. Wiley-Interscience, New York, p. 2843.

Shalaby, S. W. (1985). Fibrous materials for biomedical applications, in *High Technology Fibers* Part A, M. Lewin and J. Preston, eds. Marcel Dekker, New York.

Shalaby, S. W., and Shah, K. R. (1991). Chemical modification of natural polymers and their technological relevance, in *Water-Soluble Polymers: Chemistry and Applications*, S. W. Shalaby, G. B. Butler, and C. L. McCormick, eds., ACS Symp. Series, Amer. Chem. Soc., p. 74. Washington, DC.

Shalaby, S. W., Hoffman, A. S., Ratner, B. D. and Horbett, T. A. (1984). *Polymers as Biomaterials.* Plenum Publ., New York.

Skjak-Braek, G., and Sanford, P. A. (eds.) (1989). *Chitin and Chitosan: Sources, Chemistry, Biochemistry, Physical Properties, and Applications.* Elsevier, New York.

van der Lei, B., and Wildevuur, R. H. (1991). The use of biodegradable grafts for arterial constructs: A review of the experience at the University of Groningen, in *The Prosthetic Substitution of Blood Vessels*, H. C. Kogel, ed. Quitessenz-Verlags, München, Germany, p. 67.

von Recum, A. F. (ed.) (1986). *Handbook of Biomaterials Evaluation.* Macmillan, New York.

Tittmann, F. R., and Beach, W. F. (1980). in *Synthetic Biomedical Polymers: Concepts and Applications*, M. Szycher and W. J. Robinson, eds. Technomic Publ., Westport, CT, p. 117.

Williams, D. F., and Roaf, R. (1973). *Implants in Surgery.* Saunders, Toronto.

2.11 BIOLOGICALLY FUNCTIONAL MATERIALS

Allan S. Hoffman

Polymers are especially interesting as biomaterials because they can be readily combined physically or chemically with biomolecules (or cells) to yield biologically functional systems (Piskin and Hoffman, 1986). Biomolecules such as enzymes, antibodies, or drugs, as well as cells have been immobilized on and within polymeric systems for a wide range of therapeutic, diagnostic, and bioprocess applications. The synthesis, proper-

TABLE 1 Examples of Biologically Active Molecules That May Be Immobilized on or within Polymeric Biomaterials

Proteins/peptides	Drugs
Enzymes	Antithrombogenic agents
Antibodies	Anticancer agents
Antigens	Antibiotics
Cell adhesion molecules	Contraceptives
"Blocking" proteins	Drug antagonists
Saccharides	Peptide, protein drugs
Sugars	Ligands
Oligosaccharides	Hormone receptors
Polysaccharides	Cell surface receptors (peptides, saccharides)
Lipids	Avidin, biotin
Fatty acids	Nucleic acids, nucleotides
Phospholipids	Single or double-stranded
Glycolipids	DNA, RNA (e.g., antisense
Other	oligonucleotides)
Conjugates or mixtures of the above	

TABLE 2 Applications of Immobilized Biomolecules and Cells

Enzymes	Bioreactors (industrial, biomedical)
	Bioseparations
	Biosensors
	Diagnostic assays
	Biocompatible surfaces
Antibodies, peptides, and other affinity molecules	Biosensors
	Diagnostic assays
	Affinity separations
	Targeted drug delivery
	Cell culture
Drugs	Thrombo-resistant surfaces
	Drug delivery systems
Lipids	Thrombo-resistant surfaces
	Albuminated surfaces
Nucleic acid derivatives and nucleotides	DNA probes
	Gene Therapy
Cells	Bioreactors (industrial)
	Bioartificial organs
	Biosensors

ties, applications, and advantages and disadvantages of such systems are discussed in this chapter.

Many different biologically functional molecules can be chemically or physically immobilized on polymeric supports (Table 1) (Laskin, 1985; Tomlinson and Davis, 1986). In addition to such biomolecules, a wide variety of living cells and microorganisms may also be immobilized. Both solid polymers and soluble polymer molecules may be considered as immobilization supports for covalent binding of biomolecules. Solid supports are available in many diverse forms, such as particulates, fibers, fabrics, membranes, tubes, hollow fibers, and porous systems. When some of these solids are water swollen they become hydrogels, and biomolecules and cells may be immobilized within the aqueous pores of the polymer gel network. Examples of applications of these immobilized biological species are listed in Table 2. It can

TABLE 3 Bioreactor Supports and Designs

"Artificial cell" suspensions
(microcapsules, RBC ghosts, liposomes, reverse micelles [w/o] microspheres)
Biologic supports
(membranes and tubes of collagen, fibrin ± glycosaminoglycans)
Synthetic supports
(porous or asymmetric hollow fibers, particulates, parallel plate devices)

TABLE 4 Examples of Immobilized Enzymes in Therapeutic Bioreactors

Medical application	Substrate	Substrate action
Cancer treatment		
L-Asparaginase	Asparagine	Cancer cell nutrient
L-Glutaminase	Glutamine	Cancer cell nutrient
L-Arginase	Arginine	Cancer cell nutrient
L-Phenylalanine lyase	Phenylalanine	Toxin
Indole-3-alkane α hydroxylase	Tryptophan	Cancer cell nutrient
Cytosine deaminase	5-Fluorocytosine	Toxin
Liver failure (detoxification)		
Bilirubin oxidase	Bilirubin	Toxin
UDP-Gluceronyl transferase	Phenolics	Toxin
Other		
Heparinase	Heparin	Anticoagulant
Urease	Urea	Toxin

be seen that there are many diverse uses of such biofunctional systems in both the medical and biotechnology fields. For example, a number of immobilized enzyme supports and reactor systems (Table 3) have been developed for therapeutic uses in the clinic (Table 4) (De Myttenaere *et al.*, 1967; Kolff, 1979; Sparks *et al.*, 1969; Chang, 1972; Nose *et al.*, 1983; Schmer *et al.*, 1981; Callegaro and Denti, 1983; Lavin *et al.*, 1985; Sung *et al.*, 1986). Some of the advantages and disadvantages of immobilized biomolecules are listed in Table 5, using enzymes as an example.

TABLE 5 Some Advantages and Disadvantages of Immobilized Enzymes

Advantages
   Enhanced stability
   Can modify enzyme microenvironment
   Can separate and reuse enzyme
   Enzyme-free product
   Lower cost, higher purity product
   No immunogenic response (therapeutics)
Disadvantages
   Difficult to sterilize
   Fouling by other biomolecules
   Mass transfer resistances (substrate in and product out)
   Adverse biological responses of enzyme support surfaces (*in vivo* or *ex vivo*)
   Greater potential for product inhibition

TABLE 6 Biomolecule Immobilization Methods

Physical adsorption
   van der Waals
   Electrostatic
   Affinity
   Adsorbed and cross-linked
Physical "entrapment"
   Barrier systems
   Hydrogels
   Dispersed (matrix) systems
Covalent attachment
   Soluble polymer conjugates
   Solid surfaces
   Hydrogels

IMMOBILIZATION METHODS

There are three major methods for immobilizing biomolecules and cells (Table 6) (Stark, 1971; Zaborsky, 1973; Dunlap, 1974). It can be seen that two of them are physically based, while the third is based on covalent or "chemical" attachment to the support molecules. Thus, it is important to note that the term "immobilization" can refer either to a temporary or to a permanent localization of the biomolecule (or cell) on or within a support. In the case of a drug delivery system, the immobilized drug is supposed to be released from the support, while an immobilized enzyme (or cell) in an artificial organ is designed to remain attached to or entrapped within the support over the duration of use. Either physical or chemical immobilization can lead to "permanent" retention on or within a solid support, the former being due to the large size of the biomolecule (or cell). If the polymer support is biodegradable, then the physically or chemically immobilized biomolecule may be released as the matrix erodes or degrades away.

A large number of interesting methods have been developed for covalent binding of biomolecules to soluble or solid polymeric supports (Weetall, 1975; Carr and Bowers, 1980; Dean *et al.*, 1985; Shoemaker, *et al.*, 1987; Yang *et al.*, 1990; Park and Hoffman, 1990; Gombotz and Hoffman, 1986). Most of these are illustrated in Fig. 1. The same biomolecule may be immobilized by many different methods; specific examples of many of the most common methods are shown in Fig. 2.

For covalent binding to an inert solid polymer surface, the surface must first be chemically modified to provide reactive groups (e.g., $-OH$, $-NH_2$, $-COOH$) for the subsequent immobilization step. If the polymer support does not contain such groups, then it is necessary to modify it in order to permit covalent immobilization of biomolecules to the surface. A wide number of solid surface modification techniques have been used, including ionizing radiation graft copolymerization, plasma gas discharge, photochemical grafting, chemical modification (e.g., ozone grafting), and chemical derivation (Hoffman, 1984; Gombotz and Hoffman, 1987; Hoffman *et al.*,

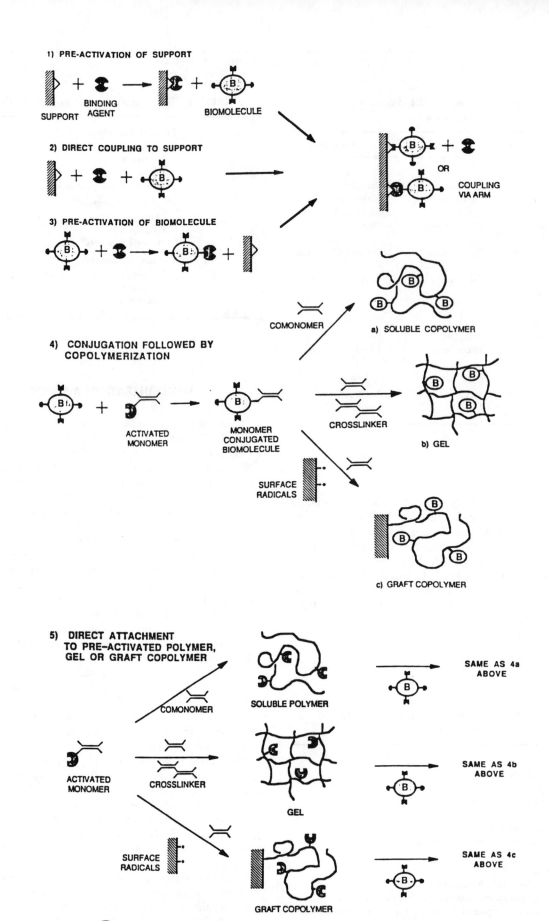

1) PRE-ACTIVATION OF SUPPORT

SUPPORT   BINDING AGENT

BIOMOLECULE

2) DIRECT COUPLING TO SUPPORT

COUPLING VIA ARM

OR

3) PRE-ACTIVATION OF BIOMOLECULE

4) CONJUGATION FOLLOWED BY COPOLYMERIZATION

ACTIVATED MONOMER

MONOMER CONJUGATED BIOMOLECULE

COMONOMER

a) SOLUBLE COPOLYMER

CROSSLINKER

b) GEL

SURFACE RADICALS

c) GRAFT COPOLYMER

5) DIRECT ATTACHMENT TO PRE-ACTIVATED POLYMER, GEL OR GRAFT COPOLYMER

ACTIVATED MONOMER

COMONOMER

SOLUBLE POLYMER

SAME AS 4a ABOVE

CROSSLINKER

GEL

SAME AS 4b ABOVE

SURFACE RADICALS

GRAFT COPOLYMER

SAME AS 4c ABOVE

NOTE: ⒷMAY BE IMMOBILIZED WITH OR WITHOUT A "TETHER" ARM IN ANY OF THE ABOVE

FIG. 1. Schematic cartoons showing various methods for covalent biomolecule immobilization.

126

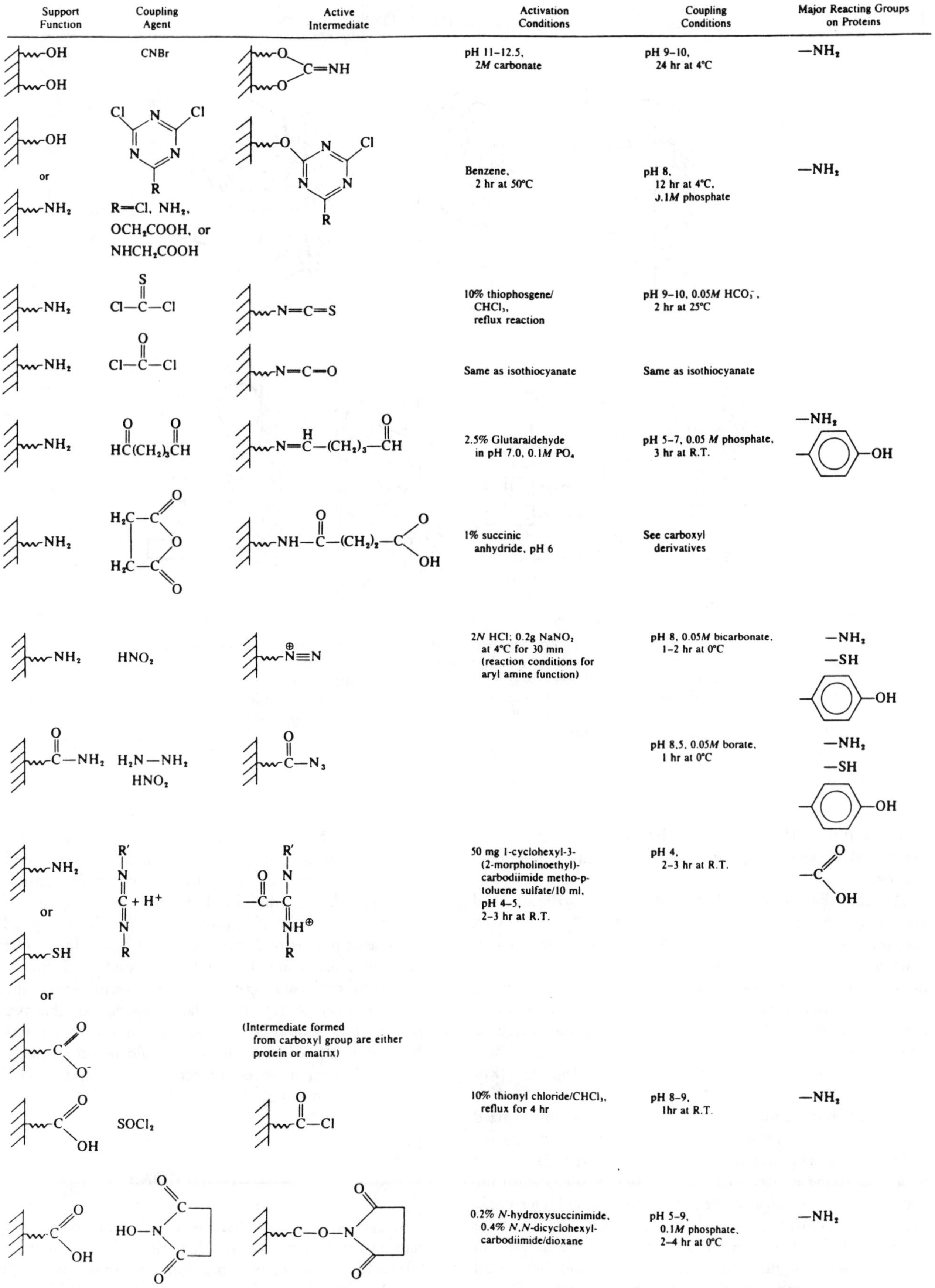

Support Function	Coupling Agent	Active Intermediate	Activation Conditions	Coupling Conditions	Major Reacting Groups on Proteins
—OH, —OH	CNBr	O, O C=NH	pH 11–12.5, 2M carbonate	pH 9–10, 24 hr at 4°C	—NH₂
—OH or —NH₂	(triazine) Cl, Cl, R; R=Cl, NH₂, OCH₂COOH, or NHCH₂COOH	—O (triazine) Cl, R	Benzene, 2 hr at 50°C	pH 8, 12 hr at 4°C, J.1M phosphate	—NH₂
—NH₂	Cl—C(S)—Cl	—N=C=S	10% thiophosgene/CHCl₃, reflux reaction	pH 9–10, 0.05M HCO₃⁻, 2 hr at 25°C	
—NH₂	Cl—C(O)—Cl	—N=C=O	Same as isothiocyanate	Same as isothiocyanate	
—NH₂	HC(CH₂)₃CH (O)(O)	—N=CH—(CH₂)₃—CH (O)	2.5% Glutaraldehyde in pH 7.0, 0.1M PO₄	pH 5–7, 0.05 M phosphate, 3 hr at R.T.	—NH₂, (phenol) OH
—NH₂	succinic anhydride	—NH—C(O)—(CH₂)₂—C(O)OH	1% succinic anhydride, pH 6	See carboxyl derivatives	
—NH₂	HNO₂	—N⁺≡N	2N HCl; 0.2g NaNO₂ at 4°C for 30 min (reaction conditions for aryl amine function)	pH 8, 0.05M bicarbonate, 1–2 hr at 0°C	—NH₂, —SH, (phenol) OH
—C(O)—NH₂	H₂N—NH₂, HNO₂	—C(O)—N₃		pH 8.5, 0.05M borate, 1 hr at 0°C	—NH₂, —SH, (phenol) OH
—NH₂ or —SH or —C(O)O⁻	R'—N=C=N—R + H⁺ (carbodiimide)	R'—N, —C(O)—C=NH⁺—R	50 mg 1-cyclohexyl-3-(2-morpholinoethyl)-carbodiimide metho-p-toluene sulfate/10 ml, pH 4–5, 2–3 hr at R.T.	pH 4, 2–3 hr at R.T.	—C(O)OH
		(Intermediate formed from carboxyl group are either protein or matrix)			
—C(O)OH	SOCl₂	—C(O)—Cl	10% thionyl chloride/CHCl₃, reflux for 4 hr	pH 8–9, 1hr at R.T.	—NH₂
—C(O)OH	HO—N (succinimide)	—C(O)—O—N (succinimide)	0.2% N-hydroxysuccinimide, 0.4% N,N-dicyclohexyl-carbodiimide/dioxane	pH 5–9, 0.1M phosphate, 2–4 hr at 0°C	—NH₂

FIG. 2. Examples of various chemical methods used to bond biomolecules directly to reactive supports. (From P. W. Carr and L. D. Bowers, *Immobilized Enzymes in Analytical and Clinical Chemistry: Fundamentals and Applications.* Wiley, 1980, pp. 172–173.)

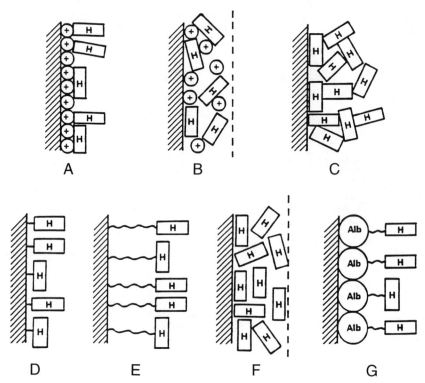

FIG. 3. Various methods for heparinization of surfaces: (A) heparin bound ionically on a positively charged surface; (B) heparin ionically complexed to a cationic polymer, physically coated on a surface; (C) heparin self-cross-linked physically coated on a surface; (D) heparin covalently linked to a surface; (E) heparin covalently immobilized via spacer arms; (F) heparin dispersed into a hydrophobic polymer; (G) heparin–albumin conjugate immobilized on a surface. (From S. W. Kim and J. Feijen, in D. Williams, ed., *Critical Reviews in Biocompatibility*. CRC Press, 1985, pp. 229–260.)

1987; Hoffman, 1987). Soluble polymers may be chemically derivitized or synthesized *de novo* with similar functional groups.

A chemically immobilized biomolecule may also be attached via a spacer group, sometimes called an "arm" or a "leash" (Cuatrecasas and Anfinsen, 1971). Most often the spacer arm reactive end groups are amine, carboxylic acid, and/or hydroxyl groups. Such spacer groups can provide greater steric freedom and thus greater specific activity for the immobilized biomolecule, especially in the case of smaller biomolecules. The spacer arm may also be biodegradable and therefore will release the immobilized biomolecule as it degrades (Kopecek, 1977).

Sometimes more than one biomolecule may be immobilized to the same support. For example, a soluble polymer designed to "target" a drug molecule may have separately conjugated to it a targeting moiety such as an antibody, along with the drug molecule, which may be attached to the polymer backbone via a biodegradable spacer group (Ringsdorff, 1975; Kopecek, 1977; Goldberg, 1983). In another example, an immunodiagnostic microtiter plate usually will be coated first with an antibody and then with albumin, each physically adsorbed to it, the latter acting to reduce nonspecific adsorption during the assay.

It is evident that there are many different ways that the same biomolecule may be immobilized to a polymeric support. Heparin and albumin are two common biomolecules which have been immobilized by a number of widely differing methods. These are illustrated schematically in Figs. 3 and 4. Some of the major features of the different immobilization techniques are compared and contrasted in Table 7. The important molecular criteria for successful immobilization of a biomolecule are that a large fraction of the available biomolecules should be immobilized, and a large fraction of those immobilized biomolecules should retain an acceptable level of bioactivity over an economically and/or clinically appropriate time period.

CONCLUSIONS

It can be seen that there is a wide and diverse range of materials and methods available for immobilization of biomolecules and cells on or within biomaterial supports. Combined with the great variety of possible biomedical and biotechnological applications, this represents a very exciting and fertile field for applied research in biomaterials.

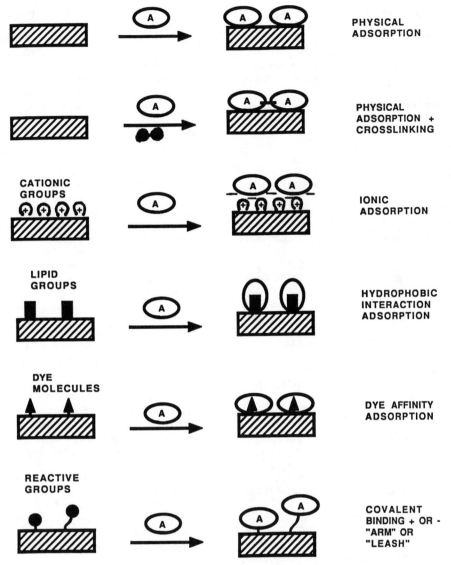

FIG. 4. Various methods for immobilization of albumin on surfaces.

TABLE 7 Comparison of Some Biomolecule Immobilization Techniques

Method	Physical and electrostatic adsorption	Cross-linking (after physical adsorption)	Entrapment	Covalent binding
Ease	High	Moderate	Moderate to low	Low
Loading level possible	Low (unless high S/V)	Low (unless high S/V)	High	(depends on S/V and site density)
Leakage (loss)	Relatively high (sens. to ΔpH salts)	Relatively low	Low to none[a]	Low to none
Cost	Low	Low to moderate	Moderate	High

[a]Except for drug delivery systems.

Bibliography

Callegaro, L., and Denti, E. (1983). Applications of bioreactors in medicine. *Internat. J. Artif. Organs* 6(Suppl 1): 107.

Carr, P. W. and Bowers, L. D. (1980). *Immobilized Enzymes in Analytical and Clinical Chemistry: Fundamentals and Applications*. Wiley, New York.

Chang, T. M. S. (1972). *Artificial Cells*. C. C. Thomas, Springfield, IL.

Cuatrecasas, P. and Anfinsen, C. B. (1971). Affinity Chromatography. *Ann Rev. Biochem.* 40: 259.

Dean, P. D. G., Johnson, W. S., and Middle, F. A., eds. (1985). *Affinity Chromatography*. IRL Press, Oxford and Washington, DC.

De Myttenaere, M. H., Maher, J., and Schreiner, G. (1967). Hemoperfusion through a charcoal column for glutethimide poisoning. *Trans. ASAIO* 13: 190.

Dunlap, B. R., ed. (1974). *Immobilized Biochemicals and Affinity Chromatography*. Plenum, New York.

Goldberg, E., ed. (1983). *Targeted Drugs*. Wiley-Interscience, New York.

Gombotz, W. R., and Hoffman, A. S. (1986). In *Hydrogels in Medicine and Pharmacy*, Vol. 1 (N. A. Peppas, ed.), pp. 95–126. CRC Press, Boca Raton, FL.

Gombotz, W. R. and Hoffman, A. S. (1987). In *Critical Reviews in Biocompatibility*, Vol. 4 (D. Williams, ed.), pp. 1–42. CRC Press, Boca Raton, FL.

Hoffman, A. S. (1984). In *Advances in Polymer Science*, Vol. 57 (K. Dusek, ed.). Springer-Verlag, Berlin.

Hoffman, A. S. (1987). Modification of material surfaces to affect how they interact with blood. *Ann. N. Y. Acad. Sci.* 516: 96–101.

Hoffman, A. S. (1988). *J. Appl. Polymer Sci. Symp.* (H. Yasuda and P. Kramer, eds.) 42: 251.

Hoffman, A. S., Leonard, E., Vroman, L., and Turitto, V., eds. (1987). *Ann. N. Y. Acad. Sci.* 516: 96–101.

Kim, S. W. and Feijen, J. (1985). In *Critical Reviews in Biocompatibility* (D. Williams, ed.), pp. 229–260. CRC Press, Boca Raton, FL.

Kolff, W. J. (1979). Artificial organs in the seventies. *Trans. ASAIO* 16: 534.

Kopecek, J. (1977). Soluble biomedical polymers. *Polymer Med* 7: 191.

Laskin, A. I., ed. (1985). *Enzymes and Immobilized Cells in Biotechnology*. Benjamin/Cummings, Menlo Park, CA.

Lavin, A., Sung, C., Klibanor, A. M., and Langer, R. (1985). Enzymatic

removal of bilirubin from blood: A potential treatment for neonatal jaundice. *Science* 230: 543.

Nose, Y., Malchesky, P. S., and Smith, J. W., eds. (1983). *Plasmapheresis: New Trends in Therapeutic Applications*. ISAO Press, Cleveland, OH.

Nose, Y., Malchesky, P. S., and Smith, J. W., eds. (1984). *Therapeutic Apheresis: A Critical Look*. ISAO Press, Cleveland, OH.

Park, T. G., and Hoffman, A. S. (1990). Immobilization of *Arthrobacter simplex* in a thermally reversible hydrogel: Effect of temperature cycling on steroid conversion. *Biotech. Bioeng.* 35: 152–159.

Piskin, E. and Hoffman, A. S., eds. (1986). *Polymeric Biomaterials*. M. Nijhoff, Dordrecht, The Netherlands.

Ringsdorf, H. (1975). Structure and properties of pharmacologically active polymers. *J. Polymer Sci.* 51: 135.

Schmer, G., Rastelli, L., Newman, M. O., Dennis, M. B., and Holcenberg, J. S. (1981). The bioartificial organ: Review and progress report. *Internat. J. Artif. Organs* 4: 96.

Shoemaker, S., Hoffman, A. S., and Priest, J. H. (1987). Synthesis and properties of vinyl monomer/enzyme conjugates: Conjugation of L-asparaginase with N-succinimidyl acrylate. *Appl. Biochem. Biotechnol.* 15: 11.

Sparks, R. E., Solemme, R. M., Meier, P. M., Litt, M. H., and Lindan, O. (1969). Removal of waste metabolites in uremia by microencapsulated reactants. *Trans. ASAIO* 15: 353.

Stark, G. R., ed. (1971). *Biochemical Aspects of Reactions on Solid Supports*. Academic Press, New York.

Sung, C., Lavin, A., Klibanov, A., and Langer, R. (1986). An immobilized enzyme reactor for the detoxification of bilrubin. *Biotech. Bioeng.* 28: 1531.

Tomlinson, E. and Davis, S. S. (1986). *Site-Specific Drug Delivery: Cell Biology, Medical and Pharmaceutical Aspects*. Wiley, New York.

Weetall, H. H. ed. (1975). *Immobilized Enzymes, Antigens, Antibodies, and Peptides: Preparation and Characterization*, Dekker, New York.

Yang, H. J., Cole, C. A., Monji, N., and Hoffman, A. S. (1990). Preparation of a thermally phase-separating copolymer, poly(N-isopropylacrylamide-co-N-acryloxysuccinimide) with a controlled number of active esters per polymer chain. *J. Polymer Sci. A. Polymer Chem.* 28: 219–220.

Zaborsky, O. (1973). *Immobilized Enzymes*. CRC Press, Cleveland, OH.

II

Biology, Biochemistry, and Medicine

3

Some Background Concepts

THOMAS A. HORBETT, BUDDY D. RATNER, JEFF M. SCHAKENRAAD, AND FREDERICK J. SCHOEN

3.1 INTRODUCTION
Buddy D. Ratner

Much of the richness of biomaterials science lies in its interdisciplinary nature. The two pillars of fundamental knowledge that support the structure that is biomaterials science are materials science, introduced in Part I, and the biological-medical sciences, introduced here. Complete introductory texts and a large body of specialized knowledge dealing with each of the chapters in this section are available. However, these chapters present sufficient background material so that a reader might reasonably follow the arguments presented later in this volume on biological interaction, biocompatibility, material performance and biological performance.

In as short a time as can be measured after implantation in a living system (<1 sec), proteins are already observed on biomaterial surfaces. In seconds to minutes, a monolayer of protein adsorbs to most surfaces. The protein adsorption event occurs well before cells arrive at the surface. Therefore, cells see primarily a protein layer, rather than the actual surface of the biomaterial. Since cells respond specifically to proteins, this interfacial protein film may be the event that controls subsequent bioreaction to implants. Protein adsorption is also of concern for biosensors, immunoassays, marine fouling, and a host of other phenomena. Protein adsorption concepts are introduced in Chapter 3.2.

After proteins adsorb, cells arrive at an implant surface propelled by diffusive, convective, or active (locomotion) mechanisms. The cells can adhere, release active compounds, recruit other cells, or grow. These processes probably occur in response to the proteins on the surface. Cell processes lead to responses (some desirable and some undesirable) that physicians and patients observe with implants. Cell processes at artificial surfaces are also integral to the unwanted buildup of marine organisms on ships, and the growth of cells in bioreactors used to manufacture biochemicals. Cells at surfaces are discussed in Chapter 3.3.

After cells arrive and attach at surfaces, they may multiply and organize into tissues. Synthetic materials can interact with or disrupt living tissues. The organization of tissues must be understood to appreciate the response to synthetic materials implanted in those tissues. Tissue structure and organization are reviewed in Chapter 3.4.

3.2 PROTEINS: STRUCTURE, PROPERTIES, AND ADSORPTION TO SURFACES
Thomas A. Horbett

The importance of proteins in biomaterials science stems primarily from their inherent tendency to deposit on surfaces as a tightly bound adsorbate, and the strong influence these deposits have on subsequent cellular interactions with the surfaces. It is thought that the particular properties of surfaces, as well as the specific properties of individual proteins, together determine the organization of the adsorbed protein layer, and that the nature of this layer in turn determines the cellular response to the adsorbed surfaces. Since the cellular responses largely determine the degree of biocompatibility of the material, the properties of proteins and their behavior at interfaces need to be understood by those interested in biomaterials. Figure 1 illustrates the interaction of a cell with an adsorbed protein layer on a solid substrate.

It is also worth noting that this subject has other important aspects, including the fact that the interfacial behavior of proteins is a fundamental, general property of proteins and enzymes that needs to be understood to fully appreciate protein behavior. In addition, phenomena at the air–water interface (e.g., interfacial coagulation and foaming), at the oil–water interface (e.g., the "receptor" proteins located in cell membranes that serve many important signalling functions), and at the solid–liquid interface in nonbiomaterial settings (e.g.,

Biomaterials Science

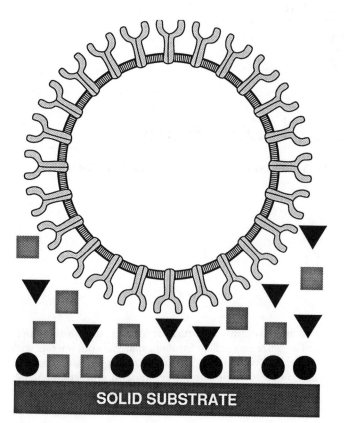

FIG. 1. Cell interaction with an adsorbed protein layer on a solid substrate. The cell is shown as a circular space with a bilayer membrane in which the adhesion receptor protein molecules (the slingshot-shaped objects) are partly embedded. The proteins in the extracellular fluid are represented by squares and triangles. The receptor proteins recognize and cause the cell to adhere to only the surface-bound form of one protein, the one represented by a solid circle. The bulk phase of this same adhesion protein is represented by a triangle, indicating that the solution and solid-phase forms of this same protein have a different biological activity. The figure is schematic and not to scale.

marine fouling, bacterial adhesion, and cell growth on surfaces in culture), are all strongly influenced by the behavior of proteins at interfaces.

A brief historical sketch of proteins at interfaces reminds us that the current biomaterials interest in this subject is just the latest in a long series of studies of this phenomenon. Thus, for example the fact that shaking of protein solutions causes the proteins to undergo "interfacial coagulation," in which the proteins are actually separated from the solution phase into the surface phase owing to their denaturation and insolubilization, was reported in 1851. Actual experimentation on proteins at interfaces began as early as 1873, when the viscosity at the air interface of protein solutions was shown to be considerably greater than in the bulk phase of the solution by observing the resistance to a magnetic field of a small magnetic needle floating at the solution interface. The behavior of proteins at the air–water interface was heavily studied in the 1920s through the early 1940s by Langmuir and others, with an emphasis on the structural changes in the proteins as judged from the rather large areas occupied per molecule when a monolayer was

formed. The subject has been revived again in more recent times owing to the interest of biomaterials scientists and those immobilizing enzymes, mostly focused on the solid–liquid interface. Most recently, the very modern technique of making variant or mutant proteins via single amino acid substitutions at a specific place in the protein chain, called "protein engineering," has been profitably applied to the further study of protein behavior at the air–water and the solid–liquid interfaces. The protein variant studies have given new insights into the molecular mechanisms of protein behavior at interfaces, since a correlation between changes in thermodynamic stability of the protein variants and the changes in surface activity was found (Horbett, 1993).

STRUCTURE AND PROPERTIES OF PROTEINS RELEVANT TO ADSORPTION

The soluble proteins present in biological fluids such as blood plasma and peritoneal exudate are the type of proteins that are primarily involved in adsorption to implanted materials. Insoluble proteins, such as collagen, which form the structural basis of tissue, are not normally free to diffuse to the implant surface, although they may be deposited in fibrous form adjacent to or actually on the implant by cells as part of the foreign body capsule formation. The soluble proteins differ from the insoluble proteins in many ways, including the fact that they are less regular in their amino acid composition and three-dimensional structure. The soluble proteins are therefore difficult to describe except in certain general terms. Fundamentally, this diversity originates in the linear sequence of amino acids that uniquely characterizes each protein. This sequence is the same for all molecules of a particular type of protein, yet both the length and sequence differ dramatically from protein to protein. For example, all albumin molecules have the same sequence, while all fibrinogen molecules have another sequence that is much longer and very different than the albumin sequence.

The role of the sequence of amino acids in generating diversity can best be appreciated by a brief review of the properties of the amino acid side chains, since these vary greatly (see Table 1). Some of the amino acids have side chains that carry no charge at any pH yet exhibit considerable polar character (serine, threonine). The ionizable side chains vary from fairly acidic ones (aspartic and glutamic acid are fully negatively charged at the physiological pH of 7.4) to more basic amino acids such as the imidazole group in histidine (which carries a partial positive charge at pH 7.4) and the still more basic amino groups in lysine and arginine that carry full charges at pH 7.4.

Another group of amino acids have no acid, base, or polar character in their side chains, instead being somewhat hydrocarbonlike in character, as attested by their generally much lower solubility in water. However, these so-called hydrophobic or "water-hating" amino acids vary considerably in this respect, depending on their specific structure. Thus, for example, the single methyl group side chain in alanine contributes only 0.5 kcal per mole to the free energy of

TABLE 1 Structure and Properties of Amino Acid Side Chains

Amino acid	Structure	Group and pK	Charge	Hydrophobicity[a]	Surface tension[b]
Isoleucine			Neutral	0.73	−15.2
Phenylalanine			Neutral	0.61	−17.3
Valine			Neutral	0.54	−3.74
Leucine			Neutral	0.53	−21.9
Tryptophane			Neutral	0.37	−9.6
Methionine	$-CH_2-CH_2-S-CH_3$		Neutral	0.26	−3.01
Alanine	$-CH_3$		Neutral	0.25	0.96
Glycine	$-H$		Neutral	0.16	1.12
Cysteine	$-CH_2SH$	—SH: 8.3	0 to −1	0.04	0.69
Tyrosine		—OH: 10.9	0 to −1	0.02	−15.1
Proline			Neutral	−0.07	−0.49
Threonine			Neutral	−0.18	0.59
Serine	$-CH_2-OH$		Neutral	−0.26	0.76
Histidine		—NH—: 6.0	0 to +1	−0.40	1.03
Glutamic acid	$-CH_2-CH_2-CO_2H$	—CO₂H: 4.3	0 to − 1	−0.62	0.86
Asparagine			Neutral	−0.64	1.17
Glutamine			Neutral	−0.69	1.21
Aspartic acid	$-CH_2-CO_2H$	—CO₂H: 3.9	0 to −1	−0.72	0.96
Lysine	$-CH_2-CH_2-CH_2-CH_2-NH_3$	—NH₂: 10.8	0 to +1	−1.1	0.92
Arginine		—NH₂: 12.5	0 to +1	−1.8	1.03

[a]A "consensus value" for the hydrophobicity of the amino acid side chains is given (Eisenberg, 1984). More positive values are more hydrophobic. According to Eisenberg, the magnitude of the values "may be considered roughly in kcal/mole for transfer from a hydrophobic to a hydrophilic phase" (e.g., from ethanol to water).

[b]The values are the surface tension lowering of water solutions of the amino acids in units of ergs/cm²/mole per liter (Bull and Breese, 1974).

transfer from water to an organic phase, whereas the double-ringed indole group in tryptophan contributes 3.4 kcal. The diverse character of the amino acid side chains, together with the variations in the proportions of the amino acids present in each particular protein, means that such physicochemical properties as their solubility and ability to interact with surfaces are also diverse.

An important consequence of the primary sequence and its inherent diversity in chemical nature is the fact that, unlike most synthetic polymers, each protein has a single, distinct, three-dimensional structure that it will assume under physiological conditions (see Table 2). These shapes are dictated by the formation of multiple noncovalent bonds formed throughout the protein's three-dimensional structure that leads to a metastable state. With soluble proteins, these shapes are often roughly spherical or globular in outline, although the important plasma protein fibrinogen violates this somewhat because it is elongated into a "three bead on a string" structure. More important here than the particular structure assumed by a protein is the fact that a unique arrangement of the amino acid sequence in three-dimensional space exists for each protein. Furthermore, the spatial arrangement results in the hydrophobic residues preferentially located "inside" the protein where they are shielded from water, while the ionized and polar residues are usually on the outside of the protein and in contact with the aqueous phase.

This spatial arrangement of the amino acids in proteins has a direct bearing on the interaction of proteins with surfaces because it means that the many residues "buried" inside the protein may not be able to participate in bond formation with the surface. For these interior residues to interact with the surface, the protein would have to unfold. Furthermore, it means that the type of residues that initially contact the interface have essentially been narrowed to the largely polar residues at the protein surface. The extremely wide range of chemical interactions theoretically possible because of the many different types of amino acid residues in the protein sequence does not necessarily come into play because there is a strong, overriding structural influence that tends to prevent the full array of possible interactions from being expressed.

The folded protein structures have densities of about 1.4 g/cm³. In comparison with water's density of 1.0, or the density of most synthetic polymers of about 1.1, this basic fact about proteins reflects their tightly folded structure. It is therefore convenient to think of proteins in a physicochemical sense as quite compact, externally charged wax droplets in water, in which the interior hydrophobic core is analogous to wax and the surface amino acids are the charged species. It is in this way that the polyelectrolyte behavior of proteins is expressed. That is, owing to their large size and corresponding large number of charged amino acid side chains of varying acidity or basicity, proteins have on them a large number of charges, both positive and negative, and these are distributed around the exterior of the protein. Therefore, depending on the pH and ionic strength of the media, a large range of charge interactions can be expected between the protein and a surface. The polyelectrolyte behavior is expressed most clearly in the degree of adsorption. Many proteins exhibit a maximum in the amount adsorbed to

neutral or slightly charged surfaces at a pH at which the net charge on the protein is minimal, i.e., near the isoelectric pH. On surfaces which themselves carry a large net charge, however, the interactions are dominated by the degree of opposition of charge on the protein and the surface, so that negatively charged proteins adsorb preferentially to positively charged surfaces and positively charged proteins adsorb preferentially to negatively charged surfaces.

The structural singularity of a particular protein generally applies only under conditions that are at least approximately physiological, i.e., in the range of 0 to 45°C, pH 5 to 8, and in aqueous solutions of about 0.15 M ionic strength. Beyond these conditions, proteins are subject to the phenomena of denaturation, a word which is meant to indicate that they lose their normal nature or structure. The natural or "native" protein can be made to undergo a transition to the denatured form simply by heating it, for example. The denatured protein is quite different than the native protein, and, in particular, generally loses the "inside/outside" nature reflected in the preferential location of polar residues on the surface and nonpolar residues on the inside of the molecule. In addition, the protein also loses the singularity of its structure when it is denatured. Instead, it will tend to become much more like a random coil that is characteristic of synthetic polymers. Denatured proteins typically lose their solubility, become much less dense, and lose biological functions such as enzyme activity. The stability of selected proteins listed in Table 2 is seen to vary greatly.

The retention of a protein's native structure upon adsorption, even under physiologic conditions, is one of the more interesting aspects of protein behavior at interfaces, for the unfolded protein, with many more exposed hydrophobic amino acid residues, is clearly capable of forming many more bonds per molecule with a surface than the native protein. The multiple bonding involved in adsorption to surfaces is a major feature that distinguishes the adsorption of proteins from the adsorption of small molecules. Generally, proteins adsorbed at the solid interface are not denatured (see the following discussion).

ADSORPTION BEHAVIOR OF PROTEINS AT SOLID–LIQUID INTERFACES

The adsorption of proteins to solid surfaces qualifies for one of the traditional definitions of adsorption in that it represents a preferential accumulation of the protein in the surface phase. The protein adsorbed to the surface does not merely reflect the retention of a thin layer of the adjacent protein solution; it is *not* sorption as might occur on a porous, absorptive matrix uch as filter paper. This can be seen from the fact that the concentration of protein in the surface phase is much higher than the bulk phase from which it came. Thus, for example, typical values for adsorption of proteins are in the range of 1 μg/cm², a plateau or monolayer value typically reached at higher bulk concentrations (see Fig. 2). To convert this two-dimensional value into an equivalent volumetric concentration unit, we can assume a monolayer of a typical protein for which a 100-Å (or 10^{-6} cm) diameter is a good approximation. Then,

TABLE 2   Properties of Selected Proteins

Protein	Function	Location	Size	Shape[a]	pI	Stability[a]	Surface activity
Albumin	Carrier	Blood	65 kDa	42 × 141 Å (1)	4.8	Denatures at 60°C (8)	Low on PE
Fibrinogen	Clotting	Blood	340 kDa	460 × 60 Å (2) trinodular string	5.8	Denatures at 56°C (9)	High on PE
IgG	Antibody	Blood	165 kDa	T-Shaped	6.5		Low on PE
Lysozyme	Bacterial lysis	Tear; hen egg	14.6 kDa	45 × 30 × 30 Å (3) Globular	11	$\Delta G_n = -14$ kcal/mol (10)	High on negatively charged surfaces
Hemoglobin	Oxygen carrier	Red cells	65 kDa	55 Å (4) Spherical A2B2 tetramer	6.87	Normal form	Very high on PE
Hemoglobin S	Oxygen carrier	Sickle red cells	65 kDa	55 Å (4) Spherical A2B2 tetramer	7.09	Lower than A form	Oxy form of HbS has much higher air–water activity than normal Hb
Myoglobin	Oxygen carrier	Muscle	16.7 kDa	45 × 35 × 25 Å (5) Spherical monomer	7.0	$\Delta G_n = -12$ kcal/mol (10)	Unknown
Collagen	Matrix factor	Tissue	285 kDa	3000 × 15 Å (6) Triple helical rod		Melts at 39°C (11)	
Bacterio-rhodopsin	Membrane protein		26 kDa	30–40 Å long (7) Seven-rod structure that self-localizes in membranes			High at cell membrane
Tryptophan synthase alpha subunit ("wild type")	Enzyme		27 kDa		5.3	$\Delta G_n = -8.8$ kcal/mole; denatures at 55°C (12)	High air–water activity compared with ovalbumin
Tryptophan synthase (glu → ileu) variant alpha subunit	Enzyme		27 kDa			$\Delta G_n = -16.8$ kcal/mole (12)	Much less active at air–water interface than wild type

[a]The numbers in parentheses refer to these references: (1) Peters, 1985, p. 176; (2) Stryer, 1981, p. 172; (3) Stryer, 1981, p. 138; (4) Stryer, 1981, p. 49; (6) Stryer, 1981, p. 188; (7) Eisenberg, 1984, p. 599; (8) Peters, 1985, p. 186; (9) Loeb and Mackey, 1972; (10) Norde and Lyklema, 1991, p. 14; (11) Stryer, 1981, p. 191; (12) Yutani et al., 1987.

a 1-cm^2 area containing 1 μg corresponds to a local protein concentration of $1/10^{-6} = 10^6$ μg/cm^3 or 1 g/cm^3. Given that the density of a pure protein is 1.4 g/cm^3, this layer is indeed tightly packed. Furthermore, the 1 g/cm^3 is equivalent to 1000 mg/cm^3, which is far higher than the bulk protein concentration of solutions from which such adsorbates form (typically 1 mg/cm^3). Thus, the surface phase is often 1000 times more concentrated than the bulk phase, corresponding to high local concentration indicative of the existence of a very different state of matter and thus truly deserving of a special term, namely, the adsorbed state.

A second major feature of the adsorbed phase is the selectivity of the process that leads to enrichment of the surface phase in one protein versus another. Here, we are speaking about adsorption as it typically occurs to biomaterial surfaces, namely from a complex mixture of proteins in the bulk phase. Since there is a limited amount of space on the surface of the solid and it can only accommodate a small fraction of the total protein typically present in the bulk phase, there is competition for the available surface sites. The monolayer of adsorbed protein is the limiting amount that can be adsorbed, resulting in competition for sites on the surface (see Fig. 2). Second, as has been discussed earlier, the proteins vary a great deal in their fundamental amino acid sequences and three-dimensional structures and therefore have very different abilities to adsorb to surfaces; their "surface activity" differs (see Table 2).

Therefore, depending on the two major driving forces for adsorption, namely, the relative bulk concentration of each protein and its intrinsic surface activity, the outcome of the competitive process of adsorption is an adsorbed layer that is richer in some proteins than others; the surface composition differs from the bulk composition. Furthermore, because the proteins have different affinities for each type of surface, the outcome of the competition is different on each surface. Our concept of the adsorbed layer formed on solid surfaces exposed to mixtures of proteins, then, is one in which variable degrees of enrichment of each of the proteins occur, so that, for example, some surfaces may be richer in albumin while others have more fibrinogen. Table 3 contains some surface enrichment data for the adsorption of plasma proteins to several surfaces.

The adsorption of proteins to solid surfaces is largely irreversible and therefore leads to the immobilization of the protein species in the surface phase since they are no longer free to diffuse away. This is somewhat different from the idea one has with gas adsorption, because there the molecules can often

A

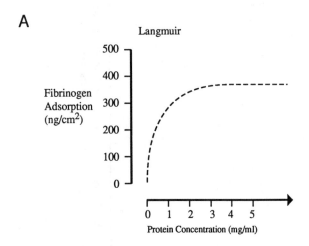

B View at plateau:

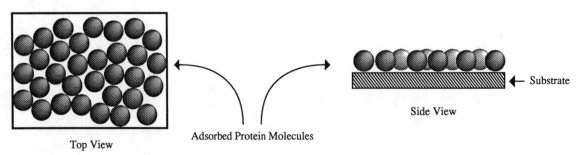

FIG. 2.   Adsorption isotherms and the monolayer concept.

TABLE 3 Enrichment of Proteins Adsorbed on Polyethylene Exposed to Blood Plasma

Protein	Enrichment[a]
Fibrinogen	1.3
γ-globulin	0.53
Albumin	0.88
Hemoglobin[b]	79

[a]Enrichment was calculated as the ratio of the surface fraction of the protein compared with the bulk fraction. The fraction in each case was calculated as the amount of each protein divided by the total amount of all of the proteins listed in the table. The data were measured in the author's laboratory using radiolabeled proteins to measure their adsorption from blood plasma.

[b]Hemoglobin is not considered a "normal" plasma protein, but is nonetheless present (typically at 0.01 mg/ml or less) in most plasma and serum preparations as a result of leakage from the red cells during preparation of the protein fraction of blood. *In vivo*, it would normally be present only in trace amounts unless a disease that caused hemolysis existed.

desorb from the surface if the pressure is lowered. When the solution phase variable corresponding to pressure is lowered— that is, when protein concentration is lowered by rinsing the surface with a protein-free medium—the protein does not readily come off of many surfaces, even after days of further rinsing. Effectively or operationally, the adsorption is irreversible unless some drastic change in the solvent is made, such as the introduction of a detergent that binds strongly to the adsorbed protein as well as the underlying substrate. The fact that the proteins are effectively immobilized by the adsorption process means that any of the processes they are normally exposed to in the bulk phase will proceed very differently in the surface phase, if only because the diffusion of one of the species is now halted. Furthermore, since transport is generally hindered near a solid interface, the approach and departure of species with which the protein may interact is greatly altered. Perhaps the most relevant consequence of the immobilization of a protein in the field of biomaterials is the often-noted phenomenon by which an adsorbed protein is able to cause a cell for which it has a receptor to adhere to the adsorbed protein–solid interface, whereas the same protein in the bulk phase is not bound by the cell (illustrated in Fig. 1). This is the case for fibrinogen and platelets; i.e., platelets adhere to adsorbed fibrinogen, but do not bind dissolved fibrinogen.[1] Table 4 contains a list of the platelet adhesion proteins present

in blood plasma (line 3c) and the receptors affected by the protein (line 3a).

The orientation of proteins in the adsorbed phase must also be considered because the proteins are not uniform in properties or structure across their surface. Indeed, the existence of regions that are enriched in one of the categories of side chains discussed earlier, e.g., a "patch" of acidic residues, is a feature that influences the functions of proteins. Similarly, particular sequences of amino acids in the protein are well known to bind specifically to a variety of agents, especially to cell-membrane bound receptors. As far as is known, proteins are not very free to rotate once adsorbed, owing to multiple bonding, and therefore a fixed portion of the surface of the protein is exposed to the bulk phase. The degree of uniformity of such orientation is, however, not known. For example, it is not known whether all the proteins are oriented the same way in order to optimize the bonding of certain favorable regions with the surface.

The reactions of proteins in the adsorbed phase may be broadly classed into noncovalent reactions represented by structural transitions, and covalent reactions, such as those that occur with protein complement C3 on hemodialyzer membranes. In the latter case, the adsorption renders this protein subject to proteolytic cleavage by other proteins in the complement system; this cleavage proceeds at only a very low rate for C3 in the bulk phase. The role of surface adsorption in this case is believed to be the formation of a covalent bond between the C3 and hydroxyl groups on the surface that somehow prevents an inhibitor that normally is present, and normally prevents further activation, from doing its job.

TABLE 4 Principles Underlying the Influence of Adsorbed Plasma Proteins on Platelet Interactions with Biomaterials

1. Synthetic foreign materials acquire bioreactivity only after first interacting with dissolved proteins. The principal means by which the transformation from an inert, nonthrombogenic polymer to a biologically active surface takes place is the interaction of the proteins with the surface, which then mediates cell adhesion.

2. Platelets are a major example of why and how adsorbed proteins are influential in cell–biomaterials interactions.

3. Sensitivity of platelets to adsorbed proteins is due to:
   a. Receptors (IIb/IIIa and Ib/IX) bind specifically to a few of the adsorbed plasma proteins, mediating adhesion.
   b. Concentrating, localizing, immobilizing effects of adsorbing proteins at the interface accentuate the receptor–adhesion protein interaction.
   c. Adhesion proteins in plasma for platelets: fibrinogen, fibronectin, vitronectin, and von Willebrand factor.

4. Principles of protein adsorption to biomaterials:
   a. Monolayer adsorption and consequent competition for available adsorption sites means that not all proteins in the plasma phase can be equally represented.
   b. Driving forces: intrinsic surface activity and bulk phase concentration.
   c. Surfaces vary in selectivity of adsorption.
   d. Biological activity of the adsorbed protein also varies on different surfaces.

[1]This applies to platelets that have not been previously activated by agonists such as ADP or thrombin. Platelets exposed to agonists do bind bulk-phase fibrinogen. See last section of 3.2 for further discussion.

TABLE 5　Enthalpy of Adsorption of Proteins to Surfaces

Protein	Surface[a]	Enthalpy of adsorption	Reference
HSA	α-Fe$_2$O$_3$	+1.9 mJ-m^{-2} at pH 5	Norde, p. 275 in
		+7.0 mJ-m^{-2} at pH 7	Andrade (1985)
RNase	PS-H	+4 mJ-m^{-2} at pH 5	Norde, p. 278 in
		−2 mJ-m^{-2} at pH 11	Andrade (1985)

[a]The α-Fe$_2$O$_3$ surface referred to is hematite. PS-H is an abbreviation for negatively charged polystyrene latex particles of high surface charge density.

Clotting factor XII also requires a surface to become active in the clotting cascade. It also undergoes cleavage, but in this case the role of the surface is thought to be to perturb the three-dimensional structure of the protein to render it active as a protease. The activation of factor XII by a conformational change induced by the adsorption process is one of the better-known cases in which a noncovalent change induced by adsorption occurs, but in general it is thought that all proteins that adsorb to solid surfaces may undergo limited conformational change. Given the metastable condition of the protein molecule, and the driving force for better adsorption through further bond formation that would arise from partial exposure of residues in the protein interior, a change in the protein's structure might seem to be likely. However, many proteins and enzymes retain at least some of their biological activity in the adsorbed state, and the solid phase immunoassay technique relies on the use of adsorbed antibody or antigen whose structure cannot be totally altered for this method to work in the binding of antigens or antibodies. Therefore, conformational changes upon adsorption seem to be limited in nature; adsorption to solid surfaces does not seem to result in a fully denatured protein.

The kinetics of adsorption of proteins to solid surfaces generally consist of a very rapid initial phase that is diffusion limited, followed by a slower phase upon approach to the steady-state value. In the initial phase, the proteins typically adsorb as quickly as they arrive at the relatively empty surface, so that plots of amount adsorbed versus time$^{1/2}$ in this regime have the linearity characteristic of a diffusion-controlled process. In the later, slower phase, it is presumably more difficult for the arriving proteins to find and fit into an empty spot on the surface.

The thermodynamics of protein adsorption are not easily characterized because the process appears to be essentially irreversible. Thus, when one exposes solid surfaces to a series of increasing concentrations of protein solutions, and then rinses away the bulk protein solution, an increasing amount of protein is retained on the surface until a plateau is reached at higher concentrations, as shown in Fig. 2, where typical adsorption isotherms are shown. Because the adsorption is irreversible, the calculation of an equilibrium binding constant from a plot of adsorption versus bulk concentration, and its conversion to a free energy value in the usual way, is not a valid method to obtain thermodynamic

information. However, direct measurements of the heat of adsorption have been made for several proteins on a variety of surfaces and under various conditions of pH and temperature (see Table 5). The enthalpy of the adsorption process has been observed to vary a great deal, even being positive in some cases. The observation of positive enthalpies upon spontaneous adsorption to certain surfaces must mean that the process is entropically driven in these cases. The net negative free energy characteristic of a spontaneous process means that $T \Delta S$ is greater than the positive ΔH term in the formula $\Delta G = \Delta H - T \Delta S$. More generally, all protein adsorption processes are thought to be strongly driven by entropic changes. The importance of entropic factors in this process can easily be envisioned to arise from changes in water binding to the surface and the protein as well as limited unfolding of the protein on the surface.

THE IMPORTANCE OF ADSORBED PROTEINS IN BIOMATERIALS

Table 4 summarizes the principles underlying the influence of adsorbed proteins in biomaterials used in contact with the blood. All of the principles listed also apply in other environments such as the extravascular spaces, albeit with other proteins and other cell types (e.g., macrophages in the peritoneum adhere via other receptors and other adhesion proteins). The platelets therefore provide a "case study," and we close this chapter by considering this case.

The sensitivity of platelet–surface interactions to adsorbed proteins is fundamentally due to the presence of adhesion receptors in the platelet membrane that bind to certain plasma proteins. There are only a few types of proteins in plasma that are bound by the adhesion receptors. The selective adsorption of these proteins to synthetic surfaces, in competition with the many nonadhesive proteins that also tend to adsorb, is thought to mediate platelet adhesion to these surfaces. However, since the dissolved, plasma-phase adhesion proteins do not bind to adhesion receptors unless the platelets are appropriately stimulated, while unstimulated platelets can adhere to adsorbed adhesion proteins, it appears that adsorption of proteins to surfaces accentuates and modulates the adhesion receptor–adhesion protein interaction. The type of surface to which the adhesion protein is adsorbed affects the ability of the protein to support platelet adhesion (Horbett, 1993). The principles that determine protein adsorption to biomaterials include monolayer adsorption, the intrinsic surface activity and bulk concentration of the protein, and the effect of different surfaces on the selectivity of adsorption.

More generally, all proteins are known to have an inherent tendency to deposit very rapidly on surfaces as a tightly bound adsorbate that strongly influences subsequent interactions of many different types of cells with the surfaces. It is therefore thought that the particular properties of surfaces, as well as the specific properties of individual proteins, together determine the organization of the adsorbed protein layer, and that the nature of this layer in turn determines the cellular response to the adsorbed surfaces.

Bibliography

Andrade, J. D. (1985). Principles of protein adsorption. in *Surface and Interfacial Aspects of Biomedical Polymers*, J. Andrade, ed. Plenum Publ., New York, pp. 1–80.

Bull, H. B., and Breese, K. (1974). Surface tension of amino acid solutions: A hydrophobicity scale of the amino acid residues. *Arch. Biochem. Biophys.* **161**: 665–670.

Eisenberg, D. (1984). Three-dimensional structure of membrane and surface proteins. *Ann. Rev. Biochem.* **53**: 595–623.

Horbett, T. A. (1982). Protein adsorption on biomaterials. in *Biomaterials: Interfacial Phenomena and Applications*, S. L. Cooper and N. A. Peppas, eds. ACS Advances in Chemistry Series, American Chemical Society, Washington, DC, Vol. 199, pp. 233–244.

Horbett, T. A. (1986). Techniques for protein adsorption studies. in *Techniques of Biocompatibility Testing*, D. F. Williams, ed., CRC Press, Boca Raton, FL, pp. 183–214.

Horbett, T. A. (1993). Principles underlying the role of adsorbed plasma proteins in blood interactions with foreign materials. *Cardiovascular Pathology.* **2**: 137S–148S.

Horbett, T. A., and Brash, J. L. (1987). Proteins at interfaces: current issues and future prospects. in *Proteins at Interfaces: Physicochemical and Biochemical Studies,* T. A. Horbett and J. L. Brash, eds. ACS Symposium Series, *343*, American Chemical Society, Washington, DC, Vol. 343, pp. 1–33.

Loeb, W. F., and Mackey, W. F. (1972). A "Cuvette Method" for the determination of plasma fibrinogen. *Bull. Amer. Soc. Vet. Clin. Path.* **1**: 5–8.

Norde, W., and Lyklema, J. (1991). Why proteins prefer interfaces. *J. Biomater. Sci.: Polymer Ed.* **2**: 183–202.

Peters, T. (1985). Serum albumin. In *Advances in Protein Chemistry*, Vol. 37 (C. B. Anfinsen, J. T. Edsall, and F. M. Richards, eds.), pp. 161–245. Academic Press, New York.

Stryer, L. (1981). *Biochemistry*, 2nd Ed. W. H. Freeman, San Francisco.

Yutani, K., Ogasahara, K., Tsujita, T., and Sugino, Y. (1987). Dependence of conformational stability on hydrophobicity of the amino acid residue in a series of variant proteins substituted at a unique position of tryptophan synthase alpha subunit. *Proc. Natl. Acad. Sci. U.S.A.* **84**: 4441–4444.

3.3 CELLS: THEIR SURFACES AND INTERACTIONS WITH MATERIALS

Jeff M. Schakenraad

THE CELL MEMBRANE

The mammalian cell is a highly organized structure. It is composed of organelles, including a nucleus containing genetic informaton (DNA, chromatin); mitochondria, the "power plants" of the cell; the Golgi apparatus, the assembly area of glycoproteins and lipoproteins; the endoplasmic reticulum (smooth and rough), involved in protein synthesis and cellular transport; and lysosomes containing proteolytic enzymes. The remainder of the cell is cytoplasm (cytosol), containing the cytoskeleton involved in cell movement.

The cell membrane surrounds the organelles and cytoplasm. Different membrane regions correspond to different functions, such as absorption, secretion, fluid transport, mechanical attachment, and communication with other cells and extracellular matrix components. The membrane is a dynamic structure composed of a double layer of phospholipids in which proteins, glycoproteins, lipoproteins, and carbohydrates "float" (Fig. 1). Intrinsic (integral) and extrinsic proteinaceous structures can be distinguished, according to their position in the membrane. During the past 60 years, several membrane models have been proposed: The Danielli–Davson model (bilayer sandwich, 1935), the unit membrane model (1959) by Robertson, and at present the fluid mosaic model, which postulates integral proteins embedded in the lipid bilayer, but free to move laterally, and possibly projecting out of the bilayer at one surface or the other.

Transport through the cell membrane (De Robertis, 1981) can occur as passive permeability, using the electrical gradient over the membrane, owing to unequal ion distribution. A transmembrane potential of about -50 to -100 mV is maintained by the exchange of Na^+ and K^+ over the membrane. Active transport can take place by using the "sodium pump," a mechanism that transports K^+ into and Na^+ out of the cell, against their respective concentration gradients, using energy provided by ATP. A third transport mechanism uses transport proteins, carrier, and fixed-pore systems.

Communication with the world outside the cell primarily takes place via receptors. Many receptors can only be activated by the binding of a specific molecule (e.g., antigen-antibody binding). This specific binding induces a response by the receptor molecule by interacting with the enzyme adenylate cyclase, which in turn activates a second messenger system on the inside of the cell by producing cyclic AMP from ATP. This second messenger takes care of further communication within the cell to create an appropriate response to the original trigger.

Microfilaments in the cytoplasm, made of actin, myosin, actinin, and tropomyosin can be connected with the cell membrane via integrin structures (Fig. 2). This microfilament network is responsible for cellular adhesion and locomotion and is called the cytoskeleton. Integrins are receptors consisting of heterodimeric proteins with two membrane-spanning subunits (Ruoslahti and Pierschbacher, 1987; Buck and Horwitz, 1987). Some integrins can bind to the RGD sequence (arginine, glycine, aspartic acid) of fibronectin or other adhesive proteins (e.g., thrombospondin, vitronectin, osteopontin). These adhesive proteins, in turn, can bind to solid substrates, extracellular matrix components, and other cells. This specific receptor is thus used to connect the cytoskeleton with extracellular adhesive sites, via the intermediate fibronectin.

The adhesive protein laminin can connect epithelial or endothelial cells to their basal lamina (Gospodarowicz *et al.*, 1981). Proteoglycans such as heparan sulfate, chondroitin sulfate and dermatan sulfate are associated with the basolateral cell membrane (Rapraeger *et al.*, 1986).

The extracellular matrix, the actual cellular glue, occurs in two forms: interstital matrix (e.g., in connective tissue) and basement membrane (e.g., in epithelium and endothelium). It has collagenous molecules, glycoproteins, elastin, proteoglycans and glycoaminoglycans as its major constituents. Its many functions include mechanical support for cellular anchorage, determination of cell orientation, control of cell growth, maintenance of cell differentiation, scaffolding for tissue renewal,

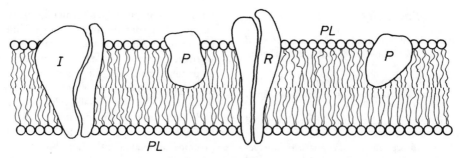

FIG. 1. Cell membrane consisting of a double layer of phospholipids (PL), in which proteins (P), lipoproteins, glycoproteins, etc. "float." Some proteinaceous structures connect the outside of the cell with the inside [e.g., integrins (I) or other specific receptor sites (R)]. These receptors can be activated by a specific trigger on the outside. A more general answer on the inside (using e.g., cyclic adenosine monophosphate (cAMP) as the second messenger) provides the communication within the cell.

establishment of tissue microenvironment, etc. As such, the extracellular matrix provides a way for one cell (type) to influence the behavior of others.

ADHESION

There are four regular adhesive sites between cells and between cells and extracellular matrix (illustrated by Fig. 3; Beck and Lloyd, 1974):

1. Gap junction (nexus): 4-nm gap; array of plaquelike connections between the plasma membranes of adjacent cells.
2. Desmosome (macula adherens): 30–50-nm gap, mechanical attachment formed by the thickened plasma membranes of two adjacent cells, containing dense material in the intercellular gap. Bundles of tonofilaments usually converge on this dense plaque.
3. Hemidesmosome: structure similar to desmosome, between cells and extracellular matrix material.

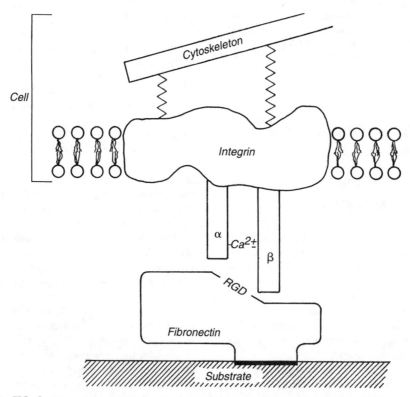

FIG. 2. Integrin, cellular receptor site for fibronectin. The α and β subunits of the integrin are only operative if calcium is present.

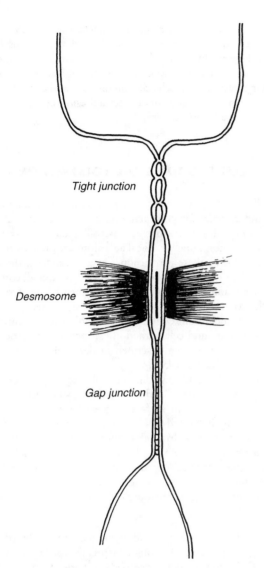

FIG. 3.   Cell–cell contact sites. The desmosome represents the strongest membrane contact. Elements of the cytoskeleton make contact with the cell membrane at the desmosome. The gap junction will allow diffusion of soluble products. The tight junction prevents transport of solubles.

4. Tight junction (zonula occludens): <5-nm gap. Structure formed when adjacent cell membranes adhere to each other, creating a barrier to diffusion.

Adhesive sites between cells and solid substrata can be described as (Culp, 1978).

1. Focal adhesion: 10–20-nm gap, often observed at the cell boundaries. It represents a very strong adhesion. Fibronectin is involved in focal adhesions.
2. Close contact; 30–50-nm gap. Often surrounding focal adhesions.
3. Extracellular matrix contacts, gap > 100 nm. Strands and cables of extracellular matrix material connect the ventral cell wall with the underlying substratum.

Some of these adhesion sites have been illustrated in Fig. 4.

In a physiological environment, protein adsorption always precedes cellular adhesion. Preadsorbed proteins, in combination with proteins produced by the cell, and depending on the substratum properties, determine the strength and type of adhesion.

From a physicochemical point of view, the kinetics of adhesion can be described as long-range interactions (Fowkes, 1964) and short-range interactions (acid-base, hydrogen bonds) (van Oss *et al.*, 1986). Others describe these forces as dispersive and polar (Schakenraad *et al.*, 1988).

The DLVO theory (Derjaguin and Landau, 1941; Verwey and Overbeek, 1948) illustrates the relationship between particle (cell) distance from the surface and repulsive (electrostatic) and attractive (e.g., van der Waals) interaction energies (Fig. 5):

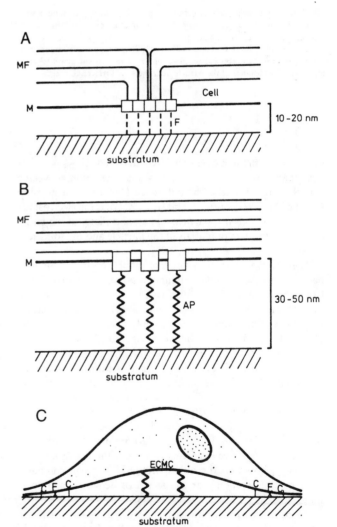

FIG. 4.   Cell–substratum contact sites. (A) Focal adhesion sites are predominantly found at the boundaries of cellular extentions. The integrin connects the cytoskeleton with the substratum via fibronectin. (B) Close contacts represent less strong adhesive sites. (C) Localization of the different adhesive sites. MF, microfilaments; M, cell membrane; AP, adhesive protein; F, fibronectin; ECMC, extracellular matrix contact; C, close contact; F, focal adhesion.

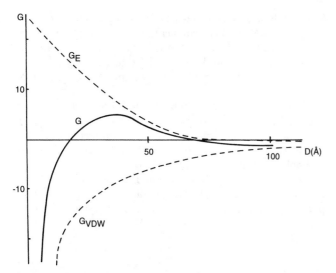

FIG. 5. Interaction energies between a particle (cell) approaching a solid surface. The total interaction energy (G, solid line) is composed of attractive van der Waals forces (G_{vdw}) and repulsive electrostatic forces (G_e). A secondary minimum can be observed at approximately 100 Å and a primary minimum at a distance <5 Å. An energy barrier can be observed at approximately 40 Å from the surface. D, distance of the particle from the solid surface; G, energy.

$$G = G_E + G_{VDW}$$

in which G is the sum of energy forces involved; G_{VDW} is $-H.a/6D$, which represents the van der Waals interaction between a particle with a radius a and a distance D to a flat substratum; H is the Hamaker constant; and $G_E(:)Z_1 \cdot Z_2$ is the electrostatic interaction between a particle and a solid plate, in which Z_1 and Z_2 are the zeta potentials of the particle and the solid substratum.

Long-range interaction forces probably result in bringing a particle into the secondary minimum at approximately 100 Å from the surface. Only little energy is needed to remove the particle from the surface again. Short-range interactions between a particle and a solid can only take place at distances <20 Å (Fig. 5). The adhesive macromolecules in cellular membranes, however, might be able to cross the energy barrier, which is necessary to reach this primary minimum, and establish acid-base or hydrogen bond interactions with the solid substratum. The energy barrier is high for low-surface free-energy substrates (e.g., polytetrafluoroethyleen) and low or absent for high-surface free-energy substrates (e.g., glass). Particles in the secondary minimum could, theoretically, freely move over the substrate without consumption of energy; however, the adhesion would be weak compared with metabolic energy or the forces of active cellular locomotion.

In contrast to this physicochemical approach, protein adsorption always precedes cellular adhesion. For example, the presence of fibronectin on either the cellular or the solid surface will exponentially increase adhesion and spreading of cells. The presence of the protein apparently circumvents the energy barrier of electrostatic repulsion.

Several authors have described cellular adhesion in a mathe-matical model, taking into account such factors as locomotion, flow rates, surface reaction rates, and diffusivity (Strong *et al.*, 1987). This modification of the Grabowski model for predicting cell adhesion (Grabowski *et al.*, 1972) allows the possibility of taking into account cell–cell interactions and can differenti-ate between the kinetics of contact adhesion and irreversible adhesion (see DLVO theory).

CELL SPREADING AND LOCOMOTION

Cell spreading is a combined process of continuing adhesion and cytoplasmic contractile meshwork activity. At first, lamellar protrusions are formed. Microfilaments can always be observed in these lamellae. Focal adhesions are often associated with the tips of these protrusions. In a later stage, the cyto-plasmic flaps in between the protrusions also expand, complet-ing cell spreading. When a cell meets the membrane of another cell, the contact inhibition process prevents further spreading. Apparently membrane contact induces inhibition of contractile protein activity and cell retraction can occur. The process of cell spreading has been described in detail elsewhere (Aber-crombie and Harris, 1958).

Cell adhesion and spreading are influenced by the physico-chemical characteristics of the underlaying solid surface (Har-ris, 1973). Substratum surface free energy is related to cell spreading, as illustrated in Fig. 6 (Schakenraad *et al.*, 1986). Poor spreading on hydrophobic substrata and good spreading on hydrophilic substrata can be observed in both the absence and presence of preadsorbed serum proteins. Apparently the substratum characteristics shine through the adsorbed proteins toward adhering and spreading cells. An explanation for this phenomenon can be:

1. Cells can reach the underlying substratum by pseudo-podia protruding through the preadsorbed protein layer.
2. Cells consume preadsorbed proteins to make direct contact.
3. The substratum characteristics are reflected in the com-position and conformation of adsorbed proteins, thus

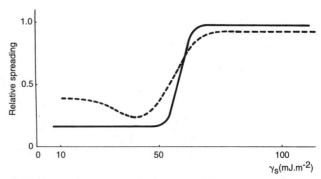

FIG. 6. Cell spreading as a function of substratum surface free energy (γ_s, wettability). Dotted line represents cell spreading in the absence of proteins. Solid line represents cell spreading in the presence of serum proteins.

presenting different molecular groups to adhering and spreading cells.

On the other hand, it was observed that adhesion and spreading decrease on very high-surface free-energy substrata (e.g., hydrogels) (van Wachem *et al.*, 1985).

Another parameter involved in cell adhesion is surface charge. Some authors report that negatively charged cells adhere with smaller contact areas to surfaces with higher negative charge (Sugimoto, 1981). Others report increased adhesion between negative cells and negatively charged substrata (Maroudas, 1975). However, these results were obtained from *in vitro* experiments. The high ionic strength of the physiological environment and the rapid establishment of ionic equilibrium indicate that surface electrical properties do not significantly influence the formation of initially adsorbing protein layers or adhering cells.

The surface topography of a biomaterial can be classified according to roughness, texture, and porosity. However, the scale of surface roughness must be kept in mind. Roughness at the level of cell adhesion (1 μm) is different from roughness at the level of protein adsorption (50 nm). It has been reported that smooth surfaces are more blood compatible than rough surfaces (Zingg *et al.*, 1982). Rough surfaces indeed promote adhesion.

Porosity is used on a large scale to promote anchorage of biomaterials to surrounding tissue. Von Recum's group described an optimal biocompatability using pore sizes of 1–2 μm. Smaller pore sizes caused poor adhesion and increased inflammatory response with little collagen formation. Larger pore sizes did allow ingrowth and anchorage, but caused a more severe foreign body reaction (Campbell and von Recum, 1989). Grooved substrata were found to induce a certain amount of cellular orientation and locomotion in the direction of the grooves (Brunette, 1986a). Applying grooved substrata will therefore induce cell contact guidance. Even shallow grooves (3 μm) result in cell orientation. If the grooves are wider than approximately 120 μm, no cell alignment is found (Brunette, 1986b).

Mechanical forces around an implant, especially in combination with a rough surface, induce abundant formation of fibrous tissue, owing to the constant irritation of the cells. Firm fixation of an implant will reduce movement and thus prevent excessive fibrous tissue formation. Rod-shaped implants with a diameter of approximately 1 mm appear to cause the least mechanical stress and thus are most biocompatible with regard to fibrous tissue formation (Matlaga *et al.*, 1976).

The texture of an implant surface and its morphology can be adapted to the clinical purpose of the biomaterial by such approaches as changing the fabrication process (e.g., woven, knitted, fibrous, grooved, veloured, smooth). It is thus obvious that often we cannot meet the theoretically optimal form, texture, and roughness, but that we have to adapt these parameters to the biomaterial's ultimate purpose.

Cellular locomotion can be directed by various gradients in the cell environment (e.g., chemotaxis, galvanotaxis, haptotaxis). For example, after implantation of a biomaterial, granulocytes are attracted by a negative oxygen gradient. Fibroblasts are attracted by agents produced by macrophages.

THERMODYNAMIC ASPECTS OF ADHESION AND SPREADING

Thermodynamically, the process of adhesion and spreading of cells from a liquid suspension onto a solid substrate can be described by (Schakenraad *et al.*, 1988):

$$\Delta F_{adh} = \gamma_{cs} - \gamma_{cl} - \gamma_{sl}$$

in which ΔF_{adh} is the interfacial free energy of adhesion, γ_{cs} is the cell–solid interfacial free energy, γ_{cl} is the cell–liquid interfacial free energy and γ_{sl} is the solid–liquid interfacial free energy. If $\Delta F_{adh} < 0$, adhesion and spreading are energetically favorable, while if $\Delta F_{adh} > 0$, adhesion and spreading are unfavorable. Figure 7 illustrates the relationship between ΔF_{adh} and substratum surface free energy (or wettability). It should be noted that hydrophobic substrata ($\gamma_s < 40$ erg $\cdot$ cm^{-2}) do not promote adhesion of fibroblasts. Extremely hydrophyllic substrata do not promote adhesion either (e.g., high-energy methacrylates or hydrogels; van Wachem, 1985; Horbett and Schway, 1988).

Only adhesion can be described in a thermodynamic way; cell spreading, a completely different phenomenon, cannot be described in a similar way. Cellular activity such as the production of adhesive proteins and cytoskeleton transport are involved in cell spreading, interfering with our thermodynamic model. Duval *et al.* (1990) have clearly demonstrated on a series of substrata that adhesion and spreading are indeed two separate phenomena. For example, the strength of adhesion to a substratum is not correlated with the area of contact (spreading). A cell spread on Teflon will weakly adhere and is easily removed, while a cell attached to, e.g., glass with one cellular extension will strongly adhere and cannot be easily removed. The type of adhesion site is crucial in this regard.

Other authors found no relation between biomaterial surface properties and cellular adhesion or growth, using

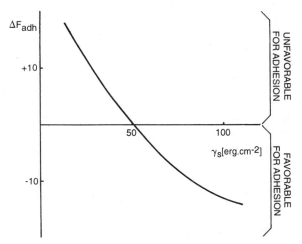

FIG. 7. Interfacial free energy of adhesion (ΔF_{adh}) as a function of substratum surface free energy (γ_s). At low surface free energy surfaces, energy is needed for adhesion; at high surface free energy surfaces, energy is available for adhesion.

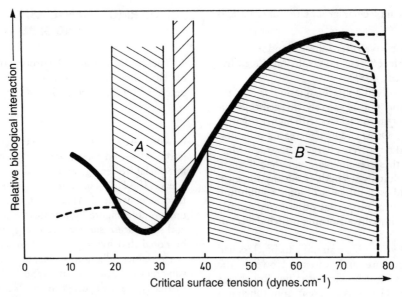

FIG. 8.   Relationship between the biological interactiveness and the critical surface tension of a biomaterial. Area "A" represents a nonadhesive zone between 20 and 30 mJ · m^{-2} critical surface tension. The area marked as "B" represents biomaterials with good adhesive properties.

metals, ceramics, and carbon (Jansen *et al.*, 1989); or found the inverse relation: better adhesion on hydrophobic surfaces (Horbett and Schway, 1988). Some of the discrepancies found can probably be explained by the fact that different parameters (adhesion, spreading, growth) were investigated, different cell types were used, or additional adhesive proteins were used.

Baier (Baier, 1975) described a related phenomenon involving a hypothetical zone of optimal biocompatibility (biological interaction) (Fig. 8). A nonadhesive zone between 20–30 mJ/m^{-2} critical surface tension (A) represents the hydrophobic substrata. The area marked as "B" represents biomaterials with good adhesive properties. It is obvious that there is a similarity with cell spreading (Fig. 6). Only materials with a minimal surface tension (or surface free energy) can provide good adhesive opportunities for cells, from a thermodynamic point of view.

Cell adhesion has also been described by the Langmuir adsorption isotherm (Eckl and Gruler, 1988), in which cellular adsorption and desorption is a rate-controlled process using a high-affinity receptor. The thermodynamic approach could not be proved valid under these circumstances.

BIOCOMPATIBILITY

A biocompatible material has been defined as a material that does not induce an acute or chronic inflammatory response and does not prevent a proper differentiation of implant-surrounding tissues (Williams, 1987) (see also Chapter 5.2 on *in vitro* and Chapter 5.3 on *in vivo* assessment of biocompatibility). It is recognized that some adverse tissue reaction around the implanted biomaterial is inevitable, owing to surgical trauma during insertion. This definition implies that biocompatibility depends on the purpose of the implant. For example, the inside of a Teflon vascular prosthesis is designed to present a blood-compatible surface, which means it should not cause adhesion or activate the clotting or complement cascade. The outside of this prosthesis, however, must firmly attach to surrounding fibrous tissue, but not induce excessive fibrous "hyperplaxia." Thus, two goals are incorporated into a single device.

Terms such as "bioinert" or "bioactive," rather than "biocompatible," more accurately describe the features required of an ideal biomaterial or device. These terms better describe the action or nonaction required from surrounding tissue and thus are related to the choice of material and material characteristics (hydrophilic/hydrophobic). In the early stages of developing a new biomaterial or device, materials scientists should take into account their demands on physicochemical properties and related cell biological requirements, in addition to mechanical properties, polymer–chemical limitations, surgical requirements, and sterilization procedures.

Bibliography

Abercrombie, M., and Harris, E. J. (1958). Interference microscopic studies of cell contacts in tissue culture. *Exp. Cell Res.* 15: 332–345.

Baier, R. E. (1975). Applied chemistry at protein interfaces. *Adv. Chem. Ser.* 145: 1.

Beck, F., and Lloyd, J. B. (1974). *The Cell in Medical Science.* Academic Press, New York, Vol. 1.

Brunette, D. M. (1986a). Spreading and orientation of epithelial cells on grooved substrata. *Exp. Cell Res.* **167**: 203–217.

Brunette, D. M. (1986b). Fibroblasts on micromachined substrata orient hierarchically to grooves of different dimensions. *Exp. Cell Res.* **16**: 11–26.

Buck, C. A. and Horwitz, A. F. (1987). Integrin, a transmembrane glycoprotein complex mediating cell-substratum adhesion. *J. Cell Sci. Suppl.* **8**: 231–250.

Campbell, C. E. and von Recum, A. F. (1989). Microtopography and soft tissue response. *J. Invest. Surg.* **2**: 51–74.

Culp, L. A. (1978). Biochemical Determinants of cell adhesion. in *Current Topics in Membranes and Transport* Vol. II, pp. 327–396, Academic Press, New York.

Curtis, A. S. G., and Pitts, J. D. (1980). *Cell Adhesion and Motility.* Cambridge University Press.

De Robertis, E. D. P., and De Robertis, E. M. F. (1981). *Essentials of Cell and Molecular Biology.* Saunders, Philadelphia.

Derjaguin, B. V. and Landau, L. D. (1941). Theory of the stability of strongly charged lyophobic sols and of the adhesion of strongly charged particles in solution of electrolytes. *Acta Physic Chimica USSR* **14**: 633–662.

Duval, J. L., Letort, M., and Sigot-Luizard, M. F. (1988). Comparative assessment of cell/substratum adhesion towards using an *in vitro* organ culture method and computerized analysis system. *Biomaterials* **9**: 155–161.

Duval, J. L., Letort, M., and Sigot-Luizard, M. F. (1990). Fundamental study of cell migration and adhesion towards different biomaterials with organotypic culture method. in *Advances in Biomaterials* Vol. 9, pp. 93–98, Elsevier, Amsterdam.

Eckl, K. and Gruler, H. (1988). Description of cell adhesion by the Langmuir adsorption isotherm. *Z. Naturforschung.* **43c**: 769–776.

Fowkes, F. M. (1964). Attractive forces at interfaces. *Ind. Eng. Chem.* **56**: 40.

Gospodarowicz, D., Greenburg, G., Foidart, J. M., and Savion, N. (1981). The production and localization of laminin in cultured vascular and corneal endothelial cells. *J. Cell. Physiol.* **107**: 171–183.

Grabowski, E. F., Friedman, L. I., and Leonard, E. F. (1972). Effects of shear rate on the diffusion and adhesion of blood platelets to a foreign surface. *Ind. Chem. Eng. Fundam,* **11**: 224–232.

Harris, A. (1973). Behaviour of cultured cells on substrata of variable adhesiveness. *Exp. Cell Res.* **77**: 285–297.

Horbett, T. A. and Schway, M. B. (1988). Correlations between mouse 3T3 cell spreading and serum fibronectin adsorption on glass and hydroxyethylmethacrylate-ethylmethacrylate copolymers. *J, Biomed. Mater. Res.* **22**: 763–793.

Jansen, J. A., van der Waerden, J. P. C. M., and de Groot, K. (1989). Effect of surface treatment on attachment and growth of epithelial cells. *Biomaterials* **10**: 604–608.

Maroudas, N. G. (1975). Adhesion and spreading of cells on charged surfaces. *J. Theor. Biol.* **49**: 417–424.

Matlaga, B. F., Yasenchak, L. P. and Salthouse, T. N. (1976). Tissue response to implanted polymers: The significance of sample shape. *J. Biomed. Mat. Res.* **10**: 391–397.

Rapraeger, A., Jalkanen, M., and Bernfield, M. (1986). Cell surface proteoglycan associates with the cytoskeleton at the basolateral cell surface of mouse mammary epithelial cells. *J. Cell Biol.* **103**: 2683–2696.

Ruoslahti, E. and Pierschbacher, M. D. (1987). New perspectives in cell adhesion: RGD and integrins. *Science* **238**: 491–497.

Schakenraad, J. M., Busscher, H. J., Wildevuur, C. R. H., and Arends, J. (1986). The influence of substratum surface free energy on growth and spreading of human fibroblasts in the pres-

ence and absence of serum proteins. *J. Biomed. Mater. Res.* **20**: 773–784.

Schakenraad, J. M., Busscher, H. J., Wildevuur, C. R. H., and Arends, J. (1988). Thermodynamic aspects of cell spreading on solid substrata. *Cell Biophysics* **13**: 75–91.

Strong, A. B., Stubley, G. D., Chang, G., and Absolom, D. R. (1987). Theoretical and experimental analysis of cellular adhesion to polymer surfaces. *J. Biomed. Mater. Res.* **21**: 1039–1055.

Sugimoto, Y. (1981). Effect on the adhesion and locomotion of mouse fibroblasts by their interacting with differently charged substrates. *Exp. Cell Res.* **135**: 39–45.

van Oss, C. J., Good, R. J., and Chaudhury, M. K. (1986). The role of van der Waals forces and hydrogen bonds in hydrophobic interactions between biopolymers and low energy surfaces. *Interface Sci.* **111**: 378–390.

van Wachem, P. B., Beugeling, T., Feijen, J., Bantjes, A., Detmers, J. P., van Aken, W. G. (1985). Interactions of cultured human endothelial cells with polymeric surfaces of different wettabilities. *Biomaterials* **6**: 403–408.

Verwey, E. J. W., and Overbeek, J. H. G. (1948). *Theory of Stability of Lyophobic Colloids.* Elsevier, Amsterdam.

Williams, D. F. (1987). *Definitions in Biomaterials.* Elsevier, Amsterdam.

Zingg, W., Neumann, A. W., Strong, A. B., Hum, O. S., and Absolom, D. R. (1982). Effect of surface roughness on platelet adhesion under static and under flow conditions. *Can. J. Surg.* **25**: 16.

3.4 TISSUES

Frederick J. Schoen

The previous chapter by Schakenraad emphasized some aspects of cell structure, function, and interaction with materials. Here, we describe how the functional structure of cells, tissues, and organs are adapted to the complex yet highly organized and regulated activities that constitute mammalian physiology. The underlying principle of organization of biological systems is that *structure is adapted to perform a specific function.* (Alberts *et al.,* 1989; Borysenko and Beringer, 1989; Darnell *et al.,* 1990; Cormack, 1987; Fawcett, 1986; Weiss, 1989; Wheater *et al.,* 1987). The key concepts of functional tissue organization include:

1. Compartmentalization of structure into intracellular and extracellular regions having well-defined purposes, and the role of membranes in providing barrier between these regions.
2. Cellular specialization, called differentiation.
3. Grouping of cell types into the basic tissues.
4. Partition and integration of specialized tissues into functional units called organs.
5. The relative capabilities of different types of cells to undergo regeneration following tissue injury.
6. Regulation and coordination of body function through communication among cells, tissues, and organs.

This chapter summarizes the concepts of structural and functional adaptation at the subcellular, cellular, tissue, and organ levels, and describes the methods by which cells and tissues are studied. The microscopic study of tissue structure, called histology, will be emphasized.

ORGANIZATION OF CELLS AND TISSUES

Cell and Tissue Compartmentalization

The structural elements of the body are cells and the extracellular matrix they secrete. Although connective tissue cells are particularly active in production of extracellular matrix,

virtually all cells secrete extracellular matrix, whose molecules are constantly remodeled.

The cell is a dynamic membranous sac with a complex set of surface structures and internal compartments that accomplish the activities of life under the direction of its genes, composed of deoxyribonucleic acid (DNA) (Fig. 1). The partitioning of the cell interior into membrane-bound units allows material to be segregated into regions of different chemical environment that can be modified by selective molecular and ionic transport. This allows regionally specialized functions. Thus, cells are composed of distinct compartments of aqueous solutions of complex proteins (including enzymes), called or-

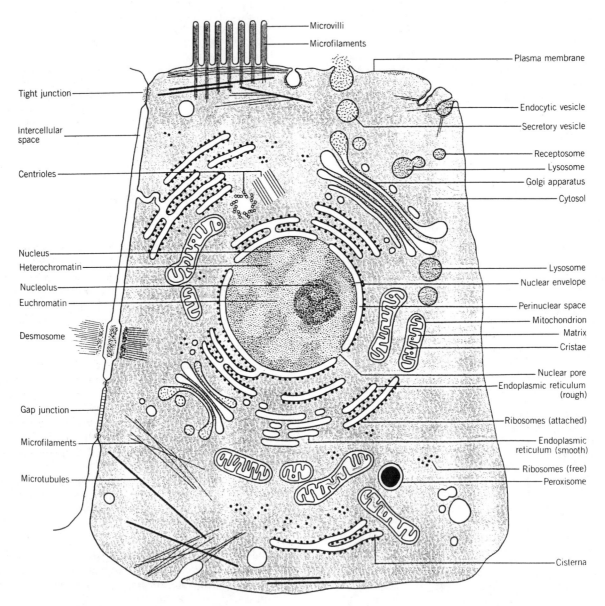

FIG. 1.   A typical cell, demonstrating organelles and their distribution, as well as several specializations of the cell membrane. (From F. Sheeler and D. E. Bianchi, *Cell and Molecular Biology*, 3rd ed. Copyright © 1987 John Wiley & Sons, Inc. Reprinted by permission of John Wiley & Sons, Inc.)

ganelles, which are surrounded by membranes of selective permeability. All cells are separated from the external environment by a limiting membrane, the plasmalemma, which acts as a selective barrier, recognizing and admitting some molecules and ions, while excluding others.

The relative amount and types of organelles reflect and support the specific functions of the cell. The most important organelles of almost all human body cells are the nucleus (the location of genetic material), mitochondria (the major sites of breakdown of macromolecules for energy generation), endoplasmic reticulum (the site of synthesis of proteins and glycoproteins used both internally and for export), the Golgi complex (which modifies synthesized compounds, especially by adding carbohydrates to proteins and packing proteins into membranes for export), and lysosomes (which usually contain proteolytic enzymes for intracellular digestion of damaged cellular components or material taken up from the environment). The remainder of the cell interior is an aqueous solution of proteins known as the cytosol. In the cytosol is the cytoskeleton, a complex network of protein filaments (including actin and myosin) that provides structural support for all cells, and in some cells is specialized for contraction (it is particularly well developed in muscle cells) or motility (in macrophages and sperm). The cytoskeleton also provides anchoring points for other cellular structures.

The plasmalemma and membranes surrounding the organelles are complex structures composed of a lipid bilayer integrated with structural and functional proteins. Although it is generally similar to organellar membranes, the plasmalemma has different lipids, proteins, and carbohydrate groups that project from the cell surface; some serve as receptors capable of selectively linking with substances outside the cell. Many important cell functions are mediated by receptors, including phagocytosis, antibody production, antigen recognition, cell–cell and cell–extracellular matrix interactions, and recognition of biochemical signal molecules (e.g., hormones or other mediators) that impinge on the cell from the external environment. Protein distribution in the plasmalemma is heterogeneous, thereby focusing membrane activities in specific sites.

The plasmalemma has morphologic specializations that include junctional complexes between cells that are in close proximity. For example, one such complex type is the gap junction, which permits the transport from cell to cell of substances of very low molecular weight, including regulatory molecules such as cyclic adenosine monophosphate (cAMP) and ions such as Ca^{2+}. Gap junctions are prominent in epithelia (especially liver), nerve cells, smooth muscle cells, and cardiac muscle cells. In contrast, tight junctions have lateral cell specializations that limit the passage of substances through the spaces separating adjacent cells lining ducts or glands, thereby sealing the lumen from the extracellular environment. Desmosomes are rigid structural anchors between adjacent cells of various types. The plasmalemma also has a glycocalyx, a coat formed by the oligosaccharide side chains of membrane glycolipids and glycoproteins.

Cell–cell contact phenomena are heterogeneous and form a continuum with cell–extracellular matrix interactions. Cell–cell and cell–matrix interactions have a high molecular specificity, requiring initial recognition, physical adhesion, electrical and chemical communication, cytoskeletal reorganization, and/or cell migration. Moreover, adhesion receptors may also act as transmembrane signaling molecules which can transmit information about the environment to the genome, thereby mediating the effects of signals initiated by growth factors or compounds controlling tissue differentiation. Moreover, in both cell–cell and cell–matrix interactions, the other cells or components of the extracellular matrix (ligands) with which cells interact are immobilized and not in solution. However, soluble (secreted) factors also modulate cell–cell communication in the normal and abnormal regulation of tissue growth and maturation. Cell surface adhesion molecules fall into several structural families, including the calcium-dependent cell–cell adhesion molecules, calcium-independent cell–cell adhesion molecules related to the immunoglobulins, the integrin adhesion receptors, and the vascular selectins.

Cell Differentiation

Basic functional attributes of nearly all cells include nutrient absorption and assimilation, respiration, synthesis of macromolecules, growth, and reproduction. However, some cells have specialized capabilities, such as irritability, conductivity, absorption, or secretion of molecules not essential to the cells that synthesize them (e.g., hormones, structural proteins, inflammatory mediators). Multicellular organisms are composed largely of individual cells with marked specialization of structure and function. These differentiated cells collectively define a division of labor in the performance and coordination of complex functions carried out in architecturally distinct and organized tissues.

Differentiated cells have developed new and usually well-defined structural and/or functional characteristics associated with increasing specialization. For example, striated muscle cells have cross-banded filaments that slide on one another, causing cellular contraction; renal tubular epithelial cells have large numbers of mitochondria that generate intracellular fuels to drive pumps performing absorption and secretory functions; and skin keratinocytes, in order to function as a protective barrier, lose virtually all their organelles to become scalelike structures filled with durable, nonliving keratin (a cytoskeletal protein). The immune system is composed of differentiated cells with specialized functional characteristics. When cells are attacked by infectious microorganisms (e.g., bacteria, parasites, and viruses), specialized cells called phagocytes detect, migrate to, and attempt to ingest and destroy these microorganisms or other foreign material. Polymorphonuclear leukocytes (also called PMNs or neutrophils) are particularly active against bacteria, and macrophages react to other types of organisms and foreign material. Lymphocytes, nonphagocytic but critical cells of the immune system, produce antibodies, attack infected or foreign cells, and regulate the immune response.

The adaptive structural changes that occur during cellular differentiation are usually irreversible. Moreover, increased structural and functional specificity is generally accompanied by not only a loss in the capacity for specialization in other ways (i.e., a loss in cell potentiality), but also a loss in the capacity of the cell to divide. For example, the newly fertilized ovum is absolutely undifferentiated and has the capacity to divide extensively, ultimately giving rise to progeny that comprise all the cells of the body. Conversely, nerve cells and heart muscle cells are highly specialized; neither can reproduce itself or regenerate following injury beyond fetal life.

Every diploid cell (i.e., having two sets of chromosomes) within a single organism or within the body has the same complement of genes, which code for all possible characteristics of all cells and extracellular substances (except ova and sperm) that comprise the body (called the genotype). Cellular differentiation involves an alteration in gene expression. The physical and biological characteristics that are expressed in a particular cell constitute the phenotype. The term "modulation" is often used to describe less dramatic functional and morphological changes that occur through differential gene expression as an adaptation to an altered environment in cells that are already differentiated (e.g., activated cells).

Cells exist in a spectrum of physiological states (with resting, or increased or decreased function) reflected by specific morphologic appearances. These changes in activity, largely regarded as normal, highlight the dynamic nature of mammalian functional structure. Cell populations respond to stimuli that call for a great increase in physiological function either by increasing their number (hyperplasia) or increasing their size (hypertrophy), or both. Greatly increased cellular activity under physiologic circumstances includes the dramatically enlarged smooth muscle cells of the myometrium that prepare the uterus for labor, the enlarged cells of the biceps muscles following exercise, the changes that breast lobular epithelium undergoes in preparation for lactation, and the transformation of lymphocytes to large blastic cells which give rise to cells that form antibodies. Nevertheless, the exaggeration of normal functional variation can overlap with certain important pathological processes. For example, markedly hypertrophied heart muscle cells can develop abnormalities in subcellular metabolism which can cause them to fail, and hyperplastic growth can sometimes result in precancerous and ultimately cancerous lesions.

Extracellular Matrix

The functions of the extracellular matrix include: (1) mechanical support for cell anchorage; (2) determination of cell orientation; (3) control of cell growth; (4) maintenance of cell differentiation; (5) scaffolding for orderly tissue renewal; (6) establishment of tissue microenvironment; and (7) sequestration, storage, and presentation of soluble regulatory molecules. Matrix components markedly influence the maintenance of cellular phenotypes, affecting cell shape, polarity, and differentiated function through receptors for specific extracellular matrix molecules on cell surfaces. Cells communicate with various matrix components using surface receptors and peripheral and integral membrane proteins; the resultant changes in cytoskeletal organization and possibly in production of other second messengers can modify gene expression. Conversely, cells produce and secrete matrix molecules, often vectorially. These functions are accomplished by reciprocal instructions between cells and matrix, an interaction termed "dynamic reciprocity."

Extracellular matrices are generally specialized for a particular function, such as strength (tendon), filtration (kidney glomerulus), or adhesion (basement membranes generally). The extracellular matrix consists of large molecules linked together into an insoluble composite. The extracellular matrix is composed of (1) fibers (collagen and elastic) and (2) a largely amorphous interfibrillary matrix (mainly proteoglycans, noncollagenous glycoproteins, solutes, and water). There are two types of matrices: the interstitial matrix and the basal lamina; the major components of each are collagen, a cell-binding adhesive glycoprotein, and proteoglycans. The interstitial matrix is produced by mesenchymal cells and contains fibrillar collagens, fibronectin, hyaluronic acid, and fibril-associated proteoglycans. The basal lamina is produced by overlying parenchymal cells. Basal laminae contain a meshlike collagen framework, laminin, and a large heparan sulfate proteoglycan. To produce additional mechanical strength, the extracellular matrix becomes calcified during the formation of bones and teeth. Although matrix turnover is quite slow in normal mature tissues, type-specific extracellular matrix components are turned over and remodeled in response to appropriate stimuli, such as tissue damage and repair.

The fibrillar components of the extracellular matrix include collagen and elastic fibers. Collagen represents a family of closely related but genetically, biochemically, and functionally distinct glycoproteins, of which at least fifteen different types have been identified. Fibrillar collagens, also called interstitial

TABLE 1 The Basic Tissues: Classification and Examples

Basic Tissues	Examples
Epithelial tissue	
Surface	Skin epidermis, gut mucosa
Glandular	Thyroid follicles, pancreatic acini
Special	Retinal or olfactory epithelium
Connective tissue	
Connective tissue proper	
Loose	Skin dermis
Dense (regular, irregular)	Pericardium, tendon
Special	Adipose tissue
Hemopoietic tissue, blood and lymph	Bone marrow, blood cells
Supportive tissue	Cartilage, bone
Muscle tissue	
Smooth	Arterial or gut smooth muscle
Skeletal	Limb musculature, diaphragm
Cardiac muscle	Heart
Nerve tissue	Brain cells, peripheral nerve

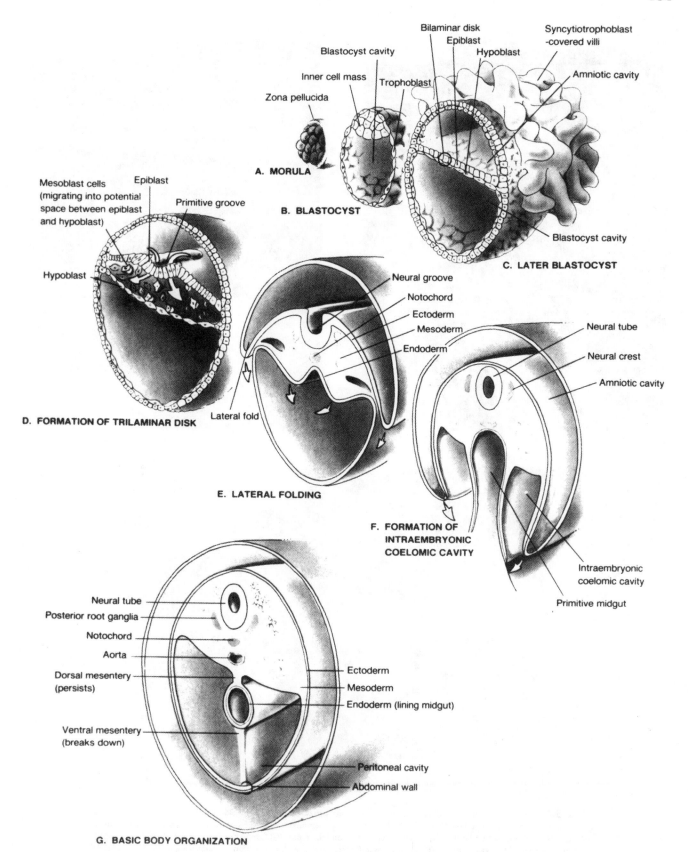

FIG. 2. Early phases of embryological development, demonstrating both essential layering that gives rise to the basic tissues and the early derivation of tubular structures. (From D. H. Cormack, *Ham's Histology*, 9th ed. Lippincott, 1987, with permission.)

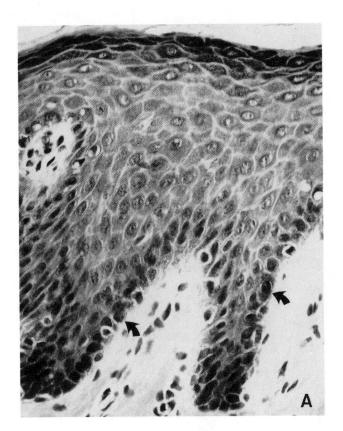

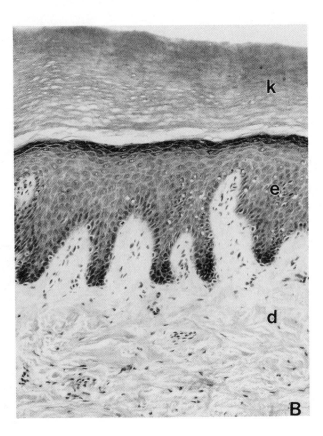

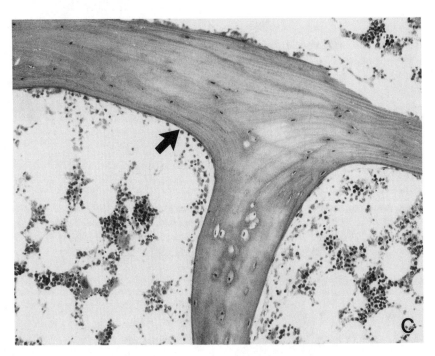

FIG. 3. Photomicrographs of histologic sections of the basic tissues. (A) Epithelium (skin epidermis, cutaneous surface at top). Maturation occurs by proliferation of basal cells (arrows) and their migration to the surface to eventually become nonliving keratin. (B) Full cross section of skin: K, keratin: e, epidermis; and d, dermis. (C) Connective tissue (bone). Vertebral trabecule (arrow) surrounded by bone marrow. (D) Muscle tissue (heart muscle), with cross-striations faintly apparent. (E) Nerve. All stained with hematoxylin and eosin (H&E); A, ×355; B, ×142; C, ×375; D and E, ×349.

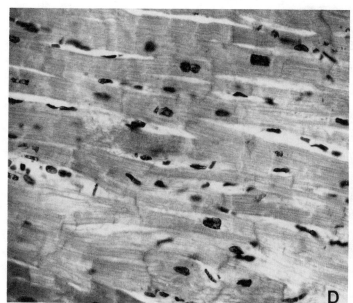

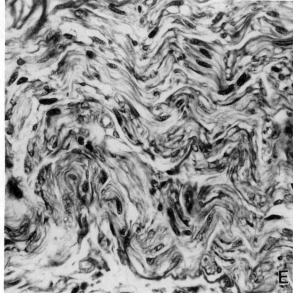

FIG. 3.—continued

collagens (types I, II, III and V), have periodic cross-striations. Type I, the most abundant, is present in most connective tissues. Type II collagen is a major component of hyaline cartilage. The fibrillar collagens provide a major component of tissue strength. Most of the remaining collagen types are nonfibrillar, the most common of which is type IV, a major constituent of all basement membranes. Elastin fibers confer an elastic flexibility to tissues.

In the amorphous matrix, glycosaminoglycans (GAGs), with the exception of hyaluronic acid, are found covalently bound to proteins (as proteoglycans). Proteoglycans serve as major structural elements of the extracellular matrix; some proteoglycans are bound to plasma membranes and appear to be involved in adhesiveness and receptor binding. A set of large noncollagenous glycoproteins is important in binding cells to the extracellular matrix, including fibronectins (the best understood of the noncollagenous glycoproteins), laminins, chondronectin, and osteonectins. Fibronectins, found almost ubiquitously in the extracellular matrix, are synthesized by many different cell types. The circulating form, plasma fibronectin, is produced mainly by hepatocytes. Fibronectin's adhesive character makes it a crucial component of blood clots and of pathways followed by migrating cells. Thus, fibronectin-rich pathways guide and promote the migration of many kinds of cells during embryonic development and wound healing.

Basement membranes provide mechanical support for resident cells, serve as semipermeable barriers between tissue compartments, and act as regulators of cellular attachment, migration, and differentiation. They consist of a discrete zone of amorphous, noncollageneous glycoprotein matrix (including laminin), proteoglycans, and type IV collagen.

Basic Tissues

The basic tissues play specific functional roles and have distinctive microscopic appearances. They have their origins in embryological development, a stage of which is the formation of a tube with three layers in its wall: (1) an outer layer of ectoderm, (2) an inner layer of endoderm, and (3) a middle layer of mesoderm (Fig. 2). The basic tissues in animals and humans have over a hundred distinctly different types of cells, which are separated into four groups: epithelium, connective tissue, muscle tissue, and nerve tissue (Table 1 and Figs. 2 and 3).

Epithelium covers the internal and external body surfaces. It provides a protective barrier (e.g., skin epidermis), and an absorptive surface (e.g., gut lining), and generates internal and external secretions (e.g., endocrine and sweat glands, respectively). Epithelium derives mostly from ectoderm and endoderm, but also from mesoderm.

Structurally heterogeneous and complex epithelia accommodate diverse functions. An epithelial surface can be (1) a protective dry, cutaneous membrane (as in skin); (2) a moist, mucous membrane, lubricated by glandular secretions (digestive and respiratory tracts); (3) a moist, serous membrane, lubricated by serous fluid that derives from blood plasma (peritoneum, pleura, pericardium), lined by mesothelium; and (4) the inner lining of the circulatory system, lubricated by blood or lymph, called endothelium. Epithelial cells play fundamental roles in the directional movement of ions, water, and macromolecules between biological compartments, including absorption, secretion, and exchange. Therefore the architectural and functional organization of epithelial cells includes structurally, biochemically, and physiologically distinct plasma membrane domains that

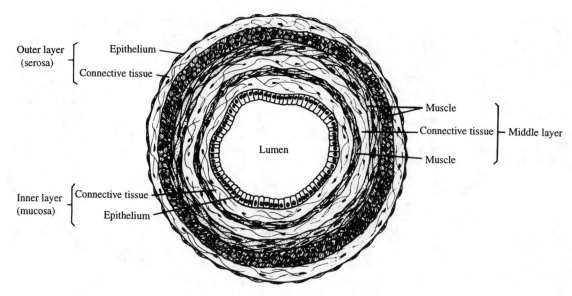

FIG. 4. Organization of tissue layers in the digestive tract (e.g., stomach or intestines). (Reproduced by permission from M. Borysenko and T. Beringer, *Functional Histology,* 3rd ed. Copyright © 1989 Little, Brown, and Co.)

contain region-specific ion channels, transport proteins, enzymes and lipids, and cell–cell junctional complexes that integrate cells into a monolayer. This forms a cellular barrier between biological compartments in organs. Epithelial specializations are best studied by transmission electron microscopy (TEM).

Rich in extracellular matrix, connective tissue supports the other tissues of the body. It develops from mesenchyme, a derivative of mesoderm. Connective tissue cells, such as fibroblasts, produce and maintain the extracellular matrix. Connective tissue also houses nerves and supports blood vessels that serve the specialized tissues. Other types of tissue with varying functions are also of mesenchymal origin, such as dense, ordinary connective tissue, adipose (fat) tissue, blood cells, blood cell-forming tissues, inflammatory cells that defend the body against infectious organisms and other foreign agents, and cartilage and bone.

Muscle cells develop from mesoderm and are highly specialized for contraction. They have the contractile proteins actin and myosin in varying amounts and configuration, depending on cell function. Muscle cells are of three types: smooth muscle, skeletal muscle, and cardiac muscle. The latter two are striated, with sarcomeres, a further specialization of the cytoskeleton. Smooth muscle cells, which have a less compact arrangement of myofilaments, are prevalent in the walls of blood vessels and the gastrointestinal tract. Their contraction regulates blood vessel caliber and proper movement of food and solid waste, respectively.

Nerve tissue, which derives from ectoderm, is highly specialized with respect to irritability and conduction. Nerve cells not only have cell membranes that generate electrical signals called action potentials, but also secrete neurotransmitters (i.e., molecules that trigger adjacent nerve or muscle cells to either transmit an impulse or to contract).

Organs

Several different types of tissues arranged into a functional unit constitute an organ. Many organs have a composite structure in which epithelial cells perform the special work while

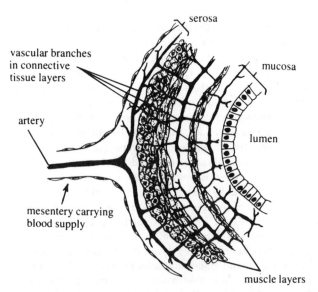

FIG. 5. Vascularization of hollow organs. (Reproduced by permission from M. Borysenko and T. Beringer, *Functional Histology,* 3rd ed. Copyright © 1989 Little, Brown, and Co.)

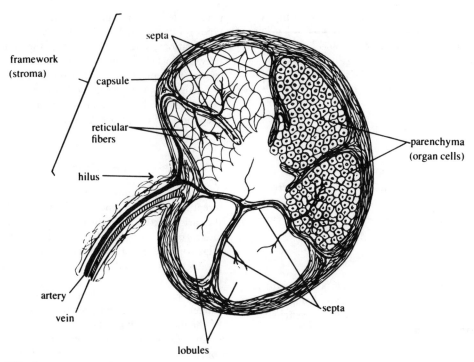

framework
(stroma)

septa

capsule

reticular
fibers

hilus

artery

vein

lobules

parenchyma
(organ cells)

septa

FIG. 6.   Organization of compact organs. (Reproduced by permission from M. Borysenko and T. Beringer, *Functional Histology,* 3rd ed. Copyright © 1989 Little, Brown, and Co.)

connective tissue and blood vessels support and provide nourishment to the epithelium. There are two basic patterns: tubular (or hollow) organs and compact (or solid) organs.

The blood vessels and the digestive, urinary–genital, and respiratory tracts have similar architectures in that each is composed of layers of tissue arranged in a specific sequence. For example, each has an inner coat consisting of an internal lining of epithelium, a middle coat consisting of layers of muscle (usually smooth muscle) and connective tissue, and an external coat consisting of connective tissue and often covered by epithelium. Specific variations reflect organ-specific functional requirements (Fig. 4).

The inner epithelial surface of a tubular organ can be either (1) a protective dry membrane (e.g., skin epidermis); (2) a moist, mucous membrane, lubricated by glandular secretions (e.g., digestive and respiratory tracts); (3) a moist, serous membrane, lubricated by serous fluid that derives from blood plasma, lined by mesothelium (e.g., peritoneum, pleura, pericardium); or (4) the inner lining of the circulatory system, lubricated by blood or lymph, whose surface is composed of endothelium. Thus, blood vessels, considered tubular organs, have an intima (primarily endothelium), an epithelium, a media (primarily smooth muscle and elastin), and an adventitia (primarily collagen). The outside epithelial lining of an organ suspended in a body cavity is called a serosa. In contrast, the outer coat of an organ that blends into surrounding structures is called the adventitia.

The blood supply of an organ comes from its outer aspect. In tubular organs, large vessels penetrate the outer coat, perpendicular to it, and give off branches that run parallel

to the tissue layers (Fig. 5). These vessels divide yet again to give off penetrating branches that course through the muscular layer, and branch again in the connective tissue parallel to the layers. The small blood vessels have junctions (anastomoses) with one another in the connective tissue. These junctions may provide collateral pathways that can allow blood to bypass obstructions. Compact, solid organs have an extensive connective tissue framework, surrounded by a dense, connective tissue capsule (Fig. 6). Such organs have a hilus or area of thicker connective tissue where blood vessels and other conduits (e.g., bronchi in the lungs) enter the organ. From the hilus, strands of connective tissue extend into the organ and may divide it into lobules. The remainder of the organ has a delicate structural framework, including supporting cells, extracellular matrix, and vasculature (essentially the "maintenance" or "service core"), which constitutes the stroma.

The dominant cells in specialized tissues comprise the parenchyma (e.g., thyroglobulin-producing epithelial cells in the thyroid, or cardiac muscle cells in the heart). Parenchyma occurs in masses (e.g., endocrine glands), cords (e.g., liver), or tubules (e.g., kidney). Parenchymal cells can be arranged uniformly in an organ, or they may be segregated into a subcapsular region (cortex) and a deeper region (medulla), each performing a distinct functional role. In compact organs, the blood supply enters the hilus and then branches repeatedly to small arteries and ultimately capillaries in the parenchyma. In both tubular and compact organs, veins and nerves generally follow the course of the arteries.

Parenchymal cells are generally less resistant than stroma

TABLE 2 Regenerative Capacity of Cells Following Injury

Category	Normal rate of replication	Response to stimulus/injury	Examples
Renewing/labile	High	Modest increase	Skin, intestinal mucosa, bone marrow
Expanding/stable	Low	Marked increase	Endothelium, fibroblasts, liver cells
Static/permanent	None	No replication; replacement by scar.	Heart muscle cells, nerves

to chemical, physical, or ischemic (i.e., low blood flow) injury. Moreover, when an organ is injured, the underlying stroma must be present to permit orderly replacement of parenchymal cells, for cells capable of regeneration.

Cell Regeneration

The turnover of cell population generally is strictly regulated. The rates of proliferation of various populations are frequently divided into three categories: (1) renewing (also called labile) cells have continuous turnover, with proliferation balancing cell loss that accrues by death or physiological depletion; (2) expanding (also called stable) cells, normally having a low rate of death and replication, retain the capacity to divide following stimulation, and (3) static (also called permanent) cells have no normal proliferation, and have lost their capacity to divide. The relative replicative and regenerative capacity of various cells following injury is summarized in Table 2.

In renewing (labile) cell populations (e.g., skin, intestinal epithelium, bone marrow), cells that are less differentiated than the cells they are meant to replace (often called stem cells) proliferate to form "daughter" cells that can become differentiated. A particular stem cell produces many daughter cells, and, in some cases, several different kinds of cells can arise from a common multipotential ancestor cell (e.g., bone marrow multipotential cells lead to several different types of blood cells). In epithelia, renewing cells are at the base of the tissue layer, away from the surface; differentiation and maturation occur as the cells move toward the surface (Fig. 3A).

In expanding (stable) populations, cells can increase their rate of replication in response to suitable stimuli. Stable cell populations include glandular epithelial cells, liver, fibroblasts, smooth muscle cells, osteoblasts and vascular endothelial cells. In contrast, permanent (static) cells have virtually no normal mitotic capacity and, in general, cannot be induced to regenerate. In labile or stable populations, cells that die are generally replaced by new ones of the same kind. However, since more specialized (i.e., permanent) cells are not easily regenerated, injury to such cells is repaired by the formation of granulation tissue, which evolves gradually into a fibrotic connective tissue

patch called a scar. The inability to regenerate certain tissue types results in a deficit that in certain cases can have clinical ramifications, since the function of the damaged tissue is irretrievably lost. For example, an area of dead heart muscle cells cannot be replaced by viable ones; the necrotic (dead) area is repaired by scar tissue, which not only has no contractile potential, but the remainder of the heart muscle must assume the workload of the lost tissue.

Cell Communication

Integrated systems coordinate and regulate the growth, differentiation, and metabolism of diverse cells and tissues so that the various activities of the body are coherent. Communication and regulation among cells can be local through direct intercellular connections or chemicals, or long range by extracellular products, such as endocrine hormones and soluble peptide mediators.

An important mechanism of communication (short or long range) is through chemical signals. There are three major types of chemical signaling: (1) Chemical mediators are secreted by many cells; some are taken up or destroyed rapidly, and therefore only act on cells in the immediate environment, other can act long range. (2) Hormones, produced and secreted by specialized and grouped endocrine cells, travel through the bloodstream to influence distant target cells. (3) Neurotransmitters are very short-range chemical mediators which act at specialized junctions between nerve cells (called synapses) or between nerve and muscle cells (called neuromuscular junctions).

Most chemical signals influence specific target cells that have receptors for the signal molecules, either by altering the properties or rates of synthesis of existing proteins, by initiating the synthesis of new proteins, or by acting to accomplish an immediate function, such as secretion, electrical depolarization, or contraction. Thus, target cells are adapted in two ways: (1) they have a distinct set of receptors capable of responding to a complementary set of chemical signals, and (2) they are programmed to respond to each signal in a characteristic way. The distinction is made among effects which are due to (1) a response to a cell's own secreted products (autocrine stimulation), (2) a response to secreted products of another cell in the vicinity (paracrine stimulation), or (3) a response to secretory products originating at a distance and travelling to the target cell(s) via the circulation (endocrine stimulation).

Factors in the extracellular medium control a cell's response through receptors. However, many hormones and other extracellular chemical signals are not soluble in lipids and therefore cannot diffuse across the cell membrane to interact with intracellular receptors. Therefore, a receptor protein on the surface of the target cell, or in its nucleus or cytosol, has a binding site with high affinity for a signaling substance that initiates a process called signal transduction. Processing of the resulting information activates an enzyme that generates a short-lived increase of intracellular signalling compound, termed a "second messenger." Second messenger molecules control the functions of an enormous variety of intracellular proteins by altering their activity (e.g., enzymes). Some receptors, however,

TABLE 3 Techniques for Studying Cells and Tissues[a]

Technique	Purpose
Gross examination	Overall specimen configuration; many diseases and processes can be diagnosed at this level
Light microscopy (LM)	Study overall microscopic tissue architecture and cellular structure; special stains for collagen, mucin, elastin, organisms, etc. are available.
Transmission electron microscopy (TEM)	Study ultrastructure (fine structure) and identify cells and their organelles and environment
Scanning electron microscopy (SEM)	Study topography and structure of surfaces
Enzyme histochemistry	Demonstrate the presence and location of enzymes in gross or microscopic tissue sections
Immunohistochemistry	Identify and locate specific molecules, usually proteins, for which a specific antibody is available
In situ hybridization	Localizes specific DNA or RNA in tissues to assess tissue identity or recognize a cell gene product
Microbiologic cultures	Diagnose the presence of infectious organisms
Morphometric studies (at gross, LM or TEM levels)	Quantitate the amounts, configuration, and distribution of specific structures
Chemical, biochemical, and spectroscopic analysis	Assess concentration of molecular or elemental constituents
Energy-dispersive X-ray analysis (EDXA)	Perform site-specific elemental analysis on surfaces
Autoradiography (at LM or TEM levels)	Locate the distribution of radioactive material in sections

[a]Modified by permission from F. J. Schoen, *Interventional and Surgical Cardiovascular Pathology: Clinical Correlations and Basic Principles*, Saunders, 1989.

do not use second messengers, but rather directly modify the activity of cytoplasmic proteins by regulating their phosphorylation and dephosphorylation. The appropriate target cells respond with great selectivity to light, transmembrane potential, carbohydrates, small amines, large proteins, and lipids. Thus, the plasma membrane translates extracellular signals into a form intelligible to the limited, conserved system of intracellular controls.

Each type of receptor includes both a binding site that recognizes its own special ligand, and a signalling domain that determines the intracellular control pathway along which its information will be sent. Closely related receptors can exert a wide variety of regulatory effects. Diverse extracellular influences thereby feed information into cells through relatively few but ubiqitous signaling pathways and intracellular messengers

that control an enormous variety of responses. Normal cell growth is controlled by the opposing effects of growth stimulators and growth inhibitors (growth factors). Some of these are polypeptides present in serum or produced by cells locally, which stimulate or retard cell growth.

Pathology

As a speciality of medicine, pathology includes not only the diagnosis of disease but also study of the mechanisms by which abnormalities at the various levels of microscopic and submicroscopic structure lead to the manifestations of macroscopic disease, thereby permitting scientifically based treatment. Disease is usually caused by environmental influences (deleterious physical or chemical stimuli), intrinsic genetic or other defects, or exaggeration of normal physiologic processes in individual cells. Cells may (1) be injured or die, (2) become hypo/hyperactive, or (3) become hyperactive with bizarre growth patterns (cancer). Lethal cell injury (cell death) is the permanent cessation of the life functions of a cell and has two forms: necrosis and apoptosis.

Necrosis comprises cell death with consequent loss of function. The body treats necrotic tissue as a foreign body and attempts to remove it by an inflammatory response. In some instances, the dead tissue is replaced by tissue of the same type by regeneration. If this cannot be accomplished, necrotic tissue is replaced by unspecialized connective tissue (fibrous repair by granulation tissue) to form a scar. In contrast, apoptosis (programmed cell death) is the chief normal mechanism of the body for eliminating unwanted cell populations. In contrast to necrosis, apoptosis induces neither an inflammatory nor a connective tissue response.

TECHNIQUES FOR ANALYSIS OF CELLS AND TISSUES

There are a number of techniques available to observe cells directly in the living state in culture systems, which can be extremely useful in investigating the structure and functions of isolated cell types. Cells in culture (*in vitro*) can often continue to perform some of the normal functions they do in the body (*in vivo*). Through measurement of changes in secreted products under different conditions, for example, culture methods can be used to study how cells respond to certain stimuli. However, since cells in culture do not have the usual intercellular organizational environment, alterations from normal physiological function can be present.

Techniques commonly used to study the structure of either normal or abnormal tissues, and the purpose of each mode of analysis, are summarized in Table 3. The most widely used technique for examining tissues is light microscopy, which is described in the following paragraphs. Details of other useful procedures are available (Schoen, 1989).

Light Microscopy

The conventional light microscopy technique involves obtaining the tissue sample, followed by fixation, paraffin embed-

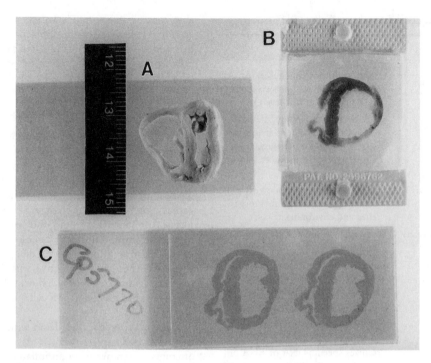

FIG. 7. Tissue processing steps for histology. (A) Tissue section. (B) Paraffin block. (C) Resulting histologic section.

TABLE 4 Stains for Light Microscopic Histology[a]

To demonstrate	Stain
Overall morphology	Hematoxylin and eosin (H&E)
Collagen	Masson's trichrome
Elastin	Verhoeff-van Gieson
Glycosoaminoglycans (GAGs)	Alcian blue
Collagen–elastic–GAGs	Movat
Bacteria	Gram
Fungi	Methenamine silver or periodic acid-Schiff (PAS)
Iron	Prussian blue
Calcium phosphates (or calcium)	von Kossa (or alizarin red)
Fibrin	Lendrum or phosphotungstic acid hematoxylin (PTAH)
Amyloid	Congo red
Inflammatory cell types	Esterases (e.g., chloroacetate esterase for neutrophils, nonspecific esterase for macrophages)

[a]Reproduced by permission from F. J. Schoen, *Interventional and Surgical Cardiovascular Pathology: Clinical Correlations and Basic Principles,* Saunders, 1989.

ding, sectioning, mounting on a glass slide, staining, and examination. Photographs of conventional tissue sections taken through a light microscope (photomicrographs) are illustrated in Fig. 3. Photographs of a tissue sample, paraffin block, and resulting tissue section on a glass slide are shown in Fig. 7. The key processing steps are summarized in the following paragraphs.

Tissue Sample

The tissue is obtained by either surgical excision (removal), biopsy (sampling), or autopsy (postmortem examination). A sharp instrument is used to remove and dissect the tissue to avoid distortion from crushing. Specimens should be placed in fixatives as soon as possible after removal.

Fixation

To preserve the structural relationships among cells, their environment, and subcellular structures in tissues, it is necessary to cross-link and preserve (i.e., fix) the tissue in a permanent state. Fixative solutions prevent degradation of the tissue when it is separated from its source of oxygen and nutrition (i.e., autolysis) by coagulating (i.e., cross-linking, denaturing, and precipitating) proteins. This prevents cellular hydrolytic enzymes, which are released when cells die, from degrading tissue components and spoiling tissues for microscopic analysis. Fixation also immobilizes fats and carbohydrates, reduces or

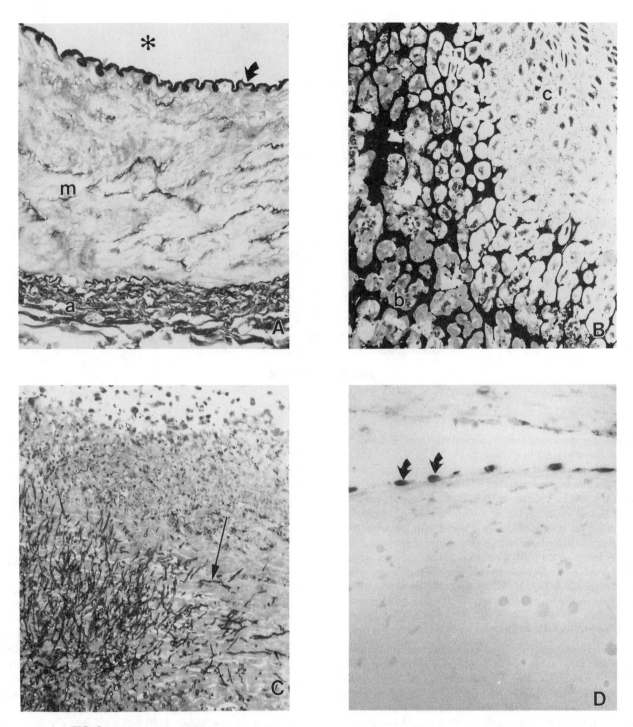

FIG. 8. Photomicrographs of tissue highlighted by special stains and other histologic techniques. (A) Elastin stain of artery [elastin (arrow) is black]. The lumen of the vessel (asterisk), media (m), and adventitia (a) are noted. (B) von Kossa stain of epiphyseal growth plate of bone (calcium phosphates are black). Cartilage (c) and calcified bony trabeculae (b) are noted. (C) Fungal stain of infected heart valve (individual fungal organism designated by arrow). (D) Immunohistochemical staining for factor VIII-related antigen (von Willebrand factor), a specific marker for endothelial cells (reaction product dark, endothelial cells designated by arrows). (E) Enzyme histochemical staining for alkaline phosphatase in bovine pericardial bioprosthetic heart valve (reaction products dark, designated by arrow).

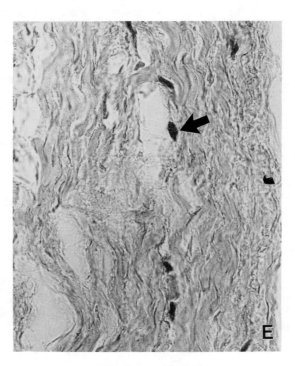

FIG. 8.—continued

eliminates enzymic and immunological reactivity, and kills microorganisms present in tissues.

A 37% solution of formaldehyde is called formalin; thus, 10% formalin is approximately 4% formaldehyde. This solution is the routine fixative in pathology for light microscopy. For TEM and scanning electron microscopy (SEM), glutaraldehyde preserves structural elements better than formalin. Adequate fixation in formalin and/or glutaraldehyde requires tissue samples less than 1.0 and 0.1 cm, respectively, in largest dimension. For adequate fixation, the volume of fixative into which a tissue sample is placed should generally be at least 5–10 times that of the tissue block.

Dehydration and Embedding

In order to support the specimen during sectioning, specimen water (approximately 70% of tissue mass) must be replaced by paraffin wax or other embedding medium, such as glycolmethacrylate. This is done through several steps, beginning with dehydration of the specimen through increasing concentrations of ethanol (eventually to absolute). However, since alcohol is not miscible with paraffin, xylol (an organic solvent) is used as an intermediate solution.

Following dehydration, the specimen is soaked in molten paraffin and placed in a mold larger than the specimen, so that tissue spaces originally containing water, as well as a surrounding cube, are filled with wax. The mold is cooled, and the resultant solid block containing the specimen can then be easily handled.

Sectioning

Tissue specimens are sectioned on a microtome, which has a blade similar to a single-edged razor blade, that is advanced through the specimen block. The shavings are floated on water and picked up on glass slides. Sections for light microscopic analysis must be thin enough to both transmit light and avoid superimposition of various tissue components. Typically sections are approximately 5 μm thick—slightly thicker than a human hair, but thinner than the diameter of most cells. If thinner sections are required (e.g., for TEM analysis, approximately 0.06 μm thick ultrathin sections are necessary), a harder supporting (embedding) medium (usually epoxy plastic) and a correspondingly harder knife are used (usually diamond). Section for TEM analysis are cut on an ultramicrotome. Because the conventional paraffin technique requires overnight

FIG. 9.   Photomicrographs taken in an electron microscope. (A) Myocardial biopsy, showing capillary lumen (*), endothelial cell (e), and heart muscle cell (hm) with sarcomeres (between open arrows). (B) Porcine bioprosthetic heart valve tissue, demonstrating fibroblast, including nucleus (nu) and cytoplasm (c), surrounded by collagen fibrils (cf). In (A) bar = 2 μm; in (B) bar = 0.7 μm.

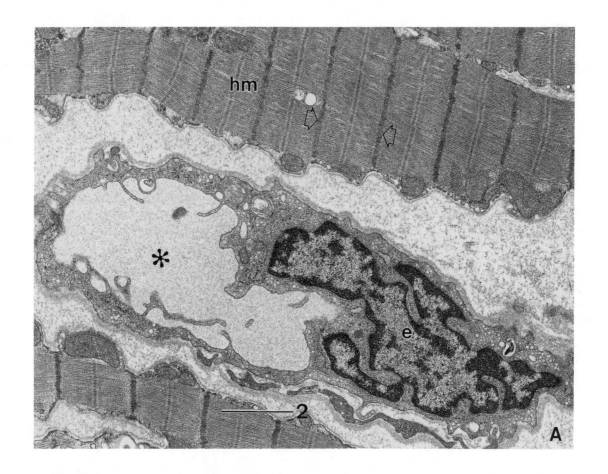

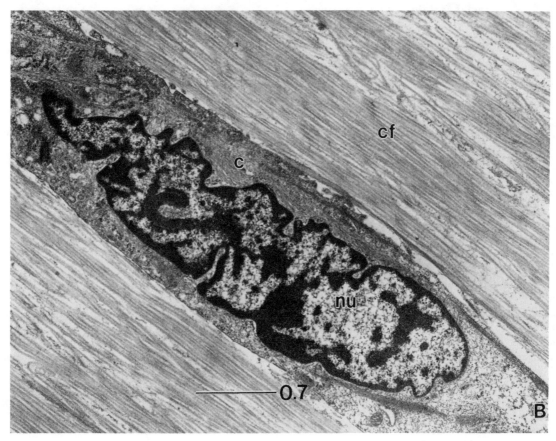

processing, frozen sections can be used to render an immediate diagnosis (e.g., during a surgical procedure that might be modified according to the diagnosis). In this method, the specimen itself is frozen, so that the solidified internal water acts as a support medium, and sections are then cut in a cryostat (i.e., a microtome in a cold chamber). Although frozen sections are extremely useful for immediate tissue examination, the quality of the appearance is inferior to that obtained by conventional fixation and embedding methods.

Staining

Tissue components have no intrinsic contrast and are of fairly uniform optical density. Therefore, in order for tissue to be visible by light microscopy, tissue elements must be distinguished by selective adsorption of dyes (Luna, 1968). Since most stains are aqueous solutions of dyes, staining requires that the paraffin in the tissue section be removed and replaced by water (rehydration). The stain used routinely in histology involves sequential incubation in the dyes hematoxylin and eosin (H&E). Hematoxylin has an alkaline (basic) pH that stains blue-purple; substances stained with hematoxylin are said to be "basophilic" (e.g., cell nuclei). In contrast, substances that stain with eosin, an acidic pigment that colors tissue components pink-red, are said to be "acidophilic" or "eosinophilic" (e.g., cell cytoplasm, collagen). The tissue sections shown in Fig. 3 were stained with hematoxylin and eosin.

Special Staining

There are special staining methods for highlighting components that do not stain well with routine stains (e.g., microorganisms) or for indicating the chemical nature or the location of a specific tissue component (e.g., collagen, elastin (Table 4). There are also special techniques for demonstrating the specific chemical activity of a compound in tissues (e.g., enzyme histochemistry). In this case, the specific substrate for the enzyme of interest is reacted with the tissue; a colored product precipitates in the tissue section at the site of the enzyme. In contrast, immunohistochemical staining takes advantage of the immunological properties (antigenicity) of a tissue component to demonstrate its nature and location by identifying sites of antibody binding. Antibodies to the particular tissue constituent are attached to a dye, usually a compound activated by a peroxidase enzyme, and reacted with a tissue section (immunoperoxidase technique), or the antibody is attached to a compound that is excited by a specific wavelength of light (immunofluorescence). Although some antigens and enzymatic activity can survive the conventional histological processing technique, both enzyme activity and immunological reactivity are often largely eliminated by routine fixation and embedding. Therefore, histochemistry and immunohistochemistry are frequently done on frozen sections, although special preservation and embedding techniques now available often allow immunological methods to be carried out on carefully preserved tissue. Special histologic techniques are illustrated in Fig. 8.

Electron Microscopy

Contrast in the electron microscope depends on relative electron densities of tissue components. Sections are stained with salts of heavy metals (osmium, lead, and uranium), which react differentially with different structures, creating patterns of electron density that reflect tissue and cellular architecture. Examples of electron photomicrographs are shown in Fig. 9.

It is often possible to derive quantitative information from routine tissue sections using various manual or computer-aided methods. Morphometric or stereologic methodology, as these techniques are called, can be extremely useful in providing an objective basis for otherwise subjective measurements (Loud and Anversa, 1984).

Three-Dimensional Interpretation

Interpretation of tissue sections depends on the reconstruction of three-dimensional information from two-dimensional observations on tissue sections that are usually thinner than a single cell. Therefore, a single section may yield an unrepresentative view of the whole. A particular structure (even a very simple one) can look very different, depending on the plane of section. Figure 10 shows how multiple sections must be examined to appreciate the actual configuration of an object or a collection of cells.

Artifacts

Artifacts are unwanted or confusing features in tissue sections that result from errors or technical difficulties in either obtaining, processing, sectioning, or staining the specimen. Recognition of artifacts avoids misinterpretation. The most frequent and important artifacts are autolysis, tissue shrinkage, separation of adjacent structures, precipitates formed by poor buffering or by degradation of fixatives or stains, folds or wrinkles, knife nicks, or rough handling (e.g., crushing) of the specimen.

CONCLUSIONS

Cells, tissues, and organs are adapted to the complex yet highly organized and regulated activities that make up body functions. Key concepts of biological structure–function correlation include compartmentalization, differentiation, the basic tissues, organs, regeneration following injury, and multicellular communication. Although a wide variety of techniques are available for observing tissue structure and function, the microscopic study of tissue slices, called histology, is the most important tool used to investigate functional tissue architecture in clinical or laboratory investigation.

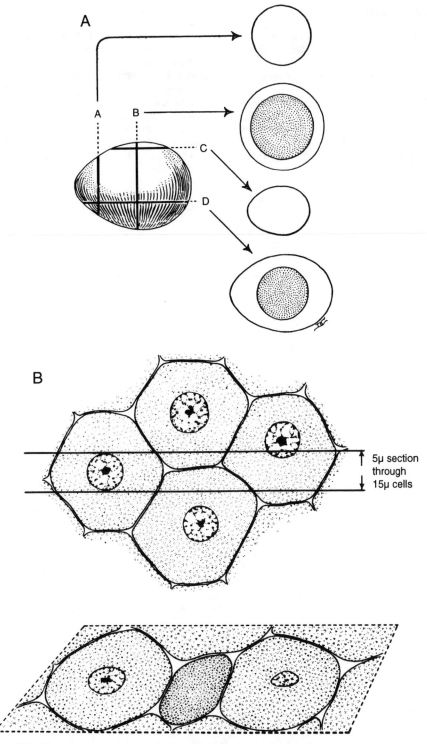

FIG. 10. Considerations for three-dimensional interpretation of two-dimensional information in relatively thin tissue sections. (A) Sections through a subject in different levels and orientations can give different impressions about its structure, here illustrated for a hard-boiled egg. (B) Sections through uniform structures can be misleading. Examination of the section indicated, shown as lower panel in B, would suggest that there were two populations of cells, one without nuclei, and even that cells of the same size had different-sized nuclei. (Reproduced by permission from A. Ham, *Histology,* 7th ed. Copyright © 1974 Lippincott.)

5μ section
through
15μ cells

Bibliography

Alberts, B., Bray, D., Lewis, L., Raff, M., Roberts, K., and Watson, J. D. (1989). *Molecular Biology of the Cell.* 2nd ed., Garland Publ., New York.

Borysenko, M., and Beringer, T. (1989). *Functional Histology,* 3rd ed. Little, Brown, Boston.

Cormack, D. H. (1987). *Ham's Histology,* 9th ed. Lippincott, Philadelphia.

Darnell, J., Lodish, H., and Baltimore, D. (1990). *Molecular Cell Biology,* 2nd ed., Scientific American Books.

Fawcett, D. W. (1986). *Bloom and Fawcett's: A Textbook of Histology.* Saunders, Philadelphia.

Loud, A. V., and Anversa, P. (1984). Morphometric analysis of biological processes. *Lab. Invest.* 50: 250–261.

Luna, M. G. (1968). *Manual of Histologic Staining Methods of the Armed Forces Institute of Pathology,* 3rd ed. McGraw-Hill, New York.

Schoen, F. J. (1989). *Interventional and Surgical Cardiovascular Pathology: Clinical Correlations and Basic Principles.* Saunders, Philadelphia.

Weiss, L. (1989). *Cell and Tissue Biology,* 6th ed. Urban and Schwarzenberg, Baltimore, MD.

Wheater, P. R., Burkitt, H. G., and Daniels, V. G. (1987). *Functional Histology: A Text and Color Atlas,* 2nd ed. Churchill Livingstone, New York.

4

Host Reactions to Biomaterials and Their Evaluation

JAMES M. ANDERSON, ANTHONY G. GRISTINA, STEPHEN R. HANSON, LAURENCE A. HARKER, RICHARD J. JOHNSON, KATHARINE MERRITT, PAUL T. NAYLOR, AND FREDERICK J. SCHOEN

4.1 INTRODUCTION
Frederick J. Schoen

Most implanted medical devices serve their recipients well for extended periods of time by alleviating the conditions for which they were implanted, improving quality of life and, with some device types, enhancing survival. However, some implants and extracorporeal devices ultimately develop complications (adverse patient–device interactions) that constitute device failure, and thereby may cause harm to or death of the patient. All implants interact to some extent with the tissue environment in which they are placed; the biomaterials–tissue interactions encountered most frequently are summarized in the table. The important complications of medical devices are largely based on biomaterials–tissue interactions that include *both* effects of the implant on the host tissues and effects of the host on the implant. Several of the most important interaction mechanisms in clinical and experimental implants and medical devices will be discussed in this section of the book, including inflammation and the "foreign body reaction," immunological sequelae, systemic toxicity, blood–surface interactions, thrombosis, device-related infection, and tumorigenesis. To a great extent, these interactions comprise aberrations of physiological

b. Toxicity
c. Modification of normal healing
    Encapsulation
    Foreign body reaction
    Pannus formation
d. Infection
e. Tumorigenesis
2. Systemic and remote
    a. Embolization
        Thrombus
        Biomaterial
    b. Hypersensitivity
    c. Elevation of implant elements in blood
    d. Lymphatic particle transport
B. Effect of the host on the implant
1. Physical–mechanical effects
    a. Abrasive wear
    b. Fatigue
    c. Stress-corrosion cracking
    d. Corrosion
    e. Degeneration and dissolution
2. Biological effects
    a. Absorption of substances from tissues
    b. Enzymatic degradation
    c. Calcification

Biomaterial–Tissue Interactions

A. Effect of the implant on the host
1. Local
    a. Blood–material interactions
        Protein absorption
        Coagulation
        Fibrinolysis
        Platelet adhesion, activation, release
        Complement activation
        Leukocyte adhesion/activation
        Hemolysis

processes that function as common host defense mechanisms (such as inflammation or thrombosis).

4.2 INFLAMMATION, WOUND HEALING, AND THE FOREIGN BODY RESPONSE
James M. Anderson

Inflammation, wound healing, and foreign body response are generally considered as parts of the tissue or cellular host

TABLE 1 Sequence of Local Events Following Implantation

Injury
Acute inflammation
Chronic inflammation
Granulation tissue
Foreign body reaction
Fibrosis

responses to injury. Table 1 lists the sequence of these events following injury. From a biomaterials perspective, placing a biomaterial in the *in vivo* environment involves injection, insertion, or surgical implantation, all of which injure the tissues or organs involved.

The placement procedure initiates a response to the injury by the body, and mechanisms are activated to maintain homeostasis. The degrees to which the homeostatic mechanisms are perturbed and the pathophysiologic conditions created and resolved are a measure of the host's reaction to the biomaterial and may ultimately determine its biocompatibility. While it is convenient to separate homeostatic mechanisms into blood–material or tissue–material interactions, it must be remembered that the various components or mechanisms involved in homeostasis are present in both blood and tissue and are a part of the physiologic continuum. Furthermore, it must be noted that host reactions can be tissue-dependent, organ-dependent, and species-dependent. Obviously, the extent of injury varies with the implantation procedure.

OVERVIEW

Inflammation is generally defined as the reaction of vascularized living tissue to local injury. Inflammation serves to contain, neutralize, dilute, or wall off the injurious agent or process. In addition, it sets into motion a series of events that may heal and reconstitute the implant site through replacement of the injured tissue by regeneration of native parenchymal cells, formation of fibroblastic scar tissue, or a combination of these two processes.

Immediately following injury, there are changes in vascular flow, caliber, and permeability. Fluid, proteins, and blood cells escape from the vascular system into the injured tissue in a process called exudation. Following changes in the vascular system, which also include changes induced in blood and its components, cellular events occur and characterize the inflammatory response. The effect of the injury and/or biomaterial *in situ* on plasma or cells can produce chemical factors that mediate many of the vascular and cellular responses of inflammation.

Regardless of the tissue or organ into which a biomaterial is implanted, the initial inflammatory response is activated by injury to vascularized connective tissue (Table 2). Since blood and its components are involved in the initial inflammatory responses, blood clot formation and/or thrombosis also occur. Blood coagulation and thrombosis are generally considered humoral responses and may be influenced by other homeostatic mechanisms such as the extrinsic and intrinsic coagulation systems, the complement system, the fibrinolytic system, the kinin-generating system, and platelets. Blood interactions with biomaterials are generally considered under the category of hematocompatibility and are discussed elsewhere in this book.

The predominant cell type present in the inflammatory response varies with the age of the injury. In general, neutrophils predominate during the first several days following injury and then are replaced by monocytes as the predominant cell type. Three factors account for this change in cell type: neutrophils are short lived and disintegrate and disappear after 24–48 hr; neutrophil emigration is of short duration; and chemotactic factors for neutrophil migration are activated early in the inflammatory response. Following emigration from the vasculature, monocytes differentiate into macrophages and these cells are very long lived (up to months). Monocyte emigration may continue for days to weeks, depending on the injury and implanted biomaterial, and chemotactic factors for monocytes are activated over longer periods of time.

The sequence of events following implantation of a biomaterial is illustrated in Fig. 1. The size, shape, and chemical and physical properties of the biomaterial may be responsible for variations in the intensity and duration of the inflammatory or wound-healing process. Thus, intensity and/or time duration of inflammatory reaction may characterize the biocompatibility of a biomaterial.

While injury initiates the inflammatory response, the chemicals released from plasma, cells, and injured tissue mediate the

TABLE 2 Cells and Components of Vascularized Connective Tissue

Intravascular (blood) cells
    Neutrophils
    Monocytes
    Eosinophils
    Lymphocytes
    Basophils
    Platelets
Connective tissue cells
    Mast cells
    Fibroblasts
    Macrophages
    Lymphocytes
Extracellular matrix components
    Collagens
    Elastin
    Proteoglycans
    Fibronectin
    Laminin

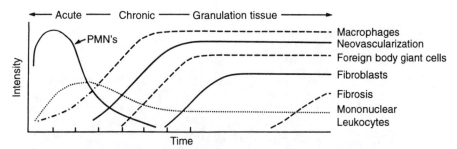

FIG. 1. The temporal variation in the acute inflammatory response, chronic inflammatory response, granulation tissue development, and foreign body reaction to implanted biomaterials. The intensity and time variables are dependent upon the extent of injury created by implantation and the size, shape, topography, and chemical and physical properties of the biomaterial.

response. Important classes of chemical mediators of inflammation are presented in Table 3. Several points must be noted in order to understand the inflammatory response and how it relates to biomaterials. First, although chemical mediators are classified on a structural or functional basis, different mediator systems interact and provide a system of checks and balances regarding their respective activities and functions. Second, chemical mediators are quickly inactivated or destroyed, suggesting that their action is predominantly local (i.e., at the implant site). Third, generally the lysosomal proteases and oxygen-derived free radicals produce the most significant damage or injury. These chemical mediators are also important in the degradation of biomaterials.

TABLE 3 Important Chemical Mediators of Inflammation Derived from Plasma, Cells, and Injured Tissue

Mediators	Examples
Vasoactive amines	Histamines, serotonin
Plasma proteases	
Kinin system	Bradykinin, kallikrein
Complement system	C3a, C5a, C3b, C5b–C9
Coagulation/fibrinolytic system	Fibrin degradation products; activated Hageman factor (FXIIA)
Arachidonic acid metabolites	
Prostaglandins	PGI_2, TxA_2
Leukotrienes	HETE, leukotriene B_4
Lysosomal proteases	Collagenase, elastase
Oxygen-derived free radicals	H_2O_2, superoxide anion
Platelet activating factors	Cell membrane lipids
Cytokines	Interleukin 1 (IL-1); tumor necrosis factor (TNF)
Growth factors	Platelet-derived growth factor (PDGF); fibroblast growth factor (FGF); transforming growth factor (TGF-α or TGF-β)

ACUTE INFLAMMATION

Acute inflammation is of relatively short duration, lasting from minutes to days, depending on the extent of injury. Its main characteristics are the exudation of fluid and plasma

FIG. 2. Acute inflammation with inflammatory exudate and polymorphonuclear leukocytes adjacent to the outer surface of an ePTFE vascular graft. Hematoxylin and eosin stain. ×140.

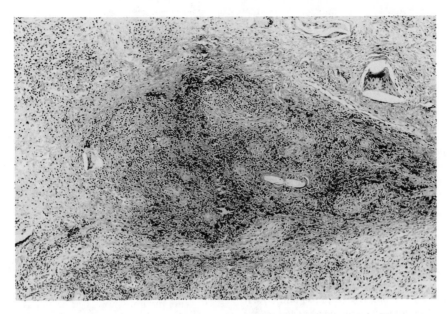

FIG. 3.   Focal chronic inflammation with the presence of lymphocytes and monocytes in the subsynovial capsule from an elbow prosthesis. Polyethylene wear debris is seen as open clefts surrounded by macrophages and foreign body giant cells. The wear debris is present but transparent to light in this histological preparation. Hematoxylin and eosin stain. ×88.

proteins (edema) and the emigration of leukocytes (predominantly neutrophils) (Fig. 2). Neutrophils and other motile white cells emigrate or move from the blood vessels to the perivascular tissues and the injury (implant) site. Leukocyte emigration is assisted by "adhesion molecules" present on leukocyte and endothelial surfaces. The surface expression of these adhesion molecules can be induced, enhanced, or altered by inflammatory agents and chemical mediators. White cell emigration is controlled, in part, by chemotaxis, which is the unidirectional migration of cells along a chemical gradient. A wide variety of exogenous and endogenous substances have been identified as chemotactic agents. Specific receptors for chemotactic agents on the cell membranes of leukocytes are important in the emigration or movement of leukocytes. These and other receptors may also play a role in the activation of leukocytes. Following localization of leukocytes at the injury (implant) site, phagocytosis and the release of enzymes occur following activation of neutrophils and macrophages. The major role of the neutrophils in acute inflammation is to phagocytose microorganisms and foreign materials. Phagocytosis is seen as a three-step process in which the injurious agent undergoes recognition and neutrophil attachment, engulfment, and killing or degradation. In regard to biomaterials, engulfment and degradation may or may not occur, depending on the properties of the biomaterial.

Although biomaterials are not generally phagocytosed by neutrophils or macrophages because of the disparity in size, (i.e., the surface of the biomaterial is greater than the size of the cell), certain events in phagocytosis may occur. The process of recognition and attachment is expedited when the injurious agent is coated by naturally occurring serum factors called "opsonins." The two major opsonins are immunoglobulin G (IgG) and the complement-activated fragment, C3b. Both of these plasma-derived proteins are known to adsorb to biomaterials, and neutrophils and macrophages have corresponding cell membrane receptors for these opsonization proteins. These receptors may also play a role in the activation of the attached neutrophil or macrophage. Owing to the disparity in size between the biomaterial surface and the attached cell, frustrated phagocytosis may occur. This process does not involve engulfment of the biomaterial but does cause the extracellular release of leukocyte products in an attempt to degrade the biomaterial.

Henson has shown that neutrophils adherent to complement-coated and immunoglobulin-coated nonphagocytosable surfaces may release enzymes by direct extrusion or exocytosis from the cell. The amount of enzyme released during this process depends on the size of the polymer particle, with larger particles inducing greater amounts of enzyme release. This suggests that the specific mode of cell activation in the inflammatory response in tissue depends upon the size of the implant and that a material in a phagocytosable form (i.e., powder or particulate) may provoke a different degree of inflammatory response than the same material in a nonphagocytosable form (i.e., film).

CHRONIC INFLAMMATION

Chronic inflammation is less uniform histologically than acute inflammation. In general, chronic inflammation is charac-

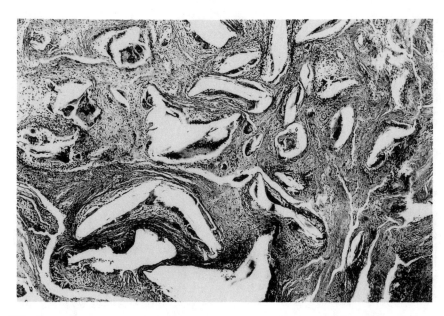

FIG. 4. Foreign body reaction to polyethylene wear debris in the subsynovial capsule of a total knee prosthesis. Foreign body giant cells and macrophages are present at the surfaces of small fragments of polyethylene wear debris and macrophages are present containing metallic wear debris. Hematoxylin and eosin stain. ×56.

terized by the presence of macrophages, monocytes, and lymphocytes, with the proliferation of blood vessels and connective tissue (Fig. 3). It must be noted that many factors can modify the course and histologic appearance of chronic inflammation.

Persistent inflammatory stimuli lead to chronic inflammation. While the chemical and physical properties of the biomaterial in themselves may lead to chronic inflammation, motion in the implant site by the biomaterial may also produce chronic inflammation. The chronic inflammatory response to biomaterials is usually of short duration and is confined to the implant site. The presence of mononuclear cells, including lymphocytes and plasma cells, is considered chronic inflammation, while the foreign body reaction with the development of granulation tissue is considered the normal wound healing response to implanted biomaterials (i.e., the normal foreign body reaction, Fig. 4).

Lymphocytes and plasma cells are involved principally in immune reactions and are key mediators of antibody production and delayed hypersensitivity responses. Their roles in nonimmunologic injuries and inflammation are largely unknown. Little is known regarding humoral immune responses and cell-mediated immunity to synthetic biomaterials. The role of macrophages must be considered in the possible development of immune responses to synthetic biomaterials. Macrophages process and present the antigen to immunocompetent cells and thus are key mediators in the development of immune reactions.

Monocytes and macrophages belong to the mononuclear phagocytic system (MPS), also known as the reticuloendothelial system (RES). These systems consist of cells in the bone narrow, peripheral blood, and specialized tissues. Table 4 lists the tissues that contain cells belonging to the MPS and RES.

The specialized cells in these tissues may be responsible for systemic effects in organs or tissues secondary to the release of components or products from implants through various tissue–material interactions (e.g., corrosion products, wear debris, degradation products) or the presence of implants (e.g., microcapsule or nanoparticle drug delivery systems).

The macrophage is probably the most important cell in chronic inflammation because of the great number of biologically active products it can produce. Important classes of products produced and secreted by macrophages include

TABLE 4 Tissues and Cells of MPS and RES

Tissues	Cells
Implant sites	Inflammatory macrophages
Liver	Kupffer cells
Lung	Alveolar macrophages
Connective tissue	Histiocytes
Bone marrow	Macrophages
Spleen and lymph nodes	Fixed and free macrophages
Serous cavities	Pleural and peritoneal macrophages
Nervous system	Microglial cells
Bone	Osteoclasts
Skin	Langerhans' cells
Lymphoid tissue	Dendritic cells

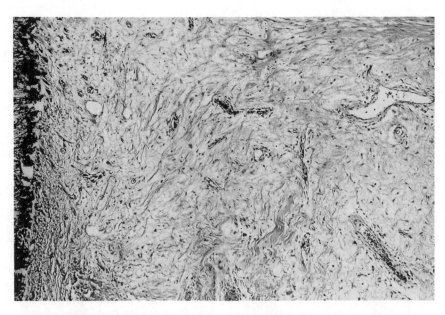

FIG. 5.   Granulation tissue with capillaries, fibroblasts, collagen, and macrophages at the outer surface of an ePTFE vascular graft. A thin, linear array of macrophages and foreign body giant cells is seen at the tissue–biomaterial interface to the left. Hematoxylin and eosin stain. ×70.

neutral proteases, chemotactic factors, arachidonic acid metabolites, reactive oxygen metabolites, complement components, coagulation factors, growth-promoting factors, and cytokines.

Growth factors such as platelet-derived growth factor (PDGF), fibroblast growth factor (FGF), transforming growth factor-β (TGF-β), TGF-α/epidermal growth factor (EGF), and interleukin-1 (IL-1) or tumor necrosis factor (TNF) are important to the growth of fibroblasts and blood vessels and the regeneration of epithelial cells. Growth factors released by activated cells can stimulate production of a wide variety of cells; initiate cell migration, differentiation, and tissue remodeling; and may be involved in various stages of wound healing.

GRANULATION TISSUE

Within 1 day following implantation of a biomaterial (i.e., injury), the healing response is initiated by the action of monocytes and macrophages. Fibroblasts and vascular endothelial cells in the implant site proliferate and begin to form granulation tissue, which is the specialized type of tissue that is the hallmark of healing inflammation. Granulation tissue derives its name from the pink, soft granular appearance on the surface of healing wounds and its characteristic histologic features include the proliferation of new small blood vessels and fibroblasts (Fig. 5). Depending on the extent of injury, granulation tissue may be seen as early as 3–5 days following implantation of a biomaterial.

The new small blood vessels are formed by budding or sprouting of preexisting vessels in a process known as neovascularization or angiogenesis. This process involves proliferation, maturation, and organization of endothelial cells into capillary tubes. Fibroblasts also proliferate in developing granulation tissue and are active in synthesizing collagen and proteoglycans. In the early stages of granulation tissue development, proteoglycans predominate but later collagen, especially type III collagen, predominates and forms the fibrous capsule. Some fibroblasts in developing granulation tissue may have the features of smooth muscle cells. These cells are called myofibroblasts and are considered to be responsible for the wound contraction seen during the development of granulation tissue. Macrophages are almost always present in granulation tissue. Other cells may also be present if chemotactic stimuli are generated.

The wound healing response is generally dependent on the extent or degree of injury or defect created by the implantation procedure. Wound healing by primary union or first intention is the healing of clean, surgical incisions in which the wound edges have been approximated by surgical sutures. Healing under these conditions occurs without significant bacterial contamination and with a minimal loss of tissue. Wound healing by secondary union or second intention occurs when there is a large tissue defect that must be filled or there is extensive loss of cells and tissue. In wound healing by secondary intention, regeneration of parenchymal cells cannot completely reconstitute the original architecture and much larger amounts of granulation tissue are formed that result in larger areas of fibrosis or scar formation.

Granulation tissue is distinctly different from granulomas,

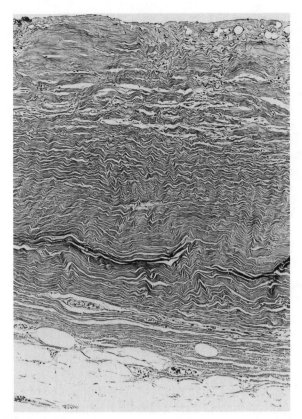

FIG. 6. Dense fibroconnective tissue capsule (fibrous capsule) at the interface of a silicone rubber mammary prosthesis. Note the lack of vessels in the fibrous capsule. The mammary prosthesis was ruptured, and small oval spaces, indicative of silicone gel, are seen at the tissue–material interface at the top of the photomicrograph. Hematoxylin and eosin stain. ×88.

which are small collections of modified macrophages called epithelioid cells that are usually surrounded by a rim of lymphocytes. Langhans' or foreign body-type giant cells may surround nonphagocytosable particulate materials in granulomas. Foreign body giant cells are formed by the fusion of monocytes and macrophages in an attempt to phagocytose the material.

FOREIGN BODY REACTION

The foreign body reaction to biomaterials is composed of foreign body giant cells and the components of granulation tissue. These consist of macrophages, fibroblasts, and capillaries in varying amounts, depending upon the form and topography of the implanted material. Relatively flat and smooth surfaces such as those found on breast prostheses have a foreign body reaction that is composed of a layer of macrophages one to two cells in thickness (Fig. 6). Relatively rough surfaces such as those found on the outer surfaces of expanded poly(tetrafluoroethylene) (ePTFE) vascular prostheses have a foreign body reaction composed of macrophages and foreign body giant cells at the surface (Fig. 7). Fabric materials generally have a surface response composed of macrophages and foreign body giant cells, with varying degrees of granulation tissue subjacent to the surface response.

As previously discussed, the form and topography of the surface of the biomaterial determines the composition of the foreign body reaction. With biocompatible materials, the composition of the foreign body reaction in the implant site may be controlled by the surface properties of the biomaterial, the form of the implant, and the relationship between the surface area of the biomaterial and the volume of the implant. For example, high surface-to-volume implants such as fabrics or porous materials will have higher ratios of macrophages and foreign body giant cells in the implant site than smooth surface implants, which will have fibrosis as a significant component of the implant site.

The foreign body reaction consisting mainly of macrophages and/or foreign body giant cells may persist at the tissue–implant interface for the lifetime of the implant (Fig. 1). Generally, fibrosis (i.e., fibrous encapsulation) surrounds the biomaterial or implant with its interfacial foreign body reaction, isolating the implant and foreign body reaction from the local tissue environment. Early in the inflammatory and wound healing response, the macrophages are activated upon adherence to the material surface.

While it is generally considered that the chemical and physical properties of the biomaterial are responsible for macrophage activation, the subsequent events regarding the activity of macrophages at the surface are not clear. Tissue macrophages, derived from circulating blood monocytes, may coalesce to form multinucleated foreign body giant cells. It is not uncommon to see very large foreign body giant cells containing large numbers of nuclei on the surface of biomaterials. While these foreign body giant cells may persist for the lifetime of the implant, it is not known if they remain activated, releasing their lysosomal constituents, or become quiescent.

FIBROSIS AND FIBROUS ENCAPSULATION

The end-stage healing response to biomaterials is generally fibrosis or fibrous encapsulation. However, there may be exceptions to this general statement (e.g., porous materials inoculated with parenchymal cells or porous materials implanted into bone). As previously stated, the tissue response to implants is in part dependent upon the extent of injury or defect created in the implantation procedure.

Repair of implant sites can involve two distinct processes: regeneration, which is the replacement of injured tissue by parenchymal cells of the same type, or replacement by connective tissue that constitutes the fibrous capsule. These processes are generally controlled by either (1) the proliferative capacity of the cells in the tissue or organ receiving the implant and the extent of injury as it relates to the destruction, or (2) persistence of the tissue framework of the implant site.

The regenerative capacity of cells allows them to be

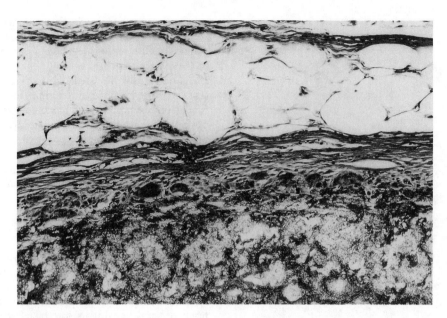

FIG. 7.   Fibrous capsule in the end-stage healing response to an ePTFE vascular graft. A normal focal foreign body reaction (i.e., macrophages and foreign body giant cells) is present at the tissue–ePTFE interface. Fibrous tissue and adipose (fat) tissue encapsulate the ePTFE vascular graft with its interfacial foreign body reaction. Hematoxylin and eosin stain. ×140.

classified into three groups: labile, stable (or expanding), and permanent (or static) cells (see Chapter 3.3). Labile cells continue to proliferate throughout life; stable cells retain this capacity but do not normally replicate; and permanent cells cannot reproduce themselves after birth. Perfect repair with restitution of normal structure can theoretically only occur in tissues consisting of stable and labile cells, while all injuries to tissues composed of permanent cells may give rise to fibrosis and fibrous capsule formation with very little restitution of the normal tissue or organ structure. Tissues composed of permanent cells (e.g., nerve cells, skeletal muscle cells, and cardiac muscle cells) most commonly undergo an organization of the inflammatory exudate, leading to fibrosis. Tissues composed of stable cells (e.g., parenchymal cells of the liver, kidney, and pancreas); mesenchymal cells (e.g., fibroblasts, smooth muscle cells, osteoblasts, and chondroblasts); and vascular endothelial and labile cells (e.g., epithelial cells and lymphoid and hematopoietic cells) may also follow this pathway to fibrosis or may undergo resolution of the inflammatory exudate, leading to restitution of the normal tissue structure.

The condition of the underlying framework or supporting stroma of the parenchymal cells following an injury plays an important role in the restoration of normal tissue structure. Retention of the framework may lead to restitution of the normal tissue structure while destruction of the framework most commonly leads to fibrosis. It is important to consider the species-dependent nature of the regenerative capacity of cells. For example, cells from the same organ or tissue but from different species may exhibit different regenerative capacities and/or connective tissue repair.

Following injury, cells may undergo adaptations of growth and differentiation. Important cellular adaptations are atrophy (decrease in cell size or function), hypertrophy (increase in cell size), hyperplasia (increase in cell number), and metaplasia (change in cell type). Other adaptations include a change by cells from producing one family of proteins to another (phenotypic change), or marked overproduction of protein. This may be the case in cells producing various types of collagens and extracellular matrix proteins in chronic inflammation and fibrosis. Causes of atrophy may include decreased workload (e.g., stress-shielding by implants), and diminished blood supply and inadequate nutrition (e.g., fibrous capsules surrounding implants).

Local and systemic factors may play a role in the wound healing response to biomaterials or implants. Local factors include the site (tissue or organ) of implantation, the adequacy of blood supply, and the potential for infection. Systemic factors may include nutrition, hematologic derangements, glucocortical steroids, and preexisting diseases such as atherosclerosis, diabetes, and infection.

Finally, the implantation of biomaterials or medical devices may be best viewed at present from the perspective that the implant provides an impediment or hindrance to appropriate tissue or organ regeneration and healing. Given our current inability to control the sequence of events following injury in the implantation procedure, restitution of normal tissue structures with function is rare. Current studies directed toward developing a better understanding of the modification of the inflammatory response, stimuli providing for appropriate proliferation of permanent and stable cells, and the appropriate application of growth factors may provide keys to the control of inflammation, wound healing, and fibrous encapsulation of biomaterials.

Bibliography

Anderson, J. M. (1988). Inflammatory response to implants. *ASAIO* **11**(2): 101–107.

Anderson, J. M. (1993). Mechanisms of inflammation and infection with implanted devices. *Cardiovasc. Pathol.* **2**(3)(Suppl.): 33S–41S.

Anderson, J. M., and Miller, K. M. (1984). Biomaterial biocompatibility and the macrophage. *Biomaterials* **5**: 5.

Clark, R. A. F., and Henson, P. M. (eds.) (1988). *The Molecular and Cellular Biology of Wound Repair*. Plenum Publ. New York.

Cotran, R. Z., Kumar, V., and Robbins, S. L. (eds.) (1994). Inflammation and repair. in *Pathologic Basis of Disease*, 5th ed. Saunders, Philadelphia, pp. 51–92.

Deuel, T. F. (1987). Polypeptide growth factors: Roles in normal and abnormal cell growth. *Ann. Rev. Cell Biol.* **3**: 443.

Gallin, J. I., Goldstein, I. M., and Snyderman, R. (eds.) (1992). *Inflammation: Basic Principles and Clinical Correlates*. Raven Press, New York.

Henson, P. M. (1971). The immunologic release of constituents from neutrophil leukocytes: II. Mechanisms of release during phagocytosis, and adherence to nonphagocytosable surfaces. *J. Immunol.* **107**: 1547.

Hunt, T. K., Heppenstall, R. B., Pines, E., and Rovee, D. (eds.) (1984). *Soft and Hard Tissue Repair*. Praeger Scientific, New York.

Johnston, R. B., Jr. (1988). Monocytes and macrophages. *N. Engl. J. Med.* **318**: 747.

Rae, T. (1986). The macrophage response to implant materials. *Crit. Rev. Biocompatibility* **2**: 97.

Spector, M., Cease, C., and Tong-Li, X. (1989). The local tissue response to biomaterials. *Crit. Rev. Biocompatibility* **5**: 269–295.

Thompson, J. A., Anderson, K. D., DiPietro, J. M., Zwiebel, J. A., Zametta, M., Anderson, W. F., and Maciag, T. (1988). Site-directed neovessel formation *in vivo*. *Science* **241**: 1349–1352.

Weissman, G., Smolen, J. E., and Korchak, H. M. (1980). Release of inflammatory mediators from stimulated neutrophils. *N. Engl. J. Med.* **303**: 27.

4.3 IMMUNOLOGY AND THE COMPLEMENT SYSTEM

Richard J. Johnson

The immune system acts to protect each of us from the constant exposure to pathogenic agents, such as bacteria, fungi, viruses, and cancerous cells, that pose a threat to our lives. The shear multitude of structures that the immune system must recognize, differentiate from "self," and mount an effective response against, has driven the evolution of this system into a complex network of proteins, cells, and distinct organs. An immune response to any foreign element involves all of these components, acting in concert, to defend the host from intrusion. Historically, the immune system has been viewed from two perspectives: cellular or humoral. This is a somewhat subjective distinction, since most humoral components (such as antibodies, complement components, and cytokines) are made by cells of the immune system and, in turn, often function to regulate the activity of these same cells. Nonetheless, we will maintain this distinction here.

The cells of the immune system arise from precursors (called stem cells) in the bone marrow (Fig. 1). These cells differ from each other in morphology, function, and the expression of cell surface antigens that have been given the name cluster determinants or CDs (e.g., CD2, CD3, CD4). However, they all share the common feature of maintaining cell surface receptors that help recognize and/or eliminate foreign material. Humoral components of the immune system are generally proteins that help facilitate various aspects of an immune response. These components include antibodies, made by B cells, complement proteins made by hepatocytes and monocytes, and cytokines made by a variety of cell types including T cells, B cells, and monocytes. The vast majority of work on immune responses elicited by biomaterials has focused on these humoral elements and particularly on complement activation. Therefore, a fairly detailed description of the humoral response to biomaterials will be given before discussing some of the cellular elements shown in Fig. 1.

COMPLEMENT

Complement is a term devised by Paul Ehrlich to refer to plasma components that were known to be necessary for antibody-mediated bactericidal activity. We now know that complement is composed of more than 20 distinct plasma proteins involving two separate pathways: classical and alternative (Ross, 1986). The complement system directly and indirectly contributes to the acute inflammatory reaction and the immune response. One of the principal functions of complement is the nonspecific recognition and elimination of foreign elements from the body. This is accomplished by coating a foreign material with complement fragments that permit phagocytosis by granulocytes, a process called opsonization. Although the system evolved to protect the host from the invasion of adventitious pathogens, the nonspecific and spontaneous nature of the alternative pathway permits activation by various biomaterial surfaces. Because complement activation can follow two distinct but interacting pathways, these pathways are treated separately.

Classical Pathway

The classical pathway (CP) is activated primarily by immune complexes (ICs) composed of antigen and specific antibody. The proteins of this pathway are C1, C2, C4, C1 inhibitor (C1-Inh), and C4 binding protein (C4bp) (Fig. 2). Their properties are summarized in Table 1.

Complement activation by the CP is illustrated in Fig. 2. This system is an example of an enzyme cascade in which each step in the series, from initiation to the final product, involves enzymatic reactions that result in some amplification. The com-

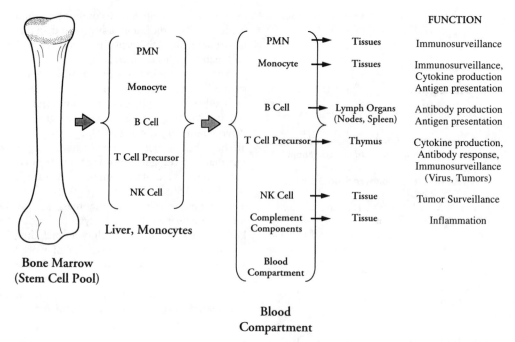

FIG. 1.   The functional elements of the immune system. Cells of the immune system are produced from stem cells in the bone marrow. These cells (PMN, monocytes, etc.) enter the blood and lymph, migrating to the various organs and tissues of the body where they perform their immunosurveillance function. Many of the cells (B cells, T cells, monocytes) secrete proteins (antibodies, cytokines, complement proteins) that are components of the humoral response. Together, these elements act to protect the host from a diverse array of pathogens.

plement cascade is initiated when C1 binds to the Fc portion of an antigen–antibody complex that has adopted a specific conformation. The C1 protein is a multimeric complex composed of three different types of subunits called C1q, C1r, and C1s. The C1q subcomponent is an arrangement of 18 separate protein chains that has been compared to a bouquet of flowers. Two or more of the C1q globular heads bind to an IC formed between an antigen and a molecule of pentameric immunoglobulin (Ig) M or several closely spaced IgG molecules. This interaction is believed to produce a conformational change in the C1q collagen-like stem which results in activation of the two C1r and C1s subcomponents. Both C1r and C1s are zymogen serine proteases, bound to the C1q stem in a calcium-dependent manner. Activation by an IC results in autocatalytic proteolysis of the two C1r enzymes, which then cleave the two C1s subunits. The proteolysis of C1s completes the activation of C1, which then proceeds to act on the next proteins in the cascade, C4 and C2.

C4 is composed of three separate chains, α, β and γ (Fig. 3), bound together by disulfide bonds. Activated C1s cleaves C4 near the amino terminus of the α chain, yielding a 77-amino acid polypeptide called C4a and a much larger (190,000 Da) C4b fragment. The C4 protein contains a unique structural element called a thioester. This reactive bond is formed by condensation of a cysteine sulfhydryl group and glutamic acid residue that are four amino acids apart in the α chain. Thioesters have only been detected in two other plasma proteins, α_2-macroglobulin and C3. Upon cleavage of C4, the buried thioester becomes exposed and available to nucleophilic attack.

In the presence of a surface containing amino or hydroxyl moieties, about 5% of the C4b molecules produced react through the thioester and become covalently attached to the surface. This represents the first amplification step in the pathway since each molecule of C1 produces a number of surface-bound C4b sites.

The C4b protein, attached to the surface, acts as a receptor for C2. After binding to C4b, C2 becomes a substrate for C1s. Cleavage of C2 yields two fragments: a 34,000-Da C2a portion diffuses into the plasma, while the 70,000-Da C2b remains bound to the C4b. The C2b protein is another serine protease that, in association with C4b, represents the classical pathway C3/C5 convertase.

As the name implies, the function of the C4b · C2b complex is to bind and cleave C3. The C3 protein sits at the juncture of the classical and alternative pathways and represents one of the critical control points. Cleavage of C3 by C2b yields a 9000-Da C3a fragment and a 175,000-dalton C3b fragment that is very similar to C4b in both structure and function. As with C4b, generation of C3b exposes a thioester that permits covalent attachment of 10–15% of the resulting C3b to the surface of the activator (Fig. 3). This is the second amplification step in the sequence since as many as 200 molecules of C3b can become attached to the surface surrounding every C4b · C2b complex. Eventually one of the C3b molecules reacts with a site on the C4b protein, creating a C3b–C4b · C2b complex that acts as a C5 convertase.

In contrast to C3, which can be cleaved in the fluid phase (see later discussion), proteolytic activation of C5 occurs only

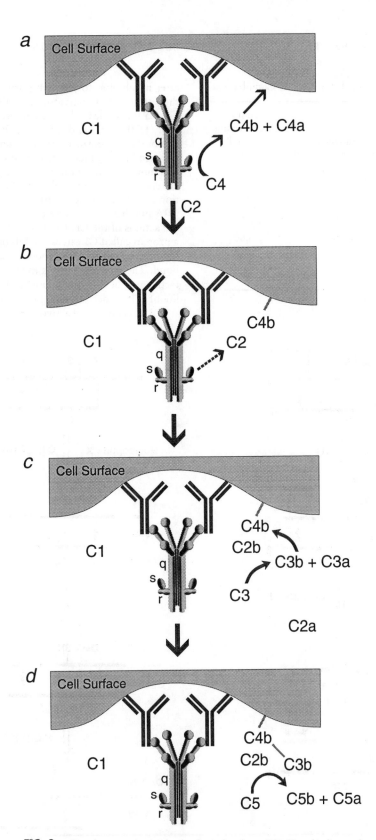

FIG. 2. Complement activation by the classical pathway (CP). Upon binding to the Fc region of an immune complex, C1 is activated and cleaves C4 (a), exposing its thioester, which permits covalent attachment of C4b to the activating surface. C2 is cleaved (b), producing C2b, which binds to C4b to form the CP C3 convertase (c). C2b is a serine protease that specifically acts on C3 to generate C3b and C3a. Attachment of C3b (through its thioester) to the C4b component of the C3 convertase alters the substrate specificity of the enzyme, making it a C5 convertase (d). Production of C5b leads to formation of the membrane attack complex (Fig. 5).

TABLE 1 Proteins of the Classical Pathway of Complement

Protein	Molecular weight	Subunits	Plasma concentration (μg/ml)
C1q	410,000	6A, 6B, 6C	70
C1r	85,000	1	35
C1s	85,000	1	35
C2	102,000	1	25
C4	200,000	α, β, γ	600
C1-Inh	104,000	1	200
C4bp	570,000	8	230

after it is bound to the C3b portion of the C5 convertase on the surface of an activator (e.g., the immune complex). Like C3, C5 is also cleaved by C2b to produce fragments designated C5a (16,000 Da) and C5b (170,000 Da). The C5b molecule combines with the proteins of the terminal components to form the membrane attack complex described later. C5a is a potent inflammatory mediator and is responsible for many of the adverse reactions normally attributed to complement activation in various clinical settings.

A number of control mechanisms have evolved to regulate the actions of the CP. Control of C1 activation is mediated by a protein called C1 esterase inhibitor (C1-Inh). C1-Inh acts by binding to activated C1r and C1s subunits, forming a covalent bond. The stoichiometry is thus C1r · C1s · (C1-Inh)$_2$. The effectiveness of this interaction is illustrated by the short half-life of C1s under physiological conditions (13 sec).

C3/C5 convertase activity is controlled in several ways. The

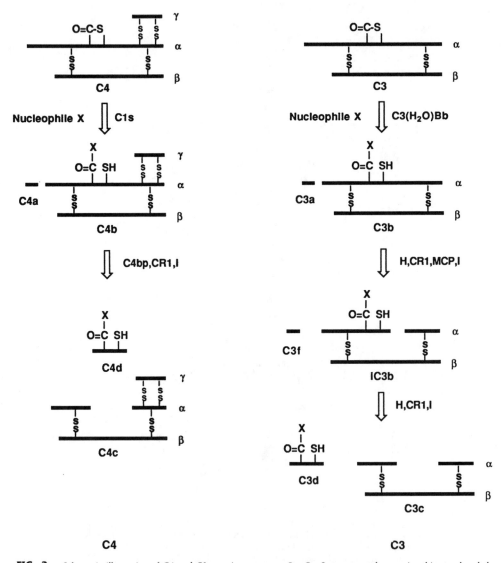

FIG. 3.   Schematic illustration of C4 and C3 protein structures. O=C—S represents the reactive thioester bond that permits covalent attachment of both C3b and C4b to surface nucleophiles. The pattern of proteolytic degradation and the resulting fragments are also shown. Although factor I is the relevant *in vivo* protease, some of these same fragments can be generated with trypsin, plasmin, and thrombin.

TABLE 2 Proteins of the Alternative Pathway
of Complement

Protein	Molecular weight	Subunits	Plasma concentration (μg/ml)
C3	185,000	α, β	1300
B	93,000	1	210
D	24,000	1	1
H	150,000	1	500
I	88,000	α, β	34
P	106,000–212,000	2–4	20

half-life of the C4b · C2b complex is about 3 min at 37°C. Thus the enzyme spontaneously decays by dissociation of the C2b catalytic subunit. This rate of dissociation is increased by C4 binding protein (C4bp), which competes with C2 for a binding site on C4b. C4bp also acts as a cofactor for another control protein called factor I, which destroys the C4b by proteolytic degradation (Fig. 3).

Alternative Pathway

The alternative pathway (AP) was originally discovered in the early 1950s by Pillemer *et al.* (1954). Pillemer's group studied the ability of a yeast cell wall preparation, called zymosan, to consume C3 without affecting the amount of C1, C2, or C4. A new protein, called properdin, was isolated and implicated in initiating C3 activation independent of the CP. This new scheme was called the properdin pathway. However, this work fell into disrepute when it was realized that plasma contains natural antibodies against zymosan, which implied CP involvement in Pillemer's experiments. The pathway was rediscovered in the late 1960s with the study of complement activation by bacterial lipopolysaccharide and with the discovery of a C4-deficient guinea pig. The 1970s witnessed the isolation and characterization of each of the proteins of this pathway until it was possible to completely reconstruct the entire AP by recombining each of the purified proteins (Schreiber *et al.*, 1978). The AP is responsible for most of the complement activation produced by biomaterials, although the CP may also be involved to some degree.

The proteins of this pathway are described in Table 2. Their actions can be conceptually divided into three phases: initiation, amplification, and regulation (Fig. 4). Initiation is a spontaneous process that is responsible for the nonselective nature of complement. During this stage, a small portion of the C3 molecules in plasma (0.005%/min) undergo a conformational change that results in hydrolysis of the thioester group, producing a form of C3 called $C3(H_2O)$.

For a short period this $C3(H_2O)$ looks like C3b and will bind to factor B in solution. This interaction requires Mg^{2+} and can be inhibited by EDTA. The $C3(H_2O) \cdot B$ complex is a substrate for factor D, which cleaves the B protein to form an alternative pathway C3 convertase: $C3(H_2O) \cdot Bb$. When it is cleaved by D to form Bb, a cryptic serine protease site is exposed on the Bb protein. This $C3(H_2O) \cdot Bb$ enzyme then can act to cleave more C3 to form C3b. The C3b can then follow several paths. First, it may attach, through its thioester, to a surface (Figs. 3 and 4) and lead to more C3 convertase (amplification), or it may react with water and become metabolized. Since the half-life of the thioester bond is about 60 μsec, most of the C3b produced is hydrolyzed and inactivated, a process that has been termed "C3 tickover."

The C3 tickover is a continuous and spontaneous process that ensures that whenever an activating surface presents itself, reactive C3b molecules will be available to mark the surface as foreign. Recognition of the C3b by factor B, cleavage by factor D, and generation of more C3 convertase leads to the amplification phase (Fig. 4). During this stage, many more C3b molecules are produced, bind to the surface, and in turn lead to additional C3b · Bb sites. Eventually, a C3b molecule attaches to one of the C3 convertase sites by direct attachment to the C3b protein component of the enzyme. This C3b–C3b · Bb complex is the alternative pathway C5 convertase and, in a manner reminiscent of the CP C5 convertase, converts C5 to C5b and C5a.

As with the classical pathway, there are a number of control mechanisms that operate to regulate this pathway. The intrinsic instability of the C3b thioester bond (half-life = 60 μsec) ensures that most of the C3b (80–90%) is inactivated in the fluid phase. Once formed, the half-life of the C3 convertase (C3b · Bb complex) is only 1.5 to 4 min and this rate of dissociation is increased by factor H. After displacement from the C3b, Bb is relatively inactive. Aside from accelerating the decay of C3 convertase activity, factor H also promotes the proteolytic degradation of C3b by factor I (Figs. 3 and 4). Factors H and I also combine to limit the amount of active $C3(H_2O)$ produced in the fluid phase.

In addition to factor H, there are several cell membrane-bound proteins that have similar activities and act to limit complement-mediated damage to autologous, bystander cells. Decay-accelerating factor, or DAF, displaces Bb from the C3 convertase and thus destroys the enzyme activity. DAF is found on all cells in the blood (bound to the plasma membrane through a unique lipid group) but is absent in a disease called proximal nocturnal hemoglobinuria (PNH), which manifests a high spontaneous rate of red blood cell lysis. In addition to DAF, there are two other cell-bound control proteins: membrane cofactor protein (MCP) and CR1 (complement receptor 1, see later discussion). MCP is found on all blood cells except erythrocytes, while CR1 is expressed on most blood cells. Both function like factor H to displace Bb from the C3 convertase and act as a cofactor for the factor I-mediated cleavage of C3b. A soluble recombinant form of CR1 (sCR1) is now being evaluated for its efficacy in mitigating complement-mediated damage in several settings of acute inflammatory injury.

Properdin, the protein originally discovered by Pillemer, functions by binding to surface-bound C3b and stabilizing the C3 and C5 convertase enzymes. The normal half-life of the C3b · Bb complex is increased to 18–40 min. Although properdin is not necessary for activation of the alternative pathway,

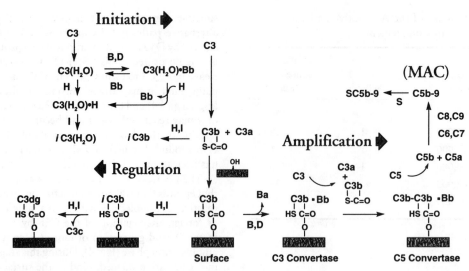

FIG. 4.   Complement activation by the alternative pathway (AP). The spontaneous conversion of C3 to C3(H₂O) permits the continuous production of C3b from C3, a process called C3 tickover. In the presence of an activating surface, the C3b is covalently bound and becomes the focal point for subsequent interactions. Binding of factor B results in C3 convertase formation and amplification of the original signal. Interaction of factors H and I lead to control of the pathway and production of C3 fragments that are recognized by specific receptors on neutrophils and monocytes.

a genetic deficiency of this protein has been associated with an increased susceptibility to meningococcal infections.

MEMBRANE ATTACK COMPLEX

Both the CP and the AP lead to a common point: cleavage of C5 to produce C5b and C5a. C5a is a potent inflammatory mediator and is discussed later in the context of receptor-mediated white blood cell activation. The production of C5b initiates the formation of a macromolecular complex of proteins called the membrane attack complex (MAC) (Table 3) that disrupts the cellular lipid bilayer, leading to cell

death. The sequence of events in MAC formation is outlined in Fig. 5.

Following cleavage of C5 by C5 convertase, the C5b remains weakly bound to C3b in an activated state in which it can bind C6 to form a stable complex called C5b6. This complex binds to C7 to form C5b67, which has ampiphilic properties that allow it to bind to, and partially insert into lipid bilayers. The C5b67 complex then binds C8 and inserts itself into the lipid bilayer. The C5b678 complex disrupts the plasma membrane and produces small pores (r = 1 nm) that permit leakage of small molecules. The final step occurs when multiple copies of C9 bind to the C5b678 complex and insert into the membrane. This enlarges the pore to about 10 nm and leads to lysis and cell death.

As with the other stages of complement activation, there are several control mechanisms that operate to limit random lysis of "bystander" blood cells. The short half-life of the activated C5b (2 min) and the propensity of the C5b67 complex to self-aggregate into a nonlytic form help limit MAC formation. In addition, a MAC inhibitor, originally called S protein and recently shown to be identical to vitronectin, binds to C5b67 (also C5b678 and C5b679) and prevents cell lysis. Recently another group of control proteins called homologous restriction factors (HRF) have been discovered. They are called HRFs because they control assembly of the MAC on autologous cells (i.e., human MAC on human cells) but do not stop heterologous interactions (e.g., guinea pig MAC on sheep red blood cells). One well-characterized member of this group is called CD59. It is widely distributed, found on erythrocytes, white blood cells, endothelial cells, epithelial cells, and hepatocytes. CD59 functions by interacting with C8 and C9, prevent-

TABLE 3   Proteins of the Membrane Attack Complex

Protein	Molecular weight	Subunits	Plasma concentration (μg/ml)
C5	190,000	α, β	70
C6	120,000	1	60
C7	105,000	1	60
C8	150,000	α, β	55
C9	75,000	1	55
S-protein	80,000	1	500

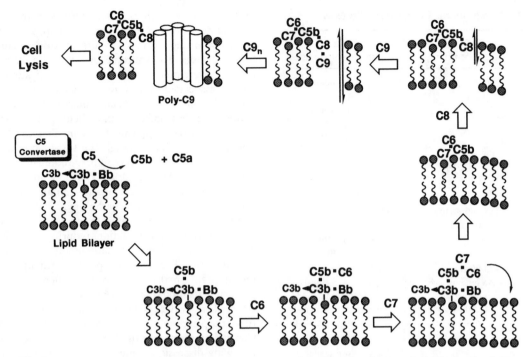

FIG. 5. Formation of the membrane attack complex (MAC). C5b, formed by either the CP or the AP, binds C6 and C7 to form a complex that associates with the plasma membrane. This C5b67 multimer then binds C8, which results in the formation of a small hole in the lipid bilayer that allows small molecules to pass through. Association of multiple C9 proteins enlarges the pore, leading to loss of membrane integrity and cell death.

ing functional expression of C56-8 and C56-9 complexes on autologous cells.

COMPLEMENT RECEPTORS

Except for the cytotoxic action of the MAC, most of the biological responses elicited by complement proteins result from ligand–receptor-mediated cellular activation. These ligands are listed in Table 4 and are discussed briefly here.

The ability of complement to function in the opsonization of foreign elements is accomplished in part by a set of receptors that recognize various C3 and C4 fragments bound to these foreign surfaces. For example, complement receptor 1 (CR1) is found on a variety of cells including erythrocytes, neutrophils, monocytes, B cells, and some T cells. CR1 recognizes a site within the C3c region of C3b (Fig. 3). On neutrophils and monocytes, activated CR1 will facilitate the phagocytosis of C3b- and C4b-coated particles. As discussed before, CR1 is also involved in the regulatory phase of complement. A second type of complement receptor, CR2, recognizes the C3d fragment. This receptor has been found on B cells and has been implicated in the process of antigen–immune complex-driven B cell proliferation. CR3 represents another complement receptor that binds to iC3b, β-glucan structures found on zymosan, and, on activated monocytes, CR3 has been shown to bind fibrinogen and factor X (of the coagulation cascade). CR3 is

a member of the β_2-integrin family of cell adhesion proteins that includes leukocyte functional antigen-1 (LFA-1) and p150,95. LFA-1, CR3, and p150,95 are routinely referred to as CD11a, CD11b, and CD11c respectively. Each of these proteins associates with a molecule of CD18 to form an α-β heterodimer that is then, transported and expressed on the cell surface. These proteins help mediate the cell–cell interactions necessary for such activities as chemotaxis and cytotoxic killing. A genetic deficiency in CR3/LFA proteins leads to recurrent life-threatening infections. Finally, CR4 is a protein found on neutrophils and platelets. CR4 also binds C3d, but is antigenically distinct from CR2. CR4 can also bind iC3b and may facilitate the accumulation of neutrophils and platelets at sites of immune complex deposition.

In contrast to the ligands discussed earlier, which remain attached to activating surfaces, C3a, C4a, and C5a are small cationic polypeptides that diffuse into the surrounding medium to activate specific cells. These peptides are called anaphylatoxins because they stimulate histamine release from mast cells and cause smooth muscle contraction, which can produce increased vascular permeability and lead to anaphylactic shock. These activities are lost when the peptides are converted to their des arg analogs (i.e., with the loss of their carboxyl terminal arginine residue). This occurs rapidly *in vivo* and is catalyzed by serum carboxypeptidase N.

In addition to its anaphylatoxic properties, C5a and C5a$_{\text{des arg}}$ bind to specific receptors found on neutrophils, monocytes, and macrophages. This C5a–receptor interaction

TABLE 4 Receptors for Complement Proteins

Receptors name/ligand	Structure	Cellular distribution/response
CR1/C3b, C4b	200,000-Da single chain	RBC, PMN, monocytes, B and T cells/ clearance of immune complexes, phagocytosis, facilitates cleavage of C3b to C3dg by Factor I
CR2/C3dg	140,000-Da single chain	B cells/regulate B cell proliferation
CR3/iC3b, β-glucan fibrinogen, factor X	185,000 α chain 95,000 β chain	PMN, monocyte/phagocytosis of microorganisms; respiratory burst activity
C5a/C5a	47,000-Da C5a binding chain may have another 40,000-Da subunit	PMN, monocytes, fibroblasts/chemotaxis, degranulation, hyperadherence, respiratory burst, IL-1 production
C3a/C3a, C4a, C5a		Mast cells/histamine release
C1q/C1q	70,000 Da	PMN, monocytes, B cells/respiratory burst activity
H/H	50,000 Da (3 chains)	B cells, monocytes/secretion of factor I, respiratory burst activity

leads to a variety of responses, including chemotaxis of these cells into an inflammatory locus; activation of the cells to release the contents of several types of secretory vesicles and produce reactive oxygen species that mediate cell killing; increased expression of CR1, CR3, and LFA-1, resulting in cellular hyperadherence; and the production of other mediators such as various arachidonic acid metabolites and cytokines, e.g., IL-1, -6, and -8. Many of the adverse reactions seen during extracorporeal therapies, such as hemodialysis, are directly attributable to C5a production.

Clinical Correlates

The normal function of complement is to mediate a localized inflammatory response to a foreign material. The complement system can become clinically relevant in situations where it either fails to function or where it is activated inappropriately. In the first instance, a lack of activity due to a genetic deficiency in one or more complement proteins has been associated with increased incidence of recurrent infections, glomerulonephritis, and other pathologies. The second instance, inappropriate activation, occurs in a variety of circumstances. Activation of the classical pathway by immune complexes occurs in various auto-

immune diseases such as systemic lupus erythematosus. Glomerular deposition of these immune complexes results in local inflammation and can contribute to kidney damage. Adult respiratory distress syndrome (ARDS) is an illness characterized by severe pulmonary edema and respiratory failure that has an associated fatality rate of 65%. The disease can be initiated by fat emboli (e.g., bone marrow from a broken bone) or severe bacterial infection and is believed to involve extensive activation of the complement system. This can lead to white blood cell activation through the C5a–receptor interaction, which may result in inadvertent tissue damage. Consistent with this hypothesis, Stevens *et al.* (1986) demonstrated that pretreatment of baboons with an anti-C5a antibody (to block its activity) attenuated ARDS pathology.

One of the major settings where complement has been implicated in adverse clinical reactions is during extracorporeal therapies (e.g., hemodialysis, cardiopulmonary bypass, and apheresis therapy). The same nonspecific mechanism that permits the alternative pathway to recognize microbes results in complement activation by the various biomaterials found in different medical devices. One of the most investigated materials (from the perspective of complement activation) is the cellulosic Cuprophan membrane used extensively for hemodialysis. The following discussion relates principally to clinical experience with Cuprophan hemodialysis membranes, but the same sequela have been noted with cardiopulmonary bypass and apheresis membranes.

Some of the adverse reactions that occur during clinical use of a Cuprophan dialyzer are listed in Table 5. In 1977, Craddock *et al.* showed that some of these same manifestations (neutropenia, leukosequestration, and pulmonary hypertension) could be reproduced in rabbits and sheep when the animals were infused with autologous plasma that had been incubated *in vitro* with either Cuprophan or zymosan. This effect could be abrogated by treatment of the plasma to inhibit complement activation (heating to 56°C or addition of EDTA), thus linking these effects with complement. The development and use of specific radioimmunoassays (RIAs) to measure C3a and C5a by Dennis Chenoweth (1984) led to the identification of these complement components in the plasma of patients during dialysis therapy.

TABLE 5 Clinical Symptoms Associated with Cuprophan-Induced Biocompatibility Reactions

Cardiopulmonary:	Pulmonary hypertension
	Hypoxemia
	Respiratory distress (Dyspnea)
	Neutropenia (pulmonary leukosequestration)
	Tachycardia
	Angina pectoris
	Cardiac arrest
Other:	Nausea, vomiting, diarrhea
	Fever, chills, malaise
	Urticaria, pruritus
	Headache

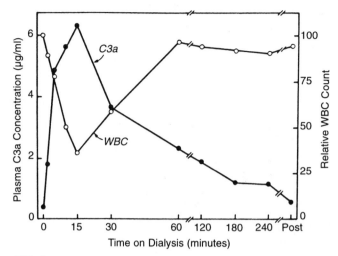

FIG. 6. A typical response pattern to dialysis with a Cuprophan membrane. Blood was taken from the venous line of a patient during dialysis and analyzed for white blood cells using a Coulter counter and for C3a antigen using RIA. The relative neutropenia (i.e., loss of neutrophils correlates with the increase in C3a, which peaks at 15 min.) (Reproduced with permission from D. E. Chenowith, Complement activation during hemodialysis: Clinical observations, proposed mechanisms, and theoretical implications, *Artif. Organs* 8: 281–287, 1984.)

A typical patient response to a Cuprophan membrane is shown in Fig. 6. The C3a (and C5a) levels rise during the first 5–12 min, peaking between 10–20 min. During this period the white blood cells become hyperadherent and are trapped in the lung, resulting in a peripheral loss of these cells (neutropenia). This is a very consistent response and many authors have noted a direct correlation between the extent of complement activation and the degree of neutropenia seen with various dialysis membranes. A variety of additional responses have been described in these patients (Fig. 7) including increased expression of CR1 and CR3 on neutrophils and monocytes and increased plasma levels of secondary mediators such as thromboxane A_2 and various interleukins.

Based on our understanding of the biochemistry of complement and its biological actions, the following scenario can be drawn (Fig. 7). Blood contact with the membrane results in deposition of C3b and consequent formation of C3 and C5 convertase enzymes. Liberation of C5a leads to receptor-mediated neutrophil and monocyte activation. This accounts for much of the pathophysiology seen clinically and outlined in Fig. 7. The critical role of C5a in mediating many of these adverse reactions was recently confirmed in experiments employing purified sheep C5a. Infusion of this isolated peptide into sheep, in a manner that would simulate exposure to this molecule during hemodialysis, produced a dose-dependent response identical to that which is seen when the sheep are subjected to dialysis (Fig. 8).

As the hemodialysis community became aware of the relationship between complement and many of the patient reactions outlined here (Table 5), membrane manufacturers began addressing the phenomena by developing new membranes (Fig. 9). These membranes tend to fall into two groups: moderately

activating modified cellulosics and low activating synthetics. Based on the foregoing discussion, the reasons for the improved biocompatibility of these membranes may be as follows.

With the exception of Hemophan, all these materials have a diminished level of surface nucleophiles. In theory, this should result in lower deposition of C3b, and in fact this has been verified experimentally. The diminished capacity to bind C3b results in lower levels of C3 and C5 convertase activity and consequently an abated production of C3a and C5a. Patient exposure to C5a is reduced even further by materials that allow for transport through the membrane to the dialysate (e.g., high flux membranes such as polysulfone will do this) or by absorbing the peptide back onto the surface (e.g., the negatively charged AN69 has been shown to have a high capacity for binding C5a). Thus, limiting C3b deposition and C5a exposure are two proven mechanisms of avoiding the clinical consequences of complement activation.

The same result can be also accomplished by facilitating the normal control of C3 convertase by factor H. Kazatchkine *et al.* (1979) have shown that heparin coupled to either zymosan or Sepharose limits the normal complement activation that occurs on these surfaces by augmenting C3b inactivation through factors H and I. Mauzac *et al.* (1985) have prepared heparinlike dextran derivatives that are extensively modified with carboxymethyl and benzylamine sulfonate groups. These researchers have shown that these modifications diminish complement activation by the dextran substrate. Recently, another advance was made with the discovery that maleic acid groups attached to a cellulose membrane (Cuprophan) inhibited the

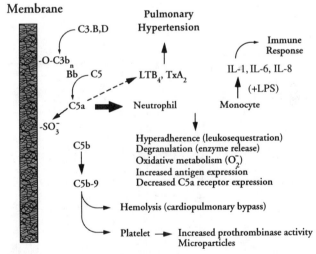

FIG. 7. The biochemical basis for complement-mediated adverse reactions during extracorporeal therapy. Production of C5a leads to receptor-dependent white blood cell activation. This results in profound neutropenia, increased concentrations of degradative enzymes, and reactive oxygen species that ultimately may lead to tissue damage and dysfunction of these important immune cells. Generation of secondary mediators, such as arachidonic acid metabolites (TxA_2, LTB_4) and cytokines can have profound consequences on whole organ systems. Finally, formation of the MAC(C5b-9) has been linked with increased hemolysis during cardiopulmonary bypass and shown to increase platelet prothrombinase activity *in vitro*. This last observation suggests that surfaces that activate complement aggressively may be more thrombogenic.

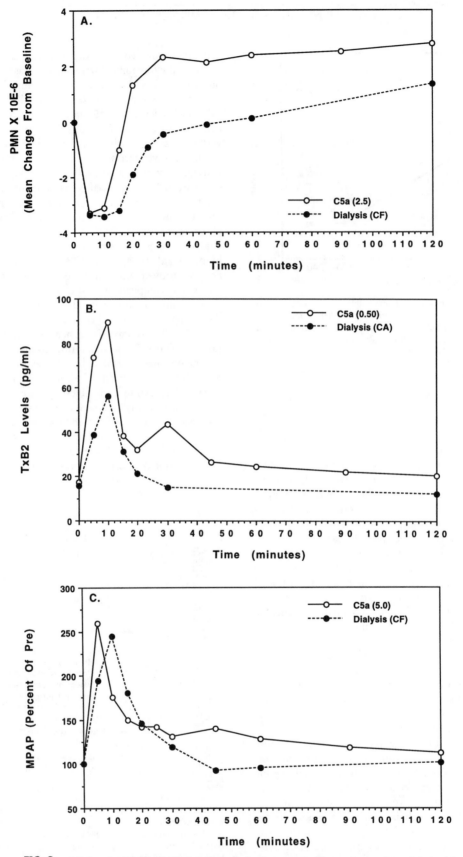

FIG. 8. Infusion of purified sheep C5a into sheep mimics the response of these animals to hemodialysis with complement-activating membranes (CF, Cuprophan; CA, cellulose acetate). C5a produces a dose-dependent neutropenia (A), thromboxane response (B), and pulmonary hypertension (C), that is identical to the responses seen during dialysis of these animals. C5a doses are shown in parentheses in units of μg/kg.

A. Cellulosics

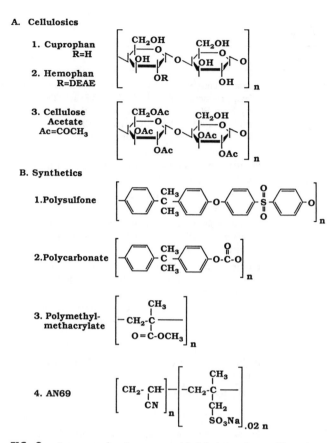

FIG. 9. Structures of various commercial dialysis membranes. Hemophan contains a low-level incorporation of DEAE (diethyl aminoethyl) groups that, for ease of illustration, has been arbitrarily placed on the 2' hydroxyl group. Most cellulose acetate membranes have a degree of substitution in the range of 2.5 to 2.7.

complement-activating potential of these materials by over 90% (Johnson *et al.*, 1990). Again, increased binding of factor H to surface-bound C3b appears to account for the improved biocompatibility of maleated cellulose. Thus materials that limit complement activation through normal regulatory mechanisms are on hand and may prove to be the next generation of complement-compatible membranes.

CYTOKINES

Cytokines are a group of proteins involved in regulating the cellular response of the immune system (Mizel, 1989). Since most of their putative functions concern communication between leukocytes, many have been given the designation interleukin (IL). However, some have retained their original names, which were based on a particular aspect of their activity; for example, interferons (IFNs) for antiviral activity and tumor necrosis factor (TNF) for cytotoxic activity against tumor cells. All cytokines are extremely potent, generally active in the picomolar range, and act through specific cell surface receptors.

They range in size from 100 to 200 amino acids and most are glycosylated. Many have multiple overlapping activities and when tested together can have synergistic or antagonistic effects.

A summary of the various known cytokines is given in Table 6, along with some of their reported activities. These activities have largely been determined with recombinant cytokines and isolated cell cultures; the relationship between *in vitro* and *in vivo* actions must still be determined in many instances. Broadly speaking, there are three areas that appear to be the principal focus of cytokine activity: (1) as growth factors for immune cells, (2) as regulators of the immune response and (3) as mediators of inflammation.

A number of cytokines help drive the process of hematopoiesis, i.e., the development of mature blood elements from primitive cells in the bone marrow (stem cells). These include IL-3, also called multi-colony-stimulating factor (CSF) because it supports the development of a number of cell types including granulocytes, macrophages, megakaryocytes, eosinophils, basophils and, especially, mast cells. IL-3 can also activate some mature blood cells such as eosinophils and mast cells. Granulocyte-macrophage colony stimulating factor (GM–CSF) has been shown to produce granulocytes and macrophages in bone marrow cultures. As a clinically used therapeutic, GM–CSF has proven effective in raising levels of neutrophils, eosinophils, and macrophages in peripheral blood. GM–CSF also has a multitude of effects on mature neutrophils and macrophages, including priming of oxidative activity, degranulation, and arachidonic acid release. It also increases the phagocytic and cytotoxic activities of these cells. G–CSF has characteristics similar to GM–CSF except its activity is largely restricted to neutrophils, while M–CSF appears to stimulate monocyte production and cytotoxic activity. IL-5 has been shown, *in vitro* and *in vivo*, to stimulate the proliferation of eosinophils and activated B cells, while IL-7 appears to induce the growth of immature B cells and thymocytes from bone marrow. Two recently discovered cytokines, IL-9 and IL-11, are also involved in hemopoietic cell development. IL-11 is a growth factor for megakaryocytes (a platelet progenitor cell) and stimulates an increase in peripheral platelet counts. Like IL-6, IL-11 also stimulates acute phase protein synthesis. IL-9 appears to act as an enhancing factor that augments the activity of other cytokines, such as IL-2, IL-3, IL-4, and erythropoietin, in promoting the growth of Th, mast, and erythroid cells.

Many cytokines are involved in regulating the immune response through their actions on T cells, B cells, and antigen-presenting cells (APC). APCs include macrophages, dendritic cells, and B cells, and, as their name implies, these cells process antigens and present them to T cells. The APCs also make IL-1 during this process. IL-1 can stimulate the T cell to make IL-2, the IL-2 receptor and interferon-γ (IFN-γ). IL-2 acts as a T-cell growth factor and is an example of autocrine stimulation, i.e., the same cell that makes the molecule also responds to its presence. IL-2 can also activate some non-T lymphocytes, most notably the lymphokine-activated killer or LAK cells. These cells become tumoricidal after exposure to IL-2, a phenomenon that serves as the basis of the LAK cell therapy developed by Steven Rosenberg and his colleagues at the National Institutes

TABLE 6 Cytokines[a]

Mediator	Source	Induced by	Actions/activities
IL-1 (13–17 kDa)[a]	Macrophages Endothelium NK cells Glial cells Keratinocytes Smooth muscle	Endotoxin, C5a, TNF/IL1	Hepatic acute phase response, leukocyte adherence to EC; production of PGI_2, PGE_2, PAF, and TF activity from EC; fibroblast collagen synthesis; neutrophilia; T, B and NK cell activation; PMNL TxA_2 release; fever (PGE_2); ACTH production
IL-2 (15.5 kDa)	T cells	IL-1 + Antigen	Proliferation and differentiation of T, B, and LAK cells; activation of NK cells
IL-3 (28 kDa)	T cells, NK cells		Hemopoietic growth factor for myeloid, erythroid, and megakaryocyte lineages
IL-4 (20 kDa)	T cells	Mitogens (Con A)	Proliferation and differentiation of B cells, isotype switching (IgE), mast and T cell proliferation, antagonistic with IFN-γ
IL-5 (45 kDa)	T cells	Mitogens (Con A)	Proliferation and differentiation of B cells and eosinophils, increased secretion of IgM and IgG from activated B cells
IL-6 (23–30 kDa)	T cells Monocytes Fibroblasts Endothelium	IL-1	Acute phase response (hepatic), proliferation and Ig secretion by activated B cells, T-cell activating factor
IL-7 (25 kDa)	Bone marrow stroma cells		Proliferation of large B progenitors, thymic maturation of T cells
IL-8 (14 kDa)	T cells, PMN Monocytes	Antigen Mitogen IL-1 TNF-α	T-cell and neutrophil chemotactic factor
IL-9	T cells		Enhancing hemopoietic growth factor for Th, mast and erythroid cells (synergizes with IL-2, IL-3 and IL-4)
IL-10 (18 kDa)	T cell, B cells Monocytes/macrophages Keratinocytes		Immunosuppression: inhibits cytokine actions and expression, stimulates IL-1ra expression, stimulates proliferation of B cells and mast cells
IL-11 (23 kDa)	Bone marrow stromal cells		Hemopoietic growth factor for megakaryocytes, myeloid progenitor cells, synergizes cytokines, stimulates acute cytotoxic activities of NK and LAK cells, growth factor for activated T cells and NK cells
L-12 (75 kDa)	B lymphoblastoid cell line		Enhances cytotoxic activities of NK and LAK cells, growth factor for activated T cells and NK cells
IL-13 (15 kDa) TNF (17 kDa)	T cells		IL-4-like activities on 15 monocytes and B cells, does not stimulate T cells, same properties as Il-1 except for T cell responses
IFN-γ (20–25 kDa)	T cells NK Cells	Viruses Bacteria IL-2	Macrophage-activating factor (antimicrobial activity); giant cell formation; acts as a cofactor in the differentiation of cytolytic T cells, B cells, and NK cells
GM–CSF	T cells Fibroblasts Endothelium	IL-1, TNF endotoxin	Proliferation and activation of granulocytes (PMN) and monocytes from bone marrow

[a]kDa, kilodaltons; NK, natural killer cells; TF, tissue factor; EC, endothelial cells.

of Health. INF-γ can activate APCs, making them more effective in antigen presentation. These various interrelationships are illustrated in Fig. 10.

Upon activation by antigen and/or IL-1, T cells can produce a number of additional cytokines that appear to play a major role in regulatory T cells, B cell activation, and antibody production. IL-4 was originally called B cell growth factor because of its ability to stimulate proliferation of activated B cells. IL-4 also induces the production of IgG1 and IgE from activated B cells and antagonizes the effects of INF-γ on B cells. IL-13, a

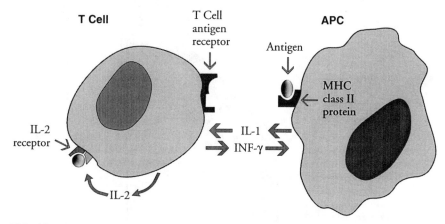

FIG. 10. Cytokines and cellular interactions involved in antigen presentation and T-cell activation. The activated T cell may then participate in immunosurveillance functions or facilitate antibody production by B cells.

newly discovered cytokine, exhibits many IL-4-like actions on B cells. IL-5, in *in vitro* assays, and in conjunction with IL-4, appears to promote the production of IgA from activated B cells. IL-6 was originally identified as B cell stimulating factor because it can induce B cell differentiation and IgM production. IL-10 has immunosuppressive activity on some immune cells. It downregulates major histocompatibility complex (MHC) class II protein on monocytes and interferes with APC activities. IL-10 inhibits the production of inflammatory cytokines (TNF-α IL-1α, IL-6, IL-8, GM–CSF) from monocytes, Th and NK cells. In contrast, IL-10 stimulates B cells, upregulating MHC class II antigens and augmenting the proliferation and differentiation of activated B cells. IL-12 stimulates the cytotoxic activities of NK and LAK cells and acts as a growth factor for activated T cells and NK cells. IL-12 is now being evaluated clinically for its efficacy as an antitumor agent.

Finally, a number of cytokines are known to contribute to inflammation, including INF-γ, GM–CSF, TNF-α, IL-6, IL-8 and IL-1. INF-γ, as noted earlier, can activate macrophages to increase their adhesive properties to promote cell–cell interactions. This cytokine also increases the microbicidal activity of these cells. GM–CSF not only helps to produce more leukocytes but also directs them to the site of inflammation by augmenting the action of other chemotaxins (e.g., C5a and IL-8). IL-8 is the most well-known member of a subfamily of cytokines, know as chemokines. (including GRO, β-thromboglobulin, platelet facter 4, IP-10, ENA-78 and RANTES). These cytokines stimulate the directed migration of neutrophils and some resting T cells to sites of inflammation. IL-8 has been shown to increase the expression of β_2-integrins (CD11b/CR3) on neutrophils, which facilitates their interaction with the endothelium. Finally, three of the cytokines listed above, TNF-α, IL-6, and IL-1, manifest many of the same activities. These three cytokines can induce the production of each other, and, when added to an assay system together, often produce synergistic (i.e., more than additive) responses.

The remaining discussion centers on IL-1 since it has been the most widely studied cytokine (Dinarello, 1988). However,

much of what follows is also true for TNF-α and IL-6. Systemic exposure to IL-1 produces a number of physiological changes, including fever, hepatic acute phase protein synthesis, increased production of the stress hormone adrenocorticotropin (ACTH) and hypotension. *In vitro*, IL-1 stimulates endothelial cells (EC) to make the vasodilators, prostaglandin I$_2$ and PGE$_2$. IL-1 induces the synthesis of adhesive receptors on ECs that promote neutrophil adherence to the endothelium. IL-1 stimulates the production of platelet-activating factor (PAF) from EC and thromboxane A$_2$ (TxA$_2$) from neutrophils. Both PAF and TxA$_2$ can promote binding and activation of platelets on the endothelium. In addition, IL-1 promotes the release of tissue factor from EC. Taken together, these activities can promote coagulation and cellular accumulation at an inflammatory site. IL-1 has also been shown to be mitogenic for fibroblasts and to induce the synthesis of types I, III, and IV collagen. This activity has implicated IL-1 in wound repair and fibrosis (Miller and Anderson, 1989).

Clinically, both TNF-α and IL-1β levels have been shown to increase in the plasma of patients during hemodialysis. This effect has been attributed to direct monocyte membrane contact, C5a production, and endotoxin exposure. IL-1 production in these patients has been associated with fever, muscle cramps, sleepiness, hypotension, and diminished T-cell responses. IL-1 has been infused into humans and induces most of these same symptoms.

Control mechanisms for these inflammatory cytokines, especially IL-1, have recently been defined. A protein called IL-1 receptor antagonist (IL1-ra) is made by the same cells that produce IL-1, usually shortly after they begin to generate IL-1. IL-1ra is highly homologous to IL-1 and binds to the same cellular receptors. However, unlike IL-1, IL1-ra does not stimulate the cells. Thus by occupying the IL-1 receptors, IL-1ra competitively inhibits IL-1 activity. In addition, many cells make two types of IL-1 receptors (RI and RII). IL-1RII binds IL-1β but does not appear to transmit a signal to activate the cell. Also IL-1RII is released from the cell and may bind to IL-1β and inhibit its activity. A similar phenomenon has been

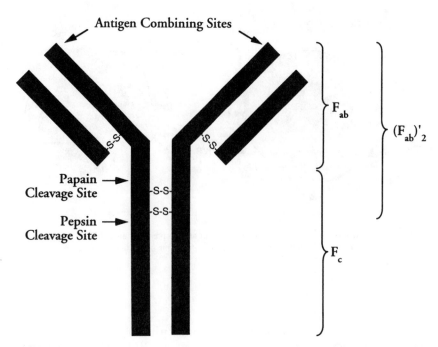

FIG. 11.   The general structure of immunoglobins. The basic structure is composed of two heavy chains (50,000 Da) and two light chains (25,000 Da) arranged in a Y shape and stabilized by multiple disulfide bonds. IgM contains five of these basic structures connected by a protein called the J protein (for joining). IgA contains two of these structures.

shown to occur with a number of other cytokines. Receptors for TNF-α, GM–CSF, G–CSF, IL-2, IL-4, and IL-6 are released by cells, bind to free cytokine, and thus potentially inhibit their activity in a competitive fashion. Thus the response elicited by various cytokines depends upon what cells they bind to, what receptors they bind to, what inhibitors may be present, and what other cytokines may be available to augment or antagonize the response.

ANTIBODIES AND ANTIBODY RECEPTORS

The final components of the humoral immune responses are the immunoglobulins. The general structure of these proteins is shown in Fig. 11. They are composed of two heavy chains (50,000 Da) and two light chains (25,000 Da) that are disulfide-bonded to each other to form a Y-shaped molecule. Different types of immunoglobulin are produced (by antigen-activated B cells) that differ in the amino acid sequence of their heavy chains. IgM has a heavy chain (called Hμ) that permits cross-linking of the five basic Ig structures shown in Fig. 10 to produce a pentamer. IgA is made with Hα heavy chains that permit cross-linking of two basic Ig structures to produce dimers. IgA is secreted by epithelia cells and provides some protection for mucosal surfaces (nose, mouth, lungs).

IgG and IgE have Hγ and Hε heavy chains and have the structure shown in Fig. 11. Papain digestion of IgG cleaves the molecule in the hinge region to produce two distinct fragments that can be separated based upon differences in their charged amino acids. One fragment, which constitutes the two arms of the Y structure, contains the antigen combining site and is therefore called Fab (for fragment-antigen binding). The remaining fragment is similar enough in most Ig preparations to be crystallized. This fragment is called the Fc portion (for fragment-crystallizable). The Fc portion of IgG and IgE contains sites that are recognized by a class of cellular receptors called FcγR or FcεR (for receptors that bind IgG or IgE, respectively). Fcγ receptors are found primarily on polymorphonuclear leukocytes (PMNs), monocytes, macrophages, and NK cells, and help (along with C3b receptors) in the phagocytosis of immune complexes and antibody-coated foreign elements. Fcε receptors are found on mast cells, basophils, and some B cells. IgE and its receptor are largely responsible for allergic reactions.

CELLULAR COMPONENTS OF THE IMMUNE SYSTEM

The cells of the immune system include leukocytes and lymphocytes that can be differentiated based upon their stem cell origin, appearance, antigen expression, and function. The most numerous immune cell is the neutrophil, also called the polymorphonuclear leukocyte. It composes 60–70% of white blood cells and after leaving the bone marrow has a life span of only 1 or 2 days. These cells contain receptors for C5a, C3b, Fcγ, and IL-1 and TNF. Upon stimulation (e.g., by C5a) these cells become hyperadherent by increasing expression of CR3 (CD11b/CD18) and other cell-adhesion proteins. This permits the cell to attach to the vascular endothelium and migrate out of the vascular space into the site of inflammation.

PMNs contain several different types of granules that are filled with proteases (e.g., elastase; cathepsins A, D, and G; and collagenase), lysozyme, myeloperoxidase, and other bactericidal proteins. They also produce (when stimulated) H_2O_2 and activated oxygen species (O_2^- and $OH\cdot$), which are extremely cytotoxic. These cells represent the first line of defense against infection, migrating to an inflammatory site to present their impressive armature against invading microbes. Defects in chemotaxis, granular constituents, or oxidative capacity have been discovered in individuals and are associated with an increased incidence of infections. In the hemodialysis setting, chronic exposure to dialysis membranes has been shown to lead to decreased chemotactic responsiveness and diminished oxidative metabolism (H_2O_2 and O_2^- production) in both PMNs and monocytes (Lewis and Van Epps, 1987). These dysfunctions have been suggested as contributing to the relatively high infection rate noted in these patients.

Monocytes originate in the bone marrow and constitute about 5% of peripheral white blood cells under normal conditions. Their average transit time in the blood is about 25 hr, after which the cells migrate into the tissues and serve as a continuous source of macrophages and histiocytes. Monocytes share many characteristics with PMNs. For example, they contain C5a, C3b, and Fc receptors and are activated in the same manner (at least *in vitro*) for directed migration (chemotaxis), phagocytosis (using C3b and Fcγ receptors), degranulation, and oxidative metabolism. However, monocytes are also a very different sort of cell. They are not terminally differentiated cells, as are neutrophils, but function as precursors for alveolar and peritoneal macrophages, Kupffer cells in the liver, and histiocytes found in many tissues. Macrophages can be activated to display different characteristics (e.g., peroxidase staining, C3b receptor density) based on their degree of activation and the type of stimulus producing the state. Thus, a single monocyte may give rise to cells with varying properties, depending upon their environment.

Monocytes and macrophages are also highly secretory, actively synthesizing and secreting a variety of molecules. Some of these are produced constitutively, such as α_2-macroglobulin, C1q, C2, C4, factors H and I, fibronectin, and lysozyme. Many more are induced upon activation and include factor B, C3, collagenase, IFNs, TNF, IL-1, and a variety of growth factors. Finally, monocytes act as antigen-presenting cells. As discussed briefly earlier, this involves processing foreign antigens internally (by proteolysis of opsonized particles), reexpressing the processed antigen on the cell surface in association with the major histocompatibility complex proteins, and finally presenting the antigen to a specific T cell. This initiates an immune response against the antigen. Individuals with a genetic deficiency of C3b receptors have been found to have an abnormal antibody response, presumably because their monocytes and B cells cannot bind and process antigen effectively.

Lymphocytes make up about 28% of white blood cells and are composed primarily of two distinct cell types called T cells and B cells. B cells arise in the bone marrow and represent about 17% of peripheral blood lymphocytes. B cells differentiate, upon stimulation by antigen, T cells and cytokines into plasma cells whose major task is the production of antibody. Mature B cells can be activated several ways. These cells express surface immunoglobulin of the IgD and IgM isotype. Each cell is specific for a given antigen based upon the specificity of the plasma membrane Ig molecule it expresses. Upon binding of these surface Igs to their respective antigen, the B cell can differentiate into an IgM-producing plasma cell. This is the most direct route to antibody production and explains why the initial humoral response to an antigen is largely of the IgM isotype. An alternative pathway involves the action of antigen-specific activated T cells that stimulate B cells by both direct interaction and by the production of cytokines (IL-4, IL-5, and IL-6). These interactions effect proliferation, differentiation, and isotype switching (i.e., IgM to IgG, IgA, or IgE isotypes).

The second major type of lymphocyte is derived from the thymus and is called a T cell. These cells represent about 24% of white blood cells, of which 60% are the helper or inducer phenotype (Th) and 30% function as cytotoxic T cells. The Th cells regulate the immune response by interacting with APC and specific B cells as described earlier. These cells are identified by a surface antigen called CD4 that facilitates the interaction of these cells with MHC class II-bearing cells (e.g., antigen-presenting monocytes and B cells). Upon binding to a Th cell through its antigen receptor, the APC (monocyte) produces Il-1, and the combination of these two signals activates the Th cell to make IL-2 and then the IL-2 receptor. This activated T cell can then proliferate and produce other cytokines (Table 6) that regulate both cell-mediated and humoral immunity (see Fig. 10).

Activated Th cells are now recognized to exist in at least two subclasses (Th-1 and Th-2) distinguished by particular patterns of cytokine production. Th-1 cells produce IL-2 and INF-γ (but not IL-4 or IL-5) and are responsible for delayed-type hypersensitivity responses, activation of macrophages, and cell-mediated cytoxicity. Th-2 type cells make IL-4, 5, 9 (and 10 in mouse Th cells) and stimulate B cell proliferation, differentiation, and isotype switching. Th-2 cells also inhibit macrophage activities, while Th-2 cell-derived cytokines IL-4 and IL-10 inhibit Th-1 cell cytokine production. Conversely, INF-γ inhibits antigen-driven growth of Th-2 cell populations. Thus these separate populations of T cells inversely control each other to balance the overall immune response. Dysregulation of this balance has been implicated in pathological responses to infection (HIV) and in autoimmune disease.

The cytotoxic T cells are recognized by the CD8 surface marker, which facilitates interaction of these cells, through their antigen receptor to MHC class I-bearing cells (MHC class I antigens are found on all cells of the body). These cells perform an immunosurveillance function by recognizing and destroying tissue grafts, tumor cells, and cells infected with either viruses or parasites.

The clinical relevance of the cellular immune response and its relationship to biocompatibility issues is complex and controversial. Most of the information available is restricted to end stage renal disease (ESDR) patients subjected to chronic hemodialysis. Many of these individuals are clearly immunosuppressed, as demonstrated by decreased cutaneous reaction to antigen stimulation, diminished ability to mount an antibody response to hepatitis B vaccine, increased incidence of infection, and prolonged survival of skin transplants. Many individuals have been found to be lymphocytopenic (decreased absolute numbers of T cells). T cells isolated from these patients have been shown to be significantly less capable of activation by mitogenic stimuli as measured by either IL-2 production, IL-2 receptor expression, or [³H]thymidine incorporation (proliferation assay). However, the relevance of these observations to complement activation, IL-1 production, or direct cell membrane interaction is, at present, not obvious. Many of the T cell dysfunctions may relate to the underlying etiology of ESRD, to the many blood transfusions these people need, or to other idiopathic causes.

SUMMARY AND FUTURE DIRECTIONS

The immune response to a biomaterial involves both humoral and cellular components. Activation of the complement cascade by either classical or alternative pathways leads to the deposition of C4b and C3b proteins. Recognition of these molecules by receptors on granulocytes can cause activation of these cells, leading to the production of degradative enzymes and destructive oxygen metabolites. Recognition of C4b or C3b by other proteins in the cascade leads to enzyme formation (C3 and C5 convertases), which amplifies the response and can lead to the production of a potent inflammatory mediator, C5a. C5a binds to specific receptors found on PMNs and monocytes. The interaction of C5a with these cells elicits a variety of responses including hyperadherence, degranulation, superoxide production, chemotaxis, and IL production. Systemic exposure to C5a during extracorporeal therapies has been associated with neutropenia and cardiopulmonary manifestations (Table 5) that can have pathologic consequences. The other portion of the C5 protein, C5b, leads to formation of a membrane attack complex that causes cytolysis and has been linked to increased hemolysis in the cardiopulmonary bypass setting.

The control of these processes is understood well enough to begin designing materials that are more biocompatible. Limiting C3b deposition (nucleophilicity), adsorbing C5a to negatively charged substituents, and facilitating the role of factors H and I are three approaches that have been shown to be effective. Translating the last mechanism into commercial materials is one of the major challenges facing the development of truly complement-compatible membranes.

The past 20 years have witnessed an explosive growth in understanding the molecular and cellular basis of the immune response. The identification and cloning of specific cytokines, their receptors, and T-cell subsets have opened up whole new research frontiers. The interaction of these components of the immune response with materials employed in medical devices has only just begun. IL-1 production during hemodialysis and T-cell dysfunction in ESRD are just two examples of where initial studies have yielded intriguing results. Considering the importance of the immune system in health and disease, an understanding of how biomaterials can both limit and elicit a given response promises a rich harvest for medical therapy.

Bibliography

Chenoweth, D. E. (1984). Complement activation during hemodialysis: clinical observations, proposed mechanisms and theoretical implications. *Artificial Org.* **8:** 231–287.

Craddock, P. R., Fehr, J., Brigham, K. L., Kronenberg, R. S., and Jacob, H. S. (1977). Complement and leukocyte-mediated pulmonary dysfunction in hemodialysis. *New Eng. J. Med.* **296:** 769–774.

Dinarello, C. A. (1988). Interleukin-1—Its multiple biological effects and its association with hemodialysis. *Blood Purif.* **6:** 164–172.

Johnson, R. J., Lelah, M. D., Sutliff, T. M., and Boggs, D. R. (1990). A modification of cellulose that facilitates the control of complement activation. *Blood Purif.* **8:** 318–328.

Kazatchkine, M., Fearon, D. T., Silbert, J. E. and Austen, K. F. (1979). Surface-associated heparin inhibits zymosan included activation of the human alternative complement pathway by augmenting the regulatory action of control proteins. *J. Exp. Med.* **150:** 1202–1215.

Ross, G. D. (1986). *Immunobiology of the Complement System.* Academic Press, New York.

Schreiber, R. D., Pangburn, M. K., Lesaure, P. H., and Muller-Eberhard, H. J. (1978). Initiation of the alternative pathway of complement: recognition of activators by bound C3b and assembly of the entire pathway from six isolated proteins. *Proc. Natl. Acad. Sci. U.S.A.* **75:** 3948–3952.

Stevens, J. H., O'Hanley, P., Shapiro, J. M., Mihm, F. G., Satoh, P. S., Collins, J. A., and Raffin, T. A. (1986). Effects of anti-C5a antibodies on the adult respiratory distress syndrome in septic primates. *J. Clin. Invest.* **77:** 1812–1816.

Lewis, S. L., and Van Epps, D. E. (1987). Neutrophil and monocyte alterations in chronic dialysis patients. *Am. J. Kidney Dis.* **9:** 381–395.

Mauzac, M., Maillet, F., Jozefonvicz, J., and Kazatchkine, M. (1985). Anticomplementary activity of dextran derivatives. *Biomaterials* **6:** 61–63.

Miller, K. M., and Anderson, J. M. (1989). *In vitro* stimulation of fibroblast activity by factors generated from human monocytes activated by biomedical polymers. *J. Biomed. Mater. Res.* **23:** 911–930.

Mizel, S. B. (1989). Interleukins. *FASEB J.* **3:** 2379–2388.

Paul, W. E. (1993). *Fundamental Immunology,* 3rd ed. Raven Press, New York.

Pillemer, L., Blum, L., Lepow, I. H., Ross, O. A., Todd, E. W., and Wardlaw, A. C. (1954). The properdin system and immunity. I. Demonstration and isolation of a new serum protein, properdin, and its role in immune phenomena. *Science* **120:** 279–285.

4.4 SYSTEMIC TOXICITY AND HYPERSENSITIVITY

Katharine Merritt

Systemic effects of biomaterials may be due to direct chemical toxicity, accumulation of products from wear, corrosion, or degradation; excess inflammatory responses (Chapters 4.2

TABLE 1 Target Organs and Signs and Symptoms in Local and Systemic Toxicity

Systemic responses

Lungs	Alteration in air exchange and breathing patterns
Kidney	Alterations in urine excretion, pain
Joints	Pain, swelling, loss of function
Liver	Alterations in blood chemistry
Lymphoid	Swelling, alteration of blood count
GI tract	Diarrhea or constipation

The following usually give local responses but may also be involved in systemic responses.

Skin	Rashes, swelling, discoloration
Eyes	Swelling, itching, watery
Nose	Itching, running, sneezing

The following usually do not give observable signs and symptoms until damage is extreme.

Brain, skeletal system, muscles

and 4.3), including the production of the various oxygen radicals (Halliwell *et al.*, 1988); generation of vasoactive products in the activation of the complement system (Chapter 4.3); or the reactions of the immune system.

NONIMMUNE SYSTEMIC TOXICITY

Systemic toxicity is broadly defined as toxicity at some distance from the site of the initial insult. The mechanisms by which substances are rendered toxic are varied and complex. The testing of biomaterials for cellular toxicity is dealt with in other chapters in this book (Chapters 5.2–5.5). Systemic toxicity following the use of biomaterials is typically caused by the accumulation, processing, and subsequent reaction of the host to degradation products and wear debris from the material.

The manifestations of toxic reactions vary depending on the site at which the response occurs. Most systemic toxic reactions are detected because damage to the target organ results in readily apparent signs and symptoms (Guyton, 1991; Cotran, 1989) (Table 1). However, some systemic responses may go undetected since the target organ does not exhibit easily observable signs and symptoms. In addition, accumulation of wear and degradation products can be substantial and yet cause no local or systemic toxicity. This is due in part to the nature of the biocompatibility of the materials and in part to the response of the individual host.

Biomaterials should be carefully evaluated and studied for toxicity *in vitro* before being implanted. Such testing must include the intact material, as well as the degradation and wear products which might be produced during function

in vivo. Systemic reactions caused by degradation and wear of biomaterials can be prevented by doing careful evaluative studies *in vitro* and in appropriate animal models. The details of the toxicity of individual drugs and manifestations of damage to target organs can be found in textbooks on pharmacology and pathology. Nonimmune systemic toxicity caused by a biomaterial is generally dose related, with higher doses giving a more severe response. With nonimmune toxicity there is usually a threshold level for each product, below which the material shows no toxicity; repeated exposures to the same substance give responses which are similar to those of the first exposure.

In contrast, systemic reactions caused by an immune response to a biomaterial and its degradation and wear products often have low threshold levels. Although the response may not be worse with increasing doses, following repeated exposure the response can occur at a lower dose or induce greater toxicity. For example, the cross-linking agents formaldehyde and glutaraldehyde induce nonimmune cytotoxicity at high doses, causing burns and cellular and protein damage. However, some individuals are markedly hypersensitive to these chemicals and show sensitivity (immune) reactions at extremely low doses.

SYSTEMIC TOXICITY DUE TO THE IMMUNE RESPONSE

The Immune Response: General Concepts

The function of the immune response is to protect the host from the onslaught of foreign substances. Healthy individuals have several mechanisms by which they combat foreign substances. The first resistance mechanism in humans is the physical barrier, especially the skin and mucous membranes. Once this barrier is breached, as occurs with surgical introduction of a biomaterial, other host defense mechanisms become involved. The internal defense mechanisms generally begin with the inflammatory response (Chapters 4.2 and 4.3), including phagocytosis. However, if this is not sufficient, and the substance is an antigen, a specific immune response may be induced. To be antigenic, a substance must be foreign and large, with a molecular weight above approximately 5000 Da. Proteins and nucleoproteins and carbohydrates are usually antigenic; lipids are weakly antigenic; and nucleic acids are not antigenic. Small molecules, called haptens, may become antigenic by binding to host cells or proteins, thereby providing the foreign antigenic site (which is called an epitope), while the host provides the large molecule. For example, drugs that bind to host cells and metal salts that bind to host components can stimulate the immune response. Thus, a seemingly innocuous chemical may become a strong antigen and be toxic at low doses. The immune response recognizes that a substance is antigenic but it cannot distinguish bad foreign substances, such as bacteria, from good foreign substances, such as the biomaterial being implanted in a therapeutic medical device.

The immune system has two effector arms: The humoral

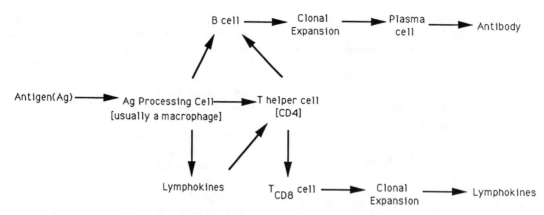

FIG. 1.   Immune pathway.

response is mediated by B cells with production of antibody by plasma cells; the other is the cell-mediated response mediated by T cells (Roitt 1989; Benjamini and Leskowitz, 1991). The pathways of the immune response are shown in Fig. 1. It is not possible to predict which arm of the immune response a previously unstudied antigen will stimulate. Most antigens are first encountered and processed by a macrophage and subsequently "presented" to T cells and B cells, small lymphocytes found in the blood and lymphoid tissues such as lymph nodes and spleen. B and T lymphocytes are morphologically identical and can be distinguished only by analysis of cell markers, specific detectable molecules that differ between the cell types. Reagents used for differentiating these cell types are available.

Recognition and presentation of an antigen stimulates the production of B or T cells reactive to the specific antigen. B cells stimulated by a processed antigen undergo cell division followed by differentiation to plasma cells, which are protein factories that produce antibodies specific to the antigen (called immunoglobulins). The basic structure of all antibody molecules is a four-chain structure with two identical light chains and two identical heavy chains as diagrammed in Fig. 2. The site that interacts with the antigen is called the Fab portion while the terminal portion of the heavy chain is called the Fc portion.

Five structurally and functionally different types of immunoglobulins are classified by their heavy chain structure (Table 2) (Roitt 1989; Benjamini and Leskowitz, 1991). The Fab portion of the antibody molecule is specific for the antigen that stimulated its production. Thus, in contrast to the nonspecific inflammatory response that is similar for all foreign substances, a specific immune response recognizes a single molecule entity. In addition, antibodies can be used as sensitive and specific tools for detecting and quantitating substances and are therefore the basis of many clinical tests for hormone levels and biomaterial–protein interactions (Benjamini and Leskowitz, 1991; Merritt, 1986). The subject of antigen–antibody reactions and their applications is exten-

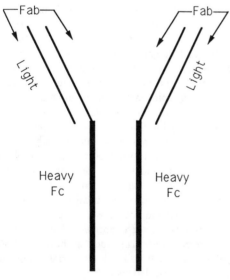

FIG. 2.   Antibody structure.

TABLE 2   Classification of Immunoglobulins

Type	Structure	Fc domains	Function
IgE	Dimer	5	Cell binding, causes allergy symptoms
IgA	Dimer	4	Secretory antibody, protective at surfaces
IgD	Dimer	4	On surface of B cells
IgG	Dimer	4	Protective antibody
IgM	Pentamer	5	Protective antibody, first class produced on surface of B cells

sive and of potential value to the biomaterials scientist (Rose and Friedman, 1986; Benjamini and Leskowitz, 1991). Radioimmunoassays (RIA), enzyme-linked immunosorbent assays (ELISA), and immunomicroscopic tests with fluorescent, enzyme, or particulate-labeled antibody have all been important in the field of biomaterials for quantification and identification of products and responses. The fluorescence-activated cell sorter (FACS) can be used to identify and collect cells with specific surface structures by tagging these structures with antibodies.

An important recent development is a method for producing large quantities of specific antibodies. A body's serum antibodies reflect its whole history of contact with foreign substances, such as vaccinations, bacteria, and viruses, and thus there are many antibodies and many specificities. However, since each plasma cell produces an antibody of a single specificity, a large quantity of specific antibody could be produced by a single plasma cell proliferating in tissue culture. The *in vitro* technique for doing this is based on the knowledge that some lymphoid cells are cancer cells (myeloma cells) and grow "forever" in culture. By chemically or electrically fusing (hybridizing) a normal antibody-producing B cell of the desired specificity with a myeloma cell, myeloma cells can be "tricked" into producing the specific antibody of the normal B cells. The cell cultures are then tested for the specific antibody being secreted by the cell. A small percentage of the fusions yield an immortal cell with the genetic information from the specifically immunized B cell (a monoclonal antibody). These cells can then be maintained in culture as a monoclonal antibody factory. Monoclonal antibodies currently are usually made using mouse cells and have provided powerful tools for immunologic testing, with expanding use into many areas. This technique allows the investigator to use the extreme sensitivity and specificity of the immune reaction to identify and quantify antigens.

A major use of the monoclonal antibody technique is in the recognition of specific subsets of T cells by markers known as cluster differentiation (CD) antigens. Three CD antigens distinct for T cells are important. CD3 is present on all mature T cells, and CD4 and CD8 are characteristic of T-helper and cytotoxic/suppressor cells, respectively. Attempts to decipher the immune response and its importance in various diseases are based on the use of monoclonal antibodies.

Antigen recognition by T cells is the initiating stimulus for T-cell activation. The receptors on T cells that are responsible for the highly specific recognition and response to antigen are composed of a complex of several integral plasma membrane proteins collectively called the T-cell receptor.

The effector phases of specific immunity are in large part mediated and regulated by protein hormones called cytokines. Cytokines are predominantly secreted by mononuclear phagocytes (macrophages), in which case they are called monokines, and lymphocytes, in which case they are called lymphokines. The functions of cytokines include (1) regulation of lymphocyte activation, growth and differentiation; (2) activation of macrophages; and (3) stimulation of immature lymphocyte growth and differentiation. Examples of cytokines are interferons (IFN), tumor necrosis factor (TNF), and interleukins (IL).

Systemic Toxicity as a Consequence of Immune Response (Hypersensitivity)

The immune response designed to protect the body against insult by foreign substances can inadvertently damage the host. Disorders that result from unusual, excessive, or uncontrolled immune reactions are called hypersensitivity reactions. Damage generally results from the release of chemicals normally confined to the internal contents of cells, or by overstimulation of the inflammatory response. It is not possible to predict systemic toxicity caused by immune reactions to a biomaterial or its degradation and wear products, since the immune response will depend on the genetics of the individual and the nature, dose, and location of release of the products. Thus, some biomaterials may release large quantities of wear and degradation products that may be spread systematically and cause no host response, while others may cause a strong reaction to those substances. Moreover, the human immune response varies from individual to individual and may be very different from that in experimental animals. Therefore, animal models can reveal some but not all potential problems. Hypersensitivity reactions have been divided into four types (Roitt, 1989). These are depicted in Fig. 3.

Type I reactions involve the interaction of antigen with antibody of the immunoglobulin class IgE, which attaches to the host cells in the skin and other tissues (mast cells, basophils, platelets, and eosinophils). An antigen encounter results in release of the cell contents, including active molecules such as histamine, heparin, serotonin, and other vasoactive substances, producing local or systemic symptoms that are manifest within minutes to a few hours following antigen-IgE interaction.

One common example of this type of hypersensitivity is hay fever, in which ragweed pollen reacts with the IgE antibody localized in the respiratory tract. If a skin test is done to determine if the patient is sensitive to ragweed, the reaction is local. However, a severe reaction in the respiratory tract may lead to systemic effects, including pulmonary problems such as asthma, and possibly to vascular collapse and death. Reports of biomaterials evoking the IgE response are rare, although IgE reactions to some components of biomaterials encountered in other applications, such as nickel and chromium salts in occupational respiratory contact, are known (Fisher 1978), and responses to silicone are controversial.

Type II reactions have a clinical course similar to type I reactions but have a different mechanism. In type II reactions, the antibody is of the IgG or IgM class and in this case it is the antigen, not the antibody, that is attached to platelets and acts as a hapten. The drug-platelet combination stimulates the immune response that makes an antibody against the drug. The reaction of antibody with the platelet activates complement and causes destruction of the platelet membrane, with release of its contents, including vasoactive substances. Documented type II reactions to biomaterials are rare.

Type III reactions fall into the category of "immune complex diseases" and are the consequences of stimulation of the inflammatory response by immune complex damage. The signs and symptoms occur days to weeks after antigen-antibody interaction. The problem arises when both antigen and anti-

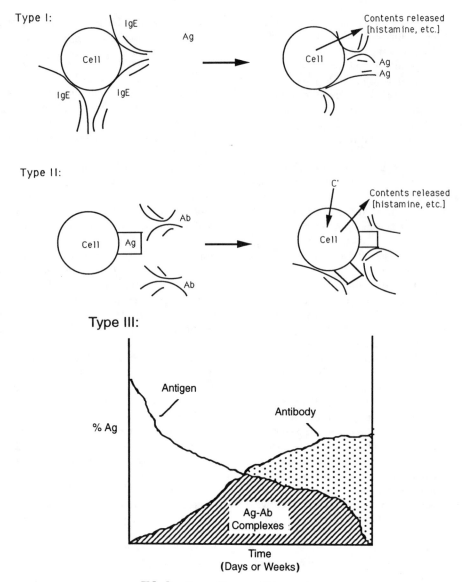

FIG. 3.   Types of hypersensitivity reactions.

body are present in the circulation at the same time, forming immune complexes that can lodge in the walls of blood vessels. These reactions are unlikely for biomaterials applications except for slowly releasing drug delivery and biodegradation systems.

Type IV hypersensitivity reactions do not involve the production of antibody, but rather involve the production of T cells that react with an antigen. This involves a complex interaction of T cells, macrophages, and soluble mediators. The common manifestation of type IV hypersensitivity is contact dermatitis, which is readily apparent as a skin rash that occurs 24–48 hr after local antigen contact, but systemic reactions are also possible. Type IV hypersensitivity occurs with plants such as poison ivy, industrial chemicals such as metal salts and photochemicals, and metal objects such as jewelry and buttons. Contact dermatitis or oral lesions have been seen with the use of

metallic biomaterials and acrylics. Deep tissue reactions of type IV hypersensitivity have been reported with the use of various biomaterials, including metals, silicones, and acrylics.

CONCLUSIONS

Systemic reactions to biomaterials that degrade or wear are possible. Toxic substances released from a biomaterial may damage a target organ. This can usually be tested for in pretrial screening of the biomaterial by analysis of the chemicals released and by tissue culture analysis of the material, leachables, and degradation and wear products. The biological response to wear products is important and controversial. Many biomaterials are composed of several components and the contribu-

Type IV:

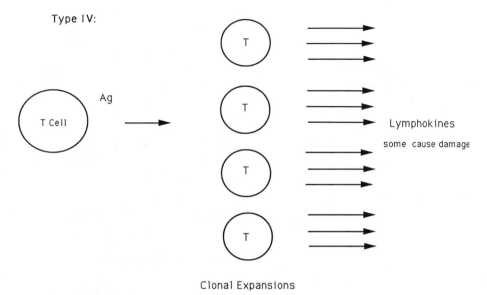

Clonal Expansions

FIG. 3—continued

tion of the wear products often cannot be attributed to only one component, such as in total joint replacements where the particles from wear can be metallic, polymeric (polyethylene and/or polymethylmethacrylate), and ceramic. Moreover, the relative contributions of particle shape, size, and chemistry are uncertain.

Systemic reactions that are a result of the immune response to the components of the biomaterial are harder to predict, difficult to test for, and need to be evaluated carefully in clinical trials. Since testing prior to release of a biomaterial for general use is extensive, there are relatively few reported clinical reactions of systemic toxicity, immune or nonimmune mediated, to biomaterials applications.

Reactions believed to be caused by hypersensitivity to biomaterials have been most thoroughly studied with metallic devices. Type IV reactions are the most common, but type I or II reactions have also been reported. Immune responses to collagen are of concern since these materials, especially when they are of non human origin, can be potent antigens. Reactions to silicone, as stated before, are controversial; type I, II, and IV responses are possible but difficult to document. Drug delivery systems or other degradative systems that slowly and continually release potentially antigenic substances into the body provide a model for the production of type III responses.

Bibliography

Benjamini, E., and Leskowitz, S. (1991). *Immunology, A Short Course.* Wiley–Liss, New York.

Cotran, R., Kumar, F., and Robbins, S. L. (1989). *Pathologic Basis of Disease.* Saunders, Philadelphia.

Guyton, A. C. (1991). *Textbook of Medical Physiology,* 8th Ed. Saunders, Philadelphia.

Halliwell, B., *et al.,* eds. (1988). *Oxygen Radicals and Tissue Injury.* Fed. Am. Soc. Exp. Biol., Bethesda, MD.

Merritt, K. (1986). Immunologic testing of biomaterials. in *Techniques of Biocompatibility Testing,* (D. F. Williams, ed., Vol. II. CRC Press, Boca Raton, FL.

Roitt, I., Brostoff, J., and Male, D. (1989). *Essential Immunology.* Blackwell Sci., Boston.

Rose, N. R., and Friedman, H. (1980). *Manual of Clinical Immunology.* Am. Soc. Microbiol., Washington, DC.

4.5 BLOOD COAGULATION AND BLOOD–MATERIALS INTERACTIONS

Stephen R. Hanson and Laurence A. Harker

The hemostatic mechanism is designed to arrest bleeding from injured blood vessels. The same process may produce adverse consequences when artificial surfaces are placed in contact with blood, and involves a complex set of interdependent reactions between (1) the surface, (2) platelets, and (3) coagulation proteins, resulting in the formation of a clot or thrombus which may subsequently undergo removal by (4) fibrinolysis. The process is localized at the surface by complicated activation and inhibition systems so that the fluidity of blood in the circulation is maintained. In this chapter, a brief overview of the hemostatic mechanism is presented. Although a great deal is known about blood responses to injured arteries and blood-contacting devices, important interrelationships are not fully defined in many instances. A more detailed discussion is given in recent reviews (Coleman *et al.,* 1994; Forbes and Courtney, 1987; Thompson and Harker, 1983).

PLATELETS

Platelets ("little plates") are non-nucleated, disk-shaped cells having a diameter of 3–4 μm, and an average volume of

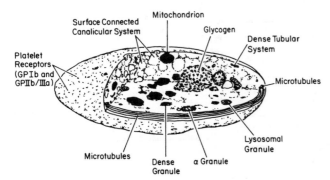

FIG. 1. Platelet structure.

10×10^{-9} mm³. Platelets are produced in the bone marrow, circulate at an average concentration of about 250,000 cells per microliter, and occupy approximately 0.3% of the total blood volume. In contrast, red cells typically circulate at 5×10^6 cells per microliter, and may comprise 40–50% of the total blood volume. As described later, platelet functions are designed to: (1) initially arrest bleeding through formation of platelet plugs, and (2) stabilize platelet plugs by catalyzing coagulation reactions, leading to the formation of fibrin.

Platelet structure provides a basis for understanding platelet function. In the normal (nonstimulated) state, the platelet discoid shape is maintained by a circumferential bundle (cytoskeleton) of microtubules (Fig. 1). The external surface coat of the platelet contains membrane-bound receptors (e.g., glycoproteins Ib and IIb/IIIa) that mediate the contact reactions of adhesion (platelet–surface) and aggregation (platelet–platelet). The membrane also provides a phospholipid surface which accelerates coagulation reactions (see later discussion), and forms a spongy, canal-like (canalicular) open network which

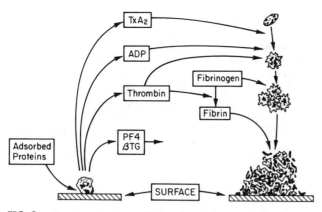

FIG. 2. Platelet reactions to artificial surfaces. Following protein adsorption to surfaces, platelets adhere and release α-granule contents, including platelet factor 4 (PF4) and β-thromboglobulin (βTG), and dense granule contents, including ADP. Thrombin is generated locally through factor XIIa and platelet procoagulant activity. Thromboxane A_2 (TxA_2) is synthesized. ADP, TxA_2, and thrombin recruit additional circulating platelets into an enlarging platelet aggregate. Thrombin-generated fibrin stabilizes the platelet mass.

represents an expanded reactive surface to which plasma factors are selectively adsorbed. Platelets contain substantial quantities of muscle protein (e.g., actin, myosin) which allow for internal contraction when platelets are activated. Platelets also contain three types of cytoplasmic storage granules: (1) α-granules, which are numerous and contain the platelet-specific proteins platelet factor 4 (PF-4) and β-thromboglobulin (β-TG), and proteins found in plasma (fibrinogen, albumin, fibronectin, coagulation factors V and VIII); (2) dense granules which contain adenosine diphosphate (ADP), calcium ions (Ca^{2+}), and serotonin; and (3) lysosomal granules containing enzymes (acid hydrolases).

Platelets are extremely sensitive cells that may respond to minimal stimulation. Activation causes platelets to become sticky and change in shape to irregular spheres with spiny pseudopods, accompanied by internal contraction and extrusion of the storage granule contents into the extracellular environment. These secreted platelet products stimulate other platelets, cause irreversible platelet aggregation, and lead to the formation of a fused platelet thrombus (Fig. 2).

Platelet Adhesion

Platelets adhere to artificial surfaces and injured blood vessels. At sites of vessel injury, the adhesion process involves the interaction of platelet glycoprotein Ib (GP Ib) and connective tissue elements which become exposed (e.g., collagen) and requires plasma von Willebrand factor (vWF) as an essential cofactor. GP Ib (~15,000 molecules per platelet) acts as the surface receptor for vWF. The hereditary absence of GP Ib or vWF results in defective platelet adhesion and serious abnormal bleeding.

Platelet adhesion to artificial surfaces may also be mediated through platelet glycoprotein IIb/IIIa, as well as through the GP Ib-vWF interaction. GP IIb/IIIa (~40,000 copies per platelet) is the platelet receptor for adhesive plasma proteins which support cell attachment, including fibrinogen, vWF, fibronectin, and vitronectin. Resting platelets do not bind these proteins, which normally occurs only after platelet activation causes a conformational change in GP IIb/IIIa. However, platelets activated near surfaces (for example, by exposure to factors released from already adherent cells) could adhere through this mechanism (e.g., to surface-adsorbed fibrinogen). Also, normally unactivated GP IIb/IIIa receptors could react with surface proteins which have undergone conformational changes as a result of the adsorption process (Chapter 3.2). The enhanced adhesiveness of platelets toward surfaces preadsorbed with fibrinogen supports this view. Following adhesion, activation, and release reactions, the expression of functionally competent GP IIb/IIIa receptors may also support tight binding and platelet spreading through multiple focal contacts with fibrinogen and other surface-adsorbed adhesive proteins.

Platelet Aggregation

Following platelet adhesion, a complex series of reactions is initiated involving: (1) the release of dense granule ADP,

(2) the formation of small amounts of thrombin (see later discussion), and (3) the activation of platelet biochemical processes to generate thromboxane A_2. The release of ADP, thrombin formation, and generation of thromboxane A_2 act in concert to recruit platelets into a growing platelet aggregate (Fig. 2). Platelet stimulation by these agonists causes the expression on the platelet surface of activated GP IIb/IIIa, which then binds plasma proteins that support platelet aggregation. In normal blood, fibrinogen, owing to its relatively high concentration (Table 1), is the most important protein supporting platelet aggregation. The platelet–platelet interaction involves Ca^{2+}-dependent bridging of adjacent platelets by fibrinogen molecules (platelets will not aggregate in the absence of fibrinogen, GP IIb/IIIa, or Ca^{2+}). Thrombin binds directly to platelets and plays a key role in platelet aggregate formation by: (1) activating platelets which then catalyze the production of more thrombin, (2) stimulating ADP release and thromboxane A_2 formation, and (3) stimulating formation of fibrin, which stabilizes the platelet thrombus.

Platelet Release Reaction

The release reaction is the secretory process by which substances stored in platelet granules are extruded from the platelet. ADP, collagen, epinephrine, and thrombin are physiologically important release-inducing agents, and interact with the platelet through specific receptors on the platelet surface. Alpha-granule contents (PF-4, β-TG, and other proteins) are readily released by relatively weak agonists such as ADP. Release of the dense granule contents (ADP, Ca^{2+}, and serotonin) requires platelet stimulation by a stronger agonist such as thrombin. Agonist binding to platelets also initiates the formation of intermediates that cause activation of the contractile-secretory apparatus, production of thromboxane A_2, and mobilization of calcium from intracellular storage sites. Elevated cytoplasmic calcium is probably the final mediator of platelet aggregation and release. As noted, substances which are released (ADP), synthesized (TxA_2), and generated (thrombin) as a result of platelet stimulation and release affect other platelets and actively promote their incorporation into growing platelet aggregates. *In vivo*, measurements of plasma levels of platelet-specific proteins (PF-4, β-TG) have been widely used as an indirect measure of platelet activation and release.

Platelet Coagulant Activity

When platelets aggregate, platelet coagulant activity is produced, including expression of membrane phospholipids which accelerate two critical steps of the blood coagulation sequence: factor X activation and the conversion of prothrombin to thrombin (see later discussion). Platelets may also promote the proteolytic activation of factors XII and XI. The surface of the aggregated platelets thus serves as a site where thrombin can form rapidly in excess of the capacity of the anticoagulant mechanisms of blood. Thrombin also activates platelets di-

rectly, and generates polymerizing fibrin, which adheres to the surface of the platelet mass.

Platelet Consumption

In man, platelets labeled with radioisotopes are cleared from circulating blood in an approximately linear fashion over time with an apparent lifespan of approximately 10 days. Platelet lifespan in experimental animals may be somewhat shorter. With the continuing thrombosis that may be produced by cardiovascular devices, platelets may be removed from circulating blood at a more rapid rate. Thus steady-state elevations in the rate of platelet destruction, as reflected in a shortening of platelet lifespan, have been used as a measure of the thrombogenicity of artificial surfaces and prosthetic devices (Hanson *et al.*, 1980).

COAGULATION

In the test tube, at least 12 plasma proteins interact in a series of reactions leading to blood clotting. Their designation as Roman numerals was made in order of discovery, often before their role in the clotting scheme was fully appreciated. Their biochemical properties are summarized in Table 1. Initiation of clotting occurs either intrinsically by surface-mediated reactions, or extrinsically through factors derived from tissues. The two systems converge upon a final common path which leads to the formation of an insoluble fibrin gel when thrombin acts on fibrinogen.

Coagulation proceeds through a "cascade" of reactions by which normally inactive factors (e.g., factor XII) become enzymatically active following surface contact, or after proteolytic cleavage by other enzymes (e.g., surface contact activates factor XII to factor XIIa). The newly activated enzymes in turn activate other normally inactive precursor molecules (e.g., factor XIIa converts factor XI to factor XIa). Because this sequence involves a series of steps, and because one enzyme molecule can activate many substrate molecules, the reactions are quickly amplified so that significant amounts of thrombin are produced, resulting in platelet activation, fibrin formation, and arrest of bleeding. The process is localized (i.e., widespread clotting does not occur) owing to dilution of activated factors by blood flow, the actions of inhibitors which are present or are generated in clotting blood, and because several reaction steps proceed at an effective rate only when catalyzed on the surface of activated platelets or at sites of tissue injury.

Figure 3 presents a scheme of the clotting factor interactions involved in both the intrinsic and extrinsic systems and their common path. Except for the contact phase, calcium is required for most reactions and is the reason why chelators of calcium (e.g., citrate) are effective anticoagulants. It is also clear that the *in vitro* interactions of clotting factors, clotting, is not identical with coagulation *in vivo*, which is triggered by artificial surfaces and by exposure of the cell-

TABLE 1 Properties of Human Clotting Factors

Clotting factor	Molecular weight (No. of chains)	Normal plasma concentration (μg/ml)	Active form
Intrinsic system			
Factor XII	80,000 (1)	30	Serine protease
Prekallikrein	80,000 (1)	50	Serine protease
High-molecular-weight kininogen	105,000 (1)	70	Cofactor
Factor XI	160,000 (2)	4	Serine protease
Factor IX	68,000 (1)	6	Serine protease
Factor VIII	265,000 (1)	0.1	Cofactor
VWF	1-15,000,000[a]	7	Cofactor for platelet adhesion
Extrinsic system			
Factor VII	47,000 (1)	0.5	Serine protease
Tissue factor	46,000 (1)	0	Cofactor
Common pathway			
Factor X	56,000 (2)	10	Serine protease
Factor V	330,000 (1)	7	Cofactor
Prothrombin	72,000 (1)	100	Serine protease
Fibrinogen	340,000 (6)	2500	Clot structure
Factor XIII	320,000 (4)	15	Transglutaminase

[a]Subunit molecular weight of factor VIII/vWF is around 220,000 with a series of multimers found in circulation.

associated protein, tissue factor. There are also interrelationships between the intrinsic and extrinsic systems such that under some conditions "crossover" or reciprocal activation reactions may be important (Coleman *et al.,* 1994; Bennett *et al.,* 1987).

MECHANISMS OF COAGULATION

In the intrinsic system, contact activation refers to reactions following adsorption of contact factors onto a negatively charged surface. Although these reactions are well understood *in vitro,* their pathologic significance remains uncertain. In hereditary disorders, only low levels of factor XI are associated with abnormal bleeding. Involved are factors XII, XI, prekallikrein, and high-molecular-weight kininogen (HMWK) (Fig. 4). All contact reactions take place in the absence of calcium. Kallikrein also participates in the fibrinolytic system and inflammation (Bennett *et al.,* 1987).

A middle phase of intrinsic clotting begins with the first calcium-dependent step, the activation of factor IX by factor XIa. Factor IXa subsequently activates factor X. Factor VIII is an essential cofactor in the intrinsic activation of factor X, and factor VIII first requires modification by an enzyme, such as thrombin, to exert its cofactor activity. In the presence of calcium, factors IXa and VIIIa form a complex on phospholipid surfaces (as expressed on the surface of activated platelets) to activate factor X. This reaction proceeds slowly in the absence of an appropriate phospholipid surface, and serves to localize the clotting reactions to the surface (vs. bulk fluid) phase.

The extrinsic system is initiated by the activation of factor VII. When factor VII interacts with tissue factor, an intracellular protein (i.e., one not found in plasma), factor VIIa becomes an active enzyme which is the extrinsic factor X activator. Tissue factor is present in many body tissues; is expressed by stimulated white cells; and becomes available when underlying vascular structures are exposed to flowing blood upon vessel injury.

The common path begins when factor X is activated by either factor VIIa-tissue factor by or the factor IXa–VIIIa complex. After formation of factor Xa, the next step involves factor V, a cofactor, which (like factor VIII) has activity after modification by another enzyme such as thrombin. Factor Xa-Va, in the presence of calcium and platelet phospholipids, then converts prothrombin (factor II) to thrombin. Like the conver-

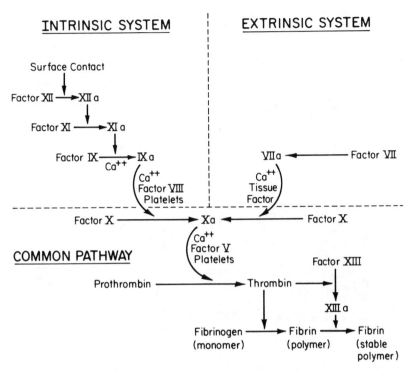

FIG. 3. Mechanisms of clotting factor interactions. Clotting is initiated by either an intrinsic or extrinsic pathway with subsequent factor interactions which converge upon a final, common path.

sion of factor X, prothrombin activation is effectively surface catalyzed. The higher plasma concentration of prothrombin (Table 1), as well as the biologic amplification of the clotting system, allows a few molecules of activated initiator to generate a large burst of thrombin activity. Thrombin, in addition to its ability to modify factors V and VIII and activate platelets, acts on two substrate: fibrinogen and factor XIII. The action of thrombin on fibrinogen releases small peptides from fibrinogen (e.g., fibrinopeptide A) which can be assayed in plasma as

evidence of thrombin activity. The fibrin monomers so formed polymerize to become a gel. Factor XIII is either trapped within the clot or provided by platelets, and is activated directly by thrombin. A tough, insoluble fibrin polymer is formed by interaction of the fibrin polymer with factor XIIIa.

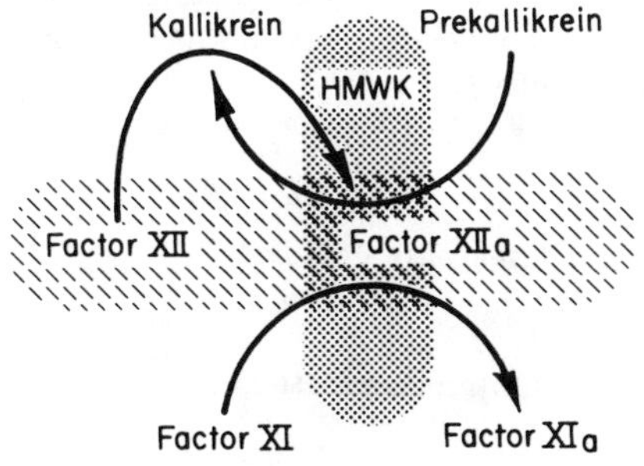

CONTROL MECHANISMS

Obviously, the body has mechanisms for avoiding massive thrombus formation once coagulation is initiated. At least four types of mechanisms may be considered. First, blood flow may reduce the localized concentration of precursors and remove activated materials by dilution into a larger volume. Also related to blood flow is the rapid removal of inactivated factors by passage through the liver. Second, the rate of several clotting reactions is fast only when the reaction is catalyzed by a surface. These reactions include the contact reactions, the activation of factor X by factor VII-tissue factor at sites of tissue injury, and reactions which are accelerated by locally deposited platelet masses (activation of factor X and prothrombin). Third, there are naturally occurring inhibitors of coagulation enzymes, such as antithrombin III, which are potent inhibitors of thrombin and other coagulation enzymes (plasma levels of thrombin–antithrombin III complex can also be assayed as a measure of thrombin production *in vivo*). Fourth, during the process of coagulation, enzymes are generated which not only activate

FIG. 4. Contact activation. The initial event *in vitro* is the adsorption of factor XII to a negatively charged surface (hatched, horizontal ovoid) where it is activated to form factor XIIa. Factor XIIa converts prekallikrein to kallikrein. Additional factor XIIa and kallikrein are then generated by reciprocal activation. Factor XIIa also activates factor XIa. Both prekallikrein and factor XI bind to a cofactor, high-molecular-weight kininogen (HMWK; dotted, vertical ovoid), which anchors them to the charged surface.

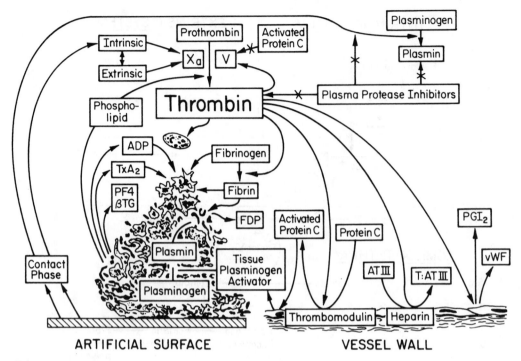

FIG. 5.   Integrated hemostatic reactions between a foreign surface and platelets, coagulation factors, the vessel endothelium, and the fibrinolytic system.

coagulation factors, but also degrade cofactors. For example, the fibrinolytic enzyme plasmin (see below) degrades fibrinogen and fibrin monomers, and can inactivate cofactors V and VIII. Thrombin is also removed when it binds to thrombomodulin, a protein found on the surface of blood vessel endothelial cells. The thrombin–thrombomodulin complex then converts another plasma protein, protein C, to an active form which can also degrade factors V and VIII. The relative importance of these inactivation pathways is only now becoming understood (Hanson *et al.,* 1993).

In summary, the platelet, coagulation, and endothelial systems interact in a number of ways which promote localized hemostasis while preventing generalized thrombosis. Figure 5 depicts some of the relationships and inhibitory pathways which apply to blood reactions following contact with both natural and artificial surfaces.

Fibrinolysis

The fibrinolytic system removes unwanted fibrin deposits to improve blood flow following thrombus formation, and to facilitate the healing process after injury and inflammation. It is a multicomponent system composed of precursors, activators, cofactors and inhibitors, and has been studied extensively (Coleman *et al.,* 1994; Forbes and Courtney, 1987; Thompson and Harker, 1983). The fibrinolytic system also interacts with the coagulation system at the level of contact activation (Ben-

nett *et al.,* 1987). A simplified scheme of the fibrinolytic pathway is shown in Fig. 6.

The most well-studied fibrinolytic enzyme is plasmin, which circulates in the inactive form as the protein plasminogen. Plasminogen adheres to a fibrin clot, being incorporated into the mesh during polymerization. Plasminogen is activated to plasmin by the actions of plasminogen activators which may be present in blood or released from tissues, or which may be administered therapeutically. Important plasminogen activators occuring naturally in man include tissue plasminogen activator (tPA) and urokinase. Following activation, plasmin digests the fibrin clot, releasing soluble fibrin–fibrinogen digestion products (FDP) into circulating blood, which may be assayed as markers of *in vivo* fibrinolysis (e.g., the fibrin D-D dimer fragment).

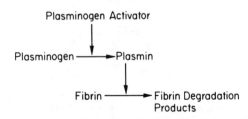

FIG. 6.   Fibrinolytic sequence. Plasminogen activators, such as tissue plasminogen activator (tPA) or urokinase, activate plasminogen to form plasmin. Plasmin enzymatically cleaves insoluble fibrin polymers into soluble degradation products (FDP), thereby effecting the removal of unnecessary fibrin clot.

Complement

The complement system is primarily designed to effect a biologic response to antigen–antibody reactions (Chapter 4.3). Like the coagulation and fibrinolytic systems, complement proteins are activated enzymatically through a complex series of reaction steps (Bennett *et al.*, 1987). Several proteins in the complement cascade function as inflammatory mediators. The end result of these activation steps is the generation of an enzymatic complex which causes irreversible damage (by lytic mechanisms) to the membrane of the antigen-carrying cell (e.g., bacteria).

There are a number of interactions between the complement, coagulation, and fibrinolytic systems which have been reviewed elsewhere (Bennett *et al.*, 1987). Thus, there has been considerable interest in the problem of complement activation by artificial surfaces, which has been prompted in particular by observations that devices having large surface areas (e.g., hemodialyzers) may cause (1) reciprocal activation reactions between complement enzymes and white cells, and (2) complement activation which may mediate both white cell and platelet adhesion to artificial surfaces. Further observations regarding the complement activation pathways involved in blood–materials interactions are likely to yield important insights.

Red Cells

Red cells are usually considered as passive participants in processes of hemostasis and thrombosis, although under some conditions (low shear or venous flows) red cells may comprise a large proportion of total thrombus mass. The concentration and motions of red cells have important mechanical effects on the diffusive transport of blood elements. Under some conditions, red cells may also contribute chemical factors that influence platelet reactivity (Turitto and Weiss, 1980). The process of direct attachment of red cells to artificial surfaces has been considered to be of minor importance, and has therefore received little attention in studies of blood–materials interactions.

White Cells

The various classes of white cells perform many functions in inflammation, infection, wound healing, and the blood response to foreign materials (Chapter 4.2). White cell interactions with artificial surfaces may proceed through as-yet poorly defined mechanisms related to activation of the complement, coagulation, fibrinolytic, and other enzyme systems, resulting in the expression by white cells of procoagulant, fibrinolytic, and inflammatory activities. For example, stimulated monocytes express tissue factor, which can initiate extrinsic coagulation. Neutrophils may contribute to clot dissolution by releasing potent fibrinolytic enzymes (e.g., neutrophil elastase). White cell interactions with devices having large surface areas may be extensive (especially with surfaces that activate complement), resulting in their marked depletion from circulating blood. Activated white cells, through their enzymatic and other activities, may produce organ dysfunction in other parts of the body. In general, the importance of white cell mechanisms of thrombosis and thrombolysis, in relation to other pathways, remains to be determined.

CONCLUSIONS

Interrelated blood systems respond to tissue injury in order to quickly minimize blood loss, and later to remove unneeded deposits after healing has occurred. When artificial surfaces are exposed, an imbalance between the processes of activation and inhibition of these systems can lead to excessive thrombus formation and an exaggerated inflammatory response. While many of the key blood cells, proteins, and reaction steps have been identified, their roles in limiting the usefulness of most cardiovascular devices is not fully understood. This will ensure future interest in studying blood reactions triggered by surface contact.

Acknowledgment

This work was supported by research grants HL-31469 and HL-31950 from the National Institutes of Health, U.S. Public Health Service.

Bibliography

Bennett, B., Booth, N. A., and Ogston, D. (1987). Potential interactions between complement, coagulation, fibrinolysis, kinin-forming, and other enzyme systems. in *Haemostasis and Thrombosis*, 2nd ed. A. L. Bloom and D. P. Thomas, eds. Churchill Livingstone, New York, pp. 267–282.

R. W. Colman, J. Hirsch, V. J. Marder, and E. W. Salzman (eds.) (1994). *Hemostasis and Thrombosis*, 3rd ed. Lippincott, New York, pp. 1–660.

Forbes, C. D., and Courtney, J. M. (1987). Thrombosis and artificial surfaces. in *Haemostasis and Thrombosis*, 2nd ed. A. L. Bloom and D. P. Thomas, eds. Churchill Livingstone, New York, pp. 902–921.

Hanson, S. R., Harker, L. A., Ratner, B. D., and Hoffman, A. S. (1980). *In vivo* evaluation of artificial surfaces using a nonhuman primate model of arterial thrombosis. *J. Lab. Clin. Med.* 95: 289–304.

Hanson, S. R., Griffin, J. H., Harker, L. A., Kelly, A. B., Esmon, C. T., and Gruber, A. (1993). Antithrombotic effects of thrombin-induced activation of endogenous Protein C in primates. *J. Clin. Invest.* 92: 2003–2012.

Thompson, A. R., and Harker, L. A. (1983). *Manual of Hemostasis and Thrombosis*, 3rd ed. F. A. Davis, Philadelphia, pp. 1–219.

Turitto, V. T., and Weiss, H. J. (1980). Red cells: Their dual role in thrombus formation. *Science* 207: 541–544.

4.6 TUMORIGENESIS AND BIOMATERIALS

Frederick J. Schoen

GENERAL CONCEPTS

Neoplasia, which literally means "new growth," is the process of excessive and uncontrolled cell proliferation (Cotran *et al.*, 1994). The new growth is called a neoplasma or tumor (i.e., a swelling, since most neoplasms are expansile, solid masses of abnormal tissue). Benign tumors do not penetrate (invade) adjacent tissues, nor do they spread to distant sites. They remain localized, and surgical excision can be curative in many cases. In contrast, malignant tumors, or cancers, not only have a propensity to invade contiguous tissues, but also have the ability to gain entrance into blood and lymph vessels. Thus, cells from a malignant neoplasm can be transported to distant sites, where subpopulations of malignant cells take up residence, grow, and again invade as satellite tumors (metastases). Most cancers do not have an identifiable cause and the general mechanisms of carcinogenesis remain largely obscure.

The primary descriptor of any tumor is its cell or tissue of origin. Benign tumors are identified by the suffix "oma," which is preceded by reference to the cell or tissue of origin (e.g., adenoma—from an endocrine gland; chondroma—from cartilage). The malignant counterparts of benign tumors carry similar names, except that the suffix carcinoma is applied to cancers derived from epithelium (e.g., squamous- or adeno-carcinoma) and sarcoma (e.g., osteo- or osteogenic or chondro-sarcoma) to those of mesenchymal origin. Neoplasms of the hematopoietic system, in which the cancerous cells are circulating in blood, are called leukemias; solid tumors of lymphoid tissue are called lymphomas. The major classes of tumors are illustrated in Fig. 1.

Cancer cells express varying degrees of resemblance to the normal precursor cells from which they derive. Thus, neoplastic growth entails both abnormal cellular proliferation and modification of the structural and functional characteristics of the cell types involved. Malignant cells are generally less differentiated than normal cells. The structural similarity of cancer cells to those of the tissue of origin enables specific diagnosis (source organ and cell type); moreover, the degree of resemblance usually predicts the biologic behavior of the cancer (which determines the expected outcome for, often called prognosis of, the patient). Therefore, poorly differentiated tumors generally are more aggressive (i.e., display more malignant behavior) than those that are better differentiated than their malignant counterparts. The degree to which a tumor mimics a normal cell or tissue type is called its grade of differentiation. The extent of spread and other effects on the host determine its stage.

Neoplastic growth is unregulated. Neoplastic cell proliferation is therefore unrelated to the physiological requirements of the tissue, and is unaffected by removal of the stimulus which initially caused it. These characteristics differentiate neoplasms from (1) normal proliferations during fetal or postnatal growth, (2) normal wound healing following an injury, and (3) hyperplastic growth to adapt to a physiological need, but which ceases when the stimulus is removed.

All tumors, benign and malignant, have two basic components: (1) proliferating neoplastic cells that constitute their parenchyma, and (2) supportive stroma made up of connective tissue and blood vessels. Although the parenchyma of neoplasms represents their underlying nature, the growth and evolution of neoplasms are critically dependent on the stroma, usually composed of blood vessels, connective tissue, and inflammatory cells. The characteristics of benign and malignant tumors are summarized in Table 1.

ASSOCIATION OF IMPLANTS WITH HUMAN AND ANIMAL TUMORS

The possibility that implant materials could cause tumors or promote tumor growth has long been a concern of surgeons and biomaterials researchers. Although cases of both human and veterinary implant-related tumors have been reported, neoplasms occurring at the site of implanted medical devices are unusual, despite the large numbers of implants used clinically over an extended period of time (Black, 1988; Pedley *et al.*, 1981; Schoen, 1987).

Whether there is a causal role for implanted medical devices in local or distant malignancy remains controversial. However, recent cohorts of patients with both total hip replacement and breast implants show no detectable increases in tumors at the implant site (Berkel *et al.*, 1992; Deapen and Brody, 1991; Mathiesen *et al.*, 1995). Indeed, a recent clinical and experimental study suggested a protective effect of silicone against breast cancer (Su *et al.*, 1995). However, one study suggested a small increase in the number of lung and vulvar cancers in patients with breast implants (Deapen and Brody, 1991).

The vast majority of malignant neoplasms induced by clinical and experimental foreign bodies in both animals and humans are sarcomas of various histologic subtypes, incuding fibrosarcoma, osteosarcoma (osteogenic sarcoma), chondrosarcoma, malignant fibrous histiocytoma, angiosarcoma, etc., and are characterized by rapid and locally infiltrative growth. Carcinomas, reported far less frequently, have usually been restricted to situations where an implant has been placed in the lumen of an epithelium-lined organ. Reported cases of human implant-related tumors include sarcomatous lesions arising adjacent to metallic orthopedic implants (including fracture fixation devices and total joint replacements) and vascular grafts. Illustrative reported cases are noted in Table 2; descriptions of other cases are available (Goodfellow, 1992; Jennings *et al.*, 1988; Jacobs *et al.*, 1992). A tumor forming adjacent to a clinical vascular graft is illustrated in Fig. 2. A distant primary tumor (gastric cancer) metastasizing to a total knee replacement has also been reported (Kolstad and Högstorp, 1990).

Caution is necessary in implicting the implant in the formation of a neoplasma; demonstration of a tumor occurring adjacent to an implant does not necessarily prove that the implant caused the tumor. Neoplasms are common in both humans and animals and can occur naturally at the sites at which biomaterials are implanted. Most clinical veterinary cases have been observed in dogs, a species with a relatively high natural frequency of osteosarcoma and other tumors at sites where

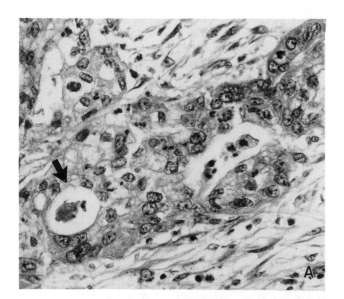

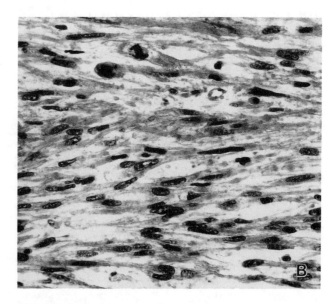

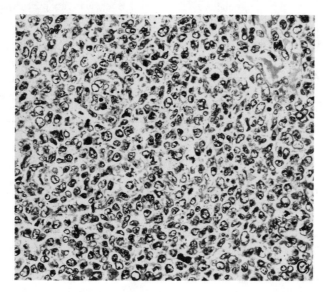

FIG. 1.   Types of malignant tumors. (A) Carcinoma, exemplified by an adenocarcinoma (gland formation noted by arrow). (B) Sarcoma (composed of spindle cells). (C) Lymphoma (composed of malignant lymphocytes). All stained with hematoxylin and eosin; all ×310.

orthopedic devices are implanted. Moreover, spontaneous human musculoskeletal tumors are not unusual. However, since sarcomas arising in the aorta and other large arteries are rare, the association of primary vascular malignancies with clinical polymeric grafts may be stronger than that with orthopedic devices. Despite the possibility that the initiating factor for orthopedic implant tumorigenesis could be metal particulates that are worn off the implant (Harris, 1994), no unequivocal case of metal particles provoking malignant transformation of tissue has been reported.

Benign but exuberant foreign body reactions may simulate neoplasms. For example, fibrohistiocytic lesions resembling malignant tumors may occur as a reaction to silica, previously injected as a soft tissue sclerosing agent (Weiss *et al.*, 1978), and foreign body reaction to silicone that has migrated from either finger joint or breast prostheses to lymph nodes could cause a mass lesion that stimulates a neoplasm on physical examination (Christie *et al.*, 1977; Hausner *et al.*, 1978).

Implant-related tumors have been reported both short and long term following implantation. The period of latency is usually relatively long, but over 25% of tumors associated with foreign bodies have developed within 15 years, and over 50% within 25 years (Brand and Brand, 1980). Cancer at foreign body sites may be mechanistically related to that which occurs in association with asbestosis (i.e., lung damage caused by chronic inhalation of asbestos), lung or liver scarring, or

TABLE 1   Characteristics of Benign and Malignant Tumors

Characteristics	Benign	Malignant
Differentiation	Well defined; structure may be typical of tissue of origin	Less differentiated with bizarre (anaplastic) cells; often atypical structure
Rate of growth	Usually progressive and slow; may come to a standstill or regress; cells in mitosis are rare	Erratic, and may be slow to rapid; mitoses may be absent to numerous and abnormal
Local invasion	Usually cohesive, expansile, well-demarcated masses that neither invade nor infiltrate the surrounding normal tissues	Locally invasive, infiltrating adjacent normal tissues
Metastasis	Absent	Frequently present; larger and more undifferentiated primary tumors are more likely to metastasize

chronic bone infections; all are diseases in which tissue fibrosis is a prominent characteristic (Brand, 1982). However, in contrast to the mesenchymal origin of most implant-related tumors, other cancers associated with scarring are generally derived from adjacent epithelial structures (e.g., mesothelioma with asebestosis). Moreover, chemical carcinogens, such as nitrosamines or those in tobacco smoke, may potentate scar-associated cancers.

PATHOBIOLOGY OF FOREIGN BODY TUMORIGENESIS

The pathogenesis of implant-induced tumors is not well understood, yet most experimental data indicate that physical effects rather than the chemical characteristics of the foreign body primarily determine tumorigenicity (Brand *et al.*, 1975). Tumors are induced experimentally by materials of any kind or composition, including those that could be considered essentially nonreactive, such as certain glasses, gold or platinum, and other relatively pure metals and polymers. Solid materials with a high surface area are most tumorigenic. Materials lose their tumorogenicity when implanted in pulverized, finely shredded, or woven form, or when surface continuity is interrupted by multiple perforations. This trend is often called the Oppenheimer effect. Thus, foreign body neoplasia is a transformation process mediated by the physical state of implants and is largely independent of their composition, so long as they are sufficiently chemically inert.

TABLE 2   Tumors Associated with Implant Sites in Humans—Representative Reports

Device (adjacent material)[a]	Tumor[b]	References	Postimplantation (years)
Fracture fixation			
Intramedullary rod (V)	L	MacDonald (1981)	17
Smith-Petersen (V)	OS	Ward *et al.* (1987)	9
Total hip			
Charnley–Mueller (UHMWPE, PMMA)	MFH	Bago-Granell *et al.* (1984)	2
Mittlemeier (Al_2O_3)	STS	Ryu *et al.* (1987)	1+
Charnley-Mueller (UHMWPE)	OS	Martin *et al.* (1988)	10
Charnley–Mueller (SS, PMMA)	SS	Lamovec *et al.* (1988)	12
Total knee			
Unknown (V)	ES	Weber (1986)	4
Vascular graft			
Abdominal aortic graft (D)	MFH	Weinberg *et al.* (1980)	1+
Abdominal aortic graft (D)	AS	Fehrenbacker *et al.* (1981)	12

[a]Materials: D, Dacron; PMMA, poly(methacrylate) bone cement; SS, stainless steel, UHMWPE, ultrahigh molecular-weight-polyethylene; V, Vitallium.

[b]Tumor types: AS, angiosarcoma; ES, epithelioid sarcoma; L, lymphoma; MFH, malignant fibrous histiocytoma; OS, osteosarcoma; SS, synovial sarcoma; STS, soft tissue sarcoma.

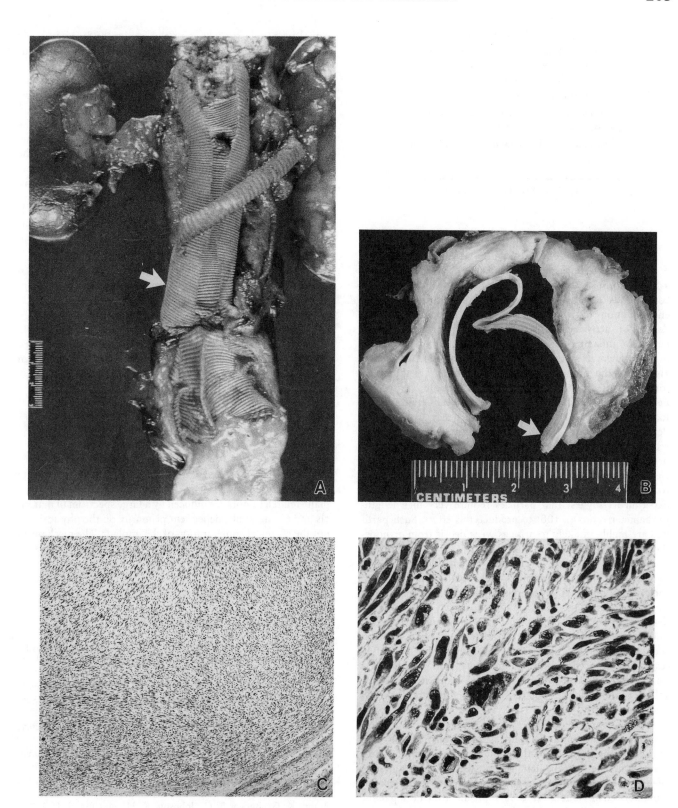

FIG. 2. Sarcoma arising 1 year following and in association with Dacron graft repair of abdominal aortic aneurysm. (A) and (B) Gross photographs (graft designated by arrow). (C) and (D) Histologic appearance of tumor. (C) and (D) Stained with hematoxylin and eosin. (C) ×49, (D) ×300. [(A), (C), and (D) reproduced by permission from D. S. Weinberg and B. S. Maini, *Cancer* **46**, 398–402, 1980.]

TABLE 3 Steps in Implant-Associated Tumorigenesis: A Hypothesis[a]

---

1. Cellular foreign body reaction
2. Fibrous capsule formation
3. Preneoplastic cells contact implant surface during quiescent tissue reaction
4. Preneoplastic cell maturation and proliferation
5. Tumor growth

---

[a]Following K. G. Brand and colleagues.

Solid-state tumorigenesis depends on the development of a bland fibrous capsule around the implant. Tumorigenicity corresponds directly to the extent and maturity of tissue encapsulation of a foreign body and inversely with the degree of active cellular inflammation. Thus, an active, persistent inflammatory response inhibits tumor formation in experimental systems. Host (especially genetic) factors also affect the propensity to form tumors as a response to foreign bodies. Humans are less susceptible to foreign body tumorigenesis than are rodents, the usual experimental model. In rodent systems, tumor frequency and latency depend on species, strain, sex, and age. Questions have recently been raised over the possibility that foreign body neoplasia can be induced by the release of needlelike particles from composites in a mechanism that is analogous to that of asbestos-related mesothelioma. However, animal experiments suggest that particles require very high length-to-diameter ratios (>100) to produce this effect. Such particles are highly unlikely to arise as wear debris from orthopedic implants.

Chemical induction of tumors is possible and merits concern. Implants of chromium, nickel, cobalt, and some of their compounds, either as foils or debris, are carcinogenic in rodents (Swierenza *et al.*, 1987). Moreover, even "nonbiodegradable" and "inert" implants have been shown to contain and/or release trace amounts of substances such as remnant monomers, catalysts, plasticizers, and antioxidants. Nevertheless, such substances injected in experimental animals at appropriate test sites (without implants) in quantities comparable to those found adjacent to implants, are generally not tumorigenic.

Foreign body tumorigenesis is characterized by a single transformation event and a long resting or latent period, during which the presence of the implant is required for tumor formation. The available data suggest the following sequence of essential developmental stages in foreign body tumorigenesis (summarized in Table 3); (1) cellular proliferation during tissue inflammation and the acute foreign body reaction (specific susceptible preneoplastic cells may be present at this stage); (2) progressive formation of a well-demarcated fibrotic tissue capsule surrounding the implant; (3) quiescence of the tissue reaction (i.e., dormancy and phagocytic inactivity of macrophages attached to the foreign body), but direct contact of clonal preneoplastic cells with the foreign body surface; (4) final maturation of preneoplastic cells; and (5) sarcomatous proliferation.

The essential hypothesis is that initial acquisition of neoplastic potential and the determination of specific tumor characteristics does not depend on direct physical or chemical reaction between susceptible cells and the foreign body, and, thus, the foreign body per se probably does not initiate the tumor. However, although the critical initial event occurs early during the foreign body reaction, the final step to neoplastic autonomy is accomplished only when preneoplastic cells attach themselves to the foreign body surface. Subsequently, maturation and proliferation of abnormal mesenchymal cells occur in this relatively quiescent microenvironment, a situation not permitted with the prolonged active inflammation associated with less inert implants.

Thus, the critical factors in sarcomas induced by foreign bodies include implant configuration, fibrous capsule development, and a period of latency long enough to allow progression to neoplasia in a susceptible host. The major role of the foreign body itself seems to be that of stimulating cell maturation and proliferation. The rarity of human foreign body-associated tumors suggests infrequent cancer-prone cells in human foreign body reactions.

CONCLUSIONS

Neoplasms associated with therapeutic clinical implants are rare; causality is difficult to demonstrate. Experimental implant-related tumors are induced by a large spectrum of materials and biomaterials, dependent primarily on the physical and not the chemical configuration of the implant. The mechanism of tumor formation, yet incompletely understood, appears related to the implant fibrous capsule.

Bibliography

Bago-Granell, J., Aguirre-Canyadell, M., Nardi, J., and Tallada, N. (1984). Malignant fibrous histiocytoma at the site of a total hip arthroplasty. *J. Bone Joint Surg.* **66B:** 38–40.

Berkel, H., Birdsell, D. C., and Jenkins, H. (1992). Breast augmentation: A risk factor for breast cancer? *N. Engl. J. Med.* **326:** 1649–1653.

Black, J. (1988). *Orthopedic Biomaterials in Research and Practice.* Churchill-Livingstone, New York, pp. 1–394.

Brand, K. G. (1982). Cancer associated with asbestosis, schistosomiasis, foreign bodies and scars. in *Cancer: A Comprehensive Treatise,* 2nd ed., F. F. Becker, ed. Plenum Publ. New York, Vol. I, pp. 661–692.

Brand, K. G., Buoen, L. C., Johnson, K. H., and Brand, I. (1975). Etiological factors, stages, and the role of the foreign body in foreign body tumorigenesis: A review. *Cancer Res.* **35:** 279–286.

Brand, K. G., and Brand, I. (1980). Risk assessment of carcinogenesis at implantation sites. *Plast. Reconstr. Surg.* **66:** 591–595.

Brittberg, M., Lindahl, A., Nilsson, A., Ohlsson, C., Isaksson, O., and Peterson, L. (1994). Treatment of deep cartilage defects in the knee with autologous chondrocyte transplantation. *N. Engl. J. Med.* **331:** 889–895.

Christie, A. J., Weinberger, K. A., and Dietrich, M. (1977). Silicone lymphadenopathy and synovitis. Complications of silicone elastomer finger joint prostheses. *JAMA* **237:** 1463–1464.

Cotran, R. S., Kumar, V., and Robbins, S. L., eds. (1994). *Robbins Pathologic Basis of Disease*, 5th Ed., Saunders, Philadelphia.

Deapen, D. M., and Brody, G. S. (1991). Augmentation mammaplasty and breast cancer: A 5-year update of the Los Angeles study. *Mammaplast Breast Cancer* **89:** 660–665.

Fehrenbacker, J. W., Bowers, W., Strate, R., and Pittman, J. (1981). Angiosarcoma of the aorta associated with a Dacron graft. *Ann. Thorac. Surg.* **32:** 297–301.

Goodfellow, J. (1992). Malignancy and joint replacement. *J. Bone Joint. Surg.* **74B:** 645.

Harris, W. H. (1994). Osteolysis and particle disease in hip replacement. *Acta Orth. Scand.* **65:** 113–123.

Hausner, R. J., Schoen, F. J., and Pierson, K. K. (1978). Foreign body reaction to silicone in axillary lymph nodes after prosthetic augmentation mammoplasty. *Plast. Reconst. Surg.* **62:**381–384.

Jacobs, J. J., Rosenbaum, D. H., Hay, R. M., Gitelis, S., and Black, J. (1992). Early sarcomatous degeneration near a cementless hip replacement. *J. Bone Joint. Surg. Br. Vol.* **74B:** 740–744.

Jennings, T. A., Peterson, L., Axiotis, C. A., Freidlander, G. E., Cooke, R. A., and Rosai, J. (1988). Angiosarcoma associated with foreign body material. A report of three cases. *Cancer* **62:** 2436–2444.

Kolstad, K., and Högstorp, H. (1990). Gastric carcinoma metastasis to a knee with a newly inserted prosthesis. *Acta Orth. Scand.* **61:** 369–370.

Lamovec, J., Zidar, A., and Cucek-Plenicar, M. (1988). Synovial sarcoma associated with total hip replacement. *J. Bone Joint Surg.* **70A:** 1558–1560.

Martin, A., Bauer, T. W., Manley, M. T., and Marks, K. H. (1988). Osteosarcoma at the site of total hip replacement. *J. Bone Joint Surg.* **70A:** 1561–1567.

Mathiesen, E. B., Ahlbom, A., Bermann, G., and Lindsgren, J. U. (1995). Total hip replacement and cancer. A cohort study. *J. Bone Joint. Surg. Br. Vol.* **77B:** 345–350.

McDonald, W. (1980). Malignant lymphoma associated with internal fixation of a fractured tibia. *Cancer* **48:** 1009–1011.

Pedley, R. B., Meachim, G., and Williams, D. F. (1981). Tumor induction by implant materials. in *Fundamental Aspects of Biocompatibility*, D. F. Williams, ed. CRC Press, Boca Raton, FL, Vol. II, pp. 175–202.

Ryu, R. K. N., Bovill, E. G., Jr, Skinner, H. B., and Murray, W. R. (1987). Soft tissue sarcoma associated with aluminum oxide ceramic total hip arthroplasty. A case report. *Clin. Orth. Rel. Res.* **216:** 207–212.

Schoen, F. J. (1987). Biomaterials-associated infection, tumorigenesis and calcification. *Trans. Am. Soc. Artif. Int. Organs* **33:** 8–18.

Su, C. W., Dreyfuss, D. A., Krizek, T. J., and Leoni, K. J. (1995). Silicone implants and the inhibition of cancer. *Plast. Reconstr. Surg.* **96:** 513–520.

Swierenza, S. H. H., Gilman, J. P. W., and McLean, J. R. (1987). Cancer risk from inorganics. *Cancer Metas. Rev.* **6:** 113–154.

Ward, J. J., Dunham, W. K., Thornbury, D. D., and Lemons, J. E. (1987). Metal-induced sarcoma. *Trans. Soc. Biomater.* **10:** 106.

Weber, P. C. (1986). Epithelioid sarcoma in association with total knee replacement. A case report. *J. Bone Joint Surg.* **68B:** 824–826.

Weinberg, D. S., and Maini, B. S. (1980). Primary sarcoma of the aorta associated with a vascular prosthesis. A case report. *Cancer* **46:** 398–402.

Weiss, S. W., Enzinger, F. M., and Johnson, F. B. (1978). Silica reaction simulating fibrous histocytoma. *Cancer* **42:** 2738–2743.

4.7 IMPLANT-ASSOCIATED INFECTION

Anthony G. Gristina and Paul T. Naylor

SIGNIFICANCE AND SCOPE

Infections involving artificial organs, synthetic vessels, joint replacements, or internal fixation devices usually require reoperation, and may result in amputation, osteomyelitis, or death. Infected cardiac, abdominal, and extremity vascular prostheses result in amputation or death in 25–50% of cases (Dougherty and Simmons, 1982; Gristina and Costerton, 1984; Gristina and Kolkin, 1989; Gristina *et al.*, 1985). Intravenous catheters, periotoneal dialysis, and urologic devices used for more than a few days frequently become infected or cause secondary tissue-sited infections. The aged or immuno-compromised are even more vulnerable to infection. The rate of infection for the total artificial heart approaches 100% when the heart is implanted for more then 90 days (DeVries, 1988; Gristina *et al.*, 1985). The use and development of implanted artificial organs is at a critical pass because of infectious complications.

HISTORY

The relevance of biomaterial and foreign bodies to infection was first suggested by Elek and Conen (1957) in a landmark experiment in 1957 when they reported that a foreign body (a silk suture) decreased the numbers of the infection-producing inoculum from 10^6 *Staphylococcus aureus* to 100 organisms. In 1963, after observing the high rate of infection associated with stainless steel implants in orthopedic surgery (10%), Gristina suggested that the internal fixation devices, biomaterials, and dead bone provided a structural framework along which microorganisms (*S. aureus* and *S. albus*) colonized and propagated (Gristina and Rovere, 1963). In 1972, Bayston and Penny described the excessive production of an mucoid substance as a possible virulence factor in the colonization of plastic neurosurgical shunts in children by *S. epidermidis* (Bayston and Penny, 1972).

Gibbons and van Houte stimulated research in 1975 when they described the attachment and complex interactions of *Streptococcus mutans* with dental pellicle and plaque (Gibbons and van Houte, 1975).

In 1976, Gristina *et al.* reported that surface effects and sequestration of ions and substances from metals, polymers, or organic substrata were factors in bacterial colonization of foreign bodies and biomaterials. In this study, some reactive metals inhibited bacterial growth, but growth was abundant adjacent to "inert" substrata such as stainless steel and polymeric sutures. Biomaterials that resisted infection were tolerated by tissues and augmented host defenses were proposed. In 1979 bacterial adhesion and biofilm-mediated polymicrobial infections were demonstrated on tissue and internal fixation devices retrieved from infected surgical wounds (Gristina *et al.*, 1980a, b). The authors hypothesized that microbial adhesion to biomaterials and compromised tissues, and production of extracapsular polysaccharides (slime) were the molecular mechanisms for and explained in part (1) the susceptibility of biomaterial sites to infection (Gristina *et al.*, 1980a, b), (2) resistance to antibiotic treatment and host defenses, (3) difficulties in identifying the organisms involved, and (4) the persistence of infection until the prosthesis was removed (Gristina *et al.*, 1980a, b).

In 1980, Beachey emphasized that bacterial adherence to tissue cells was the primary molecular mechanism for bacterial infection in man and animals. In 1980 and 1981, Gristina *et al.* (1980a, b), Costerton and Irvin (1981), Peters *et al.* (1981), and Beachey (1981) reported bacterial adhesion to metals and polymers as a cause of biomaterial-centered infection.

In 1985, Gristina (Gristina *et al.*, 1985) indicated that compromised tissues and bone were the target substrata for bacterial colonization in osteomyelitis. In 1988, cartilage and collagen were clearly identified as the specific ligands and substrata for bacterial adhesion in intraarticular sepsis (Voytek *et al.*, 1988).

THE RACE FOR THE SURFACE

A full understanding of host–environment interactions in the pathogenesis of infection also requires an appreciation of how tissue cells respond to biomaterials. Biomaterials cause significant perturbations both locally and systemically in recipient milieus. A stage is set at the site of implantation on which the players (bacteria, host cells, and organic moieties) interact and compete and for which the host is a reactive audience.

Infection around biomaterials and damaged tissues is caused by bacterial adhesion to those surfaces. Tissue integration, the adhesion (chemical bonding) of tissue cells to a biomaterial, has not been programmed by natural evolution. The processes of tissue integration and bacterial adhesion are biochemically parallel, may be competitive, and may be mutually exclusive. Surfaces well colonized by healthy tissue cells tend to be resis-

tant to infection (by virtue of cell membranes and eukaryotic extracellular polysaccharides). Bacterial colonies, if they are established on biomaterial surfaces, destroy tissue and are resistant to antibodies and host defense mechanisms.

The density of an implanted biomaterial is directed by a virtual "race for the surface" between bacteria, tissue cells, and matrix molecules. Adhesive or integrative processes for bacteria or tissue cells, respectively, are based on similar molecular mechanisms (Fig. 1).

In natural environments bacteria are surface creatures and 99% of their biomass exists on surfaces rather than in floating or planktonic forms. Exposure at surgery or from bacteremia later as a result of dental immunologic diseases may allow ubiquitous microorganisms to colonize a biomaterial first, resulting in infection and preventing tissue integration. Bacteria have adapted to colonize inanimate substrata in nature as well as specific cell surfaces in animal and plant hosts. The colonization potential of most synthetic surfaces for bacteria is high compared with tissue cells because such surfaces are acellular, inanimate, and resemble substrata in nature. Materials and damaged tissues are ideal sites of colonization; infection is thus spread to and damages adjacent tissues. It is logical that infection may be prevented by encouraging colonization (integration) of material surfaces by healthy tissue cells, which then occupy available binding sites on biomaterials and form a new layer somewhat resistant to bacterial colonization.

IMPLANT INFECTION (FOREIGN BODY EFFECT)

The features of implant-associated sepsis include (1) a biomaterial or damaged tissue substratum, (2) adhesive bacterial colonization of the substratum, (3) resistance to host defense mechanisms and antibiotic therapy, (4) characteristic bacteria such as *S. epidermidis* and *Pseudomonas aeruginosa*, (5) specificity of phenomena (material, organism, host location), (6) the transformation of nonpathogens or opportunistic pathogens into virulent organisms by the presence of a biomaterial substratum, (7) polymicrobiality, (8) persistence of infection until removal of the substratum, (9) the absence of tissue inte-

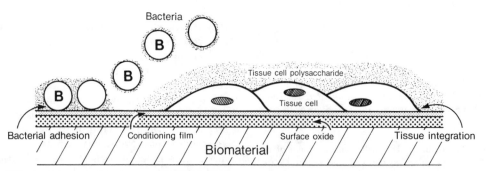

FIG. 1.   At the instant of insertion, a biomaterial represents a ready surface for colonization. The outer atomic layers of biomaterial surfaces interact instantly with the juxtaposed biologic environment. Macromolecules, bacteria, and tissue cells compete for surface domains at the reactive interface. (Reprinted with permission from *Science* 237: 1588–1595, 1987.)

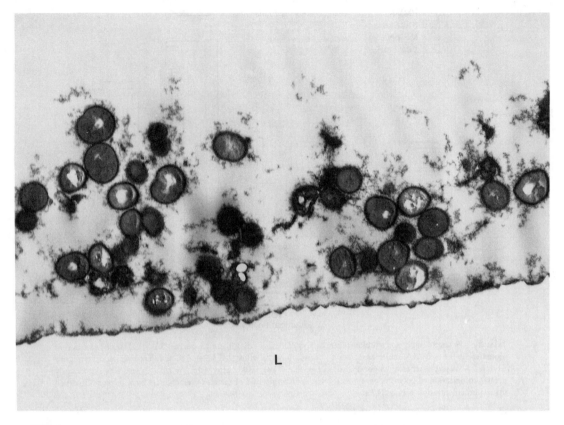

FIG. 2. Ruthenium red-stained material encompasses bacteria within clusters, and flocculent material appears to provide continuity with lens surface. (Reprinted with permission from *Science* **237**: 1588–1595, 1987.)

gration at the biomaterial–tissue interface, and (10) the presence of tissue cell damage or necrosis.

BACTERIAL PATHOGENS (CLINICAL AND LABORATORY OBSERVATIONS)

Studies of clinical retrievals indicate that a few species seem to dominate biomaterial-centered infections. *S. epidermidis* and *S. aureus* have been most frequently isolated from infected biomaterial surfaces, but *Escherichia coli, Pseudomonas aeruginosa, Proteus mirabilis,* β-hemolytic streptococci, and enterococci have also been isolated (Christensen *et al.*, 1985; Gristina *et al.*, 1987; Surgarman and Young, 1984). *Pseudomonas aeruginosa, P. maltophila,* and *P. stutzeri* in combination with *S. epidermidis* are associated with Jarvik VII infections, along with *Candida albicans* (Dobbins *et al.*, 1988).

S. epidermidis, a commensal human skin saprophyte, is a primary cause of infection of implanted polymeric biomaterials such as total artificial hearts, total joints, vascular grafts, catheters, and shunts (Gristina *et al.*, 1987).

S. aureus is frequently the major pathogen isolated from metallic-related bone, joint, and soft tissue infections. It is the most common pathogen isolated in osteomyelitis when damaged or dead bone acts as a substratum (Cierny *et al.*, 1986; Gristina *et al.*, 1985). Infection by *S. aureus*, which is a natural tissue pathogen, may be directed in part by the damaged tissue and matrix environment around implants.

Pseudomonas aeruginosa, S. epidermidis, and bacterial keratitis organisms are the primary causes of infection from extended-wear contact lenses (Slusher *et al.*, 1987; Wilson *et al.*, 1981). *Pseudomonas aeruginosa* appears to be the primary pathogen at special sites (extended-wear contact lenses; Slusher *et al.*, 1987) and the total artificial heart (Ghristina *et al.*, 1988) (Fig. 2).

Substratum-centered infections are also frequently polymicrobial. Two-thirds of osteomyelitic infections in adults are polymicrobial (Gristina *et al.*, 1985). The pathogens isolated are *S. aureus* and *S. epidermidis,* and *Pseudomonas, Enterococcus, Streptococcus, Bacillus,* and *Proteus* species.

Bacteria are divided into two major groups: gram positive and gram negative organisms. Within each of these major categories, there are a number of different genera and species that have short reproduction times, allowing for significant changes

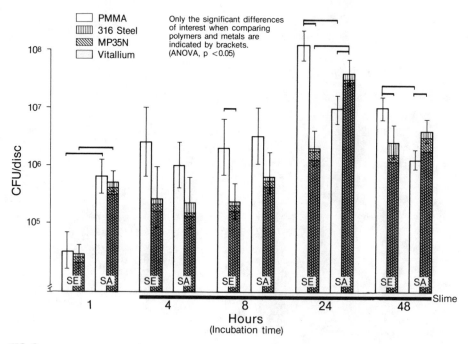

FIG. 3. *In vitro* comparison of colonization of *S. epidermidis* (SE-360) and *S. aureus* (SA-WF1) strains of different biomaterials [i.e., PMMA, multiphase cobalt chromium nickel alloy (MP35N), 316-stainless steel, and vitallium]. At 1 hr, *S. aureus* reaches a temporary maximum. Slime was visible after 4 hr. At 24 hr, there was significantly greater colonization of polymers over metals by *S. epidermidis* and of metals over polymers by *S. aureus*. Observed maximum colonization was at 24 hr.

in genetic and phenotypic expression. Thus, when studying bacterial adhesion mechanisms, great variability in expression of these mechanisms can be expected.

Reviews of the literature on bacterial adherence to biomaterials suggest that binding is very species specific. Dankert and Hogt (Dankert *et al.*, 1986) have presented data indicating that *S. epidermidis* species bind more readily to hydrophobic than to hydrophilic surfaces (polytetrafluorethylene-cohexafluoropropylene versus cellulose acetate), again noting a species specificity.

Data from our laboratory reveal that binding of *S. epidermidis* strains to stainless steel, ultrahigh-molecular-weight polyethylene (UHMWPE), and poly(methyl methacrylate) (PMMA) is influenced by the biomaterial composition and the organism's ability to elaborate an extracellular polysaccharide. Three species of *S. epidermidis* adhered more to PMMA than the other biomaterials. The production of exopolysaccharide slime increased adherence to the biomaterials by one order of magnitude.

Laboratory studies of comparative colonization of biomaterials (*S. epidermidis* and *S. aureus*) have indicated that initial attachment times for *S. epidermidis* were longer than for *S. aureus* to all materials. Slime production is evident after 4 hr of incubation for both strains. *S. epidermidis* has a significantly greater preference for polymers than metals. *S. aureus* has a greater affinity for metals than polymers) Gristina *et al.*, 1987) (Fig. 3). Knox *et al.* (1985) demonstrated that when the culture environment for a single species

of streptococcus was altered by changing the culture nutrients, pH, and surface hydrophobity, such varying degrees of adherence occurred that no correlation could be made for surface hydrophobity and adherence. The differences seen with different growth conditions in their study were attributed to the organism's ability to phenotypically change with environmental conditions. These observations are chemically relevant to treating biomaterial-adherent organisms with antibiotics since bacterial metabolism on a biomaterial surface is different that for planktonic or suspended forms, resulting in phenotypic changes in adherent organisms (Gristina *et al.*, 1989) which increase resistance to antibiotics.

Direct examination of *in situ* retrievals of biomaterials, tissues, and osteomyelitic bone from infected sites indicates that causative bacteria grow in biofilms that adhere to biomaterials and tissues. Scanning and transmission electron microscopy demonstrate that the infecting bacteria aggregate in coherent microcolonies in an extensive adherent biofilm. Accurate microbiological sampling of adhesive infections is difficult. The analysis of joint fluids, swabs of excised tissue, and prosthetic surfaces usually yields growth of only one species from what is frequently a polymicrobial population. Combined electron microscopic and multiple laboratory studies of tissue homogenates show additional species. Antibiotics chosen by conventional minimum incubation concentration (MIC) methods, that is, tests of antibiotic effects on suspension populations from inaccurate retrievals of organisms present in biofilm infections, will not be effective (Gristina *et al.*, 1989).

MOLECULAR MECHANISMS OF CELL-TO-SURFACE ADHESION

Molecular mechanisms in cell-to-surface adhesion are suggested by generalizations from colloid chemistry, microbiology, biochemistry, and physics. Bacterial particles less than 1 μm in size may be thought intially to behave as in colloid systems. The following model, in the main applicable to bacteria and in part to tissue cells, is derived from the suggestions of many authors.

Microorganisms are exposed to the surface of a biomaterial, foreign body, or tissue substrata by direct contamination, contiguous spread, or hematogenous seeding. The outer atomic layers of biomaterial surfaces are not completely saturated at the surface of a material implant and especially at grain boundaries and therefore present high energy sites and dangling bond energy profiles that act as available binding sites (Tromp *et al.*, 1986). Metallic implants have a thin but variable (100 to 200 Å) oxide layer perturbed by inconsistencies of stoichiometry that act as the true interface. Polymer surfaces also present reactive sites. The numbers of binding sites are modified by surface texture, manufacturing processes, trace chemicals, debris, and ionic and glycoproteinaceous constituents of the host environment. Surface steps and kinks, especially at grain boundaries, may act as catalytic stages for molecular and cellular activities (Gristina *et al.*, 1987; Lehninger, 1982). Profiles of biomaterial surfaces in biologic environments, including electronic state, oxidation layers, contamination level, and se-

quence of glycoprotein adsorbates, have not been elucidated. Interactions can be assumed to be specific to the host environment, the material involved, and the type of cells involved in adhesion or integration (Baier *et al.*, 1984; Dankert *et al.*, 1986).

The substratum and cell surface are usually negatively charged, repelling each other. van der Waals forces act to position a particle or bacterium near the surface (Dankert *et al.*, 1986). Hydrophobic interactions also present at most bacterial surfaces are stronger than repulsive forces and tend to attract a bacterium to within a few nanometers of the surface. Within 1 nm of the surface, short-range chemical interactions (ionic, hydrogen, and covalent bonding) are possible between bacterial extracapsular molecules and the outer atomic layers of a surface or surface adsorbates.

Initial attachment (reversible nonspecific adhesion) depends on the general physical characteristics of the bacterium, the fluid interface, and the substratum. Subsequent to attachment, specific irreversible adhesion may occur as a time-dependent chemical process involving chemical binding, hydrophobic interactions, and specific receptor–ligand interactions, as well as nonspecific exopolysaccharide polymer-to-surface interactions (Christensen, 1985; Savage, 1985) (Fig. 4).

Fimbria, pili, curli, or other receptor appendages may interact with surfaces at distances greater than 15 nm, bridging repulsive forces (Christensen, 1985; Oga *et al.*, 1988). Fimbriae may also react nonspecifically (by charge or hydrophobic interaction) with implant surfaces. Bacterial

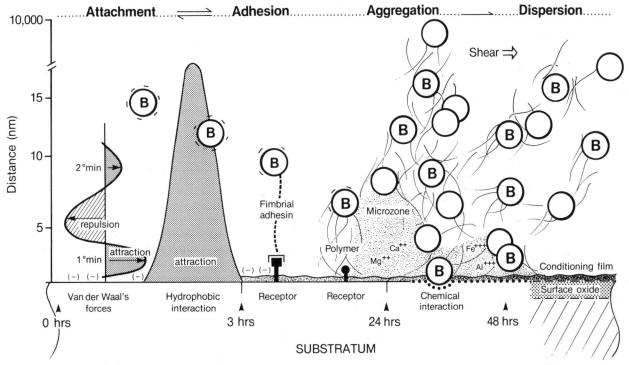

FIG. 4. Molecular sequences in bacterial (B) attachment, adhesion, and aggregation to substratum. (Reprinted with permission from *Science* **237**: 1588–1595, 1987.)

exopolysaccharides may also bind nonspecifically to surfaces or to surface adsorbates, facilitating cell-to-cell aggregation and the formation of microcolonies (Gristina *et al.*, 1988; Gibbons and van Houte, 1980; Savage and Fletcher, 1985). When environmental conditions such as temperature, nutrient substrates, antagonists, and cation balance are favorable, bacterial propagation occurs. It is unlikely that most biomaterials present a receptive substratum for tissue cells such as bone cartilage or endothelial cells.

MATRIX PROTEINS

Albumen, fibronectin, collagen, vitronectin, and other milieu proteins play an important role in specifically augmenting or blunting the adhesive tendencies of bacterial and tissue cells, depending on the concentration and the presence of integrin receptors for matrix proteins on tissue cells and a similar class of receptors on bacterial cells. The exact role of each intermediate matrix protein is assumed to be specific to the cell involved, surface denaturization, or folding interactions, as well as pH and cation concentration.

Implanted biomaterials are rapidly coated by constituents of the serum and surrounding matrix, including fibronectin, osteonectin, vitronectin, albumin, fibrinogen, laminin, collagen, and covalently bound, short-chain oligosaccharides (Gristina *et al.*, 1987; Baier *et al.*, 1984). Bacterial and tissue cells may then adhere to constituents of this film. Eukaryocytes have been shown to adhere to surfaces coated with fibronectin and collagen (Proctor, 1987; Van Wachem *et al.*, 1987). Certain strains of *S. aureus*, *S. epidermidis*, and *E. coli* have receptors or sets of receptors for fibronectin and collagen epitopes (Hermann *et al.*, 1988; Vaudaux *et al.*, 1984). Studies have shown that pathogenic bacteria have greater numbers of conditioning film protein receptors than do similar strains of nonpathogenic bacteria.

The interaction between conditioning film proteins and tissue and bacterial cells also depends on the physical state and quantity of bound proteins on the biomaterial surface. *In vitro* studies of fibronectin-coated biomaterials demonstrate that a low concentration of bound fibronectin enhances bacterial adherence, while a high concentration of bound fibronectin inhibits binding by the same bacteria (Hermann *et al.*, 1988). Bacterial adhesion may also be decreased by the presence of albumin. Thus, conditioning protein molecules may play a variety of roles in bacterial adhesion, depending on their concentration and environmental conditions (Baier *et al.*, 1984).

BIOMATERIALS—THE NATURE OF SURFACES

This section considers the general characteristics of biomaterials as substrata with special regard for their interactions with bacteria and tissue. Many implants consist of one or more metals or polymers. Biomaterials, foreign bodies, and devital-ized tissue and bone in a biological environment are passive and susceptible substrata because they are inanimate and do not resist infection. In fact, regardless of "inertness," they are physiocochemically active and may directly modulate adhesion or interact with host defenses.

Surface atoms and molecules of metal alloys and polymers are not bound on all sides by bulk phase atoms or molecules. This imbalance causes changes in the distribution of constituents of the bulk phase (segregation) and rearrangement of outer atomic layers (relaxation), depending on the ambient, nonbulk, liquid, or gas environment. Even the most inert alloys and polymers have oxide defects and reactive zones that will differentially influence adsorption and chemical interactions at their surfaces. In a like manner, surfaces of tissue cells or acellular biologic surfaces such as articular cartilage or damaged bone present specific ligands or crystalline faces for receptor binding.

Most important for each material is the surface interaction of the outer atomic layers with environmental moieties, glycoproteins, elemental constituents, or prokaryotic and eukaryotic cells. Ideally, we would like to influence interactions to promote compatibility and/or integration and resistance to infections, as well as formal host defense responses.

Titanium alloys used in dental implants have been suggested to form a direct bone (osteocyte)-implant contact (osseointegration) at the ultrastructural level (Albrektsson, 1989; Gristina *et al.*, 1985). Titanium and titanium alloy surfaces directly colonized by osteoblasts or tissue cells may be protected in part against colonization by bacterial pathogens. Porous coated surfaces used without methyl methacrylates have been reported recently as having a lower rate of infection. There has been no reported difference in infection rates between porous and smooth surfaces of the same material or alloy. The increase in surface area in these applications is less than one order of measure and not likely to increase colonization.

Studies have indicated greater bacterial adhesion to polymers than to metals for *S. epidermidis*, both *in vitro* and *in vivo* (Barth *et al.*, 1989; Gristina *et al.*, 1987). Interestingly, antibiotic resistance and host defense inhibition may be greater for infections centered on polymers (Barth *et al.*, 1989).

BIOFILMS

Microorganisms in colonies on surfaces form layers two to hundreds of organisms thick composed of cellular material, extracellular polysaccharides, environmental adsorbates, and debris. This surface composite is called the biofilm or slime. In nature, biofilms accumulate to significant thicknesses and may impede flow in hydraulic systems.

Most organisms within a mature microcolony are not attached to a two-dimensional surface but are fixed within the three-dimensional structure of the biofilm. Bacterial species in consortia have shown higher metabolic activities than those of free organisms, a characteristic that may be related to cross-feeding and the development of suitable microenvironments within the adherent biofilm (Paerl, 1985).

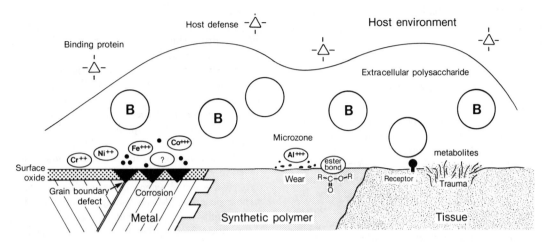

FIG. 5.   Surface disruption by wear, corrosion, trauma, or bacterial mechanisms frees metabolites or ions, which are then available to bacteria (B) within a biofilm microenvironment. At microzones, metal ions required by pathogenic bacteria are not lost by diffusion and may be shielded from host protein-binding complexes. Bacteria are also protected by biofilms and may metabolize polymer or tissue components. Interactions occur between exposed receptors on bacteria or surfaces.

Many different species of bacteria may thrive within the biofilm, with significant interaction and even interdependency between organisms (Hamilton, 1987), such as the growth and activity of anaerobic species within a biofilm that is itself located in a highly aerobic bulk environment (Gibbons and van Houte, 1975; Hamilton, 1984; Paerl, 1985). For example, the formation of dental caries involves a succession of microbes. The primary colonizing aerobic species grows on salivary glycoproteins. Facultative and anaerobic organisms than use the metabolic products of the primary colonizer and produce deleterious organic acids (Hamilton, 1987). In oil pipelines, hydrocarbon-degrading bacteria create the anaerobic conditions and produce the necessary nutrients for sulfate-reducing bacteria, which are the prinicipal cause of anaerobic corrosion (Hamilton, 1987).

WEAR AND TEAR: SUBSTRATA AND MICROZONES

Damage to substrata by wear, corrosion, toxins, viral effects, bacterial mechanisms, or biosystemic chemical degradation establishes environmental conditions that microorganisms can exploit. Surfaces provide an interface for the concentration of charged particles, molecules, and nutrients, or may themselves be metabolized (Fig. 5).

The microzone may be thought of as an environmental, metabolic microclimate at a surface created by a complex biofilm (Paerl, 1985; Stenstrom, 1985). This concept may be applied to implant surfaces on which adhesive, possibly polymicrobial, colonization creates a microclimate of favorable conditions and excludes antagonistic environmental factors.

Under normal conditions, host defense mechanisms are available to limit the nutrients necessary for bacterial propaga-

tion. For example, iron is an essential element in bacterial growth and virulence (Bullen *et al.*, 1974; Bullen, 1981). However, free ionic iron is not usually available in the extracellular matrix of the human host (Weinberg, 1974). In serum, the concentration of free iron is kept at a low level through binding by the protein transferrin (Bullen *et al.*, 1974).

Biomaterials disrupt these mechanisms (Bullen *et al.*, 1974) by providing a surface for bacterial adhesion, and corrosion of the biomaterial surface, which provides nutrients for bacterial propagation and growth.

Bacteria may use trace ions such as Mg^{2+} and Ca^{2+} to stabilize (via acidic groups) expolysaccharides in a gel, which enhances cell-to-cell and cell-to-surface adhesion and increases resistance to antagonists (Gristina *et al.*, 1987; Fletcher, 1980). Some synthetic polymers, such as polyurethane and methyl methacrylate, contain ester bonds that may be hydrolyzed by staphylococci (Ludwicka *et al.*, 1983).

The formation of slime-enclosed microcolonies allows bacteria to compete with host proteins for iron and other nutrients. Bacteria may accumulate ions such as iron by localization of metabolites and siderophores (Sriyoschati and Cox, 1986) and by sequestering them from binding by transferrin or lactoferrin. The release of free iron from the biomaterial surface may also serve to saturate the iron-binding proteins. The sequence of inflammation, corrosion, reduced pH, increased iron concentration, and saturation of iron-binding proteins may result in increased virulence and inhibition of macrophage function (Weinberg, 1974; Bullen *et al.*, 1974; Lehninger, 1982).

Biofilm development involves not only phases of attachment, growth, and polysaccharide production, but also a phase of detachment (Hamilton, 1987). Portions of the slime-enclosed microcolony may eventually be detached secondary to hemodynamic forces or trauma. Detachment may also be a normal feature of a dynamic equilibrium resulting, for example, from nutrient or oxygen depletion arising within the film (Howell and Atkinson, 1976). These pathogenic inocula may then serve as a source of secondary infection and hematogenous septic emboli.

ARTIFICIAL ORGANS

The total artificial heart and artificial organs present special problems because they are composed of many materials. They are implanted in damaged tissue cavities in immuno-compromised patients, and require compatibility between material and adjacent tissues. Complexity is increased by the requirement for proadhesive tissue system integration and antiadhesive hemodynamic system compatibility. External power sources (drive lines), traversing body cavities, mucous membranes, and skin provide a septic gateway for external pathogens.

HOST DEFENSE RESPONSE

Host defense or immune mechanisms are invariably disrupted by the implantation of biomaterials and tissue transplants. The mechanisms of this disruption are unknown.

The quality of the immune and defense responses defines susceptibility to and the specificity and outcome of infection. Spontaneous bacteremia, trauma, and surgery are common contributors to infection. For the most part, however, because of normal defense mechanisms and the use of antibiotics and antibiotic prophylaxis in surgery, sepsis is uncommon.

Certain patient groups with immune system depression or aberrations are predictably at risk. Rheumatoid patients manifest a spontaneous and somewhat cryptic sepsis in joints (Kellgren *et al.*, 1958). Diabetics, infants, children, the aged, patients with vascular disease, drug abusers, and HIV patients are at increased risk from specific organisms, as are patients with hematologic abnormalities and neoplastic disease.

Normal individuals rarely if ever contract infections due to *S. epidermidis* unless some aspect of the host defense is abrogated or compromised. The most likely component of the host defense system to become impaired or compromised in infected biomaterial implants is the phagocyte system.

Staphylococci are extracellular parasites, and therefore the major host defense mechanisms involve deployment of two classes of phagocytes: polymorphonuclear phagocytes (early) and macrophages (late). To promote deployment, chemotactic factors generated at the site of infection direct phagocytes to the implant surface–host tissue interface, the focal point of infection. However, when staphylococci colonize biomaterial surfaces, the host defense mechanisms seem to be greatly impaired.

Extracellular slime produced by *S. epidermidis* has been reported to inhibit T-cell and B-cell blastogenesis, polymorphonuclear leukocyte chemotaxis, opsonization, and chemiluminescence, and to enhance the virulence of *S. epidermidis* in mice (Peters, 1988). Johnson *et al.* (1986) showed that phagocytosis by polymorphonuclear leukocytes of *S. epidermidis* grown on a plastic surface decreased over time. Because slime production by *S. epidermidis* is time-dependent, these results

support the theory that excess extracellular slime impairs phagocytosis by polymorphonuclear leukocytes. Conversely, the presence of slime alone cannot fully explain the impairment of host defense mechanisms.

The interaction between phagocytes and a foreign body may somehow exhaust or preempt killing mechanisms. Klock and Bainton (1976) observed that granulocytes exposed to nylon fibers retained their chemotactic response but had an impaired bactericidal response. Studies have shown that neutrophils inside Teflon cages placed in peritoneal cavities exhibited decreased phagocytic and bactericidal potential and reduced superoxide production (Zimmerli *et al.*, 1984). A similar pattern was noted when normal rabbit alveolar macrophages were exposed to PMMA spheres (Myrvik *et al.*, 1989). From these results, it appears that foreign body surfaces trigger a "slow burst" and "preempt" a second burst if the phagocyte subsequently encounters an infectious organism. It is almost certain that the killing capacity of such preempted phagocytes is impaired. If a biomaterial implant surface can be found that promotes normal tissue integration rather than triggering an exhausting foreign body inflammatory response, host defense mechanisms could be expressed normally.

ANTIBIOTIC RESISTANCE

Infections involving biomaterials or damaged tissues are often resistant to antibiotic therapy and require the removal of the substrata to resolve the infection. We compared the antibiotic sensitivities of biomaterial-adherent *S. epidermidis* and *S. aureus* with the sensitivities of standard suspension cultures for nafcillin, vancomycin, gentamycin, and Daptomycin, and examined antibiotic killing kinetics on surfaces (Naylor *et al.*, 1988). The biomaterials colonized were stainless steel, PMMA, and UHMWPE. The sensitivities obtained were 2- to 250-fold higher for biomaterial-adherent bacteria than for bacteria in suspension, and were independent of slime production and specific not only for the organism but also the biomaterial. Bacteria adhering to PMMA were more resistant to antibiotics than organisms adhering to stainless steel.

Elevated minimum bacterial concentrations (MBC) independent of slime production suggest that extracapsular polysaccharides may not present a diffusion barrier to antibiotics (Naylor *et al.*, 1988; Nichols *et al.*, 1988).

FUTURE STRATEGIES

The following strategies show promising possibilities:

1. Antibiotics, systemically delivered or placed *in situ* in biomaterials, should be more effective because they are at surfaces before biomaterial-modified bacterial behavior and colonization can occur.
2. Pathogens may be predicted for each biomaterial and tissue type of infection, allowing the selection of correctly

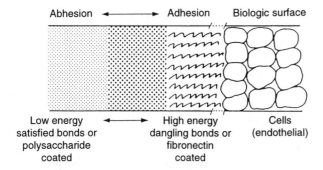

Abhesion ←————→ Adhesion   Biologic surface

Low energy ←————→ High energy   Cells
satisfied bonds or   dangling bonds or   (endothelial)
polysaccharide   fibronectin
coated   coated

FIG. 6. Biomaterial surface, at left, (e.g., artificial vessel) is programmed by (1) surface energy state, (2) polysaccharide coating, or (3) glycoprotein coating for abhesion or adhesion. At site of junction with biologic surface at right (e.g., vessel, bone), adhesion or integration by tissue cells (endothelial) is encouraged.

targeted preventive and not resistant-strain-producing antibiotics.

3. Precolonization of surfaces by healthy tissue cells (osteoblasts, fibroblasts, endothelial cells) before implantation should reduce infection and increase binding to adjacent tissue.

4. Genetic modification of eukaryotic host cells should improve tissue integration. Genetic coding for qualities of increased adhesion may be transformed to endothelial cells, for example, to allow better endothelialization of preseeded grafts.

5. Analogs such as peptides and oligosaccharides designed to fit specific ligand or receptor sites may be used to block or reverse bacterial adhesion.

6. In a similar fashion, analogs and lectins preadsorbed to biomaterial surfaces may also encourage adhesion of specific macromolecules or cells.

7. Biomaterial surfaces will be characterized and interfacial phenomena defined by advanced instrumentation such as scanning tunneling microscopy, atomic force microscopy, Auger electron spectroscopy, electron spectroscopy for chemical analysis, and other techniques being developed.

8. Biomaterial surfaces can be modified for controlled biologic response (Fig. 6) (Gristina *et al.*, 1987, 1988).

In the future, surfaces will be modified by advanced techniques that create a known interface response based on programmed surface quantum states, directing desired molecular or cellular interactions. Heavy ion implantation, chemical vapor deposition, and vacuum evaporation may be used to create a surface that "directs" tissue or macromolecular integration to build tissue and hemodynamic systems, rather than bacterial adhesion.

Bibliography

Albrektsson, T. (1989). The response of bone to titanium implants. in *Molecular Mechanisms of Microbial Adhesion*, L. Switalski, M. Hook, and E. Beachey, eds. Springer-Verlag, New York, p. 207.

Baier, R. E., Meyer, A. E., Natiella, J. R., Natiella, R. R., and Carter, J. M. (1984). Surface properties determine bioadhesive outcomes: Methods and results. *J. Biomed. Mater. Res.* **18**: 337–355.

Barth, E., Myrvik, Q. N., Wagner, W., and Gristina, A. G. (1989). *In vitro* and *in vivo* comparative colonization of *Staphylococcus aureus* and *Staphylococcus epidermidis* on orthopedic implant materials. *Biomaterials* **10**: 325–328.

Bayston, R., and Penny, S. R. (1972). Excessive production of mucoid substances in *Staphylococcus* SIIA: A possible factor in colonization of Holter shunts. *Dev. Med. Child. Neurol.* **14**: 25–28.

Beachey, E. H. (1980). *Bacterial Adherence: Receptors and Recognition, Series B.* Chapman & Hall, London.

Bullen, J. J., Rogers, H. J., and Griffiths, E. (1974). Bacterial iron metabolism in infection and immunity. in *Microbial Iron Metabolism: A Comprehensive Treatise*, J. B. Beilands, ed. Academic Press, New York, pp. 518–551.

Bullen, J. J. (1981). The significance of iron in infection. *Rev. Infect. Dis.* **3**: 1127–1138.

Christensen, G. D., Simpson, W. A., and Beachey, E. H. (1985). Microbial adherence in infection. in *Principles and Practice of Infectious Diseases*, 2nd Ed., G. L. Mandell, R. G. Douglas Jr., and J. E. Bennett, eds. Wiley, New York, pp. 6–22.

Cierny, G., Couch, L., and Mader, J. (1986). Adjunctive local antibiotics in the management of contaminated orthopaedic wounds. American Academy of Orthopaedic Surgeons Final Program of 53rd Annual Meeting, New Orleans, LA, February 20–25, 1986, Am. Acad. Orthopaedic Surgeons, Park Ridge, IL, p. 86.

Costerton, J. W. and Irvin, R. T. (1981). The bacterial glycocalyx in nature and disease. *Annu. Rev. Microbiol.* **35**: 299–324.

Dankert, J., Hogt, A. H., and Feijen, J. (1986). Biomedical polymers: Bacterial adhesion, colonization, and infection. *CRC Crit. Rev. Biocompat.* **2**: 219–301.

DeVries, W. (1988). The permanent artificial heart: Four case reports. *JAMA* **249**: 849–859.

Dobbins, J. J., Johnson, G. S., Kunin, C. M., and DeVries, W. C. (1988). *JAMA* **259**: 865–869.

Dougherty, S. H. and Simmons, R. L. (1982). Infections in bionic man: The pathobiology of infections in prosthetic devices, Part 2. *Curr. Prob. Surg.* **19**: 269–318.

Elek, S. D. and Conen, P. E. (1957). The virulence of *Staphylococcus pyogenes* for man: A study of the problems of wound infection. *Br. J. Expl. Pathol.* **38**: 573–586.

Fletcher, M. (1980). Adherence of marine microorganisms to smooth surfaces. in *Bacterial Adherence: Receptors and Recognition, Series B*, E. H. Beachey, ed. Chapman & Hall, London, Vol. 6, pp. 345–374.

Gibbons, R. J. and van Houte, J. V. (1975). Bacterial adherence in oral microbial ecology. *Annu. Rev. Microbiol.* **29**: 19–44.

Gibbons, R. J. and van Houte, J. V. (1980). Bacterial adherence and the formation of dental plaques. in *Bacterial Adherence Receptors and Recognition, Series B*, E. H. Beachey, ed. Chapman & Hall, London, Vol. 6, pp. 63–104.

Gristina, A. G. and Costerton, J. W. (1984). Bacterial adherence to biomaterials: The clinical significance of its role in sepsis. *Soc. Biomater. 1984 Trans. VII*, p. 175.

Gristina, A. G. and Kolkin, J. (1989). Current concepts review: Total joint replacement and sepsis. *J. Bone Surg.* **65A**: 128–134.

Gristina, A. G., Oga, M., Webb, L. X., and Hobgood, C. (1985). Bacterial colonization in the pathogenesis of osteomyelitis. *Science* **228**: 990–993.

Gristina, A. G., Dobbins, J. J., Giammara, B., Lewis, J. C., and DeVries, W. C. (1988). Biomaterial-centered sepsis and the total artificial heart. *JAMA* **259**: 870–874.

Gristina, A. G. and Rovere, G. D. (1963). An *in vitro* study of the effects of metals used in internal fixation on bacterial growth and dissemination. *J. Bone Joint Surg.* **45A**: 1104.

Gristina, A. G., Rovere, G. D., Shoji, H., and Nicastro, J. F. (1976). An *in vitro* study of bacterial response to inert and reactive metals and to methylmethacrylate. *J. Biomed. Mater. Res.* **10**: 273–281.

Gristina, A. G., Costerton, J. W., Leake, E., and Kolkin, J. (1980a). Bacterial colonization of biomaterials: Clinical and laboratory studies. *Orthop. Trans.* **4**: 405.

Gristina, A. G., Kolkin, J., Leake, E., Costerton, J. W., and Wright, M. J. (1980b). Bacteria and their relationship to biomaterials. in *Book of Abstracts* First World Biomaterials Conference, Vienna, April 8–12, 1980. Eur. Soc. Biomater. pp. 2–39.

Gristina, A. G., Hobgood, C. D., and Barth, E. (1987). Biomaterial specificity, molecular mechanisms and clinical relevance of *S. epidermidis* and *S. aureus* infections in surgery. in *Pathogenesis and Clinical Significance of Coagulase-Negative Staphylococci*, G. Pulverer, P. G. Quie, and G. Peters, eds. Gustav Fischer Verlag, Stuttgart, pp. 143–157.

Gristina, A. G., Jennings, R. A., Naylor, P. T., Myrvik, Q. N., and Webb, L. X. (1989). Comparative *in vitro* antibiotic resistance of surface-colonizing coagulase-negative staphylococci. *Antimicrob. Agents Chemother.* **33**: 813–816.

Hamilton, W. A. (1987). Biofilms: Microbial interactions and metabolic activities. *Symp. Soc. Gen. Microbiol.* **41**: 361–385.

Hermann, M. N., Vaudaux, P. E., Pittet, D., Auckenthaler, R., Lew, P. D., Schumacher-Perdreau, Peters, G., and Waldvogel, F. A. (1988). Fibronectin, fibrinogen, and laminin act as mediators of adherence of clinical staphylococcal isolates to foreign materials. *J. Infect. Dis.* **158**: 693–701.

Howell, J. A. and Atkinson, B. (1976). Sloughing of microbial film on trickling filters. *Water Res.* **78**: 307–315.

Johnson, G. M., Lee, D. A., Regelmann, W. E., Gray, E. D., Peters, G., and Quie, P. G. (1986). Interference with granulocyte function by *Staphylococcus epidermidis* slime. *Infect. Immun.* **54**: 13–20.

Kellgren, J. H., Ball, J., Fairbrother, R. W., and Barnes, K. L. (1958). Suppurative arthritis complicating rheumatoid arthritis. *Br. Med. J.* **1**: 1193–1200.

Klock, J. C. and Bainton, D. F. (1976). Degranulation and abnormal bactericidal function of granulocytes procured by reversible adhesion to nylon wool. *Blood* **48**: 149–161.

Knox, K. W., Hardy, L. N., Markevics, L. J., Evans, J. D., and Wicken, A. J. (1985). Comparative studies on the effect of growth conditions on adhesion, hydrophobicity and extracelular protein. Profile of *Streptococcus sanguis* 69B. *Infect. Immun.* **50**: 545–554.

Lahninger, A. L., ed. (1982). *Principles of Biochemistry.* Worth, New York.

Ludwicka, A., Locci, R., Jansen, B., Peters, G., and Pulverer, G. (1983). Microbial colonization of prosthetic devices v. attachment of coagulase-negative staphylococci and "slime"-production on chemically pure synthetic polymers. *Zentralbl. Bakteriol. Mikrobiol. Hyg. Abt. I Orig. B* **177**: 527–532.

Myrvik, Q. N., Wagner, W., Barth, E., Wood, P., and Gristina, A. G. (1989). Effects of extracellular slime produced by *Staphylococcus*

epidermidis on oxidative responses of rabbit alveolar macrophages. *J. Invest. Surg.*

Naylor, P. T., Jennings, R., Webb, L. X., and Gristina, A. G. (1988). Antibiotic sensitivity of biomaterial-adherent *Staphylococcus epidermidis*. *Orthop. Trans.* **12**: 524–525.

Nichols, W. W., Dorrington, S. M., Slack, M. P. E., and Walmsley, H. L. (1988). Inhibition of tobramycin diffusion by binding to alginate. *Antimicrob. Agents Chemother.* **32**: 518–523.

Oga, M., Sugioka, Y., Hobgood, C. D., Gristina, A. G., and Myrvik, Q. N. (1988). Surgical biomaterials and differential colonization by *Staphylococcus epidermidis*. *Biomaterials* **9**: 285–289.

Paerl, H. W. (1985). Influence of attachment on microbial metabolism and growth in aquatic ecosystems. in *Bacterial Adhesion: Mechanisms and Physiologic Significance,* D. C. Savage and M. Fletcher, eds. Plenum, New York, pp. 363–400.

Peters, G., Locci, R., and Pulverer, G. (1981). Microbial colonization of prosthetic devices. II. Scanning electron microscopy of naturally infected intravenous catheters. *Zentralbl. Bakertiol. Makrobiol. Hyg. Abt. I Orig. B* **173**: 293–299.

Proctor, R. A. (1987). Fibronectin: A brief overview of its structure, function, and physiology. *Rev. Infect. Dis. (Suppl. 4)* **9**: S317–S321.

Savage, D. C. and Fletcher, M., eds. (1985). *Bacterial Adhesion Mechanisms and Physiological Significance.* Plenum, New York.

Slusher, M. M., Myrvik, Q. N., Lewis, J. C., and Gristina, A. G. (1987). Extended-wear lenses, biofilm, and bacterial adhesion. *Arch. Ophthalmol.* **105**: 110–115.

Sugarman, B. and Young, E. J., eds. (1984). *Infections Associated with Prosthetic Devices.* CRC Press, Boca Raton, FL.

Tromp, R. M., Hamers, R. J., and Demuth, J. E. (1986). Quantum states and atomic structure of silicone surfaces. *Science* **234**: 304–390.

Van Wachem, P. B., Vreriks, C. M., Beugeling, T., Feigen, J., Bantjes, A., Detmers, J. P., and van Aken, W. G. (1987). The influence of protein adsorption of interactions of cultured human endothelial cells with polymers. *J. Biomed. Mater. Res.* **21**: 701–718.

Vaudaux, P. E., Waldvogel, F. A., Morgenthaler, J. J., and Nydegger, U. E. (1984). Adsorption of fibronectin onto polymethylmethacrylate and promotion of *Staphylococcus aureus* adherence. *Infect. Immun.* **45**: 768–774.

Voytek, A., Gristina, A. G., Barth, E., Myvik, Q., Switalski, L., Hook, M., and Speziale, P. (1988). Stephylococcal adhesion to collagen in intraarticular sepsis. *Biomaterials* **9**: 107–110.

Weinberg, E. D. (1974). Iron and susceptibility to infectious disease. *Science* **184**: 952–956.

Wilson, L. A., Schlitzer, R. L., and Ahern, D. G. (1981). Pseudomonas corneal ulcers associated with soft contact lens wear. *A. J. Ophthalmol.* **92**: 546–554.

Zimmerli, W., Lew, P. D., and Walvogel, F. A. (1984). Pathogenesis of foreign body infections: Evidence for a local granulocyte defect. *J. Clin. Invest.* **73**: 1191–1200.

5

Testing Biomaterials

STEPHEN HANSON, PEGGY A. LALOR, STEVEN M. NIEMI, SHARON J. NORTHUP, BUDDY D. RATNER,
MYRON SPECTOR, BRAD H. VALE, AND JOHN E. WILLSON

5.1 INTRODUCTION
Buddy D. Ratner

How can biomaterials be evaluated to determine if they are biocompatible and whether they function appropriately in the *in vivo* environment? This section discusses testing procedures. A few general comments will clarify key themes and ideas that are common to biological testing of all biomaterials.

Some biomaterials fulfill their intended function in seconds. Others are implanted for a lifetime (10 years? 70 years?). Are 6-month implantations useful for learning about a device intended for 3-min insertion or another device developed for lifetime implantation? These are not simple questions. However, they must be addressed and carefully considered in designing a biomaterials testing protocol.

Evaluation under *in vitro* (literally "in glass") conditions can provide rapid and inexpensive data on biological interaction. However, the question must always be raised whether the *in vitro* test is truly measuring what will occur in the much more complex environment *in vivo. In vitro* tests tests minimize the use of animals in research, a desirable goal. However, this must be tempered with an appreciation that the results of *in vitro* tests may not be relevant to the implant situation.

Animals are used in testing biomaterials to model the environment that might be encountered in humans. However, there is great range in animal anatomy, physiology, and biochemistry. Will the animal model provide data useful for predicting how a device performs in humans? Without validation through human clinical studies, it is often difficult to draw strong conclusions from performance in animals. The first step in designing animal testing procedures is to choose an animal model that offers a reasonable parallel anatomically or biochemically to the situation in humans. Experiments designed to minimize the number of animals needed, ensure that the animals are treated humanely (e.g., National Institutes of Health guidelines for the use of laboratory animals), and maximize the relevant information generated are essential.

This is a book on biomaterials, with the emphasis on materials. However, there are important differences between implanting a sheet of cellulose in an animal (a material), and evaluating the performance and biological response of the same sheet of cellulose used as a dialysis membrane in an artificial kidney (a device). The pros and cons of testing materials (a relatively low-cost procedure providing opportunities for carefully controlled experiments) versus evaluation in a device configuration (an expensive and difficult-to-control, but completely relevant, situation) must always be weighed.

Testing always leads to experimental variability, particularly tests in living systems. The more complex the system (e.g., a human versus cells in culture), the larger the variability that might be expected. In order to draw defensible conclusions from expensive testing, statistics provides an assurance that within a defined probability, the results are meaningful. Statistics should be used at two stages in testing biomaterials. Before an experiment is performed, statistical experimental design will indicate the minimum number of samples that must be evaluated to yield meaningful results. After the experiment is completed, statistics will help to extract the maximum useful information (see Chapter 9.7 for further insights on this issue).

Assistance in the design of many biomaterials tests is available through national and international standards organizations. Thus, the American Society for Testing and Materials (ASTM) and the International Standards Organization (ISO) can often provide detailed protocols for widely accepted, carefully thought out testing procedures (also see Chapter 10.2). Other testing protocols are available through government agencies (e.g., the Food and Drug Administration and the National Institutes of Health) and through commercial testing laboratories.

5.2 *IN VITRO* ASSESSMENT OF TISSUE COMPATIBILITY
Sharon J. Northup

The term "cytotoxicity" means to cause toxic effects (death, alterations in cellular membrane permeability, enzymatic inhibition, etc.) at the cellular level. It is distinctly different from physical factors that affect cellular adhesion (surface charge of a material, hydrophobicity, hydrophilicity, etc.). This chapter reviews the evaluation of biomaterials by methods that use isolated, adherent cells in culture to measure cytotoxicity and biological compatibility.

HISTORICAL OVERVIEW

Cell culture methods have been used to evaluate the biological compatibility of materials for more than two decades (Northup, 1986). Most often today the cells used for culture are from established cell lines purchased from biological suppliers or cell banks (e.g., the American Type Tissue Culture Collection, Rockville, MD). Primary cells (with the exception of erythrocytes in hemolysis assays) are seldom used because they have less assay repeatability, reproducibility, efficiency, and, in some cases, availability. Several methods have been validated for repeatability (comparable data within a given laboratory) and reproducibility (comparable data among laboratories). These methods have been incorporated into national and international standards used in the commercial development of new products. In addition, there are a wide variety of methods in the research literature that have been used in specialized applications and that are on the leading edge of scientific development. These are not discussed in this chapter. As the science of biomaterials evolves, some of these research methods may become incorporated into routine products.

BACKGROUND CONCEPTS

Toxicity

A toxic material is defined as a material that releases a chemical in sufficient quantities to kill cells either directly or indirectly through inhibition of key metabolic pathways. The number of cells that are affected is an indication of the dose and potency of the chemical. Although a variety of factors affect the toxicity of a chemical (e.g., compound, temperature, test system), the most important is the dose or amount of chemical delivered to the individual cell.

Delivered and Exposure Doses

The concept of delivered dose refers to the dose that is actually absorbed by the cell. It differs from the concept of exposure dose, which is the amount applied to a test system. For example, if an animal is exposed to an atmosphere containing a noxious substance (exposure dose), only a small portion of the inhaled substance will be absorbed and delivered to the internal organs and cells (delivered dose). Because different cells have differing susceptibilities to the toxic effects of xenobiotics (foreign substances), the cells that are most sensitive are referred to as the target cells. Taken together, these two concepts mean that cell culture methods evaluate target cell toxicity by using delivered doses of the test substance. This distinguishes cell

culture methods from whole animal studies, which evaluate the exposure dose and do not determine the target cell dose of the test substance. This difference in dosage at the cellular level accounts for a significant portion of the difference in sensitivity (i.e., quantitation range) of cell culture methods compared with whole animal toxicity data. To properly compare the sensitivity of cell culture methods with *in vivo* studies, data from local toxicity models such as dermal irritation, implantation, and direct tissue exposure should be compared. These models reduce the uncertainties of delivered dose associated with absorption, distribution, and metabolism that are inherent in systemic exposure test models.

Safety Factors

A highly sensitive test system is desirable for evaluating the potential hazards of biomaterials because the inherent characteristics of the materials often do not allow the dose to be exaggerated. There is a great deal of uncertainty in extrapolating from one test system to another, such as from animals to humans. To allow for this, toxicologists have used the concept of safety factors to take into account intra- and interspecies variation. This practice requires being able to exaggerate the anticipated human clinical dosage in the nonhuman test system. In a local toxicity model in animals, there is ample opportunity for reducing the target cell dose by distribution, diffusion, metabolism, and changes in the number of exposed cells (because of the inflammatory response). On the other hand, in cell culture models, in which the variables of metabolism, distribution, and absorption are minimized, the dosage per cell is maximized to produce a highly sensitive test system.

Solubility Characteristics

The principal components in medical devices are water-insoluble materials (polymers, metals, and ceramics), meaning that less than one part of the material is soluble in 10,000 parts of water. Other components may be incorporated into the final product to obtain the desired physical, functional, manufacturing, and sterility properties. For example, plastics may contain plasticizers, slip agents, antioxidants, fillers, mold release agents, or other additives, either as components of the formulation or trace additives from the manufacturing process. The soluble components may be differentially extracted from the insoluble material. Till *et al.* (1982) have shown that the migration of chemicals from a solid plastic material into liquid solvents is controlled by diffusional resistance within the solid, chemical concentration, time, temperature, mass transfer resistance on the solvent side, fluid turbulence at the solid–solvent interface, and the partition coefficient of the chemical in the solvent. Because of these variables, the conditions for preparing extractions of biomaterils have been carefully standardized to improve the reproducibility of the data.

Complete dissolution of biomaterials is an alternative approach for *in vitro* testing. Its main limitation is that it does not simulate the intended clinical application or may create degradation products that do not occur in the clinical application. Therefore, the actual clinical dosage or agent exposed to the cells in pharmacokinetic terms may be exaggerated because the rate of diffusion from the intact material or device may be very slow or different than that for complete dissolution.

ASSAY METHODS

Three primary cell culture assays are used for evaluating biocompatibility: direct contact, agar diffusion, and elution (also known as extract dilution). These are morphological assays, meaning that the outcome is measured by observations of changes in the morphology of the cells. The three assays differ in the manner in which the test material is exposed to the cells. As indicated by the nomenclature, the test material may be placed directly on the cells or extracted in an appropriate solution that is subsequently placed on the cells. The choice of method varies with the characteristics of the test material, the rationale for doing the test, and the application of the data for evaluating biocompatibility.

To standardize the methods and compare the results of these assays, the variables of number of cells, growth phase of the cells (period of frequent cell replication), cell type, duration of exposure, test sample size (e.g., geometry, density, shape, thickness) and surface area of test sample must be carefully controlled. This is particularly true when the amount of toxic extractables is at the threshold of detection where, for example, a small increase in sample size could change the outcome from nontoxic to moderate or severe toxicity. Below the threshold of detection, differences in these variables are not observable. Within the quantitation range of these assays, varying slopes of the dose–response curve or exposure–effect relationship (Klaassen, 1986) will occur with different toxic agents in a manner similar to that in animal bioassays.

In general, cell lines that have been developed for growth *in vitro* are preferred to primary cells that are freshly harvested from live organisms because the cell lines improve the reproducibility of the assays and reduce the variability among laboratories. That is, a cell line is the *in vitro* counterpart of inbred animal strains used for *in vivo* studies. Cell lines maintain their genetic and morphological characteristics throughout a long (sometimes called infinite) life span. This provides comparative data with the same cell line for the establishment of a database. The L-929 mouse fibroblast cell has been used most extensively for testing biomaterials. Initially, L-929 cells were selected because they were easy to maintain in culture and produced results that had a high correlation with specific animal bioassays (Northup, 1986). In addition, the fibrobast was specifically chosen for these assays because it is one of the early cells to populate a healing wound and is often the major cell in the tissues that attach to implanted medical devices. Cell lines from other tissues or species may also be used. Selection of a cell line is based upon the type of assay, the investigator's experience, measurement endpoints (viability, enzymatic activity, species specific receptors, etc.), and various other factors. It is not necessary to use human cell lines for this testing because, by definition, these cells have undergone some dedifferentiation and lost receptors and metabolic pathways in the process of becoming cell lines.

Positive and negative controls are often included in the assays to ensure the operation and suitability of the test system. The negative control of choice is a high-density polyethylene material. Certified samples may be obtained from the U.S. Pharmacopeial Convention, Inc., Rockville, MD. Several materials have been proposed as candidates for positive controls. These are low-molecular-weight organotin stabilized polyvinyl chloride, gum rubber, and dilute solutions of toxic chemicals, such as phenol and benzalkonium chloride. All of the positive controls are commercially available except for the organotin-stabilized polyvinylchloride.

The methodologies for the three primary cell culture assays are described in the U.S. Pharmacopeia, and in standards published by the American Society for Testing and Materials (ASTM), the British Standards Institute (BSI), and the International Standards Organization (ISO). There are minor variations in the methods among these sources because of the evolving changes in methodology, the time when the standards were developed, and the individual experiences of those participating in standards development. Pharmacopeial assays are legally required by the respective ministries of health in the United States (Food and Drug Administration), Europe, Japan, Australia, and other countries. It is expected that the ISO methods will replace the individual national regulations in Europe whereas the ASTM and BSI standards are voluntary, consensus standards. The basic methodologies, as described in the U.S. Pharmacopeia (1990), are described in the following paragraph.

Direct Contact Test

A near-confluent monolayer of L-929 mammalian fibroblast cells is prepared in a 35-mm-diameter cell culture plate. The culture medium is removed and replaced with 0.8 ml of fresh culture medium. Specimens of negative or positive controls and the test article are carefully placed in individually prepared cultures and incubated for 24 hr at 37 ± 1°C in a humidified incubator. The culture medium and specimens are removed and the cells are fixed and stained with a cytochemical stain such as hematoxylin blue. Dead cells lose their adherence to the culture plate and are lost during the fixation process. Live cells adhere to the culture plate and are stained by the cytochemical stain. Toxicity is evaluated by the absence of stained cells under and around the periphery of the specimen.

At the interface between the living and dead cells, microscopic evaluation will show an intermediate zone of damaged cells. The latter will have a morphological appearance that is abnormal. The change from normalcy will vary with the toxicant and may be evidenced as increased vacuolization, rounding due to decreased adherence to the culture plate, crenation, swelling, etc. For example, dying cells may roundup and detach from the culture plate before they disintegrate. Crenation and swelling are often related to osmotic or oncotic pressures. Vacuolization frequently occurs with basic substances and is due to lysosomal uptake of the toxicant and fluids. This interface area should be included in determining the toxicity rating.

Agar Diffusion Test

A near-confluent monolayer of L-929 is prepared in a 60-mm diameter plate. The culture medium is removed and replaced with a culture medium containing 2% agar. After the agar has solidified, specimens of negative and positive controls and the test article are placed on the surface of the same prepared plate and the cultures incubated for at least 24 hr at $37 \pm 1°C$ in a humidified incubator. This assay often includes neutral red vital stain in the agar mixture, which allows ready visualization of live cells. Vital stains, such as neutral red, are taken up and retained by healthy, viable cells. Dead or injured cells do not retain neutral red and remain colorless. Toxicity is evaluated by the loss of the vital stain under and around the periphery of the specimens. The interface area should be evaluated as described previously.

Selection of a proper agar for use in this assay continues to be a major problem. Agar is a generic name for a particular colloidal polymer derived from a red alga. There are many different grades of agar that are distinguished by their molecular weight and extent of cross-linking of the colloid. The mammalian tissue culture product called agar agar and agarose seem to work best. Agarose is a chemical derivative of agar that has a lower gelling temperature and is less likely to cause thermal shock. The thickness of the agar should be constant because the diffusion distance affects the cellular concentration of a toxicant. From a theoretical viewpoint, it could be expected that different chemicals will diffuse through the agar at different rates. This is true from a broad perspective, but because most toxicants are low molecular weight (less than 100 Da), the diffusion rate will not be sufficiently dissimilar within the 24-hr assay period.

Elution Test

An extract of the material is prepared by using (1) 0.9% sodium chloride or serum-free culture medium and a specified surface area of material per milliliter of extractant and (2) extraction conditions that are appropriate for the application and physical characteristics of the material. Alternatively, serum-containing culture media may be used with an extraction temperature of $37 \pm 1°C$. The choice of extractant sets an upper limit on the quantitation range of the assay in that, without added nutrients, 0.9% sodium chloride will itself be toxic to the cells after a short incubation period. The extract is placed on a prepared, near-confluent monolayer of L-929 fibroblast cells and the toxicity is evaluated after 48 hr of incubation at $37 \pm 1°C$ in a humidified incubator. Live or dead cells may be distinguished by the use of histochemical or vital stains as described earlier.

Interpretation of Results

Each assay is interpreted roughly on the basis of quartiles of affected cells. This corresponds to the customary morpholog-

ical and clinical rating scales of no, slight, moderate, and severe response grades. The terms used to describe these grades refer to the characteristics of the assays. In the direct contact and agar diffusion assays, one expects a concentration gradient of toxic chemicals, with the greatest amount appearing under the specimen and then diffusing outward in more or less concentric areas. Physical trauma from movement of the specimen in the direct contact assay is evident by patches of missing cells interspersed with normal healthy cells. This is not a concern with the agar diffusion assay because the agar cushions the cells from physical trauma. Interpretation of the elution test is based upon what happens to the total population of cells in the culture plate. That is, any toxic agent is evenly distributed in the culture plate and toxicity is evaluated on the basis of the percent of affected cells in the population. Generally, more experience in cell culture morphology is required to appropriately evaluate the elution test than is required for the other two techniques.

Table 1 lists the advantages and disadvantages of the three assays. The chief concern in each of the assays is the transfer or diffusion of some chemical(s) X from the test sample to the cells. This involves the total available amount of X in the material, the solubility limit of X in the solution phase, the equilibrium partitioning of X between the material surface and the solution, and the rate of migration of X through the bulk phase of the material to the material surface. If sufficient analytical data are available to verify that there is one and only one leachable chemical from a given material, then empirical toxicity testing *in vitro* or *in vivo* could be replaced with literature reviews and physiologically based pharmacokinetic modelling of hazard potential. Usually a mixture of chemicals migrate from materials and therefore, empirical testing of the biological effects of the mixture is necessary.

The direct contact assay mimics the clinical use of a device in a fluid path, e.g., blood path, in which the material is placed directly in the culture medium and extraction occurs in the presence of serum-containing culture media at physiological temperatures. The presence of serum presumably aids in the solubilization of leachable substances through protein binding, the *in vivo* mechanism for transporting water-insoluble substances in the blood path. The direct contact assay may be used for testing samples with a specific geometry (for example 1×1-cm^2 squares using extruded sheeting or molded plaques of material) or with indeterminate geometries (molded parts). The major difficulty with this assay is the risk of physical trauma to the cells from either movement of the sample or crushing by the weight of a high-density sample. In most direct contact assays, there will be a zone of affected cells around the periphery of a toxic sample because of a slow leaching rate from the surface and bulk matrix of the material being tested. However, if the toxicant is water soluble, the rate of leaching may be sufficient to cause a decrease in the entire cell population in the culture plate rather than only those cells closest to the sample.

The disadvantages of the direct contact assay can be avoided by using the agar diffusion assay. The layer of agar between the test sample and the cells functions as a diffusion barrier to enhance the concentration gradient of leachable toxicants while also protecting the cells from physical trauma. The test

TABLE 1 Advantages and Disadvantages of Cell Culture Methods

	Direct contact	Agar diffusion	Elution
Advantages	Eliminate extraction preparation	Eliminate extraction preparation	Separate extraction from testing
	Zone of diffusion	Zone of diffusion	Dose response effect
	Target cell contact with material	Better concentration gradient of toxicant	Extend exposure time
	Mimic physiological conditions	Can test one side of a material	Choice of extract conditions
	Standardize amount of test material or test indeterminate shapes	Independent of material density	Choice of solvents
	Can extend exposure time by adding fresh media	Use filter paper disk to test liquids or extracts	
Disadvantages	Cellular trauma if material moves	Requires flat surface	Additional time and steps
	Cellular trauma with high density materials	Solubility of toxicant in agar	
	Decreased cell population with highly soluble toxicants	Risk of thermal shock when preparing agar overlay	
		Limited exposure time	
		Risk of absorbing water from agar	

sample itself may also be tested as a diffusion barrier to the migration of inks or labeling materials through the material matrix to the cellular side of the sample. Even contact between the test sample and the agar ensures diffusion from the material surface into the agar and cell layers. That is, diffusion at the material–solution interface is much greater than that at the material–air interface. Absorbant test samples, which could remove water from the agar layer, causing dehydration of the cells below, should be hydrated prior to testing in this assay.

The elution assay separates the extraction and biological testing phases into two separate processes. This could exploit the extraction to the extent of releasing the total available pool of chemical X from the material, especially if the extraction is done at elevated temperatures that presumably enhance the rate of migration and solubility limit of chemical X in a given solvent. However, when the extractant cools to room temperature, chemical X may precipitate out of solution or partition to the material surface. In addition, elevated extraction temperatures may foster chemical reactions and create leachable chemicals that would not occur in the absence of excessive heating. For example, the polymeric backbone of polyamides and polyurethanes may be hydrolyzed when these polymers are heated in aqueous solutions. Basically, these arguments lead back to a standardized choice of solvents and extraction conditions for all samples rather than optimized solvents for each material.

As with any biological or chemical assay, these assays occasionally are affected by interferences, false negatives, and false positives. For example, a fixative chemical such as formaldehyde or glutaraldehyde will give a false negative in the direct contact but not the agar diffusion assay, which uses a vital stain. A highly absorbant material could give a false positive in the agar diffusion assay because of dehydration of the agar. Severe changes in ononicity, osmolarity, or pH can also interfere with the assays. Likewise, a chelating agent that makes an essential element such as calcium unavailable to the cells could appear as a false positive result. Thus, a judicious evalua-

tion of the test material and assay conditions is required for an appropriate interpretation of the results.

CLINICAL USE

The *in vitro* cytotoxicity assays are the primary biocompatibility screening tests for a wide variety of elastomeric, polymeric, and other materials used in medical devices. After the cytotoxicity profile of a material has been determined, then more application-specific tests are performed to assess the biocompatibility of the material. For example, a product which will be used for *in vitro* fertilization procedures would be tested initially for cytotoxicity and then application-specific tests for adverse effects on a very low cell population density would be evaluated. Similarly, a new material for use in culturing cells would be initially tested for cytotoxicity, followed by specific assays comparing the growth rates of cells in contact with the new material with those of currently marketed materials.

Current experience indicates that a material that is judged to be nontoxic *in vitro* will be nontoxic in *in vivo* assays. This does not necessarily mean that materials that are toxic *in vitro* could not be used in a given clinical application. The clinical acceptability of a material depends on many different factors, of which target cell toxicity is but one. For example, glutaraldehyde-fixed porcine valves produce adverse effects *in vitro* owing to low residues of glutaraldehyde; however, this material has the greatest clinical efficacy for its unique application.

In vitro assays are often criticized because they do not use cells with significant metabolic activity such as the P-450 drug-metabolizing enzymes. That is, the assays can only evaluate the innate toxicity of a chemical and do not test metabolic products that may have greater or lesser toxicity potential. In reality, biological effects of the actual leachable chemicals are the most relevant clinically because most medical devices are in contact with tissues having very low metabolic activity (e.g., skin, muscle, subcutaneous or epithelial tissues) or none. Metabolic products do not form at the implantation site, but rather, require transport of the leachable chemical to distant tissues which are metabolically active. In the process, there is significant dilution of concentration in the blood, tissues, and total body water to the extent that the concentration falls below the threshold of biological activity.

NEW RESEARCH DIRECTIONS

The current interest in developing alternatives to animal testing has resulted in the development and refinement of a wide variety of *in vitro* assays. Cell cultures have been used for several decades for screening anticancer drugs and evaluating genotoxicity (irreversible interaction with the nucleic acids). Babich and Borenfreund (1987) have modified the elution assay for use with microtiter plates to evaluate the dose–response cytotoxicity potential of alcohols, phenolic derivatives, and chlorinated toluenes. This system has also been modified to include a microsomal (S-9) activating system to permit drug metabolism *in vitro* when evaluating pure chemicals such as chemotherapeutic and bacteriostatic agents (Borenfreund and Puerner, 1987). The microtiter methods are likely to have increased application because they provide reproducible, semiautomatic, quantitative, spectrophotometric analyses. The major hurdle will be in identifying the appropriate benchmark or quantitation range for interpreting the data for clinical risk assessment. In earlier quantitative methods for *in vitro* biocompatibility assays, statistical differences in biocompatibility, which are attainable with quantitative assays, were not found to be biologically different (Johnson *et al.*, 1985). That is, the objective data were biologically different only when they were separated into quartiles of response similar to the subjective data. Thus, the major direction of new research will be in defining the benchmarks for application of quantitative methodology.

Bibliography

ASTM (1995a). Practice for direct contact cell culture evaluation of materials for medical devices. Annual Book of ASTM Standards, 13.01, **F813**, 233–236.

ASTM (1995b). Standard test method for agar diffusion cell culture screening for cytotoxicity. Annual Book of ASTM Standards 13.01, **F895**, 247–250.

Babich, H., and Borenfreund, E. (1987). Structure-activity relationship (SAR) models established *in vitro* with the neutral red cytotoxicity assay. *Toxicol. In Vitro* 1: 3–9.

Borenfreund, E., and Puerner, J. A. (1987). Short-term quantitative *in vitro* cytotoxicity assay involving an S-9 activating system. *Cancer Lett.* 34: 243–248.

ISO (1992). "In vitro" method of test for cytotoxicity of medical and dental materials and devices. International Standards Organization, Pforzheim, W. Germany. ISO/10993-5.

Johnson, H. J., Northup, S. J., Seagraves, P. A., Atallah, M., Garvin,

P. J., Lin, L., and Darby, T. D. (1985). Biocompatibility test procedures for materials evaluation *in vitro*. 11. Objective methods of toxicity assessment. *J. Biomed. Mater. Res.* **19**: 489–508.

Klaassen, C. D. (1986). Principles of toxicology. in *Casarett and Douell's Toxicology*, 3rd ed. C. D. Klaassen, M. O. Amdur, and J. Doull, eds. Macmillan, New York, pp. 11–32.

Northup, S. J. (1986). Mammalian cell culture models. in *Handbook of Biomaterials Evaluation: Scientific, Technical and Clinical Testing of Implant Materials,* A. F. von Recum, ed. Macmillan, New York, pp. 209–225.

Till, D. E., Reid, R. C., Schwartz, P. S., Sidman, K. R., Valentine, J. R., and Whelan, R. H. (1982). Plasticizer migration from polyvinyl chloride film to solvents and foods. *Food Chem. Toxicol.* 20: 95–104.

U.S. Pharmacopeia (1995). Biological reactivity tests, in-vitro. in *U.S. Pharmacopeia 23.* United States Pharmacopeial Convention, Inc. Rockville, MD, 1697–1699.

5.3 *IN VIVO* ASSESSMENT OF TISSUE COMPATIBILITY

Myron Spector and Peggy A. Lalor

The *in vivo* assessment of the compatibility of biomaterials and medical devices with tissue is a critical element of the development and implementation of implants for human use. While *in vitro* systems yield important fundamental information about certain elements of cellular and molecular interactions with biomaterials, they cannot replace *in vivo* evaluation. Animals models are necessary in order to account for the effects of the following biological interactions on the response to the biomaterial or medical device:

1. Interactions of various regulatory, parenchymal, stromal, and facultative cell types with the implant and with each other
2. Paracrine and endocrine factors acting on cells around the implant
3. Cell interactions with insoluble extracellular matrix components and soluble regulatory molecules that may have been altered by the presence of the implant
4. Interactions with blood-borne cells, proteins, and molecules

The principles underlying the tissue response to implants are founded in the biomedical sciences (e.g., cell and molecular biology, biochemistry, and physiology).

The biological processes comprising the tissue response are affected by implant-related factors (Fig. 1), including:

1. The "dead space" created by the implant
2. Soluble agents released by the implant (e.g., ions or polymer fragments)
3. Insoluble particulate material released from the implant (e.g., wear debris)
4. Chemical interactions of biological molecules with the implant surface
5. Alterations in the strain distribution in tissue caused by the mismatch in modulus of elasticity between the implant and surrounding tissue, and the movement of

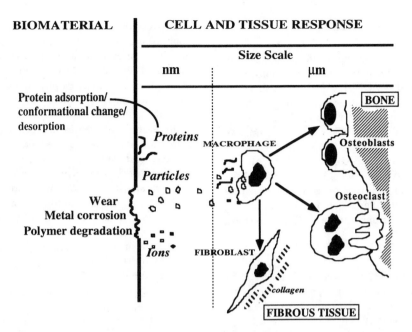

FIG. 1. This schematic is an example of a few of the molecular and cellular interactions that can comprise a tissue response to biomaterials. Cells (of micrometer size—the macrophage in this example) can respond directly to proteins interacting with a material and particles and ions (of nanometer size) released from the surface. The macrophage can release cellular mediators [cytokines and eicosanoids (arrows)] that affect cells such as fibroblasts, osteoblasts, and osteoclasts, and thereby, tissue of a millimeter size. The time scale for these interactions begins within seconds of implantation and can continue through years.

the implant relative to adjacent tissue as a result of the absence of mechanical coupling

The effects of implant-related factors on biological reactions are in the province of biomaterials science. Study of the tissue response to implants requires a methodology capable of measurements at the molecular, cellular, and tissue levels (Fig. 1). Moreover, time is an important variable owing to the criticality of the temporal relationship between the molecular and cellular protagonists of the biological reactions, and because implant-related factors act with different time constants on the biological responses. The dynamic nature of implant–tissue interactions requires that the final assessment of tissue compatibility be qualified by the time frame in which it has been evaluated.

An assessment of the biocompatibility of an implant requires that the tissue compatibility of the biomaterial and the efficacy of the medical device be determined (usually in an animal model that simulates human use). The tissue response to an implant is the cumulative physiological effect of (1) modulation of the acute wound healing response due to the surgical trauma of implantation and the presence of the implant (Fig. 2), (2) the subsequent chronic inflammatory reaction, and (3) remodeling of surrounding tissue as it adapts to the implant (Spector *et al.*, 1989). This complex sequence of biological processes cannot be modeled *in vitro*. Moreover, the healing and stress-induced adaptive remodeling responses of different tissues vary considerably. The selection of a specific *in vivo* model to assess

the tissue compatibility of a particular biomaterial or medical device should be predicated on the similarity of the healing and remodeling responses of the test site with the site that will ultimately be used for implantation in human subjects. This chapter addresses implant sites, surgical protocols, controls, and the methodology for evaluating tissue reactions to biomaterials and medical devices.

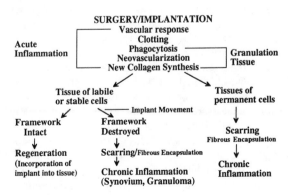

FIG. 2. Diagram showing the sequence of biological processes associated with the healing of implantation sites.

IMPLANT SITES

The primary criterion for selecting an implant site in an animal model is its similarity to the site to be employed in human use of the medical device. However, often there are limitations in the ability of certain tissues and organs in animal models to accommodate implants of a meaningful size. In assessing the appropriateness of a specific tissue as an implant site in any animal model, the healing and remodeling characteristics of the four basic types of tissue (connective tissue, muscle, epithelia, and nerve) should be considered. The characteristics of the parenchymal cells in each type of tissue can provide a rationale for selecting a particular tissue or organ as the site of implantation. In selecting an implant site, the following must be considered:

1. Vascularity
2. The nature of the parenchymal cell with respect to its capability for mitosis and migration, because these processes determine the regenerative capability of the tissue
3. The presence of regulatory cells such as macrophages and histiocytes
4. The effect of mechanical strain, associated with deformation of the extracellular matrix, on the behavior of the parenchymal cell

Surgical wounds in avascular tissue (e.g., the cornea and inner third of the meniscus) may not heal because of the limited potential for the proliferation and migration of surrounding parenchymal cells into the wound site. Gaps between an implant and surrounding avascular tissue can remain indefinitely. Implant sites in vascular tissues in which the parenchymal cell does not have the capability for mitosis (e.g., nerve tissue) heal by the formation of scar in the gap between the implant and surrounding tissue. Moreover, adjacent cells that have died as a result of the implant surgery will be replaced by fibroblasts and scar tissue.

The very presence of the implant provides a dead space in tissue that attracts macrophages to the implant–tissue interface (Silver, 1984). These macrophages, along with fibroblasts of the scar, often form a definable layer of cells that surrounds an implant. The tissue is synovium-like (Spector *et al.*, 1992), and is the chronic response to implants (unless the device is directly apposed to osseous tissue, i.e., osseointegrated). This process is often termed "fibrous encapsulation" (Coleman *et al.*, 1974; Laing *et al.*, 1967). The presence of regulatory cells such as macrophages at the implant–tissue interface can profoundly influence the host response to a device because these cells have the potential to release proinflammatory mediators (Anderson and Miller, 1984) if they are irritated by the movement of the device or substances released from the biomaterial. The inflammatory response of the tissue around implants is comparable to the inflammation that can occur in the synovium lining any bursa (e.g., bursitis); hence, the response to implants has also been termed "implant bursitis" (Spector *et al.*, 1992).

The presence of the implant can also alter the stress distribution on the extracellular matrix, and thereby reduce or increase the strains experienced by the constitutent cells. Many studies have demonstrated immobilization-induced atrophy of certain tissues resulting from a decrease in mechanical strain. Loss of bone mass around stiff femoral stems and femoral condylar prostheses of total hip and knee replacement devices has been associated with the reduced strain due to "stress shielding." Hyperplasia and hypertrophy of tissue in which mechanical strains have increased owing to the presence of an implant have also been observed.

Because of differences in tissue characteristics among implantation sites, investigators have sought to modify implantation test protocols to provide a better assessment of the biocompatibility of biomaterials targeted for specific biomedical applications. It is important to note that materials yielding acceptable tissue compatibility in one implantation site might yield unfavorable results in another site.

An important issue that must be considered when selecting an implant site is the similarity of the device used in the animal model with the device to be employed in the human subject. The size and shape of a device to be used for animal experimentation may have a surface area and stiffness that, relative to surrounding tissue, could be very different than the device intended to be used in human subjects. Such differences are often due to differences in the size and shape of anatomical structures.

Connective Tissue: Bone and Musculoskeletal Soft Tissue

Implantation protocols developed to assess the compatibility of materials for orthopedic prostheses have understandably used bone as the site of implantation. Because of its capability for regeneration, bone should be expected to appose implants in osseous sites. However, the density of bone formed around the implant depends on the site of implantation. Cortical and cancellous bone differ in their vascularity and the size of the pool of preosteoblasts that proliferate in response to the surgical trauma of implantation. When selecting cortical or cancellous sites for implantation, consideration should be given to the origin of the bone cells that are to proliferate and migrate into the implantation site (i.e., the gap between the implant and surrounding bone). In cortical bone, preosteoblasts line the Haversian canals of the osteons and the vascular channels. The periosteal lining of the cortical bone is also composed of preosteoblasts. In cancellous bone the endosteal surface of the trabeculae is covered by preosteoblasts.

The mechanical loading experienced by a biomaterial or medical device at a particular implant site is also an important determinant of the tissue response. Movement of the device relative to surrounding bone early in the wound healing process (i.e., high strains in tissue) can destroy the stroma of the regenerating osseous tissue, leading to repair with scar tissue (encapsulation). Mechanical factors can also influence the remodeling phase of the bone response to implants. The degree to which applied loads are shared between the implant and surrounding bone is a significant determinant of the stress-induced adaptive remodeling of osseous tissue. The greater the mismatch in modulus of elasticity and stiffness between the implant and host bone, the greater the alteration of the normal strain distribution in the osseous tissue and the greater the nonanatomic remodeling because of atrophy (due to stress shielding) or hyperplasia (due to abnormally high strains). The profound effects of mechanical factors on bone remodeling have been recognized for more than a hundred years. The fact that other connective tissues might similarly be influenced by mechanical strains has become evident only recently.

Certain site-selection criteria have been implemented in studies investigating the mechanical (Galante *et al.*, 1971) and chemical (Geesink *et al.*, 1987) bonding of bone to biomaterials. In these studies, cylindrical specimens were implanted in holes drilled in the diaphyseal cortex of dogs. Smaller animals do not have sufficiently thick cortices to allow enough contact between cortical bone and the implant to yield meaningful data. Considering that most orthopedic implants are apposed to cancellous bone, other more meaningful sites of implantation are the metaphyses of the proximal and distal femur and proximal tibia of the dog. In the rabbit, it is only the femoral condyle that has a sufficient amount of cancellous bone to accommodate cylindrical specimens a few millimeters in diameter.

The musculoskeletal soft tissues, including tendons, ligaments, menisci, and articular cartilage have also been implantation sites to test materials to be used in devices to treat defects in those tissues. In many respects less is known about the response of the musculoskeletal soft tissues to implants than about the response to materials in bone. Questions remain about the relative contribution of intrinsic versus extrinsic factors in the healing of defects in soft tissues. Limitations in our understanding of the proliferative capabilities of parenchymal cells and the vascularity of musculoskeletal soft tissues can make assessments of the response of these tissues to implants difficult.

Each of the musculoskeletal soft tissues offers a special challenge, particularly in preparing the subject to receive an implant so that the mechanical forces applied during function of the tissue will not bias the assessment of tissue compatibility. Control of mechanical loading during the initial healing phase is particularly important. In human subjects, the load applied to medical devices can be controlled to some extent during the healing phase by appropriate immobilization or mobilization (e.g., continuous passive motion) protocols. However, similar approaches are more difficult to implement in animal models.

Connective Tissue: Subcutaneous and Cutaneous Tissue

Subcutaneous sites are frequently chosen to assess biocompatibility of implant material. These are readily accessible sites that allow a generalized assessment of biological responses to biomaterials. Tissue reactions to inflammatory agents released by an implant material can be readily detected using this approach. The thickness of the fibrous capsule around implants placed subcutaneously has been used as a measure of the "biocompatibility" of biomaterials for more than 25 years (Laing *et al.*, 1967).

The behavior of this peri-implant tissue is regulated by the macrophages that are sensitive to agents released by the bioma-

terial. The "activation" of this regulatory cell by soluble (e.g., metal ions) and insoluble (e.g., particulate wear debris) agents from an implant can lead to production and release of proinflammatory agents known to influence fibroblast behavior (e.g., fibroblast growth factor and platelet-derived growth factor). These agents can result in thickening of the fibrous capsule. Macrophages can also produce fibrogenic agents that result in thickening of the capsule in response to relative motion at the implant site, thus confounding the interpretation of the tissue response to the material itself.

Surgical defects in dermis have been used in specific investigations as implantation sites for materials developed to facilitate the regeneration of the dermal component of skin (for epidermal regeneration, see "Epithelia" in this Chapter). An important consideration in using this implantation site is the contraction that occurs during the healing process in certain animal models (e.g., the guinea pig). In some studies, the delay in the time course of skin contraction has been employed as a quantitative measure of the effectiveness of certain substances in facilitating regeneration (Yannas, 1989).

Muscle

The paravertebral muscle of rats, rabbits, and dogs has been used as a standardized implantation site to detect problematic constituents released from an implant. Because of the relative motion between the implant and surrounding muscle and the limited capability of skeletal muscle for regeneration, scar tissue forms around the implant. Macrophages can also be found adhering to the biomaterial surface. As with the subcutaneous site of implantation, the relative motion can lead to thickening of the fibrous capsule. Specific studies have compared tissue responses to implants placed in muscles that exposed test coupons to various degrees of relative motion. Thicker scar capsules were found around implants placed in muscles undergoing larger displacements (Black, 1981).

Epithelia

The many disorders affecting the epithelial layer of organs (e.g., epidermis and endothelium) have prompted investigations on the effects of biomaterials in facilitating regeneration of epithelial tissues. Considerable attention has been given to substances that might be used as temporary covering materials to facilitate reepithelization of skin wounds (Rovee, 1991). Epidermal wounds have been produced experimentally by heat, chemical agents, and excision of tissue. The nature of the healing of such wounds is related in large part to the extent to which the wound is desiccated during the healing phase (Winter and Scales, 1963). The beneficial effects of certain biomaterial coverings can be attributed, in large part, to their role in preventing such desiccation, thus allowing for moist wound healing. The rapidity with which epidermis regenerates when desiccation is prevented makes it difficult to assess the role of candidate biomaterials in accelerating the healing process in partial-thickness wounds of skin. Epidermal cells do not come only from the periphery of the wound; the epidermal cells lining the hair follicles also rapidly proliferate to resurface a wound (Winter and Scales, 1963).

Materials to be used in fabricating vascular prostheses have generally been evaluated for their blood compatibility as replacement segments in selected vessels in various animal models; these include carotid-jugular and femoral arteriovenous (Hanson *et al.*, 1980) shunts. It is important to be aware that there is a large difference in the reendothelialization of vascular prostheses in different animal models (Keough *et al.*, 1984). One important consideration is that components of the blood, such as platelets and fibronectin, can profoundly affect the migration of endothelial cells onto a biomaterial surface.

Nerve

Studies in recent years have shown the capability of nerve cells for "regenerating" severed axons (Yannas, 1989). Nerve cells do not have the capability for division. However, the elongation of severed axons allows a degree of electrophysiological continuity to be reestablished across defect sites. Certain matrices facilitate the elongation of such axons, thereby accelerating the regeneration of the nerve and restoration of some function. Biomaterials to be used in nerve regeneration experiments have been contained in silicone elastomer tubes implanted in segmental defects produced in the peripheral nerve of rats.

In other studies, materials to be employed in fabricating electrodes to be used in the neural system have been implanted in brain tissue in the subdural space. Fibrous encapsulation has also been noted in this site.

SURGICAL PROTOCOL AND FORM OF THE IMPLANT

The surgical protocol implemented to assess compatibility of biomaterials and devices in tissues depends on the size and shape of the implant and certain characteristics of the tissue. The surgical trauma associated with implantation will initiate a wound healing response that has a profound effect on the subsequent assessment of the compatibility of the implant in the tissue. Therefore, surgical techniques that minimize trauma to the tissue should be employed. The greater the amount of maceration of the tissue, the more extensive the formation of scar tissue. Power tools with high-speed burrs, drills, and blades that are used to produce implant sites in bone can generate excessive heat that devitalizes adjacent tissue; show-speed instruments with irrigation should be employed to minimize this effect.

When implementing surgical protocols to assess the tissue compatibility of devices, there is a need for adherence to strict sterile technique. Stringent control of sterility is required for implant surgery because of the propensity of bacteria to colonize the implant surface. Care needs be exercised to prevent the introduction of extraneous material, such as talc from surgical gloves, into the implant site.

Studies have indicated that conditioning of an implant in aqueous solution prior to its introduction into the body could be important. Air nuclei on the surface of implants have been found to increase the activation of complement molecules and therefore could influence the tissue response.

The size of the implant is particularly important. Substances in particulate form, of a size that can be phagocytosed, are likely to elicit an inflammatory response because of the release of proinflammatory cytokines and eicosanoids by phagocytes (namely, macrophages) during phagocytosis. Implants (and particulate debris) of a size larger than the mononuclear phagocyte might be covered by the same cell type but do not provoke as much of an inflammatory response because the cells are not capable of phagocytosing the implant. Macrophage activity on the surface of these larger implants generally leads to fusion of the cells to form multinucleated foreign body giant cells (Behling and Spector, 1986).

There can be cases where the form of the implant in a particular animal model elicits an unfavorable response that is not found in implants of the same type in human subjects. Films of a wide range of materials implanted subcutaneously in rats were found to cause malignant tumors at the site of implantation (Oppenheimer *et al.*, 1958). This process has been related to the uninterrupted plane surface area of the implant in this particular *in vivo* model (Brand *et al.*, 1975), and has not been observed in human subjects.

Several methods have been employed for exposing biomaterials to tissues *in vivo*. Generally the biomaterial is placed in direct contact with host tissue as it is implanted into the surgical site produced by an open procedure. In some cases the material has been contained within a stainless steel cage (Marchant *et al.*, 1983) or gelatin capsule in order to investigate the biological response to biomaterials *in vivo* without allowing direct contact of tissue with the implant, which could affect the response. Because these methods introduce a second foreign body into the implant site, controls comprising the carrier implant alone are critical. Materials in particle form or in the form of small-diameter rods, able to fit through the needle of a syringe, have been injected percutaneously into tissue sites. The advantage of the percutaneous route is the lesser degree of surgical trauma which can confound interpretation of the response to the material.

One of the most important variables of a surgical protocol is the "dose" of the biomaterial. Studies of the pharmacologic response to drugs and biologics are not considered complete without an assessment of dose response. In the case of soluble substances (namely, drugs), dose is generally determined by weight (and in some cases associated "activity units"). However, dose-response analyses of biomaterials are rarely performed because generally the dose of a biomaterial is dependent on several variables that could affect the response: weight, surface area, bulk size, number of implants (particles), topography. These parameters are interrelated, confounding experimental designs to determine the dose response; changing the dose according to one of these parameters also changes the values of the other parameters in a way that could alter the tissue reaction. In investigations of the biological response to agents that might be released from an implant, surface area might be considered the controlling variable. Surface area is also an important variable in studies assaying the interactions of biological molecules and/or cells with the biomaterial surface. However, because surface area is strongly influenced by rugosity (at the nano- and microstructural levels as well as macroscopically), methods should be implemented to quantify surface area.

CONTROLS

Controls for *in vivo* investigations of tissue compatibility can include (1) contralateral intact tissues as anatomic controls; (2) sham-operated controls (e.g., surgical incision only); (3) unfilled surgical implant sites; and (4) material and device controls. The nature of the controls employed will depend on the endpoint (i.e., outcome variables) of the experiment. Because of the important effect of surgical trauma on wound healing and tissue response to implants, unfilled surgical implant site controls can be particularly valuable. For example, if a surgical site is prepared in a particular tissue and no implant is inserted, the amount of scar formed could be a valuable guide in assessing the significance of a fibrous capsule found around an implant at the test site. A sham-operated limb can display the effects of altered load bearing on the recipient tissue, and thus serve as control for the effects of altered load bearing on the limb in which the implant has been placed.

Material and device controls must be assessed for their chemistry, mechanical properties, and topography, relative to the experimental specimens, prior to their use as positive or negative controls. In some cases the electrical characteristics of the materials might also have to be considered. For functional devices (e.g., femoral stems of total hip replacement prostheses), the test and control devices should be of identical shape and size.

EVALUATION OF TISSUE REACTION

The method used to evaluate the tissue reaction to any biomaterial should consider the intended use of the material and should be incorporated in the experimental design. Local responses to implanted materials can be assessed qualitatively and quantitatively by several methods, depending on the objective of the experiment. Systemic responses are also important in assessing *in vivo* compatibility, and are discussed elsewhere in this volume. The methods of evaluation addressed in this section are histology and histochemistry, immunohistochemistry, transmission electron microscopy (TEM), scanning electron microscopy (SEM), biochemistry, and mechanical testing.

Histology, SEM, and TEM are tools used to determine the morphological features of the implant material and surrounding tissue at the tissue, cellular, and molecular levels. The morphology and staining patterns of the cells and extracellular matrix provide the means for identifying these components within the surrounding tissues. The condition of cellular organelles observed in the TEM can provide insights into the functional capabilities of the constituent cells. Cell phenotype and activity can also be determined by analyzing for regulatory cytokines and eicosanoids (e.g., prostaglandin and leuko-

trienes) and for extracellular matrix components using immunohistochemistry and immuno-TEM; these methods allow biochemistry to be correlated with morphology. Because of the dynamic nature of the implant site, sampling times are critical, as is subsequent handling of the tissue to preserve morphology and/or biological activity.

Histology and Histochemistry

Histology is the tool used most often to assess the tissue compatibility of an implant material. Qualitative determination of the relative numbers of various cell types and the amount of extracellular matrix components around implants has been the foundation of most *in vivo* protocols. A measurement commonly used to grade the tissue response to a biomaterial is the thickness of the fibrous capsule around an implant. Recently, quantitative analysis of histological sections has been facilitated by computerized image analysis.

Histochemical methods have been used to identify certain types of cells and extracellular matrix components. Localization of lysosomal enzymes has facilitated identification of macrophages in histological sections. Tartrate-resistant acid phosphatase localization has been found to be of value in distinguishing osteoclasts from multinucleated foreign body giant cells, and stains for alkaline phosphatase to assist in the identification of osteoblasts. Other histochemical studies have used a panel of stains to reveal certain components of extracellular matrix (e.g., collagen, elastin, and glycosaminoglycans). A particular challenge in employing any histological method is the preservation of the tissue. Fixation and dehydration of tissues can be achieved by a number of methods, all of which affect tissue morphology to varying degrees. The investigator must be aware of these effects in order to interpret results correctly. Also, care must be exercised to ensure that the molecule of interest is not leached from the section during processing.

Studies focusing on cells and extracellular material at the implant–tissue interface can be problematic. The investigator is challenged to prepare histological sections in which the implant remains *in situ*. Removal of the implant disturbs the interface of interest, often removing critical interfacial cells with the device. The difference in the mechanical properties (namely, hardness) among the implant and surrounding tissue and the conventional embedding medium of paraffin makes microtomy difficult. Nonconventional histological methods involving the embedding of tissue in polymeric media such as glycol methacrylate, poly(methyl methacrylate), or epoxy resins have been developed for preparation of tissue for sectioning with heavy-duty microtomes or by sawing and grinding techniques. Challenges associated with polymer embedding media relate to the inability of many stains to penetrate the medium, and the added difficulty of uniformly depolymerizing sections processed in this way. The relatively thick sections (50 μm and greater) produced by ground section histology often limit the detail that can be clearly discerned because of the projection of overlapping structures when the section is viewed with conventional transmission light microscopy.

An important consideration in the processing of implant-containing tissue for histological methods is the effect of the dehydrating agents on the implant material, particularly when it is a polymer. Solvents such as xylene, acetone, and propylene oxide, commonly used in histological procedures, can lead to softening or dissolution of certain polymeric biomaterials [e.g., poly(methyl methacrylate) and polysulfone]. In this case, investigators have had to develop special processing protocols to limit the alteration of the implant materials.

Histological evaluation alone is not usually sufficient to determine if a material is biocompatible. This method is limited since not all cell types can be determined by light microscopy of histological sections, even with enzyme localization. For example, activated lymphocytes can often be mistaken for monocytes and macrophages for fibroblasts. However, if it is used as a tool in conjunction with other methods, histology can be very useful. Analysis of histological sections can answer many questions on the nature of the tissue surrounding an implanted material.

Immunohistochemistry

Immunohistochemical techniques are a more recent addition to the protocols for evaluating tissue responses to implant materials. Utilization of immunohistochemistry allows specific cell types and extracellular matrix components around an implant to be identified. For this technique, monoclonal or polyclonal antibodies specific for membrane, intracellular, and extracellular molecules (antigenic determinants) are applied to tissue sections. This antibody, or secondary antibody, specific for the primary antibody, is substrate conjugated and can, therefore, be localized by a chemical reaction to produce a colored or fluorescent precipitate at the site of the antigen. Alternatively, several secondary antibodies can be used prior to a substrate-conjugated final antibody, all of which are specific for the preceding antibody. Colored or fluorescent precipitates can be seen by light microscopy and can be assessed both qualitatively and quantitatively. Because monoclonal antibodies are antigen specific, double and even triple labeling can be performed on the same tissue section to evaluate two or more parameters simultaneously. However, before employing multiple labels, consideration should be given to the fact that steric hindrance could affect interpretation of results if the antigens are in close proximity.

Despite the specificity and scope of analyses offered by immunohistochemistry, few investigators have utilized this technique in assessing tissue responses to implant materials. This is due, in part, to the fact that there have not been many commercially available antibodies specific for molecules in certain animal models apart from the mouse, rat, and guinea pig. This has changed rapidly, however, in the past few years, due to increased demand. One application of the immunolabeling technique was the determination of the population of Ia-presenting cells (macrophages) and T lymphocytes and B lymphocytes in rats which had been paraspinally implanted with commercially pure titanium disks. It was shown that Ia-presenting cells were the main contributor to the inflammatory reaction (which decreased after 50 days). T lymphocytes of both the

helper and suppresser subtypes were present in fewer numbers, and decreased in number at an earlier stage than the Ia-presenting cells, indicating that a relatively low T-lymphocyte-mediated response was initially adjacent to this material after implantation (Lalor and Revell, 1993).

Specimens for immunohistochemistry examination must be prepared in such a way as to preserve the antigenicity of the molecules of interest in the tissues. Equally important is a working knowledge of where the antigenic determinant can be found within the tissue under investigation. In some cases, an antigen may be masked by another protein in the tissue or by the process used to embed the tissue. For example, glycol methacrylate, which is used as an embedding medium, often masks antigenic determinants. In cases in which a protein is masking the molecule of interest, enzyme digestion might reveal the antigen.

Immunohistochemically labeled sections have generally been evaluated qualitatively. Recent emergence of digitized image analysis can facilitate quantitation of the labeled elements; for example, the amounts of T lymphocytes and B lymphocytes adjacent to metal implants in a fibrocapsule surrounding an implant in the paraspinal musculature of rats can be evaluated in this way.

A method of investigation which employs labeling principles similar to those used in immunohistochemistry is in situ hybridization. In situ hybridization is generally used to provide evidence that certain proteins are being synthesized by cells by demonstrating the presence of messenger RNA (mRNA). This method utilizes labeled (either with a substrate or a radio isotope) RNA or DNA probes that have nucleotide sequences complementary to mRNA. mRNA production can therefore be demonstrated either by autoradiography or by an enzyme–substrate reaction. Either method enables protein production to be quantified by computerized image analysis. This method is not yet routinely used to evaluate biomaterials in vivo. However, it is likely that this method will become important in the future as more investigators focus on mechanisms involved in biological responses to biomaterials.

Transmission Electron Microscopy

TEM allows cells at the interface of an implant to be examined at the ultrastructural level. However, the difficulty in preparing the ultrathin sections (100 nm or less) has limited implementation of this technique. Recent investigations using ion-thinning methods and special ultramicrotomy techniques have begun to reveal important features of certain biomaterial–tissue interfaces. Immunolocalization of certain proteins can be performed at the TEM level using antibodies conjugated to gold particles or enzyme substrates.

One area that has benefited greatly from TEM is the investigation of the bone-bonding behavior of certain calcium-containing materials. TEM studies have demonstrated the precipitation of biological apatites on synthetic calcium phosphate materials implanted in soft tissue and bone sites. Ultrathin sections were produced by ultramicrotomy and ion-thinning methods. An additional benefit of transmission electron microscopy for these studies was the ability to identify specific crystalline phases using selected area electron diffraction. In addition, the relationship of various crystalline materials at the ultrastructural level was facilitated using high-resolution electron microscopy to image the crystalline lattices of individual crystallites.

The particle size of wear debris from implants (e.g., orthopedic protheses) can be below the resolution limit of light microscopy. In these types of investigations TEM can be used to determine the cellular response to the particles of debris. It can be used to identify the condition of certain cellular organelles that reflect the metabolic health of individual cells.

TEM can offer further advantages if methods of elemental analyses are linked to the microscope. Energy-dispersive X-ray microanalysis (EDX) and electron energy loss spectroscopy (EELS) allow elemental analysis of the tissue surrounding an implant. They provide the means to: (1) identify particulate substances, (2) determine which elements have leached from component materials, and (3) determine the location of the particles or elements within the cell or the extracellular matrix.

Scanning Electron Microscopy

Scanning electron microscopy has been used to image tissue around implants in many *in vivo* models. The implant–tissue interface has been prepared for SEM by cutting cross sections with saws or knives or by freeze-fracture techniques. The integrity of the implant–tissue interface depends on the nature of the tissue and the fixation and dehydration protocols employed. Shrinkage of the tissue due to processing can be advantageous in determining features of the implant–tissue interface by revealing the surface of the tissue apposed to the implant. However, in some instances shrinkage may destroy a tissue-implant interface.

Conventional SEM utilizing imaging with secondary electrons is of limited value in identifying cellular and extracellular elements because such identification must rely on morphological criteria that generally lack specificity. However, in some cases elemental analysis using energy-dispersive X-ray microanalysis in the SEM can provide complementary information to facilitate identification of certain features in the implant and surrounding tissue (e.g., mineral deposits and particulate debris).

Recent investigations have demonstrated the value of backscattered electron imaging (BSE) in revealing certain features of mineralized tissues and ceramic implants (namely, calcium phosphate substances). The difference in mineral density allows for the distinction of implant and bone, and facilitates digitized, quantitative area analysis of bone around certain types of implants. This method has generally replaced microradiography for imaging the bone response to implants.

Biochemistry

The metabolic characteristics of tissue around implants, particularly with respect to inflammatory reactions, have been effectively evaluated using biochemical methods. The levels of certain inflammatory mediators such as prostaglandin E_2 and certain interleukins (e.g., IL-1) in the tissue around implants can be determined using radioimmunoassay techniques. However, a limitation of this method is the amount of tissue required in situations where the levels of these eicosanoids and inflammatory cytokines are relatively low. A complementary method is to isolate cells from the peri-implant tissue in culture. The levels of inflammatory mediators in the conditioned medium can be assessed after various culture times. This method has the advantage of providing information about the metabolic behavior of the cells in the peri-implant tissue. However, the handling required to isolate and culture the cells can activate them to produce and release more of these mediators than they might have while *in situ*.

Biochemical evaluation of tissue surrounding an implant is not routinely used to determine tissue biocompatibility. It can be a valuable tool in investigating the production of regulatory cytokines and eicosanoids by cells in response to the presence of a biomaterial. However, manipulation of tissues at and after explantation can dramatically alter the production and release of cellular mediators.

Mechanical Testing

It is useful to test a material for certain mechanical properties if the intended use will subject the implant to mechanical loading. These tests have been particularly valuable in determining the strength and compliance of certain implant–bone interfaces to evaluate the efficacy of certain coatings as attachment vehicles for orthopedic prostheses and dental implants. For screening purposes, cylindrical rods of test materials have been implanted transcortically through the lateral aspect of the canine femur. Push-out tests are utilized to compute the interfacial shear strength by dividing the load to failure by the cortical surface area apposing the implant. However, this approach has certain limitations. Most orthopedic implants reside in cancellous rather than cortical bone. Cancellous bone has lower density, strength, and modulus than cortical bone. In addition, because push-out testing produces a nonuniform shear stress distribution in bone, the method of computing interfacial shear strength is problematic. Alternative investigations have implanted cylindrical specimens in the cancellous bone of dogs in the proximal and distal femur. Pull-out tests from these sites lead to a more uniform distribution of stress in surrounding bone, and might model the clinical use of the implants more closely. Stress analyses indicate that torsion tests produce the most uniform distribution of stress in surrounding bone and therefore might be the best method to use if results are to be compared from one investigation to another.

Most studies are designed to perform mechanical testing on explanted specimens within hours of sacrifice of the animal or after the specimens have been frozen and thawed. Any type of fixation by chemical methods will affect the mechanical properties of biological tissues, and in many cases, the biomaterial.

CRITERIA FOR ASSESSING ACCEPTABILITY OF THE TISSUE RESPONSE

The *in vivo* assessment of tissue compatibility requires that certain criteria be used for determining the acceptability of the tissue response relative to the intended application of a material. The device should be considered biocompatible only in the context of the criteria used to assess the acceptability of the tissue response. In this regard, every study involving the *in vivo* assessment of tissue compatibility should provide a working definition of biocompatibility. Biomaterials and devices implanted into bone can become apposed by the regenerating osseous tissue, and thus be considered compatible with bone regeneration. However, altered bone remodeling around a device, with a net loss of bone mass (i.e., osteopenia), could lead to the assessment that a material or device is not compatible with normal bone remodeling. It situations in which an implant is surrounded by fibrous tissue, the macrophages on the surface of the material are the expected response to the dead space produced by the implant. The synovium-like tissue thus produced might be considered an acceptable response relative to the chemical compatibility of the material. Utilization of the thickness of the scar capsule around an implant alone as a measure of biocompatibility is problematic because the formation of scar tissue can be influenced by movement of the tissue at the site relative to the implant.

The cellular and molecular makeup of tissue and the interactions between these components are complex. Therefore, criteria for assessing certain features of the biocompatibility of biomaterials or devices should focus on specific aspects of the biological response. The tissue compatibility of materials should be assessed specifically in the context of the effects of the material or device on certain aspects of the response.

It is difficult to assess the acceptability of an implant without knowledge of the etiology of tissue responses. Suitable negative controls are vital for comparison. Controls which have been proven to produce minimal adverse response in the host provide adequate comparisons.

Extreme care should be used in extrapolating the results obtained in animal models to human tissues. Animal models do not always accurately model the tissue response that will occur in humans because of the diverse biology and anatomy of different species.

Bibliography

Anderson, J. M., and Miller, K. M. (1984). Biomaterial biocompatibility and the macrophage. *Biomaterials* 5: 5–10.

Behling, C. A., and Spector, M. (1986). Quantitative characterization of cells at the interface of long-term implants of selected polymers. *J. Biomed. Mater. Res.* 20: 653–666.

Black, J. (1981). *Biological Performance of Materials*, pp. 208–209. Dekker, New York.

Brand, E. G., Buoan, L. C., Johnson, K. H., and Brand, I. (1975). Etiological factors stages, and the role of the foreign body in foreign body tumorigenesis: A review. *Cancer Res.* 35: 279–286.

Coleman, D. L., King, R. N., and Andrade, J. D. (1974). The foreign body reaction: a chronic inflammatory response. *J. Biomed. Mater. Res.* 8: 199–211.

Galante, J., Rostoker, W., Lueck, R., and Ray, R. D. (1971). Sintered fiber metal composites as a basis for attachment of implants to bone. *J. Bone Joint Surf.* 53A: 101–115.

Geesink, R. G. T., deGroot, K., and Klein, C. (1987). Chemical implant fixation using hydroxy-apatite coatings. *Clin. Orthop. Rel. Res.* 225: 147–170.

Hanson, S. R., Harker, L. A., Ratner, B. D., and Hoffman, A. S. (1980). *In vivo* evaluation of artificial surfaces with a non-human primate model of arterial thrombosis. *J. Lab. Clin. Invest.* 95: 289.

Keough, E. M., Callow, A. D., Connoly, R. J., Weinberg, K. S., Aalberg, J. J., and O'Donnell, T. F. (1984). Healing pattern of small caliber Dacron grafts in the baboon: an animal model for the study of vascular prostheses. *J. Biomed. Mater. Res.* 18: 281–292.

Laing, P. G., Ferguson, A. B., and Hodge, E. S. (1967). Tissue reaction in rabbit muscle exposed to metallic implants. *J. Biomed. Mater. Res.* 1: 135–149.

Lalor, P. A, and Revell, P. A. (1993). T-lymphocytes and titanium–aluminium–vanadium (TiAlV) alloy: Evidence for immunological events associated with debris deposition. *Clin. Mater.* 12: 57–62.

Marchant, R., Hiltner, A., Hamlin, C., Rabinovitch, A., Slododkin, R., and Anderson, J. (1983). *In vivo* biocompatibility studies. I. The cage implant system and a biodegradable hydrogel. *J. Biomed. Mater. Res.* 17: 301–325.

Oppenheimer, B. S., Oppenheimer, E. T., Stout, A. P., Willhite, M., and Danishefsky, I. (1958). The latent period in carcinogenesis by plastics in rats and its relation to the presarcomatous stage. *Cancer* 11: 204–213.

Rovee, D. T. (1991). Evolution of wound dressings and their effects on the healing process. *Clin. Mater.* 8: 183–188.

Silver, I. A. (1984). The physiology of wound healing. in *Wound Healing and Wound Infection*, T. K. Hunt, ed. Appleton-Century-Crofts, New York, p. 11.

Spector, M., Cease, C., and Xia, T.-L. (1989). The local tissue response to biomaterials. *CRC Crit. Rev. Biocompat.*, 5: 269–295.

Spector, M., Shortkroff, S., Hsu, H.-P., Taylor-Zapatka, S., Lane, N., Sledge, C. B., and Thornhill, T. S. (1992). Synovium-like tissue from loose joint replacement prostheses: comparison of human material with a canine model. *Semin. Arthr. Rheum.* 21: 335–344.

Winter, G. D., and Scales, J. T. (1963). Biology: Effect of air drying and dressings on the surface of a wound. *Nature* 91: 2.

Yannas, I. V. (1989). Regeneration of skin and nerves by use of collagen templates. in *Collagen* Vol. III: *Biotechnology*, M. Nimni, ed. CRC Press, Boca Raton, FL, pp. 87–115.

5.4 TESTING OF BLOOD–MATERIALS INTERACTIONS

Stephen Hanson and Buddy D. Ratner

Every day, thousands of devices made from synthetic materials or processed natural materials are interfaced with blood (see Chapters 7.2 and 8.2). How can the biomaterials engineer know which materials might be best used in fabricating a blood-contacting device? This chapter outlines some methods and concerns in evaluating the blood compatibility of biomaterials, and the blood compatibility of medical devices. It does not automatically follow that if the materials comprising a device are blood compatible, a device fabricated from those materials will also be blood compatible. This important point should be clear upon completion of this chapter. Before considering the evaluation of materials and devices, the reader should be familiar with the protein and cellular reactions of blood coagulation, platelet responses, and fibrinolysis as discussed in Chapter 4.5.

BLOOD COMPATIBILITY

A discussion of the nature of blood compatibility would be straightforward if there were a list of standard tests that might be performed to evaluate blood compatibility. By simply performing the tests outlined in such a list, a material could be rated "blood compatible" or "not blood compatible." Unfortunately, no widely recognized, standard list of blood compatibility tests exists. Owing to the complexity of blood–materials interactions (BMI), there is a basic body of ideas that must be mastered in order to appreciate what blood interaction tests actually measure. This section introduces the rationale for BMI testing.

"Blood compatibility" can be defined as the property of a material or device that permits it to function in contact with blood without inducing adverse reactions. Unfortunately, this simple definition offers little insight into what a blood-compatible material is. More useful definitions become increasingly complex. This is because there are many mechanisms that the body has available to respond to insults to the blood. A material that will not trigger one response mechanism may be highly active in triggering another mechanism. The mechanisms by which blood responds to synthetic materials were discussed in Chapter 4.5. This chapter discusses how one assesses the blood compatibility of materials in light of these response mechanisms.

We can also view blood compatibility from a different perspective by considering a material that is *not* blood compatible, i.e., a thrombogenic material. Such a material would produce specific adverse reactions when placed in contact with blood: formation of clot or thrombus composed of various blood elements; shedding or nucleation of emboli (detached thrombus); the destruction of circulating blood components; activation of the complement system and other immunologic pathways (Salzman and Merrill, 1987). Clearly, in designing blood-contacting materials and devices, the aim is to minimize these obviously undesirable blood reactions.

Many devices and materials are used in humans to treat or to facilitate treatment of various disease states. Such devices include the extracorporeal pump-oxygenator (heart-lung machine) used in many surgical procedures, hollow fiber hemodialyzers for treatment of kidney failure, catheters for blood access and blood vessel manipulation (e.g., angioplasty), heart assist devices, stents and permanently implanted devices to replace diseased heart valves (prosthetic heart valves) and arteries (vascular grafts). Since these and other blood-contacting devices

have been successfully used in patients for many years, and are judged to be therapeutically beneficial, it is reasonable to ask: (1) Is there a continued need for assessing BMI? and (2) Are there important problems that remain to be addressed? The answer in both cases is clearly "yes." For example, many existing devices are frequently modified by incorporation of new design features or synthetic materials primarily intended to improve durability and physical and mechanical characteristics; that is, devices may be modified to improve characteristics other than BMI. However, since these changes may also affect blood responses, and since BMIs are not entirely predictable based on knowledge of device composition and configuration, blood compatibility testing is nearly always required to document safety.

The performance of many existing devices is also less than optimal (Salzman and Merrill, 1987; Williams, 1987; McIntire *et al.*, 1985). For example, prolonged cardiopulmonary bypass and membrane oxygenation can produce a tendency to severe bleeding. Mechanical heart valves occasionally shed emboli to the brain, producing stroke. Synthetic vascular grafts perform less well than grafts derived from natural arteries or veins; graft failure due to thrombosis can lead to ischemia (lack of oxygen) and death of downstream tissue beds; small-diameter vascular grafts (<4 mm i.d.) cannot be made. Thus, while performance characteristics have been judged to be acceptable in many instances (i.e., the benefit/risk ratio is high), certain existing devices could be improved to extend their period of safe operation (e.g., oxygenators), and to reduce adverse BMI long term (e.g., heart valves). Furthermore, many devices are only "safe" when anticoagulating drugs are used (e.g., oxygenators, heart valves, hemodialyzers). Device improvements that would reduce adverse BMI and thereby eliminate the need for anticoagulant therapy would have important implications both for health (fewer bleeding complications from drug effects) and cost (complications can be expensive to treat). The reusability of devices which can undergo repeated blood exposure in individual patients (e.g., dialyzers) is also an important economic consideration.

For certain applications there are no devices available at present that perform adequately (owing to adverse BMI) even when antithrombotic drugs are used. Thus there is a need for devices which could provide long-term oxygenation for respiratory failure, cardiac support (total artificial heart or ventricular assist device), and intravascular measurement of physiologic parameters (O_2, CO_2, pH), as well as for small-diameter vascular grafts (<5 mm internal diameter) and other conduits (e.g., stents) for reconstruction of diseased arteries and veins. Overall, there is a compelling need for continued and improved methods for evaluating BMI.

THROMBOGENICITY

The thrombogenic responses induced by a material or device can be categorized into two groups. First, as the term implies,

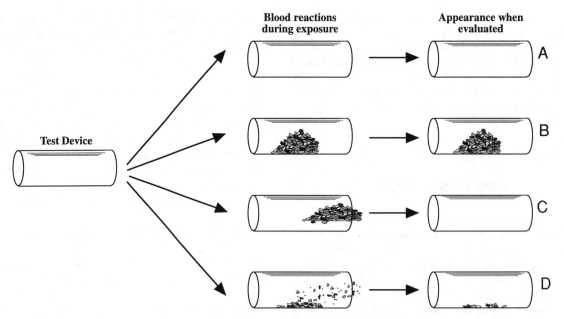

FIG. 1. Possible scenarios for blood–materials interactions, and limitations of evaluating only local thrombus formation at fixed time points. (A) Device remains free of thrombus. (B) Large thrombus forms and remains attached. (C) Large thrombus forms but detaches (embolizes). (D) Surface is highly reactive toward blood but deposited material is quickly removed through microembolism and/or lysis. Inspection of devices (C) and (D) could lead to the incorrect conclusion that these surfaces are blood compatible.

a thrombogenic device may cause the accumulation of various blood elements which are preferentially concentrated locally relative to their concentrations in circulating blood (thrombus formation). Cardiovascular devices may also exhibit regions of disturbed flow or stasis which lead to the formation of blood clots (coagulated whole blood). These *local* effects may compromise device functions, such as the delivery of blood through artificial blood vessels, the mechanical motions of heart valves, gas exchange through oxygenators, and the removal of metabolic waste products through dialysis membranes. These local blood reactions may also produce effects in other parts of the host organism, i.e., *systemic* effects. Thus thrombi may detach from a surface (embolize) and be carried downstream, eventually occluding a blood vessel of comparable size and impairing blood flow distal to the site of occlusion. Chronic devices may produce steady-state destruction or "consumption" of circulating blood elements, thereby lowering their concentration in blood (e.g., mechanical destruction of red cells by heart prostheses, producing anemia, or removal of platelets as a result of continuing thrombus formation), with a concomitant rise in plasma levels of factors released from those blood elements (e.g., plasma hemoglobin, platelet factor 4). Mediators of inflammatory responses and vessel tone may also be produced or released from cells (e.g., platelets, white cells, the complement pathway) following blood–surface interactions which can affect hemodynamics and organ functions at other sites. Thus "thrombogenicity" may be broadly defined as the extent to which a device, when employed in its intended configuration, induces the adverse responses outlined here.

While all artificial surfaces interact with blood, an acceptably nonthrombogenic device can be defined as one which would produce neither local nor systemic deleterious effects of consequence to the health of the host organism.

With "thrombogenicity" now defined globally in terms of adverse outcomes associated with device use, the obvious goal is to design and improve devices using materials which are blood compatible (nonthrombogenic) for specific applications. Ideally, the biomaterials engineer should be able to consult a handbook for a list of materials useful in fabricating a device. Unfortunately, there is little consensus as to what materials are blood compatible. Because of this lack of consensus, there is no reliable or "official" list of blood-compatible materials. As a result, an individual interested in learning which materials might be suitable for construction of a new blood-contacting device generally consults published studies, or directly studies candidate materials.

Despite intensive efforts, the blood compatibility of specific materials for particular device applications is not well established because: (1) the types of devices used are numerous, may exhibit complex flow geometries, and are continuously evolving; (2) the possible blood responses are numerous, complex, dynamic, and not fully understood; and (3) it is difficult and expensive to measure device thrombogenicity (clinically significant local thrombosis or systemic effects) in an extensive and systematic way in either experimental animals or humans (2). Most tests purported to measure blood compatibility really evaluate certain blood–materials interactions. Figure 1 illustrates how alternative interpretations can be applied to data from "blood compatibility" tests. This concept is further ex-

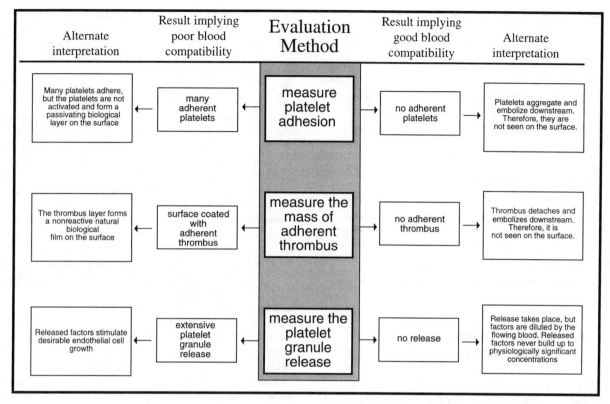

FIG. 2. Alternative scenarios that can be applied for interpreting results of blood–materials interaction assays.

panded upon in Fig. 2. These alternative interpretations often invalidate the conclusions one would like to draw from such tests. For accuracy, the term "BMI assessment" will be used for the remainder of this chapter instead of "blood compatibility test." Based upon the characteristics of the evaluation method (i.e., what is really being measured), the biomaterials scientist must relate the significance of the events being observed (the BMI) to the blood compatibility of the material or device. A solid understanding of the physical and biological mechanisms of blood–materials interactions is required to make this connection in a rational way.

In more specific terms, BMIs are the interactions (reversible and irreversible) between surfaces and blood solutes, proteins, and cells (adsorption, absorption, adhesion, denaturation, activation, spreading) which occur under defined conditions of exposure time, blood composition, and blood flow. Since each of these variables influences BMI, we cannot generally: (1) extrapolate results obtained under one set of test conditions to another set of conditions; (2) use short-term testing to predict long-term results; and (3) predict *in vivo* performance of a device based on BMI testing of materials per se in idealized flow geometries. Nonetheless, such tests have provided important insights into the mechanisms of thrombus formation in general, and the relationships between BMI and blood compatibility. These studies also permit some general guidelines for constructing devices and, to a limited

extent, may allow prediction of the performance of devices in humans. These points are addressed in subsequent sections of this chapter.

The above considerations suggest that no material may be simply "blood compatible" or "nonthrombogenic" since this assessment will depend strongly on details of the test system or usage configuration. In fact, under conditions of sluggish (low shear) blood flow or stasis, most if not all polymeric materials may become associated with localized blood clotting and thus be considered "thrombogenic." This is because synthetic materials, unlike the endothelium (which lines all blood vessels), cannot actively inhibit thrombosis and clotting by directly producing and releasing inhibitors or by inactivating procoagulant substances. The possibility that there may be no "biomaterials solution" to certain problems, or that the performance of a device could be improved by emulating strategies found in nature, has led some investigators to consider coating devices with endothelial cells, antithrombotic drugs, or anticoagulating enzymes. While there is no evidence that these methods have solved the problem of biomaterial thrombogenicity for any device, the approaches appear promising and are being widely explored. As for conventional synthetic materials and devices, establishing the usefulness of biologic surfaces and drug delivery devices requires appropriate methods for evaluating their blood interactions.

TABLE 1 Factors Important in the Acquisition and
Handling of Blood for BMI Experiments

Species of the blood donor

Health of the blood donor

Blood reactivity of the donor (individual physiological differences)

Time interval between blood draw and BMI experiment

Care with which the puncture for the blood draw was made

Temperature (for blood storage and testing)

Anticoagulation

Drugs and anesthetics present in the blood

Blood damage caused by centrifugation and separation operations

Blood damage caused by contact with foreign surfaces prior to the
  BMI experiment (syringe, needle, blood bag, bottles, tubing, etc.)

Blood damage caused by the air–blood interface

Blood damage caused by pumping and recirculation

IMPORTANT CONSIDERATIONS FOR
BMI ASSESSMENT

In 1856 Rudolph Virchow proposed that the three factors
that contribute to the coagulation of the blood are the blood
chemistry, the blood-contacting surface, and the flow regime
(commonly referred to as Virchow's triad).

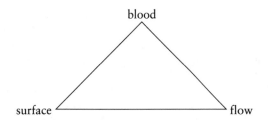

This assessment is still valid, and provides a framework
for more formally introducing the variables important in any
system intended to evaluate BMI. As described here, these
variables may each profoundly influence the results and interpretation of BMI testing, and the ranking of blood compatibility of materials according to that test.

FACTORS AFFECTING THE PROPERTIES OF BLOOD

The source and methods for handling blood can have important effects on BMI. Blood obtained from humans and
various animal species has been employed *in vitro* and *in vivo*,
both in the presence and absence of anticoagulants. Blood
reactivity is also influenced by the extent and period of manipulation *in vitro*, the surface-to-volume ratio of blood placed
in extracorporeal circuits, and the use of pumps for blood
recirculation. These aspects are discussed in the following paragraphs and are summarized in Table 1.

The blood chemistry of each animal species is different (see
Chapter 5.5). In particular, blood may vary with respect to
the concentrations and functions of blood proteins and cells
which participate in coagulation, thrombosis, and fibrinolysis
(Chapter 4.5). The size of blood-formed elements may also
differ. A comparison of blood chemistry between man and
the more commonly used animal species has been published
(McIntire *et al.*, 1985). While human blood is obviously prefer-
able for BMI, it is often impossible to use human blood in
certain experiments. In addition, there are significant health
concerns in experimenting with human blood, and animals are
commonly used for both *in vitro* and *in vivo* studies. Unfortunately, most investigations have employed a single animal species or blood source. There have been few comparisons between
human and animal blood responses for evaluating BMI in any
particular test situation. In many instances, differences between
human and animal blood responses are likely to be large, and
must be borne in mind when interpreting experimental results.
For example, the initial adhesiveness of blood platelets for
artificial surfaces appears to be low in man and some primates,
and high in the dog, rat, and rabbit (Grabowski *et al.*, 1976).
Following the implantation of chronic blood-contacting devices (e.g., vascular grafts), there may be large differences between man and all other animal species in terms of device
healing and incorporation of surrounding tissues, which will be
reflected as differences in the *time course* of BMI. Furthermore,
while animal groups serving as blood donors for BMI experiments may represent a relatively homogeneous population in
terms of age, health status, blood responses, etc., the human
recipients of blood-contacting devices may vary considerably
in terms of these parameters. Thus, the results obtained with
any animal species must be viewed with great caution if conclusions are to be drawn as to the significance of the results
for humans.

Despite these limitations, animal testing has been particularly helpful in defining *mechanisms* of BMI and thrombus
formation, and the interdependence of blood biochemical pathways, the nature of the surface, and the blood flow regime. In
addition, while results of animal testing may not quantitatively
predict results in man, in many cases results are likely to be
qualitatively similar. These aspects are discussed further later.

In general, studies in lower animal species such as the rabbit,
rat, and guinea pig may be useful to screen for profound differences between materials which might be achieved, for example,
by incorporation of an antithrombotic drug delivery system
into an otherwise thrombogenic device. Short-term screening
to identify markedly reactive materials, and longer term studies
to evaluate the effects of healing phenomena on BMI can also
be performed in other species such as the dog and pig. When
differences in BMI are likely to be modest (for example, as a
consequence of subtle changes in surface chemistry or device
configuration), the ranking of materials based on tests with
lower animal species may be unrelated to results that would
be obtained in man; studies with primates, which are hematologically similar to man, are more likely to provide results
which are clinically relevant. However, the relationships between human and primate BMIs are also not well established
in many models and applications, and should therefore be
interpreted with caution.

In vitro testing generally requires anticoagulation of the blood, which can have a profound effect on BMI. *In vivo* testing and the use of extracorporeal circuits are also commonly performed with anticoagulants. Two anticoagulants are most frequently used: sodium citrate, which chelates calcium ions that are required for certain reactions of platelets and coagulation proteins, and heparin, a natural polysaccharide used to block the action of the coagulation protease thrombin (Chapter 4.5). Both can markedly affect BMI. In particular, the removal of calcium ions may profoundly depress platelet–surface reactions and the capacity of platelets to form aggregates and thrombi. Thus the relationship between BMI in the presence of citrate anticoagulant and "thrombogenicity" in the absence of anticoagulant is questionable. Similar concerns apply to heparin anticoagulation. Although this agent is less likely to interfere with the earliest platelet reactions, platelet thrombus formation may be impaired by inhibition of thrombin activity. The use of heparin is appropriate for evaluating devices whose use normally requires heparinization *in vivo* (e.g., oxygenators, dialyzers). In general, results with anticoagulated systems cannot be used to predict performance in the absence of anticoagulants. Anticoagulants in the context of BMI were discussed by McIntire *et al.* (1985).

Blood is a fragile tissue that begins to change from the moment it is removed from the body. It may become more active (activated) or less active (refractory) in several ways, even with effective anticoagulation. Thus, BMI evaluations with blood aged more than a few hours are probably meaningless. If purified blood components or cells are used (e.g., platelets, fibrinogen), studies must be performed to ensure that they remain functionally normal. In most cases the volume of blood used, relative to test surface area, should be large. Similarly, the area of nontest surfaces, including exposure to air interfaces, should be minimized. Changes in blood or test surface temperature, or exposure to intense light sources, can also produce artifactual results. When blood is pumped, the recirculation rate (fraction of total blood volume pumped per unit time) should be minimized since blood pumping alone can induce platelet and red cell cell damage, platelet release reactions, and platelet refractoriness.

EFFECTS OF BLOOD FLOW

Blood flow controls the rate of transport (by diffusion and convection) of cells and proteins in the vicinity of artificial surfaces and thrombi. This subject was reviewed by Leonard (1987) and Turitto and Baumgartner (1987). While physiologically encountered blood shear forces probably do not damage or activate platelets directly, such forces can dislodge platelet aggregates and thrombi, which may attach farther downstream or be carried away (embolize) to distal circulatory beds. Platelet diffusion in flowing blood and early platelet attachment to surfaces may be increased 50- to 100-fold by the presence of red blood cells, which greatly enhance the movement of platelets across parallel streamlines. At higher shear forces, red cells may also contribute chemical factors that enhance platelet reactivity (Turitto and Baumgartner, 1987).

A number of studies using well-characterized flow geometries have suggested that the initial attachment of platelets to artificial surfaces increases with increasing blood flow, or, more specifically, with increasing wall shear rate (slope of the velocity profile at the surface). Under conditions of low wall shear rate (less than ~ 1000 sec^{-1}) early platelet adhesion (over the first minutes of exposure) may depend more upon the platelet arrival rate (i.e., platelet availability) than on substrate surface properties (Friedman *et al.*, 1970). Under these conditions the platelet–surface reaction rate is said to be diffusion controlled. At higher shear rates, platelet adhesion may depend upon both the rate of platelet transport as well as surface properties; thus, studies designed to assess the role of surface properties are best performed under flow conditions where platelet transport is not limiting. Following initial platelet adhesion, subsequent processes of platelet aggregation and *in vivo* thrombus formation (over minutes to hours) may be partly reaction controlled. For example, platelet accumulation on highly thrombogenic artificial surfaces (e.g., fabric vascular grafts) or biologic surfaces (e.g., collagen) may be quite rapid and depends on both the substrate reactivity and factors influencing platelet availability (shear rate, hematocrit, and the platelet content of blood) (Harker *et al.*, (1991). Under other circumstances, the rate of platelet–surface interactions may be almost entirely reaction controlled. For example, with smooth-walled artificial surfaces which cause repeated embolization of small platelet aggregates continuously over days, the overall rate of platelet destruction depends strongly on material properties but not on blood flow rate or circulating platelet numbers over wide ranges of these variables (Hanson *et al.*, 1980).

It has been observed that under arterial flow conditions (high shear rate), a thrombus that forms *in vivo* may be largely composed of platelets ("white thrombus"), while a thrombus that forms under venous flow conditions (low shear rate) may contain mostly red cells entrapped in a fibrin mesh ("red thrombus"). The process of platelet thrombus formation may not be affected by administration of heparin in normal anticoagulating amounts (i.e., arterial thrombosis may be heparin resistant), while venous thrombosis is effectively treated with heparin. This lack of effect against platelet reactions is somewhat surprising since the procoagulant enzyme, thrombin, one of the most potent activators of platelets, is strongly inhibited by heparin. These observations have been incorrectly interpreted to mean that arterial and venous thromboses are separable processes, with the former depending only on platelet reactions and the latter depending only on coagulation-related events. However, while platelet-dependent (arterial) thrombosis may be little affected by heparin, it is blocked quite effectively by other inhibitors of thrombin (Hanson and Harker, 1988; Wagner and Hubbell, 1990), indicating that heparin is limited in its capacity to block the enzyme when it is produced locally in high concentrations through reactions which may be catalyzed on the platelet surface (Chapter 4.5).

The formation of fibrin, owing to the action of thrombin on fibrinogen, is also important for the formation and stabilization of a thrombus since: (1) fibrinolytic enzymes can reduce platelet thrombus formation, and (2) arterial thrombi are often composed of alternating layers of platelets and fibrin. Thus, in most circumstances, thrombin is a key promoter of local platelet and fibrin accumulation (on surfaces), under both high shear and low shear conditions. Thrombi may differ in appearance because under high flow conditions, thrombin and precursor procoagulant enzymes (e.g., factor Xa) may be diluted sufficiently to prevent bulk phase clotting and trapping of red cells.

In summary, the formation of a thrombus requires the transport by flow of platelets and coagulation proteins to surfaces. Fibrin polymerization, as well as local platelet activation and recruitment into growing thrombi, requires conversion of prothrombin to thrombin, the end product of a sequential series of coagulation reactions which are also catalyzed by platelets, and may be amplified or inhibited by various feedback mechanisms (Chapter 4.5). Blood flow regulates each reaction step such that under low (venous) flow conditions fibrin formation is abundant; thrombi may resemble coagulated whole blood with many entrapped red cells. Under high (arterial) flows, platelets, stabilized by much smaller amounts of fibrin, may comprise the greater proportion of the total thrombus mass.

PROPERTIES OF BIOMATERIALS AND DEVICES

Many different artificial surfaces, in various device applications, are used in contact with blood. As discussed subsequently, tests designed to assess certain blood–materials interactions have shown that the surface physicochemical properties of materials and devices can have important effects on early events, for example, on protein adsorption and platelet adhesion, yet how these effects relate to subsequent thrombus formation remains uncertain.

When placed in contact with blood, most if not all artificial surfaces first acquire a layer of adsorbed blood proteins whose composition and mass may vary with time in a complex manner, depending on substrate surface type (Chapter 3.2). This layer mediates the subsequent attachment of platelets and other blood cells that can lead to the development of platelet aggregates and thrombi. The relationship between material properties, the protein layer, and the propensity of a material or device to accumulate thrombi is not well understood because: (1) protein–surface reactions involve complex, dynamic processes of competitive adsorption, denaturation, and activation; (2) cell–surface interactions may modify the protein layer, i.e., cells may deposit lipid and protein "footprints" derived from the cell membrane; (3) the importance of specific adsorbed proteins for subsequent cell interactions is not well defined; (4) there have been few relevant tests in which both protein adsorption and later thrombus formation has been assessed. Under conditions of low blood shear, the capacity of negatively charged surfaces (such as glass) to activate intrinsic coagulation (via factor XII) can lead to thrombin production with subsequent platelet deposition and fibrin clot formation. Under other circumstances, the availability on surfaces of adhesive plasma proteins, such as fibrinogen, may be important for regulating platelet attachment (Horbett et al., 1986).

With anticoagulated blood, initial platelet attachment to a variety of surfaces may be comparable and limited to a partial platelet monolayer, suggesting that surface properties may be "inconsequential" for early platelet adhesion (Friedman et al., 1970). In the absence of anticoagulants, initial platelet attachment may vary, but no general relationship to substrate surface properties has been demonstrated. In attempts to establish such relationships, thrombus formation has been studied using devices implanted in animals and composed of various materials, including polymers, metals, carbons, charged surfaces, and hydrogels. Correlations have been sought between the blood response and surface properties such as charge (anionic-cationic), hydrophilicity, hydrophobicity, polarity, contact angle, wettability, and critical surface tension (Salzman and Merrill, 1987; Williams, 1987; McIntire et al., 1985). These parameters have not proven satisfactory for predicting device performance even in idealized test situations, reflecting the complexity of the phenomena being investigated, the limitations of animal modeling (e.g., Fig. 1), and, in some cases, inadequate characterization of material surface properties (see Chapters 1.3 and 9.7 and Ratner, 1993a).

In many cases, material properties are constrained by the intended application of the device. For example, vascular grafts and the sewing ring of prosthetic heart valves are composed of fabric or porous materials to permit healing and tissue anchoring. Other materials must be permeable to blood solutes and gases (dialysis and oxygenator membranes) or distensible (pump ventricles, balloon catheters), and may necessarily exhibit complex flow geometries. In general, devices with flow geometries which cause regions of flow recirculation and stasis tend to produce localized clotting in the absence of heparin anticoagulant. On a microscopic scale, surface imperfections, cracks, and trapped air bubbles may serve as a focus to initiate thrombus formation. While surface smoothness is usually desirable, many devices having a fabric or microporous surface (e.g., vascular grafts) function well if the layer of thrombus that forms is not thick enough to interfere with the function of the device (Salzman, 1987).

INTERACTION TIME

Different events may occur at short and long BMI times. A test in which blood contacts a device for seconds or minutes may yield a result that will have no meaning for devices used for hours or days, or which may be implanted chronically. Thus, measurements of protein adsorption may not predict levels of platelet adhesion. Platelet adhesion alone is not an adequate measure of thrombogenicity, and does not predict the extent of local or systemic thrombogenic effects which could be harmful to the host organism. However, several studies indicate that an early maximum in platelet thrombus accumulation may be seen within hours of device exposure, which can be sufficient to produce device failure (e.g., small-diameter vascular graft occlusion) (Harker et al., 1991). Therefore, short-term testing (over hours) may be appropriate for predicting the clinical usefulness of devices which can produce an acute, severe thrombotic response. In general, the nature and extent of BMI may change continuously over the entire period of device exposure. An exception to this rule may be chronic implants which are not covered by tissue (e.g., heart valve struts, arteriovenous shunts), and which may interact with blood elements at a constant rate as shown, for example, by steady-state rates of platelet consumption (Hanson *et al.*, 1980).

EVALUATION OF BMI

In this section we provide a general summary and interpretation of more commonly used *in vitro* and *in vivo* animal testing procedures used to evaluate BMI. It is emphasized that a thorough characterization of material properties is critical for the interpretation of these tests (Chapters 1.2 and 1.3 and Ratner, 1993a).

In Vitro Tests

In vitro BMI tests involve placing blood or plasma in a container composed of a test material or recirculating blood through a flow chamber in which test materials contact blood under well-defined flow regimes which simulate physiologic flow conditions. Many flow geometries have been studied, including tubes, parallel plates, packed beds, annular flows, rotating probes, and spinning disks. These studies have been reviewed (McIntire *et al.*, 1985; Turitto and Baumgartner, 1987), and have yielded considerable insight into how proteins and platelets are transported to, and react with, artificial surfaces. They also provide a wealth of morphologic information at the cellular level on details of platelet–surface and platelet–platelet interactions (Sakariassen *et al.*, 1989). However, as discussed earlier, these tests are usually of short duration and are strongly influenced by the blood source, handling methods, and the use of anticoagulants. Thus, *in vitro* test results generally cannot be used to predict longer term BMI and *in vivo* outcome events, or to guide the selection of materials for particular devices.

However, *in vitro* tests may be of some usefulness in screening materials that may be highly reactive toward blood. Tests of the whole blood clotting time and variations thereof involve placing nonanticoagulated whole blood (or blood anticoagulated with sodium citrate which is then recalcified) into containers of test material and measuring the time for a visible clot to form. Materials which quickly activate intrinsic coagulation and cause blood to clot within a few minutes (such as negatively charged glass) are probably unsuitable for use in devices with low shear blood flow, or in the absence of anticoagulants.

Recirculation of heparinized blood through tubular devices and materials may lead to deposition of platelets onto highly thrombogenic materials with the appearance in plasma of proteins released from platelets (Kottke-Marchant *et al.*, 1985). Thus, this and similar methods may identify materials that might cause rapid platelet accumulation *in vivo* over short time periods, and therefore be unsuitable for certain applications such as small-diameter vascular grafts or blood conduits. Both this test and *in vitro* clotting assays can be considered for preliminary screening and identification of materials that could be highly thrombogenic. Most artificial surfaces in common use would probably "pass" these tests. Since small differences in test results are likely to be meaningless for predicting material performance in actual use, these tests are not appropriate for optimizing or refining material properties. *In vivo* testing is therefore required.

In Vivo Tests of BMI

Many studies have been performed in which test materials, in the form of rings, tubes, and patches, are inserted for short or long periods into the arteries or veins of experimental animals (Salzman and Merrill, 1987; Williams, 1987; McIntire *et al.*, 1985). For the following reasons, most of these tests are of questionable relevance to the use of biomaterials in man: (1) The timing and type of measurements may be such that important blood responses are unrecognized. In particular, the measurement of gross thrombus formation at a single point may lead to incorrect conclusions about local thrombus formation (e.g., Fig. 1), and does not provide assessment of systemic effects of thrombosis such as embolization and blood element consumption; (2) With more commonly used animal species (e.g., dogs), blood responses may differ from humans both quantitatively and qualitatively; (3) The hemodynamics (blood flow conditions) of the model may not be controlled or measured; (4) There may be variable blood vessel trauma and tissue injury which can cause local thrombus formation through the extrinsic pathway of blood coagulation (Chapter 4.5). Thus, *in vivo* testing of materials in idealized flow geometries (rather than actual device configurations) may provide few insights into the selection of materials for use in man.

Evaluations of BMI may be performed in animals having arteriovenous (A-V) or arterioarterial (A-A) shunts, i.e., tubular blood conduits placed between an artery and vein, or between an artery and artery. A-V shunts have been studied in a variety of animals, including baboons, dogs, pigs, and rabbits (3). An A-V shunt system is illustrated in Fig. 3. Once established, shunts may remain patent (not occluded) for long periods of time (months) without the use of anticoagulants. Test materials or devices are simply inserted as extension segments or between inlet and outlet portions of the chronic shunt. These systems have the advantages that: (1) blood flow is easily controlled and measured; (2) native or anticoagulated blood can be employed; and (3) both short-term and long-term BMI, including local and systemic effects, can be evaluated. For example, in baboons, which are hematologically similar to man, the blood responses to tubular biomaterials and vascular grafts have been quantitatively compared with respect to: (1) localized thrombus accumulation; (2) consumption of circulating platelets and fibrinogen; (3) plasma levels of factors released by platelets and coagulation proteins during thrombosis; and (4) embolization of microthrombi to downstream circulatory beds (Harker *et al.*, 1991). These studies in primates are consistent with observations in man that certain commonly used polymers [e.g., polytetrafluoroethylene, polyethylene, poly(vinyl chloride), silicone rubbers] and some vascular grafts [e.g., polytetra-

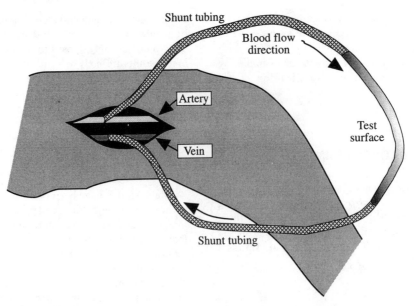

FIG. 3.   Illustration of an arteriovenous (A-V) shunt placed between the femoral artery and vein (in the leg) of an experimental animal. Materials to be tested (in this case a tubular device) are interposed between inlet and outlet segments of the shunt.

fluoroethylene] are relatively nonthrombogenic in extracorporeal circuits and arteries. Thus, results with shunt models, particularly in higher animal species, may predict BMI in humans when employed under comparable flow conditions (laminar unidirectional flow with arterial shear rates). Since extracorporeal shunts exclude the modulating effects of blood vessel cells and tissue injury, results with these models may be less relevant to the behavior of devices which are placed surgically or the responses which may be mediated by interactions with the vessel wall as well as the blood (e.g., heart valves, grafts, indwelling catheters and sensors).

In Vivo Evaluation of Devices

Since the blood response to devices is complex and not well predicted by testing materials per se in idealized configurations, animal testing, and ultimately clinical testing, of functioning devices is required to establish safety and efficacy. Broad guidelines, based on the type of device being considered, are given in the following paragraphs and apply to both animal and clinical testing. A summary of *in vivo* blood responses to devices, and of commonly used methods which have proven useful for evaluating those responses, is given in Table 2.

Devices which have relatively small surface areas and are exposed for short periods of time (hours to days) include catheters, guidewires, sensors, and some components of extracorporeal circuits. With these devices the primary concern is the formation of significant thrombi that could interfere with their function (e.g., cause blood sensor malfunction), occlude vessels (catheters), and embolize either spontaneously or be stripped from the device surface when it is removed from the body (e.g.,

during catheter withdrawal through a vessel insertion site), producing occlusion of distal vessels and tissue ischemia. Devices exposed for short periods which have large surface areas (dialyzers) and complex circuitries (pump-oxygenators) may, in addition, produce (1) a marked depletion of circulating blood cells and proteins (e.g., platelets and coagulation factors); (2) an immune or inflammatory response through activation of complement proteins and white cells; and (3) organ dysfunction mediated by hemodynamic, hematologic, and inflammatory reactions. Mechanical devices which are used for long periods of time (heart assist devices, extracorporeal membrane oxygenators) may produce profound systemic effects and organ dysfunction such that their use in man remains problematic. Other long-term implants (grafts, heart valves, stents) might be further improved by extending their period of safe operation and patency, and by reducing the frequency of embolic phenomena (e.g., stroke) and requirements for the concurrent use of antithrombotic drugs.

With both long-term and short-term devices, thrombus formation can be assessed directly and indirectly. Important indirect assessments include depletion from circulating blood of cells and proteins consumed in thrombus formation, and the appearance in plasma of proteins generated by thrombus formation (e.g., fibrinopeptide A, platelet factor 4). Direct assessment of blood flow rate, flow geometry, and extent of flow channel occlusion can in many cases be achieved using sophisticated methods, including angiography, ultrasound imaging, and magnetic resonance imaging. Devices which are removed from the circulation should be visually inspected to assess whether a thrombus has formed at particular sites or on certain materials. Emboli in flowing blood may be detected using ultrasound and laser-based techniques, although these methods are

TABLE 2 Blood–Materials Responses and Their Evaluation

System	Blood response	Assessment[a]
Device/material	Thrombosis	Direct visual and histologic evaluation; noninvasive imaging (angiography, ultrasound, radioisotope, magnetic resonance); evidence of device dysfunction.
	Thromboembolism	Emboli detection (ultrasound, laser); evidence of organ/limb ischemia, stroke.
Platelets	Consumption	Increased removal of radioisotopically labeled cells; reduced blood platelet count.
	Dysfunction[b]	Reduced platelet aggregation *in vitro*; prolonged bleeding time.
	Activation	Increased plasma levels of platelet factor 4 and β-thromboglobulin; platelet membrane alterations (e.g., by flow cytometry).
Red cells[b]	Destruction	Decreased red cell count; increased plasma hemoglobin.
White cells[b]	Consumption/activation	Decreased counts of white cell populations; increased white cell plasma enzymes (e.g., neutrophil elastase).
Coagulation factors	Consumption[b]	Reduced plasma fibrinogen, factor V, factor VIII.
	Thrombin generation	Increased plasma levels of prothrombin fragment 1.2 and thrombin : antithrombin III complex.
	Fibrin formation	Increased plasma level of fibrinopeptide A.
	Dysfunction[b]	Prolonged plasma clotting times.
Fibrinolytic proteins	Consumption[b]	Reduced plasma plasminogen level.
	Plasmin generation	Increased plasma level of plasmin : antiplasmin complex.
	Fibrinolysis	Increased plasma level of fibrin D-dimer fragment.
Complement proteins[b]	Activation	Increased plasma levels of complement proteins C5a and C3b.

[a]Radioimmunoassays (RIA) and enzyme-linked immunoassays (ELISA) may not be available for detection of nonhuman proteins.
[b]Tests which may be particularly important with long-term and/or large surface area devices.

not used widely at present. Thrombus formation and rates of platelet destruction by both acutely placed and chronically implanted devices can be determined quantitatively by measurements of platelet lifespan and scintillation camera imaging of radioisotopically labeled blood elements (McIntire *et al.*, 1985; Hanson *et al.*, 1990).

Finally, it is important to emphasize that thrombosis occurs dynamically, so that thrombi continuously undergo both formation and dissolution. Device failure represents the imbalance of these processes. Older thrombi may also be reorganized considerably by the enzymatic and lytic mechanisms of white cells. While the initial consequences of implanting surgical devices include tissue injury, thrombosis due to tissue injury, and foreign body reactions, the flow surface of long-term implants may become covered with a stable lining of cells (e.g., ingrowth of vascular wall endothelial and smooth muscle cells onto and into vascular grafts), or blood-derived materials (e.g., compacted fibrin). Certain reactions of blood elements (e.g., platelets, thrombin) may also stimulate the healing response. Ultimately, long-term devices, such as the small-caliber graft, may fail as a result of excessive tissue ingrowth which could be largely unrelated to biomaterials properties.

While many applications of devices described here, as well as laboratory and clinical methods for evaluating their biologic responses, will be unfamiliar to the bioengineer, it is important to appreciate that: (1) each device may elicit a unique set of blood responses, both short-term and long-term; (2) methods are available to assess systemic changes in the blood and host organism which indirectly reflect thrombus formation; and (3) localized thrombus formation can usually be measured directly

and quantitatively. Whenever possible, serial and dynamic studies should be performed to establish the time course of thrombus formation and dissolution. These measurements will ultimately predict device performance, and allow for the rational selection of biomaterials which will minimize adverse blood–device interactions.

CONCLUSIONS

The most blood-compatible material known is the natural, healthy, living lining of our blood vessels. This "material" functions well by a combination of appropriate surface chemistries, good blood flow characteristics, and active biochemical processes involving removal of prothrombotic substances and secretion of natural anticoagulants. It seems unlikely that we will ever match this performance in a synthetic material or device, although attempts to imitate aspects of the natural system represent a promising strategy for developing a new generation of blood-compatible devices (Chapters 2.11, 7.2, and 9.3). At present, however, synthetic materials that perform less well than the vessel wall, but still satisfactorily, will be needed. This chapter provides only a brief outline of the issues involved in evaluating materials and devices to find those that are minimally damaging or activating toward blood. The subject of blood compatibility testing is complex, and advanced study is required before considering experiments intended to elucidate basic mechanisms or improve human health. Further discussion elaborating upon the complexity of the issues involved in BMI testing can be found in Ratner (1984, 1993b). Detailed discussion on the characterization of materials for biomaterials application and on BMI testing can also be found in a recent publication coordinated by the Device and Technology Branch of the National Heart, Lung and Blood Institute, National Institutes of Health (Harker *et al.*, 1993).

Acknowledgments

The authors acknowledge generous support from research grants HL-31469, HL25951 and RR01296 from the National Institutes of Health, U.S. Public Health Service, during the preparation of this chapter and for some of the studies described here.

Bibliography

Friedman, L. I., Liem, H., Grabowski, E. F., Leonard, E. F., and McCord, C. W. (1970). Inconsequentiality of surface properties for initial platelet adhesion. *Trans. Am. Soc. Artif. Intern. Organs* **16**: 63–70.

Grabowski, E. F., Herther, K. K., and Didisheim, P. (1976). Human versus dog platelet adhesion to cuprophane under controlled conditions of whole blood flow. *J. Lab. Clin. Med.* **88**: 368–373.

Hanson, S. R., and Harker, L. A. (1988). Interruption of acute platelet-dependent thrombosis by the synthetic antithrombin D-phenylalanyl-L-prolyl-L-arginyl chloromethylketone. *Proc. Natl. Acad. Sci. U.S.A.* **85**: 3184–3188.

Hanson, S. R., Harker, L. A., Ratner, B. D., and Hoffman, A. S. (1980). *In vivo* evaluation of artificial surfaces using a nonhuman primate model of arterial thrombosis. *J. Lab. Clin. Med.* **95**: 289–304.

Hanson, S. R., Kotz'e, H. F., Pieters, H., and Heyns, A. D. (1990). Analysis of 111-indium platelet kinetics and imaging in patients with aortic aneurysms and abdominal aortic grafts. *Arteriosclerosis* **10**: 1037–1044.

Harker, L. A., Ratner, B. D., and Didisheim, P. (1993). Cardiovascular biomaterials and biocompatibility: A guide to the study of blood-material interaction. *Cardiovasc. Pathol.* **2** (Suppl. (3): 1993.

Harker, L. A., Kelly, A. B., and Hanson, S. R. (1991). Experimental arterial thrombosis in non-human primates. *Circulation* **83**(6 Suppl.): IV41–55.

Horbett, T. A., Cheng, C. M., Ratner, B. D., Hoffman, A. S., and Hanson, S. R. (1986). The kinetics of baboon fibrinogen absorption to polymers; *in vitro* and *in vivo* studies. *J. Biomed. Mater. Res.* **20**: 739–772.

Kottke-Marchant, K., Anderson, J. M., Rabinowitch, A., Huskey, R. A., and Herzig, R. (1985). The effect of heparin vs. citrate on the interaction of platelets with vascular graft materials. *Thromb. Haemost.* **54**: 842–849.

Leonard, E. F. (1987). Rheology of thrombosis. in *Hemostasis and Thrombosis*, 2nd ed. R. W. Coleman, J. Hirsh, V. J. Marder, and E. W. Salzman, eds. Lippincott, Philadelphia, pp. 1111–1122.

McIntire, L. V., Addonizio, V. P., Coleman, D. L., Eskin, S. G., Harker, L. A., Kardos, J. L., Ratner, B. D., Schoen, F. J., Sefton, M. V., and Pitlick, F. A. (1985). Guidelines for Blood–Material Interactions—Devices and Technology Branch, Division of Heart and Vascular Diseases, National Heart, Lung and Blood Institute, NIH Publication No. 85-2185, revised July 1985, U.S. Department of Health and Human Services, Washington, DC.

Ratner, B. D. (1984). Evaluation of the blood compatibility of synthetic polymers: consensus and significance, in *Contemporary Biomaterials: Materials and Host Response, Clinical Applications, New Tech-*

nology and Legal Aspects, J. W. Boretos and M. Eden, eds., Noyes Publ., Park Ridge, NJ, pp. 193–204.

Ratner, B. D. (1993a). Characterization of biomaterial surfaces. *Cardiovasc. Pathol.* **2** Suppl. (3), 87S–100S.

Ratner, B. D. (1993b). The blood compatibility catastrophe, *J. Biomed. Mater. Res.* **27**: 283–287.

Sakariassen, K. S., Muggli, R., and Baumgartner, H. R. (1989). Measurements of platelet interaction with components of the vessel wall in flowing blood. *Methods Enzymol.* **169**: 37–70.

Salzman, E. W., and Merrill, E. D. (1987). Interaction of blood with artificial surfaces. *Hemostasis and Thrombosis*, 2nd ed., R. W. Coleman, J. Hirsh, V. J. Marder, and E. W. Salzman, eds. Lippincott, Philadelphia, pp. 1335–1347.

Turitto, V. T., and Baumgartner, H. R. (1987). Platelet–surface interactions. in *Hemostasis and Thrombosis*, 2nd ed., R. W. Coleman, J. Hirsch, V. J. Marder, and E. W. Salzman, eds. Lippincott, Philadelphia, pp. 555–571.

Wagner, W. R., and Hubbell, J. A. (1990). Local thrombin synthesis and fibrin formation in an *in vitro* thrombosis model result in platelet recruitment and thrombus stabilization on collagen in heparinized blood. *J. Lab. Clin. Med.* **116**: 636–650.

Williams, D. (ed.) (1987). *Blood Compatibility*. CRC Press, Boca Raton, FL.

5.5 ANIMAL MODELS

Brad H. Vale, John E. Willson, and Steven M. Niemi

Others chapters in this book address the variety and relative utility of research techniques used to study biomaterials in contact with live tissue, either *in vitro* or *in vivo*. Key underlying issues in the selection of and success with these techniques include:

Legal and humane aspects of laboratory animal procurement and use

Animal species differences for various physiological traits and responses

Regulatory and scientific considerations in the hierarchy of preclinical testing that precedes clinical testing in human patients.

These three topics are addressed in this chapter.

RESPONSIBLE USE OF ANIMALS

When considering animal models for biomaterials research, the investigator and sponsoring institution should first assess their responsibilities with respect to federal, state, and local laws and regulations. In addition, and given the growing public and political sensitivity to the use of resarch animals, all laboratories using animals must ensure that their programs recognize and comply with current definitions of humane animal care and use. The importance of this issue is reflected in a resolution recently adopted (1990) by the American Association for the Advancement of Science (see Appendix 1). This section briefly presents legal obligations and provides practical guidelines for responsible use of laboratory animals.

At the national level, laboratory animal care and use is

covered by regulations established under the Animal Welfare Act (Public Law 89-544, as amended). The regulations [Title 9 of the *Code of Federal Regulations (CFR)*, Chapter 1, subchapter A—animal Welfare, Parts 1, 2, and 3] are enforced by the Animal and Plant Health Inspection Service (APHIS), U.S. Department of Agriculture. The Act covers dogs, cats, nonhuman primates, guinea pigs, hamsters, rabbits, and "any other warmblooded animal, which is being used, or is intended for use for research, teaching, testing, experimentation, . . ." (9 *CFR* subchapter A, Part 1, Section 1.1). Species specifically exempt from the Act include birds, rats and mice bred for use in research, and horses and livestock species used in agricultural research. Institutions using those species covered by the Act must be registered by APHHIS. Continued registration is dependent on submission of annual reports to APHIS by the institution, as well as satisfactory inspection of the institution's animal facility during unannounced site visits by APHIS inspectors. The Act provides specifications for animal procurement (i.e., from licensed suppliers), husbandry, and veterinary care which are used to determine compliance with the Act. The annual report supplied to APHIS contains a list of all species and numbers of animals used by the institution for the previous year. All animals listed in the annual report must be categorized by the level of discomfort or pain they were thought to experience in the course of the research.

The 1985 amendments to the Act (7 U.S.C. 2131, *et seq.*) were implemented and became effective on October 30, 1989, August 15, 1990, and March 18, 1991, and extend the Act to cover the institution's administrative review and control of its animal research program (*Federal Register*, Vol. 54, No. 168, pp. 36112–36163; Vol. 55, No.36, 28879–28881; Vol. 56, No. 32, pp. 6426–6505). Specifically, the new regulations require that all animal research protocols be reviewed and approved by an Institutional Animal Care and Use Committee (IACUC) before those protocols are initiated. Furthermore, the submitted protocols must state in writing that less harmful alternatives were considered but are not available, and that the proposed research is not unnecessarily duplicative. Additional requirements increase the scope of husbandry requirements for laboratory dogs and nonhuman primates.

The other relevant federal body is the National Institutes of Health (NIH). The Health Research Extension Act of 1985 (Public Law 99-158) required the director of NIH to establish guidelines for the proper care of laboratory animals and IACUC oversight of that care. Broad policy is described in the *Public Health Service Policy on Humane Care and Use of Laboratory Animals* (NIH, 1986), which identifies the *Guide for the Care and Use of Laboratory Animals* (NIH, 1985a) as the reference document for compliance. This policy applies to all activities involving animals either conducted by or supported by the U.S. Public Health Service (PHS). Before an institution can receive animal research funding from NIH or any other PHS agency, that institution must file a statement of assurance with the PHS Office for Protection from Research Risks (OPRR) that it is complying with these guidelines.

Similar to the Animal Welfare Act, the PHS policy requires annual updates on animal research use and IACUC review of animal protocols and facilities. Unlike the Act, the PHS policy covers all vertebrate species, does not include an enforcement arm for routine inspections (but does provide for inspection in cases of alleged misconduct), and penalizes noncompliant institutions by withdrawing funding support.

State and local laws and regulations also may affect animal research programs. These mainly involve restrictions on availability of cats and dogs from municipal shelters. Within the past decade several states have also enacted registration and inspection statutes similar to the Animal Welfare Act (e.g., Massachusetts). In several states, court rulings based on open meeting laws have required IACUC reports and deliberations at state-supported institutions to be conducted in public (e.g., Florida, Massachusetts, North Carolina, Washington).

Articles that document the benefits and concepts of responsible animal use are listed in the Suggested Reading section. Those concepts can be outlined as follows:

1. Use abiotic or *in vitro* models whenever possible. In those instances where live animals are appropriate, ensure that the minimum number of animals necessary is used and in a manner that minimizes animal discomfort, in the context of the aim of their intended use.
2. Treat and maintain all laboratory animals humanely and in full accordance with applicable laws and regulations. This includes appropriate methods of physical restraint, anesthesia, analgesia, and euthanasia that do not interfere with experimental results.
3. Maintain animal facilities and programs so that they meet or surpass all applicable regulatory and accreditation agency standards. This includes training of personnel and formation of internal review groups to monitor compliance with these standards.
4. Actively survey current technical and scientific literature in order to prevent unnecessary duplication and to ensure that the most appropriate (animal or nonanimal) models are used for the corresponding research goal.

There are many organizations dedicated to providing for the humane care and use of laboratory animals, and that serve as useful sources or information on regulations, animal husbandry, personnel training, and experimental procedures. Appendix 2 lists some of these organizations and their addresses.

ANIMAL MODELS AND SPECIES CONSIDERATIONS

A multitude of issues can be raised regarding the selection of specific animal models to address a hypothesis on a testing need. Within the general area of testing cardiovascular devices, for instance, the specific question of variation of blood–materials, interactions across species has generated many publications, including the *Guidelines for Blood–Material Interactions* (NIH, 1985b). An entire chapter in that publication is devoted to issues of species differences in such parameters as platelet function and fibrinogen level. Other comparative studies of differences in blood coagulation between species and between breeds or inbred lines within the same species have been published (Didisheim, 1985; Clemmons and Meyers, 1984; Dodds, 1988). It is most appropriate in this brief chapter to simply raise the concern in the biomedical researcher's mind about species variability and then suggest a bibliography for followup reading on specific species and testing regimens.

Some of the currently used sources of information in this area are listed in the Suggested Reading section. The recent volume VII of Gay's series on Methods of Animal Experimentation has a section dedicated to the use of telemetry in animal research. It is clear that the biomedical researcher, when selecting appropriate models and designing experiments, can refer to an ever-increasing body of knowledge that resides both in texts and in the support personnel at each vivarium.

TESTING HIERARCHIES

After a biomaterial or medical device is determined to be efficacious under circumstances not involving human subjects (i.e., *in vitro*, *ex vivo*, or in animals), a program of safety evaluation follows that involves several stages of increasing complexity. For example, the NIH *Guidelines for the Physico-chemical Characterization of Biomaterials* (NIH, 1980) contains a list of assays for testing the safety of blood-contacting devices. A hierarchy of testing, starting with *in vitro* systems and progressing through functionality implants *in situ*, is implied:

Cell culture cytotoxicity (mouse L929 cell line)
Hemolysis (rabbit or human blood)
Mutagenicity (human or other mammalian cells or Ames test (bacterial)
Systemic injection acute toxicity (mouse)
Sensitization (guinea pig)
Pyrogenicity (limulus amebocyte lysate [LAL] or rabbit)
Intracutaneous irritation (rabbit)
Intramuscular implant (rat, rabbit)
Blood compatibility (rat, dog, primate, etc.)
Long-term implant (rat)

These and similar assays are conducted under the Good Laboratory Practices, as described and regulated by the U.S. Food and Drug Administration (21 *CFR* Part 58; see *Federal Register*, Vol. 43, No. 27). In addition to *in vitro* and *in vivo* evaluations of the device, there are other parameters that the investigator or manufacturer should consider. Data must be provided on the characterization of raw materials, manufacturing and sterilization processes, packaging, storage, and stability of the final product prior to implantation. Final versions of the device should be used in the regulatory assays and other testing described earlier. This ensures that sample devices or device components that are submitted for regulatory (safety) testing can be evaluated with a minimum number of animals in a controlled and statistically reliable manner, thereby hastening the regulatory approval process.

Vale has addressed separately some of the unique issues specific to blood compatibility testing (Vale 1987). The *Tripartite Biocompatibility Guidance for Medical Devices* document is currently in use by the FDA (West, 1988). It lists a collection of preclinical tests currently recommended to support submission of a device for regulatory approval. The Association for the Advancement of Medical Instrumentation (AAMI) has recently published *Biological Evaluation of Medical Devices*, which is volume 4 of the AAMI Standards and Recommended Practices (AAMI, 1994). This book provides guidance on the selection of tests; animal welfare requirements; tests for genotoxicity; selection of tests for interactions with blood; tests for cytotoxicity; and tests for systemic toxicity. Future publications of the AAMI will deal with other relevant aspects of biological testing such as tests for local effects after implantation and tests for irritation and sensitization.

It should be appreciated that the purpose of safety testing new devices and biomaterials in animals is to detect adverse effects. Thus, toxic responses provide helpful data to regulatory authorities so they can anticipate problems in clinical trials. On the other hand, there is no value to increasing the parameter of animal exposure (e.g., size of medical device used in the animal, duration of treatment, number of animals) to extreme levels or similarly, evaluating a new device in an unnatural situation simply to generate toxicity since this may be an irrelevant endpoint.

In light of the interagency and international variation in safety and functionality testing protocols, plus the appearance of new technologies and applications, the investigator or sponsoring firm should contact the relevant regulatory agency as soon as possible during the development process if the device is novel or otherwise unusual. Early dialogue with regulatory scientists can save time, money, and animals by avoiding unnecessary or inappropriate assays.

In some cases a firm may take the initiative and propose certain (unconventional) tests or strategies if circumstances warrant. Such tests must be justifiable physiologically or pharmacologically; financial cost is usually not an acceptable criterion for proposing alternative assays. One benefit of this approach may be reducing the number of animals or discomfort to animals used in preclinical evaluation for a particular product. This is noteworthy since there is considerable international concern within regulatory agencies to reduce or refine animal testing, and justifiable suggestions to this end may meet with approval. The regulatory approval process may also be hastened if the manufacturer realizes that additional studies may be required *after* clinical trials begin. This avoids the situation where possibly nonessential animal assays are performed prior to requesting approval to begin clinical trials.

APPENDIX 1

American Association for the Advancement of Science Resolution on the Use of Animals in Research, Testing, and Education

Whereas society as a whole, and the scientific community in particular, supports and encourages research that will improve the well-being of humans and animals, and that will lead to the cure or prevention of disease; and

Whereas the use of animals has been and continues to be essential not only in applied research with direct clinical applications in humans and animals, but also in research that furthers the understanding of biological processes; and

Whereas the American Association for the Advancement of Science supports appropriate regulations and adequate funding to promote the welfare of animals in laboratory or field situations and deplores any violations of those regulations; and

Whereas the American Association for the Advancement of Science deplores harassment of scientists and technical personnel engaged in animal research, as well as destruction of animal laboratory faciliites; and

Whereas in order to protect the public, both consumer and medical products must be tested for safety, and such testing may in some cases require the use of animals; and

Whereas the American Association for the Advancement of Science has long acknowledged the importance and endorsed the use of animal experimentation in promoting human and animal welfare and in advancing scientific knowledge;

Be it further resolved that scientists bear several responsibilities for the conduct of research with animals; (1) to treat their subjects with proper care and sensitivity to their pain and discomfort, consistent with the requirements of the particular study and research objectives; (2) to be informed about and adhere to relevant laws and regulations pertaining to animal research; and (3) to communicate respect for animal subjects to employees, students, and colleagues; and

Be it further resolved that the development and use of complementary or alternative research or testing, methodologies, such as computer models, tissue, or cell cultures, be encouraged where applicable and efficacious; and

Be it further resolved that the use of animals by students can be an important component of science education as long as it is supervised by teachers who are properly trained in the welfare and use of animals in laboratory or field settings and is conducted by institutions capable of providing proper oversight; and

Be it further resolved that scientists support the efforts to improve animal welfare that do not include policies or regulations that would compromise scientific research; and

Be it further resolved that the American Association for the Advancement of Science encourages its affiliated societies and research institutions to support this resolution.

Joint Resolution Adopted by the
AAAS Board and Council, February 19, 1990.
Sponsored by the AAAS Committee on Scientific
Freedom and Responsibility.

APPENDIX 2

Selected Agencies and Organizations Providing Information on Laboratory Animal Care and Use

American Association for Accreditation of Laboratory
Animal Care
9650 Rockville Pike
Bethesda, MD 20814

American Association for Laboratory Animal Science
70 Timber Creek Drive
Suite #5
Cordova, TN 38018

American College of Laboratory Animal Medicine
Dr. Charles McPherson
Secretary-Treasurer
200 Summerwinds Drive
Cary, NC 27511

American Society of Laboratory Animal Practitioners
Dr. Bradford S. Goodwin, Jr.
Secretary-Treasurer
University of Texas Medical School
6431 Vannin St., Room 1.132
Houston, TX 77030-1501

Animal Welfare Information Center
National Agricultural Library
United States Department of Agriculture
Beltsville, MD 20705

Canadian Council on Animal Care
315-350 Albert
Ottawa, Ontario K1R 1B1

National Center for Research Resources
National Institutes of Health
5333 Westbard Ave.
Westwood Bldg, Rm 875
Bethesda, MD 20892

Institute of Laboratory Animal Resources
National Research Council
2101 Constitution Ave., NW
Washington, D.C. 20418

National Association for Biomedical Research
818 Connecticut Avenue, N.W.
Suite 303
Washington, D.C. 20006

Scientists Center for Animal Welfare
Golden Triangle Building One
7833 Walker Drive, Suite 340
Greenbelt, MD 20770

Suggested Reading

Selected Literature on Responsible Animal Use in Biomedical Research

Benefits of Animal Research to Human Health

Council on Scientific Affairs, American Medical Association, Animals in Research. Summary of medical accomplishments based on animal use in biomedical research. *Journal of the American Medical Association,* **261**: 3602–3606, 1989.

Foundation for Biomedical Research. Role of animals in research and testing. In *The Use of Animals in Biomedical Research and Testing,* Washington, DC. June 1988.

National Academy of Sciences and the Institute of Medicine, *Science, Medicine, and Animals* NAS/NIM, Washington, DC, 1991.

NIH Conference on Modeling in Biomedical Research. Animal models and new developments in cardiovascular/pulmonary function and diabetes. In An Assessment of Current and Potential Approaches—Summary Statement, Bethesda, MD, May 1989.

W. I. Gay. Contribution of Animal Research to human health. In *Health Benefits of Animal Research*. Foundation for Biomedical Research, Washington, DC.

U.S. Congress, Office of Technology Assessment. Report on Alternative Technologies in Biomedical and Behavioral Research, Toxicity Testing, and Education. Alternatives to Animal Use in Research, Testing, and Education, Washington, D.C. U.S. Government Printing Office, OTA-BA-273, February 1986.

Humane Principles of Animal Research

U.S. Department of Health and Human Services, *Guide for the Care and Use of Laboratory Animals*. HHS, Washington, DC, 1985. This guide is also used as the American Association for Accreditation of Laboratory Animal Care (AAALAC) standard.

The New York Academy of Sciences, Interdisciplinary Principles and Guidelines for the Use of Animals in Research, Testing, and Education. Position paper on animal use with ethical considerations for animal use, experimental design, and euthanasia. New York, 1988.

National Research Council. Report of the Committee on the Use of Laboratory Animals in Biomedical and Behavior Research. *Use of Laboratory Animals in Biomedical and Behavioral Research*, National Academy Press, Washington, DC, 1988.

Foundation for Biomedical Research. *The Biomedical Investigator's Handbook for Researchers Using Animal Models*. Washington, DC, 1987.

National Institutes of Health, Office of Protection from Research Risks. *Public Health Service Policy on Humane Care and Use of Laboratory Animals*. NIH, Bethesda, MD, 1986.

W. M. S. Russell and R. L. Burch. *Principles of Humane Experimental Technique*. Methuen, London, 1959.

American Veterinary Medical Association (AVMA). 1993 Report of the AVMA Panel on Euthanasia, *Journal American Veterinary Medical Association*. **202**: 229–249, 1993.

S. M. Niemi and J. W. Willson, eds. *Refinement and Reduction in Animal Testing*. scientists Center for Animal Welfare, Bethesda, MD, 1993.

Selected Literature on Animal Models in Biomedical Research

Edward J. Calabrese. *Principles of Animal Extrapolation*. Wiley, New York, 1983.

William I. Gay. *Methods of Animal Experimentation*, Vols. I–VI; Vol. VII: Part A, Patient Care, Vascular Access, and Telemetry; Part B, Surgical Approaches to Organ Systems; Part C, Research Surgery and Care of the Research Animal. Academic Press, San Diego, CA, 1965–1989.

P. A. Flecknell. *Laboratory Animal Anaesthesia*. Academic Press, San Diego, CA, 1988.

Edwin J. Andrews, Billy C. Ward, and Norman H. Altman, eds. *Spontaneous Animal Models of Human Disease*, Vol. I and Vol. II, Academic Press, San Diego, CA, 1979, 1980.

Steven H. Weisbroth, Ronald E. Flatt, and Alan L. Krauss, eds. *The Biology of the Laboratory Rabbit*. Academic Press, Orlando, FL, 1974.

Joseph E. Wagner and Patrick J. Manning, eds. *The Biology of the Guinea Pig*. Academic Press, Orlando, FL, 1976.

G. L. VanHoosier, Jr. and Charles W. McPherson, eds. *Laboratory Hamsters*. Academic Press, San Diego, CA, 1987.

H. L. Foster, J. D. Small, and J. G. Fox, eds. *The Mouse in Biomedical Research* (Vols. I–IV. Academic Press, San Diego, CA, 1981–1983.

James G. Fox, Bennett J. Cohen, and Franklin M. Loew, eds. *Laboratory Animal Medicine*. Academic Press, San Diego, CA, 1984.

Hubert C. Stanton and Harry J. Meisman, eds. *Swine in Cardiovascular Research*, Vols. I and II. Academic Press, San Diego, CA, 1984.

H. J. Baker and J. R. Lindsey, eds. *The Laboratory Rat*, Vol. I and Vol. II. Academic Press, Orlando, FL, 1980.

E. C. Melby, Jr. and N. H. Altman, eds. *Handbook of Laboratory Animal Science*, Vols. I–III. CRC Press, Boca Raton, FL, 1974–1976.

Bibliography

AAMI (1994). Standards and Recommended Practices, Volume 4: Biological Evaluation of Medical Devices. Association for the Advancement of Medical Instrumentation. Arlington, VA.

Clemmons, R. M. and K. M. Meyers (1984). Acquisition and aggregation of canine blood platelets: basic mechanisms of functions and differences because of breed origin. *Am. J. Vet. Res.* **45**: 137–144.

Didisheim, P. (1985). Comparative hematology in the human, calf, sheep and goat: relevance to implantable blood pump evaluation. *ASAIO J.* **8**: 123–127.

Dodds, W. J. (1988). Third International Registry of Animal Models of Thrombosis and Hemorrhagic Diseases, *Institute of Laboratory Animal Resources News* 30: R1–R32.

NIH (1980). Guidelines for Physiochemical Characterization of Biomaterials, Pub. 80-2186, U.S. Department of Health and Human Services, Washington, DC.

NIH (1985a). Guide for the Care and Use of Laboratory Animals, Pub. No. 86-23, Department of Health and Human Services, Washington, DC.

NIH (1985b). Guidelines for Blood–Material Interactions (Report of the National Heart, Lung, and Blood Institute Working Group) Pub. No. 85–2185 (revised September 1985), National Institutes of Health, Bethesda, MD.

NIH (1986). Public Health Service Policy on Humane Care and Use of Laboratory Animals, Office of Protection from Research Risks, National Institutes of Health, Bethesda, MD.

Vale, B. H. (1987). Current concepts for assessing blood compatibility: Small diameter vascular prostheses, *J. Biomater. Appl.* **2**: 149–159.

West, D. L. (1988). User of toxicology data for medical devices: The FDA perspective, *J. Am. Coll. of Toxicol.* **7**: 499–507.

6

Degradation of Materials in the Biological Environment

ARTHUR J. COURY, ROBERT J. LEVY, CARL R. MCMILLIN, YASHWANT PATHAK, BUDDY D. RATNER, FREDERICK J. SCHOEN, DAVID F. WILLIAMS, AND RACHEL L. WILLIAMS

6.1 INTRODUCTION
Buddy D. Ratner

The biological environment is surprisingly harsh and can lead to rapid or gradual breakdown of many materials. Superficially, one might think that the neutral pH, low salt content, and modest temperature of the body would constitute a mild environment. However, many special mechanisms are brought to bear on implants to break them down. These are mechanisms that have evolved over millennia specifically to rid the living organism of invading foreign substances and they now attack contemporary biomaterials.

First, consider that, along with the continuous or cyclic stress many biomaterials are exposed to, abrasion and flexure may also take place. This occurs in an aqueous, ionic environment that can be electrochemically active to metals and plasticizing (softening) to polymers. Then, specific biological mechanisms are invoked. Proteins adsorb to the material and can enhance the corrosion rate of metals. Cells secrete powerful oxidizing agents and enzymes that are directed at digesting the material. The potent degradative agents are concentrated between the cell and the material where they act, undiluted by the surrounding aqueous medium.

To understand the biological degradation of implant materials, synergistic pathways should be considered. For example, the cracks associated with stress crazing open up fresh surface area to reaction. Swelling and water uptake can similarly increase the number of sites for reaction. Degradation products can alter the local pH, stimulating further reaction. Hydrolysis of polymers can generate more hydrophilic species, leading to polymer swelling and entry of degrading species into the bulk of the polymer. Cracks might also serve as sites for the initiation of calcification.

Biodegradation is a term that is used in many contexts. It can be used for reactions that occur over minutes or over years. It can be engineered to happen at a specific time after implantation, or it can be an unexpected long-term conse-
quence of the severity of the biological environment. Implant materials can solubilize, crumble, become rubbery, or become rigid with time. The products of degradation may be toxic to the body, or they may be designed to perform a pharmacologic function. Degradation is seen with metals, polymers, ceramics, and composites. Thus, biodegradation as a subject is broad in scope, and rightfully should command considerable attention for the biomaterials scientist. This section introduces biodegradation issues for a number of classes of materials, and provides a basis for further study on this complex but critical subject.

6.2 CHEMICAL AND BIOCHEMICAL DEGRADATION OF POLYMERS
Arthur J. Coury

Biodegradation is the chemical breakdown of materials by the action of living organisms which leads to changes in physical properties. It is a concept of vast scope, ranging from decomposition of environmental waste involving microorganisms to host-induced deterioration of biomaterials in implanted medical devices. Yet it is a precise term, implying that specific biological processes are required to effect such changes (Williams, 1989). This chapter, while grounded in biodegradation, addresses other processes that contribute to the often complex mechanisms of polymer degradation. Its focus is the unintended chemical breakdown, in the body, of synthetic solid-phase polymers. (See Chapter 2.5 for a description of systems engineered to break down in the body.)

POLYMER DEGRADATION PROCESSES

Polymeric components of implantable devices are generally reliable for their intended lifetimes. Careful selection and exten-

Biomaterials Science
Copyright © 1996 by Academic Press, Inc.

TABLE 1 Typical Operations on an Injection-Moldable Polymer Biomaterial

---

Polymer: Synthesis, extrusion, pelletizing

Pellets: Packaging, storage, transfer, drying

Components: Injection molding, post-mold finishing, cleaning, inspecting, packaging, storage

Device: Fabrication, storage (presterilization), cleaning, inspecting, packaging, storage (packaged), sterilization, storage (sterile), shipment, storage (preimplant), implantation, operation in body

---

sive preclinical testing of the compositions, fabricated components, and devices usually establish functionality and durability. However, with chronic, indwelling devices, it is infeasible during qualification to match all implant conditions in real time for years or decades of use. The accelerated aging, animal implants, and statistical projections employed cannot expose all of the variables which may cause premature deterioration of performance. The ultimate measure of the acceptability of a material for a medical device is its functionality for the device's intended lifetime as ascertained in human postimplant surveillance.

No polymer is totally impervious to the chemical processes and mechanical action of the body. Generally, polymeric biomaterials degrade because body constituents attack the biomaterials directly or through other device components or the intervention of external factors.

Numerous operations are performed on a polymer from the time of its synthesis to its use in the body (see, e.g., Table 1). Table 2 lists mechanisms of physical and chemical deterioration, which may occur alone or in concert at various stages of a polymer's history. Moreover, a material's treatment prior to implant may predispose it to stable or unstable end-use behavior (Brauman *et al.*, 1981; Greisser *et al.*, 1994). A prominent example of biomaterial degradation caused by preimplant processing is the gamma irradiation sterilization of ultrahigh-molecular-weight polyethylene used in total joint prostheses. The process generates free radicals within the material which react with oxygen to produce undesirable oxidation products. Chain oxidation and scission can occur for periods of months to years, causing loss of strength and embrittlement with limited shelf life (McKellop *et al.*, 1995; Furman and Li, 1995; Weaver *et al.*, 1995). It is crucially important, therefore, that appropriate and rigorous processing and characterization protocols be followed for all operations (Coury *et al.*, 1988).

After a device has been implanted, adsorption and absorption processes occur. Polymeric surfaces in contact with body fluids immediately adsorb proteinaceous components, and the bulk begins to absorb soluble components such as water, proteins, and lipids. Cellular elements subsequently attach to the surfaces and initiate chemical processes. With biostable components, this complex interplay of factors is of little functional consequence. At equilibrium fluid absorption, there may be some polymer plasticization, causing

dimensional and mechanical property changes (Coury *et al.*, 1988). On the surface, a powerful acute attack by cells and many chemical agents, including enzymes, will have been substantially withstood. With the resolution of this acute inflammatory phase, a fibrous capsule will likely have formed over the device, and the rate of release of powerful chemicals from activated cells will have markedly decreased.

For those polymers subjected to chemical degradation *in vivo*, few if any reports have comprehensively described the multistep processes and interactions that comprise each mechanism. Rather, explant analysis and occasionally metabolite evaluation is used to infer reaction pathways. The analysis of chemically degraded polymers has almost always implicated either hydrolysis or oxidation as an essential component of the process.

HYDROLYTIC BIODEGRADATION

Structures of Hydrolyzable Polymers

Hydrolysis is the scission of susceptible molecular functional groups by reaction with water. It may be catalyzed by acids, bases, salts, or enzymes. It is a single-step process in which the rate of chain scission is directly proportional to the rate of initiation of the reaction (Schnabel, 1981). A polymer's susceptibility to hydrolysis is the result of its chemical structure, its morphology, its dimensions, and the body's environment.

In a commonly used category of hydrolyzable polymeric

TABLE 2 Mechanisms Leading to Degradation of Polymer Properties[a]

Physical	Chemical
Sorption	Thermolysis
Swelling	Radical scission
Softening	Depolymerization
Dissolution	Oxidation
Mineralization	Chemical
	Thermooxidative
Extraction	Solvolysis
Crystallization	Hydrolysis
Decrystallization	Alcoholysis
	Aminolysis, etc.
Stress cracking	Photolysis
Fatigue fracture	Visible
Impact fracture	Ultraviolet
	Radiolysis
	Gamma rays
	X-rays
	Electron beam
	Fracture-induced radical reactions

---

[a]Some degradation processes may involve combinations of two or more individual mechanisms.

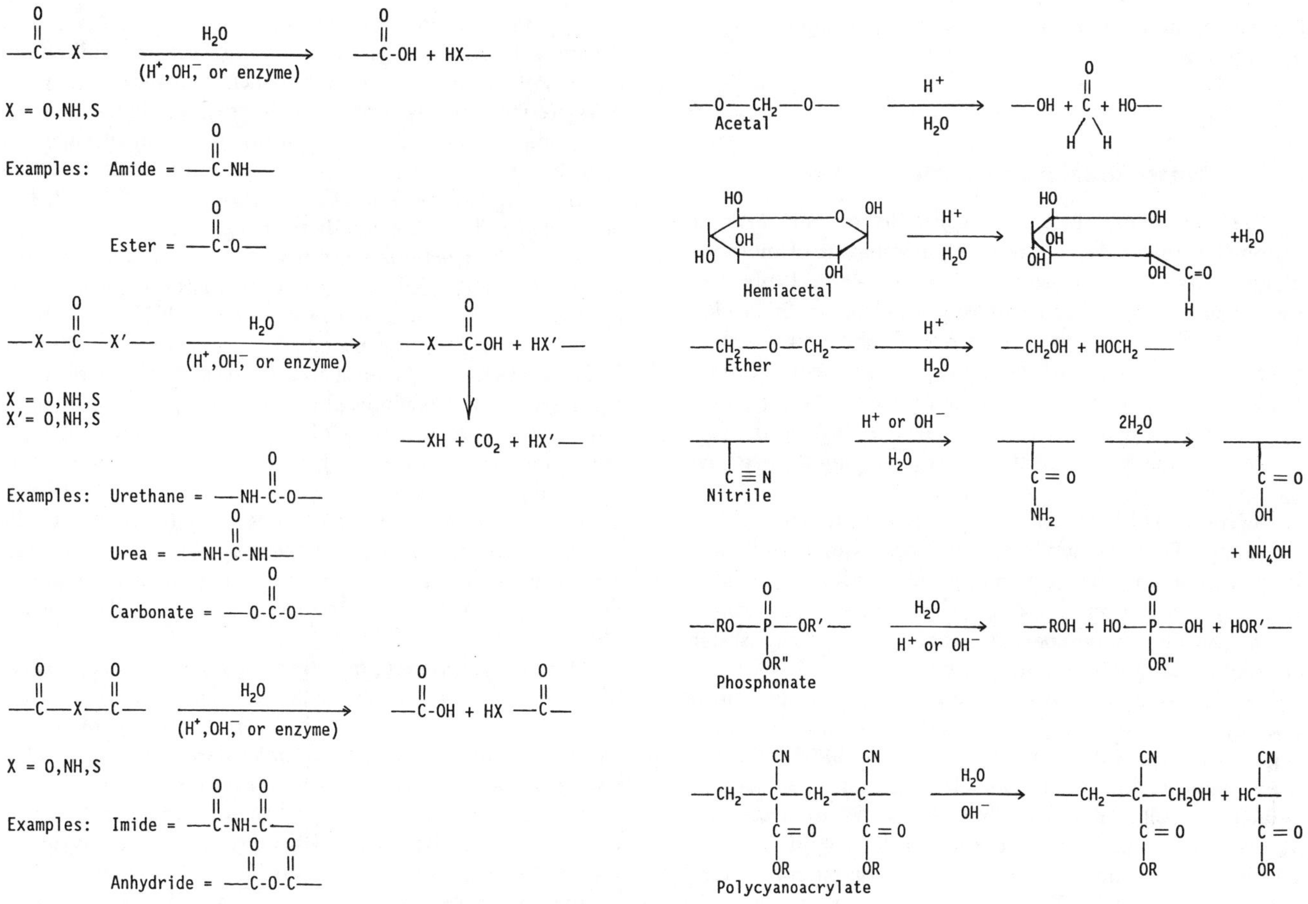

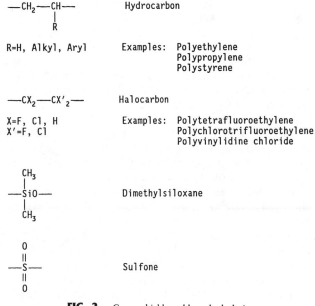

FIG. 1. Hydrolyzable groups in polymer biomaterials.

biomaterials, functional groups consist of carbonyls bonded to heterochain elements (O, N, S). Examples include esters, amides, and carbonates (Fig. 1). Other polymers containing groups such as ether, acetal, nitrile, phosphonate, or active methylenes, hydrolyze under certain conditions (Fig. 1). Groups that are normally very stable to hydrolysis are indicated in Fig. 2.

The rate of hydrolysis tends to increase with a high proportion of hydrolyzable groups in the main or side chain, other polar groups which enhance hydrophilicity, low crystallinity, low or negligible cross-link density, a high ratio of exposed surface area to volume and, very likely, mechanical stress. Porous hydrolyzable structures undergo especially rapid property loss because of their large surface area. Factors that tend to suppress hydrolysis kinetics include hydrophobic moieties (e.g., hydrocarbon or fluorocarbon), cross-linking, high crystallinity due to chain order, thermal annealing or orientation, low stress, and compact shape. While the molecular weight of linear polymers per se may not have a great effect on degradation rate, physical property losses may be retarded for a given number of chain cleavage events with relatively high-molecular-weight polymers. Property loss caused by chain cleavage

FIG. 2. Groups highly stable to hydrolysis.

is more pronounced in polymers with weak intermolecular bonding forces.

Host-Induced Hydrolytic Processes

The body is normally a highly controlled reaction medium. Through homeostasis, the normal environment of most implants is maintained at isothermal (37°C), neutral (pH 7.4), aseptic, and photoprotected aqueous steady state. By *in vitro* standards, these conditions appear mild. However, complex interactions of humoral and cellular components of body fluids involving activators, receptors, inhibitors, etc. produce aggressive responses to any foreign bodies through the processes of adhesion, chemical reaction, and particulate transport.

Several scenarios leading to hydrolysis in the host can be considered. First, essentially neutral water is capable of hydrolyzing certain polymers (e.g., polyglycolic acid) at a significant rate (Chapter 2.5 and Zaikov, 1985). However, this simple mechanism is unlikely to be significant in polymer compositions selected for long-term *in vivo* biostability.

Next, ion-catalyzed hydrolysis offers a likely scenario in body fluids. Extracellular fluids contain ions such as: H^+, OH^-, Na^+, Cl^-, HCO_3^-, PO_4^{-3}, K^+ Mg^{2+}, Ca^{2+} and SO_4^{2-}. Organic acids, proteins, lipids, lipoproteins, etc. also circulate as soluble or colloidal components. It has been shown that certain ions (e.g., PO_4^{3-}) are effective hydrolysis catalysts, enhancing, for example, reaction rates of polyesters by several orders of magnitude (Zaikov, 1985). Ion catalysis may be a surface effect or a combined surface-bulk effect, depending on the hydrophilicity of the polymer. Very hydrophobic polymers (e.g., those containing <2% water of saturation) absorb negligible concentrations of ions. Hydrogels, on the other hand, which can absorb large amounts of water (>15% by weight) are essentially "sieves," allowing significant levels of ions to be absorbed with consequent bulk hydrolysis via acid, base, or salt catalysis.

Localized pH changes in the vicinity of the implanted device, which usually occur during acute inflammation or infection, can cause catalytic rate enhancement of hydrolysis (Zaikov, 1985). Organic components, such as lipoproteins, circulating in the bloodstream or in extracellular fluid, appear to be capable of transporting catalytic inorganic ions into the polymer bulk by poorly defined mechanisms.

Enzymes generally serve a classic catalytic function, altering reaction rate (via ion or charge transfer) without being consumed, by modifying activation energy but not thermodynamic equilibrium. While enzymes function in extracellular fluids, they are most effectively transferred onto target substrates by direct cell contact (e.g., during phagocytosis). Hydrolytic enzymes or hydrolases (e.g., proteases, esterases, lipases, glycosidases) are named for the molecular structures they affect. They are cell-derived proteins which act as highly specific catalysts for the scission of water-labile functional groups.

Enzymes contain molecular chain structures and develop conformations that allow "recognition" of chain sequences (receptors) on biopolymers. Complexes form between segments of the enzyme and the biopolymer substrate which result in enhanced bond cleavage rates. Lacking the recognition sequences of susceptible natural polymers, most synthetic polymers are more resistant to enzymatic degradation. Nevertheless, comparative studies have shown some enhancement of hydrolysis rates by enzymes, particularly with synthetic polyesters and polyamides (Zaikov, 1985; Smith *et al.*, 1987; Kopecek *et al.*, 1983). Apparently the enzymes can recognize and interact with structural segments of the polymers, or more accurately, of the polymers coated with serum proteins, to initiate their catalytic action *in vivo* (Pitt, 1992).

Enzymes with demonstrated effects on hydrolysis rates can be quite selective in the presence of several hydrolyzable functional groups. For example, polyether urethane ureas and polyester urethane ureas exposed to hydrolytic enzymes (an esterase, cholesterol esterase, and a protease, elastase) were observed for rate and site of hydrolysis. Enzyme catalysis was clearly observed for the ester groups while the hydrolytically susceptible urea, urethane, and ether groups did not show significant hydrolysis as indicated by release of radiolabeled degradation products (Santerre *et al.*, 1994; Labow *et al.*, 1995).

Many enzymes exert predominantly a surface effect because of their great molecular size, which prevents absorption. Even hydrogels [e.g., poly(acrylamide)], which are capable of absorbing certain proteins, have molecular weight cutoffs for absorption well below those of such enzymes. However, as the degrading surface becomes roughened or fragmented, enzymatic action may be enhanced as a result of increased surface area if the substrates remain accessible to phagocytic cells that contain the active enzymes. Implanted devices that are in continuous motion relative to neighboring tissue can provoke inflammation, stimulating enzyme release.

Hydrolysis: Preclinical and Clinical Experience

A discussion of *in vivo* responses of several prominent polymer compositions known to be susceptible to hydrolysis follows. The structures of these polymers are described in Chapter 2.3.

Polyesters

Poly(ethylene terephthalate) (PET), in woven, velour, or knitted fiber configurations, remains a primary choice of cardiovascular surgeons for large-diameter vascular prostheses, arterial patches, valve sewing rings, etc. It is a strong, flexible, polymer, stabilized by high crystallinity as a result of chain rigidity and orientation and is often considered to be biostable. Yet, over several decades, there have been numerous reports of long-term degeneration of devices *in vivo*, owing to breakage of fibers and device dilation. The proposed causes have been structural defects, processing techniques, handling procedures, and hydrolytic degradation (Cardia *et al.*, 1989).

Systematic studies of PET implants in healthy dogs have shown slow degradation rates, which were estimated to be

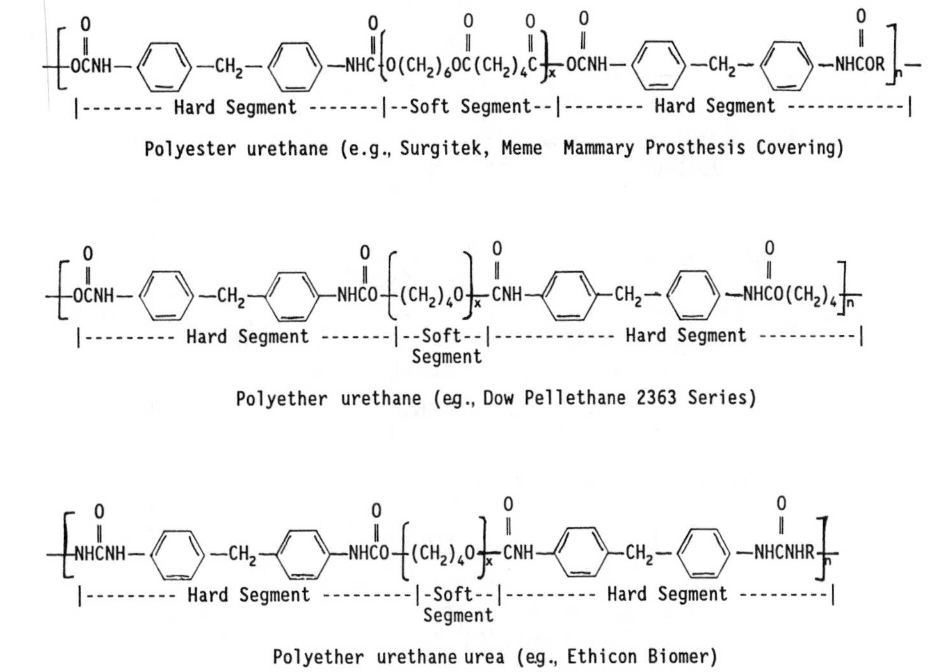

Polyester urethane (e.g., Surgitek, Meme Mammary Prosthesis Covering)

Polyether urethane (eg., Dow Pellethane 2363 Series)

Polyether urethane urea (eg., Ethicon Biomer)

FIG. 3. Structure of implantable polyester urethane, polyether urethane, and polyether urethane urea.

equivalent to those in humans. For woven patches implanted subcutaneously, a mean total absorption time by the body of 30 ± 7 years, with 50% deterioration of fiber strength in 10 ± 2 years was projected. In infected dogs, however, where localized pH dropped to as low as 4.8, degradation was enhanced exponentially, with complete loss of properties within a few months (Zaikov, 1985). Human implant retrieval studies have shown significant evidence of graft infection (Vinard *et al.*, 1991). Besides the obvious pathological consequences of infection, the enhanced risk of polymer degradation is a cause for concern.

Aliphatic polyesters are most often intended for use as biodegradable polymers, with polycaprolactone, for example, undergoing a significant decrease in molecular weight as indicated by a drop of 80–90% in relative viscosity within 120 weeks of implant (Kopecek *et al.*, 1983).

Polyester Urethanes

The earliest reported implants of polyurethanes, dating back to the 1950s, were cross-linked, aromatic polyester urethane foam compositions (Blais, 1990; Bloch *et al.*, 1972). Their use in plastic and orthopedic reconstructive surgery initially yielded promising results. Acute inflammation was low. Tissue ingrowth promoted thin fibrous capsules. However, within months, they were degraded and fragmented, producing untoward chronic effects (Bloch *et al.*, 1972). Foci of initial degradation are generally considered to be the polyadipate ester soft segments which undergo hydrolysis (Fig. 3). By comparison, corresponding polyether urethanes

are very resistant to hydrolysis, although more susceptible to oxidation (see the section on oxidative biodegradation). Whether such hydrolytically degraded polyester urethanes subsequently produce meaningful levels of aromatic amines (suspected carcinogens) by hydrolysis of urethane functions *in vivo* is currently an unresolved subject of considerable debate (Szycher *et al.*, 1991; Blais, 1990).

It is noteworthy that polyester urethane foam-coated silicone mammary implants have survived as commercial products until recently (Blais, 1990), despite their known tendency to degrade. Apparently the type of fibrous capsules formed on devices containing degradable foam were favored by some clinicians over those caused by smooth-walled silicone implants. In large devices, unstabilized by tissue ingrowth, the frictional effects of sliding may cause increased capsule thickness and contraction (Snow *et al.*, 1981) along with extensive chronic inflammation.

Polyamides

Nylon 6 (polycaproamide) and nylon 6,6 [poly(hexamethylene adipamide)] contain a hydrolyzable amide connecting group, as do proteins. These synthetic polymers can absorb 9–11% water, by weight, at saturation. It is predictable, then, that they degrade by ion-catalyzed surface and bulk hydrolysis (Fig. 1). In addition, hydrolysis due to enzymatic catalysis leads to surface erosion (Zaikov, 1985). Quantitatively, nylon 6,6 lost 25% of its tensile strength after 89 days, and 83% after 726 days in dogs (Kopecek, 1983). An example of polyamide degradation of particular consequence involved the *in vivo*

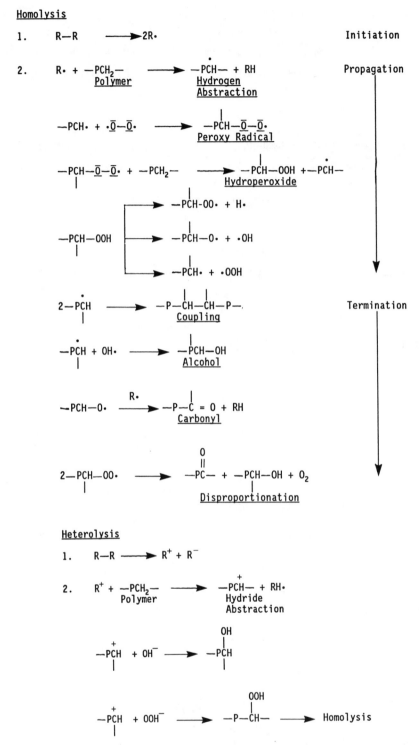

FIG. 4.   Proposed homolytic chain reaction and heterolytic oxidation mechanisms.

fragmentation of the nylon 6 tail string of an intrauterine contraceptive device. This string consisted of nylon 6-coated nylon 6 multifilaments. The combination of fluid absorption (>10%) and hydrolysis was claimed to produce environmental stress cracking. The cracked coating allegedly provided a path-

way for bacteria to travel from the vagina into the uterus, resulting in significant pelvic inflammatory disease (Hudson and Crugnola, 1987).

Degradation of a polyarylamide intended for orthopedic use (the fiber-reinforced polyamide from *m*-xylylene diamine and adipic acid) was also shown in a rabbit implant study. [Although the material provoked a foreign body reaction comparable to a polyethylene control, surface pitting associated with resolving macrophages was noted at 4 weeks and became more pronounced by 12 weeks. This result was not predicted since polyarylamides are very resistant to solvents and heat (Finck *et al.*, 1995).]

Polyamides with long aliphatic hydrocarbon chain segments (e.g., polydodecanamide) are more hydrolytically stable than shorter chain nylons and correspondingly degrade slower *in vivo*.

Polyalkylcyanoacrylates

This class of polymers, used as tissue adhesives, is noteworthy as a rare case in which carbon–carbon bonds are cleaved by hydrolysis (Fig. 1). This occurs because the methylene (—CH$_2$—) hydrogen in the polymer is highly activated inductively by electron-withdrawing neighboring groups. Formation of the polymer adhesive from monomers is base catalyzed, with adsorbed water on the adherend being basic enough to initiate the reaction.

Catalysts for equilibrium reactions affect the reverse, as well as the forward reaction. Therefore, water associated with tissue can induce polycyanoacrylate hydrolysis by a "reverse Knoevenagel" reaction (Fig. 1). More basic conditions and (as suggested by *in vitro* cell culture or implant studies) enzymatic processes are much more effective. In chick embryo liver culture (a rich source of a variety of enzymes), methyl cyanoacrylate degraded much faster than in culture medium alone. In animal implants, methyl cyanoacrylate was extensively degraded within 4–6 months (Kopecek, 1983). Higher alkyl (e.g., butyl) homologs degraded slower than the methyl homolog and were less cytotoxic (Hegyeli, 1973).

OXIDATIVE BIODEGRADATION

Oxidation Reaction Mechanisms and Polymer Structures

While much is known about the structures and reaction products of polymers susceptible to oxidative biodegradation, confirmation of the individual reaction steps has not yet been demonstrated analytically. Still, mechanistic inferences are possible from extensive knowledge of physiological oxidation processes and polymer oxidation *in vitro*.

The polymer oxidation processes to be discussed may be consistent with a homolytic chain reaction or a heterolytic mechanism. Species such as carbonyl, hydroxyl, and chain scission products are detectable. Classic initiation, propagation, and termination events for homolysis and ionic heterolytic processes are detailed in Fig. 4.

Except for the nature of susceptible functional groups, the hydrolysis resistance principles stated in the section on the structures of hydrolyzable polymers are valid for predicting relative oxidation resistance of polymers. Sites favored for ini-

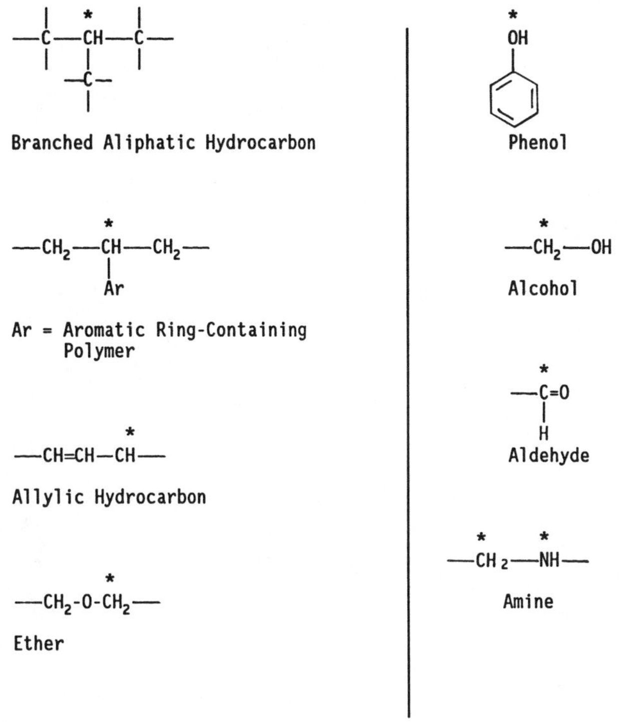

FIG. 5. Readily oxidizable functional groups (* is site of homolysis or heterolysis).

tial oxidative attack, consistent with a homolytic or heterolytic pathway, are those that allow abstraction of an atom or ion and provide resonance stabilization of the resultant radical or ion. Figure 5 provides a selection of readily oxidized groups and the atom at which initial attack occurs. In Fig. 6, examples of radical and ion stabilization by resonance in ether and branched hydrocarbon structures are provided. Peroxy, carbonyl, and other radical intermediates are stabilized by similar resonance delocalization of electrons from the elements C, O, H, or N.

Two general categories of oxidative biodegradation, based on the source of initiation of the process, are direct oxidation by the host and device or external environment-mediated oxidation.

Direct Oxidation by Host

In these circumstances, host-generated molecular species effect or potentiate oxidative processes directly on the polymer. Current thinking, based on solid analytical evidence, is that such reactive molecules are derived from activated phagocytic cells responding to the injury and properties of the foreign body at the implant site (Zhao *et al.*, 1991). These cells, which originate in the bone marrow and populate the circulatory system and connective tissues, are manifest as two types, the neutrophils (polymorphonuclear leukocytes, PMNs) and the monocytes. The latter can differentiate into macrophage and foreign body giant cell (FBGC) phenotypes.

Much work is under way to elucidate the sequence of events

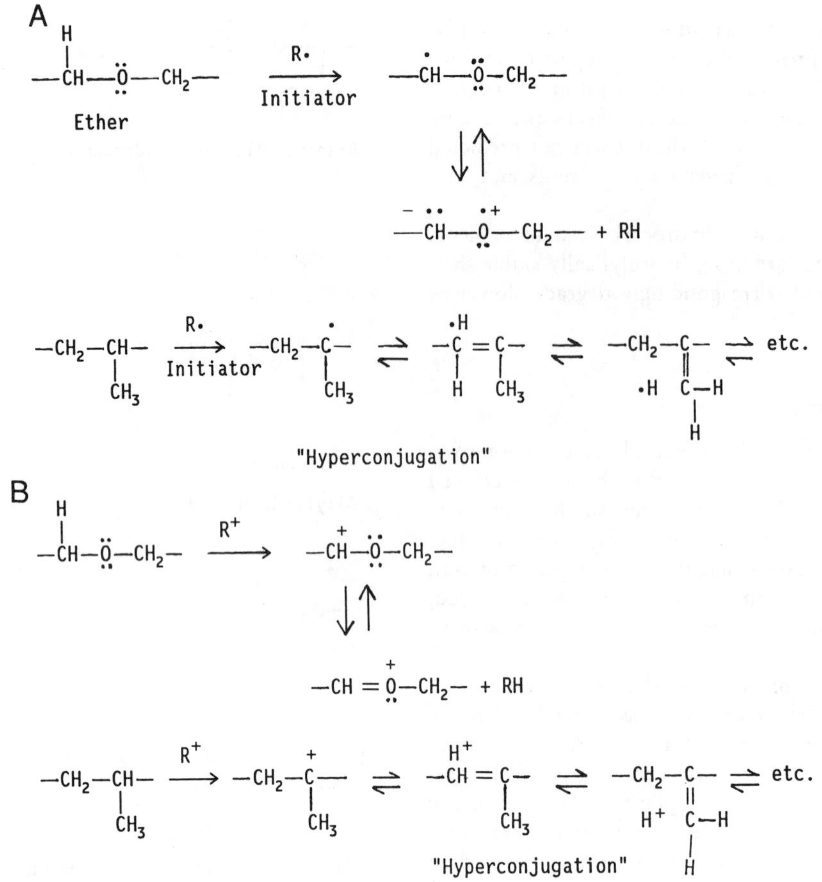

FIG. 6. (A) Resonance stabilization of ether and hydrocarbon radicals. (B) Resonance stabilization of ether and hydrocarbon cations.

leading to phagocytic oxidation of biomaterials. Certain important processes of wound healing in the presence of biologically derived foreign bodies such as bacteria and parasites, are showing some relevance to biomaterial implants (Northrup, 1987).

Neutrophils, responding to chemical mediators at the wound site, mount a powerful but transient chemical attack within the first few days of injury (Northrup, 1987; Test and Weiss, 1986). Chemically susceptible biomaterials may be affected if they are in close apposition to the wound site (Sutherland *et al.*, 1993). Activated macrophages subsequently multiply and subside within days at a benign wound site or in weeks if stimulants such as toxins or particulates are released at the site. Their fusion products, foreign body giant cells, can survive for months to years on the implant surface. Macrophages also remain resident in collagenous capsules for extended periods.

While we recognize that the mechanism of cellular attack and oxidation of biomaterials is as yet unconfirmed, the following discussion attempts to provide logical biological pathways to powerful oxidants capable of producing known degradation products.

Both PMNs and macrophages metabolize oxygen to form a superoxide anion ($O_2^{\bar{\cdot}}$). This intermediate can undergo trans-

$$\text{Neutrophil or Macrophage} + O_2 + e^- \xrightarrow{\text{Activating Factors}} \underset{\text{Superoxide Anion}}{O_2^{\bar{\cdot}}}$$

$$2O_2^{\bar{\cdot}} + 2H^+ \xrightarrow{\text{SOD}} O_2 + \underset{\text{Hydrogen Peroxide}}{H_2O_2}$$

$$O_2^{\bar{\cdot}} + \underset{\text{Ferric Ion}}{Fe^{+3}} \longrightarrow O_2 + Fe^{+2}$$

$$O_2^{\bar{\cdot}} + H^+ \longrightarrow HO_2\cdot$$

$$H_2O_2 + Fe^{+2} \longrightarrow Fe^{+3} + \underset{\text{Hydroxyl Radical}}{OH\cdot} + OH^-$$

$$H_2O_2 + Cl^- + H^+ \xrightarrow{\text{MPO}} \underset{\text{Hypochlorous Acid}}{HOCl} + H_2O$$

$$HOCl + R_2NH \rightleftharpoons \underset{\text{Chloramine}}{R_2N\text{-}Cl} + H_2O$$

$$\underset{\text{Nitric Oxide}}{NO\cdot} + O_2^{\bar{\cdot}} \longrightarrow \underset{\text{Peroxynitrite Anion}}{ONOO^-}$$

FIG. 7. Generation of potential oxidants by phagocytic processes.

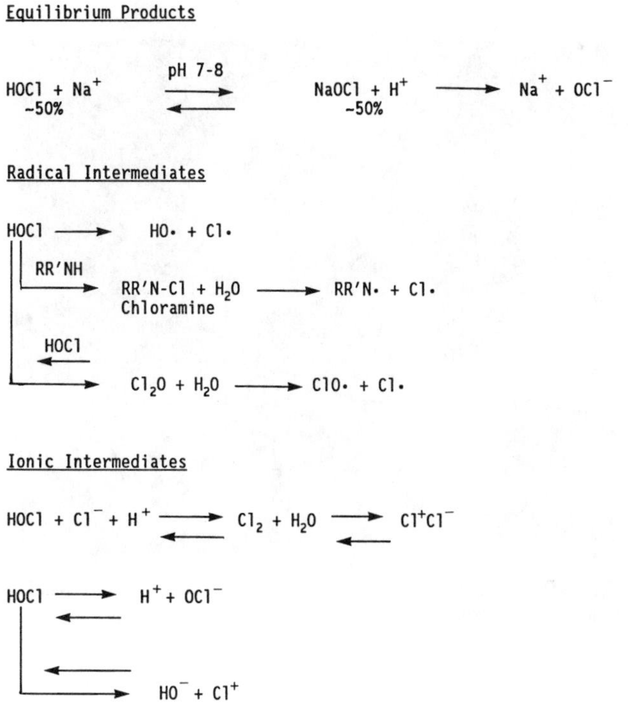

Equilibrium Products

$$HOCl + Na^+ \underset{\sim 50\%}{\overset{pH\ 7-8}{\rightleftharpoons}} NaOCl + H^+ \longrightarrow Na^+ + OCl^-$$
$$\sim 50\%$$

Radical Intermediates

$$HOCl \longrightarrow HO\cdot + Cl\cdot$$

$$\downarrow RR'NH$$

$$RR'N-Cl + H_2O \longrightarrow RR'N\cdot + Cl\cdot$$
$$Chloramine$$

$$\downarrow HOCl$$

$$Cl_2O + H_2O \longrightarrow ClO\cdot + Cl\cdot$$

Ionic Intermediates

$$HOCl + Cl^- + H^+ \rightleftharpoons Cl_2 + H_2O \rightleftharpoons Cl^+Cl^-$$

$$HOCl \longrightarrow H^+ + OCl^-$$

$$\longrightarrow HO^- + Cl^+$$

FIG. 8. Hypochlorous acid: formation and potential reaction intermediates.

formation to more powerful oxidants, or, conceivably, can initiate homolytic reactions on the polymer. Superoxide dismutase (SOD), a ubiquitous peroxidase enzyme, can catalyze the conversion of superoxide to hydrogen peroxide, which, in the presence of myeloperoxidase (MPO), derived from PMNs, is converted to hypochlorous acid (HOCl). A potent biomaterial oxidant in its own right (Coury *et al.*, 1987), hypochlorite can oxidize free amine functionality (e.g., in proteins) to chloramines that can perform as long-lived sources of chlorine oxidant (Test and Weiss, 1986, Figs. 7, 8). Hypochlorite can oxidize other substituted nitrogen functional groups (amides, ureas, urethanes, etc.) with potential chain cleavage of these groups.

The following paragraphs describe potential cooperative reactions involving acquired peroxidase and free ferrous ions. Macrophages contain essentially no MPO, so their hydrogen peroxide is not normally converted to HOCl. However, PMN-derived MPO can bind securely to foreign body surfaces (Locksley *et al.*, 1982), and serve as a catalyst reservoir for macrophage- or FBGC-derived HOCl production. If free ferrous ion, which is normally present in negligible quantities in the host, is released to the implant site by hemolysis or other injury, it can catalyze the formation of the powerfully oxidizing hydroxyl radical via the Haber–Weiss cycle (Klebanoff, 1982; Fig. 7).

Figure 8 shows radical and ionic intermediates of HOCl that may initiate biomaterial oxidation. Figure 9 is a diagram showing a leukocyte phagocytic process which employs endogenous MPO catalysis of HOCl formation. In a more general sense, the MPO may come from within or outside of the cell.

The foregoing discussion of sources of direct oxidation focused primarily on acute implant periods in which bursts of PMN activity followed by macrophage activity normally resolve within weeks. However, since the foreign body subsequently remains implanted, a sustained, if futile attempt to phagocytose an implanted device provides a prolonged release of chemicals onto the biomaterial. This phenomenon, called

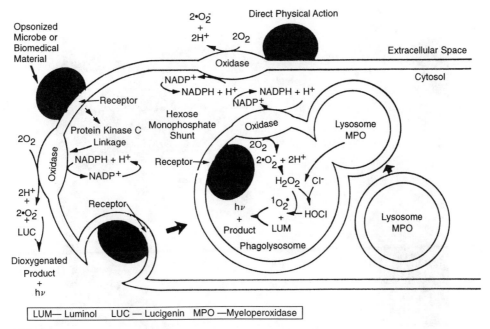

LUM— Luminol   LUC — Lucigenin   MPO —Myeloperoxidase

FIG. 9. Activation of phagocyte redox metabolism: chemiluminigenic probing with luminol and lucigenin. From R. C. Allen, personal communication.

TABLE 3   Characteristics of Polyether Urethanes That Cracked *in Vivo*

Components contained residual processing and/or applied mechanical stresses/strains

Components were exposed to a medium of viable cellular and extracellular body constituents

Polymers had oxidatively susceptible (aliphatic ether) groups

Analysis of polymers showed surface oxidation products

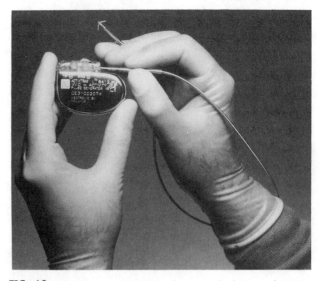

FIG. 10.   Cardiac pacemaker with polyurethane lead, tine, and connector. Courtesy of Medtronic, Inc.

exocytosis, occurs over months to possibly years (Zhao *et al.*, 1990) and results primarily from the macrophage-FBGC line. It can contribute to long-term chemical degradation of the polymer.

The oxidation processes induced by phagocytes are the result of oxidants produced by general foreign body responses, not direct receptor–ligand catalysis by oxidase enzymes. Attempts to degrade oxidatively susceptible polymers by direct contact with oxidase enzymes have generally been unsuccessful (Santerre *et al.*, 1994; Sutherland *et al.*, 1993).

Macrophages mediate other processes, such as fibrous capsule formation around the device. Their release of cellular regulatory factors stimulates fibroblasts to populate the implant site and produce the collagenous sheath. Any knowledge of the effects of such factors as fibroblasts or fibrous capsules on rates and mechanisms of polymer degradation is, at this time, extremely rudimentary.

Stress Cracking

An important category of host-induced biodegradation with an oxidative component is stress cracking as manifest in polyether methane elastomers. It differs from classic environmental stress cracking (ESC), which involves a susceptible material at a critical level of stress in a medium which may permeate, but does not dissolve, the polymer. Classic ESC is not accompanied by significant chemical degradation (Stokes, 1988). Instead,

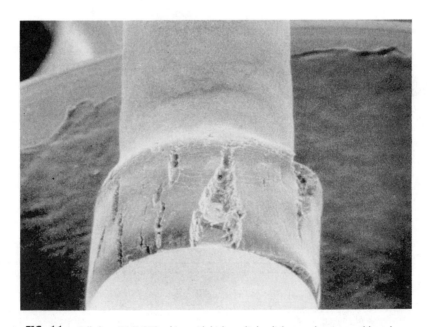

FIG. 11.   Pellethane 2463-80A tubing with high applied radial stress showing total breach.

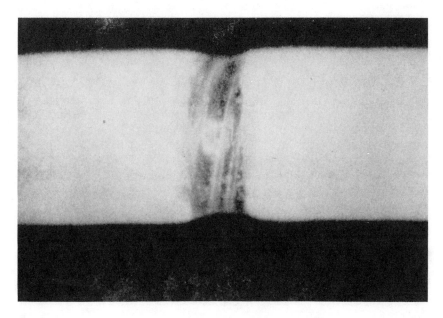

FIG. 12. Pellethane 2363-80A tubing showing "frosting" due to stress from tight ligature.

stress cracking of polyurethanes is characterized by surface attack of the polymer and by chemical changes induced by relatively specific *in vivo* or *in vitro* oxidizing conditions. Conditions relevant to stress cracking of certain polyether urethane compositions are stated in Table 3.

Recent information on the stress cracking of polyether urethanes and polyether urethane ureas (e.g., Fig. 3) has provided

insights that may be valid for these and other compositions that can be oxidized (e.g., polypropylene; Altman *et al.*, 1986; polyethylenes, Wasserbauer *et al.*, 1990; Zhao *et al.*, 1995).

Polyether urethanes, which are resistant to hydrolysis *in vivo*, are used as connectors, insulators, tines, and adhesives for cardiac pacemakers and neurological stimulators (Fig. 10). They have performed with high reliability in chronic clinical

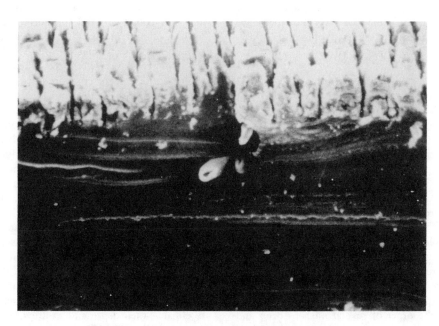

FIG. 13. Stress crack pattern (frosting) near tight ligature. ×14.

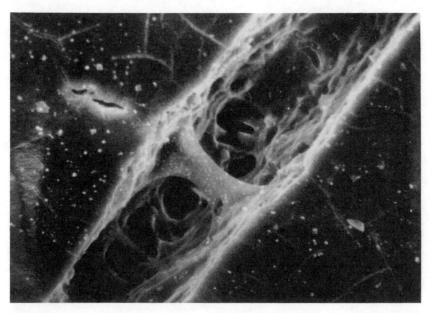

FIG. 14.   Single stress crack with rough walls and "tie fibers" indicative of ductile fracture. ×700.

applications since 1975. Certain polyether urethane pacing leads have displayed surface cracks in their insulation after residence times *in vivo* of months to years. These cracks are directly related in frequency and depth to the amount of residual stress (Figs. 11, 12) and the ether (soft segment) content of the polyurethane (Coury *et al.*, 1987).

Morphologically, the cracks display regular patterns predominately normal to the force vectors, with very rough walls, occasionally with "tie fibers" bridging the gaps, indicative of ductile rather than brittle fracture (Figs. 13, 14). Infrared analysis indicates that oxidation does not take place detectably in the bulk, only on the surface, where extensive loss of ether functionality (1110 cm^{-1}) and enhanced absorption in the hydroxyl and carbonyl regions are observed (Stokes *et al.*, 1987). Possible mechanisms for the oxidative degradation of ethers are presented in Fig. 15.

In a seminal study, Zhao *et al.* (1990) placed polyurethane tubing in cages permeable to fluids and cells under strain (therefore under high initial stress, which was subject to subsequent stress relaxation), and implanted them in rats. In certain cases, antiinflammatory steroids or cytotoxic polymers were coimplanted in the cages. Implants of up to 15 weeks were retrieved. The only prestressed samples to crack were those that did not reside in the cages with the coimplants. The authors concluded that adherent cells caused the stress cracking, and cell necrosis or deactivation inhibited crack induction.

Subsequently, viable phagocytic cells were implicated as a cause of crack initiation *in vivo* (Zhao *et al.*, 1991). By removing adherent foreign body giant cells after implantation of a curved polyether urethane urea film in a wire cage for up to 10 weeks, exposed "footprints" showed localized surface cracking, on the order of several microns deep and wide. Adjacent areas of polymer, which were devoid of

attached cells, were not cracked. Owing to relatively low stresses in the implanted film, deep crack propagation was not observed.

In vitro studies of strained (Stokes, 1988) and unstrained polyether urethane films (Phua *et al.*, 1987; Bouvier *et al.*, 1991; Ratner *et al.*, 1988) using oxidants, enzymes, etc., have sought to duplicate *in vivo* stress cracking. Although some surface chemical degradation with products similar to those seen *in vivo* was demonstrated, stress crack morphology was not closely matched *in vitro* until recently, in two studies. A test which involves immersing stressed polyether urethane tubing in a medium of glass wool, hydrogen peroxide, and cobalt chloride produces cracks which duplicate those produced *in vivo* but with rate acceleration of up to seven times (Zhao *et al.*, 1995). In another study, comparable crack patterns were produced when specimens of stressed tubing in rats were compared with those incubated with PMNs in culture (Sutherland *et al.*, 1993). Moreover, this study revealed a difference in chemical degradation products with time of implant which correlated with products from oxidants generated primarily by PMNs (HOCl) and macrophages (ONOO$^-$). Early implant times, activated PMNs, and HOCl caused preferential decrease in the urethane oxygen stretch peak while longer implant times and ONOO$^-$ caused selective loss of the aliphatic ether stretch peak (by infrared spectroscopy).

Taken together, the foregoing observations are consistent with a two-step mechanism for stress cracking *in vivo*. This hypothesis, as yet unproven, is under investigation. In the first step, surface oxidation induces very shallow, brittle microcracks. The second step involves propagation of the cracks in which specific body fluid components act on the formed cracks to enhance their depth and width without inducing major detectable bulk chemical reactions. Should this hypothesis prove

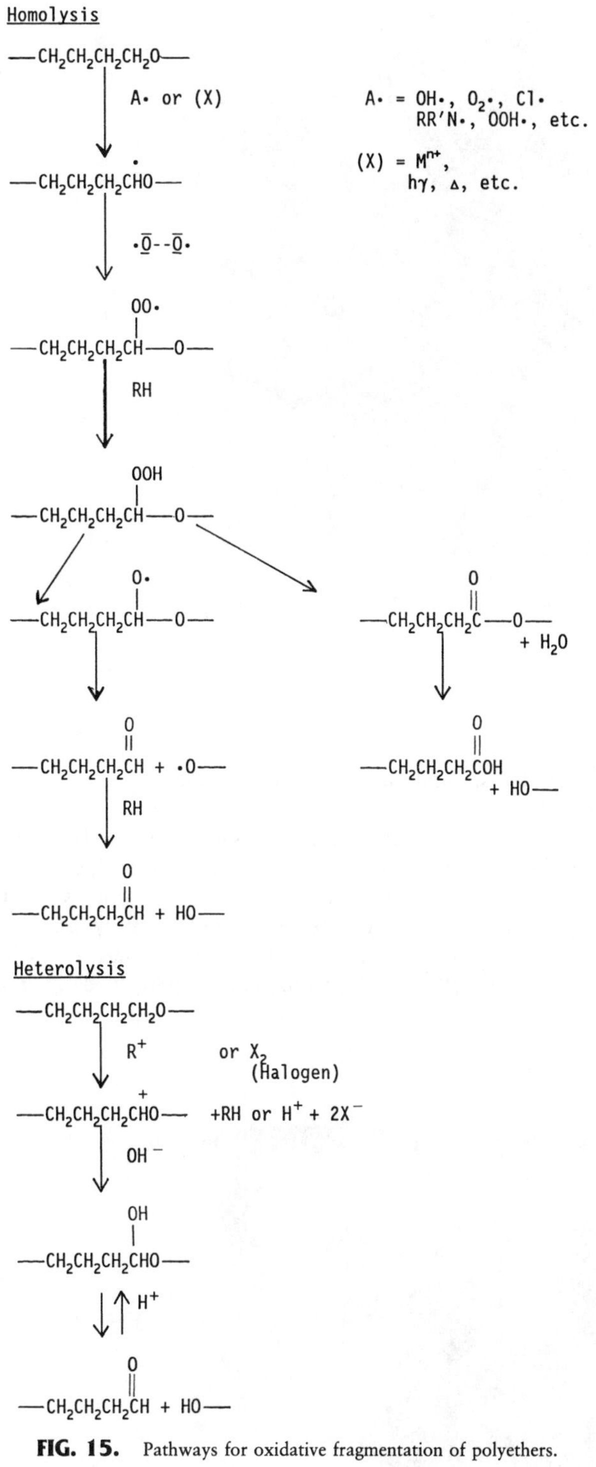

<u>Homolysis</u>

$$—CH_2CH_2CH_2CH_2O—$$

$$\downarrow \quad A\cdot \text{ or } (X)$$

A$\cdot$ = OH$\cdot$, O$_2\cdot$, Cl$\cdot$
RR'N$\cdot$, $^-$OOH$\cdot$, etc.

(X) = M^{n+},
hγ, Δ, etc.

$$—CH_2CH_2CH_2CHO—$$

$$\downarrow \quad \cdot\overline{O}--\overline{O}\cdot$$

$$\begin{array}{c} OO\cdot \\ | \\ —CH_2CH_2CH_2CH—O— \end{array}$$

$$\downarrow \quad RH$$

$$\begin{array}{c} OOH \\ | \\ —CH_2CH_2CH_2CH—O— \end{array}$$

$$\begin{array}{c} O\cdot \\ | \\ —CH_2CH_2CH_2CH—O— \end{array} \qquad \begin{array}{c} O \\ \| \\ —CH_2CH_2CH_2C—O— \\ + H_2O \end{array}$$

$$\begin{array}{c} O \\ \| \\ —CH_2CH_2CH_2CH + \cdot O— \end{array} \qquad \begin{array}{c} O \\ \| \\ —CH_2CH_2CH_2COH \\ + HO— \end{array}$$

$$\downarrow \quad RH$$

$$\begin{array}{c} O \\ \| \\ —CH_2CH_2CH_2CH + HO— \end{array}$$

<u>Heterolysis</u>

$$—CH_2CH_2CH_2CH_2O—$$

$$\downarrow \quad R^+ \qquad \text{or } X_2$$
$$\qquad \qquad \text{(Halogen)}$$

$$\begin{array}{c} + \\ —CH_2CH_2CH_2CHO— \end{array} \quad +RH \text{ or } H^+ + 2X^-$$

$$\downarrow \quad OH^-$$

$$\begin{array}{c} OH \\ | \\ —CH_2CH_2CH_2CHO— \end{array}$$

$$\downarrow\uparrow \quad H^+$$

$$\begin{array}{c} O \\ \| \\ —CH_2CH_2CH_2CH + HO— \end{array}$$

FIG. 15. Pathways for oxidative fragmentation of polyethers.

correct, the term "oxidation-initiated stress cracking" would be reasonably descriptive.

Stress cracking has been controlled by reducing residual stress, isolating the polymer from cell contact (Tang *et al.*, 1994), protecting the polymer from stress cracking media, or using stress crack-resistant (in the case of urethanes, ether-free)

polymers (Takahara *et al.*, 1994; Coury *et al.*, 1990). Stress cracking is next compared with another type of degradation, metal ion-induced oxidation.

Device or Environment-Mediated Oxidation

Metal Ion-Induced Oxidation

A process of oxidative degradation that has, thus far, only been reported clinically for polyether urethane pacemaker leads, requires, as does stress cracking, a very specific set of conditions. The enabling variables and fracture morphology are quite different from stress cracking, although oxidative degradation products are similar. Biodegradation of implanted devices through stress cracking always occurs on polymer surfaces exposed to cells and provides characteristic rough-walled fissures (indicative of ductile fracture) oriented perpendicular to the stress vector (Figs. 11–14). Metal ion-induced oxidation takes place on the enclosed inner surfaces of pacing lead insulation near corroded metallic components and their entrapped corrosion products. Smooth crack walls and microscopically random crack orientation is indicative of brittle fracture (Figs. 16, 17). Macroscopically, crack patterns that track metal component configurations may be present (Fig. 18). Degradation products which may be found deeper in the bulk than with stress cracking, again are indicative of brittle fracture.

This phenomenon, called metal ion-induced oxidation, has been confirmed by *in vitro* studies in which polyether urethanes were aged in metal ion solutions of different standard oxidation potentials. Above an oxidation potential of about +0.77, chemical degradation was severe. Below that oxidation potential, changes in the polymer that are characteristic of simple plasticization were seen (Coury *et al.*, 1987; Table 4). This technique also showed that metal ion-induced oxidation was proportional to the ether content of the polyurethane (Coury *et al.*, 1987; Table 5).

The effect of various metals on oxidation *in vitro* and *in vivo* has also been studied. Different metallic components of pacing lead conductors were sealed in polyether urethane (Dow Pellethane 2363-80A) lead tubing and immersed in 3% hydrogen peroxide at 37°C for up to 6 months (Stokes *et al.*, 1987) or implanted in rabbits for up to 2 years (Stokes *et al.*, 1990). Both techniques resulted in corroded metals and degraded tubing lumen surfaces, under certain conditions, within 30 days. *In vivo,* the interaction of body fluids with cobalt and its alloys, in particular, resulted in oxidative cracking of the polymer.

The metal ion-induced oxidation process clearly involves corrosion of metallic elements to their ions and subsequent oxidation of the polymer. In operating devices, the metal ion may be formed by solvation, galvanic corrosion, or chemical or biochemical oxidation (Fig. 19). In turn, these metal ions develop oxidation potentials that may well be enhanced in body fluids over their standard half-cell potentials. As strong oxidants, they produce intermediates or attack the polymer to initiate the chain reaction (Fig. 20). Metal ion-induced oxidation is, therefore, the result of a highly complex interaction of the device, the polymer, and the body.

FIG. 16. Random crack pattern of Pellethane 2363-80A lead insulation caused by metal ion-induced oxidation. ×480.

Should metal ion-induced oxidation be a possibility in an implanted device, several approaches are available to control this problem. They are not universally applicable, however, and should be incorporated only if functionality and biocompatibility are retained. Potentially useful techniques include using corrosion-resistant metals, "flushing" corrosive ions away from the susceptible polymer, isolating the metals and polymer from electrolyte solutions, incorporating appropriate antioxidants, and using oxidation-resistant polymers if available.

Recently, polyurethane elastomers with enhanced oxidation stability have been developed. They are segmented, ether- and ester-free polymers with unconventional soft segments, including, for example, hydrogenated polybutadiene, polydimethylsiloxane, polycarbonate, and dimerized fat acid derivatives (Takahara *et al.*, 1991, 1994; Coury *et al.*, 1990; Pinchuk *et al.*,

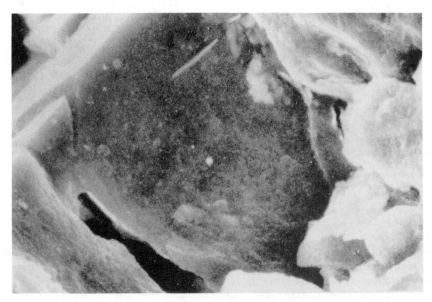

FIG. 17. Smooth crack wall indicative of brittle fracture caused by metal ion-induced oxidation. ×830.

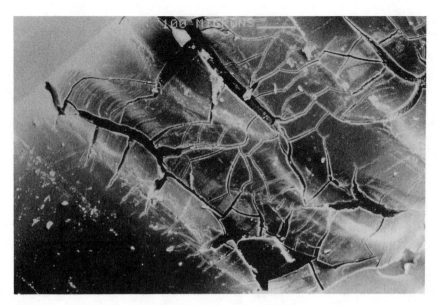

FIG. 18. Crack pattern on inner lumen of polyether urethane lead insulation tracking coil indicative of metal ion-induced oxidation. ×100.

1991; Kato *et al.*, 1995; Ward *et al.*, 1995). In implant tests, they have shown reduced tendency to stress crack, and some of them have shown high resistance to metal ion oxidants *in vitro*.

Oxidative Degradation Induced by External Environment

Under very limited circumstances the body can transmit electromagnetic radiation that may affect the integrity of implanted polymers. For example, the cornea and vitreous humor of the eye as well as superficial skin layers allow the passage of long-wave (320–400 nanometer) "ultraviolet A" radiation. Absorption of ultraviolet radiation causes electron excitation that can lead to photo-oxidative degradation. This process has been suggested in the breakdown of polypropylene components of intraocular lens (Altman *et al.*, 1986).

In maxillofacial exo- and, very likely, endoprostheses, elastomers may undergo undesirable changes in color and physical properties as a consequence of exposure to natural sunlight-frequency radiation (Craig *et al.*, 1980). Photo-oxidation mechanisms involving the urethane function of aromatic polyether- or polyester urethanes are shown in Fig. 21. Antioxidants and ultraviolet absorbers provide limited protection for these materials.

TABLE 4 Effect of Metal Ion Oxidation Potential on Properties of Polyetherurethane (Pellethane 2363-80A)[a]

Aqueous solution	Standard oxidation potential	Change in tensile strength (%)	Change in elongation (%)
PtCl$_2$	Ca +1.2	−87	−77
AgNO$_3$	+0.799	−54	−42
FeCl$_3$	+0.771	−79	−10
Cu$_2$Cl$_2$	+0.521	−6	+11
Cu$_2$(OAc)$_2$	+0.153	−11	+22
Ni(OAc)$_2$	−0.250	−5	+13
Co(OAc)$_2$	−0.277	+1	+13

[a]Conditions: 0.1 *M* solutions/90°C/35 days vs. controls aged in deionized water; ASTM (D-1708) microtensile specimens; specimens were tested wet.

TABLE 5 Effect of Ether Content of Polyether Urethane on Susceptibility to Metal Ion-Induced Oxidation[a]

Polyetherurethane	Polyether content	Change in tensile strength (%)	Change in elongation (%)
Pellethane 2363-80A	High	−54	−42
Pellethane 2363-55D	Low	−23	−10
Model segmented polyurethane	None	+9	+3

[a]Conditions: 0.1 *M* AgNO$_3$/90°C/35 days vs. controls aged in deionized water; ASTM (D-1708) microtensile specimens.

$$M^\circ \xrightarrow{\text{Electrolysis}} M^{+n} + ne^-$$

$$2M^\circ + 2H^+ \longrightarrow 2M^+ + H_2$$

$$4M^\circ + O_2 + 2H_2O \longrightarrow 4M^+ + 4OH^-$$

$$M^\circ + HOOH \longrightarrow M^+ + HO^- + HO\cdot$$

$$M^\circ + HOCl \longrightarrow M^+ + HO\cdot + Cl^-$$

etc.

FIG. 19. Formation of metal ion from metal.

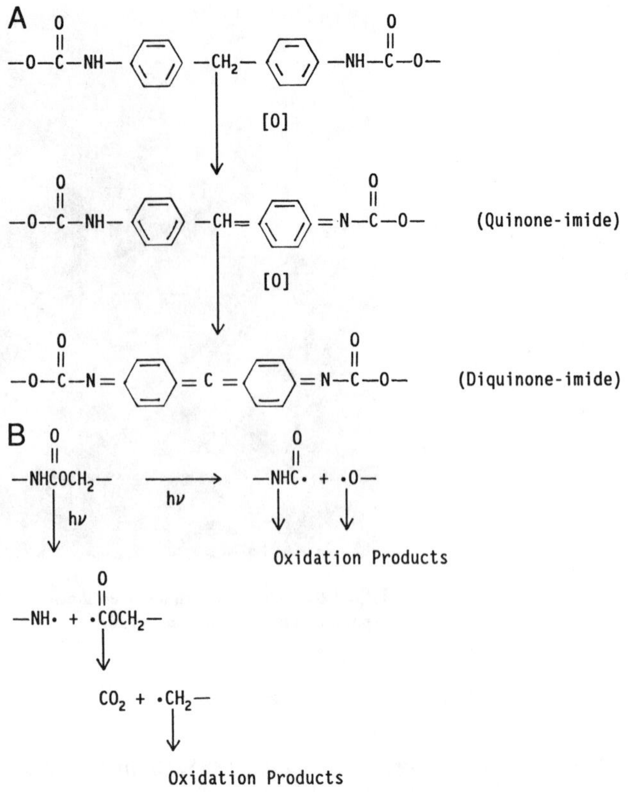

CONCLUSION

Polymers that are carefully chosen for use in implanted devices generally serve effectively for their intended lifetimes if they are properly processed and device–material–host interactions are adequately addressed. In certain limited circumstances, unintended hydrolytic or oxidative biodegradation occurs. This may be induced by direct attack by the host or via the intermediacy of the device or the outside environment. With susceptible polymers, protective measures can be taken to ensure extended efficacy, although new, biodegradation-resistant polymers, which are on the horizon, will require less protection. Knowledge of biodegradation mechanisms and the employment of appropriate countermeasures will promote the continued growth in compositions and uses of polymers as implantable biomaterials.

Acknowledgments

The author is very grateful to Dr. R. C. Allen for providing the drawing on activated phagocyte redox metabolism. For their technical advice and contributions, I sincerely thank Drs. James Anderson, John Eaton, John Mahoney, Maurice Kreevoy, Grace Picciolo and Buddy Ratner. For the preparation of this manuscript, I am deeply indebted to my "computer wizard," Mrs. Jayne McCaughey.

$$M^{+n} + H_2O \longrightarrow M^{+(n-1)} + HO\cdot + H^+$$

$$M^{+n} + -PH- \longrightarrow M^{+(n-1)} + -P\cdot + H^+$$
(PH = Polymer)

$$M^{+n} + O_2 \longrightarrow M^{+(n+1)} + O_2^-$$

$$M^{+n} + HOCl \longrightarrow M^{+(n+1)} + HO\cdot + Cl^-$$

$$M^{+n} + -PH- \longrightarrow M^{+(n-1)}H + -P^+-$$

$$M^{+n} + H_2O_2 \longrightarrow M^{+(n+1)} + OH^- + HO\cdot$$

$$M^{+n} + H_2O_2 \longrightarrow M^{+(n-1)} + HO_2\cdot + H^+$$

FIG. 20. Initiation of oxidation pathways by metal ion.

FIG. 21. Photo-oxidative reactions of aromatic polyurethanes. (A) Formation of quinone-imide from aromatic polyurethane. (From A. J. Coury *et al.*, *J. Biomater. Appl.* 3, 1988.) (B) Photolytic cleavage of urethane link. (From S. K. Brauman *et al.*, *Ann. Biomed. Eng.* 9, 1981.)

Bibliography

Allen, R. C. (1991). Activation of phagocyte redox metabolism: chemiluminigenic probing with luminol and lucigenin, Drawing Provided.

Altman, J. J., Gorn, R. A., Craft, J., Albert, D. M. (1986). The breakdown of polypropylene in the human eye: Is it clinically significant? *Ann. Ophthalmol.* 18: 182–185.

Blais, P. (1990). Letter to the editor. *J. Appl. Biomater.* 1: 197.

Bloch, B., and Hastings, G. (1972). *Plastics Materials in Surgery*, 2nd ed. Charles C. Thomas, Springfield, IL, pp. 97–98.

Bouvier, M., Chawla, A. S., and Hinberg, L., (1991). *In vitro* degradation of a poly (ether urethane) by trypsin. *J. Biomed. Mater. Res.* 25: 773–789.

Brauman, S. K., Mayorga, G. D., and Heller, J. (1981). Light stability and discoloration of segmented polyether urethanes. *Ann. Biomed. Eng.* 9: 45–58.

Cardia, G., and Regina, G. (1989). Degenerative Dacron graft changes: Is there a biological component in this textile defect?—A case report. *Vasc. Surg.* 23(3): 245–247.

Coury, A. J., Slaikeu, P. C., Cahalan, P. T. and Stokes, K. B. (1987). Medical applications of implantable polyurethanes: Current issues. *Prog. Rubber Plastics Tech.* 3(4): 24–37.

Coury, A. J., Slaikeu, P. C., Cahalan, P. T., Stokes, K. B., and Hobot, C. M. (1988). Factors and interactions affecting the performance of polyurethane elastomers in medical devices. *J. Biomater. Appl.* 3: 130–179.

Coury, A. J., Hobot, C. M., Slaikeu, P. C., Stokes, K. B. and Cahalan, P. T., (1990). A new family of implantable biostable polyurethanes. *Trans. 16th Annual Meeting Soc. for Biomater.* May 20–23, 158.

Craig, R. G., Koran, A., and Yus, R. (1980). Elastomers for maxillofacial applications. *Biomaterials* 1(Apr.): 112–117.

Finck, K. M., Grosse-Siestrup, C., Bisson, S., Rinck, M., and Gross, U. (1994). Experimental *in vivo* degradation of polyarylamide. *Trans. 20th Annual Meeting Soc. for Biomater.* April 5–9, p. 210.

Furman, B., and Li, S. (1995). The effect of long-term shelf life aging of ultra high molecular weight polyethylene. *Trans. 21st Annual Meeting Soc. for Biomater.* March 18–22, p. 114.

Greisser, H. J., Gengenbach, T. R., and Chatelier, R. C. (1994). Long-term changes in the surface composition of polymers intended for biomedical applications. *Trans. 20th Annual Meeting Soc. for Biomater.* April 5–9, p. 19.

Hegyeli, A. (1973). Use of organ cultures to evaluate biodegradation of polymer implant materials. *J. Biomed. Mater. Res.* 7: 205–214.

Hudson, J., and Crugnola, A. (1987). The *in vivo* biodegradation of nylon 6 utilized in a particular IUD. *J. Biomater. Appl.* 1: 487–501.

Kato, Y. P., Dereume, J. P., Kontges, H., Frid, N., Martin, J. B., MacGregor, D. C., and Pinchuk, L. (1995). Preliminary mechanical evaluation of a novel endoluminal graft. *Trans. 21st Annual Meeting Soc. for Biomater.* March 18–22, p. 81.

Klebanoff, S. (1982). Iodination catalyzed by the xanthine oxidase system: Role of hydroxyl radicals. *Biochemistry* 21: 4110–4116.

Kopecek, J., and Ulbrich, K. (1983). Biodegradation of biomedical polymers. *Prog. Polym. Sci.* 9: 1–58.

Labow, R. S., Erfle, D. J., and Santerre, J. P. (1995). Neutrophil-mediated degradation of segmented polyurethanes. *Biomaterials* 16: 51–59.

Locksley, R., Wilson, C., and Klebanoff, S. (1982). Role of endogenous and acquired peroxidase in the toxoplasmacidal activity of murine and human mononuclear phagocytes. *J. Clin. Invest.* 69(May): 1099–1111.

McKellop, H., Yeom, B., Campbell, P., and Salovey, R. (1995). Radiation induced oxidation of machined or molded UHMWPE after seventeen years. *Trans. 21st Annual Meeting Soc. for Biomater.* March 18–22, p. 54.

Northrup, S. (1987). Strategies for biological testing of biomaterials. *J. Biomater. Appl.* 2: 132–147.

Phua, S. K., Castillo, E., Anderson, J. M., and Hiltner, A. (1987). Biodegradation of a polyurethane *in vitro*. *J. Biomed. Mater. Res.* 21: 231–246.

Pinchuk, L., Esquivel, M. C., Martin, J. B., and Wilson, G. J. (1991). Corethane: A new replacement for polyether urethanes for long-term implant applications. *Trans. 17th Annual Meeting of the Soc. for Biomater.*, May 1–5, p. 98.

Pitt, C. G. (1992). Non-microbial degradation of polyesters: Mechanisms and modifications. in *Biodegradable Polymers and Plastics,* M. Vert, J. Feijin, A. Albertson, G. Scott, and E. Chiellini, eds. R. Soc. Chem., Cambridge, UK, pp. 1–19.

Ratner, B. D., Gladhill, K. W., and Horbett, T. A. (1988). Analysis of *in vitro* enzymatic and oxidative degradation of polyurethanes. *J. Biomed. Mater. Res.* 22: 509–527.

Santerre, J. P., Labow, R. S., Duguay, D. G., Erfle, D., and Adams, G. A. (1994). Biodegradation evaluation of polyether- and polyesterurethanes with oxidative and hydrolytic enzymes. *J. Biomed. Mater. Res.* 28: 1187–1199.

Schnabel, W. (1981). *Polymer Degradation Principles and Practical Applications,* Macmillan, New York, pp. 15–17, 179–185.

Smith, R., Oliver, C., and Williams, D. F. (1987). The enzymatic degradation of polymers *in vitro*. *J. Biomed. Mater. Res.* 21: 991–1003.

Snow, J., Harasaki, H., Kasick, J., Whalen, R., Kiraly, R. and Nosè, Y. (1981). Promising results with a new textured surface intrathoracic variable volume device for LVAS. *Trans. Am. Soc. Artif. Intern. Organs* XXVII: 485–489.

Stokes, K. (1988). Polyether polyurethanes: Biostable or not? *J. Biomater. Appl.* 3(Oct.): 228–259.

Stokes, K., Coury, A., and Urbanski, P. (1987). Autooxidative degradation of implanted polyether polyurethane devices. *J. Biomater. Appl.* 1(Apr.): 412–448.

Stokes, K., Urbanski, P., and Upton, J., (1990). The *in vivo* autooxidation of polyether polyurethane by metal ions. *J. Biomater. Sci., Polymer Edn.* 1(3): 207–230.

Sutherland, K., Mahoney, J. R., II, Coury, A. J., and Eaton, J. W. (1993). Degradation of biomaterials by phagocyte-derived oxidants. *J. Clin. Invest.* 92: 2360–2367.

Szycher, M., and Siciliano, A. (1991). An assessment of 2,4-TDA formation from Surgitek polyurethane foam under stimulated physiological conditions. *J. Biomater. Appl.* 5: 323–336.

Takahara, A., Coury, A. J., Hergenrother, R. W., and Cooper, S. L. (1991). Effect of soft segment chemistry on the biostability of segmented polyurethanes. I. *In vitro* oxidation. *J. Biomed. Mater. Res.* 25: 341–356.

Takahara, A., Coury, A. J., and Cooper, S. L. (1994). Molecular design of biologically stable polyurethanes. *Trans. 20th Annual Meeting Soc. for Biomater.* April 5–9, p. 44.

Tang, W. W., Santerre, J. P., Labow, R. S., Waghray, G., and Taylor, D. (1994). The use of surface modifying macromolecules to inhibit biodegradation of segmented polyurethanes. *Trans. 20th Annual Meeting Soc. for Biomater.* April 5–9, p. 62.

Test, S., and Weiss, S. (1986). The generation of utilization of chlorinated oxidants by human neutrophils. *Adv. Free Radical Biol. Med.* 2: 91–116.

Vinard, E., Eloy, R., Descotes, J., Brudon, J. R., Giudicelli, H., Patra, P., Streichenberger, R., and David, M. (1991). Human vascular graft failure and frequency of infection. *J. Biomed. Mater. Res.* 25: 499–513.

Ward, R. S., White, K. A., Gill, R. S., and Wolcott, C. A. (1995). Development of biostable thermoplastic polyurethanes with oligomeric polydimethylsiloxane end groups. *Trans. 21st Annual Meeting Soc. for Biomater.* March 18–22, p. 268.

Wasserbauer, R., Beranova, M., Vancurova, D., and Dolezel, B. (1990). Biodegradation of polyethylene foils by bacterial and liver homogenates. *Biomaterials* 11(Jan.): 36–40.

Weaver, K. D., Sauer, W. L., and Beals, N.B. (1995). Sterilization induced effects on UHMWPE oxidation and fatigue strength. *Trans. 21st Annual Meeting Soc. for Biomater.* March 18–22, p. 114.

Williams, D. F. (1989). *Definitions in Biomaterials.* Elsevier, Amsterdam.

Zaikov, G. E. (1985). Quantitative aspects of polymer degradation in the living body. *JMS-Rev. Macromol. Chem. Phys.* C25(4): 551–597.

Zhao, Q., Topham, N., Anderson, J. M., Hiltner, A., Lodoen, G., and Payet, C. R. (1991). Foreign-body giant cells and polyurethane biostability: *In vivo* correlation of cell adhesion and surface cracking. *J. Biomed. Mater. Res.* 25: 177–183.

Zhao, Q., Agger, M., Fitzpatrick, M., Anderson, J., Hiltner, A., Stokes, P., and Urbanski, P. (1990). Cellular interactions with biomaterials: *In vivo* cracking of pre-stressed pellethane 2363-80A. *J. Biomed. Mater. Res.* 24: 621–637.

Zhao, Q., Casas-Bejar, C., Urbanski, P., and Stokes, K. (1995). Glass wool–H_2O_2/$CoCl_2$ for *in vitro* evaluation of biodegradative stress cracking in polyurethane elastomers. *J. Biomed. Mater. Res.* **29**: 467–475.

Ziats, N., Miller, K., and Anderson, J. (1988). *In vitro* and *in vivo* interactions of cells with biomaterials. 9(Jan.): 5–13.

6.3 DEGRADATIVE EFFECTS OF THE BIOLOGICAL ENVIRONMENT ON METALS AND CERAMICS

David F. Williams and Rachel L. Williams

The environment to which biomaterials are exposed during prolonged use (i.e., the internal milieu of the body) can be described as an aqueous medium containing various anions, cations, organic substances, and dissolved oxygen. The anions are mainly chloride, phosphate, and bicarbonate ions. The principal cations are Na^+, K^+, Ca^{2+}, and Mg^{2+}, but with smaller amounts of many others. The organic substances include low-molecular-weight species as well as relatively high-molecular-weight polymeric components. The pH in this well-buffered system is around 7.4 and the temperature remains constant around 37°C.

On the basis of existing knowledge of the stability of materials in various environments, we should predict that metals, as a generic group, should be relatively susceptible to corrosion in this biological environment, whereas ceramics should display a varying susceptibility, depending on solubility. This correlates fairly well with experimental observations and clinical experience, since it is well known that all but the most corrosion-resistant metals will suffer significant and destructive attack upon prolonged implantation. Also, even the most noble of metals and those that are most strongly passivated (i.e., naturally protected by their own oxide layer) will still show some degree of interaction. There are some ceramics that have a combination of very strong partially ionic, partially covalent bonds that are sufficiently stable to resist breakdown within this environment, such as the pure simple oxide ceramics, and others in which certain of the bonds are readily destroyed in an aqueous medium so that the material essentially dissolves, for example certain calcium phosphates.

With these general statements in mind, we have to consider the following questions in relation to the corrosion and degradation of metals and ceramics:

1. Within these groups, how does the susceptibility to corrosion and degradation vary; by what precise mechanisms do the interfacial reactions take place; and how is material selection (and treatment) governed by this knowledge?
2. Are there variables within this biological environment other than those described above that can influence these processes?
3. What are the consequences of such corrosion and degradation phenomena?

We review each of these questions in this chapter. It is particularly important to bear in mind some general points as these questions are discussed.

1. Material selection cannot be governed solely by considerations of stability, and mechanical and physical properties especially may be of considerable importance. Since corrosion is a surface phenomenon, however, it may be possible to optimize corrosion resistance by attention to or treatment of the surface rather than by manipulation of the bulk chemistry. This offers the possibility of developing sufficient corrosion resistance in materials of excellent bulk mechanical and physical properties. Thus, noble metals such as gold and platinum are rarely used for structural applications (apart from dental restoration) because of their inferior mechanical properties, even though they have excellent corrosion resistance; instead, base metal alloys with passivated or protected surfaces offer better all-around properties.
2. Medical devices are not necessarily used in mechanically stress-free conditions and indeed the vast majority of those using metals or ceramics are structurally loaded. It is well known that mechanical stress plays a very important role in the corrosion and degradation process, both potentiating existing effects and initiating others. This has to be taken into consideration.
3. We cannot expect the biological environment to be constant. Within the overall characteristics described earlier there are variations (with time, location, activity, health status, etc.) in, for example, oxygen levels, availability of free radicals, and cellular activity, all of which may cause variations in the corrosive nature of the environment. Most important, corrosion is not necessarily a progressive homogeneous reaction with zero-order kinetics. Corrosion processes can be quiescent but then become activated, or they can be active but then become passivated and localized, with transient fluctuations in conditions playing a part in these variations.
4. The effects of corrosion or degradation may be twofold. First, and in the conventional metallurgical sense, the most obvious, the problem can lead to loss of structural integrity of the material, volume, and function. This may be undesirable, as in the case of many long-term prostheses, or desirable, as in devices intended for short-term function (e.g., ceramics for drug delivery systems) or where the material is replaced by tissue during the degradation process, as with ceramic bone substitution. In addition to this, however, and usually of much greater significance with biomaterials, when released into the tissue, the corrosion or degradation products can have a significant and controlling effect on that tissue. Indeed, it is likely that the corrosion process is the most important mediation of the tissue response to materials. It is therefore important that we know both the nature of the reaction products and their rate of generation.

METALLIC CORROSION

Basic Principles

The most pertinent form of corrosion related to metallic biomaterials is aqueous corrosion. This occurs when electrochemical reactions take place on a metallic surface in an aqueous electrolyte. There are always two reactions that occur: the anodic reaction, which yields metallic ions, for example, involving the oxidation of the metal to its salt:

$$M \rightarrow M(n+) + n(\text{electrons}) \qquad (1)$$

and the cathodic reaction, in which the electrons so generated are consumed. The precise cathodic reaction will depend on the nature of the electrolyte, but two of the most important in aqueous environments are the reduction of hydrogen:

$$2H^+ + 2e^- \rightarrow H_2 \qquad (2)$$

and the reduction of dissolved oxygen:

$$O_2 + 4H^+ + 4e^- \rightarrow 2H_2O \qquad (3)$$

in acidic solutions or:

$$O_2 + 2H_2O + 4e^- \rightarrow 4OH^- \qquad (4)$$

in neutral or basic solutions.

In all corrosion processes, the rate of the anodic or oxidation reaction must equal the rate of the cathodic or reduction reaction. This is a basic principle of electrochemically based metallic corrosion. It also explains how variations in the local environment can affect the overall rate of corrosion by influencing either the anodic or cathodic reactions. The whole corrosion process can be arrested by preventing either of these reactions.

From a thermodynamic point of view, first consider the anodic dissolution of a pure metal isolated in a solution of its salt. The metal consists of positive ions closely surrounded by free electrons. When the metal is placed in a solution, there will be a net dissolution of metal ions since the Gibbs free energy (ΔG) for the dissolution reaction is less than for the replacement reaction. This leaves the metal with a net negative charge, thus making it harder for the positive ions to leave the surface and increasing the ΔG for the dissolution reaction. There will come a point when the ΔG for the dissolution reaction will equal the ΔG for the replacement reaction. At this point, a dynamic equilibrium is reached and a potential difference will be set up across the charged double layer surrounding the metal. The potential difference will be characteristic of the metal and can be measured against a standard reference electrode. When this is done against a standard hydrogen electrode in a 1 N solution of its salt at 25°C, it is defined as the standard electrode

TABLE 1 Electrochemical Series

Metal	Potential (V)
Gold	1.43
Platinum	1.20
Mercury	0.80
Silver	0.79
Copper	0.34
Hydrogen	0.00
Lead	−0.13
Tin	−0.14
Molybdenum	−0.20
Nickel	−0.25
Cobalt	−0.28
Cadmium	−0.40
Iron	−0.44
Chromium	−0.73
Zinc	−0.76
Aluminum	−1.33
Titanium	−1.63
Magnesium	−2.03
Sodium	−2.71
Lithium	−3.05

potential for that metal (Table 1). The position of a metal in the electrochemical series primarily indicates the order with which metals displace each other from compounds, but it also gives a general guide to reactivity in aqueous solutions. Those at the top are the noble, relatively unreactive metals, whereas those at the bottom are the more reactive. This is the first guide to corrosion resistance, but, as we shall see, there are major difficulties related to the use and interpretation of reactions from this simple analysis.

Now consider a system in which the metal is in an aqueous solution that does not contain its ions. In this situation, the electrode potential at equilibrium (i.e., when the rate of the anodic reaction equals the rate of the cathodic reaction) will be shifted from the standard electrode potential and can be defined by the Nernst equation:

$$E = E_0 + (RT/nF \ln(a_{\text{anod}}/a_{\text{cath}}) \qquad (5)$$

where E_o is the standard electrode potential, RT/F is a constant, n is the number of electrons transferred, and a is the activity of the anodic and cathodic reactants. At low concentrations, the activity can be approximated to the concentration. In this situation, there is a net dissolution of the metal and a current will flow. At equilibrium, the rate of the metal dissolution is equal to the rate of the cathodic reaction, and the rate of the reaction is directly proportional to the current density by Faraday's law; therefore:

$$i_{\text{anodic}} = i_{\text{cathodic}} = i_{\text{corrosion}} \qquad (6)$$

and the Nernst equation can be rewritten:

$$E - E_{\text{o}} = \pm B \ln(i_{\text{corr}}/i_0) \qquad (7)$$

where B is a constant and i_0 is the exchange current density, which is defined as the anodic (or cathodic) current density at the standard electrode potential. Current density is the current, measured in amperes, normalized to the surface area of the metal.

These conditions represent convenient models for the basic mechanisms of corrosion, but they are hardly realistic. Indeed, in this situation of a homogeneous pure metal existing within an unchanging environment, an equilibrium is reached in which no further net movement of ions takes place. In other words, the corrosion process takes place only transiently, but is effectively stopped once this equilibrium is reached.

In reality, we usually have neither entirely homogeneous surfaces or solutions, nor complete isolation of the metal from other parts of the environment, and this equilibrium is easily upset. If the conditions are such that the equilibrium is displaced, the metal is said to be polarized. There are several ways in which this can happen. Two main factors control the behavior of metals in this respect and determine the extent of corrosion in practice. The first concerns the driving force for continued corrosion (i.e., the reasons why the equilibrium is upset and the nature of the polarization), and the second concerns the ability of the metal to respond to this driving force.

It is self-evident that if either the accumulating positive metal ions in the surrounding media or the accumulating electrons in the metal are removed, the net balance between the dissolution and the replacement of the ions will be disturbed. The equilibrium is established precisely because of the imbalance of charge, so that if the latter is disturbed, so is the former. The result will be continued dissolution as the system attempts to achieve this equilibrium—in other words, sustained corrosion. An electron sink in contact with the metal or a dynamic medium will achieve this.

The process of galvanic corrosion may be used to demonstrate this effect. Consider a single homogeneous pure metal, A, existing within an electrolyte (Fig. 1). The metal will develop its own potential, V_A, with respect to the electrolyte. If a different metal electrode, B, is placed into the same electrolyte, but without contacting A, it will develop its own potential V_B. If V_A is not equal to V_B, there will be a difference in the numbers of excess free electrons in each. This is of no consequence if A and B are isolated from each other, but should they be placed in electrical contact, electrons will flow from that metal with the greater potential in an attempt to make the two electrodes equipotential. This upsets the equilibrium and causes continued and accelerated corrosion of the more active metal (anodic dissolution) and protects the less active (cathodic protection).

Galvanic corrosion may be seen whenever two different metals are placed in contact in an electrolyte. It has been frequently observed with complex, multicomponent surgical im-

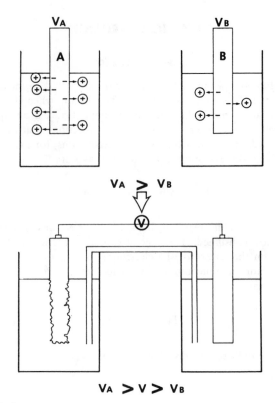

FIG. 1. When electrical contact is made between electrodes A and B, electrode B acts as an electron sink, thus upsetting the equilibrium and causing continued dissolution of A.

plants. It is not necessary for the components to be macroscopic, monolithic electrodes for this to happen, and the same effect can be seen when there are different microstructural features within one alloy. In practice, it is the regional variations in electrode potential over an alloy surface that are responsible for much of the generalized surface corrosion that takes place in metallic components.

Many of the commonly used surgical alloys contain highly reactive metals (i.e., with high negative electrode potentials), such as titanium, aluminum, and chromium. Because of this high reactivity, they will adsorb oxygen upon initial exposure to the atmosphere. This initial oxidation stage leaves an impervious oxide layer firmly adherent to the metal surface; thus all other forms of corrosion may then be stifled because the oxide layer acts as a protective barrier, passivating the metal. It should also be noted that it is possible to enhance the oxide layer artificially to provide better corrosion resistance.

In summary, the basic principles of corrosion determine that:

1. In theory, corrosion resistance can be predicted from standard electrode potentials. This explains the nobility of some metals and the considerable reactivity of others, but is not useful for predicting the occurrence of corrosion of most alloy systems in practice.

2. Irrespective of standard electrode potentials, the corro-

sion resistance of many materials is determined by their ability to become passivated by an oxide layer that protects the underlying metal.

3. Corrosion processes in practice are influenced by variations in surface microstructural features and in the environment that disrupt the charge transfer equilibrium.

INFLUENCE OF THE BIOLOGICAL ENVIRONMENT

It is reasonable to assume that the presence of biological macromolecules will not cause a completely new corrosion mechanism. However, they can influence the rate of corrosion by interfering in some way with the anodic or cathodic reactions discussed earlier. Four ways in which this could occur are discussed below:

1. The biological molecules could upset the equilibrium of the corrosion reactions by consuming one or other of the products of the anodic or cathodic reaction. For example, proteins can bind to metal ions and transport them away from the implant surface. This will upset the equilibrium across the charged double layer and allow further dissolution of the metal; in other words, it will decrease ΔG for the dissolution reaction.

2. The stability of the oxide layer depends on the electrode potential and the pH of the solution. Proteins often have electron-carrying roles and can thus affect the electrode potential, and bacteria can alter the pH of the local environment through the generation of acidic metabolic products.

3. The stability of the oxide layer is also dependent on the availability of oxygen. The adsorption of proteins onto the surface of materials could limit the diffusion of oxygen to certain regions of the surface. This could cause preferential corrosion of the oxygen-deficient regions and lead to the breakdown of the passive layer.

4. The cathodic reaction often results in the formation of hydrogen, as shown earlier. In a confined locality, the buildup of hydrogen tends to inhibit the cathodic reaction and thus restricts the corrosion process. If the hydrogen can be eliminated, then the active corrosion can proceed. It is possible that bacteria in the vicinity of an implant could utilize the hydrogen and thus play a crucial role in the corrosion process.

There is sufficient evidence to support the premise that the presence of proteins can influence the rate of corrosion of some metals. Studies have examined these interactions electrochemically and have found very few differences in many of the parameters measured (e.g., electrode potential, polarization behavior, and current density at a fixed potential). However, analysis of the amount of corrosion through weight loss or chemical analysis of the electrolyte has shown significant effects from the presence of relatively low concentrations of proteins. These effects have varied from severalfold increases for some metals

under certain conditions, to slight decreases under other conditions.

It has been shown that proteins adsorb onto metal surfaces and that the amount adsorbed appears to be different on a range of metals. Similarly, proteins have been shown to bind to proteins and it is suggested that they are transported away from the local site as a protein–metal complex and distributed systemically in the body. It is therefore likely that proteins will influence the corrosion reactions that occur when a metal is implanted, although there is no direct evidence to explain the mechanism of the interaction at this time.

CORROSION AND CORROSION CONTROL IN THE BIOLOGICAL ENVIRONMENT

The need to ensure minimal corrosion has been the major determining factor in the selection of metals and alloys for use in the body. Two broad approaches have been adopted. The first has involved the use of noble metals, that is, those metals and their alloys for which the electrochemical series indicates excellent corrosion resistance. Examples are gold, silver, and the platinum group of metals. Because of cost and relatively poor mechanical properties, these are not used for major structural applications, although it should be noted that gold and its alloys are extensively used in dentistry; silver is sometimes used for its antibacterial activity; and platinum group metals (Pt, Pd, Ir, Rh) are used in electrodes.

The second approach involves the use of the passivated metals. Of the three elements that are strongly passivated (i.e., aluminum, chromium, and titanium), aluminum cannot be used for biomedical purposes because of toxicity problems. Chromium is very effectively protected but cannot be used in bulk. It is, however, widely used in alloys, especially in stainless steels and in the cobalt–chromium-based alloys, where it is normally considered that a level of above 12% gives good corrosion resistance and about 18% provides excellent resistance. Titanium is the best in this respect, and is used as a pure metal or as the major constituent of alloys.

Although these metals and alloys have been selected for their corrosion resistance, corrosion will still take place when they are implanted in the body. Two important points have to be remembered. First, whether noble or passivated, all metals will suffer a slow removal of ions from the surface, largely because of local and temporal variations in microstructure and environment. This need not necessarily be continuous and the rate may either increase or decrease with time, but metal ions will be released into that environment. This is particularly important with biomaterials, since it is the effect of these potentially toxic or irritant ions that is the most important consequence of their use. Even with a strongly passivated metal, there will be a finite rate of diffusion of ions through the oxide layer, and possibly a dissolution of the layer itself. It is well known that titanium is steadily released into the tissue from titanium implants. Second, some specific mechanisms of corrosion may be superimposed on

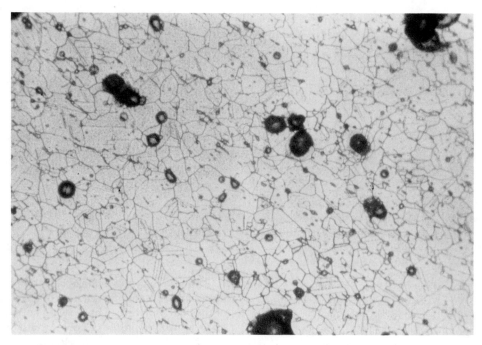

FIG. 2.   This etched metallographic micrograph demonstrates the pitting corrosion of stainless steel.

this general behavior; some examples are given in the next section.

Pitting Corrosion

The stainless steels used in implantable devices are passivated by the chromium oxide that forms on the surface. It has been shown, however, that in a physiological saline environment, the driving force for repassivation of the surface is not high. Thus, if the passive layer is broken down, it will not repassivate and active corrosion can occur.

Localized corrosion can occur as a result of imperfections in the oxide layer, producing small areas in which the protective surface is removed. These localized spots will actively corrode and pits will form in the surface of the material. This can result in a large degree of localized damage because the small areas of active corrosion become the anode and the entire remaining surface becomes the cathode. Since the rate of the anodic and cathodic reactions must be equal, it follows that a relatively large amount of metal dissolution will be initiated by a small area of the surface, and large pits may form (Fig. 2).

Fretting Corrosion

The passive layer may be removed by a mechanical process. This can be a scratch that does not repassivate, resulting in the formation of a pit, or a continuous cyclic process in which any reformed passive layer is removed.

This is known as fretting corrosion, and it is suggested that this can contribute to the corrosion observed between a fracture fixation plate and the bone screws attaching the plate to the bone. There are three reasons why fretting can affect the corrosion rate. The first is due to the removal of the oxide film as just discussed. The second is due to plastic deformation of the contact area; this can subject the area to high strain fatigue and may cause fatigue corrosion. The third is due to stirring of the electrolyte, which can increase the limited current density of the cathodic reaction.

Crevice Corrosion

The area between the head of the bone screw and countersink on the fracture fixation plate can also be influenced by the crevice conditions that the geometry creates (Fig. 3). Accelerated corrosion can be initiated in a crevice by restricted diffusion of oxygen into the crevice. Initially, the anodic and cathodic reactions occur uniformly over the surface, including within the crevice. As the crevice becomes depleted of oxygen, the reaction is limited to metal oxidation balanced by the cathodic reaction on the remainder of the surface. In an aqueous sodium chloride solution, the buildup of metal ions within the crevice causes the influx of chloride ions to balance the charge by forming the metal chloride. In the presence of water, the chloride will dissociate to its insoluble hydroxide and acid. This is a rapidly accelerating process since the decrease in pH causes further metal oxidation.

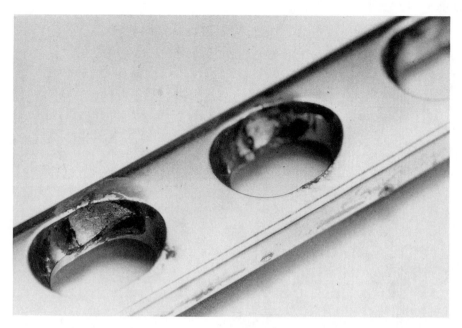

FIG. 3. Crevice corrosion is evident in the screw hole in this fracture fixation plate.

Intergranular Corrosion

As mentioned earlier, stainless steels rely on the formation of chromium oxides to passivate the surface. If some areas of the alloy become depleted in chromium, as can happen if carbides are formed at the grain boundaries, the regions adjacent to the grain boundaries become depleted in chromium. The passivity of the surface in these regions is therefore affected and preferential corrosion can occur (Fig. 4). Although this problem can easily be overcome by heat treating the alloys, it has been observed on retrieved implants, and can cause severe problems since once initiated it will proceed rapidly and may well cause fracture of the implant and the release of large quantities of corrosion products into the tissue.

Stress Corrosion Cracking

Stress corrosion cracking is an insidious form of corrosion since an applied stress and a corrosive environment can work together and cause complete failure of a component, when neither the stress nor the environment would be a problem on their own. The stress level may be very low, possibly only residual, and the corrosion may be initiated at a microscopic crack tip that does not repassivate rapidly. Incremental crack growth may then occur, resulting in fracture of the implant. Industrial uses of stainless steels in saline environments have shown susceptibility to stress corrosion cracking and therefore it is a potential source of failure for implanted devices.

Galvanic Corrosion

If two metals are independently placed within the same solution, each will establish its own electrode potential with respect to the solution. If these two metals are placed in electrical contact, then a potential difference will be established between them, electrons passing from the more anodic to the more cathodic metal. Thus equilibrium is upset and a continuous process of dissolution from the more anodic metal will take place. This accelerated corrosion process is galvanic corrosion. It is important if two different alloys are used in an implantable device when the more reactive may corrode freely.

Whenever stainless steel is coupled with another alloy, it will suffer from galvanic corrosion. If both alloys remain within their passive region when coupled in this way, the additional corrosion may be minimal. Some modular orthopedic systems are made of titanium alloys and cobalt-based alloys on the basis that both should remain passive. Galvanic corrosion may also take place on a microscopic scale in multiphase alloys where phases are of considerably different electronegativity. In dentistry, some amalgams may show extensive corrosion because of this mechanism.

CERAMIC DEGRADATION

The rate of degradation of ceramics within the body can vary considerably from that of metals in that they can be either highly corrosion resistant or highly soluble. As a

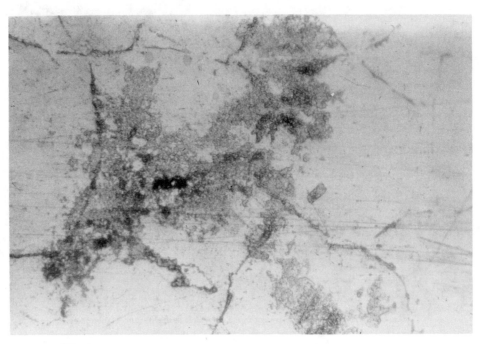

FIG. 4.   Intergranular corrosion is demonstrated on this etched stainless steel specimen.

general rule, we should expect to see a very significant resistance to degradation with ceramics and glasses. Since the corrosion process in metals is one of a conversion of a metal to ceramic structure (i.e., metal to a metal oxide, hydroxide, chloride, etc.) we must intuitively conclude that the ceramic structure repesents a lower energy state in which there would be less driving force for further structural degradation. The interatomic bonds in a ceramic, being largely ionic but partly covalent, are strong directional bonds and large amounts of energy are required for their disruption. As extraction metallurgists know, it takes a great deal of energy to extract aluminum metal from the ore aluminum oxide, but as we have seen, the reverse process takes place readily by surface oxidation. Thus, we should expect ceramics such as Al_2O_3, TiO_2, SiO_2, and TiN to be stable under normal conditions. This is what is observed in clinical practice. There is limited evidence to show that some of these ceramics (e.g., polycrystalline Al_2O_3) do show "aging" phenomena, with reductions in some mechanical properties, but the significance of this is unclear.

Alternatively, there will be many ceramic structures that, although stable in the air, will dissolve in aqueous environments. Consideration of the classic fully ionic ceramic structure NaCl and its dissolution in water demonstrates this point. It is possible, therefore, on the basis of the chemical structure, to identify ceramics that will dissolve or degrade in the body, and the opportunity exists for the production of structural materials with controlled degradation.

Since any material that degrades in the body will release its constituents into the tissue, it is necessary to select anions and cations that are readily and harmlessly incorporated into metabolic processes and utilized or eliminated. For this reason, it is compounds of sodium, and especially calcium, including calcium phosphates and calcium carbonates, that are primarily used.

The degradation of such compounds will depend on chemical composition and microstructure. For example, tricalcium phosphate $[Ca_3(PO_4)_2]$ is degraded fairly rapidly while calcium hydroxyapatite $[Ca_{10}(PO_4)_6(OH)_2]$ is relatively stable. Within this general behavior, however, porosity will influence the rates so that a fully dense material will degrade slowly, while a microporous material will be susceptible to more rapid degradation.

In general, dissolution rates of these ceramics *in vivo* can be predicted from behavior in simple aqueous solution. However, there will be some differences in detail within the body, especially with variations in degradation rate seen with different implantation sites. It is possible that cellular activity, either by phagocytosis or the release of free radicals, could be responsible for such variations.

In between the extremes of stability and intentional degradability lie a small group of materials in which there may be limited activity. This is particularly seen with a number of glasses and glass ceramics, based on Ca, Si, Na, P, and O, in which there is selective dissolution on the surface involving the release of Ca and P, but in which the reaction then ceases because of the stable SiO_2-rich layer that remains on the surface. This is of considerable interest because of the ability of such surfaces to bond to bone, and this subject is dealt with elsewhere in this book.

On the basis of this behavior, bioceramics are normally classified under three headings:

Inert, or "nearly inert" ceramics
Resorbable ceramics
Ceramics of controlled surface reactivity

SUMMARY

This chapter has attempted to demonstrate that metals are inherently susceptible to corrosion and that the greatest care is needed in using them within the human body. In general, ceramics have much less tendency to degrade, but care still has to be taken over aging phenomena. The human body is very aggressive toward all of these materials.

Bibliography

Cook, S. D., Tomas, K. A., Harding, A. F., Collins, C. L., Haddad, R. J., Milicic, M., and Fischer, W. L. (1987). The *in vivo* performance of 250 internal fixation devices; a follow up study. *Biomaterials* **8**: 177–184.

Dalgleish, B. J. and Rawlings, R. D. (1981). A comparison of the mechanical behaviour of aluminas in air and simulated body environments. *J. Biomed. Mater. Res.* **15**: 527–542.

de Groot, K. (1983). *Bioceramics of Calcium Phosphate.* CRC Press, Boca Raton, FL.

Fraker, A. C., and Griffith, C. D., eds. (1985). *Corrosion and Degradation of Implant Materials.* ASTM S.T.P. No. 859, American Society for Testing and Materials, Philadelphia.

Marek, M. (1990). Corrosion of dental materials. in *Encyclopedia of Medical and Dental Materials,* D. F. Williams, ed. Pergamon, Oxford, pp. 121–126.

Williams, D. F. (1981). Electrochemical aspects of corrosion in the physiological environment. in *Fundamental Aspects of Biocompatibility,* D. F. Williams, ed. CRC Press, Boca Raton, FL, Vol. I, pp. 11–42.

Williams, D. F. (1985). Physiological and microbiological corrosion. *Crit. Rev. Biocompat.* **1**(1): 1–24.

6.4 MECHANICAL BREAKDOWN IN THE BIOLOGICAL ENVIRONMENT

C. R. McMillin

Contact with biological or physiological solutions causes polymer degradation, ceramic dissolution, and metallic corrosion and this effect can be amplified in the presence of mechanical stresses. This chapter discusses effects commonly seen in metals, ceramics, and polymers.

METALS AND CERAMICS

There are a number of mechanisms by which a combination of the biological environment and mechanical stress can cause increased mechanical breakdown through corrosion, including galvanic, fatigue, wear, microbiological, and stress-enhanced degradation and corrosion. The combined effect of environment and these stresses is much greater than either one by itself.

Galvanic Corrosion

A bent metallic rod or plate will have different electrochemical potentials on opposing sides of the bend. The tensile side may be anodic with respect to the compression side, leading to increased galvanic corrosion. This same phenomenon occurs at stress risers in loaded devices, including design flaws such as sharp corners; surface flaws such as cuts, nicks, and scratches; and internal material flaws, imperfections, and gain boundaries. The higher stressed area of the material will corrode or degrade at a higher rate than the surrounding, less stressed area.

Galvanic corrosion can be increased by residual stresses in metal parts caused by processing. If the high hardness or strength of cold-worked materials is not needed, manufactured parts can be annealed to relieve internal stresses.

Fatigue and Wear Corrosion

A passive and insoluble film may be produced on the surface of a number of biologically stable metals that makes them more resistant to corrosion. Bending, strain from fatigue, abrasion, or wear may continuously rupture this passive film, exposing the underlying metal. When this happens, the material loses much of its corrosion resistance. Coupled with the loss of the passive film is anodic dissolution at a growing crack tip, if one is present. When these mechanisms result in increased corrosion in a fatigue environment, it is called corrosion fatigue. Static stress can cause similarly increased corrosion and cracking, which is generally called stress-induced corrosion cracking. These effects are substantially greater at elevated temperatures.

The effect of dynamic stress on the corrosion of surgical implant alloys was studied by Luedemann and Bundy (1989). They found that even the relatively low loads typical of normal walking had a marked effect on the corrosion characteristics of implant alloys tested in Ringer's solution. The amounts of ions released from the surfaces of cobalt–chromium alloy and 316L stainless steel were up to 26 and 38 times those released from unstressed polished surfaces in the same solution. With metal fatigue in the presence of an aggressive environment that causes corrosion, there is no limiting stress (endurance limit) below which corrosion-induced cracking will not occur (Williams, 1981).

Microbiological Corrosion

Microorganisms can influence the corrosion of some metals by a variety of mechanisms, most of which are enhanced by the presence of mechanical stresses (Williams, 1981). They can alter the pH and the oxygen content of the local environment, can release corrosive metabolic products, and can cause depolarization phenomena. For example, *Thiobacillus*, a genus of sulfur-oxidizing bacteria, produces acids, including sulfuric acid. The bacteria themselves can easily withstand strongly acidic conditions and are thus able to colonize a sulfur-containing site and cause increased corrosion of metals in the vicinity. Similarly, the fungal microorganisms *Clad osporium resinae* colonize aircraft fuel tanks, producing oxygen concentration gradients and thereby causing corrosion. The sulfate-reducing bacteria *De sulphovibro* utilize hydrogen in the reduction of sulfates. If the cathodic reaction of corrosion is one of hydrogen reduction, the utilization of the hydrogen by the bacteria can cause cathodic depolarization and allow corrosion to continue. Although not considered to be a significant issue for medical implant devices, some dental implant devices can be influenced by microbiological corrosion phenomena.

Stress-Enhanced Degradation and Corrosion

Stressing a material puts it in a higher energy state and makes it easier to degrade or corrode. This is true for metals, polymers, and ceramics. For example, the material in the immediate vicinity of vacancies, slip bands, dislocations, and grain boundaries in metals can be more highly stressed than the homogeneous material within the grains. If metal samples are etched, the metal is preferentially etched from the higher energy areas around these imperfections, making pits that can be visually observed. Under stress, some dislocations will move. If a sequence of stress applications and etchings is used, the movement of the dislocations can be followed by the resulting sequential pits formed at dislocation sites developed by the etching (Hayden *et al.*, 1965). This type of phenomenon is also likely to be related to the environmental stress cracking seen in polymers.

If a ceramic material contains both amorphous or glassy regions, as well as either crystalline regions or embedded crystallites, energetic favor the dissolution of the amorphous regions in preference to the crystalline regions.

Hench has said that all ceramic materials exhibit an environmentally sensitive fatigue behavior (Hench, 1981). Partially because of this loss of strength, he concluded that "bulk bioglass and bioglass-ceramics are likely to have insufficient long-term mechanical reliability for load-bearing prosthetic applications unless improvements in physical properties beyond those now accessible are achieved." This concept is further discussed in Chapter 2.6.

Effects of Metallic Corrosion

In addition to the obvious loss of mechanical strength of implants caused by corrosion, the corrosion products can have a number of other adverse physiological effects. Metal sensitivity and allergic reactions from metallic corrosion products and wear debris are of concern, particularly with metals containing nickel and chromium (Bruner *et al.*, 1993). When allergic responses are seen, they can lead to localized problems such as inhibiting fusion or to systemic problems such as dermatitis. Both of these conditions usually require removal of the implant. Concern has also been expressed that the localized release of certain metal ions may cause cancer (Sharkness *et al.*, 1993). However, the relatively few cancers found associated with me-

tallic implants compared with the large number that have been implanted for long periods of time suggest that this effect, if any, is small. Metallic corrosion and wear debris particulates may cause other nonspecific effects that are covered in more detail in the next section of this chapter.

POLYMERS

There are several ways in which a biological environment increases the rate of mechanical breakdown of polymers. These include increased wear, swelling or leaching, environmental stress crazing and cracking, and mechanical effects on blood compatibility and calcification.

Increased Wear in Biological Environments

Generally, friction is reduced but wear may be increased for wet sliding wear compared with dry sliding wear for unfilled polymers (Lhymn, 1987). The wear volume also increases in carbon fiber and glass-reinforced composites compared with unfilled polymers. The presence of a liquid can remove wear debris immediately, thus enhancing the cutting/plowing/cracking mechanism of abrasive particles against the polymer or polymer composite material (Lhymn, 1987). In a physiological environment, metallic, ceramic, or polymeric wear particles may be trapped between two moving surfaces, causing three-body wear, which generally causes a significantly higher wear rate than two-body wear. Other mechanisms for increased *in vivo* wear include environmental stress cracking, polymer degradation, microstructural imperfections, and creep.

The analysis of artificial joints demonstrates the increase of wear rates under *in vivo* versus *in vitro* conditions. In one study, dry *in vitro* tests of hip joints predicted an average wear rate of 0.02 mm a year, whereas the average clinical penetration rate was found to be 0.07 mm a year over the first 9 years of use (Dowling, 1983). This wear may cause loosening of the prostheses by the resulting poor mechanical fit between the ball and socket of the hip. If the wear becomes extensive, the artificial hip may need to be replaced.

Of much greater concern, however, is the release of wear debris particles. Generation of wear debris is an important factor both because of the potential for wear debris to migrate to distant organs, particularly the lymph nodes where accumulation of particle-containing macrophages causes enlargement and chronic lymphadenitis (Benz *et al.*, 1994) and because of local physiological responses such as inflammatory, cytotoxic, and osteolytic reactions.

It is now apparent that large quantities of micron and submicron wear debris from most materials cause nonspecific osteolysis of local bone. For example, a major cause of late failures of artificial hips has been loosening between the hip stem and the partially resorbed surrounding boone, causing pain and requiring hip replacement. Sometimes femoral bone weakened by osteolysis can fracture under high loading conditions and require replacement. It was thought that this condition was the result of osteolysis of the bone caused by bone cement particles released by wear and fatigue microfractures (Willert *et al.*, 1990a), and it was occasionally called bone cement disease.

A new generation of artificial hips has been developed and introduced which uses bioingrowth of bone into porous surfaces of the hip stem so that cement is not required; thus it was hoped to eliminate bone cement disease. Unfortunately, the newer bioingrowth artificial hips have continued to fail by the same osteolytic mechanism, in the absence of any bone cement! Further research has shown that polyethylene (Willert *et al.*, 1990b), titanium, and ceramic wear debris can cause the same osteolytic problem associated with bone cement debris. Both micron-sized particulates seen in earlier histological slides, as well as submicron particulates now found using digestion and scanning electron microscopic techniques, have been implicated in osteolytic destruction of the bone–implant interface. Assuming a million steps per year and 0.03 to 0.08 mm per year wear into polyethylene acetabular cups, one recent calculation estimates that on an average as many as 470,000 polyethylene particles of 0.5-micron size could be released to the surrounding tissues at each step (McKellop *et al.*, 1993).

Other examples of nonspecific osteolysis presumed to be caused by wear debris of devices include bone resorption around failed elastomeric finger, toe, and wrist prostheses and around failed poly(tetrafluoroethylene)-carbon fiber temporomandibular joint (jaw) prostheses. Therefore, an important criterion in the selection of materials for use in medical implants is the quantity, size, shape, and composition of wear debris that may be released *in vivo*.

Mechanical Failure by Swelling and Leaching

In a biological environment, polymers both absorb components from the surrounding media and also leach components from the polymer into the media. The absorption of the chemicals can result in a physical change in shape of the device, which can then lead to failure through a variety of mechanisms.

One of the most important historical examples of failure by swelling was in the original Star–Edwards and Smeloff–Cutter caged ball heart valve prostheses. Lipid infiltration and subsequent swelling, along with ball wear, caused some device-related deaths. It was found that an incomplete silicone rubber cure may have contributed to the high lipid uptake and later failure of these heart valves.

When fiber-filled composites adsorb solvents and chemicals (including water, lipids, and proteins), it can lead to degradation of the polymer–fiber interface, with resulting reduced strength and stiffness (Wilfong, 1989).

Under *in vivo* conditions, the leaching of plasticizers, stabilizers, antioxidants, pigments, lubricants, fillers, residual monomers, polymerization catalysts, and other chemical additives from plastics or rubbers both change the properties of the polymer and cause local or systemic toxicological problems.

When plasticizers migrate from flexible poly(vinyl chloride) parts, the parts become brittle. This is typically why automobile vinyl seat covers become brittle as the di-2-ethylhexyl phthalate plasticizer slowly migrates from the plastic, leaving the rigid poly(vinyl chloride). The leaching of this same plasticizer from blood bags into stored blood has been a cause for concern from the bioaccumulation standpoint for patients receiving many units of stored blood (e.g., leukemia patients).

Both leaching of mineral oil and absorption of lipids and fluids have been reported to contribute to changes in the properties of a modified copolymer of styrene–ethylene/butylene–styrene after subcutaneous implantation (Hawkins *et al.*, 1993).

Environmental Stress Crazing, Cracking, and Rupture of Polymers

The combination of stresses and environmental conditions can act synergistically to cause failure in polymers. In these cases, the polymers generally undergo what is called environmental stress crazing, cracking, and rupture.

Crazing can be described as the formation of internal voids with load-bearing fibers that stretch across the voids. Crazes can look like cracks, but are differentiated from cracks by the load-bearing nature of the fibers that extend across the craze. As the craze grows, some of the fibers in the widest area of the craze break, and a nonload-bearing crack is formed. The crack can then grow and progress across the material, leading to rupture or failure of the part.

Craze formation is dependent on the molecular weight of the polymer. Generally, high-molecular-weight polymers form crazes that can progress to cracks. For lower molecular weight polymers, crack formation occurs directly, without the formation of craze material. A mixture of high- and low-molecular-weight polymers forms crazes more readily than an equivalent medium-molecular-weight polymer.

The sequence of environmental stress crazing, cracking, and rupture occurs at low stress levels for polymers sensitive to these effects when they are in contact with aggressive environments (which are specific for each polymer). The stresses can either be externally applied or the result of residual internal stresses from molding or processing.

Thus, polymers can be resistant to high stress *in vitro*, and can be implanted *in vivo* for long periods of time under low or no stress conditions with little or no degradation of properties. When stressed in the same environment, however, they can actually fall apart as a result of environmental stress cracking.

For example, Daniels *et al.* (1989) evaluated the dynamic cyclic bending of a poly(ortho ester) in different media. They showed that 25,000 cycles in air had little effect on the mechanical properties. Specimens in contact with simulated body fluid under no load also showed very little effect, while the cyclically loaded specimens in the test fluid showed a 75% decrease in the flexural strength and modulus.

Environmental Stress Cracking—Polyurethane Pacemaker Leads

Some polyurethane cardiac pacemaker leads have been found to fail prematurely. Damaged insulation has resulted in rapid drainage of battery power and erratic pacing, resulting in the forced replacement of the pacing system. Medtronic has reported 240 cases of cracked insulation in 185,000 poly(ether urethane)-insulated cardiac pacing leads distributed between 1977 and 1983, with 104 of these affecting

TABLE 1 Potential Factors Involved with
Pacemaker Lead Failure

Stress Component
  Residual processing stresses
  Bends or kinks during insertion
  Postimplant zones of sheath shrinkage
  In vivo flexing during operation
Environmental Component
  Voltage (current) used in pacing
  Oxidation
  Enzymes
  Chloride anion
  Cholesterol and lipids
  Hydrolytic cleavage
  Extraction of low molecular weight material
  Metal ion-catalyzed degradation
  Auto-oxidation involving metals
Polymer Sensitivity
  Ether links attacked
  Poly(tetramethylene oxide) blocks attacked
  Silicone added to resist attack

the performance of the device (Stokes, 1984). Much greater degradation was seen in flexion points of the leads compared with areas not under the same mechanical stresses (Chawla *et al.*, 1988). However, the great majority of the same pacing leads survived for long periods of time without significant degradation. Although the failure of the leads has clearly been attributable to environmental stress cracking, there has been considerable debate about what aspect of the environment causes the failures, as indicated in Table 1 and discussed in the following paragraphs.

During the production of pacemaker leads, residual stresses are created that are tensile on the outer surface and compressive in the core. These stresses, combined with externally applied stress, are thought to be a major mechanical factor in the environmental stress cracking of the implanted leads (Hayshi, 1987).

The effect of applied voltage on environmental stress cracking in Pellethane (Dow Chemical Co.)-coated pacing leads has been studied *in vitro* using a balanced salt solution (Hank's solution) (Sung and Fraker, 1987). It was found that current leakage occurred in only a few areas of the lead after exposure to the solution for 14 months with a 5-V to 20-V potential. Although the outer surface of the leads revealed some cracks, the checkerboard-type cracking reported in *in vivo* studies was not observed in this unstressed *in vitro* test.

Another environmental factor that has been examined is the effect of enzymes on polyurethanes. In one study, papain and urease were found to degrade the polyurethane Biomer (Ethicon, Inc.) (Ratner *et al.*, 1988). However, degradation was not noted in the postimplantation examination of Biomer used in Jarvik-7 artificial heart dia-

phragms (Benson and Wong, 1988). In another study, papain, chymotrypsin, and leucine aminopeptidase were all found to degrade some polyurethanes (Phua *et al.*, 1987). In a third study, papain, esterase, bromelain, ficin, chymotrypsin, trypsin, and cathepsin C (but not collagenase or the oxidative enzymes xanthine oxidase and cytochrome C oxidase) were found to degrade a radiolabeled polyurethane (Smith *et al.*, 1987). The mixed enzymes found in rabbit liver homogenate also had a deleterious effect on the polyurethane used in this study.

Polyurethane-insulated Co–Ni–Cr–Mo (MP35N) pacemaker leads have been used for cardiac pacing since 1977. Tests were therefore conducted to determine if metal ions were involved in the degradation of the polyurethane insulation. In one study, exposure to 0.1 M silver nitrate at 90°C for 35 days oxidized the ether linkages of polyurethanes in increasing severity for Pellethane, Tecoflex EF (Thermedics, Inc.), Biomer and Cardiothane-51 (Kontron Instruments) (Coury *et al.*, 1988). In another study, cobalt was found to cause bulk degradation of Pellethane by what was described as an auto-oxidative mechanism (Stokes *et al.*, 1988).

The role of anions in the degradation of polyurethanes continues to be studied. It has been reported that the chloride anion has much more effect on polyurethane degradation than the acetate anion (Thoma *et al.*, 1988). Cholesterol and lipids may also contribute to polyurethane degradation as evidenced by the decrease in fatigue life of polyurethanes in solutions of these chemicals (Hayashi, 1988).

Thus, environmental stress cracking of polyurethanes is a continuing research issue. The relative contribution of all the above-mentioned factors has not been determined and is likely to be different with differing polyurethanes and environments.

Environmental Stress Cracking—Polysulfone

Another polymer sensitive to environmental stress cracking is polysulfone, a high-strength engineering thermoplastic. In the medical field, polysulfone is used in the manufacture of respirators, nebulizers, prosthesis packing, dental tools, and sterilizer trays. It has also been proposed for use in *in vivo* biomedical applications, including use as a coating and matrix material for prototype composite hip implants, as a porous coating on prototype metallic orthopedic implants, and as the rigid housing for artificial hearts.

An engineering design package from a supplier of medical-grade polysulfone contains 15 pages of tables on the environmental stress rupture of polysulfone in contact with many different substances.[1] Their general conclusion is that, if possible, polysulfone should be used in contact only with nonsolvents. When polysulfone must be in contact with a partial solvent, the polysulfone object must be virtually free of both

---

[1]Udel Design Engineering Data, Section 5—Chemical and Solvent Resistance, Union Carbide Engineering Polymers, Danbury, CT. Note: Udel polysulfone is currently being marketed by Amoco Performance Products, Ridgefield, CT.

external and internal stresses at the surface to avoid environmental stress cracking.

Asgian *et al.* (1989) reported that polysulfone test pieces stressed to 5000 to 9000 psi failed or crazed when in contact with bone marrow from the long bones of sheep or the femur of a human. On analysis, the bone marrows were found to contain 82% and 57% lipids, respectively. Control specimens of polysulfone loaded to the same stress displayed no crazing.

Residual Stresses and Stress Relief of Devices

For polymers such as polysulfone, the effect of crazing or cracking caused by environmental stress can be minimized by reducing the residual stresses that are created in a part by the molding process. The parts can be stress relieved by annealing them at a temperature below the melting temperature of the polymer.

A by-product of the annealing process can be an increase in the crystallinity of the polymer. Increased crystallinity can be either beneficial or detrimental to the properties of the final product. For example, when poly(ether ether ketone) (PEEK) is annealed, the increase in crystallinity makes the product more resistant to the uptake of solvents and increases its strength, both of which are good for most aerospace applications. However, when poly(ethylene terephthalate) (e.g., Dacron, Mylar, Cleartuf) becomes more crystalline, it loses its clarity and much of its ultimate elongation, either of which could be important for some applications.

The stress-relieving effects of annealing can sometimes be obtained by the action of solvents that promote an increase in at least the surface crystallinity of a polymer. However, residual solvents can also be very detrimental to the properties of polymers, as in the case of polysulfone where small quantities of residual methylene chloride will initiate environmental stress crazing.

Effect of Flexing on Blood Compatibility and Calcification

Calcification of artificial heart pump diaphragms is known to occur primarily in the flexing regions of the diaphragms (Dew *et al.*, 1984). Calcified areas of polyurethane and biological tissue heart valves occur *in vivo* at the stressed areas within the cusps and along the commissures (Hennig *et al.*, 1988). Nonflexing areas of these devices show little or no calcification.

As calcified areas of the heart valves become stiffer, they can fail either by (1) loss of function (failure of the valve to open or close because of the calcified deposits); (2) mechanical failure of the elastomer due to the higher local stresses, abrasion, surface and polymer degradation caused by the calcification; or (3) nucleation of thrombosis formation by the calcified deposits. *In vivo* flexing of biomaterials can also have a direct adverse effect on the thromboresistance of polymers (McMillin *et al.*, 1988).

Bibliography

Asgian, C. M., Gilbertson, L. N., Blessing, E. E., and Crowninshield, R. D. (1989). Environmentally induced fracture of polysulfone in lipids. *Trans. Soc. Biomater.* XII: 17.

Benson, R. S., and Wong, R. P. (1988). The dynamic mechanical properties of the blood diaphragm of the JARVIK-7 TAH, in *Trans. 3rd World Biomaterials Cong.*, Vol. XI, Kyoto, Japan, Business Center for Academic Societies Japan, Tokyo, p.124.

Benz, E., Sherburne, B., Hayek, J., Falchuk, K., Godleski, J. J., Sledge, C. B., and Spector, M. (1990). Migration of polyethylene wear debris to lymph nodes and other organs in total joint replacement patients. *Trans. Soc. Biomater.* XVII: 83.

Bruner, R. J., Merritt, K., Brown, S. A., and Kraay, M. J. (1993). Metal sensitivity in patients undergoing total joint revision. *Trans. Soc. Biomater.* XVI: 219.

Chawla, A. S., Blais, P., Hinberg, I., and Johnson, D. (1988). tion of explanted polyurethane cardiac pacing leads and of polyurethane. *Biomater. Artif. Cells. Artif. Org.* 16(4): 785–800.

Coury, A., Cahalan, P., Halverson, E., Slaikeu, P., and Stokes, K. (1988). Oxidation resistance of implantable polyurethanes, in *Trans. 3rd World Biomaterials Cong.*, Vol. XI, Kyoto, Japan, Business Center for Academic Societies Japan, Tokyo, p. 432.

Daniels, A. U., Smutz, W. P., Andriano, K. P., Chang, M. K. O., and Heller, J. (1989). Dynamic environmental exposure testing of biodegradable polymers. *Trans. Soc. Biomater.* XII: 74.

Dew, P. A., Olsen, D. B., Kessler, T. R., Coleman, D. L., and Kolff, W. J. (1984). Mechanical failures in *in vivo* and *in vitro* studies of pneumatic total artificial hearts. *Trans. Am. Soc. Artif. Internal. Organs.* XXX: 112–116.

Dowling, J. M. (1983). Wear analysis of retrieved prostheses, in *Biocompatible Polymers, Metals, and Composites,* M. Szycher, ed. Technomic Publ., Lancaster, PA. p. 407.

Hawkins, M. V., Zimmerman, M. C., Parsons, J. R., Langrana, N. A., and Lee, C. K. (1993). Environmental effect on a tic elastomer (TPE) for use in a composite intervertebral disc spacer. in *Composite Materials for Implant Applications in the Human Body,* R. D. Jamison and L. N. Gilbertson, eds., ASTM STP 1178, American Society for Testing and Materials, Philadelphia, pp. 17–26.

Hayashi, K. (1987). Biodegradation of implant materials. *JSME ternat.* 30:(268) 1517–1525.

Hayashi, K. (1988). Effects of environmental temperature and cyclic rate on the fatigue properties of segmented polyether polyurethanes. in *Trans. 3rd World Biomaterials Cong.*, Vol. XI, Kyoto, Japan, Business Center for Academic Societies Japan, Tokyo, p. 123.

Hayden, H. W., Moffatt, W. G., and Wulff, J. (1965). *The Structure and Properties of Materials: Mechanical Behavior.* Wiley, New York.

Hench, L. L. (1981). Stability of ceramics in the physiological ment. in *Fundamental Aspects of Biocompatibility.* D. F. Williams, ed. CRC Press, Boca Raton, FL, Vol. 1, pp. 67–85.

Hennig, E., John, A., Zartnack, F., Lemm, W., Bucherl, E. S., Wick, G., and Gerlach, K. (1988). Biostability of polyurethanes. *Int. J. Artif. Organs* 11(6): 416–427.

Lhymn, C. (1987). Effect of environment on the two-body abrasion of polyetheretherketone (PEEK)/carbon fiber composites. *ASLE Trans.* 30 3: 324–327.

Luedemann, R. E., and Bundy, K. J. (1989). The effect of dynamic stress on the corrosion characteristics of surgical implant alloys. *Trans. Soc. Biomater.* XII: 56.

McKellop, H. A., Schmalzried, T., Park, S. H., and Campbell, P. (1993). Evidence for the generation of sub-micron polyethylene wear particles by mirco-adhesive wear in acetabular cups. *Trans. Soc. Biomater.* XVI: 184.

McMillin, C. R., Malladi, M. R., Ott, D. W., Evancho, M. M., and Schmidt, S. P. (1988). Cellular morphology and distribution on a stretching blood-material interface. *J. Biomed. Mater. Res.* 22: 339–351.

Phua, S. K., Castillo, E., Anderson, J. M., and Hiltner, A. (1987). Biodegradation of a polyurethane *in vitro*. *J. Biomed. Mater. Res.* 21: 231–246.

Ratner, B. D., Gladhill, K. W., and Horbett, T. A. (1988). Analysis of *in vitro* enzymatic and oxidative degradation of polyurethanes. *J. Biomed. Mater. Res.* 22: 509–527.

Sharkness, C. M., Acosta, S. K., More, R. M. Jr., Hamburger, S., and Gross, T. P. (1993). Metallic orthopedic implants and their possible association with cancer. *J. Long-Term Effects of Medical Implants* 3(3): 237–249.

Smith, R., Williams, D. F., and Oliver, C. (1987). The biodegradation of poly(ether urethanes). *J. Biomed. Mater. Res.* 21: 1149–1166.

Stokes, K. (1984). Environmental stress cracking in implanted polyether polyurethanes. in *Polyurethanes in Biomedical Engineering*, H. Planck, G. Egbers, and I. Syrè, eds. Elsevier, Amsterdam, pp. 243–255.

Stokes, K., Urbanski, P., and Coury, A. (1988). *In vivo* autooxidation of polyester polyurethanes in the presence of metals: Interim results. in *Trans. 3rd World Biomaterials Cong.*, Vol. XI, Kyoto, Japan, Business Center for Academic Societies Japan, Tokyo, p. 434.

Sung, P. and Fraker, A. C. (1987). Corrosion and degradation of a polyurethane/Co-Ni-Cr-Mo pacemaker lead. *J. Biomed. Mater. Res.* 21(3A): 287–297.

Thoma, R. J., Phillips, R. E., and Tan, F. R. (1988). In-vitro studies of polyurethane degradation, in *Trans. 3rd World Biomaterials Cong.*, Vol. XI, Kyoto, Japan, Business Center for Academic Societies Japan, Tokyo, p. 430.

Wilfong, D. L. (1989). Effects of sorbed caprolactam on the crystallinity, morphology, and deformation behavior of polyetheretherketone and poly(phenylene sulfide). *Polymer Composites* 10(2): 92–97.

Willert, H. G., Bertram, H., and Buchhorn, G. H. (1990a). Osteolysis in alloarthroplasty of the hip—The role of bone cement fragmentation. *Clin. Orthop. Rel. Res.* No. 258, 108–121.

Willert, H. G., Bertram, H., and Buchhorn, G. H. (1990b). Osteolysis in alloarthroplasty of the hip—The role of ultra-high molecular weight particles. *Clin. Orthop. Rel. Res.* No. 258, 95–107.

Williams, D. F. (1981). Electrochemical aspects of corrosion in the physiological environment. in *Fundamental Aspects of Biocompatibility*, D. F. Williams, ed. CRC Press, Boca Raton, FL, Vol. 1, pp. 11–42.

6.5   PATHOLOGIC CALCIFICATION OF BIOMATERIALS

Yashwant Pathak, Frederick J. Schoen, and Robert J. Levy

The failure of certain clinical devices, particularly in the cardiovascular system, is frequently caused by the formation of nodular deposits of calcium phosphate or other calcium-containing compounds, a process known as calcification or mineralization. Although deposition of mineral salts of calcium occurs as a normal process in bones and teeth, the biomaterials that comprise medical devices are not intended to calcify since mineral deposits can interfere with function. Therefore, calcification of biomaterials is abnormal or pathologic. Pathologic calcification can be either dystrophic or metastatic. Dystrophic calcification is the deposition of calcium salts in damaged or diseased tissues or biomaterials in individuals with normal calcium metabolism. In contrast, metastatic calcification is the deposition of calcium salts, in previously normal tissues, as a result of deranged mineral metabolism (usually elevated blood calcium levels). Dystrophic and metastatic calcification can be synergistic; in the presence of abnormal mineral metabolism, calcification associated with biomaterials or injured tissues is enhanced.

Calcification of biomaterials can affect a variety of prostheses implanted into the circulatory system, within connective tissues, or at other sites (Table 1). For example, dystrophic calcification has been encountered as degeneration of bioprosthetic or homograft cardiac valve replacements, calcification in blood pumps used as cardiac assist devices, mineralization of intrauterine contraceptive devices, encrustation of urinary prostheses, and mineral deposition within soft contact lenses.

Calcification has been associated with both synthetic and biologically derived biomaterials. The mature mineral phase of most biomaterial calcifications is a poorly crystalline calcium phosphate, known as apatite, which is related to calcium hydroxyapatite, the mineral that provides the structural rigidity for bone and has the chemical formula $Ca_{10}(PO_4)_6(OH)_2$. In general, the determinants of biomaterial mineralization includes factors related to both host metabolism and implant structure and chemistry (Fig. 1). Furthermore, mineralization of a biomaterial is generally enhanced at the sites of intense mechanical deformations, such as the flexing points in circulatory devices. Calcification can be potentiated in the presence of implant infection. Of additional importance is the fact that calcification may occur on the surface of an implant (extrinsic calcification), where it is often associated with attached tissue or cells, or within the structural components (intrinsic calcification). Finally, the underlying mechanism of pathologic calcification shares many similarities with that of normal bone mineralization. In fact, some investigators have attempted to exploit the calcification of biomaterial implants to create new hard tissue.

CALCIFICATION OF PROSTHESES AND DEVICES

Calcification of Bioprosthetic Heart Valves

Calcific degeneration of glutaraldehyde-pretreated bioprosthetic heart valves (Fig. 2) is the most important example of a clinically significant dysfunction of a medical device due to biomaterial calcification. Glutaraldehyde-preserved (i.e., cross-linked) stent-mounted bioprostheses fabricated from porcine aortic valve, or related devices fabricated from bovine pericardial tissue, have been implanted in hundreds of thousands of patients since 1971. However, more than half of such valves implanted in patients fail within 12–15 years. Nearly all bioprosthetic valves retrieved at reoperation have tears or stiffening, or both, resulting from intrinsic calcification. Furthermore, accelerated mineral accumulation leading to valve failure in less

TABLE 1 Prostheses and Devices Affected by Calcification of Biomaterials

Configuration	Biomaterial	Result
Cardiac valve prostheses	Glutaraldehyde-pretreated porcine aortic valve or bovine pericardium Polyurethane	Valve obstruction or incompetency
Cardiac ventricular assist system bladders	Polyurethane	Dysfunction by stiffening or leaking
Vascular grafts	Aortic homografts Dacron grafts	Graft obstruction or stiffening
Soft contact lens	Hydrogels	Opacification
Intrauterine contraceptive devices	Silicone rubber, or polyurethane containing Cu^{2+} or other agents	Birth control failure by dysfunction or expulsion
Urinary prostheses	Silicone rubber or polyurethane	Incontinence and/or infection

than 4 years is almost uniform in adolescent and preadolescent children with bioprostheses. Calcification is most pronounced at the flexure regions of the cusps, the points of greatest functional valve stresses.

Homograft valves are valves that are removed from a person who has died, and then are sterilized, often further preserved, and implanted in another individual. Human aortic homograft valves have been in widespread clinical use for the past 20 years in reconstructive surgery for congenital cardiac defects or acquired aortic valve disease, or as vascular conduits with the valve removed. A homograft aortic valve is surrounded by a sleeve of aorta. Many homografts, whether containing aortic valves or nonvalved (to replace a large blood vessel), undergo calcification and/or degeneration in the aortic wall. Homograft calcification has been noted regardless of the type of sterilization procedure used. Aortic homograft calcification can lead to prosthesis failure as a result of either valve dysfunction or deterioration of the aortic wall due to mineral deposits.

Calcification of Polymeric Bladders in Blood Pumps

The deposition of calcific crystals on flexing bladder surfaces limits the functional longevity of blood pumps composed of polyurethane and used as ventricular assist systems or total artificial hearts. Thus far, calcification has only been noted in experimental animals. Rigidity caused by the mineral deposits can result in deterioration of pump or valve performance through loss of bladder pliability or the initiation of tears, or both. Blood pump calcification, regardless of the type of polyurethane used, generally predominates along the flexing areas of the diaphragm, either at the diaphragm housing junction or at flexing areas of the pusher plate-type or cylindrical diaphragm, indicating that mechanical factors play an important potentiating role.

Calcification of blood pump components can be intrinsic or extrinsic. The calcific deposits are frequently associated with microscopic surface defects, perhaps originating during bladder fabrication or resulting from environmental stress or mechanical effects, leading to cracking. It is hypothesized that these surface defects and subsurface voids can serve as preferred sites for calcific deposits in smooth polymer surfaces, but this is not yet proven. If degraded blood cells and their subcellular components trapped in these microscopic cavities become calcified, it is termed "extrinsic calcification." However, some calcific deposits occur below the surface in the absence of defects (clearly intrinsic) or within the adherent layer of deposited proteins and cells (pseudointima) on the blood-contacting surface (extrinsic).

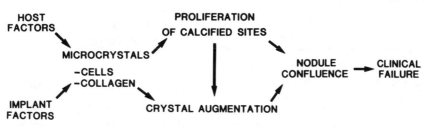

FIG. 1. Hypothesis for calcification of clinical bioprosthetic heart valves emphasizing relationships among host and implant factors, nucleation and growth of calcific nodules, and clinical failure of the device. (Reproduced with permission from F. J. Schoen *et al.*, *Lab. Invest.* 52: 531, 1985.)

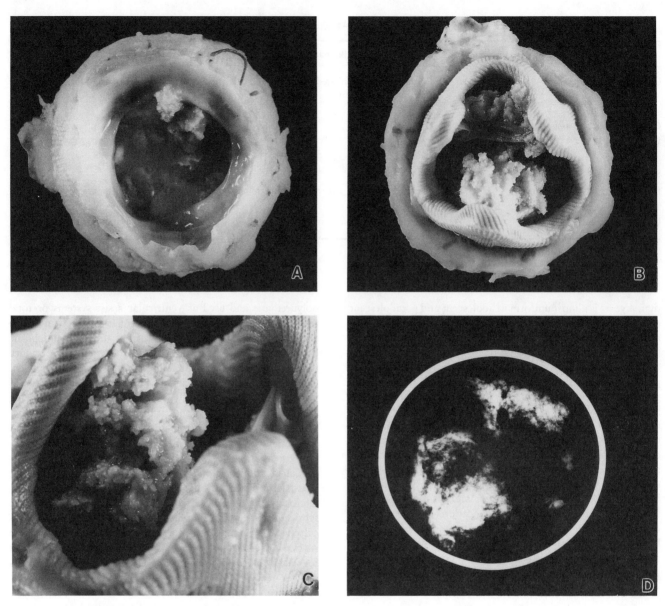

FIG. 2.   Calcified clinical porcine bioprosthetic valve removed because of extensive calcification, causing stenosis. (A) Inflow surface of valve. (B) Outflow surface of valve. (C) Closeup of large calcific nodule ulcerated through cuspal surface. (D) Radiograph of valve illustrating radio-opaque, dense calcific deposits.

Calcification on Contraceptive Intrauterine Devices

Calcification of contraceptive intrauterine devices (IUD) has been suspected of causing dysfunction leading to either detachment of the device or contraceptive failure. Accumulation of calcific plaque on copper-containing IUDs can prevent the release of ionic copper, thereby altering the effectiveness of the devices, since copper release is needed to prevent conception. Similarly, calcific deposits that interfere with the release of an active agent may also be a problem with hormone-releasing IUD systems. Studies of explanted IUDs, using transmission and scanning electron microscopy coupled with X-ray micro-

probe analysis, have shown that virtually all devices examined have varying amounts of surface calcium deposition. IUDs that function for longer periods have greater amounts of deposition, although there is lesser accumulation on devices that release conception-preventing hormones.

Calcification on Soft Contact Lenses

Calcium phosphate deposits lead to the opacification of soft contact lenses, which are composed of poly-2-hydroxyethy-methylacrylate (HEMA). Calcium from tears is considered to be the source of the deposits found on HEMA contact lenses.

TABLE 2 Methods for Assessing Calcification

Technique	Sample preparation	Analytical results
Morphologic procedures		
Gross examination	Gross specimen	Overall morphology
Radiographs	Gross specimen	Calcific distribution
Light microscopy–von Kossa or alizarin red	Formalin or glutaraldehyde fixed	Microscopic calcium distribution
Transmission electron microscopy	Glutaraldehyde fixed	Ultrastructural mineral
Scanning electron microscopy with electron microprobe	Glutaraldehyde fixed	Elemental localization
Electron energy loss spectroscopy	Glutaraldehyde fixed or rapidly frozen	Elemental localization (highest sensitivity)
Chemical procedures		
Atomic absorption	Ash or acid hydrolyzate	Bulk Ca^{2+}
Colorimetric phosphate analysis	Ash or acid hydrolyzate	Bulk PO_4^{3-}
X-ray diffraction	Powder	Crystal phase
Infrared spectroscopy	Powder	Carbonate mineral phase

Calcific deposits grow progressively larger with time and are virtually impossible to remove without destroying the lens. Calcification of soft lens is often associated with elevated tear calcium levels. However, at present, there is an incomplete understanding of the causes of this calcification. No soft lens material is yet available which is not susceptible to this problem, although the polyglycerylmethyl methacrylate lens has been shown to fare better than the HEMA soft lens. Similarly, improved assessment of systemic and ocular conditions may also help in identifying patients at risk of early lens failure, and improvements in lens materials may result in materials less likely to calcify.

Encrustation of Urinary Prosthesis

The increased use of polymeric prostheses to alleviate urinary obstruction or incontinence has drawn attention to the frequent formation of mineral crusts on the surfaces of these devices. This problem has been observed in prostheses used in either male or female urethral implants, as well as in artificial ureters. Encrustation can lead to obstruction of urinary devices and thus has been associated with device failure. The mineral crust consists of either hydroxyapatite or struvite, an ammonium- and magnesium-containing phosphate mineral. Although there is some evidence that encrustation may be enhanced by the initial presence of a bacterial infection, this has not been established. Conversely, encrustation may predispose the implant to bacterial infection.

ASSESSING CALCIFICATION OF BIOMATERIALS

Calcific deposits are investigated by using a combination of chemical and morphologic techniques. Morphologic techniques help to delineate the microscopic and ultrastructural distribution of the calcific deposits (Table 2), and chemical techniques serve to identify and quantitate both the bulk mineral constituents and crystalline mineral phase. Some morphologic techniques also allow site-specific quantitation.

Morphologic Evaluation

The morphologic distribution of calcification has been analyzed with a number of readily available and well-established techniques that range from direct gross examination or X-rays of explanted prostheses to sophisticated electron energy loss spectroscopy. Each technique has unique advantages; by combining several approaches it is possible to attain a comprehensive understanding of the structure and composition of each type of calcification.

Following careful gross examination, often under a dissecting (low power) microscope, explant X-rays (Fig. 2) have been used to assess the distribution of mineral in explanted bioprosthetic heart valves and ventricular assist systems. The technique utilized typically involves placing the explanted prostheses on an X-ray film plate and exposing the specimen to an X-ray beam at an energy level of 35 keV for 1 min in a special X-ray device used for small samples. Mineralization is evident on the roentgenograms as bright radiolucent images.

Light microscopy of calcified tissues is widely used. Identification of the mineral is facilitated through the use of either calcium- or phosphorus-specific stains, such as alizarin red (calcium) or von Kossa (phosphates) (Fig. 3). Both of these histologic stains are readily available and can be easily applied to either paraffin or plastic-embedded tissue sections. These stains are most useful for confirming and characterizing suspected calcified areas noted by routine hematoxylin and eosin staining techniques.

Electron microscopic techniques, which involve the bombardment of the specimen with a highly focused electron beam, have much to offer in the characterization of cardiovascular calcifications. Scanning electron microscopy, which images by backscattering of the electron beam, can be coupled with elemental localization by energy-dispersive X-ray analysis

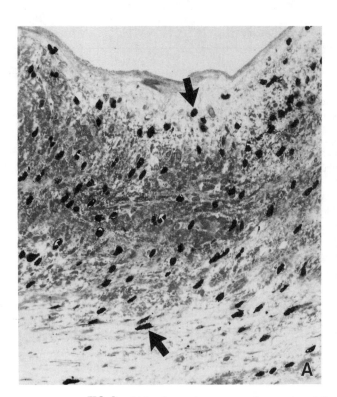

FIG. 3. Light microscopic appearance of progressive calcification of experimental porcine aortic heart valve tissue implanted subcutaneously in 3-week-old rats, demonstrated by specific staining. (A) 72-hr implantation illustrating initial discrete deposits (arrows). (B) 21-day implant demonstrating early nodule formation (arrow). Both stained with von Kossa stain (calcium phosphates black). (A) ×356; (B) ×190. (Reproduced with permission from F. J. Schoen *et al.*, *Lab. Invest.* 52: 526, 1985.)

(EDXA), allowing a semiquantitative evaluation of the local progression of calcium and phosphate deposition in a site-specific manner. Transmission electron microscopy analysis (Fig. 4) of the ultrastructure of calcifications also facilitates the understanding of the initial sites of calcific crystals. This has been widely used to investigate calcification of bioprosthetic heart valves. The electron microprobe or, more recently, electron energy loss spectroscopy (EELS) are techniques that couple transmission electron microscopy with highly sensitive elemental analyses. These techniques provide the most powerful means for localizing the ultrastructural nucleation sites of calcium phosphate deposits. In general, the more highly sensitive and sophisticated morphologic techniques require more demanding, careful, and expensive preparation of specimens to avoid artifacts.

Chemical Assessment

It is important to quantitate calcium and phosphorus in biomaterial calcifications in order to make relevant comparisons in terms of severity of deposition and the effectiveness of preventive measures. The techniques that have been used for assaying bulk mineral as well as mineral phase development in bone may be useful for biomaterials. Specimens for mineral analyses are best prepared in a uniform powder by either milling them in liquid nitrogen or finely mincing freeze-dried speci-

mens. Calcium has been quantitated by using atomic absorption spectroscopy of acid-hydrolyzed tissues. Ashing samples in a muffle furnace is also another acceptable means for preparing explanted materials for calcium and phosphate analyses. Phosphorus, as phosphate, is usually quantitated using a molybdate complexation technique with spectrophotometric detection.

As mentioned earlier, crystalline calcium phosphates, such as hydroxyapatite, make up the actual mineral deposits. The type of crystalline form of calcium phosphate (mineral phase) can be determined by X-ray diffraction. Powdered specimens are also useful for mineral phase analyses by X-ray diffraction. Powdered analyses, using a Debye–Scherer camera, require minimal amounts of material and provide an X-ray diffraction pattern produced by rotating the specimen in a capillary tube through an X-ray beam in the presence of a photographic emulsion. In addition, samples may also be analyzed for carbonate-containing mineral phases using infrared spectroscopy.

OVERVIEW OF EXPERIMENTAL MODELS AND PATHOPHYSIOLOGY

General Considerations

Common aspects of the various biomaterial-associated calcifications include such features as the mineralization of preex-

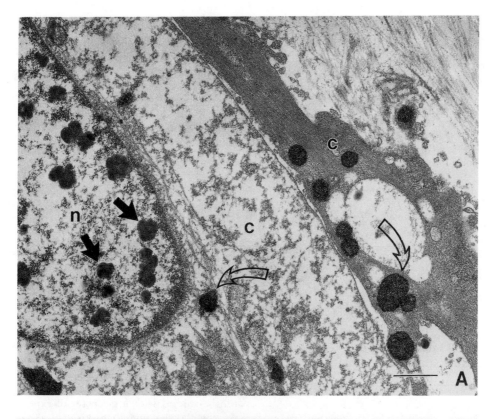

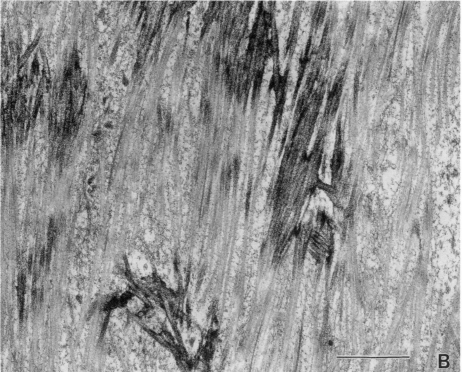

FIG. 4. Transmission electron microscopy of calcification of experimental porcine aortic heart valve implanted subcutaneously in 3-week-old rats. (A) 48-hr implant demonstrating focal calcific deposits in nucleus of one cell (closed arrows) and cytoplasm of two cells (open arrows), n, nucleus; c, cytoplasm. (B) 21-day implant demonstrating collagen calcification. Bar = $2\mu m$. Ultrathin sections stained with uranyl acetate and lead citrate. (Figure 4A reproduced with permission from F. J. Schoen *et al., Lab. Invest.* **52:** 521, 1985.)

TABLE 3 Experimental Models of Calcification

Type	System	Duration
Calcification of bioprosthetic heart valve	Rat subdermal implant	3 weeks
	Calf or sheep orthotopic valve replacement	5 months
Calcification of polyurethane	Calf or sheep artificial heart implant	5 months
	Trileaflet polymeric valve implant in calf or sheep	5 months
	Rat subdermal implants	2 months
Calcification of hydrogel	Rat subdermal implant	3 weeks
Calcification of collagen	Rat subdermal implant	3 weeks
Urinary encrustation	*In vitro* incubation	Hours to days
	In vivo bladder implants (rats and rabbits)	10 weeks

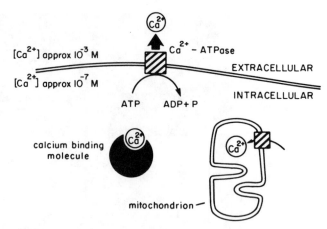

FIG. 5. Maintenance of the physiologic gradient of free calcium across the cell membrane is depicted as an energy-dependent process. With cell death or membrane dysfunction, calcium phosphate formation can ensue within cellular structures. (Reproduced with permission from Schoen, F. J. *et al.*, *J. Appl. Biomater.* **2**: 29, 1988.)

isting, adsorbed, or imbibed devitalized cell membranes; the associated deposition of various calcium-binding proteins; and the apparent enhancement of calcification by mechanical stress. All of these features are also comparable with essential aspects of bone mineralization. Cell and membrane-oriented calcification may be due in part to the high concentrations of phosphorus normally present in cellular membranes and nucleic acids. In addition, cell membranes contain high levels of the enzyme, alkaline phosphatase, which is thought to play a role in initiating normal skeletal mineralization. Calcium-binding proteins containing calcium-binding amino acids, such as γ-carboxyglutamic acid or aminomalonic acid, are present in both mineralized bone and pathologic calcifications and may also play a role in initiating mineral deposition. Stress-enhanced deposition may occur as a result of enhanced cell injury and membrane disruption at high stress sites.

Cardiovascular Calcification

Animal models have been developed for the investigation of the calcification of bioprosthetic heart valves, aortic homografts, cardiac assist devices, and trileaflet polymeric valves (Table 3). Experimental circulatory implants of each type develop calcification. Expense and technical complexity pose important limitations to all the circulatory models, since these model systems require the use of large animals such as sheep or calves, and implantation must be carried out by using complex techniques such as cardiopulmonary bypass. However, calcification of experimental bioprosthetic heart valves has also been investigated with subdermal (under the skin) implant models in a variety of species, including rats, rabbits, and mice. Subder-

mal bioprosthetic implants have demonstrated calcification pathology comparable to that seen in circulatory explants, with the advantages that (1) calcification is markedly accelerated; (2) the model is sufficiently economical so that many specimens can be studied with a given set of experimental conditions, thereby allowing quantitative statistical comparisons; and (3) the specimens are rapidly retrieved from the experimental animals, thereby allowing the rapid processing required for sophisticated morphological analyses. Polyurethane calcification has also been studied with subdermal implants in rats. These implants demonstrate calcific deposits comparable to those noted in the bloodstream.

Pathophysiology of Calcification of Experimental Bioprosthetic Heart Valves

Subdermal and circulatory implants of bioprosthetic heart valve tissue in animal models have elucidated the pathophysiology of calcification that occurs in an important clinical implant system. This work has aided our understanding of pathologic calcification in general. Calcification of the two principal types of biomaterials used in bioprostheses, namely, glutaraldehyde-pretreated porcine aortic valves or glutaraldehyde-pretreated bovine pericardium, is virtually identical. Calcification in bioprosthetic heart valves proceeds as an interaction of host and implant factors. The most important host factor is young age, with more rapid calcification taking place in immature animals. The most important implant factor is the glutaraldehyde pretreatment of the bioprosthetic tissue. It has been hypothesized that the cross-linking agent, glutaraldehyde, stabilizes and perhaps modifies phosphorous-rich calcifiable structures in the bioprosthetic tissue. These sites seem to be capable of mineralization upon implantation when exposed to the high calcium levels of extracellular fluid. Thus, mineralization depends on a susceptible substrate in a permissive environment.

These initial calcification sites are dead cells or cell membrane fragments (Fig. 5). Living cells normally maintain an intracellular calcium level that is 10,000 times less than that of the extracellular fluid. Cells may calcify after glutaraldehyde pretreatment because this cross-linking agent stabilizes all the phosphorous stores, but the normal mechanisms for calcium exclusion are not available in glutaraldehyde-pretreated tissue. Later in the calcification process, the structural protein, collagen, which is widespread throughout bioprosthetic tissue, calcifies as well. Initial ultrastructural calcification deposits eventually enlarge and coalesce, resulting in grossly mineralized nodules that cause the prosthesis to malfunction.

Mineralization of Urinary Implants: In Vitro and in Vivo Calcification Model Systems

Experimental systems that mimic the encrustation of urinary prosthetic implants use either an *in vitro* system with a synthetic fluid having a pH and mineral composition reasonably close to that of human urine, or bladder implants in experimental animals. Polymer surfaces to be tested are incubated in the circulating synthetic urine at 37°C and are observed for the formation of a mineral crust. This system has been particularly useful for screening possible improvements in prosthetic materials that may prevent urinary encrustation. Although this system is technically advantageous because of convenience and low cost, its questionable relevance to *in vivo* phenomena is an important disadvantage. In addition, urinary bladder implants of polymeric biomaterials in rats or rabbits, such as silicone elastomers, become encrusted. Although this *in vivo* system has not been extensively investigated, it may eventually prove to be more useful than the *in vitro* system just described.

Calcification of Hydrogels

A subdermal rat model has also been successfully used for studying the calcification of polyhydroxyethymethacrylate hydrogels used in soft contact lenses. As in bioprosthetic heart valves, young age potentiates mineralization of this system. Significant differences were observed in the rate and amount of hydrogel calcification between growing (6 weeks old) and adult rats (>12 weeks old). The onset of calcification in the adult rats was significantly delayed, and the young rats had deposits that were significantly larger than those in mature animals after the same implant period. Most important, the distribution of the calcific deposits was the same as that noted in contact lens calcification. However, little is known of the pathogenesis of this disorder.

Calcification of Collagen

Collagen-containing implants are widely used in various surgical applications, such as tendon prostheses and surgical absorptive sponges. Calcification of these types of biomaterial implants also compromises their usefulness owing to stiffening of tendons caused by calcium phosphate deposits. Calcification of collagen implants has been studied using a rat subdermal model. Pretreatment of collagen sponge implants made of purified collagen with either glutaraldehyde or formaldehyde pro-

motes their calcification, but the extent of calcification does not correlate with the degree of cross-linking.

PREVENTION OF CALCIFICATION

Three strategies for preventing calcification of biomaterial implants have been investigated: (1) systemic therapy with anticalcification agents; (2) local therapy with implantable drug delivery devices; and (3) biomaterial modifications, either by removal of a calcifiable component, addition of an exogenous agent, or chemical alteration.

Investigations of an anticalcification strategy must demonstrate not only the effectiveness of the therapy but also the absence of adverse effects. Adverse effects in this setting could include systemic or local toxicity, tendency toward thrombosis, induction of immunological effects or degradation, with either immediate progressive loss of mechanical properties or stability. A disadvantage of the systemic use of anticalcification agents for preventing pathologic calcification is side effects on bone, leading to insufficient amounts of skeletal calcium phosphates. To avoid this, recent approaches for preventing the calcification of bioprosthetic heart valves have been based on either modifying the biomaterial or coimplanting a drug delivery system adjacent to the prosthesis. While none of these methodologies is yet clinically applicable, several show promise and may be of conceptual importance (Table 4).

Detergent pretreatment of bioprosthetic heart valve leaflets prevents calcification of both rat subdermal implants and sheep circulatory implants. The mechanism of action of this type of mitigation has not been established, but it may be due to disruption or dissociation of the cells and cell-derived debris that contribute to the development of calcific lesions. Cell-membrane phospholipids, thought to be the major calcifiable component, are nearly completely removed by moderate detergent pretreatment. Nevertheless, when studied experimentally, detergent pretreatments have been shown to be only transiently effective in the circulation, perhaps due to reabsorption of phospholipids from the blood, and thus apparently only delay the onset of calcification.

The use of specific anticalcification drugs is another important approach for preventing calcification of biomaterials. Bisphosphonates are synthetic compounds that inhibit hydroxyapatite formation *in vitro* and *in vivo*. However, effective systemic therapy with bisphosphonates always results in side effects on bone. This occurs because of the similar mechanisms of calcification of bioprosthetic tissue and bone. However, systemic adverse effects caused by bisphosphonate are completely avoided by using either controlled release polymers as coimplants or by covalently linking these agents, such as aminobisphosphonate, to residual aldehyde functions on glutaraldehyde-cross-linked bioprosthetic tissue. Controlled release of ethanehydroxybisphosphonate (EHDP) using a polymer-based drug delivery system has been effective in a subdermal setting for preventing calcification of bioprosthetic heart valves with local (directly into the heart valve) drug administration, but has not yet been shown to be effective in the circulation. Controlled-release EHDP can be sustained for extrapolated dura-

TABLE 4 Prevention of Calcification

Strategy	Application	Mechanism
Detergent pretreatment	Glutaraldehyde-pretreated biomaterials	Charge modifications or phospholipid extraction
Al^{3+} or Fe^{3+} pretreatment	Glutaraldehyde-pretreated biomaterials	Restrict growth of mineral phase
Aminodiphosphonate or ethanehydroxydiphosphonate covalent bonding	Glutaraldehyde-pretreated biomaterials and polyurethane	Restrict growth of mineral phase
Diphosphonates (controlled release)	All implants	Restrict growth of mineral phase
Al^{3+} (controlled release)	All implants	Restrict growth of mineral phase
Protamine pretreatment	Glutaraldehyde-pretreated biomaterials	Charge modification

tions of more than 30 years by using formulations containing the slightly soluble calcium salt of EHDP in the matrices. Diphosphonates covalently bonded to polyurethane also prevent implant calcification.

Another potentially useful approach for preventing calcification involves the use of specific metal ions that are calcification inhibitors. Pretreatment of bioprosthetic heart valve cusps with Al^{3+} has also been shown to prevent implant mineralization without generalized side effects. This approach was suggested by the observations that patients with kidney disease, who are exposed to trace-level Al^{3+}, have insufficiently calcified bones. The mechanism of action of Al^{3+} inhibition of calcification has been shown to be due, at least in part, to the inhibition of hydroxyapatite crystal growth. Other metallic cations that have been shown to inhibit hydroxyapatite crystallization *in vitro*, such as Fe^{3+}, Ga^{3+}, and Cd^{2+}, may also be useful in preventing calcification.

Charge modification of glutaraldehyde-pretreated bovine pericardium with covalent binding of protamine has also inhibited calcification of this biomaterial experimentally. Protamine is a highly positively charged protein that may inhibit calcification by repelling calcium ions, thus blocking the formation of calcium phosphate.

CONCLUSIONS

Calcification of biomaterial implants is an important pathologic process affecting a variety of tissue-derived biomaterials as well as synthetic polymers in various functional configurations. The pathophysiology has been partially characterized with a number of useful animal models; a key common feature is the involvement of devitalized cells and cellular debris. Although no clinically useful preventive approach is yet available, several strategies based on either modifying biomaterials or local drug administration appear to be promising.

Acknowledgments

The authors thank Mrs. Catherine Wongstrom for preparing this manuscript. This work was supported in part by NIH Grants HL38118 and HL36574.

Bibliography

General Review

Anderson, H. C. (1988). Mechanisms of pathologic calcification. *Rheum. Dis. Clin. N. Am.* **14**: 303–319.

Anderson, H. C. (1989). Mechanism of mineral formation in bone. *Lab. Invest.* **60**: 320–330.

Bonucci, E. (1987). Is there a calcification factor common to all calcifying matrices? *Scanning Microscopy* **1**: 1089–1102.

Cheng, P. T. (1988). Pathologic calcium phosphate deposition in model systems. *Rheum. Dis. Clin. N. Am.* **14**: 341–351.

Levy, R. J., Schoen, F. J., and Golomb, G. (1986). Bioprosthetic heart valve calcification: Clinical features, pathobiology and prospects for prevention. *CRC Crit. Rev. Biocompatibil.* **2**: 147–187.

Mandel, N., and Mandel, G. (1988). Calcium pyrophosphate crystal deposition in model systems. *Rheum. Dis. Clin. N. Am.* **14**: 321–339.

Nicholls, D. G. (1986). Intracellular calcium homostasis. *Brit. Med. Bull.* **42**: 353–358.

Russel, R. G. G., Coswell, A. M., Hern, P. R., and Sharrard, R. M. (1986). Calcium in mineralized tissues and pathological calcification. *Brit. Med. Bull.* **42**: 435–446.

Schoen, F. J., and Levy, R. J. (1984). Bioprosthetic heart valve failure: Pathology and pathogenesis. *Cardiol. Clin.* **2**: 717–739.

Schoen, F. J., Harasaki, H., Kim, K. H., Anderson, H. C., and Levy, R. J. (1988). Biomaterials-associated calcification: pathology, mechanisms, and strategies for prevention. *J. Biomed. Mater. Res.: Appl. Biomater.* **22** A1: 11–36.

Examples of Calcification Studies

Goldfarb, R. A., Neerhut, G. J., Lederer, E. (1989). Management of acute hydronephrosis of pregnancy by uretheral stenting: Risk of stone formation. *J. Urol.* **141**(4): 921–922.

Hilbert, S. L., Ferrans, V. J., Tomita, Y., Eidbo, E. E., and Jones, M. (1987). Evaluation of explanted polyurethane trileaflet cardiac valve prostheses. *J. Thorac. Cardiovasc. Surg.* **94**: 419–429.

Khan, S. R., and Wilkinson, E. J. (1985). Scanning electron microscopy, x-ray diffraction, and electron microprobe analysis of calcific deposits on intrauterine contraceptive devices. *Human Pathology* **16**: 732–738.

Lentz, D. L., Pollock, E. M., Olsen, D. B., and Andrews, E. J. (1982). Prevention of intrinsic calcification in porcine and bovine xenograft materials. *Trans. Am. Soc. Artif. Intern. Organs* **28**: 494–497.

Levy, R. J., Schoen, R. J., Levy, J. T., Nelson, A. C., Howard, S. L., and Oshry, L. J. (1983). Biologic determinants of dystrophic calcification and osteocalcin deposition in glutaraldehyde-preserved porcine aortic valve leaflets implanted subcutaneously in rats. *Am. J. Pathol.* **113**: 142–155.

Levy, R. J., Wolfrum, J., Schoen, F. J., Hawley, M. A., Lund, S. A., and Langer, R. (1985). Inhibition of calcification of bioprosthetic heart valves by local controlled release diphosphonate. *Science* **227**: 190–192.

Schoen, F. J., Levy, R. J., Nelson, A. C., Bernhard, W. F., Nashef, A., and Hawley, M. (1985). Onset and progression of experimental bioprosthetic heart valve calcification. *Lab. Invest.* **52**: 523–532.

Schoen, F. J., Tsao, J. W., and Levy, R. J. (1986). Calcification of bovine pericardium used in cardiac valve bioprostheses. Implications for mechanisms of bioprosthetic tissue mineralization. *Am. J. Pathol.* **123**: 143–154.

Schoen, F. J., Kujovich, J. L., Webb, C. L., and Levy, R. J. (1987). Chemically determined mineral content of explanted porcine aortic valve bioprostheses: correlation with radiographic assessment of calcification and clinical data. *Circulation* **76**: 1061–1066.

Webb, C. L., Benedict, J. J., Schoen, F. J., Linden, J. A., and Levy, R. J. (1987). Inhibition of bioprosthetic heart valve calcification with covalently bound aminopropanehydroxydiphosphonate. *Trans. Am. Soc. Artif. Intern. Organs* **33**: 592–595.

7

Application of Materials in Medicine and Dentistry

JOHN F. BURKE, PAUL DIDISHEIM, DENNIS GOUPIL, JORGE HELLER, JEFFREY B. KANE, J. LAWRENCE KATZ, SUNG WAN KIM, JACK E. LEMONS, MIGUEL F. REFOJO, LOIS S. ROBBLEE, DENNIS C. SMITH, JAMES D. SWEENEY, RONALD G. TOMPKINS, JOHN T. WATSON, PAUL YAGER, AND MARTIN L. YARMUSH

7.1 INTRODUCTION

Jack E. Lemons

Synthetic biomaterials have been evaluated and used for a wide range of medical and dental applications. From the earliest uses (~1000 B.C.) of gold strands as soft tissue sutures for hernia repairs, silver and gold as artificial crowns, and gemstones as tooth replacements (inserted into bone and extending into the oral cavity), biomaterials have evolved to standardized formulations. Since the late 1930s, high-technology polymeric and ceramic substrates have played a central role in expending the application of biomaterial devices.

Most students enter the biomaterials discipline with a strong interest in applications. Critical to understanding these applications is the degree of success and failure and, most important, what can be learned from a careful evaluation of this history. The following chapters present topics across the spectrum of applications, ranging from blood contact and cardiovascular devices to drug delivery and sensors for diagnostic purposes. A central emphasis is the correlation of application limits with the basic properties of the various biomaterials and devices and how it might be possible to extend and improve existing applications. One goal for future applications of devices is to extend functional longevities by a factor of four (to 80 or more years) so that the need for revisions and replacements will be minimized.

More extended literature on applications can be found in the numerous books that have been written by professionals within the various fields, and more recently, by the edited versions of conferences that are available through the professional societies and government agencies. To obtain this literature, the reader is referred to the computer-based (MedLine or equivalent) methods for initial surveys within discipline areas. This will provide an extensive list of books and standard reference materials that should complement the basic information contained here.

7.2 CARDIOVASCULAR APPLICATIONS

Paul Didisheim and John T. Watson

Over the past 30 years, major advances have been made in using biomaterials to develop cardiovascular devices. Table 1 lists some of these devices and their annual use. However, the materials used in the fabrication of these devices have been primarily designed for nonmedical applications rather than specifically synthesized for medical purposes. In addition to having the required mechanical properties, both the materials and the devices made from them must be biocompatible (Didisheim and Watson, 1989; Webster, 1988). (Chapters 3.3.2., 3.3.3.), i.e., they must not provoke undesirable responses or complications during use.

Blood–Material Interactions

Blood–material interactions include any interaction between a material or device and blood or any component of blood, resulting in effects on the device or on the blood or on any organ or tissue. The effects commonly occur in various combinations, since there is considerable synergism among them. Such effects may or may not have clinically significant or undesirable consequences.

Classification of Blood–Material Interactions

A. Those which primarily affect the material or device and which may or may not have an undesirable effect on the subject.

TABLE 1 Estimated Annual Use of Some Cardiovascular Devices in the United States

Device	Implants
Heart valves	
Mechanical	32,000[a]
Bioprosthetic	20,000[b]
Pacemakers	144,000[a]
Vascular grafts	160,000[c]
Oxygenators (Cardiopulmonary bypass)	260,000[a]
Heart assist systems	
Intra-aortic balloon pumps	31,300[a]
Ventricular assist devices (pulsatile and nonpulsatile)	400[d]
Total artificial hearts	17[d]

[a]National Center for Health Statistics (1989). Vital and Health Statistics. Detailed diagnosis and procedures, National Hospital Discharge Survey, 1987. Series 13: Data from the National Health Survey No. 100. *DHHS Publ.* (PHS) 89-1761.
[b]Biomedical Business International, Tustin, CA.
[c]Frost and Sullivan, Inc., New York.
[d]W. E. Pae, C. A. Miller, and W. S. Pierce, *J. Heart Transpl.* 8: 277–280, 1989.

1. Adsorption of plasma proteins, lipids, calcium (see Chapter 3.4.4.), or other substances from the blood onto the surface of the device, or absorption of such substances into the device.
2. Adhesion of platelets, leukocytes, or erythrocytes onto the surface of the device or absorption of their components into the device.
3. Formation of pseudointima on the luminal (inner) surface of the device or formation of tissue capsule on the outer surface of the device.
4. Alterations in mechanical or other properties of the device.

B. Those which have a potentially undesirable effect on the subject.
1. Activation of platelets or leukocytes or the coagulation, fibrinolytic, immunologic, or other pathways involving blood components, including immunotoxicity (see Chapter 4.3).
2. Formation of thrombi on the device surface.
3. Embolization of thrombotic or other material from the device's luminal surface to another site within the circulation.
4. Injury to circulating blood cells resulting in anemia, hemolysis, leukopenia, thrombocytopenia, or altered function of blood cells.
5. Injury to cells and tissues adjacent to the device.
6. Intimal hyperplasia on or adjacent to the device, resulting in reduced flow or affecting other functions of the device.

In the case of cardiovascular devices, one must consider interactions not only with blood, but also with adjacent and surrounding tissues, and in addition adhesion and growth of bacteria or other infectious agents to or near the device. This chapter describes some of the cardiovascular devices currently being used or developed, the materials from which they are made, the clinical benefits, and the problems observed with their use.

CARDIOPULMONARY BYPASS

Also known as the extracorporeal circulation or heart-lung machine, the cardiopulmonary bypass (CPB) system is designed to pump unoxygenated venous blood from the right side of the heart through a synthetic oxygenator rather than through the lungs and return oxygenated blood to the systemic arterial circulation (Didisheim and Watson, 1989; Didisheim, 1993; Webster, 1988; Galletti and Brecher, 1962; Sabiston, 1991) (Fig. 1). Developed in the 1950s, CPB has

FIG. 1. Maxima hollow-fiber membrane oxygenator. Two units of different size are shown. (Medtronic, Inc., Minneapolis, MN. Courtesy Arthur J. Coury.)

enabled the extraordinary advances in open-heart surgery that have been made over the past 40 years. By using CPB, surgeons can now repair complex congenital or acquired defects involving the walls of the heart's chambers, the septa dividing the left and right sides of the heart, and the valves or vessels entering or leaving the heart. The most common of these procedures is coronary artery bypass grafting (CABG). CPB is also useful for: (1) extracorporeal membrane oxygenation (ECMO) to assist in exchange of respiratory gases in patients with pulmonary diseases, especially neonatal respiratory distress syndrome (RDS), and (2) for sustaining life after severe damage to the heart or lungs (e.g., myocardial infarctions, trauma, or pulmonary embolism). Percutaneous CPB is a recently developed technique by which, within a few minutes, partial, life-sustaining circulatory support may be initiated and maintained for several hours while the patient receives adjuvant therapy (Shawl *et al.*, 1990).

The usual components of a CPB unit include a gas exchange device (oxygenator), blood pumps, heat exchanger, ventricular vent, pericardial suction line, tubing, filters, and reservoirs. Occlusive roller pumps are commonly used because they are reliable and cause little trauma to blood. Current oxygenators are of two types: bubble and membrane. The membranes (across which oxygen diffuses and enters red blood cells) may be in the form of plates, coils, or hollow fibers.

Figure 1 shows two hollow-fiber membrane oxygenators; the membranes consist of microporous polypropylene (Celanese Celgard). Blood passes along the outside surface of the fibers, which are embedded in a cast two-part polyurethane (Biothane), and emerge with open lumens from the casting of the case. Inside the container, also in contact with blood, is a series of epoxy resin-coated aluminum tubes which act as heat exchangers.

Membrane oxygenators cause less trauma to blood than bubble oxygenators. This results in less hemolysis and protein denaturation, and better preservation of platelet count and platelet functions. Consequently, fewer transfusions of blood and blood components are required. The disadvantages of membrane oxygenators are their greater cost and complexity of operation. With bubble oxygenators, respiratory gases are mixed directly with blood, requiring a defoamer and bubble trap as integral components. Complement activation of a similar degree occurs with both types of oxygenators. Activation of the contact, coagulation, and fibrinolytic pathways (Chapter 4.5) also occurs with both systems. Clinically, the most significant hemostatic disorder caused by CPB is an abnormality in platelet functions (Chapter 4.5), which is believed to contribute to increased bleeding and the need for blood transfusions (Harker *et al.*, 1980).

The "postperfusion lung syndrome," characterized by respiratory distress following CPB, is believed to be caused by circulatory platelet or leukocyte aggregates blocking the pulmonary microcirculation. This process may be mediated by activated complement. A similar mechanism in the cerebral circulation has been suggested for the temporary changes in behavior and personality that may occur in patients following CPB. Microbubbles may also contribute. The incidence of these complications increases with prolonged duration of CPB, and microfilters may be effective in reducing their occurrence. The neutropenia commonly observed during the first 15 min of CPB may be caused by complement activation (Chapter 4.3). Certain materials such as polyacrylonitrile, polysulfone, and poly(methyl methacrylate) cause less activation of the complement system than do some of the materials commonly used in CPB components and other cardiovascular devices (Hakim, 1993; Johnson, 1989). In a related condition, called "first-use syndrome," the complement activation observed with hemodialysis using previously unused Cuprophan membranes (see Chapter 7.9) may be responsible for the respiratory distress and pulmonary leukocyte aggregation sometimes seen. In both conditions, toxic reactive oxygen species produced by activated polymorphonuclear leukocytes (PMNLs) may play a role in the cellular injury that can occur.

Silicone elastomer has been widely used as a material for fabricating oxygenator membranes because of its high permeability to respiratory gases. Microporous polypropylene is also used although its advantages with regard to improved hemocompatibility have not been established. Other materials used in various components of CPB systems include polyester (mesh in filter), acrylonitrile–styrene poolymers, polyurethane, polycarbonate, and stainless steel. Because no synthetic surface known is completely thromboresistant, systemic anticoagulation is a requirement of CPB use (as it is with hemodialysis). Although heparin in the high doses used completely prevents coagulation in the CPB device, it may have deleterious effects on platelet functions and induce their aggregation, further compromising hemostasis. Various techniques of binding heparin to surfaces have been investigated. However, heparin-coated oxygenator membranes have not been evaluated sufficiently to determine their efficacy. Prostacyclin (prostaglandin I_2, PGI_2) has been administered instead of heparin during CPB because of its ability to prevent platelet adhesion and aggregation. However, although it prevents coagulation, prostacyclin's marked hypotensive effect in the doses required has precluded its general use. Analogs of prostacyclin (e.g., Iloprost) may provide the benefits of prostacyclin while minimizing its risks of hypotension.

HEART VALVES

Prosthetic (mechanical or tissue) heart valves (Giddens *et al.*, 1993) are used to replace the natural heart valves (mitral, tricuspid, aortic, and pulmonic) when these no longer perform their normal functions because of disease. Prosthetic heart valves were first used in humans in 1960. Dramatic functional improvement is usually observed with replacement of abnormal heart valves.

Mechanical Valves

Mechanical valves are of two types: caged ball and tilting disk. The materials most widely used are silicone elastomer, cobalt–chrome-based alloys, titanium, and pyrolytic carbon. The moving component or "occluder" (caged ball or tilting

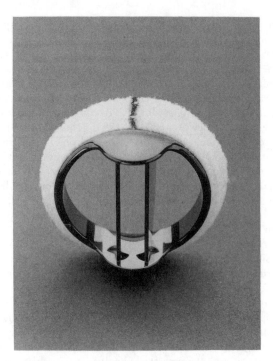

FIG. 2. St. Jude Medical bileaflet mechanical heart valve. (St. Jude Medical, Inc., St. Paul, MN. Courtesy John H. Wang.)

disk) responds passively to changes in pressure and flow within the heart. The incidence of thromboembolism is 2–5% per patient year even with adequate anticoagulant treatment. Continuous anticoagulant therapy is required, resulting in a significant risk of bleeding. The Starr–Edwards caged ball valve, in which the sewing ring is cloth covered to reduce the incidence of thromboembolism, has a record of excellent durability for up to 20 years of use. Silicone elastomer has been used for the poppet or ball in caged valves.

The Bjork–Shiley valve was the first clinically available tilting-disk valve and has undergone several significant design changes since its introduction in 1969. The struts (disk retainers) and orifice ring are made of cobalt–chrome-based alloy, the occluder (disk) of pyrolytic carbon, and the sewing (outer) ring of expanded poly(tetrafluoroethylene) (ePTFE) fabric. Other valves that have been marketed in the United States include the Medtronic–Hall tilting-disk valve, the Omniscience tilting-disk valve, the St. Jude Medical bileaflet valve (Fig. 2), and the Duromedics bileaflet valve. Pyrolytic carbon is used for the occluder and orifice ring of the bileaflet valves and for the occluder of the other valves. Dense, isotropic carbon deposition at temperatures between 900°C and 1500°C has been found useful for various medical applications. For mechanical heart valves, silicon is usually codeposited with carbon to increase strength and wear resistance.

Some valve designs allow the cage–strut structure and orifice to move as a unit within the surgical sewing ring. This feature enables the surgeon to provide the patient with the most advantageous occluder orientation. Spontaneous rotation

or perivalvular leaks are rare complications of this aspect of the design. Strut fracture is a rare but potentially catastrophic complication of any strut-bearing valve.

Tissue Valves

Tissue valves (bioprosthetic valves) are made entirely or partly of materials of biologic origin and are classified as homografts (prepared from human cadavers—allografts) or heterografts (prepared from tissues of another species—xenografts). Homograft valves are obtained at autopsy from patients without valvular disease, and are then stored in antibiotic-containing solutions or freeze-dried. Cryopreserved homograft valves have had fewer complications than mechanical or xenograft valves. However, they have not been widely used because of limited commercial availability, difficulties in obtaining valves of the size required, and the complexity of the surgical implant technique compared with xenograft or mechanical valves. The human dura mater valve is prepared from cadaver tissue that is mounted on a polyester fabric-covered stent or mold, as are pericardial xenografts. Despite the low incidence of thromboembolism and lack of endocarditis, this valve is also rarely used because it is not commercially available.

Xenografts are made of either porcine aortic valve cusps or bovine pericardium. First used in 1965, xenografts have proved less durable than mechanical valves. Deterioration is commonly observed after several years of implantation, and functional failure occurs in about 40% of patients by 10 years. Calcification is believed to be accelerated by high bending strains and leaflet deformation. Tissue valves are not normally used in children because they tend to calcify more than mechanical valves. On the other hand, these valves are less prone to thromboembolic problems (0.1–2.0% per patient year) in all age groups, and therefore anticoagulant therapy is generally not necessary. After unsatisfactory results with a variety of preservatives, glutaraldehyde has been found to yield a stable product. Several manufacturers now commercially distribute glutaraldehyde-fixed porcine valves (e.g., Carpentier–Edwards, Hancock) (Fig. 3). Glutaraldehyde appears to promote the stability of collagen cross-links and renders the valves virtually nonantigenic. The Hancock valve is mounted on a Dacron poly(ethylene terephthalate) cloth-covered flexible polypropylene strut. The Carpentier–Edwards valve is mounted on an ePTFE-covered alloy strut. Xenograft valves prepared from fascia lata to form an unstented valve or attached to cloth-covered stents were found to be prone to deterioration and are not widely used. However, stented xenografts using glutaraldehyde-cross-linked bovine pericardium (Ionescu–Shiley valves) appear to have minimal antigenicity, good hemodynamics, and low thrombogenicity. The three leaflets are mounted on a Dacron-covered titanium frame. Tethering and pressure fixation are means of preserving the tensile viscoelastic properties of aldehyde-fixed bovine pericardium for use as a xenograft.

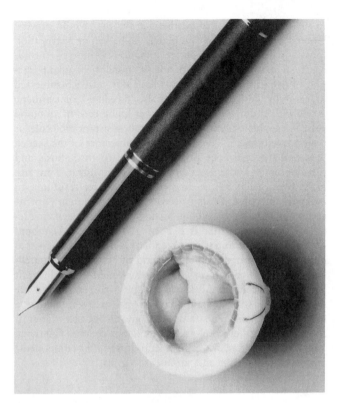

FIG. 3. Medtronic Hancock bioprosthetic (tissue) valve. (Medtronic, Inc., Minneapolis, MN. Courtesy Arthur J. Coury.)

VASCULAR GRAFTS

Biologic Grafts

Vascular grafts, like heart valves, are of either biologic or synthetic origin (Sabiston, 1991; Clowes, 1993; Kambic *et al.*, 1986). Venous autografts, introduced in 1949, are obtained from an autologous saphenous (ASV; vein (the patient's own calf vein). In terms of duration of patency and freedom from complications, they are more satisfactory than any synthetic vascular grafts, Although 20–30% of patients requiring bypass grafting do not have a suitable saphenous vein either because it is too small, unusable due to phlebitis (inflammation of a vein) or varicosities (permanently distended areas of a vein), or has already been removed, each year 300,000 patients in the United States receive a venous autograft to bypass an occluded or narrowed coronary or peripheral (usually leg) artery. Autologous saphenous vein (ASV) yields the best results in reconstructive arterial surgery below the knee, where homografts, xenografts, and synthetic grafts are unsatisfactory.

ASV has been used successfully to replace or bypass coronary, visceral, renal, and peripheral arteries; patency has been superior to that obtained with any other vessel of either tissue or synthetic origin except internal mammary artery, which is discussed later. Causes of failure of ASV are thrombosis in the first few days of grafting or neointimal hyperplasia subsequently.

Hyperplasia results from the proliferation of normally quiescent smooth muscle cells (SMC) (Chapter 4.2) (Clowes, 1993). Factors postulated to be responsible include endothelial injury associated with harvesting and handling the ASV, and the increased shear forces to which the ASV is subjected when placed in the arterial circulation. In the process of harvesting and implanting the autograft, injury and denudation of endothelial cells are difficult to avoid and result in (1) removal of the endothelial source of heparin and the related glycosaminoglycans that normally inhibit SMC proliferation, and (2) circulating blood platelets contacting and adhering to subendothelial collagen fibers and releasing intracellular components, including platelet-derived growth factor (PDGF), known to stimulate SMC proliferation. However, SMC proliferation also occurs in the absence of endothelial denudation, probably as a result of increased PDGF release from the SMCs themselves. The increased shear forces to which the ASV is exposed when implanted in the arterial system may stimulate PDGF secretion from SMC.

Internal mammary arteries, located on the anterior thoracic wall, yield even more favorable results than saphenous veins for CABG, perhaps in part because of their higher blood flow velocity owing to their smaller diameter and their higher content of nitric oxide (NO) endothelial-derived relaxing factor, EDRF. However, their usefulness is diminished by their limited accessibility and the location and number of lesions to be bypassed. The superiority of saphenous veins and internal mammary arteries to any prosthetic grafts for small-diameter arterial reconstruction is related to the veins' and arteries' lining of endothelial cells which synthesize molecules (e.g., prostacyclin, heparin, tissue plasminogen activator, NO) that inhibit thrombosis, vasoconstriction, and intimal hyperplasia (see Chapter 4.1) (Libby and Schoen, 1993).

Venous homografts were used in the 1950s but were abandoned in the early 1970s because of a high failure rate. They are antigenic and sensitize the recipient. Umbilical vein homografts were first used clinically in the 1960s. Glutaraldelyde tanning improved clinical results and significantly reduced their antigenicity. The grafts were enclosed in a polyester mesh to increase their wall strength. As femoropopliteal grafts [between femoral (thigh) and popliteal (knee) artery of the leg], they have yielded patency rates similar to those reported for ASV. However, in the femorotibial position (the distal anastomosis being below the knee), the patency rate is distinctly inferior to that of ASV. Occasional formation of aneurysms (localized dilatations) and separation of the inner lining with resultant luminal occlusion have been reported.

Arterial xenografts were used widely in the 1950s, but are rarely used now because of the high incidence of thrombosis and rupture. Subsequent modifications improved the efficacy of these grafts: this involved treatment of bovine carotid arteries with the proteolytic enzyme, ficin, which leaves a tube of collagen that is then tanned with dialdehyde. The use of this graft in humans results in a moderate fibroblastic reaction, minimal inflammation, and infrequent thrombus formation. In time, these xenografts develop a thin inner lining of host-derived fibroblasts. These grafts perform well in the aortic and iliac positions, but not below the inguinal (groin) ligament, where aneurysms form in 3–6% of grafts. Infection is also a significant problem. Patency when used as femoropopliteal bypass is about 40%, which is clearly inferior to that achieved with ASV or synthetic grafts.

Synthetic Grafts

The rationale for developing synthetic vascular grafts has been based on: (1) the need to replace or bypass large vessels such as the aorta that become weakened, aneurysmal, stenosed, or occluded; (2) the unsatisfactory performance of homografts or xenografts for this purpose; and (3) the unavailability of a suitable autologous vein for coronary and other small-diameter artery grafting.

Of the many polymers tested, ePTFE has proved to be the most satisfactory in terms of requisite tensile strength and low incidence of occlusion caused by thrombosis or excessive neointimal hyperplasia (see Chapters 4.2 and 4.5). However, these events do occur, especially with grafts of 4 mm internal diameter or less. Therefore, prosthetic vascular grafts are not suitable for use in bypass or reconstructive procedures in the coronary arterial circulation or in arteries below the knee. Possible reasons for this failure include (1) increased surface-to-volume ratio in small-diameter vessels, resulting in increased rate of activation of blood coagulation, complement, platelet, and other pathways by a surface which is less thromboresistant than autologous vein; and (2) reduced flow in small-diameter vessels, resulting in increased contact time of blood components with surfaces capable of activating these pathways. Rarely, anastomotic aneurysm or graft infection may occur.

Knitted Dacron is the most widely used prosthetic arterial graft for medium and large-diameter (>4 mm) sites. When used for aortofemoral bypass, the patency rate at 5 years is about 90%. Grafts are available in many types of weaves and knits; however, the contribution of these factors to graft patency has not been clearly established. To prevent excessive bleeding through the graft during and following implantation, knitted and woven grafts must be preclotted by forcing a small volume of the patient's blood by syringe into and through the graft's interstices prior to implantation. The clot that results is replaced by ingrowth of fibroblasts and collagen from the host. To eliminate the need for preclotting, manufacturers have impregnated the interstices with albumin or collagen. Collagen is chemotactic to fibroblasts, macrophages, and granulocytes (Chapter 4.1) and thus may enhance tissue repair.

The healing process in humans (Chapter 4.2) involves the formation of a fibrous capsule around the graft and a pseudointima (not a true intima because of a lack of endothelial cells) on the luminal surface. In contrast to other species, humans have a very limited ability to produce a true intima. Luminal healing composed of endothelial cells and SMC occurs only near suture lines. The remainder of the graft is lined primarily with fibrin, which may later be replaced or covered by fibroblasts or macrophages. Healing of a graft is aided by porosity, which permits the ingrowth of cells that take part in the healing process and of capillaries that nurture the pseudointima (Clowes, 1993).

Numerous attempts have been made to bind molecules to the luminal surface of synthetic grafts that would prevent blood coagulation (heparin), platelet adhesion or aggregation (prostacyclin or dipyridamole), enhance fibrinolysis (urokinase), or inhibit intimal hyperplasia (nonanticoagulant fraction of heparin). These attempts have not yet been subjected to carefully designed clinical trials. "Seeding" of synthetic grafts with endothelial cells (the inner lining cells of all blood vessels) capable of synthesizing thromboresistant molecules is under development. The functional competence of the seeded cells and the resistance of seeded grafts to thrombosis and pseudointimal hyperplasia have been demonstrated in animals. Endothelial cell seeding may also confer resistance to bacterial infection. Trials of seeded grafts in human patients are under way. All blood-conducting conduits have a characteristic compliance that is defined as the rate and amount of distension resulting from a standard increment of increased pressure. The role of compliance and of "compliance-matching" between the natural and the implanted vessel in graft patency is uncertain. All synthetic grafts currently used are significantly less compliant than normal arteries, so that a mismatch exists at the site of anastomosis between the graft and artery. Furthermore, graft compliance changes following implantation as various cellular and noncellular components are deposited within and around the graft.

VASCULAR ACCESS

Some patients who receive chronic intravenous therapy e.g., hemodialysis require continuous vascular access or a conduit between the skin surface and the implanted device. This represents a special challenge because the skin organ system functions to expel foreign objects by epithelial downgrowth. The epithelium forms an extensive sinus tract around the implant until it is able to reattach, thus expelling the implant. For success, a stable interface must be established between the epithelium and the device surface.

One approach is to encourage collagen ingrowth near the epithelium to provide a contact inhibition layer. This can be seen in Fig. 4 where the concept for a "skin button" has been adapted to a peritoneal dialysis catheter. The textured thermoplastic, aliphatic polyurethane (Tecoflex), used on the implant neck has carefully sized interstices (pores) for collagen ingrowth. The collagen secures the implant in place and also significantly inhibits the epithelial downgrowth by contact inhibition. The reaction products for this device, which are typical for polyurethanes, include methylene bis(cyclohexyl) diisocyanate, poly(tetramethylene)ether glycol, and 1,4 butane diol chain extender. Aliphatic polyurethanes usually do not discolor with ultraviolet light, oxygen, or age as do aromatic formulations.

DRUG ADMINISTRATION SYSTEMS

The Medtronic drug administration system consists of an implantable programmable drug pump and delivery catheter.

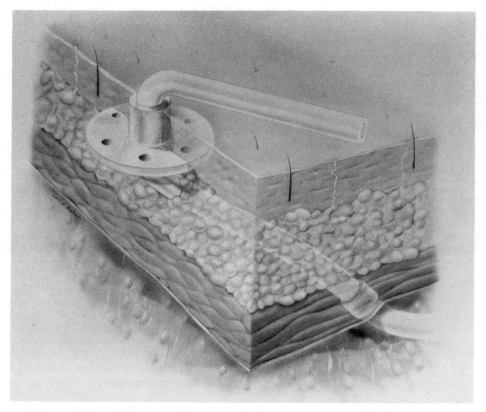

FIG. 4. "Skin button" adapted to a peritoneal dialysis catheter. (TCI Thermo Cardiosystems Inc., Woburn, MA. Courtesy Victor L. Poirier.)

The pump provides peristaltic drug delivery from a bellows-design reservoir powered by the vapor pressure of a fluid-vapor equilibrium. In addition to the titanium case (ASTM Grade 1), components include a refill septum (ETR silicone—Dow), a suture pad (MDX—Dow), and silicone elastomer catheters. When implanted, the device is sutured in place using a Dacron mesh pouch.

STENTS

A stent is a device made of inert materials and designed to serve as a temporary or permanent internal scaffold to maintain or increase the lumen of a vessel (Schatz, 1989). First introduced in the 1960s, this device has received increased attention in recent years because of the high frequency of restenosis (30–50%) following percutaneous transluminal coronary angioplasty (PTCA, balloon angioplasty). Four designs are being investigated: (1) a springlike design with an initially small diameter that expands to a predetermined dimension when a constraint is removed, (2) thermal memory stents in which the "memory metal" Nitinol is used and changes shape upon heating, and (3) balloon-expandable stents, which operate on the principle of plastic deformation of a metal beyond its elastic limit (Fig. 5), and stents made of biodegradable polymers, which serve as scaffolds for a finite period and may in addition serve as drug delivery devices as the drug-containing polymers are degraded.

Essential stent features include radial and torsional flexibility, biocompatibility, visibility by X-ray, and reliable expandability. Concerns with any stent include the injury to the vessel wall that is caused by its insertion, and the potential consequences of injury, which include acute localized thrombus formation or intimal hyperplasia (leading to restenosis); both these complications have been observed. Results of clinical trials suggest that the incidence of failure caused by thrombosis or restenosis, is lower with stents than with angioplasty.

CATHETERS AND CANNULAS

Catheters, cannulas, and related intravascular tubing are placed in virtually every portion of the arterial and venous circulation of patients. They are used for administering fluids (e.g., blood and blood products, nutrients, isotonic saline, glucose, medications, contrast medium for angiography) as well as for obtaining data (e.g., pulmonary artery pressure, gases) and withdrawing blood specimens for chemical analysis. The

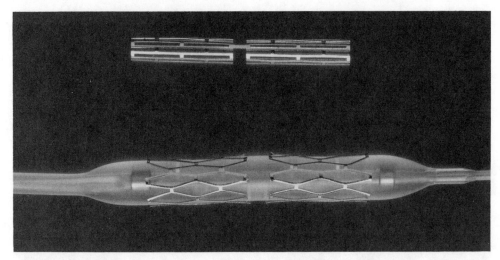

FIG. 5.   Balloon-expandable coronary artery stent. (Courtesy Richard A. Schatz, Arizona Heart Institute Foundation, Phoenix, AZ.)

large majority of these procedures are performed without incident. Rarely, catheter-related thrombosis may occur, creating a potential source for thromboembolism.

Catheterization of arteries or veins in the newborn, which is performed with increasing frequency in neonatology, is associated with an especially high risk of thrombosis and vascular occlusion, in part because of the very small diameter of the catheters and consequently the high surface-to-volume ratio, which favors thrombosis. Other factors include the flexibility and chemical composition of the catheter, activation of blood components by the toxic effects of the delivered solution on the vessel wall, and the duration of catheter placement. Polyurethane and silicone elastomer are preferred materials, but controlled clinical studies of the relative incidence of complications with different catheter materials are not available and are difficult to conduct. Injury to the vascular endothelium caused by the catheter or its contents probably plays a major role in initiating thrombosis. Several methods have been developed for binding anticoagulants, especially heparin, to the luminal surface of catheters, directly or by using a spacer between the anticoagulant and the surface to enhance function of the bound molecule. The efficacy of these methods in reducing the risk of thrombosis is not clear.

PACEMAKERS

Pacemakers (Webster, 1988; Sabiston, 1991) were developed to overcome abnormalities in heart rhythm. Approximately 500,000 persons in the United States are living with these devices. A pacemaker system consists of a pulse generator with lead connector, electrodes, and a lead wire connecting the two (Fig. 6). The pulse generator contains a hermetically sealed battery and a electronic module that communicates via a feedthrough to an external connector. It is encased in ASTM grade 1 titanium. The pulse generator is implanted beneath the skin, below the clavicle in the case of endocardial pacemakers (on the inner lining of the heart) or over the upper abdomen for epicardial pacemakers (on the outer surface of the heart). The electrodes are made of platinum, silver, titanium, stainless steel, or cobalt alloys. Most pacemakers use a lithium–iodine battery. The lack of gas evolution in these batteries (compared with previous mercury–zinc batteries) allows the pulse generator to be encased without requiring a vent. Failures from premature battery exhaustion or moisture infiltration are rare.

The lead connector is a complex subunit providing a body fluid seal for the pulse generator while permitting continuous communication between the electronics and the pacing lead. A typical connector is injection molded of polyether urethane and composed of polytetramethylene ether glycol (methylene-bis-p-phenyl isocyanate) and 1,4-butanediol. The encapsulated metal connector block is grade 2 titanium. The set screws, made of $TiAL_6V_4$ alloy with feedthrough, are insulated with single-crystal (sapphire) or polycrystalline alumina. The lead must be durable, flexible, noncorrosive, and a good electrical conductor. It must withstand repeated flexing by cardiac contractions (30–40 million cycles per year) in a warm, corrosive, salinelike medium. The material used must be as inert as possible, not subject to fatigue, and nontoxic. The bipolar and unipolar leads are secured in place by tines, anchoring sleeves, and electrode pads of polyurethane or silicone. Usually leads are composites with metal alloy conductor coils whose composition includes cobalt, chromium, molybdenum, and nickel. As an electrode, platinum meets these requirements but lacks structural strength; when alloyed 10% with iridium, it becomes stronger than most steels. Pure silver is even softer than platinum, but when alloyed 10% with palladium, it acquires the appropriate mechanical properties while retaining suitable electrochemical properties.

A steroid-loaded electrode tip contains dexamethasone so-

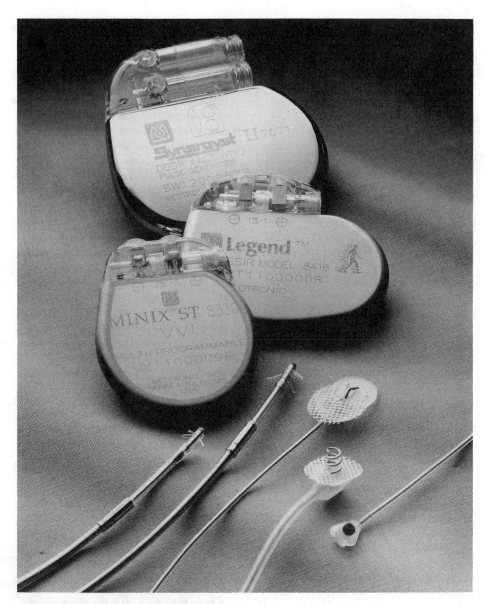

FIG. 6. Two bipolar pulse generators: the dual chamber Synergist II, the rate-responsive Legend ST miniature ventricular device. Left to right are two bipolar urethane-insulated tined endocardial leads and three myocardial leads; a stab-in, urethane-insulated lead with silicone sewing pad and Dacram mesh disk; a screw-in lead, insulated with silicone; and an epicardial urethane-insulated lead with a silicone sewing pad. (Medtronic, Inc., Minneapolis, MN. Courtesy Arthur J. Coury.)

dium phosphate in a polydimethylsiloxane matrix. The steroid acts to prevent an increase in stimulation threshold, which is caused by cell attachment and growth on the tip. Early lead sheaths were made of polyethylene; modern leads are made of more resilient materials such as silicone rubbers or, more recently, polyurethanes (Bruck and Muller, 1988). Silicone elastomer is easily damaged by a scalpel, forceps, or too tight a suture. Polyurethanes are stronger, enabling a thinner walled sheath and thinner lead to be used. Polyurethanes also have the advantage of having a very low coefficient of friction when wet. Strain relief of the polyurethane sheath (and other materi-

als) to prevent stress corrosion cracking is usually required (Bruck and Muller, 1988). Another mechanism of degradation, metal ion-induced oxidation, affects polyurethanes but not silicones. Biomaterial-related complications of pacemaker use are rare, especially when they are implanted epicardially and are not in contact with flowing blood. However, 90% of pacemaker leads are implanted endocardially, by passage up an arm vein through a transvenous approach so that the lead tip is lodged against the endocardial surface. Rarely, thromboembolic problems may occur as a result of localized endocardial injury by the lead tip.

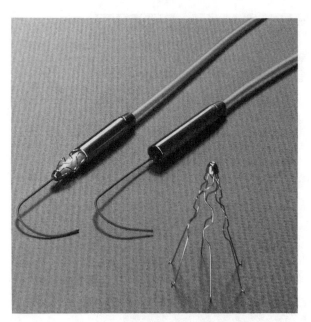

FIG. 7. Greenfield inferior vena cava filter. Left, filter loaded in catheter. Center, catheter. Right, cone-shaped filter with wires that hook into vein wall. (Courtesy Boston Scientific Corp., Medi-Tech Div., Watertown, MA.)

INFERIOR VENA CAVA FILTERS

Filters are introduced into the inferior vena cava of patients who have had or are at risk of having pulmonary embolism (Sabiston, 1991), but in whom anticoagulant therapy to prevent recurrence is contraindicated. The "umbrella filter" is made of silicone elastomer-coated stainless steel and is passed like a closed umbrella, under local anesthesia, by way of the jugular (neck) vein by means of a catheter, then opened in the inferior vena cava just below the renal vessels. Rarely, this filter has migrated to the right ventricle and pulmonary artery, causing catastrophic obstructive symptoms. Greenfield and associates' cone-shaped stainless steel filter (Fig. 7) is inserted into the inferior vena cava by way of the femoral (leg) or jugular vein and is designed to trap emboli without significantly reducing venous flow. Fixation is achieved by hooks that grasp the wall of the vein. Rare complications include recurrent pulmonary embolism from thrombi forming on the filter surface, distal migration of the filter down to the bifurcation of the inferior vena cava, and protrusion of the struts through the venous wall. The "bird's nest" filter (Raehm *et al.*, 1986) is inserted and fixed in a similar fashion; its unique feature is the dense network of interconnected, thin 302 and 304 stainless steel wires designed to trap flowing thrombi. The filter is secured to the vascular wall by hooks silver-soldered to the wires.

INTRA-AORTIC BALLOON PUMP

The intra-aortic balloon pump (IABP), introduced into clinical medicine by Kantrowitz in 1967, is the most commonly used form of mechanical circulatory assistance (Webster, 1988; Sabiston, 1991). It operates on the principle of counterpulsation: pressure pulses are cyclically applied from an external pressure source, out of phase with the heart's systolic pulses. The primary objectives are to reduce the workload of the abnormal heart and to increase coronary artery perfusion. The device consists of a cylindrical polyurethane balloon that is placed in the descending thoracic aorta by way of a femoral artery and connected by a catheter to an external cyclical pump (Fig. 8). The balloon is rapidly inflated during diastole (the period of relaxation and dilatation of the heart) in order to pump blood both up and down the aorta from the site of the device. The

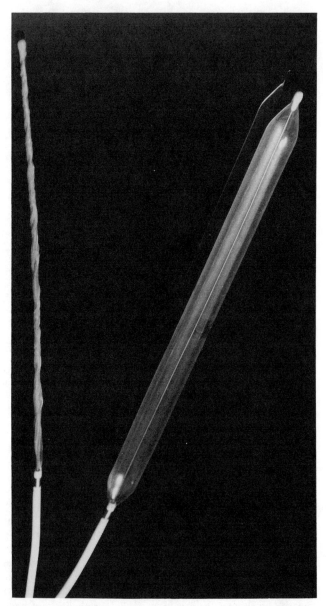

FIG. 8. Percutaneous intra-aortic balloon pump. Left, balloon deflated for insertion. Right, balloon inflated. (Datascope Corp., Oakland, NJ. Courtesy S. Wolvek.)

balloon rapidly deflates before systole (the period of contraction of the heart), reducing the resistance to ejection and allowing the heart to work more efficiently at a smaller end-systolic volume (lower wall stress). For a number of clinical situations, this reduction in workload allows recovery of normal heart function and the ability to support the patient's circulation without the balloon. It is useful in the treatment of cardiogenic shock and other cardiac disorders and in the support of patients undergoing coronary artery bypass surgery. Currently, it is used primarily to treat cardiac failure developing after cardiac surgery.

Complications following IABP use are primarily due to surgical accidents associated with the procedure rather than to device failure or bioincompatibility of materials used. While the blood-contacting surface is large, the high blood flow velocity in the aorta prevents the accumulation of critical concentrations of procoagulants that might lead to thromboembolic complications. Carbon dioxide or helium is used to inflate the balloon, and leakage of both these gases by diffusion through the balloon necessitates periodic refilling. Patients have been supported for up to a year with the IABP, although the duration is more commonly a few days.

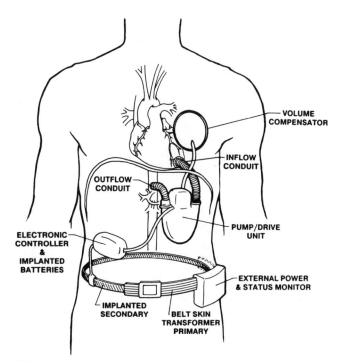

FIG. 9. Totally implantable left ventricular assist device (LVAD). (Novacor Division, Baxter Healthcare Corp., Oakland, CA. Courtesy P. M. Portner.)

VENTRICULAR ASSIST DEVICES AND TOTAL ARTIFICIAL HEARTS

Heart transplantation has become an effective form of therapy for patients with intractable heart failure; at present 1- and 5-year survival rates are approximately 80% and 70%, respectively. However, the number of hearts available for transplant, about 2000 per year in the United States, falls far short of the number of patients who develop intractable heart failure, which is estimated to be between 17,000 and 35,000 per year in the United States. This discrepancy between supply and need has stimulated research and development of both ventricular assist devices (VADs) and total artificial hearts (TAHs).

VADs and TAHs are used to replace the mechanical functions of part or all of the heart when those functions have failed irreversibly and heart transplantation is not possible. It was demonstrated in the early 1980s that patients with irreversible heart disease could survive for several months to over a year with a totally implantable (Jarvik-7) artificial heart replacing their own heart. However, these patients all succumbed to thromboembolic complications (e.g., stroke), infections in and around the device, and the failure of other organs (lungs, kidneys) (Didisheim *et al.*, 1989; DeVries, 1988). For these reasons, the clinical approach has shifted away from chronic implantation of a TAH to replace the heart, to shorter term (days to weeks) implantation of a TAH or VAD to support the patient until a donor heart becomes available for transplant ("bridge to transplant"). Improved design and better modes of antithrombotic and antibacterial therapy may allow these devices to be used for more prolonged periods. Meanwhile, research is supporting the development of improved, totally implantable TAHs intended for long-term clinical use. It is projected that these devices will be fully developed and tested during the next decade. A substantial fraction of their design features will

evolve from the clinical studies that are planned for implantable VADs.

Beyond the totally implantable artificial heart, an implantable ventricular assist system (VAS) (Fig. 9) is the most complex cardiovascular device to be implanted in humans. Table 2 lists typical biomaterials used in a VAS. The design and enabling biomaterials of the blood pumping chamber represent a substantial challenge to the bioengineer. The blood pumping must (1) be biocompatible (an objective not yet fully realized), (2) possess necessary mechanical properties (structure, flex pattern), (3) be impermeable to water, (4) allow gas transfer for barometric and altitude equilibration, (5) prevent bacteria adhesion (not yet achieved), and (6) not degrade during the useful life of the implant.

Other design factors consist of possible local tissue response to biomaterial extractables and erosion caused by device motion or vibration. Continuous relative motion between the VAS and the surrounding tissue produces inflammation and a thick fibrous encapsulation of the implant. This can lead to a site of pocket infection that is untreatable because oral antibiotics are walled off from the implant. This process must be minimized so that only a thin, stable capsule is formed. Integration of the VAS with surrounding tissue is desirable but has not yet been shown in animals or humans.

VASs designed for long-term use (months, years) have now been implanted in hundreds of patients as a bridge to cardiac transplant. Patients are routinely supported for 3–4 months, with a few VASs functioning for nearly 18 months before transplantation. These results are significant when compared to the early experience of the 112 days Dr. Barney Clark was maintained using the Jarvik-7 TAH (DeVries, 1988). These

TABLE 2   Novacor Implantable Ventricular Assist System

Device component	Device subcomponent	Biomaterial
Blood-contacting materials (Inner surface)		
Pump/drive unit	Blood pump sac	Segmented polyether, polyurethane (Biomer)
	Inflow and outflow valves	
		Porcine valve (with silicone flange)
Inflow and outflow conduits	Luminal surface	Urethane elastomer (Adiprene L-100)
		Dacron vascular graft
Tissue-contacting materials (Outer surface)		
Pump/drive unit	Encapsulation shell	Titanium (CP-1)
		Medical grade adhesive A (silicone)
		Expoxy (polyamine)
Inflow and outflow conduits	Outer surface (Graft)	Dacron vascular graft
Variable volume compensator	External reinforcement	Polypropylene
	Flexing diaphragm, connecting tube	Segmented polyether, polyurethane (Biomer), Dacron velour fabric
	Rigid housing	Titanium (6A1, 4 V)
Energy control and power unit	Hermetic encapsulation shell	Titanium (CP-1)
Belt skin transformer	Belt body	
		Silicone
		Medical grade adhesive A (silicone), Silver contacts
Interconnecting leads	Outer encapsulation	Silicone, Medical grade Adhesive A (silicone)
Special structural materials		
Pump/drive unit	Solenoid energy Converter	
		Titanium decoupling Spring (6A1, 4 V), Vanadium permendur, magnetic core, Copper coils
	Blood pump	Lightweight structural composite
Variable volume compensator	Gas-filled (Replenishable) reservoir	
Energy control and power unit	Hybrids	
	Application specific-integrated circuits, rechargeable batteries	Ni–Cad
Belt skin transformer (Secondary)	Multistrand wire	Silver, copper

patients are ambulatory and exercise regularly, and preexisting organ dysfunction is often resolved. They perform moderately in sports shooting a basketball and playing volleyball. One interesting case involved a blues musician who continued playing his guitar at concerts, while singing and dancing. These patients seem to have an excellent quality of life, manage their own care, and rapidly recover from their subsequent heart transplant operation.

The Heartmate VAS manufactured by Thermo Cardiosystems, Inc. (Woburn, MA) has demonstrated an interesting biocompatibility concept (Dasse *et al.*, 1987; Meconi *et al.*, 1995). The blood pump's blood-contacting surface is lined with sintered titanium alloy of 75- to 100-micron microspheres and the flexing silicon elastomer diaphragm with 25 × 300-micron polyurethane fibrils that form an interdigitated matrix. The design rationale is to create a surface that stimulates the development of a firmly attached fibrin layer that in time matures

into a thin, collagenous, fibroblast-lined, blood-compatible lining. Both the rigid and the flexing blood surfaces provide the anchoring function.

The Heartmate was the first implantable, vented VAS approved by the Food and Drug Administration for clinical use in the United States. A vented VAS uses a small percutaneous conduit which provides volume compensation during the pulsatile pumping cycle. The patient must protect this conduit from fluids and other contaminants which could cause blockage or infection. This VAS is also approved in Europe in a battery-powered version as an alternative to cardiac transplantation.

The future for implantable ventricular assist and artificial heart systems appears quite bright. Clinical trials will provide data on patient quality of life, function, survival, and cost. Extending device lifetimes from 2 to 5 years is a reasonable extension of existing knowledge. Newer innovative designs use electrically transformed autologous skeletal muscle which is fatigue resistant to actuate the pumping chamber. Rotary blood pumps are much smaller devices which will anatomically fit in nearly all adults and many children. A rotary or continuous-flow blood pump is equivalent to a pulsatile blood pump but with a very small stroke volume and a very high beat rate. Those new concepts are under study in a new program of the National Heart, Lung, and Blood Institute (Kraft *et al.*, 1994).

BLOOD SUBSTITUTES

Because normal human blood is dependent on the availability of healthy, willing donors, its supply has not always been adequate to meet the demand. Furthermore, the recognition that blood transfusions can be a means of transmitting diseases such as hepatitis or AIDS (current donor tests make this possibility unlikely) has heightened the interest in developing blood substitutes which would not carry this risk (Cheng, 1993). Although blood contains many separable therapeutic components (erythrocytes, leukocytes, platelets, and plasma proteins including gamma globulin, albumin, fibrinogen, and other coagulation factors), the term "blood substitutes" usually refers to oxygen carriers to substitute for erythrocytes (red blood cells). One group of compounds that have shown promise are perfluorochemicals, some of which can carry as much oxygen as hemoglobin. Fluosol, first developed in Japan, is an emulsion of perfluorodecalin and perfluorotripropylamine; it is undergoing clinical trials. Fluosol has been approved for use during percutaneous transluminal coronary angioplasty (PTCA) for high-risk patients. It is also being evaluated as an adjunct to cancer therapy and in the treatment of myocardial infarction in conjunction with thrombolytic therapy.

Another approach has been to encapsulate or cross-link hemoglobin following its separation from stroma. Compared with erythrocytes, microencapsulated hemoglobin has a markedly shortened survival in the circulation and is rapidly taken up by the reticuloendothelial system. Hemoglobin is cross linked to polymers to prolong its survival in the circulation. These red blood cell substitutes are effective in the treatment of acute experimental hemorrhagic shock. However nonoxygen-carrying solutions such as isotonic saline and Ringer's lactate may be equally effective for replacement of 30–40% loss in blood volume. With larger losses, cross-linked hemoglobin is more effective. Safety and efficacy in clinical trials have not yet been determined for microencapsulated or cross-linked human hemoglobin.

SUMMARY

Most of the devices described in this chapter have advanced from the animal experiment stage to the demonstration of major clinical value within the past 40 years. TAHs and VADs have not yet attained acceptance for long-term therapy. Problems remaining to be solved with these devices are primarily related to biological events at the interface, especially thromboembolism and infection. The small-diameter prosthetic vascular graft requires further development. In this case, the principal problems are thrombosis and intimal hyperplasia, both of which can result in narrowing or occlusion, distal ischemia (reduced blood flow), and tissue injury.

Current Research

Investigative approaches being pursued to overcome the problem of thrombosis and intimal hyperplasia on device surfaces include

1. Binding and controlled release of heparin and other antithrombotic compounds; both heparin and nonanticoagulant fractions of heparin inhibit smooth muscle cell replication and intimal hyperplasia; binding of other growth factor inhibitors.
2. Seeding of surfaces with endothelial cells, fetal fibroblasts, or other cells (fetal cells have the advantage over adult cells of inducing a diminished antigenic response).
3. Binding of synthetic peptides that stimulate endothelial cell adhesion and spreading.
4. Using the techniques of genetic engineering, inserting into endothelial or other cells the gene for tPA enhancing their capacity to lyse clots; seeding such cells on the luminal surface of devices, or producing an "organoid" (a hybrid organ composed of a network of biomaterial fibers lined by such cells) capable of producing a systemic clot-inhibitory effect. Genes controlling the expression of other thrombus-inhibiting or hyperplasia-inhibiting molecules could also be used.
5. Development of bioresorbable materials from which to fabricate small-diameter vascular grafts that become replaced by tissues of suitable strength as the polymers are resorbed; development of bioresorbable stents whose stimulus for thrombosis or intimal hyperplasia diminishes as they are resorbed or which incorporate inhibitors of these processes.
6. Improved hemodynamic design aimed at eliminating zones of stasis, stagnation, recirculation, flow separation, turbulence, or excessively high shear stress (Slack and Turitto, 1993).

Approaches to the problem of infection include:

1. Study and modification of receptors to bacterial adherence and colonization on polymeric and metallic surfaces.
2. Binding and controlled release of antibacterial agents.
3. Seeding of surfaces with endothelial or other cells known to have antibacterial properties.
4. Genetic engineering of endothelial or other cells to increase their antibacterial properties; seeding of such cells onto the luminal and outer surface of implanted devices.

Future Directions

In the future, there will be less emphasis on solving problems related to short-term exposure of devices to blood and tissues, since the complications can usually be tolerated or pharmacologically suppressed to acceptable levels. Instead, attention will be focused on interfacial events occurring after long-term implantation of devices, an area of increasing interest in device development. In the cardiovascular field, the approach in long-term implants is shifting away from attempts to develop materials that are increasingly inert, and towards the development of a cardiovascular implant which in time becomes integrated with its biological environment. Such a device would contain on its luminal surface or release molecules similar to the organ it replaces, encourage entry and organization of tissue cells within its structure, and become partially or completely replaced (by bioresorption) by the host's cells that provide the mechanical and functional properties possessed by the organ it replaces. This strategy is an application of the evolving discipline called tissue engineering. Tissue engineering is the application of engineering principles to create devices for the study, restoration, modification, and assembly of functional tissues from native or synthetic sources (Anderson *et al.*, 1995).

Acknowledgments

The authors thank the following for assistance and suggestions in preparing this manuscript: A. S. Berson, Ph.D.; M. J. Domanski, M.D.; G. Nemo, Ph.D.; M. L. Offen, M.D.; R. M. Hakim, M.D., Ph.D.; R. J. Turner, Ph.D.; and Ms. Kendra Brown and Ms. Gloria Dean for typing the manuscript.

Glossary of Terms

Angioplasty Surgical reconstruction of a blood vessel.

Anticoagulation The administration of a drug (anticoagulant) such as heparin or coumadin, which prevents or delays blood coagulation.

Balloon angioplasty The process whereby the narrowed segment of a vessel is dilated by inflating a balloon directed to the site of narrowing by a catheter.

Bifurcation Division into two branches.

Cardiogenic shock Shock or failure of the circulation resulting from diminution of cardiac output in heart disease, as in myocardial infarction.

Descending thoracic aorta The portion of the aorta in the chest distal to the ascending or proximal segment that arises from the left ventricle and distal to the arch of the aorta.

Dura mater Outermost tough fibrous membrane encasing the brain and spinal cord.

Endothelial-derived relaxing factor (EDRF) A labile factor produced by and released from endothelial cells and having the properties of dilating blood vessels and inhibiting platelet aggregation.

Extracorporeal membrane oxygenation (ECMO) A system whereby unoxygenated blood from a peripheral (arm or leg) blood vessel is oxygenated by flowing it through a synthetic membrane oxygenator outside the body and returning it to the circulation.

Fascia lata A sheet of fibrous tissue that encases the muscles of the thigh.

Fibrinolysis The dissolution of fibrin by an enzyme, such as tPA, urokinase, or streptokinase.

Fibroblast A connective tissue cell that plays a role in supporting and binding tissues.

Hemocompatibility Compatibility of a material with blood to which it is exposed.

Hemolysis Destruction of red blood cells.

Hemostasis The mechanisms whereby blood coagulation, platelet functions, vascular integrity, and vasoconstriction (constriction of blood vessels) control excessive bleeding following vascular injury.

Heparin An acid mucopolysaccharide present in many tissues and having the property of preventing blood from coagulating. Clinically it is used to prevent thrombosis.

Hyperplasia Abnormal increase in number of cells in normal arrangement in a tissue.

Hypotensive Causing a drop in blood pressure; having a low blood pressure.

Intimal Pertaining to the innermost layer of arteries or veins.

Inferior vena cava The major vein in the abdominal cavity draining the legs, pelvic, and abdominal organs.

Leukocyte White blood cell.

Myocardial infarction Heart attack caused by necrosis (death) of a portion of heart muscle.

Neointimal Pertaining to a newly formed inner layer of a blood vessel.

Neutropenia A decrease in the concentration of polymorphonuclear leukocytes (PMNLs) to a subnormal level.

Perianastomotic In the vicinity of the anastomosis or point where two blood vessels, or a natural tissue (i.e., blood vessel) and a prosthesis (i.e., synthetic vascular graft), are sutured together.

Pericardium Fibrous sac that encases the heart.

Polymorphonuclear leukocyte (PMNL) Neutrophil; one of several types of white blood cells. Plays important roles in defense against infection and in immune response.

Pulmonary artery (left and right) Major artery carrying blood from the heart to the lungs.

Pulmonary embolism Obstruction of an artery in the lung by thrombotic material that lodged there after flowing from a distant site such as the heart or a vein.

Restenosis Renarrowing of the lumen of a vessel (usually by neointimal hyperplasia) after the original narrowed segment has been reopened, as by angioplasty.

Thrombocytopenia A decrease in the concentration of blood platelets to a subnormal level, commonly associated with bleeding.

Thromboembolism The dislodgment of thrombotic material from its site of formation resulting in blockage of blood flow at a downstream site.

Thrombolytic therapy Therapy aimed at lysing clots. See fibrinolysis, tPA, and urokinase.

Thromboresistant Resistant to the initiation or formation of thrombus.

Thrombosis The formation of a clot in flowing blood in the heart or a blood vessel.

Thrombus A blood clot forming in flowing blood in the heart or a blood vessel.

tPA Tissue-type plasminogen activator, an enzyme found in various tissues, capable of activating plasminogen to plasmin and therefore of lysing clots.

Urokinase An enzyme found in the urine of mammals, including humans, capable of activating plasminogen to plasmin, a fibrinolytic enzyme, which dissolves clots.

Vasoconstriction Diminution of caliber of vessels caused by contraction of cells in the vessel wall, resulting in decreased flow.

Bibliography

Altieri, F. D., Watson, J. T., and Taylor, K. D. (1986). Mechanical support for the failing heart. *J. Biomater. Applic.* **1**: 106–156.

Anderson, J. M., Cima, L. G., Eskin, S. G., *et al.* (1995). Tissue engineering in cardiovascular disease: A report. *J. Biomed. Mater. Res.* **29**: 1473–1475.

Bruck, S. D. and Muller, E. P. (1988). Material aspects of implantable cardiac pacemaker leads. *Med. Prog. through Technol.* **13**: 149–160.

Chang, T. M. S., ed. (1993). *Blood Substitutes and Oxygen Carriers.* Marcel Dekker, New York.

Clowes, A. W. (1993). Intimal hyperplasia and graft failure. in *Cardiovascular Biomaterials and Biocompatibility,* L. A. Harker, B. D. Ratner, and P. Didisheim, eds. Special Supplement to Vol. 2, No. 3, *Cardiovascular Pathology,* pp. 179S–186S.

Dasse, K. A., Chipman, S. D., Sherman, C. N., Levine, A. H., and Frazier, O. H. (1987). Clinical experience with textured blood contacting surfaces in ventricular assist devices. *Trans. Am. Soc. Artif. Intern. Organs* **33**: 418–425.

DeVries, W. C. (1988). The permanent artificial heart; four case reports. *J. Am. Med. Assn.* **259**: 849–859.

Didisheim, P. (1993). Introduction: An approach to biocompatibility. in *Cardiovascular Biomaterials and Biocompatibility,* L. A. Harker, B. D. Ratner, and P. Didisheim, eds. Special Supplement to Vol. 2, No. 3, *Cardiovascular Pathology,* pp. 1S–2S, July–September, 1993.

Didisheim, P. and Watson, J. T. (1989). Thromboembolic complications of cardiovascular devices and artificial surfaces. in *Clinical Thrombosis,* H. C. Kwaan and M. M. Samama, eds. CRC Press, Boca Raton, FL, pp. 275–284.

Didisheim, P., Olsen, D. B., Farrar, D. J., Portner, P. M., Griffith, B. P., Pennington, D. G., Joist, J. H., Schoen, F. J., Gristina, A. G., and Anderson, J. M. (1989). Infections and thromboembolism with implanted cardiovascular devices. *Trans. Am. Soc. Artif. Int. Organs* **35**: 54–70.

Galletti, P. M. and Brecher, G. A. (1962). *Heart-Lung Bypass: Principles and Techniques of Extracorporeal Circulation.* Grune and Stratton, New York.

Giddens, D. P., Yoganathan, A. P., and Schoen, F. J. (1993). Prosthetic cardiac valves. in *Cardiovascular Biomaterials and Biocompatibility,* L. A. Harker, B. D. Ratner, and P. Didisheim, eds. Special Supplement to Vol. 2, No. 3, *Cardiovascular Pathology,* pp. 167S–177S.

Hakim, R. M. (1993). Complement activation by biomaterials. in *Cardiovascular Biomaterials and Biocompatibility,* L. A. Harker, B. D. Ratner, and P. Didisheim, eds. Special Supplement to Vol. 2, No. 3, *Cardiovascular Pathology,* pp. 187S–197S.

Harker, L. A., Malpass, T. W., Branson, H. E., Hessel, E. A., and Slichter, S. J. (1980). Mechanism of abnormal bleeding in patients undergoing cardiopulmonary bypass: Acquired transient platelet dysfunction associated with selective α-granule release. *Blood* **56**: 824–834.

Johnson, R. (1989). A simple modification of cuprophane that dramatically limits complement activation. *Trans. Soc. Biomater. 15th Ann. Mtg.,* **12**: 187.

Kambic, H. E., Kantrowitz, A., and Sung, P. (eds.) (1986). *Vascular Graft Update: Safety and Performance.* ASTM Special Publication 898. American Society for Testing and Materials, Philadelphia.

Kraft, S. M., Berson, A. S., and Watson, J. T. (1994). Request for proposals No. NHLBI-HV-94-25: Innovative Ventricular Assist System (IVAS).

Libby, P. and Schoen, F. J. (1993). Vascular lesion formation. in *Cardiovascular Biomaterials and Biocompatibility,* L. A. Harker, B. D.

Ratner, and P. Didisheim, eds. Special Supplement to Vol. 2, No. 3, *Cardiovascular Pathology,* pp. 43S–52S.

Meconi, M. J., Pockwinse, S., Owen, T. A., Dasse, K. A., Stein, G. S., and Lian, J. B. (1995). Properties of blood-contacting surfaces of clinically implanted cardiac assist devices: Gene expression, matrix composition, and ultrastructural characterization of cellular linings. *J. Cell. Biochem.* **57**: 557–573.

Roehm, M. D. Jr., John, D. F. *et al.* (1986). The bird's nest inferior vena cava filter. *Sem. Intervent. Radiol.* **3**: 205–213.

Sabiston, D. C. (ed.) (1991). *Textbook of Surgery. The Biological Basis of Modern Surgical Practice.* 14th ed. Saunders, Philadelphia. See chapters on pulmonary embolism (D. C. Sabiston, Jr.), arterial substitutes (G. L. Moneta, Jr. and J. M. Porter), cardiac pacemakers (J. E. Lowe), mitral and tricuspid valve disease (J. S. Rankin), acquired disorders of the aortic valve (G. J. R. Whitman and A. H. Harken), cardiopulmonary bypass for cardiac surgery (W. L. Holman and J. K. Kirklin), intra-aortic balloon counterpulsation (W. R. Chitwood, Jr.), and total artificial heart (W. E. Richenbacker, D. B. Olsen, and W. A. Gay).

Schatz, R. A. (1989). A view of vascular stents. *J. Am. Coll. Cardiol.* **13**: 445–457.

Shawl, F. A., Domanski, M. J., Davis, M., and Wish, M. H. (1990). Percutaneous cardiopulmonary bypass support in the catheterization laboratory: Technique and complications. *Am. Heart J.* **120**: 195–203.

Slack, S. M., and Turitto, V. T. (1993). Fluid dynamic and hemorheologic considerations. *Cardiovascular Biomaterials and Biocompatibility,* L. A. Harker, B. D. Ratner, and P. Didisheim, eds. Special Supplement to Vol. 2, No. 3, *Cardiovascular Pathology,* pp. 11S–21S.

Webster, J. G. (ed.) (1988). *Encyclopedia of Medical Devices and Instrumentation.* Wiley, New York. See chapters of heart-lung machine (W. J. Dorson and J. B. Loria IV), heart valve prostheses (H. S. Shim and J. A. Lenker), and total artificial heart (S. F. Yared and W. C. DeVries).

7.3 NONTHROMBOGENIC TREATMENTS AND STRATEGIES

Sung Wan Kim

OVERVIEW FOR THE DESIGN OF NONTHROMBOGENIC SURFACES

The initial event after a foreign material is exposed to the blood is protein adsorption onto the surface, followed by a complicated sequences of events, including activation of the intrinsic clotting system (Chapter 4.5), activation of platelets (Chapter 4.5), thrombolysis (Chapter 4.5), and activation of the complement system (Chapter 4.3). This chapter describes strategies to modify the surface of polymeric prosthetic devices (such as vascular grafts and artificial hearts) to minimize surface-induced thrombosis and device failure.

Over the years, a substantial amount of research has been performed to improve the biocompatibility of polymeric materials in contact with blood. However, the precise relationship between the nature of the surface, blood compatibility, and the mechanism of surface-induced thrombosis has not yet been completely elucidated.

Major factors influencing blood interaction at polymer interfaces are determined by the composition of the surface and the physical and chemical properties that the surface may encounter within the biological environment. One approach taken by many groups is to synthesize nonthrombogenic polymers with surfaces tailored to minimize specific blood interactions, such as thrombus formation and platelet interactions. General methods of surface modification are described in Chapter 2.9, while specific examples of surface modification using pharmacologically active compounds are summarized in this chapter and also in Chapter 2.11.

It is generally accepted that a hydrophilic environment at the blood–polymer interface is beneficial in reducing platelet adhesion and thrombus formation. Modification of surface hydrophilicity was attempted using albumin and poly(ethylene oxide) by different research groups. Polymer surfaces were designed by Mason et al. (1971) with selective affinity toward albumin. They confirmed that platelets showed little adherence to albumin-coated surfaces, while fibrinogen- or gamma globulin-coated surfaces exhibited an increased number of adhering platelets. Kim et al. (1974) and Lee and Kim (1979) studied the interactions of adsorbed plasma proteins with platelets and postulated mechanisms of cell adhesion involving glycoprotein terminal sugar groups. Munro et al. (1981) coupled alkyl chains of 16- and 18-carbon residues (C16, C18) to polymers to enhance albumin binding to the surface and thereby improve blood compatibility. They demonstrated that albumin binding was significantly enhanced with C16 and C18 alkylation, while fibrinogen adsorption was inhibited by C18 alkylation. Plate and Matrosovich (1976) also confirmed that C16- and C18-alkylated surfaces increased albumin binding, thereby improving blood compatibility.

Poly(ethylene oxide) (PEO) has also received much attention for applications as a biomaterial interface. Detailed studies utilizing PEO can be found in Chapters 2.3 and 2.4 and in a reference by Harris (1992). The main benefits of PEO are its unlimited solubility in water, favorable chain conformation in water that allows for high mobility, and large excluded volume to repel protein and cell interactions. Furthermore, the simple chemical structure of PEO allows for sophisticated and quantitative end-group coupling to enhance the chemical reactivity for surface immobilization (Harris, 1992). These properties of PEO, especially the dynamic motion and excluded volume, were studied as surface-grafted polymers. Nagaoka et al. (1982) immobilized PEO on hydrophobic surfaces and demonstrated that the water content, surface mobility, and volume restriction of the hydrophilic interface are critical factors influencing blood interaction. Gregonis et al. (1984) showed that PEO covalently bounded to a quartz interface decreased protein adsorption as a function of molecular weight.

In addition to the benefits of increased surface hydrophilicity documented with PEO, many groups have incorporated anticoagulant drugs, such as heparin, directly into or upon the polymer surface. These drugs interact directly with the coagulation cascade and prevent fibrin formation. Furthermore, the drug is concentrated directly at the site that initiates thrombus generation: the polymer surface in contact with blood. A more in-depth discussion of blood hemostasis can be found in Chapter 4.5.

EVALUATION METHODS OF BLOOD COMPATIBILITY FOR DESIGNED POLYMER SURFACES

The ability of drugs to maintain their biological effectiveness when incorporated with polymeric materials is essential for the development of blood-contacting surfaces. A variety of unfavorable conditions, such as drug or protein denaturation, complexation, or a variety of chemical or physical instabilities may be encountered when combining bioactive agents with polymeric materials. For these reasons, the bioactivity of the drug must be determined to evaluate the effectiveness of the fabrication process. In many instances, the measurement of biological activity of a particular drug associated with a polymeric device may require analytic techniques different from those documented to measure drug in the solution state.

Blood clotting and thrombus formation are controlled by a naturally occurring polysaccharide drug, heparin. Heparin prevents thrombus formation by binding to antithrombin III (ATIII) and catalyzing the action of ATIII to neutralize thrombin, thereby preventing the catalysis and cleavage of fibrinogen to fibrin. Currently, three models exist to account for heparin interaction with ATIII and thrombin, as shown in Fig. 1. The widely accepted model is that heparin first binds to ATIII, thereby potentiating the binding of ATIII to thrombin. In the second model, heparin binds first to thrombin, before binding as a complex to ATIII. The third model proposes that heparin can bind both ATIII and thrombin simultaneously, enhancing their interaction and rate of inactivation by increasing their proximity in the complex.

Clinically, heparin is administered via subcutaneous or intravenous administration for conditions such as deep vein thrombosis. To prevent thrombosis on the surfaces of vascular grafts, many groups are investigating the incorporation of heparin directly with the polymer surface. Therefore, determining the bioactivity of the polymer-associated drug is essential for the successful design of nonthrombogenic materials.

A well-written protocol was published by the National Heart, Lung, and Blood Institute, National Institutes of Health (NIH, 1985), to act as a standard for those evaluating blood–polymer interactions. In addition, common laboratory procedures to evaluate the biological response should include in vitro tests of blood coagulation, chromogenic assays to measure heparin bioactivity, and ex vivo and in vivo implant experiments. The interaction of platelets with foreign surfaces is significant and the adhesion, aggregation, and activation properties of platelets should be determined. A general summary of methods to evaluate blood compatibility are described in Chapter 4.5 and in a reference by Brown (1984). A brief description of common techniques used by the author will be presented.

Activated-Partial Thromboplastin Time Assay

The bioactivity of heparin-containing polymers coated onto surfaces can be measured using the activated-partial thromboplastin times assay (APTT), modified to account for heparin immobilized onto a surface, rather than in solution (Kim et al., 1987; Park et al., 1992). In general, the heparinized poly-

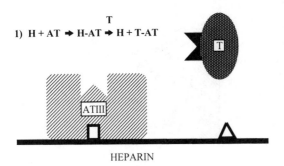

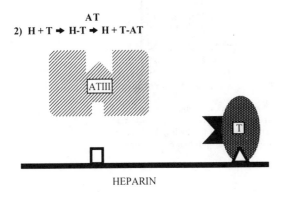

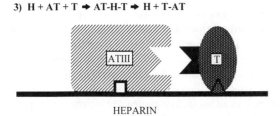

FIG. 1. Hypothesized mechanism of heparin interacting with antithrombin III and thrombin. (Top) Heparin first binding with ATIII, accelerating binding with thrombin. (Middle) Heparin first binding with thrombin, followed by ATIII binding. (Bottom) Simultaneous binding of ATIII and thrombin with heparin.

mers are coated onto glass beads, which serve as a physical support and also as a means to increase the surface area of the test sample. The heparinized polymer-coated beads are incubated with platelet-poor plasma and activated with thrombin at 37°C for 2 min. A calcium chloride solution (0.02 M) is added, and the time for a clot to form is measured with a mechanical endpoint fibrometer. The bioactivity of surface-immobilized heparin is obtained by comparing the APTT endpoints generated for surface-immobilized heparin with heparin solutions of known concentrations.

Anti-Factor Xa and Thrombin Chromogenic Assay

The bioactivity of heparinized polymers coated onto glass beads can also be determined using chromogenic assays devel-

oped by Teien *et al.* (1976). In general, these methods rely on the enzymatic activity of Factor Xa or thrombin to bind and cleave chromogenic substrates, such as S2222 (Factor Xa) or S2238 (thrombin). Thus, these methods determine the catalytic binding of heparin to ATIII. Once bound to heparin, ATIII then binds to Factor Xa or thrombin, which relates to increased heparin–ATIII binding to Factor Xa or thrombin. This binding decreases the fraction of free Factor Xa or thrombin which is necessary to bind and cleave the chromogenic substrate. This method is a direct measure of the biological activity of heparin to bind and interact with ATIII.

Adsorption of ATIII and Thrombin onto Heparin-Containing Surfaces

A two-step enzyme-linked immunosorbent assay (ELISA) is used to determine the adsorption of ATIII onto polymers, either containing heparin or without heparin, which are coated onto a glass bead surface. The adsorption of ATIII is carried out either from phosphate-buffered saline (PBS), pH 7.4, solutions or from plasma. Experimental details of this two-step ELISA are given in the literature (Hennink *et al.*, 1984).

A method to evaluate the kinetics of thrombin and ATIII binding to immobilized heparin under flow conditions was recently described by Byun *et al.* (1995). This *in vitro* model investigated the binding of ATIII to immobilized heparin, followed by thrombin. The experiments monitored the generation of the thrombin–ATIII complex, also using an ELISA under different flows.

Ex Vivo A-A Shunt Model

This is a simple and effective method to evaluate surface-induced thrombosis under low blood flow conditions in an *ex vivo* experiment (Nojiri, *et al.*, 1987). Polyurethane tubing (1.5 mm i.d. × 30 cm in length) is first coated with the various heparin-containing polymers. The tubing is equilibrated overnight in PBS (pH 7.4) and then implanted into the clamped, ligated carotid artery of a rabbit. At time $t = 0$, the clamp is removed and the shunt flow is started, maintaining the flow rate at 2.5 ml/min (shear rate = 126/sec) using a suture-tourniquet and a clamp. The experiment continues until the flow rate within the shunt stops, indicating that a thrombus large enough to occlude the tubing was produced. This is a sensitive and reliable method to evaluate different polymer or heparinized-polymer surfaces in a biological environment.

HEPARIN-CONTAINING BIOACTIVE SURFACES

Current methods to create biocompatible surfaces using heparin are discussed in this section. There are several techniques available to utilize the anticoagulant activity of heparin in conjunction with blood-contacting materials, including: (1) heparin-releasing surfaces, (2) heparin-immobilized surfaces, and (3) coating surfaces with copolymers of heparin. These methods are illustrated in Fig. 2.

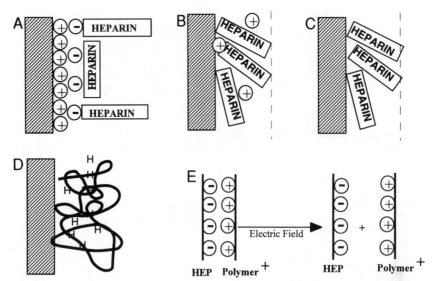

FIG. 2.   Heparin-releasing polymers. (A) Heparin (negatively charged) ionically bound onto a positively charged surface. (B) Heparin ionically bound to positively charged gel and coated onto a polymer substrate. (C) Heparin-dispersed polymers (diffusion release mechanism). (D) Thermosensitive hydrogel-grafted surface loading heparin in low temperature. (E) Heparin (negatively charged)–positive charged polymer complex (heparin released under electric current).

Heparin-Releasing Systems

Heparin-dispersed polymers release heparin primarily by diffusion from the bulk polymer, through the surface of the material, into the blood stream. Heparin-dispersed systems were initially prepared by mixing heparin with a hydrophobic polymer solution and casting to form the device or applying as a coating onto another substrate material. A summary of heparin-dispersed polymers is presented in Table 1. Limitations on the use of dispersed systems include the loading capacity of heparin which prevents long-term release, optimal control of hydrophilic heparin release rates due to the hydrophobic polymer, and the heterogeneity of heparin within the polymer matrix.

Heparin was dispersed in silicone rubber or silicone rubber/graphite and fabricated as canine arterial grafts (Hufnagel et al., 1968). This device was able to prevent thrombus formation in the animal for a minimum of 2 hr. Heparin was also incorporated into a poly(hydroxyethyl methacrylate) monolithic device and the release of heparin was detected for ~10 hr (Ebert et al., 1980). In addition, low-molecular-weight heparin fractions were released more rapidly than polydispersed heparin from polyurethane and polyvinyl chloride membranes, demonstrating a rapid initial release, followed by a more gradual release (Ebert and Kim, 1984). Heparin-releasing polyurethane catheters were also implanted in canine femoral and jugular veins, exhibiting significant reduction of thrombus formation after 1 hr (Heyman et al., 1985).

Lin et al. (1987) and Nojiri et al. (1987) coated the lumen of polyurethane tubing with a Biomer–heparin solution and implanted the device as an A-A shunt in rabbits. The heparin release profile demonstrated an initial burst effect, followed by first-order release (see Chapter 7.8). The systemic effects of heparin released from shunts were determined by APTT, platelet number, and platelet aggregation assays. Without rinsing the heparin monolithic coating, heparin released from the device caused changes in systemic blood parameters for 30 min, and extended APTT values were observed for 15 min after insertion of the shunt. Using a 15-min rinse protocol, the release of heparin did not affect systemic blood hemostasis, and the APTT values were within control values (Fig. 3). There was a consistent decrease in platelet numbers 15 min after the shunts were inserted, which subsequently rebounded to normal control values. Platelet aggregation with adenosine diphosphate

TABLE 1   Heparin-Dispersed Surfaces

Investigators	Method
Hufnagel et al. (1968)	Heparin dispersed in silicone rubber and silicone rubber plus graphite.
Holland et al. (1978)	Heparin blended into epoxy resins.
Ebert et al. (1980)	Heparin dispersed into poly(hydroxy ethyl methacrylate) monolithic devices.
Jacobs et al. (1985)	Heparin–PGE$_1$ conjugate-coated releasing surface.
Lin et al. (1989)	Heparin-releasing surfaces via polyurethane heparin solution coating.
Gutowska et al. (1992, 1994, 1995)	Heparin loading and release from temperature-sensitive polymers.

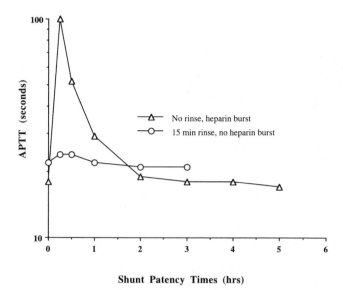

FIG. 3. APTT measurements during A-A shunt patency.

(ADP) remained relatively constant when adjusted for platelet number (Fig. 4). The occlusion time for the control, nonheparinized polyurethane was 30 ± 4 min. This compares well with Biomer shunts, which occluded in 45 ± 7 min. The heparin-releasing matrix tubing had prolonged patency times ranging from 1 to 5 hr, decreasing with increased rinse times. Thus, heparin released from the shunt was effective in improving the blood compatibility of the device.

The *ex vivo* heparin level at the time of shunt occlusion was estimated from parallel elution of [^{14}C]heparin. For experiments in which the shunts were prerinsed, the heparin release level to prevent shunt occlusion was consistently around 4×10^{-9} g/cm^2/min. This release rate was termed the minimum critical heparin release rate for nonthrombogen-

icity of the heparin-releasing matrix (Lin *et al.*, 1987). Tanzawa *et al.* (1973) studied heparin elution from cationic hydrophilic copolymers composed of vinyl chloride, ethylene, and vinyl acetate that bind heparin by coulombic interactions. The *in vivo* test of these heparin-releasing polymers showed that a constant heparin elution rate of 4×10^{-8} g/cm^2/min will keep an inferior vena cava indwelling catheter thrombus free for 2 weeks. Basmadjian and Sefton (1983) concurred with this value by using a theoretical flow model to show that heparin release will maintain the interfacial heparin concentration above the minimum concentration for prophylactic minidose heparin therapy.

The synergism between platelet aggregation and fibrin formation *in vivo* stimulated the synthesis of a covalently bonded conjugate of commercial-grade heparin and PGE$_1$. This dual-acting drug conjugate was evaluated for use as a controlled release system for blood-contacting surfaces in order to improve the blood compatibility of polymer surfaces. The compound was synthesized using a modified mixed carbonic anhydride method (Meienhofer, 1972) of amide bond formation between the carboxylic acid moiety of PGE$_1$ and a primary amine group on heparin. Bioactivity tests on PGE$_1$–heparin conjugates (Jacobs and Kim, 1986) (APTT and platelet aggregation) confirmed that the antithrombic activity of heparin was maintained. However, PGE$_1$ bioactivity, as measured by ADP-induced platelet aggregation, decreased, but was still active. Rabbit A-A shunt experiments revealed that heparin–PGE$_1$ released from the polyurethane device prevented both fibrin formation and platelet activation (Jacobs *et al.*, 1989). This approach is promising owing to the dual activity of the released compound which prevents both platelet activation and fibrin formation at the blood–polymer interface.

A new approach to create heparin-releasing systems was presented by Gutowska *et al.* (1992). They incorporated heparin into thermosensitive hydrogels (TSH), hydrogels that exhibit higher swelling at low temperature than at body temperature, as illustrated in Fig. 5. TSH were fabricated with *N*-isopropyl acrylamide (NiPAAm) and copolymerized with butyl methacrylate or acrylic acid. These NiPAAm/TSH were combined with polyurethane to form a novel interpenetrating polymer network (IPN) (Gutowska *et al.*, 1992), and heparin loading and release was studied from this device. Equilibrium swelling studies showed that modification of NiPAAm gel with polyurethane via a semi-IPN formation did not affect the gel collapse point, but resulted in decreased thermosensitivity and lower swelling levels. It was hypothesized that NiPAAm/polyurethane semi-IPNs, in which the polyurethane network is not cross-linked and therefore more susceptible to phase separation, will form a microporous structure due to the enhanced phase separation of components in the hydrated state (Gutowsaka *et al.*, 1994). This TSH–heparin–IPN was coated onto polyurethane catheters followed by heparin loading in low temperature solution and implanted as vascular access catheters in dogs. The device was removed after 1 hr to evaluate surface-induced thrombosis. The amount of thrombus attached to control surfaces (bare polyurethane or IPN without heparin) was significantly greater than to heparin-releasing surfaces (Gutowska *et al.*, 1995). Therefore, the TSH material was able to absorb heparin at low temperature and release sufficient

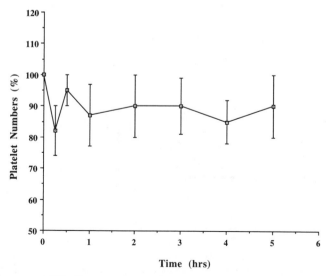

FIG. 4. Platelet counts during A-A shunt implantation.

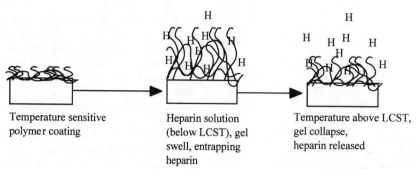

FIG. 5. Surface-grafted thermosensitive polymer release mechanism. Polymer swelling and heparin loading at low temperature; polymer collapse and heparin release at body temperature.

amounts at body temperature to minimize surface-induced thrombus formation *in vivo*.

The fact that heparin is a highly negatively charged polysaccharide led researchers to bind heparin onto a cationic surface through ionic binding. A summary of ionically bound heparin surfaces is presented in Table 2. However, the usefulness of ionically bound heparin surfaces is limited, primarily due to the short-term release of heparin.

As another approach, Kwon *et al.* (1995) prepared and investigated an electroerodible polyelectrolyte complex for the pulsatile release of heparin. An insoluble polyelectrolyte complex was formed by combining two water-soluble polymers, poly(allylamine) and heparin. Upon the application of an electric current, a rapid structural change of the complex occurred, dissolving the polymer matrix in proportion to the intensity of an applied electric current. The disruption of ionic bonds in the polymer matrix attached to the cathode and subsequent release of heparin was due to the locally increased pH near the cathode from hydroxyl ion production. Thus, the release pattern of heparin followed the applied electric current, primarily due to surface erosion of the polymer matrix, as illustrated in Fig. 6.

TABLE 2 Ionically Bound Heparin Surfaces

Investigators	Method
Gott *et al.* (1966)	Prior adsorption of cationic surfactant onto graphite-coated surfaces, followed by ionic heparin binding.
Leninger *et al.* (1966)	Formation of quaternary amine groups on the surface of polymers, followed by ionic heparin binding.
Grode *et al.* (1972)	Prior adsorption of tridodecyl methyl ammonium chloride onto the preswollen material, followed by heparin binding.
Tanzawa *et al.* (1973, 1978)	Formation of quaternary amine groups in the bulk copolymer preparation, followed by ionic binding of heparin.
Barbucci *et al.* (1984)	Synthesis of polymers containing amido and tertiary amine groups or premodification of polymer by grafting on poly(amido amines), followed by ionic binding of heparin.
Tanzi and Levi (1989)	Synthesis of polyurethane-containing poly(amido amine) blocks.
Kwon *et al.* (1994)	Heparin binding to cationic polymer to form insoluble complex; complex dissolution and heparin release under electric current.

Heparin Immobilized onto Polyurethane

The main alternative to a heparin-release device is the covalent coupling of heparin or other anticoagulant drugs directly onto the polymer surface. This procedure effectively

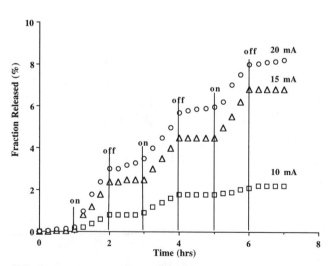

FIG. 6. Heparin release from polymer complex as a function of "on–off" electric current.

TABLE 3 Covalently Bound Heparin surfaces

Investigators	Method
Merrill *et al.* (1970) Sefton *et al.* (1983)	Gluteraldehyde and/or formaldehyde cross-linking of heparin to polyvinyl alcohol.
Labarre *et al.* (1974)	Free radical graft copolymerization of methyl methacrylate onto heparin effected by cerium salts.
Schmer *et al.* (1976) Danishefsky *et al.* (1974)	Heparin coupled directly to cyanogen bromide-activated agarose or via a spacer arm to amino-substituted and carboxylic acid-substituted agarose.
Ebert and Kim (1982)	Heparin-bound polyurethane using carbodiimide and on diamino alkyl agarose by means of Woodward K reagent. The biological activity of heparin increased as a function of alkyl chain length.
Jacobs and Kim (1989)	PGE$_1$–heparin conjugates: synthesis, characterization, and application.
Grainger *et al.* (1988) Vulic *et al.* (1987)	Heparin block copolymers, heparin bound to PEO/PS and PEO/PDMS, hydrophilic/hydrophobic block copolymer microdomains.
Park *et al.* (1988)	Heparin covalently bound onto PU via PEO spacer groups.
Lin *et al.* (1991) Piao *et al.* (1992)	Increased heparin surface concentration using branched, hydrophilic spacer groups.
Byun *et al.* (1994)	Binding kinetics for thrombin and ATIII with surface–PEO–heparin reported.
Nojiri *et al.* (1992)	Quantitation and identification of distinct proteins adsorbing onto heparin–PEO surfaces after *in vivo* implantation.
Park *et al.* (1991, 1992)	Synthesis, characterization, and application of polyurethane–PEO–heparin graft copolymers used as coating for existing surfaces.

maintains a constant concentration of heparin on the surface to interact with ATIII. Thus, compared with dispersed or ionic systems, chemically coupled surfaces should show thromboresistance for longer times than the release systems. A list of different techniques taken by various research groups to immobilize heparin onto polymer surfaces is summarized in Table 3.

Heparin was immobilized onto the surface of Biomer (a polyurethane) using hydrophilic spacer groups, specifically PEO, with molecular weights of 200, 1000, and 4000 (B–PEO–HEP). The synthetic scheme (Fig. 7) involves coupling a telechelic diisocyanate-derivatized PEO to Biomer through an allophanate–biuret reaction (i.e., the reaction between isocya-

nate and urethane NH or urea NH, respectively). The free isocyanate remaining on the spacer group is then coupled through a condensation reaction to functional groups (–OH, –NH$_2$) on heparin (Park *et al.*, 1988).

The rationale for immobilizing heparin via PEO spacer groups is that the increased hydrophilicity and dynamic motion of PEO may reduce subsequent platelet adhesion and adsorption. Heparin immobilized through PEO spacers demonstrated increased bioactivity compared with hydrophobic C6 alkyl spacer. In addition, heparin bioactivity increased with increasing PEO spacer length. The B–PEO 4000–HEP surface maintained the highest bioactivity at approximately 1.06 ($\pm$0.02) $\times$ 10^{-2} IU/cm^2 (19%), even though it contained the least amount of heparin on the surface, as detected by toluidine blue assay. APTT bioactivity assay showed behavior consistent with the Factor Xa assay, even though APTT measurements showed less bioactivity than Factor Xa complexation. These results suggest that the increasingly mobile nature of the longer hydrophilic spacer chains increases the observed bioactivity of immobilized heparin by providing a more bulk-like environment for heparin.

Correlations between platelet adhesion and the amount of PEO spacers on the Biomer surfaces (surface density) are shown in Fig. 8. Platelet adhesion onto PEO 1000-grafted surfaces decreased markedly with increasing amounts of PEO; however, this was not seen with PEO 200- or 4000-grafted surfaces. To examine the PEO chain length effect on platelet adhesion, the total amount of PEO on the surface was fixed at 12 μg/cm^2 and the value for PEO 200 was extrapolated.

The results also demonstrated the relationship between

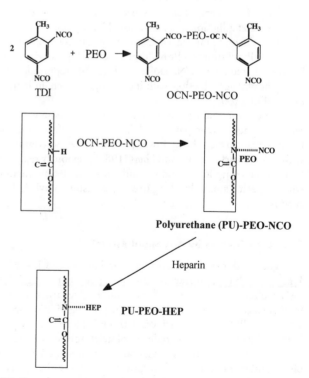

FIG. 7. Synthetic scheme for the immobilization of heparin onto polyurethane through PEO spacer groups.

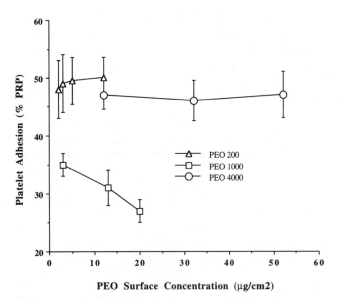

FIG. 8. Platelet adhesion onto PEO-grafted surfaces as a function of PEO chain length and PEO surface density.

platelet adhesion and PEO spacer length for B–PEO surfaces at 12 μg/cm² PEO. Relative to Biomer controls, PEO-grafted surfaces showed a decrease in platelet adhesion with minimum platelet adhesion observed with PEO 1000 surfaces. Compared with the alkyl spacer system, PEO spacer systems demonstrated consistently lower platelet adhesion. These results correlate with the interfacial free energy concept in which a decrease in interfacial free energy decreases protein adsorption, causing lower platelet adhesion and activation. Hydrophilic PEO spacers can reduce the interfacial free energy between the surface and the bulk solution, while hydrophobic alkyl spacers increase interfacial free energy. Nagaoka *et al.* (1984) reported a decrease in platelet adhesion with increasing PEO chain length, up to PEO 2500.

Furthermore, heparin-immobilized surfaces show lower platelet adhesion compared with a Biomer surface and the B–C6–HEP surface shows lower platelet adhesion compared with a B–C6 surface. Kim and Ebert (1982) previously reported that spacer group length did not influence platelet interaction with heparin-immobilized agarose gels using alkyl spacer groups.

Ex Vivo Evaluation by A-A Shunt Model

Figure 9 shows occlusion times for Biomer and B–PEO–HEP (immobilized surfaces) using different chain lengths of PEO evaluated in a rabbit A-A shunt model. Heparin-immobilized surfaces demonstrated significant prolongation of occlusion times. PEO-grafted surfaces without heparin did not prolong occlusion times, but reduced platelet adhesion *in vitro*. These results suggest that surface-induced coagulation in whole blood under conditions of low flow rate and low shear forms stabilized thrombus masses on the surface as a result of synergistic or cooperative actions of fibrin net formation together with platelet aggregation. Heparinized surfaces using PEO spacers appear to be effective in suppressing both fibrin and platelet deposition for longer patency. It has been known that PEO-grafted surfaces are less thrombogenic owing to decreased protein adsorption and platelet adhesion as a result of high dynamic motion and low binding affinities of grafted PEO chains. There is the probability that thrombus was removed from the surfaces as microemboli under the flow conditions present in these systems. Weak cohesive forces between a forming thrombus and the PEO surface would result in weak binding interactions and subsequent embolization. By contrast, PEO-grafted surfaces have generated extensive blood coagulation under low flow conditions in A-A shunt experiments. More complex thrombogenic events occurring at the interface may promote thrombus formation in excess of that by simple surface PEO mobility. These effects may include contact activation, fibrin formation, platelet adhesion, and stable thrombus formation.

B–PEO–HEP surfaces demonstrate longer occlusion times than Biomer, as shown in Fig. 9. A significant hydrophilic spacer effect is observed, that is, PEO spacers prolong patency. The nonthrombogenic activities observed *ex vivo* for spacer length systems of different lengths utilized for heparin immobilization are attributed primarily to differing effects of spacer lengths on the whole blood bioactivity of heparin.

It was obvious in these studies that the combination of hydrophilic environment produced by PEO surface grafting and the anticoagulant properties of immobilized heparin were needed to reduce surface-induced thrombosis. PEO alone was effective *in vitro* to prevent platelet adhesion and protein adsorption, but failed *in vivo* to enhance biocompatibility. Similarly, heparin coupled to surfaces with no spacer or with hydrophobic spacer did not enhance the biological activity. The hydrophilic nature of the optimum chain length of PEO coupled to heparin was necessary to optimize the biological response *in vivo*.

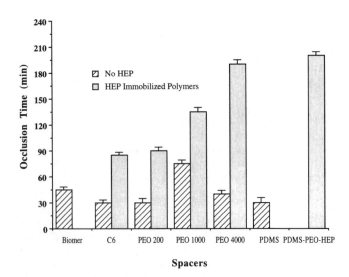

FIG. 9. A-A shunt occlusion times for PU–PEO–HEP and PDMS–PEO–HEP surfaces.

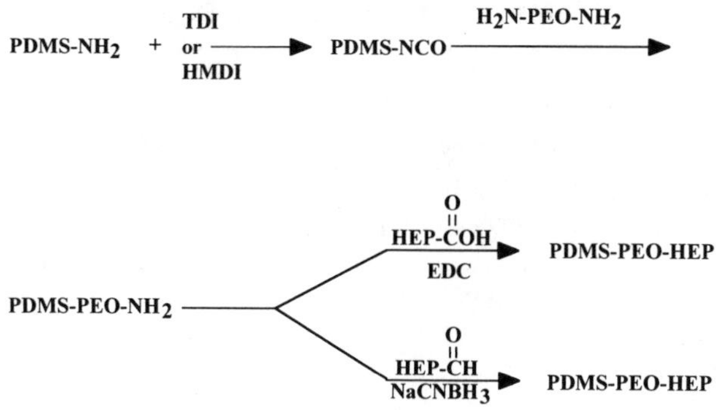

PDMS-NH₂ + TDI or HMDI → PDMS-NCO → (H₂N-PEO-NH₂)

$$PDMS\text{-}PEO\text{-}NH_2 \longrightarrow \begin{cases} \overset{\overset{O}{\|}}{HEP\text{-}COH}, EDC \rightarrow PDMS\text{-}PEO\text{-}HEP \\ \overset{\overset{O}{\|}}{HEP\text{-}CH}, NaCNBH_3 \rightarrow PDMS\text{-}PEO\text{-}HEP \end{cases}$$

FIG. 10. Synthetic scheme for the synthesis of PDMS–PEO–HEP co-polymers.

HEP–PEO–Polymer Coatings

Another method to fabricate heparin-immobilized surfaces, in contrast to chemical immobilization onto preformed surfaces, is that of synthesizing soluble heparin-grafted polymers. These polymers were synthesized in solution and coated onto the surfaces. This provides a distinct advantage over direct (Grainger *et al.*, 1988) immobilization, including scale-up synthesis and the ability to coat existing devices.

Several triblock and pentablock copolymers and comb-type polymers consisting of a hydrophobic portion to anchor and interact with the surface, a hydrophilic segment (PEO), and heparin have been synthesized. HEP–PEO–silicone rubber (polydimethyl siloxane—PDMS) and HEP–PEO polystyrene triblock copolymers were synthesized (see Chapter 2.3 for a discussion of block copolymers). Figure 10 shows the schematic

Reaction Scheme: Polyurethane-PEO-HEP

Polyurethane (PU)   OCN(CH₂)₆NCO   HMDI

PU-NCO

HO-PEO-OH

HMDI

HEPARIN

PU-PEO-HEP

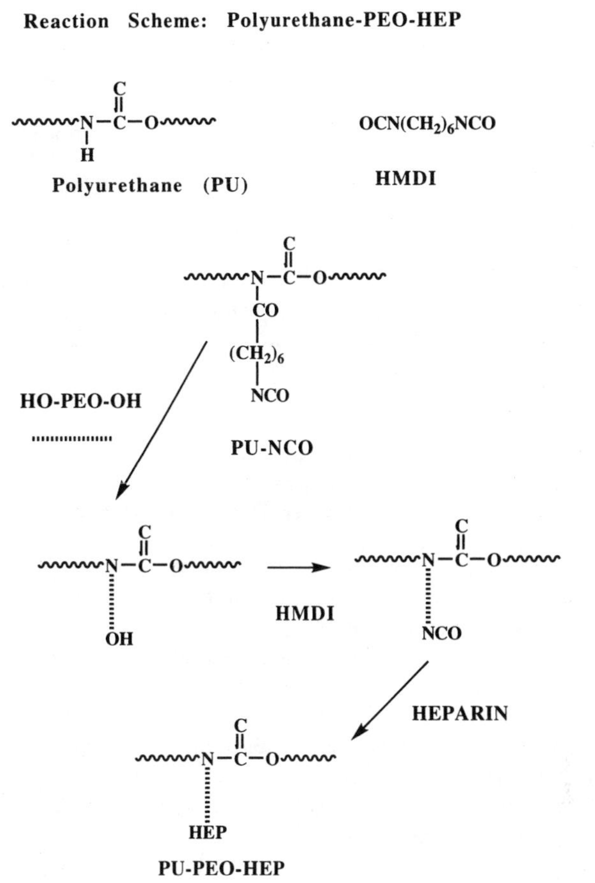

FIG. 12. Synthetic scheme for the synthesis of PU–PEO–HEP soluble copolymers.

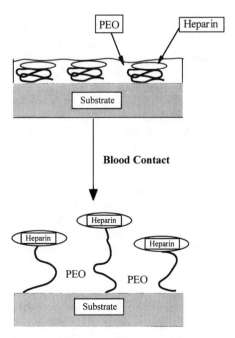

FIG. 11. Surface reorientation of HEP–PEO-grafted polymer coated onto a base polymer in contact with blood.

for the synthesis of HEP–PEO–PDMS. Ideally, the hydrophobic block will be coated on a hydrophobic substrate (Fig. 11), wherein the hydrophobic portion anchors to the substrate, and the hydrophilic PEO and heparin are extended into the aqueous environment of blood. Both triblock and pentablock–PDMS–PEO–heparin polymers were coated onto polyurethane and demonstrated significant improvement in blood compatibility compared with native polyurethane surfaces (Grainger *et al.*, 1988; Piao *et al.*, 1990; Vulic *et al.*, 1989; Grainger, 1988). Occlusion times in the A-A shunt model for triblock HEP–PEO–PDMS-coated systems and controls are shown in Fig. 9. Control surfaces (Biomer, PDMS, PEO) perform poorly in the shunt, all demonstrating occlusion times of less than 50 min. HEP–PEO–PDMS remained patent, even under these extreme flow conditions, for over 200 min. Based on the low levels of platelet adhesion detected *in vitro*, it is likely that patency is a function of low platelet adhesion and heparin bioactivity.

Park (1990) extended this research to synthesize soluble polyurethane (PU) polymers grafted with PEO and heparin (PU–PEO–HEP), as illustrated in Fig. 12. A condensation reaction between diisocyanate molecules (hexamethylene diisocyanate—HMDI) and the urethane protons on the polyurethane chain was first performed. The PU–HMDI intermediate was then coupled to a terminal hydroxyl group on PEO to form

TABLE 4 Biological Activity of Polyurethane–PEO–Heparin-Grafted Copolymer Surfaces

Surface	Surface heparin concentration (μg/cm^2)	FXa assay ($\times 10^{-2}$ IU/cm^2)	Bioactivity (%) (FXa/heparin)
(PEO 1000)–HEP	0.47 ± 0.08	0.96 ± 0.03	11.54
(PEO 4000)–HEP	0.28 ± 0.05	1.24 ± 0.02	22.17
(PEO 7500)–HEP	0.24 ± 0.04	0.081 ± 0.04	18.25

PU–PEO. Again, HMDI was used to activate the terminal hydroxyl on PEO for subsequent coupling to heparin.

This well-characterized heparinized polymer was coated onto a polyurethane surface and evaluated for *in vitro* and *in vivo* blood compatibility. As shown in Fig. 11, coated graft copolymers will reorient hydrophilic PEO and heparin when in contact with blood to reduce surface-induced thrombosis.

As investigated for *in situ* surface modification, the molecular weight of PEO (1000, 4000, and 7500) was studied to maximize the bioactivity of heparin. PU–PEO and PU–PEO–HEP graft copolymers were coated onto a glass surface and the *in vitro* bioactivity of heparin as a function of PEO chain length was determined. As shown in Table 4, the PU–(PEO 4000)–HEP copolymer-coated surface showed the highest bioactivity and correlated with results obtained by the *in situ* immobilization method. In addition, PU–PEO–HEP surfaces and the PU–PEO adsorbed less platelets than the bare polyurethane surface.

The high bioactivity of the heparin-immobilized copolymer surfaces was comparable with those achieved by the *in situ* modification techniques. The convenience of the copolymer synthesis, its ability to be used as a coating onto other polymer substrates, and the bioactivity of immobilized heparin confirmed that this procedure can improve the blood compatibility of blood-contacting surfaces.

In Vivo Arterial Graft Model

Heparin–PEO–Biomer-coated grafts were implanted in the abdominal aorta of dogs and evaluated for graft patency and

TABLE 5 Protein Surface Layer Thickness onto *in Vivo* Vascular Grafts

Polymer	Protein layer (Å)
Polyurethane (PU)	980 ± 56
PU–(PEO 4000)	1860 ± 203
PU–(PEO 4000)–heparin	326 ± 29

Note. 3 weeks–1 month after implantation; n = 3; values are means ± SD.

TABLE 6 Graft Patency in Dogs

Surface	Days
Polyurethane (PU)	4–28
PU–(PEO 4000)	22–30
PU–(PEO 4000)–heparin	>90

Note. n = 3–5.

protein adsorption (Nojiri *et al.*, 1989; Nojiri *et al.*, 1990). As listed in Table 6, B–PEO–HEP grafts demonstrated significantly longer patency when compared with Biomer and B–PEO controls. In addition (Table 5), the Biomer and B–PEO grafts demonstrated a thick (1000 to 2000 Å) protein multilayer that consisted mainly of fibrinogen and immunoglobulin G (IgG) and was low in albumin. In contrast, B–PEO–HEP displayed a relatively thin (200–300 Å) protein layer with a high concentration of albumin and IgG and which was low in fibrinogen. Biomer and B–PEO grafts may lead to denaturation of adsorbed proteins, resulting in multilayer protein deposition patterns. B–PEO–HEP may passivate the initial protein layer, preventing denaturation and producing a stable monolayer of proteins.

SUMMARY

Heparinized materials, including heparin-releasing devices and immobilized systems, demonstrate advantages over currently available blood-contacting materials. The effectiveness of heparin released from a hydrophobic matrix was dependent on the hydrophobicity of the bulk material, while rapid and controlled release patterns were evident from thermosensitive polymer coatings. Heparin released from an ionic polymer complex proved to be a unique new approach for controlled release of the anticoagulant agent.

Much practical information regarding the interaction of surfaces with blood has been gained by studying the effect of a hydrophilic spacer group (PEO) in combination with anticoagulant drugs. During *in vitro* studies, PEO surfaces demonstrated decreased protein and platelet adhesion, while the same system *in vivo* was not effective in preventing surface-induced thrombosis. However, during *in vivo* studies, heparinized surfaces did not show decreased protein and platelet interactions, while heparin immobilized directly onto a surface showed minimal *in vivo* effectiveness in preventing surface-induced thrombosis. The most effective combination, therefore, was the chemical coupling of heparin onto a surface using hydrophilic spacer groups. This method afforded heparin the most efficient biological activity, while demonstrating decreased protein and platelet interactions. The combination of the hydrophilic attributes of PEO and the retention of biological activity of heparin immobilized via PEO proved to be a viable approach to synthesize bioactive polymer surfaces. The bioactivity (antithrombin activity) of immobilized heparin was shown to be a function of PEO spacer length, while immobilized heparin demonstrated no chain-length effect on platelet adhesion. In *ex vivo* shunt experiments under low-flow, low-shear conditions, and in *in vivo* vascular graft experiments, all heparinized surfaces exhibit prolonged occlusion times compared with controls, indicating the ability of immobilized heparin to inhibit thrombosis in whole blood.

Acknowledgments

The author thanks the National Institutes of Health for supporting this research in bioactive polymer surfaces (HL 20251 and HL 17623). The author also thanks Dr. Ki Dong Park for his assistance in the preparation of this chapter.

Bibliography

Basmadjian, D., and Sefton, M. V. (1983). Relationship between release rate and surface concentration for heparinized materials. *J. Biomed. Mater. Res.* **17**: 509–518.

Brown, B. A. (1984). Coagulation. in *Hematology: Principles and Procedures.* Lea & Fegiger, Philadelphia, pp. 179–240.

Byun, Y. R., Jacobs, H., and Kim, S. W. (1994). Heparin surface immobilization through hydrophilic spacers: Thrombin and antithrombin III binding kinetics. *J. Biomat. Sci.* **6**: 1.

Byun, Y. R., Jacobs, H., and Kim, S. W. (1996). Binding of antithrombin III and thrombin to immobilized heparin under flow conditions. *Biotech. Prog.*, in press.

Casini, G., Tempesti, F., Barbucci, R., Mastacchi, R. and Sarret, M. (1984). Heparinizable materials, III: Heparin retention power of a poly(amido amine) either as crosslinked resin or surface grated on PVC. *Biomaterials* **5**: 234–237.

Danishefsky, I., and Tzeng, F. (1974). Preparation of heparin-linked agarose and its interaction with plasma. *Thromb. Res.* **4**: 237–246.

Ebert, C. D., and Kim, S. W. (1982). Immobilized heparin: Spacer arm effects on biological interactions. *Thromb. Res.* **26**: 43–57.

Ebert, C. D., and Kim, S. W. (1984). Heparin polymers for the prevention of surface thrombosis. *Med. Appl. Controlled Release* **2**: 77.

Ebert, C. D., McRea, J. C., and Kim, S. W. (1980). Controlled release of antithrombotic agents from polymer matrices. in *Controlled Release of Bioactive Materials,* R. Baker, ed. Academic Press, New York, pp. 107–122.

Goosen, M. F., and Sefton, M. V. (1983). Properties of a heparin–poly(vinyl alcohol) hydrogel coating. *J. Biomed. Mater. Res.* **17**: 359–373.

Gott, V. L., Whiffen, J. D., and Datton, R. C. (1966). Heparin bonding on colloidal graphite surfaces. *Science* **152**: 1297–1298.

Grainger, D. (1988). Blood compatibility of amphiphilic block copolymers, Ph.D. dissertation, University of Utah, Salt Lake City.

Grainger, D., Feijen, J., and Kim, S. W. (1988). Poly(dimethylsiloxane)–poly(ethylene oxide)–heparin block copolymers, I: Synthesis and characterization. *J. Biomed. Mater. Res.* **22**: 231–242.

Grainger, D., Nojiri, C., Okano, T., and Kim, S. W. (1989). *In vitro* and *ex vivo* platelet interactions with hydrophilic/hydrophobic PEO–PST multiblock copolymers. *J. Biomed. Mater. Res.* **23**: 979–1005.

Gregonis, D., Van Wagonen, R., and Andrade, J. D. (1984). Poly(ethylene glycol) surfaces to minimize protein adsorption. *Trans. Sec. World Congr. Biomater.* **7**: 266.

Grode, G. A., Falb, R. D., and Crowley, J. P. (1972). Biocompatible materials for use in the vascular system. *J. Biomed. Mater. Res.* **3**: 77–84.

Gutowska, A., Bae, Y. H., Feijen, J., and Kim, S. W. (1992). Heparin release from thermosensitive hydrogels. *J. Controlled Release* **22**: 95.

Gutowska, A., Bae, Y. H., Jacobs, H., Feijen, J., and Kim, S. W. (1994). Heparin release from thermosensitive interpenetrating polymer networks: Synthesis, characterization, and macromolecular release. *Macromolecules* **27**: 4167.

Gutowska, A., Bae, Y. H., Jacobs, H., Feijen, J., and Kim, S. W. (1995). Heparin release from thermosensitive polymer coating: *In vivo* studies. *J. Biomed. Mater. Res.* **29**: 811.

Harris, J. M. (1992). Introduction to biotechnical and biomedical applications of poly(ethylene glycol). in *Poly(ethylene Glycol) Chemistry: Biotechnical and Biomedical Application,* J. M. Harris, ed. Plenum, New York, pp. 127–136.

Hennink, W. E., Ebert, C. D., Kim, S. W., Breemhar, W., Bantjes, A., and Feijen, J. (1984). Interaction of antithrombin III with preadsorbed albumin–heparin conjugates. *Biomaterials* **5**: 264–268.

Heyman, P. W., Cho, C. S., McRea, J. C., Olsen, D. B., and Kim, S. W. (1985). Heparinized polyurethanes: *In vitro* and *in vivo* studies. *J. Biomed. Mater. Res.* **19**: 419.

Holland, F. F., Gidden, H. E., Mason, R. G., and Klein, E. (1978). Thrombogenicity of heparin-bound DEAE cellulose hemodialysis membranes. *Artif. Organs* **1**: 24–36.

Hufnagel, C. A., Conrad, P. W., Gillespie, J. F., Pifarre, R., Ilano, A., and Yokoyama, T. (1968). Characteristics of materials for intervascular applications. *Ann. N. Y. Acad. Sci.* **146**: 262–270.

Jacobs, H., and Kim, S. W. (1986). *In vitro* bioactivity of a synthesized PGE₁–heparin conjugate. *J. Pharm. Sci.* **75**: 172–175.

Jacobs, H., Okano, T., and Kim, S. W. (1989). Antithrombogenic surfaces: Characterization and bioactivity of surface immobilized PGE₁–heparin conjugate. *J. Biomed. Mater. Res.* **23**: 611–630.

Jacobs, H., Okano, T., Lin, J. Y., and Kim, S. W. (1985). PGE₁–heparin conjugate releasing polymers. *J. Controlled Release* **2**: 313–319.

Kim, S. W., Jacobs, H., Lin, J. Y., Nojiri, C., and Okano, T. (1987). Nonthrombogenic bioactive surfaces. *Ann. N. Y. Acad. Sci.* **516**: 116–130.

Kim, S. W., Lee, R. G., Oster, H., Coleman, D., Andrade, J. D., Lentz, D., and Olsen, D. (1974). Platelet adhesion to polymer surfaces. *Trans. Am. Soc. Artif. Intern. Organs* **20**: 449–455.

Kwon, I. C., Bae, Y. H., and Kim, S. W. (1994). Heparin release from polymer complex. *J. Controlled Release* **30**: 155.

Labarre, D., Boffa, M. C., and Jozefowicz, M. (1974). Preparation and properties of heparin–poly(methylacrylate) copolymers. *J. Polym. Sci.* **47**: 131–137.

Lee, E. S., and Kim, S. W. (1979). Adsorbed glycoproteins in platelet adhesion onto polymer surfaces: Significance of terminal galactose units. *Trans. Am. Soc. Artif. Intern. Organs* **25**: 124–131.

Leninger, R. I., Cooper, C. W., Falb, R. D., and Grode, G. A. (1966). Nonthrombogenic plastic surfaces. *Science* **152**: 1625–1626.

Lin, J. Y., Jacobs, H., Nojiri, C., Okano, T., and Kim, S. W. (1987). Minimum heparin release rate for nonthrombogenicity. *Trans. Am. Soc. Artif. Intern. Organs* **33**: 602–605.

Lin, S. C., Jacobs, H., and Kim, S. W. (1991). Heparin immobilization increased through chemical amplification. *J. Biomed. Mater. Res.* **25**: 792.

Mason, R. G., Read, M. S., and Brinkhaus, K. M. (1971). Effect of fibrinogen concentration on platelet adhesion to glass. *Proc. Soc. Exp. Biol. Med.* **137**: 680–682.

Meienhofer, J. (1972). The mixed carbonic anhydride method of peptide synthesis. in *The Peptides,* E. Gross and J. Meienhofer, eds. Academic Press, New York, pp. 264–309.

Merrill, E. W., Salzman, E. W., Wong, P. L., Ashford, T. P., Brown,

A. D., and Austen, W. G. (1970). Poly(vinyl alcohol)–heparin hydrogel "G". *J. Appl. Physiol.* **29**: 723–730.

Mori, Y., Nagaoka, S., Masubuchi, Y., Itoga, M., Tanzawa, H., Kiucki, T., Yamada, Y., Yonaha, T., Wantabe, H., and Idezuke, T. (1978). The effect of released heparin from the heparinized hydrophilic polymers (HRSD) on the process of thrombus formation. *Trans. Am. Soc. Artif. Intern. Organs* **24**: 736–745.

Munro, M. S., Quattrone, A. J., Ellworth, S. R., Kulkarni, P., and Eberhart, R. C. (1981). Alkyl substituted polymers with enhanced albumin affinity. *Trans. Am. Soc. Artif. Intern. Organs* **27**: 499–503.

Nagaoka, S., Mori, Y., Takiuchi, H., Yokota, K., Tanzawa, H., and Nishiumi, S. (1982). The influence of heparinized polymers on the retention of platelet aggregability during storage. *J. Biomed. Mater. Res.* **16**: 209–230.

Nagaoka, S., Mori, Y., Takiuchi, H., Yokota, K., Tanzawa, H., and Nishiumi, S. (1984). Interaction between blood components and hydrogels with poly(oxyethylene) chains. in *Polymers as Biomaterials*, S. W. Shalaby, A. Hoffman, B. D. Ratner, and T. A. Horbet, eds. Plenum, New York, pp. 361–374.

National Institutes of Health (1985). Guidelines for blood–material interactions. NIH No. 85-2185, National Heart, Lung, and Blood Institute, Bethesda, MD.

Nojiri, C., Okano, T., and Kim, S. W. (1987). Evaluation of nonthrombogenic polymers in a new rabbit A-A shunt model. *Trans. ASAIO* **33**: 596–601.

Nojiri, C., Park, K. D., Okano, T., and Kim, S. W. (1989). *In vivo* protein adsorption on polymers: Transmission electron microscopic (TEM) study. *Trans. ASAIO* **35**: 357–361.

Nojiri, C., Park, K. D., Grainger, D., Jacobs, H., Okano, T., and Kim, S. W. (1990). *In vivo* nonthrombogenicity of heparin immobilized polymer surfaces. *Trans. ASAIO* **36**: 168–172.

Nojiri, C., Okano, T., Koyanagi, T., Park, K. D., and Kim, S. W. (1992). *In vivo* protein adsorption on polymers: Visualization of adsorbed proteins on vascular implants in dogs. *J. Biomater. Sci.* **4**: 75.

Park, K. D., Okano, T., Nojiri, C., and Kim, S. W. (1988). Heparin immobilized onto segmented polyurethaneureas surfaces: Effects of hydrophilic spacers. *J. Biomed. Mater. Res.* **22**: 977.

Park, K. D., Piao, A. Z., Jacobs, H., Okano, T., and Kim, S. W. (1991). Synthesis and characterization of SPUU–PEO–heparin grafted copolymers. *J. Polym. Sci. A* **29**: 1725.

Park, K. D., Kim, W. G., Jacobs, H., Okano, T., and Kim, S. W. (1992). Blood compatibility of SPUU–PEO–heparin grafted copolymers. *J. Biomed. Mater. Res.* **26**: 739.

Piao, A. Z., Nojiri, C., Park, K. D., Jacobs, H., Feijen, J., and Kim, S. W. (1990). Synthesis and characterization of poly(dimethylsiloxane)–poly(ethylene oxide)–heparin block copolymers. *J. Biomed. Mater. Res.* **24**: 403.

Piao, A. Z., Jacobs, H., and Kim, S. W. (1992). Heparin immobilization via surface amplification. *ASAIO J.* **38**: M638.

Plate, N. A., and Matrosovich, M. M. (1976). Affinity chromatography of serum albumin for adsorption on synthetic polymers. *Akad. Nauk (USSR)* **220**: 496–503.

Plate, N. A., and Valuev, L. I. (1986). Heparin-containing polymeric materials. *Adv. Polym. Sci.* **79**: 95–133.

Rosenberg, R. D. (1983). The function and structure of the heparin molecule. in *Heparin*, I. Witt, ed. de Gruyter, Berlin, pp. 2–33.

Schmer, G., Teng, L. M. L., Cole, J. J., Vizzo, H. E., Francisco, M. M., and Schribner, B. H. (1976). Successful use of a totally heparin grafted hemodialysis system in sheep. *Trans. Am. Soc. Artif. Intern. Organs* **22**: 654–663.

Sefton, M. V., and Zingg, W. (1981). Patency of heparinized SBS shunts at high shear rates. *Biomater. Med. Dev. Artif. Organs* **9**: 127–142.

Tanzawa, H., Mori, Y., Harumiya, N., Miyama, H., Hori, M., Ohshima, N., and Idezuki, Y. (1973). Preparation and evaluation of a new antithrombogenic heparinized hydrophilic polymer for use in cardiovascular systems. *Trans. Am. Soc. Artif. Intern. Organs* **19**: 188–194.

Teien, A. N., Lie, M., and Abildgaard, N. (1976). Assay of heparin in plasma using a chromogenic substrate for activated factor X. *Thromb. Res.* **8**: 413–416.

Vulic, I. (1989). Heparin containing block copolymers, Ph.D. dissertation, Twente University of Technology, Enschede, The Netherlands.

Wilson, J. E. (1981). Heparinized polymers as thromboresistant biomaterials. *Plast. Technol. Eng.* **16**: 119–208.

7.4 DENTAL IMPLANTS

Jack E. Lemons

HISTORICAL OVERVIEW

In the early history of tooth replacement, naturally occurring minerals or parts from dead animals were used to replace lost or diseased dentition (Phillips, 1973). Most were fabricated or shaped to resemble the lost region or component, and a common practice during the seventeenth century was to implant teeth extracting from living persons or from cadavers. In most cases, the component was similar to the missing part. Little attention was given to the basic properties of the implant material or the best possible form for interfacing the implant with tissue. Interestingly, some of the early fabricated tooth replacements showed evidence of wear facets that could have occurred only from an extended period of oral function (mastication of food). Historical records and skulls in museums demonstrate that semiprecious and precious gemstones, gold, ivory, bone, and tooth substitutes provided some degree of oral function.

During the eighteenth and nineteenth centuries, design and synthetic material concepts evolved to a point where systematic investigations were being conducted to determine the physical, mechanical, chemical, electrical, and biological properties of surgical dental implant systems (see Chapter 9.3). The scientific and technological advances of the twentieth century subsequently provided the basis for in-depth considerations of both designs and biomaterials (Williams and Roaf, 1973), (Smith and Williams, 1982). This chapter concentrates on the biomaterials for dental implant systems that have found extensive clinical applications since the late 1940s.

Designs

Dental implant designs can be separated into two categories called endosteal (endosseous), which enter the bone tissue, or subperiosteal systems, which contact the exterior bone surfaces (Lemons, 1988a). These are shown schematically in Fig. 1 and Fig. 2. The endosteal forms,[1] such as root forms (cylinders,

---

[1]The various implant designs will be shown in more detail later in this chapter.

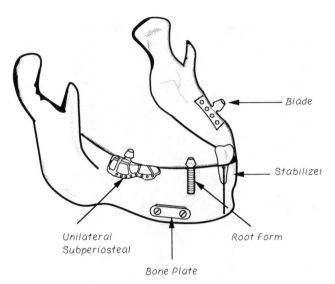

FIG. 1. Schematic representation of different dental implant designs.

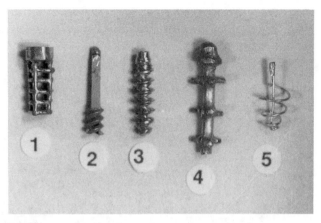

FIG. 3. Examples of some lost wax casting or bent wire endosteal implants: (1) Cast design after Greenfield, Au, Pt–Ir or Co–Cr–Mo. (2) Cast design after Chercheve, Co–Cr–Mo. (3) Cast screw design after Strock, Cr–Co–Mo. (4) Cast tree-like design after Lew, Co–Cr–Mo. (5) Bent wire design after Formiggini, Au wire.

screws, plateaus, etc.), blades (plates), transosseous or staples, ramus frames, and endodontic stabilizers, are placed into the bone as shown. In contrast, the subperiosteal devices are fitted to the bone surface as customized shapes while bone plates are placed onto the bone under the periosteum and fixed with endosteal screws.

Modern dental implants began with the endosteal systems fabricated from bent wires or lost wax castings (Phillips, 1973). Some examples are shown in Fig. 3. These early implants incorporated many of the structural features that continue to be used today. The blade or plate devices were introduced to treat the edentulous spaces where the buccallingual (thickness) dimensions of the bone were limited (see Figs. 4A–B and 5A–D). This situation, or the loss of significant bone, has been one of the adverse sequelae of utilizing an ill-fitting "free-end distal partial denture" over the long term. Transosseous, staple, and ramus frame systems (shown in later figures) were designed to treat the fully edentulous mandible, with the bone pathways located along the anterior

symphysis and ascending ramus (ramus frame) locations. Endodontic stabilizers (shown in Fig. 1) were designed to pass through the distal pump chamber regions of endodontically treated teeth, the rationale being to improve the crown-to-root ratio for compromised dentitions.

The subperiosteal system evolved much later than the early root form devices (Rizzo, 1988). These implants are custom-made devices that are designed and fabricated for each patient. The cast framework is fitted to the bone using an initial working model that is obtained from a surgical impression of the bone or a CAT-scan image. Although the materials of construction have remained essentially unchanged (cobalt alloy or titanium), the geometric forms of the bone framework regions and the intraoral denture connections have changed over time. Many of these changes have taken place since 1980 (James *et al.*, 1991).

The types of dental implants available are numerous, with experience and a history of peaks and valleys of interest influencing the introduction of different designs. In many situations, the dentist who invented an implant (Linkow,

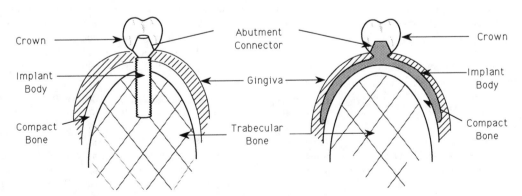

FIG. 2. Schematic representation of the interfacial regions of dental implant systems. Idealized cross sections through mandibular bone.

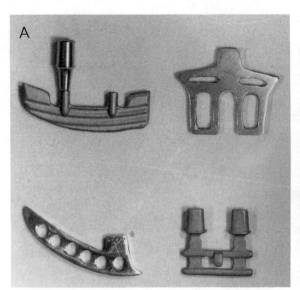

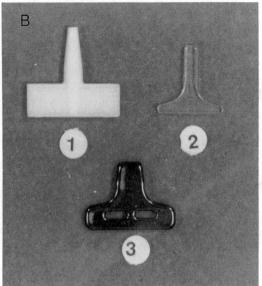

FIG. 4. Examples of blade (plate) endosteal dental implants. (A) Custom Ti, two-stage with anodized surface, upper left. Custom, 316SS design after Roberts, upper right. Ramus blade, 316SS, lower left. Fagan custom blade, Ti, Lower right. (B) Miter, Alumina (1); Bioceram, sapphire (2); Carbomedics, carbon–silicon (3).

1983), (Roberts, 1971), (Weiss, 1986), (Branemark *et al.*, 1977), (Kirsch, 1986) is also the principal clinician using the device. This has resulted in many individuals and groups that are staunch proponents of specific designs that often carry their names. This development has been viewed as offering both strengths and weaknesses. The strengths are a result of more than a casual or indirect interest in "solving all problems" and attempting to satisfy the patient's needs over both the short and long terms. The weaknesses are related to attempts by some to treat all clinical conditions with one design. Reporting may sometimes be biased, and patient, clinical, and company interests (commercial, financial, and marketing) become intertwined. However there are no simple solutions to such complex issues.

Materials and Biomaterials

Synthetic materials for root form devices were originally fabricated from precious metals such as gold, platinum, iridium, and palladium. Cost and strength considerations soon resulted in the development and use of alloys and reactive group metals such as tantalum, titanium, and zirconium, and the inert cast and wrought cobalt- (Stellites) and iron- (stainless steel) based alloy systems. Since the 1950s, materials formerly used only for general industrial applications (automotive, aircraft, aerospace, etc.) have been reconstituted to include a wide range of metals and alloys, ceramics and carbons, polymers, combinations, and composites (ASTM, 1992). Some commonly used biomaterials for dental implants are listed in Table 1.

The more commonly used biomaterials for endosteal implants are titanium and titanium alloy, aluminum oxide, and surface coatings of calcium phosphate ceramics. The cobalt-based casting alloys are most often used for the subperiosteal implant devices (McKinney, 1991). Since the properties of these biomaterials are quite different from one another, specific criteria must be applied to each device design and clinical application. These details are discussed in the following sections and other chapters.

CLINICAL ENVIRONMENT

The body of the dental implant contacts the bone and soft tissue interfaces within the submucosal regions. These contact zones within or along the bone surface provide the areas for mechanical force transfer (Bidez *et al.* 1986). Therefore, the implant-to-tissue interface becomes a critical area for force transfer and thereby the focal point for quality and stability of intraoral function. The implant body is directly connected to the transmucosal (or transgingival) post, which provides the base for the dental bridge abutment. The abutment is either cemented or screwed to the intraoral dental crown, bridge, or denture prosthesis. This arrangement is schematically shown in Fig. 2.

The dental implant device provides a percutaneous connection from the intraoral to the tissue sites results in a very complex set of conditions (Meffert, 1988). The oral cavity represents a multivariant external environment with a wide range of circumstances (e.g., foods and abrasion, pH, temperatures from 5 to 55°C, high magnitude forces, bacteria, etc.). Stability requires some type of seal between the external and internal environments, which of course depends upon both the device and the patient. Fortunately, the oral environment and

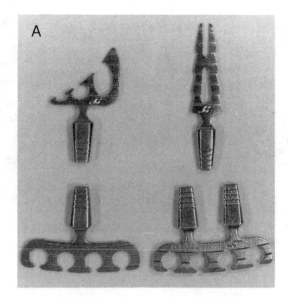

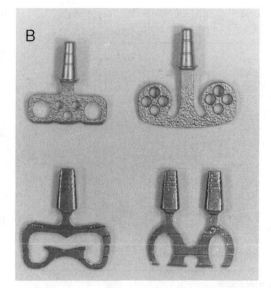

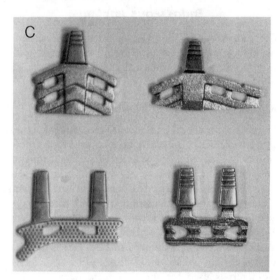

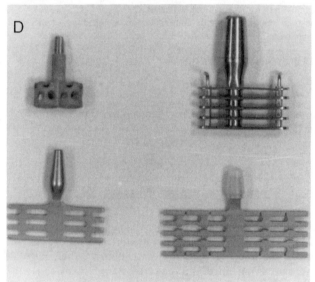

FIG. 5. Examples of endosteal blade (plate), basket, and plateau design dental implants. (A) Oratronics (Ti). (B) Park dental (Ti), upper and Oratronics (Ti), lower. (C) Oratronics (Ti, ion implanted), Upper. Bioceram (Ti, nitrided), lower left and Oratronics (Ti), lower right. (D) Colmed basket (Ti with porous Ti coating), upper left. Omni "Sinus" plateau (Ti-6Al-4V), upper right. Miter (Ti-6Al-4V), lower left, and Feigel (Ti-6Al-4V), lower right.

the associated tissues are known to be very tolerant and resistant to external challenges.

CONCEPTS OF STAGING AND OSTEOINTEGRATION

In contrast to most other types of musculoskeletal implants, where some function and a relatively high magnitude force transfer exist immediately after surgery, dental implant treatments can be staged in ways to directly influence the healing and restorative processes (Branemark *et al.*, 1977). Some de-

vices are placed into the bone for several months prior to attaching the abutment head and intraoral prosthesis. Healing takes place under conditions of no imposed oral function and thereby with minimal mechanical force transfer along the biomaterial-to-tissue interface.

Studies within the dental field have demonstrated the modeling and remodeling of bone along biomaterial surfaces here bone tissue is directly adjacent to the synthetic materials (Albrektsson and Zarb, 1989; Rizzo, 1988). This condition has been called osseo- or osteointegration. The staging concept delaying the final treatment enhances the opportunity to obtain these types of interfaces, in that bone is allowed to form along the implant during the healing process and

TABLE 1   General Summary of Synthetic Biomaterials
for Dental Implants[a]

Metals and Alloys
   Titanium and 90[b] titanium-6 aluminum-4 vanadium
   Zirconium and tantalum (99% purity)
   68 cobalt-27 chromium-5 molybdenum
   68 Iron-18 chromium-14 nickel

Ceramics and Carbons
   Aluminum or zirconium oxide (99.9[+]%)
   Calcium phosphates (<50 ppm impurities)
   Glass and glass ceramics (high-purity compositions)
   Carbon (99.9[+]%) and 85 carbon-15 silicon

Polymers
   Polyethylene (ultra high molecular weight)
   Poly(tetrafluoroethylene)
   Polysulfone
   Poly(ethylene terephthalate)

[a]Detailed properties are available in Vol. 13.01, *Medical Implants*, American Society for Testing and Materials, 1992.
[b]The numbers given provide nominal weight percentages of the various elements.

in the absence of significant imposed motion between the implant and the tissues.

DENTAL IMPLANT SYSTEMS

Dental implant systems can be separated into those that are directly in contact with tissues and those that replace the crown portions of the teeth. In addition, as mentioned earlier, the implants can be separated into the categories of endosseous and subperiosteal designs. These are discussed separately in the following sections.

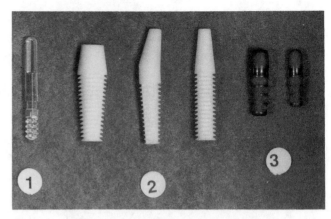

FIG. 7.   Examples of root form one-stage endosteal implants. (1) Bioceram, sapphire (Al_2O_3). (2) Synthodonts, alumina (Al_2O_3). (3) Carbomedics, carbon-silicon (C-Si).

Endosseous Implants

The endosseous implant designs include the various root forms, blades, staples, frames, bone plates, screws, and wires. Each group is reviewed separately.

Root Forms and Endodontic Stabilizers

The root form dental implants are recommended for placement at anatomical positions where natural tooth roots have previously existed. Most have external surface configurations that resemble helical spirals, screw threads, cylinders, baskets, plateaus, rods, pins, cones, structures with surface vents, porous or solid coatings, or combinations of these shapes and surface features. Published reviews of root form implants and their designs list more than 40 that are being utilized clinically (English, 1988). Some examples of both one- and two-stage root form devices are shown in Figs. 6–9.

Biomaterials used for the foot form devices include the

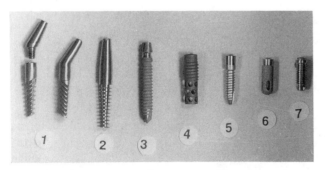

FIG. 6.   Examples of root form implants. (1) and (2) Omni R Series (Ti-6Al-4V), two-stage. (3) TPS (Ti with porous Ti coating), one-stage. (4) Core vent (Ti-6Al-4V), two-stage. (5) Sterioss (Ti), two-stage. (6) IMZ (Ti with porous Ti coating), two-stage. (7) Collagen-OTC (Ti), two-stage.

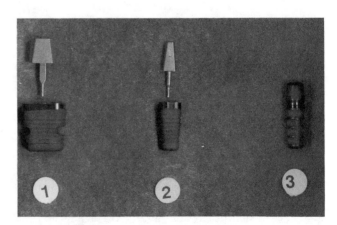

FIG. 8.   Examples of carbon and carbon-silicon endosteal root form implants. (1) ande (2) Vitredent (C), two-stage. (3) Carbomedics (C-Si), one-stage.

FIG. 9. Examples of porous endosteal root form implants and corrosion test specimens. (1) Porous Ti and Ti–6Al–4V endosteal implants. (2) Porous Co–Cr–Mo corrosion test specimen. (3) Porous Ti and Ti-6Al-4V corrosion test specimen.

metals and alloys, ceramics and carbons, and some polymers or porous surfaces for bone and soft tissue interfaces. Titanium and titanium alloy have been used for most devices, although some continue to be made from cobalt alloy. Designs fabricated from aluminum oxide as alumina (polycrystalline) or sapphire (single crystalline) have been more popular in Asia and Europe, although many have been used in North America. Some devices have been fabricated from titania and zirconia ceramics, but a limited number have been utilized clinically. Also, carbon, carbon-silicon, and polymers [e.g., poly(methymethacrylate), poly(tetrafluoroethylene), polysulfone, and poly(dimethylsiloxane)] have been used in limited numbers during the past few years. Surface configurations include porous metallic plasma sprays, sintered wire compacts, or complex geometric features along the endosteal region of the device. Most designs are available with calcium phosphate ceramic surface coatings. (These subjects are covered in Chapter 2.6.)

The intraoral abutment connector for root form implants utilizes several different types of attachment systems. These include screws or cap screws, slots or tapers for cement lutings, taper or friction lock configurations, or combinations of shapes. This multiplicity of design is can be seen by examining the upper portions (abutments) of the devices shown in Figs. 6–9. The denture attachment includes the configurations listed plus connectors with magnets, ball or undercut slots with O-rings, snap clips, nuts, or standard dental preparation shapes for cementation to a crown or bridge. This subject has been explained in detail by English (1988). These include both fixed and removable connector systems, depending upon the type of intraoral prosthesis that is required.

Endodontic stabilizers are long pins or screws that are of a small enough diameter to be passed through a tooth root canal. Examples of endodontic stabilizers and pins are shown in Fig. 10. A schematic of clinical placement was shown in Fig. 1. These designs of implants are sometimes placed into teeth that have minimal bone and periodontal ligament support. The stabilizer is intended to improve the mechanical force transfer from the tooth to the bone, to increase the tooth stability, and thereby to enhance long-term usefulness. These devices are used with endodontically treated teeth, and a relatively stable seal at the tooth root apex area is critical to long-term stability.

Stabilizers are fabricated with a mechanically rolled or swagged surface configuration to enhance mechanical strength. (See Chapter 2.2) These are usually nonalloyed titanium or titanium-6 aluminum-4 vanadium (Ti-6Al-4V), although some designs utilize cobalt alloy. Smooth surface designs are also available in single crystal sapphire (aluminum oxide) ceramic.

Blades (Plates)

The blade (plate) form endosseous implants (shown in Figs. 4A–B and 5A–D) were introduced to treat a special bone situation encountered when patients have been edentulous for extended periods (Roberts, 1971). This device was first applied to areas of bone resorption and residual dimensions of limited thickness and height. Results were such that the clinical use was expanded to include a wide range of treatments. The term "blade" was used because the shapes of several early designs resembled a wedge or knife blade. Because of the length dimension, compared with width and depth, the devices appeared to be a "blade" when viewed on two-dimensional radiographs. Design changes now include many configurations that are not wedge-shaped and thus the term "plate" has evolved.

Metallic dental blades (plates) are fabricated mostly from titanium or titanium alloy, although 316 type stainless steel continues to be used for the ramus blade design (shown in Fig. 4A), and a number of designs are cast from cobalt alloy. Aluminum oxide ceramics and some zirconia ceramic plate forms are also utilized clinically. These are also available with porous surfaces of plasma-sprayed metals and alloys or with

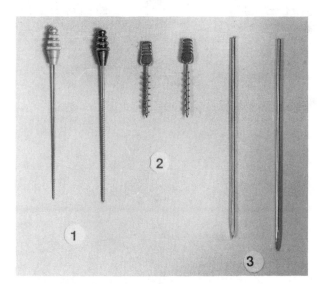

FIG. 10. Examples of endodontic, spiral, and pin endosteal implants. (1) Endodontic stabilizers (Ti). (2) Spiral root forms (Ti). (3) Pins (316LSS, Ti or Co–Cr–Mo alloy).

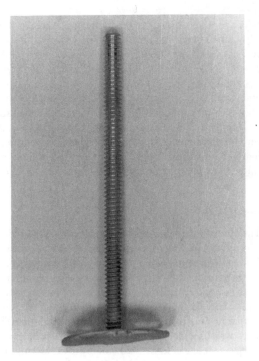

FIG. 11. Examples of a transteal bone implant cast from Co–Cr–Mo alloy.

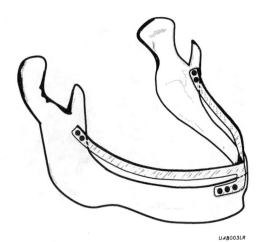

FIG. 12. Schematic of a ramus frame placed in an edulous mandible. The lower denture mounts on the intraoral bar.

calcium phosphate ceramic or carbon coatings. Most devices have relatively smooth surfaces without porosity or undercut configurations. Almost all designs contain holes or vents through the endosseous bodies of the devices to provide increased surface areas for force transfer and a cross-communication for tissue continuity.

Transosseous, Staples, and Frames

Early transosseous screw or post designs were fabricated from cast cobalt alloy, as shown in Fig. 11. These were placed through the anterior region of the mandible and protruded through the gingiva into the oral cavity (Cranin, 1970). These protrusions or abutments (usually bilateral at the cuspid location) were used to stabilize standard type, soft tissue-supported lower dentures. An extension of this clinical concept was the endosteal staple (Small, 1980) which incorporated a plate that interconnected the two transosteal threaded pins. Additional pins were placed along the plate to penetrate the inferior border

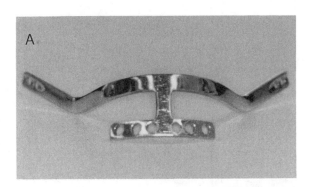

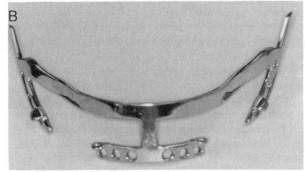

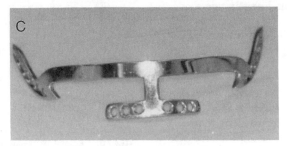

FIG. 13. Examples of ramus frame dental implants. (A) M2, H. Roberts design (316SS). (B) Ra-2, R. Roberts design (316SS). (C) M-25, H. Roberts design (316SS).

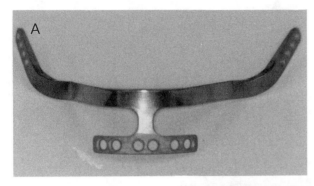

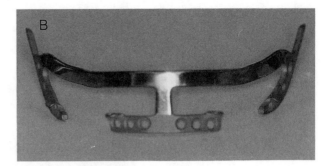

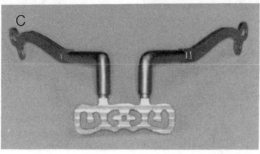

FIG. 14. Examples of ramus frame dental implants. (A) Omni (Ti). (B) RA-2, R. Roberts design (Ti). (C) R₂S₅ Linkow five-piece design (Ti).

of the mandible and thereby provide endosteal support. These designs were fabricated from titanium alloy. More recent versions now include bone screw stabilizers and calcium phosphate coatings to improve mechanical stabilities within the bone.

The ramus frame implants are also intended for treatment of the endentulous mandibular arch. This type of device application is shown schematically in Fig. 12. Several commercially marketed ramus frame designs are shown in Figs. 13A–C and 14A–C. The anterior endosteal portion of these devices is placed into the central symphysis of the mandible, while the posterior extensions or blades are implanted into the ascending ramus or distal mandibular regions. The extensions along the posterior (distal) regions, and additional surface tabs were added as design changes in an attempt to enhance resistance to bone resorption and to progressive settling into the bone of the mandible during long-term use. These devices are utilized as one-stage restorative systems in patient treatments. Many of these devices are placed in patients who are completely endentulous and the patient starts to use the denture the first day after surgery. Thus, the frame-supported lower denture, stabilized by the bone insertions, articulates with a tissue-supported upper denture.

Plates, Screws, and Wires

Bone plates, screws, and wires used in oral and maxillofacial surgery are very similar to those used in orthopedic surgery. On a relative basis, more of the oral surgical implants are fabricated from titanium compared with orthopedic devices, and, in general, are smaller in dimension. A wide range of

dental design configurations exists that is intended to stabilize both simple and complex craniofacial bone lesions. All commonly used plates, screws, and wires are constructed from metals and alloys.

Subperiosteal Implants

The subperiosteal implant designs are placed along bone surfaces under the periosteum. Each metallic framework to be placed along the mandibular or maxillary bone regions is cast to specifically fit each patient's bone structure. The bone is exposed surgically to make a polysulfide, polyether, or silicone rubber polymeric impression replica; or a radiographic three-dimensional replica is constructed from serial tomogram (CAT-scan) images (James *et al.*, 1991). The framework is cast from either cobalt alloy or titanium, with most devices made from cobalt alloy.

Examples of subperiosteal implant designs commonly used during the 1960s are shown in Figs. 15A and B. These devices include four independent abutment posts that extended across the soft tissues into the oral cavity. The denture was processed onto a metallic suprastructure framework for placement onto the abutment posts. The suprastructure and denture were removable for cleaning and overnight periods. This design and system have been used extensively for restoration of edentulous conditions.

Design Evolutions

The four-post subperiosteals with a relatively simple "over the bone" configuration have been redesigned to include a

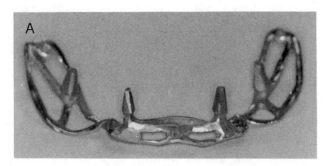
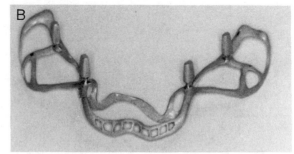

FIG. 15.   (A) and (B) Examples of mandibular subperiosteal designs cast from Co–Cr–Mo alloys.

number of structural and material modifications (Steflic and McKinney, 1991), (James, 1982, 1990). The bone surface regions of these devices now include a wide range of open, elongated vents. Some mandibular devices provide a posterior segment that incorporates an implant region along the lateral border of the ascending ramus portion of the bone. Also, the intraoral connector regions have been changed from four independent posts to connected bars, bars with additions of rings, undercuts or balls for O-rings, or removable connectors for use with a two-stage clinical restorative sequence. Examples of mandibular and maxillary subperiosteal frames are shown in Figs. 16A and B. These devices were cast from cobalt-based alloy and demonstrate a calcium phosphate coating on the maxillary device (Fig. 16A) and a two-stage mandibular design (Fig. 16B).

The subperiosteal frames are also cast from titanium and some have been coated with porous metals, carbon, calcium phosphate ceramics, or combinations of geometric forms and materials. These various surface treatments are quite involved and the student is referred to Chapters 2.2 and 9.4 and ASTM (1992) for related material. Since many of these devices are cast in dental laboratories (some owned by dentists providing the clinical treatments), procedures for adequate quality control and assurance have been incorporated into the laboratory-based manufacturing processes. A number of independent dental laboratories now provide these castings for implant applications (Root, 1990).

Mucosal Inserts

One design listed as a dental implant is the mucosal implant series. These are small, mushroom-shaped buttons that are fabricated into the denture base and protrude from the denture surface. The buttons are located using a surgical procedure, so that they penetrate into the adjacent mucosa. Epithelial cells line the crypt around the buttons during healing; therefore these systems are not true implants. However, since they are often placed into freshly prepared soft tissue lesions, implant-quality materials and surface conditions are required (e.g., most are passivated stainless steel or titanium). As mentioned, most inserts are now fabricated from type 316 stainless steel and titanium, with some made from aluminum oxide ceramics. The denture base is constructed from poly(methyl methacrylate), which also is used as an implant biomaterial for orthopedic (bone cement) and dental (bone substitute) applications.

Crown, Bridge, and Denture Restorations

Intraoral prostheses involve a wide range of metals, ceramics, polymers, and composites. Since many of these intraoral devices are mechanically or chemically (cemented) attached to the dental implants, biomaterial interactions become a part of the overall restorative treatment. Therefore, each implant modality must be evaluated to ensure that adverse biomaterial or biomechanical conditions are not introduced through incor-

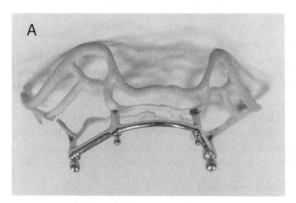
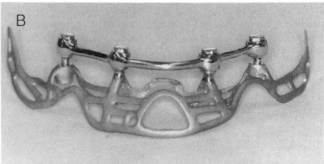

FIG. 16.   Examples of subperiosteal designs. (A) Maxillary device cast from Co–Cr–Mo alloy and coated with calcium–phosphate ceramic. (B) Mandibular device cast from Co–Cr–Mo alloy with removable suprastructure.

rect selections (Lemons, 1988). For detailed information, the reader is referred to dental material textbooks (Phillips, 1973; Craig, 1985; Leinfelder and Lemons, 1988).

Fixed Restorations

Most root form and blade (plate) systems use restorative treatments in which the implant is rigidly attached to the intraoral prosthesis. Some systems provide polyethylene or polyoxymethylene spacers within the core of the implant to enhance mobility through the interposition of lower modulus materials, while others require that occlusal surfaces of the teeth be fabricated from polymer-based restorative materials (Kirsch, 1986; Branemark, 1977). These variations result in a wide range of treatments being utilized clinically. In this regard, device retrieval analyses have shown some situations where galvanic coupling of dissimilar alloys (see Chapter 6.3) or inadequate quality castings that contained porosities and high carbon content have influenced device longevities (Lemons, 1988). At this time, most implant device-related treatments include a prescription or recommendation for the design and biomaterial of the intraoral prosthesis.

Removable Dentures

The subperiosteal, transosseous, staple, ramus frame, and mucosal insert implant designs are associated with removable denture prostheses. The dentures are fastened to the various intraoral connectors or abutments shown in previous figures. The dentures can be attached by a variety of connector types that range from magnets to simple recesses into one of the components. Dentists and dental laboratory personnel have been ingenious in developing these connector systems. In some situations, interlocking devices have evolved to provide what is called "fixed-removable" prostheses.

TISSUE INTERFACES

Occlusal forces from the mastication of food are transferred from the intraoral prosthesis through the implant abutment and neck (connector) and into the implant-to-tissue interface region. These forces are dissipated through the associated tissues, and the quality of functional stability can be correlated with the relative interfacial stability (micromotion) over time (Brunski, 1988). Two types of interfacial conditions have been described for functional dental implants: fibrous tissue integration (called pseudo-ligament or fibro-osteal integration) (Weiss, 1986) and bone tissue integration (called osseo- or osteointegration) (Branemark, 1977).

Fibro-osteal Integration

The fibrous tissue interfaces with blade (plate), subperiosteal, and insert surfaces have been described in detail in laboratory animal and human retrieval studies (James, 1982; Lemons, 1988). In general, the functional zone is described as a relatively acellular region of fibrous collagen and fibroblast tissues that is oriented along and around the endosteal portion of the implant body (Feigel and Makek, 1989). Descriptions of oriented collagen fiber networks that provide a hammock or shock absorber zone have been discussed and supported as advantageous (Weiss, 1986). Proponents of this relatively low modulus region with regard to motion profiles emphasize that dynamic movement would be more like normal teeth (with periodontal ligaments), as opposed to directly adjacent regions of osseous tissues.

Single-stage restorative treatments place many of these implant designs into mechanical function during the bone healing period. The interfacial motion could be the primary etiology of this fibrous tissue zone. These types of interfaces have been shown for a variety of designs and materials. Critical to longevity for any material or design is the dimensional stability of this fibrous tissue zone over time (5 to 20 years); and an increase in the thickness of this region is one predictor of clinical failure.

Osteointegration

The concept of osseo- or osteointegration was initially associated with root form implant designs, two-stage restorative treatments, and a titanium oxide implant surface condition (Branemark *et al.*, 1977). This is described as a direct bone-to-biomaterial interface (without fibrous tissue) for a functioning implant at the optical microscopy limits of resolution ($0.5~\mu$m). A wide range of implant biomaterials and designs have been shown to exhibit this type of interface for functional dental systems (Rizzo, 1988). Proponents support the concepts that: (1) greater force can be transferred along bone interfaces; (2) direct association or attachment eliminates or minimizes interfacial movement (slip); and (3) periodontal soft tissue regions are more biomechanically stable.

Interestingly, both osseous and fibrous tissue interfaces have provided functional situations for dental implant devices over 15 or more years (Rizzo, 1988). Clinical trials that have been developed using prospective analysis protocols should provide answers to the questions related to "which is the best interface."

Calcium phosphate ceramic, glass ceramics, porous layers added to metals, and other biomaterial surface modifications have been introduced and developed to enhance bone adaptation to endosteal and subperiosteal implant surfaces (Ducheyne and Lemons, 1988). These surface modifications could result in a more stable implant-to-tissue interface. Clinical data are inadequate to evaluate this hypothesis at this time.

TRENDS IN RESEARCH AND DEVELOPMENT

The dental implant field has provided a significant contribution to our overall understanding of basic biomaterial and biomechanical properties. The oral cavity provides 28 to 32 candidate implant sites, a necessity for crossing the epithelium to provide functional support, a wide range of anatomical shapes, and an even wider range of environmental conditions. The oral cavity and associated tissues provide a very severe environment for implant biomaterials and designs, and it is interesting to note the trends in research and development within this field.

Conservative Treatment Modalities

In the 1970s, dental implants were judged to be in a longitudinal research and development phase. Typically, they were only used in clinical situations where all other treatments had failed and where bone and soft tissue health was compromised for implant-based restorations.

As a result of the reported clinical longevities and the functionalities of several dental implant designs and biomaterials, opinions now reflect the use of implants on a routine basis where they are indicated for dental restoration of normal function. Dental implant-based treatments have moved from the longitudinal research phase in 1970 to a recognized conservative modality in 1990. Important to this change is the quality and quantity of bone and soft tissue available at the time of implant placement, i.e., proper oral diagnosis, plus the significant improvements in dental implant-based treatment systems. This trend is expected to continue to provide ever improved device function and longevity.

Computer Modeling and Surface Modifications

The recent introduction of computer-based finite element modeling and analysis (FEM and FEA) to the dental implant area is significantly influencing design concepts (Brunski, 1988). Three-dimensional models that are now processed on supercomputer systems are providing opportunities to optimize shape and material combinations (Bidez et al., 1986). These analyses should lead to improved implant designs and to more objective rationales for device selection and use. One aspect of these analyses is the interfacial condition of slip (fibro-osteal integration) or nonslip (osteointegration) (Bidez, 1987). Device shapes to better utilize coatings for bonding to bone and soft tissues should evolve. To avoid the adverse sequelae of stress (or strain) shielding within bone, the concepts need to be further developed and utilized. The addition of a calcium phosphate ceramic to an existing dental implant design may or may not be an acceptable modification. The tolerance of the host tissues may be forgiving enough to accept both types of surfaces.

Combined Delivery Systems

To provide mechanically and chemically clean implants, many manufacturers and suppliers have moved to prepackaged and presterilized dental implant delivery systems. This helps to avoid contamination or altering of implant surfaces during shipping, presurgical handling, and surgical placement.

Bibliography

Albrektsson, T., and Zarb, G. A. (eds.) (1989). The Branemark Osseointegrated Implant. Quintessence Pub. Co., Chicago.

ASTM (1992). Annual Book of ASTM Standards. Vol. 13.01, *Medical Implants,* ASTM Publ., Philadelphia.

Bidez, M. W. (1987). Stress distribution within endosseous blade implant systems as a function of interfacial boundary conditions. Ph.D. dissertation, University of Alabama at Birmingham.

Bidez, M. W. (1991). Biomechanics literature review. *Int. J. Oral Impl.* **8**: 95–100.

Bidez, M. W., Lemons, J. E., and Isenburg, B. P. (1986). Displacements of precious and nonprecious dental bridges utilizing endosseous implants as distal abutments. *J. Biomed. Mater. Res.* **20**: 785–797.

Branemark, P. I., Hansson, B. O., Adell, R., Breine, U., Lindstrom, J., Hallen, O. and Ohman, A. (1977). *Osteointegrated Implants in the Treatment of the Endentulous Jaw.* Almquist and Wiksell Int., Stockholm.

Brunski, J. B. (1988). Biomechanics of oral implants: Future research directions. *J. Dent. Ed.* **52**: 775–778.

Craig, R. G. (1985). *Restorative Dental Materials.* 7th ed., C. V. Mosby, St. Louis.

Cranin, A. N. (1970). *Oral Implantology.* Charles C. Thomas, Springfield, IL.

Ducheyne, P., and Lemons, J. E. (1988). Bioceramics: material characteristics versus *in vivo* behavior. *Ann. New York Acad. Sci.* **523**: 000–000.

English, C. E. (1988). cylindrical implant. *Calif. Dent. J.* **16**(1): 17–40.

Feigel, A., and Makek, M. (1989). The significance of sinus elevation for blade implantology—report of an autopsy case. *J. Oral Impl.* **15**: 237–249.

James, R. A. (1982). Host response to dental implants. in *Biocompatibility of Dental Materials.* CRC Press, Boca Raton, FL, pp. 163–197.

James, R. A. (1990). Implant prosthodontics. in *Prosthetic Management on Subperiosteal Implants.* Year Book Med. Publ., Chicago, Ch. II.

James, R. A., Lozada, J. L., and Truitt, H. P. (1991). Computer tomography (CT) applications in implant dentistry. *J. Oral Impl.* **17**: 10–16.

Kirsch, A. (1986). Plasma-sprayed titanium IMZ implant. *J. Oral Impl.* **12**: 494–498.

Leinfelder, K. F., and Lemons, J. E. (1988). *Clinical Restorative Materials and Techniques,* Lea and Febiger, Philadelphia.

Lemons, J. E. (1988a). Dental implant retrieval analyses. *J. Dent. Ed.* **52**: 748–757.

Lemons, J. E. (1988b). Quantitative characterization and performance of porous implants for hard tissue applications. ASTM Publication, STP 953, American Society for Testing and Materials, Philadelphia.

Linkow, L. I. (1983). *Dental Implants.* R. Speller, New York.

McKinney, R. V., Jr. (1991). *Endosteal Dental Implants,* Mosby, St. Louis.

Meffert, R. M. (1988). The soft tissue interface in dental implantology. *J. Dent. Ed.* **52**: 810–812.

Phillips, R. W. (1973). *Science of Dental Materials.* Saunders, Philadelphia.

Rizzo, A. A. (ed.) (1988). Proceedings of the consensus development conference on dental implants. *J. Dent. Ed.* **52**: 678–827.

Roberts, H. D. (1971). *The Ramus, Single Tooth and Ramus Frame Implants.* Loma Linda Univ. Press, Loma Linda, CA.

Root, D. (1990). Current laboratory observation. *Int. J. Oral Impl.* **6**(2): 39–44.

Small, I. A. (1980). Benefit and risk of mandibular staple bone plates. in *Dental Implants: Benefit and Risk,* PHS Pub. 81-1531, 139-152, U.S. Public Health Service, Washington, DC.

Smith, D.C., and Williams, D. F. (1982). *Biocompatibility of Dental Materials.* CRC Press, Boca Raton, FL.

Steflic, D. E., and McKinney, R. V., Jr. (1991). History of implantology.

in *Endosteal Dental Implants,* Mosby Year Book, St. Louis, pp. 8–19.

Weiss, C. M. (1986). Tissue integration of dental endosseous implants: Description and comparative analysis of fibro-osseous and osseous integration systems. *J. Oral Impl.* **12:** 169–215.

Williams, D. F. (1982). *Biocompatibility of Orthopaedic Implants.* CRC Press, Boca Raton, FL, Vol. 1.

Williams, D. F., and Roaf, R. (1973). *Implants in Surgery.* Saunders, London.

7.5 ADHESIVES AND SEALANTS

Dennis C. Smith

According to a definition of the American Society for Testing and Materials, an adhesive is a substance capable of holding materials together by surface attachment. Inherent in the concept of adhesion is the fact that a bond that resists separation is formed between the substrates or surfaces (adherends) comprising the joint and work is required to separate them.

"Adhesive" is a general term that covers designations such as cement, glue, paste, fixative and bonding agent used in various areas of adhesive technology. Adhesive systems may comprise one- or two-part organic and/or inorganic formulations that set or harden by several mechanisms.

Commercial adhesive systems are often designed to result in only a thin layer of adhesive for efficient bonding of the two surfaces since thick layers may contain weakening defects such as air voids or contaminants. Such systems may be low-viscosity liquids. In other situations where, for example, the surfaces to be joined are irregular, gap-filling qualities are required of the bonding agent. These systems may be solid–liquid (filled) adhesives or viscous liquids and are usually referred to as cements, glues, or sealants. Thus, the term "sealant" implies not only that good bonding and gap-filling characteristics are present in the material, but also that the bonded join is impervious, for example, to penetration by water. Since most adhesives, including biomaterials, are used to joint dissimilar materials that are subjected to a variety of physical, mechanical, and chemical stresses, good resistance to environmental degradative processes, including biodegradation, is essential.

The applications of adhesive biomaterials range from soft (connective) tissue adhesives used both externally to temporarily fix adjunct devices such as colostomy bags and internally for wound closure and sealing, to hard (calcified) tissue adhesives used to bond prosthetic materials to teeth and bone on a more permanent basis. All of these biological environments are hostile, and a major problem in the formulation of medical and dental adhesives is to develop a material that will be easy to manipulate, interact intimately with the tissue to form a strong bond, and also be biocompatible. Over the past two decades, more success at a clinical level has been achieved in bonding to hard tissues than to soft tissues.

More details on the background of adhesion and adhesives can be found in recent texts (Kinloch, 1987; Skeist, 1990; Lee, 1991a,b; Pizzi and Mittal, 1994).

HISTORICAL OVERVIEW

Wound closure by means of sutures extends back many centuries. The idea of using an adhesive is more recent but dates back to at least 1787 when it was noted "that many workmen glue their wounds with solid glue dissolved in water" (Haring, 1972). Hide glue is similar to gelatin, which itself derives from collagen. Other biological adhesives such as blood and egg white have also been known for centuries; however, first attempts to develop adhesives with specific chemical structures began in the late 1940s and 1950s.

Natural materials such as cross-linked gelatin and thrombin-plasma were investigated, but a major stimulus was provided by the discovery in 1951 of methyl 2-cyanoacrylate by Coover *et al.* (1972). This clear liquid monomer and its higher homologs (ethyl, butyl, octyl, etc.) were found to polymerize rapidly in the presence of moisture or blood, giving rapid hemostasis and highly adherent films. Extensive clinical and laboratory investigations on the cyanoacrylates took place in the 1960s and 1970s (Matsumoto, 1972), but problems of manipulation and biocompatibility, including reports of cancer in laboratory animals, have limited their current use to surface applications on oral mucosa and life-threatening arteriovenous situations.

The discovery of the adhesive properties of the cyanoacrylates prompted numerous studies on synthetic adhesive systems designed to interact with tissue protein side chain groups to achieve chemical bonding (Cooper *et al.*, 1972). Few systems have been found to possess the requisite combination of biocompatibility, ease of manipulation, and effectiveness. As a result of this experience and the more strictly controlled regulatory situation of today, little new research is being done on novel tissue adhesives. Work has been reported on synthetic polymerizable systems containing the reactive cyanoacrylate or isocyanate groups, but attention has been more focused clinically on materials on a natural basis. Some studies still continue on the gelatin–resorcinol–formaldehyde (GRF) combination (Cooper *et al.*, 1972; Chopin *et al.*, 1989; Nakayama *et al.*, 1994) but the main emphasis has been on fibrin glues derived from a fibrinogen–thrombin combination (Schlag and Redl, 1987).

As with soft tissues, interest in adhesive bonding to calcified tissues as a replacement for, or supplementation to, gross mechanical fixation such as screws has developed mainly in this century and particularly in the past 30 years. Fixation of orthopedic joint components by a cement dates back at least to Gluck (1891), and retention of metal or ceramic inlays and crowns on teeth by dental cements to about 1880. The development of acrylic room temperature polymerizing (cold-curing) systems for dental filling applications in the 1950s led to their use as dental cements and later to their application for fixation of hip joint components by Charnley and Smith (Charnley, 1970; Smith, 1971). These situations involved bonding by mechanical interlocking into surface irregularities. In the case of tooth restorations, leakage along the bonded interface developed. This so-called microleakage led to an intensive effort over the past 30 years to develop adhesive dental cements and filling (restorative) materials (Phillips and Ryge, 1961; Smith, 1991, 1994).

Bonding materials and techniques are now a major component of clinical dentistry (Neuse and Mizrahi, 1994). Effective clinical bonding of polymerizable fluid dimethacrylate monomers and composite formulations to dental enamel, the most highly calcified (98%) tissue in the body, has been achieved by using phosphoric acid etching of the surface (the "acid-etch" technique). Bonding to tooth dentin is currently achieved by using acidic primer monomeric systems containing functional groups such as polycarboxylate or polyphosphate and hydrophilic monomers such as hydroxy ethyl methacrylate (Asmussen and Hansen, 1993; Johnson *et al.*, 1991; Vanherle *et al.*, 1993). Similar materials have been investigated for adhesion to bone, which is compositionally similar to dentin (Lee and Brauer, 1989).

BACKGROUND CONCEPTS

As indicated previously, significant advances in adhesive biomaterials have occurred over the past 30–40 years as real progress has taken place in the science and technology of adhesion and adhesives. This development is continuing since the fundamental aspects of the formation of adhesive bonds at interfaces are not yet fully understood even though successful application of adhesives in technically demanding situations has been achieved.

Experience and, to some extent, theory have shown that severe hostile environments such as biological milieu may require specific surface pretreatments for the surfaces being joined in addition to selection of an adhesive with appropriate characteristics. Such surface pretreatments may involve cleaning or etching processes designed to remove contaminants and expose wettable surfaces and may require the application of primers to achieve specific chemical reactivity at the surface. These procedures are a reflection of the need for intimate interfacial contact between the bonding agent and the adherends in order to form adhesive bonds across the interface. These adhesive forces must hold the materials together throughout the required service life of the joint. However, it must also be appreciated that the factors of the design of an adhesive joint, the applied loads, and the service environment it must withstand will all affect its mechanical performance and life expectancy (Kinloch, 1987).

The establishment of intimate molecular contact between the adhesive and adherend requires, ideally, the adhesive and/or primer to (1) exhibit a zero or near zero contact angle when liquid (2) have a low viscosity during bonding and (3) be able to displace air and contaminants during application. As discussed elsewhere in this volume (Chapters 1.3 and 9.7) surface wetting to achieve these requirements involves an understanding of wetting equilibria on clean, high-energy surfaces, the kinetics of spreading of the adhesive, and the minimization of surface contaminants, including moisture, during the bonding process.

Four main mechanisms of adhesion at the molecular level have been proposed: (1) mechanical interlocking, (2) adsorp-

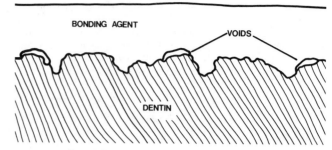

FIG. 1. Diagrammatic representation of mechanical interlocking by a cement to tooth dentin. Note voids at interface due to imperfect adaption.

tion (including chemical bonding), (3) diffusion theory, and (4) electronic theory. More complex interpretations have been proposed (Schulz and Nardin, 1994) but the validity of each theory is influenced by the system under consideration.

Mechanical Interlocking

This adhesion involves the penetration of the bonding agent into surface irregularities or porosity in the substrate surface. Gross examples of this mechanism include the retention of dental filling materials in mechanically prepared tooth cavities and of crowns by dental cements on teeth (Fig. 1) and the fixation of artificial joint components by acrylic bone cement (Fig. 2). Even apparently smooth surfaces are pitted and rough at the microscopic level, and strong bonding can arise with an adhesive that can penetrate at this level. The use of primers (chemical pretreatments) can create surface irregularities or porosities at the microscopic level, or can deposit porous layers that similarly provide effective micromechanical interlocking. Examples include the etching of dental enamel by 35–40% phosphoric acid (Fig. 3) and primer treatment of tooth dentin with acidic agents (Fig. 4). In each case the unpolymerized bonding agent penetrates 5–50 μm into the surface, creating numerous resin "tags" that provide a strong bond.

Adsorption Theory

This theory postulates that if intimate interfacial molecular contact is achieved, interatomic and intermolecular forces will establish a strong joint. Such forces include van der Waals and hydrogen bonds, donor–acceptor bonds involving acid–base interactions, and primary bond (ionic, covalent, metallic) formation (chemisorption). Numerous studies have suggested that secondary bonds (van der Waals and hydrogen bonds) alone are sufficient to establish strong bonding. However, where environmental attack is severe (e.g., by water as in biological systems), the formation of primary bonds across the interface seems to be essential.

Evidence of such bond formation has been found for commercial adhesives, particularly as a result of the use of chemical primers (e.g., silane coupling agents) on ceramics (Kinlock,

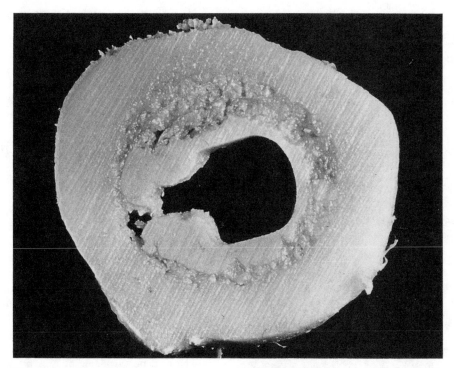

FIG. 2. Section through femur after removal of stem of hip prosthesis showing mechanical interlocking by bone cement into cancellous bone. (After J. Charnley, personal communication.)

FIG. 3. Dental enamel etched by 30-sec treatment with 35% phosphoric acid showing prismatic structure. Prisms are about 5 μm in diameter.

FIG. 4. Treatment of dentin surface by acidic primer showing demineralized collagen fibers in surface zone.

Diffusion Theory

This theory states that the intrinsic adhesion of polymers to substrates and each other involves mutual diffusion of polymer molecules or segments across the interface. This can occur only when sufficient chain mobility is present. The application of this theory is limited to specific situations. Diffusion of polymers into intimate contact with metallic or ceramic surfaces may in fact result in enhanced adsorption or even micromechanical interlocking as a source of improved bonding. This concept (and others) has led to the idea of an "interphase" that is formed between adhesive and substrate which influences bonding behavior.

Electronic Theory

Electronic theory postulates that electronic transfer between adhesive and adherent may lead to electrostatic forces that result in high intrinsic adhesion. Such interactions may arise in certain specialized situations, but for typical adhesive–substrate interfaces, any electrical double layer generated does not contribute significantly to the observed adhesion (Kinlock 1987; Schulz and Nardin, 1994).

The evidence available at present suggests that for most biological adhesives, adsorption phenomena or micromechanical interlocking account for the bond formation and behavior observed. Since few practical surfaces, especially tissues, are completely smooth and nonporous, it is likely that both mechanisms exist in practical clinical situations, with one or the other predominating according to the type of adhesive system, surface preparation technique, and bonding environment.

1987; Schulz and Nardin, 1994). In the biomedical field, the polyacrylic acid-based dental cements (zinc polycarboxylate and glass ionomer cements) (Fig. 5) have been shown to undergo carboxylate bonding with Ca in enamel and dentin (Smith, 1994). Silane primers are used also to form bonds between dental resin adhesives and dental ceramics. Primary bond formation has also been postulated in several reactive dentin bonding systems (Asmussen and Hansen, 1993), but no unequivocal evidence for this has yet been presented (Eliades, 1993).

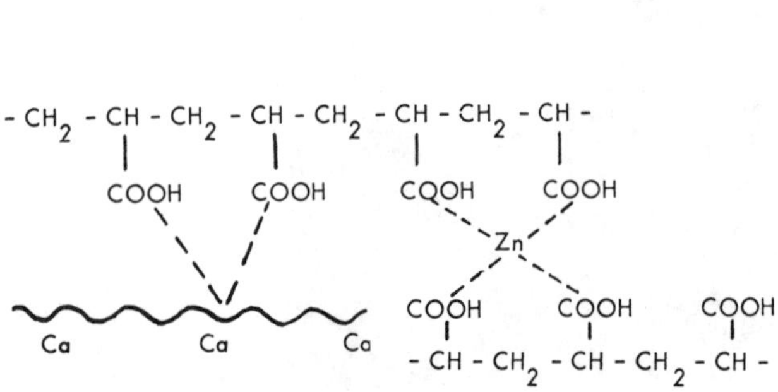

$$- CH_2 - CH - CH_2 - CH - CH_2 - CH - CH_2 - CH -$$

FIG. 5. Diagrammatic representation of setting of zinc polyacrylate and bonding to calcific surface.

COMPOSITION AND CHARACTERISTICS OF ADHESIVE BIOMATERIALS

Soft Tissue Adhesives

Most soft tissue adhesives are intended to be temporary, that is, they are removed or degrade when wound healing is sufficiently advanced for the tissue to maintain its integrity. Effective adhesion can be obtained on dry skin or wound surfaces by using wound dressing strips with acrylate-based adhesives. However, on wound surfaces that are wet with tissue fluid or blood, the adhesive must be able to be spread on such wet surfaces, provide adequate working time, develop and maintain adhesion, desirably provide hemostasis, facilitate wound healing, and maintain biocompatibility. Positive antimicrobial action would be an additional advantage.

Few, if any, systems comply with all these requirements. Currently, there are two principal systems in clinical use—cyanoacrylate esters and fibrin sealants. Another glue based on a gelatin–resorcinol–formaldehyde combination still receives limited use. An interesting but still experimental system based on polypeptides from marine organisms (mussel adhesive) does not seem to have developed into practical use.

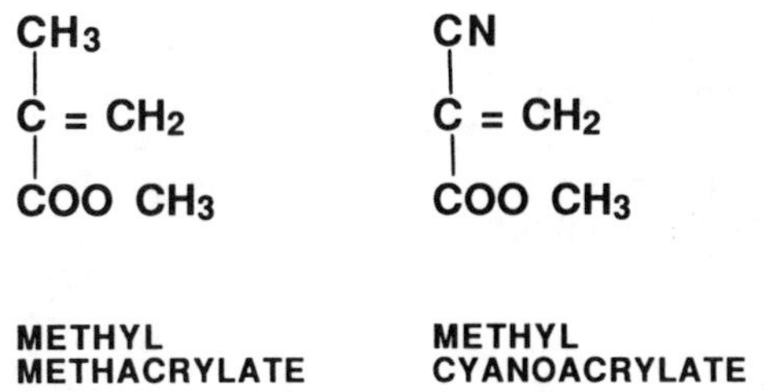

CH₃ structure and methyl cyanoacrylate / methyl methacrylate figures:

$$CH_3 \atop | \atop C=CH_2 \atop | \atop COO\ CH_3$$

**METHYL
METHACRYLATE**

$$CN \atop | \atop C=CH_2 \atop | \atop COO\ CH_3$$

**METHYL
CYANOACRYLATE**

FIG. 6. Structure of methyl cyanoacrylate and methyl methacrylate.

Cyanoacrylate Esters

These esters are fluid, water-white monomers that polymerize rapidly by an anionic mechanism in the presence of weak bases such as water or NH_2 groups. Initially, methyl cyanoacrylate (Fig. 6) was used but in the past decade isobutyl and *n*-butyl cyanoacrylate have been found more acceptable. The higher cyanoacrylates spread more rapidly on wound surfaces and polymerize more rapidly in the presence of blood. Furthermore, they degrade more slowly over several weeks, in contrast to the methyl ester, which hydrolyzes rapidly, yielding formaldehyde that results in an acute inflammatory response.

These materials achieve rapid hemostasis as well as a strong bond to tissue. However, the polymer film is somewhat brittle and can be dislodged on mobile tissue, and the materials can be difficult to apply on large wounds. Because of adverse tissue response and production of tumors in laboratory animals, cyanoacrylates are not approved for routine clinical use in the United States although a commercial material based on *n*-butyl cyanoacrylate (Histoacryl blue) is approved by several other countries.

The current uses are as a surface wound dressing in dental surgery, especially in periodontics, and in life-threatening applications such as brain arteriovenous malformations. Reports of sarcomas in laboratory animals (Reiter, 1987) late complications after dura surgery (Chilla, 1987), and evidence of *in vitro* cytotoxicity (Ciapetti *et al.*, 1994) appear likely to restrict their further use in spite of work on synthesis of new types of cyanoacrylate.

Fibrin Sealants

Fibrin sealants involve the production of a synthetic fibrin clot as an adhesive and wound-covering agent. The concept of using fibrin dates back to 1909 but was placed on a specific basis by Matras *et al.* in 1972 (Schlag and Redl, 1987). The commercial materials first available (Tisseel, Tissucol, Fibrin-Kleber) consisted of two solutions that are mixed immediately before application to provide a controlled fibrin deposition. More recently a "ready to use" formulation (Tisseel Duo) has been introduced.

The essential components of these solutions are as follows:

Solution A	Solution B
Fibrinogen	Thrombin
Aprotinin	CaCl₂

The fibrinogen is at a much higher concentration than that in human plasma. On mixing the two solutions, a reaction similar to that of the final stages of blood clotting occurs in that polymerization of the fibrinogen to fibrin monomers and a white fibrin clot are initiated under the action of thrombin and $CaCl_2$.

Fibrinogen for commercial material is manufactured from the pooled plasma of selected donors. The material is subjected to in-process virus inactivation and routinely screened for hepatitis virus and HIV (Schlag and Redl, 1987). To minimize these risks, recent processes produce the fibrinogen in a "closed" blood bank or from the patient's own blood (Silberstein *et al.*, 1988; Lerner and Binar, 1990). Autologous fibrin glue now appears to be the approach of choice (Tawes *et al.*, 1994).

Fibrin sealant has four main advantages: (1) it is hemostatic, (2) it adheres to connective tissue, (3) it promotes wound healing, and (4) it is biodegradable, with excellent tissue tolerance (Schlag and Redl, 1987). The adhesive strength is not as high as cyanoacrylates but is adequate for many clinical situations. Possible complications include formation of antibodies and thrombin inhibitors. The material has been used in a wide variety of surgical techniques that are reviewed in a seven-volume report by Schlag and Redl (1987) and in numerous papers in the recent literature (Lerner and Binar, 1990; Tawes *et al.*, 1994). The composition may be adjusted to promote hemostasis, for example, or to minimize persistence of the clot to avoid fibrosis. The use of fibrin sealant alone or in admixture with bone chips, tricalcium phosphate, and antibiotics in orthopedic surgery has been reviewed (Schlag and Redl, 1987). More recently, the material has been used as a drug release vehicle for local sites.

Gelatin–Resorcinol–Formaldehyde Glue

This glue was developed in the 1960s by Falb and co-workers (Falb and Cooper, 1966; Cooper *et al.*, 1972) as a less toxic material than methyl cyanoacrylate. The material is fabricated by warming a 3 : 1 mixture of gelatin and resorcinol and adding an 18% formaldehyde solution. Cross-linking takes place in about 30 sec.

This glue was used in a variety of soft tissue applications but technical problems and toxicity have limited its application in recent years to aortic dissection (Nakayama *et al.*, 1994). In attempts to overcome the toxicity and potential mutagenicity/carcinogenicity of the formaldehyde component, modified formulations have been developed in which other aldehydes such as glutaraldehyde and glyoxal (Ennker *et al.*, 1994a,b) are substituted for the formaldehyde. Favorable results with this material (GR-DIAL) have been reported (Ennker *et al.*, 1994a,b).

Bioadhesives

Bioadhesives are involved in cell-to-cell adhesion, adhesion between living and nonliving parts of an organism, and adhesion between an organism and foreign surfaces. Adhesives produced by marine organisms such as the barnacle and the mussel have been extensively investigated over the past 20 years be-

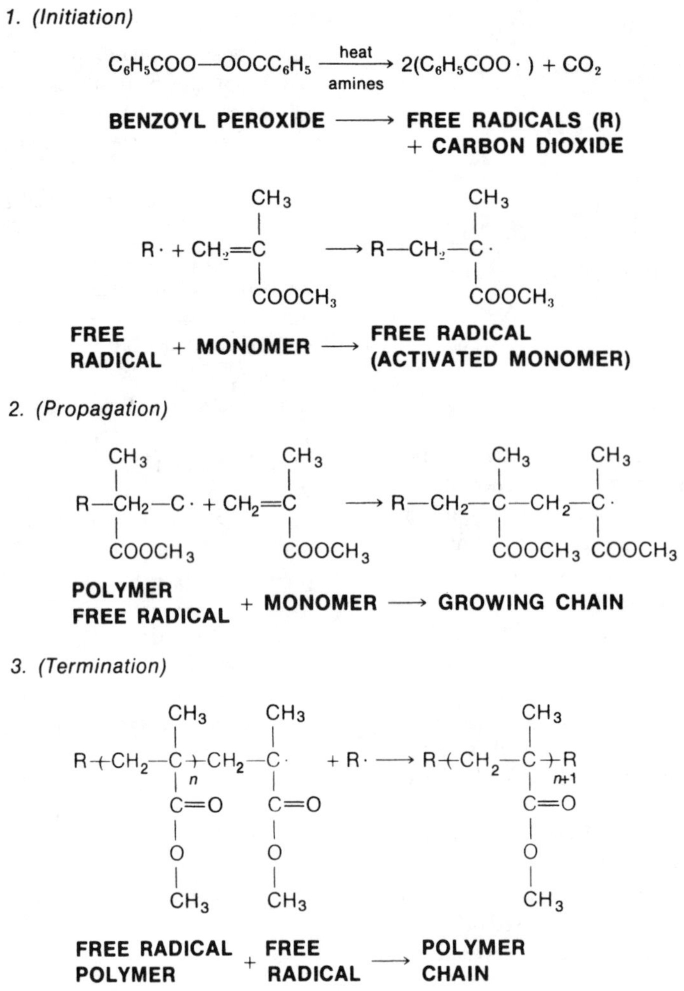

1. (Initiation)

$$C_6H_5COO—OOCC_6H_5 \xrightarrow[\text{amines}]{\text{heat}} 2(C_6H_5COO\cdot) + CO_2$$

BENZOYL PEROXIDE ⟶ **FREE RADICALS (R)**
+ CARBON DIOXIDE

$$R\cdot + CH_2{=}\underset{\underset{COOCH_3}{|}}{\overset{\overset{CH_3}{|}}{C}} \longrightarrow R—CH_2—\underset{\underset{COOCH_3}{|}}{\overset{\overset{CH_3}{|}}{C}}\cdot$$

FREE
RADICAL + **MONOMER** ⟶ **FREE RADICAL**
(ACTIVATED MONOMER)

2. (Propagation)

$$R—CH_2—\underset{\underset{COOCH_3}{|}}{\overset{\overset{CH_3}{|}}{C}}\cdot + CH_2{=}\underset{\underset{COOCH_3}{|}}{\overset{\overset{CH_3}{|}}{C}} \longrightarrow R—CH_2—\underset{\underset{COOCH_3}{|}}{\overset{\overset{CH_3}{|}}{C}}—CH_2—\underset{\underset{COOCH_3}{|}}{\overset{\overset{CH_3}{|}}{C}}\cdot$$

POLYMER
FREE RADICAL + **MONOMER** ⟶ **GROWING CHAIN**

3. (Termination)

$$R{+}CH_2—\underset{\underset{\underset{CH_3}{|}}{\underset{O}{|}}}{\overset{\overset{CH_3}{|}}{\underset{C{=}O}{C}}}{+}_n CH_2—\underset{\underset{\underset{CH_3}{|}}{\underset{O}{|}}}{\overset{\overset{CH_3}{|}}{\underset{C{=}O}{C}}}\cdot + R\cdot \longrightarrow R{+}CH_2—\underset{\underset{\underset{CH_3}{|}}{\underset{O}{|}}}{\overset{\overset{CH_3}{|}}{\underset{C{=}O}{C}}}{+}_{n+1} R$$

FREE RADICAL
POLYMER + **FREE**
RADICAL ⟶ **POLYMER**
CHAIN

FIG. 7. Auto-polymerizing methyl methacrylate systems as used in dental resins and acrylic bone cement. (From R. Roydhouse (1989), in *Dental Materials Properties and Selection*, W. J. O'Brien, ed., p. 129. Quintessence Books, Chicago, with permission.)

cause of their apparent stable adhesion to a variety of surfaces under adverse aqueous conditions. These studies have shown that these organisms secrete a liquid acidic protein adhesive that is cross-linked by a simultaneously secreted enzyme system. The bonding probably involves hydrogen and ionic bonding from the acidic groups (Waite *et al.*, 1989).

The adhesive from the mussel has been identified as a polyphenolic protein, molecular weight about 130,000 Da, which is cross-linked by a catechol oxidase system in about 3 min. A limiting factor in the practical use of this material is the difficulty of extracting it from the natural source. The basic unit of the polyphenolic protein has been identified as a specific decapeptide. Recombinant DNA technology and peptide synthesis have been used in attempts to produce an affordable adhesive with superior properties. Little information has been reported on the performance of these materials.

Hard Tissue Adhesives

As previously noted, prostheses can be attached to calcified tissues (bone, tooth enamel, dentin) by gross mechanical interlocking to machined surfaces. Thus, room temperature-polymerizing methyl methacrylate (Fig. 7) systems are used to fix orthopedic implants (e.g., acrylic bone cement, see Chapter 7.7) and for dental restorations. The former, in a closed system, has been relatively successful. However, conditions are much more stringent in the mouth because of the changing environment, thermomechanical stresses on the bond, and the presence of oral bacteria that result in renewed tooth decay. Thus, considerable development of new dental cements and adhesive systems has occurred in recent years in attempts to provide a leakproof bond to attach fillings, crowns, and veneers to the tooth (Fig. 8).

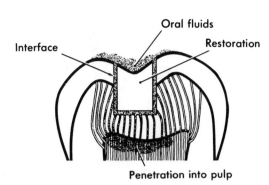

Oral fluids

Interface

Restoration

Penetration into pulp

FIG. 8. Leakage of oral fluids and bacteria around dental filling material in tooth crown. (From R. W. Phillips (1991). *Science of Dental Materials*, 9th Ed., p. 62. W. B. Saunders, Philadelphia, with permission.)

Dental cements are traditionally fast-setting pastes obtained by mixing solid–liquid components. Most of these materials set by an acid–base reaction and more recent resin cements set by polymerization (Smith, 1988, 1991).

Zinc phosphate cement is the traditional standard. This material is composed primarily of zinc oxide powder and a 50% phosphoric acid solution containing Al and Zn. The mixed material sets to a hard, rigid cement (Table 1) by forming an amorphous zinc phosphate binder. Although the cement is gradually soluble in oral fluids and can irritate pulp, it is clinically effective over 10–20-year periods. The bonding arises entirely from penetration into mechanically produced irregularities on the surface of the prepared tooth and the fabricated restorative material. Some interfacial leakage occurs because of cement porosity and imperfect adaptation (Fig. 1), but this is usually acceptable since the film thickness is generally below 100 μm.

Polycarboxylic acid cements were developed in 1968 (Smith 1988, 1994) to provide materials with properties comparable to phosphate cements but that would adhere to calcified tissues. Zinc polyacrylate (polycarboxylate) cements are formed from zinc oxide and a polyacrylic acid solution. The metal ion cross links the polymer structure via carboxyl groups, and other

carboxyl groups form a complex to Ca ions in the surface of the tissue (Fig. 5). The zinc polycarboxylate cements have adequate physical properties, excellent biocompatibility in the tooth, and proven adhesion to enamel and dentin (Smith, 1988).

The glass ionomer cements are also based on polyacrylic acid or its copolymers with itaconic or maleic acids, but utilize a calcium aluminosilicate glass powder instead of zinc oxide (Smith, 1988, 1994). In this case, the cements set by cross-linking of the polyacid with Ca and Al ions from the glass. The set structure and the residual glass particles yield a stronger, more rigid cement (Table 1) but with adhesive properties similar to the zinc polyacrylate cements. In recent materials the polyacid molecule contains both ionic carboxylate and polymerizable methacrylate groups and is induced to set both by an acid–base reaction and visible light polymerization. These cements are widely used clinically. Adhesive bonding but not complete sealing is obtained because of imperfect adaptation to the bonded surfaces under practical conditions.

Resin cements are fluid or pastelike monomer systems based on aromatic or urethane dimethacrylates (Fig. 9). Silanated ceramic fillers may be present to yield a composite composition. The two-component materials polymerize on mixing through a two-part organic peroxide–tertiary amine initiator–activator system in about 3 min. More recent are one-component materials containing diketone polymerization initiators that achieve polymerization in about 30 sec by exposure to visible (blue) light energy. These set materials are strong, hard, rigid, insoluble, cross-linked polymers (Table 1). Bonding is achieved by mechanical interlocking to surface roughness. In recent materials, reactive adhesive monomers may also be present (see the following discussion), conferring also a presumed chemisorption mechanism.

Enamel and dentin bonding systems are composite polymer–ceramic formulations similar to resin cements but are more complex systems containing reactive monomers. Their use usually involves an acidic pretreatment of the tooth surface, an unfilled monomer bonding (primer resin) composition to achieve good wetting of the tooth surface, and a filled bonding agent for the bulk of the bond. These materials are used to

TABLE 1 Properties of Dental Cements and Sealants

Material	Strength (MPa)		Modulus of elasticity (GPa)	Fracture toughness K_1C MN$^{-3/2}$
	Compressive	Tensile		
Zinc phosphate	80–100	5–7	13	~0.2
Zinc polycarboxylate	55–85	8–12	5–6	0.4–0.5
Glass ionomer	70–200	6–7	7–8	0.3–0.4
Resin sealant unfilled	90–100	20–25	2	0.5
Resin sealant filled	150	30	5	
Resin cement	100–200	30–40	4–6	
Composite resin filling material	350–400	45–70	15–20	1.6

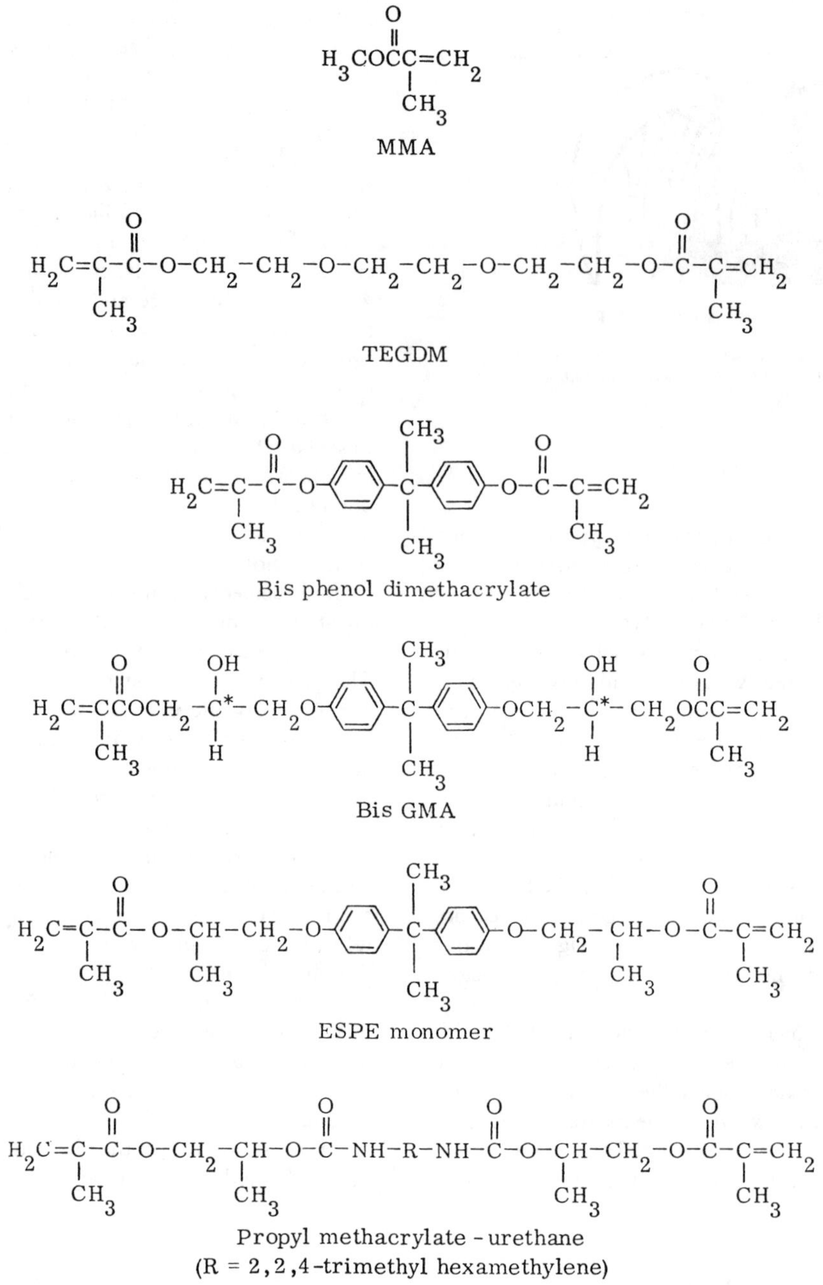

$$
\underset{\text{MMA}}{\text{H}_3\text{COCC}=\text{CH}_2}
$$

$$
\overset{\text{O}}{\underset{\text{CH}_3}{\overset{\|}{\text{H}_3\text{COCC}=\text{CH}_2}}}
$$

MMA

$$
\text{H}_2\text{C}=\text{C}-\text{C}-\text{O}-\text{CH}_2-\text{CH}_2-\text{O}-\text{CH}_2-\text{CH}_2-\text{O}-\text{CH}_2-\text{CH}_2-\text{O}-\text{C}-\text{C}=\text{CH}_2
$$

TEGDM

Bis phenol dimethacrylate

Bis GMA

ESPE monomer

Propyl methacrylate - urethane
(R = 2,2,4-trimethyl hexamethylene)

FIG. 9.   Structures of some dimethacrylate monomers used in dental composite filling and bonding systems.

attach composite resin restorative materials, ceramic veneers, and orthodontic metal and ceramic brackets to enamel and dentin surfaces. Because of the different composition and physical properties of enamel and dentin (see Chapter 3.4), more complex and greater demands are placed on multipurpose adhesive systems intended for both tissues.

Bonding to enamel is achieved by pretreating the surface with 35–50% phosphoric acid for 30–60 sec as described earlier (Fig. 3). This resulting washed and dried surface is readily wettable and penetratable by resin cements and bond-

ing agents. The resulting resin tags (5–50 μm long) in the surface of the tissue result in efficient micromechanical interlocking with a potential tensile bond strength of about 20 MPa, which is equivalent to cohesive failure in the resin or in the enamel.

Bonding to dentin currently involves pretreatment of the prepared (machined) surface with acidic solutions (phosphoric, nitric, maleic acids) or ethylenediaminetetracetic acid (EDTA) to remove cutting debris (the smear layer). This procedure opens the orifices of the cut dentinal tubules and creates micro-

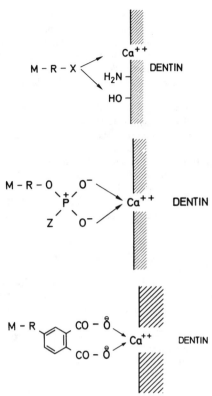

FIG. 10. Reactive monomer structures for bonding to calcific tissues. M-R, monomer portion of molecule. (After Asmussen, E., Aranjo, P. A., and Pentsfeld, A. 1989, *Trans. Acad. Dent. Mater.* 2: 59.)

porosity in the surface (Fig. 4). A primer treatment is then applied that comprises a reactive monomer system (Fig. 10) containing a carboxylate or a polyphosphate function, depending on the type of product. These primers also contain hydrophilic monomers, such as hydroxyethyl methacrylate (HEMA), and may also contain water.

The function of the primer is to penetrate the demineralized dentin surface and facilitate wetting by an unfilled dimethacrylate bonding resin which is subsequently applied. Polymerization of this treatment layer by visible light activation results in the formation of micromechanical bonds by penetration into the dentin and surface tubules, forming the so-called hybrid layer (Nakabayashi *et al.*, 1991) or resin-interdiffusion zone (Van Meerbeck *et al.*, 1992). Chemical interaction with the hydroxyapatite and/or proteinaceous phases of the dentin surface may also occur (Asmussen and Hansen, 1993). However, direct chemical evidence has not been provided yet for the postulated interactions (Eliades, 1993). Under the best conditions, initial tensile bond strengths of 15–25 MPa can be obtained depending on test conditions. The long-term durability of these bonds under oral conditions is being investigated.

NEW RESEARCH DIRECTIONS

As a result of the experience of the past two decades, the problems involved in developing an adhesive system for both soft and hard tissues have been addressed. Nevertheless, it is difficult to reconcile short- and long-term biocompatibility needs with chemical adhesion mechanisms that use reactive monomer systems.

Where relatively temporary adhesion is required, as in wound healing, systems based on natural models that allow biodegradation of the adhesive and interface and subsequent normal tissue remodeling appear to merit further development. For longer term durability in both soft and hard tissues, hydrophilic monomers and polymers of low toxicity which can both diffuse into the tissue surface and form ionic bonds across the interface seem to be the most promising approaches. Evidence has been obtained of the need for hydrophobic–hydrophilic balance in adhesive monomer systems (Nakabayashi *et al.*, 1991). The use of hydrophilic monomers such as hydroxyethyl methacrylate in commercial materials has facilitated surface penetration.

On calcified surfaces, the use of hydrophilic electrolytes such as the polycarboxylates has demonstrated that proven ionic bonding *in vitro* can also be achieved *in vivo*. An advantage of such systems is that surface molecular reorientations can improve bonding with time. Encouraging preliminary results have been obtained with such glass ionomer cements in orthopedics and there is considerable scope for the future development of such polyelectrolyte cements.

A practical limitation in many systems is ease of manipulation and application. For example, the effectiveness of the fibrin sealant is critically dependent on proper mixing of the ingredients and uniform application. Further technology transfer could improve this often-neglected area of adhesive development. For example, the visible-light polymerization technology developed in dentistry that allows extended working time and curing "on demand" in a few seconds could usefully be applied to medical applications. Laser activation of fibrin sealants has received an initial trial.

The development of more efficient adhesives and sealants that, in addition to enhancing the durability of current applications, would permit new applications such as osteogenic bone space fillers, percutaneous and permucosal seals, and functional attachment of prostheses is a challenging problem for the future.

Bibliography

Asmussen, E., de Aranjo, P. A., Peutzfeldt, A. (1989). *In vitro* bonding of resins to enamel and dentin: an update. *Trans. Acad. Dent. Mater.* 2: 36–63.

Asmussen, E. and Hansen, E. K. (1993). Dentine bonding systems. in *State of the Art on Direct Posterior Filling Materials and Dentine Bonding*, G. Vanherle, M. Degrange, G. Willems, eds. Cavex Holland BV, Haarlem, pp. 33–47.

Charnley, J. (1970). *Acrylic Cement in Orthopaedic Surgery.* E. S. Livingstone, Edinburgh.

Chilla, R. (1987). Histoacryl-induzierte Spätkomplikationen nach Duraplastiken an der Fronto- und Otobasis. *HNO* 35: 250–251.

Chopin, D. K., Abbou, C., Lottiman, H. B., Topoz, P., Popov, Z., Lang, T. R., Buisson, C. L., Belghiti, D., Colombel, M., and Auvert, J. M. (1989). Conservative treatment of renal allograft rupture with polyglactin 910 mesh and gelatin resorcinol formaldehyde glue. *J. Urol.* **142**: 363–365.

Ciapetti, G., Stea, S., Cenni, E., Sudanese, A., Marraro, D., Toni, A., and Pizzoferrato, A. (1984). Toxicity of cyanoacrylates *in vitro* using extract dilution assay on cell cultures. *Biomater.* **15**: 92–96.

Cooper, C. W., Grode, G. A., and Falb, R. D. (1972). The chemistry of bonding-alternative approaches to the joining of tissues. in *Tissue Adhesives in Surgery*, T. Matsumoto, ed., Medical Examination Publ., New York, pp. 189–210.

Coover, H. W., Jr., and McIntire, J. M. (1972). The chemistry of cyanoacrylate adhesives. in *Tissue Adhesives in Surgery*, T. Matsumoto, ed., Medical Examination Publ., New York, pp. 154–188.

Eliades, G. C. (1993). Dentin bonding systems. in *State of the Art on Direct Posterior Filling Materials and Dentine Bonding*, G. Vanherle, M. Degrange, and G. Willems, eds. Cavex Holland BV, Haarlem, pp. 49–74.

Ennker, J., Ennker, I. C., Schoon, D., Schoon, H. A., Dorge, S., Messler, M., Rimpler, M., and Hetzer, R. (1994a). The impact of gelatin-resorcinol-formaldehyde glue on aortic tissue: a histomorphologic examination. *J. Vascular Surg.* **20**: 34–43.

Ennker, I. C., Ennker, J., Schoon, D., Schoon, H. A., Rimpler, M., and Hetzer, R. (1994b). Formaldehyde free collagen glue in experimental lung gluing. *Ann. Thoracic Surg.* **57**: 1622–1627.

Falb, R. D., and Cooper, C. W. (1966). Adhesives in surgery. *New Scientist* 308–309.

Gluck, T. (1891). Referat uber die durch das moderne chirugische Experiment gewonnen positiven Resultate, betrifind die Naht und den Ersatz von defecten hoherer Gewebe sowie uber die Verwerthung resorbirbarer und lebendiger Tampons in der Chirurgie. *Langenbecks Archiv fur Klinische Chirugie* **41**: 187–239.

Haring, R. (1972). Current status of tissue adhesives in Germany. in *Tissue Adhesives in Surgery*, T. Matsumoto, ed., Medical Examination Publ., New York, pp. 430.

Johnson, G., Powell, V., and Gordon, G. (1991). Dentin bonding agents: a review. *J. Am. Dent. Assn.* **122**: 34–41.

Kinlock, A. J. (1987). *Adhesion and Adhesives.* Chapman and Hall, London.

Lee, C. H., and Grauer, G. M. (1989). Oligomers with pendant isocyanate groups as adhesives for dentin and other tissues. *J. Dent. Res.* **68**: 484–488.

Lee, L-H. (ed.) (1991a). *Fundamentals of Adhesion.* Plenum Publ., New York.

Lee, L-H. (1991b). *Adhesive Bonding.* Plenum Publ., New York.

Lerner, R., and Binar, N. S. (1990). Current status of surgical adhesives. *J. Surg. Res.* **48**: 165–181.

Matsumoto, T. (1972). *Tissue Adhesives in Surgery.* Medical Examination Publ., New York.

Nakabayashi, N., Nakamura, M., and Yasuda, N. (1991). Hybrid layer as a dentin bonding mechanism. *J. Esthet. Dent.* **3**: 133–135.

Nakayama, Y., Kitamura, S., Kawachi, K., Inoue, K., Tomiguchi, S., Fukutomi, M., Kobayashi, S., Kawata, T., and Yoshida, T. (1994). Efficacy of GRF glue on surgery for Type A aortic dissection. *J. Jpn. Assn. Thoracic Surg.* **42**: 1021–1026.

Neuse, E. W., and Mizrahi, E. (1994). Bonding materials and techniques in dentistry. in *Handbook of Adhesive Technology.* A. Pizzi and K. L. Mittal, eds., pp. 629–656, Marcel Dekker, New York.

Phillips, R. W., and Ryge, G. (eds.) (1961). *Adhesive Restorative Dental Materials*, National Institutes of Health, U.S. Public Health Service, Washington.

Pizzi, A., and Mittal, K. L. (eds.) (1994). *Handbook of Adhesive Technology.* Marcel Dekker, New York.

Schlag, G., and Redl, H. (1987). Fibrin Sealant in Operative Medicine, Vol. 4. *Plastic Surgery—Maxillofacial and Dental Surgery*, Springer-Verlag, Berlin.

Schulz, J., and Nardin, M. (1994). Theories and Mechanisms of Adhesion. in *Handbook of Adhesive Technology,* A. Pizzi and K. L. Mittal, eds., Marcel Dekker, New York, pp. 19–33.

Silberstein, L. E., Williams, L. J., Hughlett, M. A., Magee, D. A., and Weisman, R. A. (1988). An autologous fibrinogen-based adhesive for use in otologic surgery. *Transfusion* **28**: 319–321.

Skeist, I. (1990). *Handbook of Adhesives*, 3rd ed. Van Nostrand Reinhold, New York.

Smith, D. C. (1971). Medical and dental applications of cements. *J. Biomed. Mater. Res. Symp.* **1**: 189–205.

Smith, D. C. (1988). Dental cements. *Adv. Dent. Res.* **2**(1): 134–141.

Smith, D. C. (1991). Dental cements. *Curr. Opin. Dent.* **1**: 228–234.

Smith, D. C. (1994). Development of glass ionomer cement systems. in *Glass Ionomers: The Next Generation*, P. Hunt, ed., International Symposia in Dentology, Philadelphia, pp. 1–12.

Tawes, R. L., Jr., Sydorak, G. R., and DuVall, T. B. (1994). Autologous fibrin glue: the last step in operative haemostasis. *Am. J. Surg.* **168**: 120–122.

Vanherle, G., Degrange, M., and Willems, G. (eds.) (1993). *State of the Art on Direct Posterior Filling Materials and Dentine Bonding.* Cavex Holland BV, Haarlem.

Van Meerbeck, B., Inokoshi, S., Braem, M., Lambrechts, P., and Vanherle, G. (1992). Morphological aspects of resin-dentin interdiffusion zone with different dentin adhesive systems. *J. Dent. Res.* **71**: 1530–1540.

Waite, J. H. (1989). The glue protein of ribbed mussels (*Genkenska denissa*): a natural adhesive with some features of collagen. *J. Comp. Physiol.* [B] **159**(5): 517–525.

7.6 OPHTHALMOLOGIC APPLICATIONS

Miguel F. Refojo

Light that penetrates into the eye is partially refracted in the cornea, passes through the aqueous humor and the pupil (the opening in the center of the iris), is further refracted in the crystalline lens, passes through the vitreous humor, and converges on the retina (Fig. 1). Diverse polymeric devices, such as spectacles, contact lenses, and intraocular implants, are used to correct the optical function of the eye. The materials used in spectacle lenses are outside the scope of this chapter. Contact lenses, however, being in intimate contact with the tissues of the eye, are subject to the same regulations that govern the use of implant materials, and they are included in this chapter with other biomaterials used to preserve and to restore vision, such as intraocular implants.

CONTACT LENSES

General Properties

Contact lenses are optical devices that must have good transmission of visible light. Pigments and dyes are added to some contact lenses for cosmetic effect. Contact lenses also may have ultraviolet (UV) light-absorbing additives, usually copolymerized in the contact lens material, to protect the eye from the harmful effects of UV light. UV light absorbed by the normal crystalline lens is harmful to the retina and also may contribute to the clouding of the lens (cataract) (Miller, 1987).

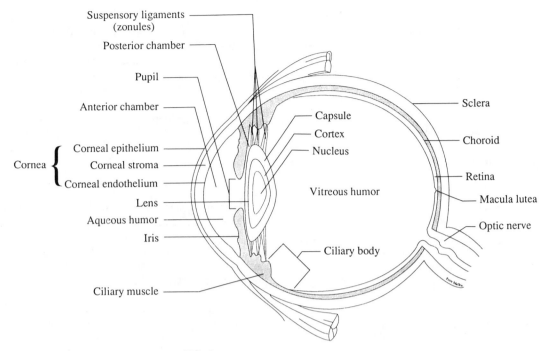

FIG. 1. Schematic representation of the eye.

The principal properties sought in contact lens materials, in addition to the required optical properties, chemical stability, and amenability to manufacture at reasonable cost, are high oxygen transmissibility (to meet the metabolic requirements of the cornea), tear film wettability (for comfort), and resistance to accumulation on the lens surfaces of mucus/protein/lipid deposits from the tear film and other external sources. Contact lenses also must be easy to clean and disinfect (Kastl, 1995).

Most of the available contact lenses were developed with the important property of oxygen permeability in mind. The oxygen permeability coefficient, P, is a property characteristic of a material. [$P = Dk$, where D is the diffusivity, in cm^2/sec, and k is the Henry's law solubility coefficient, in cm^3 (STP)/ cm^3 mm Hg.] For a given contact lens, its oxygen transmissibility (Dk/L) is more important than its permeability; oxygen transmissibility is defined as the oxygen permeability coefficient of the material divided by the average thickness of the lens (L, in cm) (Holden *et al.*, 1990).

For oxygen permeability, the ideal contact lens would be made of poly(dimethyl siloxane). For better mechanical properties and manufacture, most silicone elastomeric lenses have been made of diverse poly(methyl phenyl vinyl siloxanes). Because of its hydrophobic character, to be useful, a silicone rubber lens must be treated in an RF-plasma reactor or other suitable procedure to make its surface hydrophilic and tolerated on the eye. Nevertheless, the silicone rubber lenses have not been very successful for general cosmetic use, not only because of surface problems and comfort, but principally because they have a strong tendency to adhere to the cornea.

There are currently a wide variety of contact lens materials with diverse physical properties that determine the fitting characteristics of the lens on the eye (Kastl, 1995).

Soft Hydrogel Contact Lenses

The soft hydrogel contact lenses (SCL) are supple and fit snugly on the corneal surface. Because there is little tear exchange under these lenses, most of the oxygen that reaches the cornea must permeate through the lens. The oxygen permeability coefficient of hydrogel materials increases exponentially with the water content.

The hydrogel lenses are made of slightly cross-linked hydrophilic polymers and copolymers. The original hydrogel contact lens material was poly(2-hydroxyethyl methacrylate) (PHEMA) (Wichterle and Lim, 1960); at equilibrium swelling in physiological saline solution, it contains about 40% water of hydration. (Hydration of hydrogel contact lenses is customarily given as a percentage of water by weight, on a wet basis.) The oxygen transmissibility of the original rather thick PHEMA hydrogel contact lenses was found to be insufficient for normal corneal metabolism. New hydrogel contact lenses were soon developed with higher water content or with a water content similar to that of PHEMA but more amenable to fabrication in an ultrathin modality. New fabrication techniques were also developed to make ultrathin PHEMA lenses. This takes advantage of the law of diffusion which, applied to contact lenses, will guarantee that for any lens type under the same conditions of wear, the oxygen flux through the lens will double when the thickness is halved.

Other hydrogel contact lens materials include HEMA copolymers with other monomers such as methacrylic acid, acetone acrylamide, and vinyl pyrrolidone. Commonly used also are copolymers of vinyl pyrrolidone and methyl methacrylate, and of glyceryl methacrylate and methyl methacrylate. A variety of other monomers as well as a variety of cross-linking agents

TABLE 1 Chemical Composition of Some Hydrogel
Contact Lenses

Polymer	USAN[a]	% H$_2$O
2-hydroxyethyl methacrylate (HEMA) with ethyleneglycol dimethacrylate (EGDM)	Polymacon	38
HEMA with methacrylic acid (MAA) and EGDM	Ocufilcon A	44
	Ocufilcon C	55
HEMA with sodium methacrylate and 2-ethyl-2-(hydroxymethyl)-1,3-propanediol trimethacrylate	Etafilcon A	58
HEMA with divinyl benzene, methyl methacrylate (MMA) and 1-vinyl-2-pyrrolidone (VP)	Tetrafilcon A	43
HEMA with VP and MAA	Perfilcon A	71
HEMA with N-(1,1-dimethyl-3-oxobutyl) acrylamide and 2-ethyl-2-(hydroxymethyl)-1,3-propanediol trimethacrylate	Bufilcon A	45
	Bufilcon B	55
2,3-Dihydroxypropyl methacrylate with MMA	Crofilcon A	39
VP with MMA, allyl methacrylate and EGDM	Lidofilcon A	70
	Lidofilcon B	79
MAA with HEMA, VP and EGDM	Vifilcon A	55

[a]U.S. adopted name.

are used as minor ingredients in hydrogel contact lenses (Refojo, 1979) (Table 1).

Hydrogel lenses have been classified by the U.S. Food and Drug Administration (FDA) into four general groups: low water (<50% H$_2$O), nonionic; high water (>50% H$_2$O), nonionic; low water, ionic; and high water, ionic. The ionic character is usually due to the presence of methacrylic acid, which is responsible for higher surface protein binding to the contact lenses. High water of hydration is a desirable property for good oxygen permeability, but it carries some disadvantages, such as friability and protein penetration into the polymer network. Physiologically and optically, ultrathin low-water-content contact lenses can perform very well as daily-wear lenses.

As a result of temperature changes and water evaporation, all hydrogel contact lenses dehydrate to some degree on the eye. Higher-water-content lenses dehydrate more than low-water-content lenses, and thin lenses dehydrate more easily than thick lenses (Refojo, 1991). A drawback of high-water-content, thin hydrogel contact lenses is that as they dehydrate on the eye, they induce corneal epithelium injuries by a mechanism still unclear. Therefore, the ideal hydrogel contact lens would be ultrathin, resistant to mechanical damage, made of a nonionic polymer, and retain a high water content (i.e., >70% H$_2$O) on the eye.

Flexible Fluoropolymer Lenses

The flexible fluoropolymer (FFP) lens was made from a copolymer of a telechelic perfluoropolyether (which imparts high oxygen permeability) with vinyl pyrrolidone (which imparts wettability) and methyl methacrylate (which imparts rigidity). This flexible, nonhydrated contact lens, made by the molding procedure, had a high oxygen permeability and, owing to its high fluorine content, was claimed to be more resistant to coating by tear proteins than other contact lens materials. At this time, the FFP lenses are no longer commercially available.

Rigid Contact Lenses

The rigid contact lenses, as well as the FFP lenses, fit loosely on the cornea and move with the blink more or less freely over the tear film that separates the lens from the corneal surface. The mechanical properties of rigid and FFP contact lenses must be such that any flex on the lens provoked by the blink must recover instantaneously at the end of the blink.

The first widely available contact lenses were made of poly-(methyl methacrylate), which is an excellent optical biomaterial in almost all respects except for its virtual impermeability to oxygen. Several materials that were specially developed for the manufacture of rigid gas-permeable (RGP) contact lenses are copolymers of methyl methacrylate with siloxanylalkyl methacrylates (Refojo and Dabezies, 1984). To compensate for the hydrophobic character imparted to the polymer by the high siloxane content of these copolymers (required for oxygen permeability), the copolymer also contains some hydrophilic comonomers. The most commonly used hydrophilic comonomer in rigid lenses is methacrylic acid. There are also minor ingredients and cross-linking agents. A diversity of RGP contact lenses, consisting of different but closely related comonomers used in a variety of proportions to obtain the most desirable properties, are commercially available (Table 2). However, any subtle change in the chemistry of a contact lens material might strongly affect its clinical performance. As a general rule, the oxygen permeability coefficient of the siloxanylalkyl methacrylate contact lens materials is inversely proportional to the density.

The development of the fluorine-containing contact lenses and the realization that the fluoroderivatives may improve oxygen permeability and resistance to deposit formation caused contact lens chemists to include a fluoroalkyl methacrylate or a similar fluorine-content monomer as an additional ingredient in the siloxanylalkyl methacrylate-comethyl methacrylate RGP contact lens materials. These perfluoroalkyl-siloxanylalkyl-methyl methacrylate contact lenses have high oxygen permeability and, supposedly, better surface properties than the non-fluorine-containing rigid contact lenses.

Cellulose acetate butyrate (CAB) is also used as a rigid oxygen-permeable contact lens material. However, CAB not only has relatively low oxygen permeability compared with the siloxanylalkyl methacrylate copolymers but also has low scratch resistance and tends to warp with humidity changes.

Other copolymers useful as contact lens materials are isobutyl and isopropyl styrene, with hydrophilic comonomers of the HEMA or vinyl pyrrolidone type.

TABLE 2 Composition of Some Rigid Gas-Permeable
Contact Lenses

Polymer	USAN[a]
Cellulose acetate dibutyrate	Porofocon Cabufocon
3-[3,3,5,5,5-pentamethyl-1,1-bis[pentamethyldisiloxanyl)oxy] trisiloxanyl]propyl methacrylate, with methyl methacrylate (MMA), methacrylic acid (MAA) and tetraethyleneglycol dimethacrylate (TEGDMA)	Silafocon
MMA with MAA, EGDMA, 3-[3,3,3,-trimethyl-1,1-bis(trimethylsiloxy)disiloxanyl] propyl methacrylate (TRIS) and N-(1,1-dimethyl-3-oxybutyl)acrylamide.	Nefocon
VP with HEMA, TRIS, allyl methacrylate and α-methacryloyl-ω-(methacryloxy) poly(oxyethylene-co-oxy(dimethylsilylene)-co-oxyethylene.	Mesifilcon
TRIS with MMA, dimethyl itaconate, MAA and TEGDMA.	Itafocon
TRIS with 2,2,2,-trifluoro-1-(trifluoromethyl) ethyl methacrylate, 1-vinyl-2-pyrrolidone (VP), MAA and ethyleneglycol dimethacrylate (EGDMA).	Melafocon
TRIS with 2,2,2-trifluoroethyl methacrylate, MAA, MMA, VP with EGDMA.	Paflufocon

[a]U.S. adopted name.

CORNEAL IMPLANTS

The cornea is an avascular tissue that consists of three principal layers (Fig. 1). The outermost layer, which itself consists of about five cellular layers, is the epithelium. The central and main portion of the cornea is the stroma, a collagenous connective tissue that is 78% hydrated in its normal state. Normal corneal hydration is disrupted by injury to the limiting epithelial and endothelial membranes. The endothelium is the innermost monocellular layer, which by means of a "pump-leak" mechanism, is mostly responsible for maintaining normal corneal hydration. Swelling, tissue proliferation, and vascularization may compromise the transparency of the cornea. There are several types of corneal implants (Refojo, 1986a; Abel, 1988) that replace all or part of the cornea.

Epikeratophakia and Artificial Epithelium

To correct the optics of the eye after cataract extraction, the surgeon may perform an epikeratophakia procedure which consists of transplanting a slice of donor cornea. The transplanted tissue heals into a groove carved into the recipient corneal surface and is reepithelialized with the recipient corneal epithelium (Werblin *et al.*, 1987). A modification of this technique attempts to obtain similar results with an artificial material that would heal into the donor cornea and be able to grow the epithelium of the donor cornea on its surface (Fig. 2).

An epithelium that has become irregular through swelling and proliferation has been replaced by an artificial epithelium made of a hard plastic contact lens glued with a cyanoacrylate adhesive to the corneal stroma (Fig. 2). This procedure has not been successful mainly because of failure of the glue to maintain a tight attachment of the prosthesis to the corneal stroma and also because of epithelial penetration between the prosthesis and the cornea.

Artificial Corneas

Corneal transplants from donor eyes are usually highly successful. In the rare instance of transplant failure, an opaque cornea can be replaced with an artificial cornea (keratoprosthesis) (Barber, 1988). These are usually through-and-through corneal implants, consisting of a central optical portion and some modality of skirt that fixes the prosthesis to the recipient cornea (Fig. 3). The main problem with through-and-through keratoprostheses is common to all kinds of implants that are not fully buried in the recipient tissue: faulty tissue–prosthesis interface, epithelium downgrowth, and tissue ulceration and infection around the prosthesis. The most feasible solution to these problems would be the development of a material for the optical portion of the keratoprosthesis that would accept growth and attachment of transparent epithelium on its surface. Also needed are biomaterials that would heal into the recipient corneal tissue.

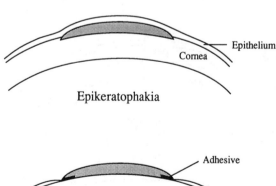

FIG. 2. Schematic representation of superficial corneal implants. (Top) In the epikeratophakia procedure, the corneal epithelium is removed before the implant is placed on the stromal surface and epithelium grows over the implant. (Bottom) The artificial epithelium or epikeratoprosthesis is a contact lens glued to a deepithelialized cornea. Ideally, the epithelium should not grow over or under the glued-on lens.

Artificial Endothelium

The corneal endothelium has been replaced, but not very successfully in the long term, by a silicone rubber membrane that passively controls corneal hydration (Fig. 3). Unfortunately, the membrane serves as a barrier not only to water but also to the nutrients that the cornea normally receives from the aqueous humor.

Intracorneal Implants

Ophthalmic surgeons may use diverse polymeric devices to correct the optical function of the eye. Thus, intracorneal implants can be used instead of spectacles or contact lenses to correct nearsightedness and farsightedness (Fig. 4). The intracorneal implants most likely to succeed are made of hydrogel materials tailored to have high permeability to metabolites and able to correct severe myopia (McCarey *et al.*, 1989). The stromal cells (keratocytes) and the epithelium receive their nutrients from the aqueous humor and also release waste products in the same direction. Therefore, some previously used intrastromal implants, such as poly(methyl methacrylate) and polysulfone, which are impermeable to metabolites, will result in the ulceration and vascularization of the overlying stroma. A more recent development is the intrastromal ring made of poly(methyl methacrylate) or silicone rubber, which may change the corneal curvature and, hence, the eye's optical power. These rings can make the corneal curvature steeper, increasing the refractive power, or flatter, decreasing the refractive power.

IMPLANTS FOR GLAUCOMA

Polymeric devices are used to control abnormally high intraocular pressure in otherwise intractable glaucoma (Krupin

Refractive Keratoplasty

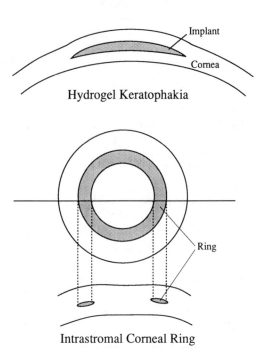

Hydrogel Keratophakia

Intrastromal Corneal Ring

FIG. 4. Schematic representation of intracorneal implants used to change the curvature of the cornea in refractive keratoplasty. (Top) An intrastromal hydrogel intracorneal implant. (Bottom) An intrastromal corneal ring.

et al., 1988). These devices consist essentially of tiny tubes that transport the aqueous humor—which normally maintains the physiological intraocular pressure and flows in and out of the eye in a well-regulated manner—from the anterior chamber to some artificially created space between the sclera and the other tissues that surround the eyeball; then the aqueous humor is absorbed into the blood circulation. The main problem with glaucoma implants is tissue proliferation around the outlet of the plastic device. Tissue proliferation, or capsule formation, takes place in and around all implanted biomaterials and may retard or even stop the outflow of aqueous humor from the eye.

INTRAOCULAR LENS IMPLANTS

Intraocular lenses (IOLs) are used after cataract extraction to replace the opaque crystalline lens of the eye (Apple *et al.*, 1984). IOLs consist of an optical portion and haptics that support the optical portion in its proper place in the eye (Fig. 5). IOLs may be placed in the anterior chamber, in the pupil, and in the posterior chamber. The last type are most commonly used at this time; they are usually placed within the posterior capsule of the crystalline lens, which remains in the eye after the lens contents have been removed surgically (Fig. 5). A large variety of IOL designs and shapes are available; the choice does not necessarily depend on need but rather on the preferences of surgeons and manufacturers.

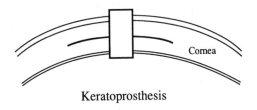

Keratoprosthesis

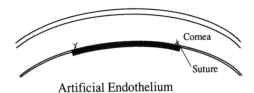

Artificial Endothelium

FIG. 3. (Top) Schematic representation of a through-and-through artificial cornea (kerastoprosthesis) that consists of an optical cylinder that penetrates the opaque tissue. The prosthesis has an intrastromal rim that holds the prosthesis in the cornea. (Bottom) Schematic representation of the artificial corneal endothelium that consists of a transparent membrane sutured to the posterior part of the cornea denuded of its endothelium. This membrane acts as a barrier to the inflow of aqueous fluid into the corneal stroma.

The requirements of IOL materials are good optical properties and biocompatibility with the surrounding tissues. Although oxygen or metabolite permeability is irrelevant for IOLs, one may have to be concerned with the potential absorption in the IOL material of aqueous humor proteins or a topical or systemic drug given to a patient wearing an IOL, particularly if the IOL material is a hydrogel or silicone rubber.

Most IOLs are made of poly(methyl methacrylate), and the haptics are often made of the same material or polypropylene fiber (Apple *et al.*, 1984). Filtration of UV light by UV-absorbing moieties polymerized into the IOL is desirable to protect the retina (Miller, 1987).

Corneal astigmatism may result from tissue distortions occurring as a consequence of the uneven healing of the wound made when the IOL was implanted. There is currently a strong interest in developing soft IOLs, which can be inserted in the eye through smaller surgical incisions that are required for implanting rigid lenses. A smaller incision may result in a lower incidence of astigmatism. Soft IOLs have been made of HEMA or other hydrogels, which can be inserted into the eye fully hydrated or in the dehydrated state; in the latter case they will swell *in situ* to their equilibrium hydration (Barrett *et al.*, 1986). Flexible IOLs are made also of silicone rubber and of alkyl acrylate copolymers.

Biopolymers in the form of a viscoelastic solution are also used in IOL implantation. The corneal endothelium is an extremely delicate cell layer and can be irreversibly damaged upon contact with an IOL, during or after insertion. The surgeon must be extremely careful not to touch the corneal endothelium with the IOL or with any instrument used during surgery. Highly viscous, and preferably viscoelastic, solutions of biopolymers such as sodium hyaluronate, chondroitin sulfate, or hydroxypropyl methylcelluose are useful adjuncts in IOL implant surgery for maintaining anterior chamber depth

during introduction of the implant and for preserving the corneal endothelium (Fernandez-Vigo *et al.*, 1989). Other important developments may be the surface modification of IOLs with permanent hydrophilic or hydrophobic coatings.

IMPLANTS FOR RETINAL DETACHMENT SURGERY

A retina detached from its source of nutrition in the choroidal circulation ceases to be sensitive to light. The choroid is the vascular layer between the retina and the sclera (Fig. 1). In cases of retinal detachment, the surgeon must reattach the retina to restore vision. Retinal detachment could result from traction of a retracting vitreous humor or from seepage of liquefied vitreous through a retinal hole between the retina and the choroid. Retina surgeons can often restore vision to these eyes with vitreous implants and scleral buckling materials (Refojo, 1986b) (Fig. 6).

Vitreous Implants

Vitreous implants are desirable in certain difficult cases of retinal detachment surgery (Refojo, 1986b). Physiological saline solution, air and other gases, as well as a sodium hyaluronate solution frequently are injected into the vitreous cavity during virtreoretinal surgery. These fluids may perform well as a short-term vitreous substitute. For long-term vitreous replacement, however, the only substance used at this time, and with variable results, is silicone oil of high viscosity (1,000 to 12,500 centistokes) and for short-term vitreous replacement, perfluorocarbon compounds of low viscosity. The main problem with long-lasting intravitreous implants is tolerance. Retinal toxicity, oil emulsification, glaucoma, and corneal clouding are some of the complications of permanent vitreous implants. These complications may be avoided by removing the implant after choroidal-retinal adhesion has been achieved, but implant removal involves further surgery, is difficult to achieve completely, and carries the risk of recurrence of the retinal detachment.

Scleral Buckling Materials

Scleral buckling materials for retinal detachment surgery must be soft and elastic. Solid silicone rubber and silicone sponge have been used successfully. More recently an acrylic hydrogel made of a copolymer of 2-hydroxyethyl acrylate with methyl acrylate has become available. It may improve the already relatively small rate of infection resulting from the use of the sponge and the potential for long-term pressure necrosis of the more rigid solid silicone rubber implants (Refojo, 1986b).

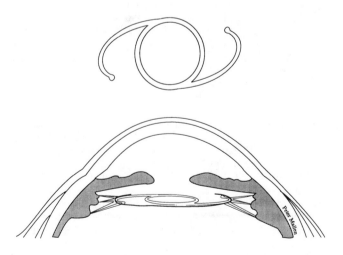

Intraocular Lens

FIG. 5. (Top) Schematic representation of a typical intraocular lens implant with a central optical portion and the haptics or side-arms that hold the lens in the eye. (Bottom) A schematic representation of the anterior segment of the eye with an intraocular lens placed into the empty crystalline lens bag.

SURGICAL ADHESIVES

As in most applications of polymers as biomedical implants, in ophthalmology any polymeric device must be as free as

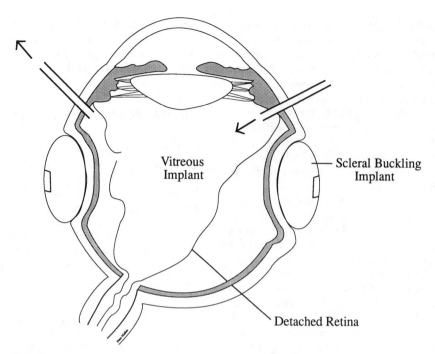

FIG. 6. Schematic representation of an eye with a detached retina. The retina can be pushed back into its normal place by injecting a fluid in the vitreous cavity (inside arrow) while the subretinal fluid is drained (outside arrow). A scleral buckling implant (the drawing represents an encircling implant) is placed over retinal tears to counteract the traction on the retina of a shrinking vitreous and to reapproximate the retina to the underlying tissues.

possible of residual monomer. However, in the unique case of the cyanoacrylate surgical adhesives, the monomers are applied directly to the tissues and almost instantaneously polymerize and adhere tenaciously to the tissues. The cyanoacrylate adhesives have been used in many diverse applications in the eye but have been particularly useful in corneal perforation and ulcers as well as in gluing artificial epithelium to the corneal surface and repairing retinal detachments (Refojo *et al.*, 1971).

Bibliography

Abel, R., Jr. (1988). Development of an artificial cornea: I. History and materials. in *The Cornea: Transactions of the World Congress on the Cornea III,* H. D, Cavanagh, ed. Raven Press, New York, pp. 225–230.

Apple, D. J., Loftfield, K., Mamalis, N., Normal, D. K.-V., Brady, S. E., and Olson, R. J. (1984). Biocompatibility of implant materials: A review and scanning electron microscopic study. *Am. Intraocular Implant Soc. J.* 10: 53–66.

Barber, J. C. (1988). Keratoprostheses: past and present. *Int. Ophthalmol. Clin.* 28: 103–109.

Barrett, G. D., Constable, I. J., and Stewart, A. D. (1986). Clinical results of hydrogel lens implantation. *J. Cataract Refract. Surg.* 12: 623–631.

Fernandez-Vigo, J., Refojo, M. F., and Jumblatt, M. (1989). Elimination of hydroxypropyl methylcellulose from the anterior chamber of the rabbit. *J. Cataract Refract. Surg.* 15: 191–195.

Holden, B. A., Newton-Homes, J., Winterton, L., Fatt, I., Hamano, H., La Hood, D., Brennan, N. A., and Efron, N. (1990). The Dk project: An interlaboratory comparison of Dk/L measurements. *Optom. Vis. Sci.* 67: 476–481.

Kastl, P. R. (ed.). (1995). *Contact Lenses: The CLAO Guide to Basic Science and Clinical Practice.* Kendall/Hunt Publishing Co., Dubuque, IA.

Krupin, T., Ritch, R., Camras, C. B., Brucker, A. J., Muldoon, T. O., Serle, J., Podos, S. M., and Sinclair, S. H. (1988). A long Krupin-Denver valve implant attached to a 180° scleral explant for glaucoma surgery *Ophthalmology* 95: 1174–1180.

McCarey, B. E., McDonald, M. B., van Rij, G., Salmeron, B., Pettit, D. K., and Knight, P. M. (1989). Refractive results of hyperopic hydrogel intracorneal lenses in primate eyes. *Arch. Ophthalmol.* 107: 724–730.

Miller, D. (ed.). (1987). *Clinical Light Damage to the Eye.* Springer Verlag, New York.

Refojo, M. F. (1979). Contact lenses. in *Kirk-Othmer: Encyclopedia of Chemical Technology.* 3rd ed. Wiley, New York, vol. 6, 720–742.

Refojo, M. F. (1986a). Current status of biomaterials in ophthalmology. in *Biological and Biomechanical Performance of Biomaterials,* P. Christel, A.Meunier, and A. J. C. Lee, eds. Elsevier, Amsterdam, pp. 159–170.

Refojo, M. F. (1986b). Biomedical materials to repair retinal detachments. in *Biomedical Materials,* J. M. Williams, M. F. Nichols, and W. Zingg, eds. Materials Research Society Symposia Proc., Materials Research Society, Pittsburgh, Vol. 55, pp. 55–61.

Refojo, M. F. (1991). Tear evaporation considerations and contact lens wear. in *Considerations in Contact Lens Use under Adverse Conditions,* P. E. Flattau, ed., pp. 38–43. National Academy Press, Washington, DC.

Refojo, M. F., and Dabezies, O. H., Jr. (1984). Classification of the types of material used for construction of contact lenses. in *Contact*

Lenses: The CLAO Guide to Basic Science and Clinical Practice, O. H. Dabezies, Jr., ed. Grune & Stratton, Orlando, FL, pp. 11.1–11.12.

Refojo, M. F., Dohlman, C. H., and Koliopoulos, J. (1971). Adhesives in ophthalmology: A review. *Surv. Ophthalmol.* **15**: 217–236.

Werblin, T. P., Peiffer, R. L., and Patel, A. S. (1987). Synthetic keratophakia for the correction of aphakia. *Ophthalmology* **94**: 926–934.

Wichterle, O., and Lim, D. (1990). Hydrophilic gels for biological use. *Nature* **185**: 117–118.

7.7 ORTHOPEDIC APPLICATIONS

J. Lawrence Katz

The developer of biomaterials for orthopedic purposes faces the same duality of concerns present in all other implant use: (1) the material must not affect adversely its biological environment and (2) the material in return must not be adversely affected by the surrounding host tissues and fluids. A starting point to address this issue is the understanding of the interrelationship between the structure and properties of the natural tissues that are being replaced. An appreciation of the "form–function" relationship in calcified tissues will help provide insight into factors determining implant design as well as determining which are the materials of choice to meet a specific orthopedic need.

STRUCTURE AND PROPERTIES OF CALCIFIED TISSUES

Structure

There are several different calcified tissues in the human body and several different ways of categorizing them. All calcified tissues have one thing in common: in addition to the principal protein component, collagen,[1] and small amounts of other organic phases, they all have an inorganic component hydroxyapatite (abbreviated OHAp) or $Ca_{10}(PO_4)_6(OH)_2$.[2] In the case of long bones such as the tibia or femur, an understanding of the organization of these two principal components is the beginning phase of the characterization of bone structure according to scale, (i.e., the level of the observation technique). It has been convenient to treat the structure of compact cortical bone (e.g., the dense bone tissue found in the shafts of long bones) on four levels of organization (Fig. 1) (Katz, 1980a; Park, 1984).

The collagen triple helical structure (tropocollagen) and OHAp crystallography, comprise the initial or molecular level. The structure of OHAp is shown in Fig. 2 (Young, 1975). It

forms a hexagonal unit cell with space group symmetry P 6_3/m and lattice constants a = 9.880 Å and c = 6.418 Å, containing two molecular units, $Ca_5(PO_4)_3OH$, per unit cell. How this mineral phase is produced by cells and whether it is the first calcium phosphate laid down are subjects of considerable research at present. Because of its small crystallite size in bone (approximately $2 \times 20 \times 40$ nm), the X-ray diffraction pattern of bone exhibits considerable line broadening, compounding the difficulty of identifying additional phases. As we will see later, the fact that a Ca-bearing inorganic compound is one of the components of calcified tissues has led to the development of a whole class of ceramic and glass-ceramic materials that are osteophilic within the body (i.e., they present surfaces that bone chemically attaches to).

To appreciate the way in which collagen contributes to the hierarchical structural levels of bone, it is necessary to discuss the structure of collagen. However, this would require an introduction to amino acids, how they link to form polypeptides, and how these lead to proteins, which the constraints of space will not allow. A simplified picture of how the alpha chain is formed, how three chains coil together to form the tropocollagen molecule, and how molecules form a protofibril is shown in Fig. 3. For further reading, see Glimcher and Krane (1968).

As yet, we do not know fully how the two components, collagen and OHAp, are arranged or what forms hold them together at this molecular level. Whatever the arrangement, when it is interfered with, as is apparently the case in certain bone pathologies in which the collagen structure is altered during formation, the result is a bone that is formed with seriously compromised physical properties.

The second, or ultrastructural, level may be loosely defined as the structural level observed with transmission electron microscopy (TEM) or high-magnification scanning electron microscopy (SEM) (Fig. 1). Here too, we have not yet achieved a full understanding of the collagen–OHAp organization. It appears that the OHAp can be found both inter- and intrafibrillarly within the collagen. As we shall see later, at this level, it appears that we can model the elastic properties of this essentially two-component system by resorting to some sort of linear superposition of the elastic modules of each component, weighted by the percent volume concentration of each.

These fibrillar composites form larger structures, fibers, and fiber bundles, which then pack into lamellar-type units that can be observed with both SEM and optical microscopy. This is the third, or microstructural level of organization. Figure 4 shows two such types of lamellar organizations (5). Figure 4a illustrates the circular (or nearly circular) lamellar units forming the secondary osteons (Haversian systems) found in mature human bone. Figure 4b shows the straight lamellar units forming the plexiform (lamellar) bone found generally in young quadruped animals the size of cats and larger. This is the structural level that is being described when the term "bone tissue" is used or when histology is generally being discussed. At this level, composite analysis can also be introduced to model the elastic properties of the tissue, thus providing an understanding of the macroscopic properties of bone (i.e., those associated with the behavior of the whole bone, or fourth level of structure). Unfortunately, this modeling is much more

---

[1]There is one exception to this, enamel, which is found in the outer sheath of teeth; enamel has a small amount of another protein, enamelin.

[2]In actuality, this stoichiometric formula is not achieved in biological apatites. There are carbonate and other ions present as well in bones and teeth.

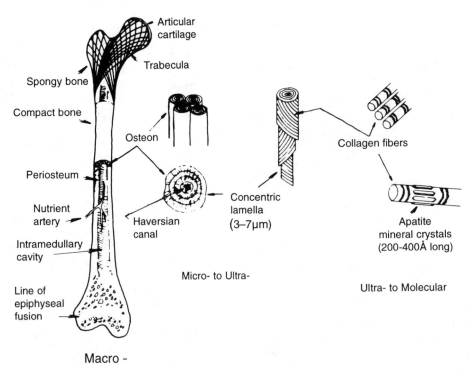

FIG. 1. Hierarchical levels of structural organization in a human long femur. (Adapted with permission from J. B. Park, *Biomaterials: An Introduction,* Plenum Publ., 1979, p. 105.)

complex than that cited earlier so that a complete description lies beyond the scope of this chapter. Interested readers are referred to some of the original sources (Katz, 1980a,b, 1981). A short account of some of this modeling is provided below in the section on composites.

Since a significant portion of bone is composed of collagen, it is not surprising to find that in addition to being anisotropic and inhomogeneous, bone is also viscoelastic like all other biological tissues. Clearly, duplicating such properties with a synthetic material is a forbidding if not hopeless task. However,

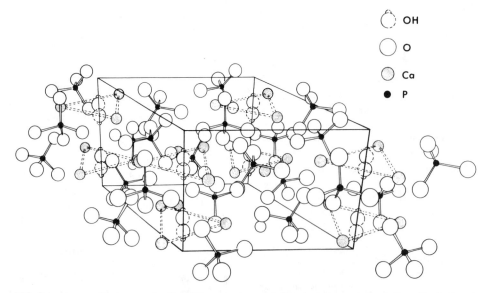

FIG. 2. Hexagonal hydroxyapatite. Each OH position is statistically only 50% occupied, as indicated by the dotted outline of half of each OH group. (Reproduced courtesy of the Centre National de la Recherche Scientific, Paris, from R. A. Young in *Physico-chimie et Christallographie des Apatites d'Interest Biologique,* CNSR Publ. No. 230, 1975.)

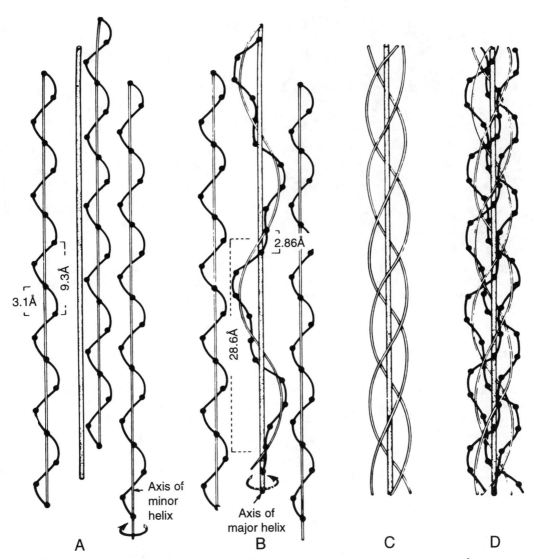

FIG. 3. The tropocollagen unit. (A) The α-helix is left handed. Each spot is a residue and repeat length is 3.1 Å. (B) the α-helix is twisted into a right-handed helix which reduces repeat length between residues to 28 Å and the repeat of the supercoil is 10 times the residue length or 28 Å. (C) and (D) Three α-helices are threaded to form a single tropocollagen (TC) unit. (Adapted from M. J. Glimcher and S. M. Krane, in *Treatise on Collagen*, Vol. IIB, *Biology of Collagen*, G. N. Ranchandran and B. S. Gould, eds., Academic Press, 1968.)

a description of the various properties does provide some insight into potential starting points for the design of implant materials.

Elastic Properties

Although bone is viscoelastic, a considerable amount of understanding is obtained by examining its elastic properties. Most of the original and subsequent studies of such properties have been performed using mechanical testing instruments at quasi-static strain rates. These experiments resulted in stress-strain curves, providing data such as Young's modulus, Poisson's ratio, yield stress, fracture stress and elastic and plastic strain in tension, compression, torsion, shear, and fatigue. A comprehensive survey of such efforts through the beginning of the 1970s will be found in F. Gaynor Evans' 1973 book, *Mechanical Properties of Bone* (Evans, 1973). This book provides an instructive overview of the large number of factors, some intrinsic and some extrinsic, that affect the mechanical properties in one way or another.

There have been sufficient measurements of bone properties to make it quite clear that a profound relationship exists between the microstructure of bone and its material properties. By analyzing results of a number of such experiments, it has

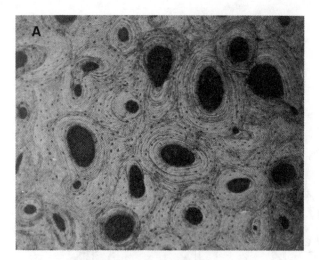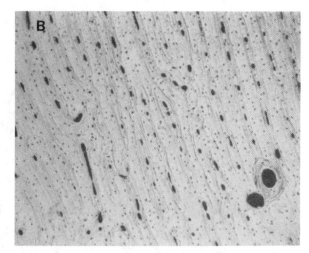

FIG. 4. Optical micrographs of transverse cross section showing the microstructure of human and bovine femora. (A) Human Haversian bone. (B) Bovine plexiform bone. (From H. S. Yoon and J. L. Katz, *J. Biomech.* **9:** 408, 1976. Copyright 1976 Pergamon Press, Ltd. Reprinted with permission.)

been possible to show analytically that bone has either transverse isotropic (hexagonal) symmetry requiring either five elastic compliances or stiffness coefficients to describe its elastic properties, or orthotropic (orthorhombic) symmetry requiring nine such constants.

This identification of the elastic anisotropy and its relation to bone microstructure has become the hallmark of the recent studies of elastic properties. One of the principal driving forces is the desire to know what kinds of changes in bone due to resorption and remodeling are going on at the bone–implant biomaterial interface. For the general Hooke's law:

$$\sigma_{ij} = c_{ijkl}\varepsilon_{kl} \tag{1}$$

where σ_{ij} and ε_{kl} are the stress and strain tensors, respectively, and c_{ijkl} are the fourth-rank elastic stiffness coefficients. The matrix of the latter for the transverse isotropic case is given in the usual reduced notation ($11 \rightarrow 1$, $22 \rightarrow 2$, $33 \rightarrow 3$, $23 \rightarrow 4$, $13 \rightarrow 5$, $12 \rightarrow 6$) by:

$$[c_{ij}] = \begin{bmatrix} c_{11} & c_{12} & c_{13} & 0 & 0 & 0 \\ c_{12} & c_{11} & c_{13} & 0 & 0 & 0 \\ c_{13} & c_{13} & c_{33} & 0 & 0 & 0 \\ 0 & 0 & 0 & c_{44} & 0 & 0 \\ 0 & 0 & 0 & 0 & c_{44} & 0 \\ 0 & 0 & 0 & 0 & 0 & c_{66} \end{bmatrix}, \tag{2}$$

where $c_{66} = 1/2\,[c_{11} - c_{12}]$.

In cases where bone resorption leads to porosity, there is a gradient in the structure so that measurements of properties in the radial and tangential directions are significantly different. This is also observed in the plexiform lamellar bone found in young mammals. For such bone, orthotropic symmetry is the

proper description. In that instance, the matrix of stiffness coefficients is given by:

$$[c_{ij}] = \begin{bmatrix} c_{11} & c_{12} & c_{13} & 0 & 0 & 0 \\ c_{12} & c_{22} & c_{23} & 0 & 0 & 0 \\ c_{13} & c_{23} & c_{33} & 0 & 0 & 0 \\ 0 & 0 & 0 & c_{44} & 0 & 0 \\ 0 & 0 & 0 & 0 & c_{55} & 0 \\ 0 & 0 & 0 & 0 & 0 & c_{66} \end{bmatrix}. \tag{3}$$

There are also forms of bone that are isotropic both in microstructure and in measurements. While elastic properties are a significant aid in understanding the magnitude and anisotropy present in bones, a complete description requires knowledge of the viscoelastic properties of bone as well.

Viscoelastic Properties

Interest in the viscoelastic properties of bone arose during the nineteenth century at about the same time as interest in elastic properties. In more recent times, a number of investigators have measured various aspects of the viscoelastic properties of bones and related calcified tissues such as dentin. There are measurements of creep, stress relaxation, strain rate dependence, and even some dynamic measurements. Recent experiments have shown that bone as a material is nonlinear and thermorheologically complex (i.e., time-temperature superposition cannot be used to obtain its properties) (Lakes *et al.*, 1979). A relaxation spectrum for compact human bone is shown in Fig. 5. A calculation of the amplitudes and frequencies of various mechanisms likely to occur in bone is presented in Fig. 6. Comparing these calculations with the relaxation

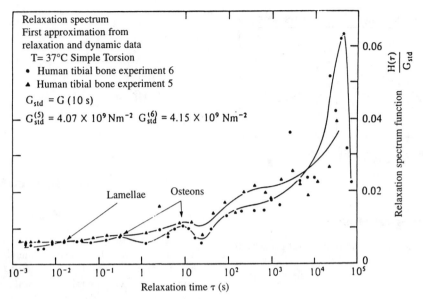

FIG. 5. The relaxation spectra for two specimens of wet human tibial bone in simple torsion. (From R. S. Lakes and J. L. Katz, *J. Biomech.* **12:** 685, 1979.)

spectrum shows that the fit is not unreasonable, with most of the mechanisms appearing to occur at frequencies within the physiological range.

Also, it should be noted that although bone is nonlinear, the nonlinearity occurs outside the physiological range of interest. Thus, for most physiological purposes we can use the Boltzmann superposition integral as the viscoelastic constitutive relationship for bone, which in general is given by:

$$\sigma_{ij}(t) = \int_0^t c_{ijkl}(t-\tau)\frac{d\varepsilon_{kl}(\tau)}{d\tau}\,d\tau, \tag{4}$$

where the tensors here are now the time-dependent analogs to those in Hooke's law. As a result, we can use simple linear rheological models to describe creep, stress relaxation, etc., not only for the behavior of bone but also for the soft connective tissues, collagen, tendon, ligaments, etc.

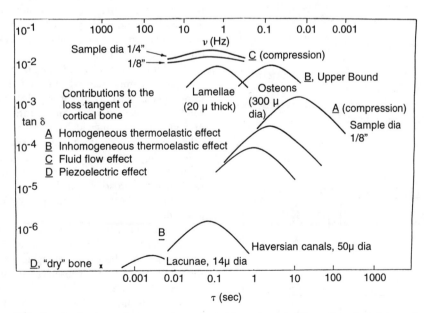

FIG. 6. Predicted contributions to the mechanical losses in cortical bone. (From R. S. Lakes and J. L. Katz, *J. Biomech.* **12:** 684, 1979.)

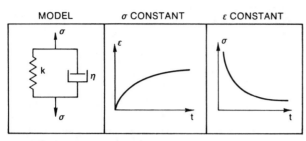

FIG. 7.   Kelvin-Voight rheological model (creep compliance).

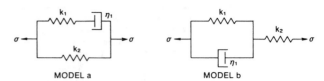

FIG. 9.   Three-element or general linear models.

Rheological Models

Figure 7 describes the Kelvin–Voigt model that is useful in portraying creep:

$$\sigma = \sigma_k + \sigma_\eta$$

$$\dot{\varepsilon} + \frac{k}{\eta}\varepsilon = \frac{1}{\eta}\sigma. \qquad (5)$$

$$\varepsilon = \varepsilon_k = \varepsilon_\eta$$

On the other hand, the Maxwell system provides a good model for stress relaxation (Fig. 8).

$$\varepsilon = \varepsilon_k + \varepsilon_\eta$$

$$\dot{\sigma} + \frac{k}{\eta}\sigma = E\dot{\varepsilon}. \qquad (6)$$

$$\sigma = \sigma_k = \sigma_\eta$$

Of course, to provide a better description for bone or other tissues, a three-element model or general linear system should be used (e.g., either Fig. 9a or 9b), which provides equivalent results. The solution to the dynamic experiment can be obtained using the general linear system by substituting

$$\varepsilon = \varepsilon_0 e^{i\omega t} \quad \sigma = \sigma_0 e^{i(\omega t + \delta)},$$

where ω is the applied frequency and δ is the phase angle between stress and strain. With these descriptions of bone properties, the stage is set for introducing the various categories of materials available for possible use as implants.

BIOMATERIALS

Metals

Since the principal function of the long bones of the lower body is to act as load-bearing members, it was reasonable that the initial materials introduced to replace joints, such as artificial hips, were metals. Both stainless steel, such as 316L, and Co–Cr alloys became the early materials of choice, because of their relatively good corrosion resistance and reasonable fatigue life within the human body. Of course, their stiffness, rigidity, strength, exceeded those of bone considerably. However, in certain applications, owing to size restrictions and design limitations (e.g., in rods used to straighten the spine in scoliosis), fatigue failures did occur.

The hip prosthesis has been the most active area of joint replacement research (Fig. 10) (Williams, 1984). A dramatic improvement in the efficacy of the hip implant occurred with the introduction by English orthopedic surgeon John Charnley (later knighted for this innovation) of his total hip arthroplasty consisting of a metal femoral prosthesis that was held in place by poly(methyl methacrylate) (PMMA), with the acetabulum component made of ultrahigh-molecular-weight linear polyethylene (UHMWPE), also cemented in place with PMMA (Fig. 11) (Charnley, 1972). This system has seen many variations over the years, but is still a significant factor in modern joint replacement surgery. Further discussion of the properties of polymers and the function of each is covered in the section on polymers.

As alluded to above, a number of properties of the implant materials are important when selecting suitable candidates for joint replacements in current use. Because of the success of the total hip, similar material configurations have been used with an appropriate geometric design for the knee and almost every other large joint in the body. In addition to the 316L and Co–Cr alloys mentioned above, titanium and the so-called aviation alloy Ti–6A1–4V are now also used for the femoral portion of hip prostheses. Table 1 lists the relevant mechanical properties of these metals along with those of bone for comparison.

In addition, considerations of corrosion and fatigue life are obviously of prime importance. Also, the metals must of necessity not be toxic, mutagenic, or carcinogenic within the body, either of themselves or due to the release of chemical components as a result of interactions with the various body fluids. More complete descriptions of the behavior and properties of metals used for implants are available (Park, 1984; Boretos and Eden, 1984; Mears, 1979).

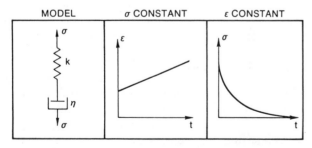

FIG. 8.   Maxwell rheological model (stress relaxation).

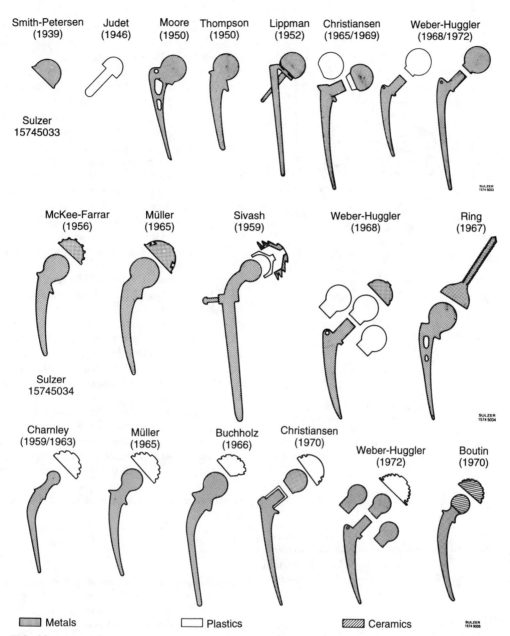

Smith-Petersen (1939) · Judet (1946) · Moore (1950) · Thompson (1950) · Lippman (1952) · Christiansen (1965/1969) · Weber-Huggler (1968/1972)

Sulzer 15745033

McKee-Farrar (1956) · Müller (1965) · Sivash (1959) · Weber-Huggler (1968) · Ring (1967)

Sulzer 15745034

Charnley (1959/1963) · Müller (1965) · Buchholz (1966) · Christiansen (1970) · Weber-Huggler (1972) · Boutin (1970)

▨ Metals   ▢ Plastics   ▨ Ceramics   SULZER 1574 5035

FIG. 10.   The stages in the development of arthroplastic devices for the hip joint are shown schematically. (Reproduced with permission of Sulzer Bros., Ltd., Winterthur, Switzerland, in *Biocompatibility of Orthopedic Implants*, D. F. Williams, ed., CRC Press, 1984.)

There is also an extended series of books published by CRC Press on various biomaterials, their biocompatibility and properties, not only for use in orthopedics but as implants for other tissues and organs as well. Owing to the importance of these considerations, the ASTM Committee F-4 on Surgical Implant Materials has written standard specifications for these metals as well as many other materials used for implants. At present, there are five metallic systems used as implants: (1) stainless steel, (316L), (2) cast cobalt chromium-molybdenum alloy, (3) wrought cobalt–chromium–tungsten–nickel alloy, (4) commercially pure titanium and (5) titanium–6A1–4V alloy. However, recently there has been concern over the wear of the load-bearing surfaces of titanium hip implants.

There is still considerable research activity in this area, with new developments being introduced regularly. Because of various concerns with the efficacy of the PMMA fixation of implants, alternative methods of fixation have been introduced. These include the use of porous outer coatings on the metals in the case of titanium because of its low crevice corrosion. In

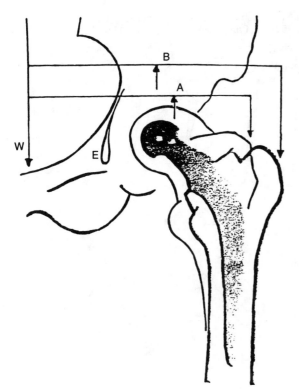

FIG. 11.   The moment arms about the hip joint are shown before operation, A, and after reconstruction, with a small-diameter femoral head replacement, B (22 mm). The arthroplastic hip shows medial displacement of the center of rotation and lateral projection of the greater trochanter. (Reproduced with permission of J. Charnley, *Acrylic Cement in Orthopaedic Surgery.* Churchill Livingstone, 1972.)

addition, either solid or porous claddings of polymers, composites, and ceramics on metallic shafts are all being tried. Descriptions of the properties of these nonmetallic materials will be discussed later.

Polymers

There are two specific applications in orthopedics in which polymers have proved useful. First, they are used for one of the articulating surface components in joint prostheses. Thus, they must have a low coefficient of friction and low wear rate when they are in contact with the opposing surface, which is usually made of metal or ceramic. Initially, Charnley used Teflon for the acetabular component of his total hip arthroplasty. However, its accelerated creep and poor stress corrosion (for the material he used) caused it to fail *in vivo*, requiring replacement with his ultimate choice, UHMWPE. Thus, the twofold demands of "not doing damage" and "not being damaged" were preeminent once again.

Second, polymers are used for fixation as a structural interface between the implant component and bone tissue.

Clearly in this case, the appropriate mechanical properties of the polymer are of major importance. There are two different ways in which this fixation can be produced and each requires a different type of material with substantially different properties. The first type of use was again Charnley's. He used PMMA as a grouting material to fix both the stem of the femoral component and the acetabular component in place, and distribute the loads more uniformly from the implants to the bone.

Since high interfacial stresses result from the accommodation of a high-modulus prosthesis within the much lower modulus bone, the use of a lower modulus interpositional material has been introduced as an alternative to PMMA fixation. Thus recently, polymers such as polysulfone have been used as porous coatings on the implant's metallic core to permit mechanical interlocking through bone and/or soft tissue ingrowth into the pores. This requires that the polymers have surfaces that resist creep under the stresses found in clinical situations and have high enough yield strength to minimize plastic deformation.

As indicated earlier, the mechanical properties of concern in polymer applications are yield stress, creep resistance, and wear rate. These factors are controlled by such polymer parameters as molecular chain structure, molecular weight, and degree of branching or (conversely) of chain linearity.

Polyethylene is available commercially in three different grades: low-density, high density, and UHMWPE, the form used in orthopedic implants. The better packing of linear chains with the resulting increased crystallinity in the last form provides the improved mechanical properties required for orthopedic use even though there is a decrease in both ductility and fracture toughness. The properties of the three grades are compared in Table 2.

TABLE 1   Mechanical Properties of Cortical Bone, 316L Stainless Steel, Cobalt–Chromium Alloy, Titanium and Titanium-6-Aluminum-4-Vanadium

Material	Young's modulus (GPa)	Compressive strength (GPa)	Tensile strength (GPa)
Bone			
(wet at low strain rate)	15.2	0.15	0.090
(wet at high strain rate)	40.7	0.27–0.40	—
316L stainless steel	193	—	0.54
Co–Cr (cast)	214	—	0.48
Ti			
0% porosity	110	—	0.40
40% porosity	24[a]	—	0.076
Ti–6Al–4V			
0% porosity	124	—	0.94
40% porosity	27[a]	—	0.14

[a] Calculated based on assuming $E = E (1 - V)^3$.

TABLE 2 Properties of Polyethylene[a]

	Low density	High density	Low density	UHMWPE[b]
Molecular weight (g/mol)		$3\ 4 \times 10^3$	5×10^5	2×10^6
Density (g/ml)		0.90–0.92	0.92–0.96	0.93–0.944
Tensile strength (MPa)		7.6	23–40	3
Elongation (%)		150	400–500	200–250
Modulus of elasticity (MPa)		96–260	410–1240	—

[a]Reprinted with permission from J. B. Park, *Biomaterials Science and Engineering*, Plenum Publ., New York, 1984.
[b]Data from ASTM F 648.

Ceramics and Glasses

In recent years, ceramics and glass ceramics have played an increasingly important role in implants. This has occurred because of two quite disparate uses. First, there is the use associated with improved properties such as resistance to further oxidation (implying inertness within the body), high stiffness, and low friction and wear as articulating surfaces. This requires the use of full-density, controlled, small, uniform grain size (usually less than 5 μm) materials. The small grain size and full density are important since these are the two principal bulk parameters controlling the ceramic's mechanical properties. Clearly, any void within the ceramic's body will increase stress, degrading the mechanical properties. Grain size controls the magnitude of the internal stresses produced by thermal contraction during cooling. In ceramics, such thermal contraction stresses are critical because they cannot be dissipated as in ductile materials via plastic deformation. Thus full-density, small grain size alumina (Al_2O_3) has proven quite successful as the material in matched pairs of femoral head and acetabular components in total hip arthoplasty. Some of alumina's chemical and mechanical properties are listed in Table 3. Other ceramics either in use or under consideration for such use include TiO_2 and Si_3N_4.

The second, and scientifically more exciting application, takes advantage of the osteophilic surface of certain ceramics and glass ceramics. These materials provide an interface of such biological compatibility with osteoblasts (bone-forming cells) that these cells lay down bone in direct apposition to the material in some form of direct chemicophysical bond. The initial suggestion was the use of special compositions of glass ceramics, which were termed bioglasses, for implant applications in orthopedics. The bioglasses have nominal compositions in the ranges: 40–50% SiO_2; 5–31% Na_2O; 12–35% CaO; 0–15% P_2O_5; plus in some cases MgO or B_2O_3 or CaF_2, etc. Table 4 lists the compositions of several of the more relevant materials (Hench and Ethridge, 1982).

Several of these compositions have seen wide use as implant

materials. The model proposed for the "chemical" bond formed between glass and bone is that the former undergoes a controlled surface degradation, producing an SiO_2 rich layer and a Ca, P-rich layer at the interface. Originally amorphous, the Ca, P-rich layer eventually crystallizes as a mixed hydroxy-carbonate apatite structurally integrated with collagen, which permits subsequent bonding by newly formed mineralized tissues.

There is still an entirely different series of inorganic compounds that also have been shown to be osteophilic. These include OHAp, which, is the form of the naturally occurring inorganic component of calcified tissues, and calcite, $CaCO_3$, and its Mg analog, dolomite, among others being studied. The most extensive applications in both orthopedics and dentistry have involved OHAp. This has been used as a cladding for metal prostheses for the former, and in dense, particulate form for the latter. The elastic properties of OHAp and related compounds are compared with those of bone, dentin, and enamel in Table 5.

The use of both OHAp and the glass ceramics as claddings on the metallic stems of hip prostheses is still another method of providing fixation instead of using PMMA. In these cases, the fixation is via the direct bonding of bone to the cladding surface. Several of these systems are undergoing clinical trials.

Composites

It is evident from the description of bone properties and a comparison with the corresponding properties of the various metals, polymers, and ceramics, that there is a considerable

TABLE 3 Chemical and Mechanical Properties of Alumina for Implants[a]

Property	Value
Chemical	
Al_2O_3	99.7%
MgO	0.23% (max)
SiO_2 and alkali metal oxides	0.1%
Physical and Mechanical	
Density	3.4 g cm^3
Grain size	4 μm
Fracture strength (compression)	580.0 ksi or 4000.0 N mm^{-2}
Flexure strength (MOR)	58.0 ksi or 400.0 N mm^{-2}
Young's modulus	55.000 ksi or 380,000 N mm^2
Coefficient of friction: water lubricant	0.05
Wear rate (50 lb/in.^{2}; 400 mm sec^{-1})	3×10^{-9} mg/mm

[a]Adapted with permission from Gibbons (1984).

TABLE 4   Compositions of Surface-Active Glasses and Glass Ceramics[a,b]

Material	SiO₂	P₂O₅	CaO	Na₂O	Other
45S5	45.0	6.0	24.5	24.5	—
45S5-F	45.0	6.0	12.25	24.5	Caf₂, 12.15
45S5-B5	40.0	6.0	24.5	24.5	B₂O₃, 5
45S5-OP	45.0	0	24.5	30.5	—
45S5-M	48.3	6.4	—	26.4	MgO, 18.5
Alkali-rich glass-ceramic	42–47	5–7.5	20–25	20–25	—
Ceravital	40–50	10–15	30–35	5–10	K₂), 0.5–3.0 MgO, 2.5–5.0
Composition C	42.4	11.2	22.0	24.4	—

[a]Adapted with permission from L. L. Hench and E. C. Ethridge, *Biomaterials: An Interfacial Approach*, Academic Press, New York, 1982.
[b]In weight percent.

mismatch among the relative properties; thus, the present interest in composite materials.

As indicated in the introductory discussion of bone as a composite, the starting point used in modeling composite behavior is the introduction of the simple superposition relationships for a pair of disparate materials. When two materials with different elastic moduli but identical Poisson's ratios are bonded together, there are two distinct limiting cases on the elastic properties of the composite. (When the Poisson's ratios are not equal, a correction term must be introduced into the equations.) Figure 12a illustrates the behavior when the strains are distributed uniformly across the materials' interface, the so-called parallel or Voigt model. This results in the Voigt modulus, E^V, being the linear superposition of the respective moduli of the components, E_1 and E_2, weighted by their respective volume fractions V_1 and V_2 (rule of mixtures):

$$E^V = E_1 V_1 + E_2 V_2 \text{ where } V_1 + V_2 = 1. \tag{7}$$

Similarly, Fig. 12b illustrates the behavior of the isostress distribution, the so-called series or Reuss model. In this case, the superposition is over the compliances (inverses of the moduli), again weighted by their respective volume fractions, yielding as the Reuss modulus, E_R, the value:

$$\frac{1}{E_R} = \frac{V_1}{E_1} + \frac{V_2}{E_2} \text{ or } E_R = \frac{E_1 E_2}{E_1 V_2 + E_2 V_1}, \tag{8}$$

where again $V_1 + V_2 = 1$.

TABLE 5   Elastic Properties of Mineral and Synthetic Apatites and Various Calcium-Bearing Minerals from Ultrasonic Measurements Compared with Bone and Teeth

Material	Bulk Modulus (10¹⁰ N m⁻²)	Shear Modulus (10¹⁰ N m⁻²)	Young's Modulus (10¹⁰ N m⁻²)	Poisson's ratio
FAp (mineral)	9.40	4.64	12.0	0.26
OHAp (mineral)	8.90	4.45	11.4	0.27
OHAp (synthetic)	8.80	4.55	11.7	0.28
ClAp (synthetic	6.85	3.71	9.43	0.27
Co₃Ap (synthetic, type B)	8.17	4.26	10.9	0.28
CaF₂	7.74	4.70	11.8	0.26
Dicalcium phosphate dihydrate	5.50	2.40	6.33	0.31
Bone (human femur, powdered)	2.01	0.800	2.11	0.33
Dentine (bovine, powdered)	3.22	1.12	3.02	0.35
Enamel (bovine, powdered)	6.31	2.93	7.69	0.32

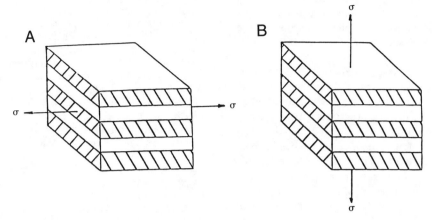

FIG. 12. (A) Parallel or uniform strain model of two phases (also call the Voight model); the stress axis is parallel with the laminae. (B) Series or uniform stress model of two phases (also called the Reuss model); the stress axis is perpendicular to the laminae.

While neither of these expressions is rigorously correct, the Voigt and Reuss models do provide the upper and lower bounds respectively on the behavior of such simple composites. When the two moduli differ by no more than a factor of two or three, the simple linear superposition of the Voigt and Reuss values provides a good approximation. Of course, when one component is now a low modulus matrix, E_m, and the second component is composed of high modulus, E_f, unidirectional fibers oriented at some angle ϕ with respect to the stress direction, then a simple suitable modification of Eq. 8 by introducing a fiber reinforcement factor that depends on ϕ can provide for such angular-dependent behavior:

$$E = E_m V_m + E_f V_f \cos^4 \phi, \tag{9}$$

whereas before $V_m + V_f = 1$.

Such a disparity in moduli means that there will be a corresponding disparity in the composite modulus as a function of angle. When the reinforcing fibers line up with the stress axis, the composite modulus is determined predominantly by the fiber modulus. However, for specimens with fibers oriented at right angles ($\phi = 90°$) to the stress axis, the trigonometric term causes the value of the composite modulus to be reduced to that of the matrix alone, which is usually that of a compliant material. Thus, depending on the fiber orientation, the desired modulus can be designed for the composite. Actually, the equations provided are simplified ones based on limiting cases. For specific cases of interest, it is clear that the rich literature on composites must be utilized.

The number of composite applications in orthopedics has been rising in recent years. There have been attempts to improve the properties of the UHMWPE used in total joint replacement by including short, chopped, carbon fibers. However, in a series of laboratory experiments at the Hospital for Special Surgery in New York City on total knee replacements using such components, it was found that the increased elastic modulus of the fiber-reinforced UHMWPE was responsible for increasing the maximum contact stress, the maximum shear stress, the maximum principal stress, and the range of the maximum principal stress (Wright *et al.*, 1985). All of these suggested that the surface damage in the carbon-reinforced components would be at least as considerable as in the ordinary component, if not actually even more severe. Collateral fracture toughness and fatigue crack propagation tests also showed rather interesting results. There was no significant difference between the composite and ordinary polyethylenes in the former tests, while in the latter, the fatigue crack growth rates were nearly eight times higher for the carbon-reinforced component. Poor bonding between carbon fibers and the polyethylene matrix was found to be responsible for this poor fracture resistance. Clinical studies by the same group on retrieved components from total knee replacements confirmed the similar nature of the surface damage modes observed on both the carbon-reinforced and unreinforced polyethylene components. Thus in this use, the composite system provided no improvement, and possibly even a reduction in the functioning level of the component.

An area where composites appear to provide an interesting and important application is in bone plates. These are plates placed across fracture sites to provide fixation during the initial stages of healing. However, the rigidity and stiffness of the metallic plates generally used are much greater than the adjacent bone under the plate. The plate continues to carry a major portion of the load even after healing has been initiated, thus effectively "stress shielding" the underlying bone and delaying, even possibly retarding, full healing. Eventually a second operation is required to remove the plate. A number of composite plates of various compositions have been designed and tested, including carbon-reinforced epoxy resin with the fibers oriented at various angles to the longitudinal plate axis. In some instances, thin layers with different fiber orientations are laminated together to form a "super" composite. Carbon fiber-reinforced polysulfone has been used as well as metal-reinforced polyacetal resin plates. Animal studies comparing these with comparable-sized stainless steel plates showed that both the degree of bone resorption and the change in elastic modulus of the plated bone were greater the greater the stiffness of the plates. However, even with such plates a second operation is required for retrieval.

This situation is one where composites can provide a unique solution to the problem. There are mateials that can be biodegraded within the body, often with the timescale governed by the manner in which the material is processed. Thus, in principle, it might be possible to design a plate that would remain stiff and rigid enough to support the loads of physical activity until the bone itself could oblige. Then, instead of continuing to bypass part of the load away from the underlying bone, as the plate biodegraded, more and more of the load would transfer onto the bone, thus providing the proper loading required for the full healing dynamics to take place. Totally bioresorbable composite plates have been made from poly(glycolic acid) (PGA) fibers embedded in a poly(lactic acid) (PLA) matrix. The time evolution of the properties of the PLA can be controlled by making stereo copolymers of D- and L-lactic acid (denoted by PLA X, where X is the percentage of L-lactic acid units); the number of D- units apparently controls the degradation rate. These composites are bioresorbable, exhibit good biocompatibility with the tissues in which they are implanted, and have adjustable resorption rates, depending on the relative amounts of L- and D-lactic acid units as well as on the quantities of GA and LA repeating units. This concept of bioresorption or biodegradability is an important one in many areas of bioimplants (e.g., time-released drug delivery) (Boretos and Eden, 1984).

Still another use of composites in orthopedics is in replacement of damaged tendons and ligaments. Both of these behave as tensile systems within the body with a tendon providing a connection between a muscle and a bone, and a ligament serving the same purpose between a pair of bones. Thus, carbon fiber-reinforced polymer systems have been introduced in this area. In this case, biodegradable polymers such as PLA have been used to coat high-strength uniaxial filamentous carbon to form a composite ribbon. This system is based on the concept that the implant acts as a scaffold, allowing ingrowth of fibrous collagenous tissue in an attempt to reform the connective tissue itself.

The systems described have by no means exhaust the composites being used nor the areas in orthopedics in which they are utilized. The leading biomedical materials journals regularly publish research articles describing some new composite, ceramic, alloy, or polymer for implant use.

Bibliography

Boretos, J. W., and Eden, M. (eds.) (1984). *Contemporary Biomaterials: Material and Host Response, Clinical Applications. New Technology and Legal Aspects,* Noyes Med. Publ., Park Ridge, NJ.

Charnley, J. (1972). *Acrylic Cement in Orthopaedic Surgery,* Churchill Livingstone, London.

Evans, F. E. (1973). *Mechanical Properties of Bone.* Charles C. Thomas, Springfield, IL.

Gibbons, D. F. (1984). Materials for orthopedic implants, in *Biocompatibility of Orthopedic Implants,* D. F. Williams, ed. CRC Press, Boca Raton, FL, Vol. 1, pp. 112–137.

Glimcher, M. J., and Krane, S. M. (1968). The organization and structure of bone and the mechanism of calcification, in *Treatise on Collagen,* Vol. 11B, *Biology of Collagen,* G. N. Ranachandran and B. S. Gould, eds., Acad. Press, New York, pp. 68–251.

Hench, L. L., and Ethridge, E. C. (1982). *Biomaterials: An Interfacial Approach,* Academic Press, New York.

Katz, J. L. (1980a). The structure and biomechanics of bone, in *The Mechanical Properties of Biological Materials,* J. F. V. Vincent and J. D. Currey, eds. Cambridge Univ. Press, London and New York, pp. 137–168.

Katz, J. L. (1980b). Anisotropy of Young's modulus of bone. *Nature* **283**: 106–107.

Katz, J. L. (1981). Composite material models for cortical bone, in *Mechanical Properties of Bone,* S. C. Cowin, Ed. A.S.M.E., New York, AMD Vol. 45, pp. 171–184.

Lakes, R. S., and Katz, J. L. (1979). Viscoelastic properties and behavior of cortical bone; Part II: Relaxation mechanisms. *J. Biomech.* **12**: 679–687.

Lakes, R. S., Katz, J. L., and Sternstein, S. S. (1979). Viscoelastic properties of wet cortical bone: I, Torsional and biaxial studies. *J. Biomech.* **12**: 657–678.

Mears, D. C. (1979). *Materials and Orthopaedic Surgery, Williams and Wilkins,* Baltimore, MD.

Park, J. B. (1989). *Biomaterials Science and Engineering.* Plenum Publ., New York.

Wright, T. M., Bartel, D. L., and Rimnac, C. M. (1985). Carbon fiber-reinforced UHMWPE for total joint replacement components, in *Proc. 1st International Conf.: Composites in Bio-Medical Engineering,* Plastics and Rubber Institute, London, pp. 21/1-4.

Yoon, H. S., and Katz, J. L. (1976). Ultrasonic wave propagation in human cortical bone—I. Theoretical considerations of hexagonal symmetry. *J. Biomech.* **9**: 407–412.

Young, R. A. (1975). Some aspects of crystal structural modeling of biological apatites, in *Physico-chimie et Cristallographie des Apatites d'Interet Biologique.* CNSR Publication #230, Editions du CNRS, Paris.

7.8 DRUG DELIVERY SYSTEMS
Jorge Heller

The major advantage in developing systems that release drugs in a controlled manner can be appreciated by examining Fig. 1, which illustrates changes in blood plasma levels following a single dose of a therapeutic agent. As shown, the blood plasma level rapidly rises and then exponentially decays as the drug is excreted and/or metabolized. The figure also shows concentrations above which the drug produces undesirable side effects and below which it is not therapeutically effective. The difference between these two levels is known as the therapeutic index.

Using a single dose, the time during which the concentration of the drug is above the minimum effective level can only be extended by increasing the size of the dose. However, by doing so, blood plasma concentrations extend into the

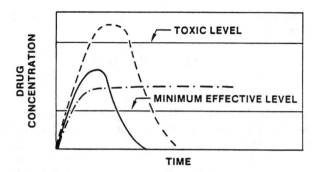

FIG. 1. Drug concentration following absorption of therapeutic agent as a function of time. (——) safe dose, (———) unsafe dose, (—·—) controlled release. (Reprinted from T. J. Roseman and S. H. Yalkowsky, in *Controlled Release Polymeric Formulations*, D. R. Paul and F. W. Harris, eds., ACS Symposium Series 33, American Chemical Society, 1976, pp. 33–52.)

TABLE 1 Classification of Controlled Release Systems

Type of system	Rate-control mechanism
Diffusion controlled	
Reservoir devices	Diffusion through membrane
Monolithic devices	Diffusion through bulk polymer
Water penetration controlled	
Osmotic systems	Osmotic transport of water through semipermeable membrane
Swelling systems	Water penetration into glassy polymer
Chemically controlled	
Monolithic systems	Either pure polymer erosion (surface erosion) or combination of erosion and diffusion (bulk erosion)
Pendant chain systems	Combination of hydrolysis of pendant group and diffusion from bulk polymer
Regulated systems	
Magnetic or ultrasound	External application of magnetic field or ultrasound to device
Chemical	Use of competitive desorption or enzyme-substrate reactions. Rate control is built into device

toxic level, an undesirable situation. For this reason, developing controlled-release devices that can maintain a desired blood plasma level for long periods without reaching a toxic level or dropping below the minimum effective level is of great interest.

Aside from the clear therapeutic advantage of controlled-release products, there are also compelling business reasons for the development of such devices. Because of increasingly more stringent U.S. Food and Drug Administration (FDA) regulations, the cost for introducing new drug entities has escalated to an average of $150 million for each drug and it is not uncommon for development to require more than 10 years of research and development work. Thus, it is reasonable for pharmaceutical companies to attempt to maximize their financial return for each drug by research aimed at extending the patented lifetime of the drug. One means of doing this is to develop a patented controlled-release formulation. However, the commercial feasibility of such a strategy is predicated on a demonstration that the controlled-release formulation is indeed superior to the unformulated drug and most important, that the cost of the controlled-release formulation is low enough to ensure a reasonable market penetration.

Successful efforts to produce pharmaceutical formulations that would prolong the action of therapeutic agents, thus minimizing the frequency between dosing, go back to the late 1940s and early 1950s with the introduction of the first commercial product known as Spansules. This product was designed to increase the duration of orally administered drugs and consisted of small spheres with a soluble coating. By using coatings having varying thicknesses, dissolution times could be varied, prolonging the action of the therapeutic agent. Such formulations are now known as sustained-release or prolonged-release products.

However, the functionality of such products depends on the external environment which varies greatly from patient to patient. For this reason, a major effort has been under way since the late 1960s to develop products which are capable of releasing drugs by reproducible and predictable kinetics. Ideally, such products are not significantly affected by the exter-

nal environment so that patient-to-patient variability is greatly reduced. Such devices are known as controlled-release products. Controlled-release polymeric systems can be classified according to the mechanism which controls the release of the therapeutic agent and are shown in Table 1.

DIFFUSION-CONTROLLED DELIVERY SYSTEMS

Two fundamentally different devices in which the rate of drug release is controlled by diffusion can be constructed. These are monolithic devices and membrane-controlled devices.

Monolithic Devices

In a monolithic device, the therapeutic agent is dispersed in a polymer matrix and its release is controlled by diffusion from the matrix. Mathematical treatment of diffusion depends on the solubility of the agent in the polymer and two cases must be considered. In one case, the agent is present below its solubility limit and is dissolved in the polymer. In the other case, the agent is present well above its solubility limit and is dispersed in the polymer.

The release of an agent that is dissolved in the polymer can be calculated by two equations, known as early-time approximation (Eq. 1), and late-time approximation (Eq. 2) (Baker and Lonsdale, 1974).

$$\frac{dM_t}{dt} = 2M_x \left[\frac{D}{\pi \, l^2 \, t} \right]^{1/2} \qquad (1)$$

$$\frac{dM_t}{dt} = \frac{8DM_x}{l^2} \, exp \, \frac{\pi^2 Dt}{l^2} \qquad (2)$$

These equations predict the rate of release from a slab of thickness l, where D is the diffusion coefficient, M_x is the total amount of agent dissolved in the polymer, and M_t is the amount released at time t. According to Eq. 1, which is valid over the first 60% of the release time, the rate decreases as the square root of time. During the latter 40% of the release time, the rate decays exponentially, as shown by Eq. 2. Plots of these two approximations are shown in Fig. 2.

When the agent is dispersed in the polymer, release kinetics can be calculated by the Higuchi equation (Higuchi, 1961) (Eq. 3).

$$\frac{dM_t}{dt} = \frac{A}{2} \left[\frac{2DC_sC_o}{t} \right]^{1/2} \qquad (3)$$

In this equation, A is the slab area, C_s is the solubility of the agent in the matrix, and C_o is the total concentration of dissolved and dispersed agent in the matrix. Clearly, in this particular case, release rate decreases as the square root of time over the major portion of the delivery time and deviates only after the concentration of the active agent remaining in the matrix falls below the saturation value C_s.

Membrane-Controlled Devices

In this particular delivery system, the active agent is contained in a core that is surrounded by a thin polymer membrane, and release to the surrounding environment occurs by diffusion through the rate-limiting membrane.

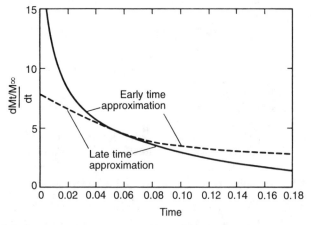

FIG. 2. Release rate of drug initially dissolved in a slab as a function of time. (Reprinted from R. W. Baker and H. K. Lonsdale, in *Controlled Release of Biologically Active Agents*, A. C. Tankaray and R. E. Lacey, eds., Plenum Publ., 1974, p. 43.)

When the membrane is nonporous, diffusion can be described by Fick's first law

$$J = -D \frac{dC_m}{dx} \qquad (4)$$

Where J is the flux in g/cm^2-sec, C_m is the concentration of the agent in the membrane in g/cm^3, dC_m/dx is the concentration gradient, and D is the diffusion coefficient of the agent in the membrane in cm^2/sec.

Because the concentration of the agent in the membrane cannot be readily determined, Eq. 4 can be rewritten using partition coefficients which describe the equilibrium ratio of the saturation concentration of the agent in the membrane to that in the surrounding medium.

$$J = \frac{DK \, \Delta C}{l} \qquad (5)$$

where ΔC is the difference in concentration between the solutions on either side of the membrane, K is the partition coefficient, and l is the thickness of the membrane.

Reservoir devices can also be constructed with membranes that have well-defined pores connecting the two sides of the membrane. Diffusion in such microporous membranes occurs principally by diffusion through the liquid-filled pores. In this system, the flux is described by Eq. 6.

$$J = \frac{EDK \, \Delta C}{\tau l} \qquad (6)$$

where E is the porosity (number of pores per unit area) of the membrane and τ is the tortuosity (average length of channel traversing the membrane).

One of the most important consequences of Eqs. 5 and 6 is that the flux J will remain constant provided that the membrane material does not change with time so that D, K, E, and τ remain constant. However, another very important requirement is that ΔC also remains constant. The practical consequences of this requirement is that the concentration of the agent in the core cannot change with time and that the agent released from the device must be rapidly removed. Constant agent concentration in the core can be readily achieved by dispersing the agent in a medium in which it has a very low solubility. However, it is not always possible to ensure that the concentration of an agent does not increase around the device, particularly with agents having very low water solubility and deviations from zero-order kinetics, known as boundary-layer effects.

Another factor that contributes to deviations from zero-order release kinetics is migration of agent from the core into the membrane on storage. Then, when the device is placed in a release medium, initial release is rapid because the agent diffuses from the saturated membrane. This nonlinear portion of the release is known as the burst effect.

Applications

Decisions as to which type of device is most appropriate for an intended application must consider the need for constant

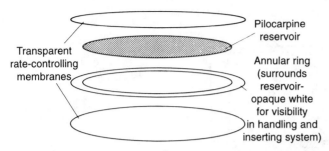

FIG. 3. Schematic diagram of Ocusert. (Reprinted from S. K. Chandrase-karan, H. Benson, and J. Urquhart, in *Sustained and Controlled Release Drug Delivery Systems,* J. R. Robinson, ed., Marcel Dekker, 1978, p. 571.)

drug release, manufacturing cost, and safety. Although the long-term zero-order drug release achievable with reservoir devices is highly desirable, it is expensive to manufacture such devices and in many applications where cost is important, inexpensive matrix-type devices are used even though their release rate declines with time. The utilization of such devices is very common in the veterinary or agricultural fields where low cost is of paramount importance.

When cost is not an overriding consideration, as is the case in human therapeutics, then reservoir-type devices are an excellent choice and a number of such devices are currently available. These are the Ocusert, which is inserted in the lower cul-de-sac of the eye to control glaucoma, the intrauterine contraceptive device Progestasert, and the contraceptive implant Norplant. The complete Norplant system consists of six silicone tubes, 20 × 2 mm, and contains the contraceptive steroid levonorgestrel in the core of the devices, which are designed for implantation in the upper arm. They are capable of maintaining a therapeutically effective concentration of levonorgestrel for as long as 5 years. The construction of the Ocusert device is shown in Fig. 3. Even though reservoir-type devices can yield zero-order kinetics for long periods of time, the safety of the device is of concern because rupture of the membrane can rapidly release the entire contents of the core. Thus, the amount of drug in the core and its toxicity must be considered.

To date, the most commercially successful use of diffusional systems is in transdermal applications. In these devices, a polymeric delivery system is held on the skin by an adhesive. The device contains the drug either in a reservoir with a rate-controlling membrane, or dispersed in a polymer matrix. A schematic of a membrane-controlled device is shown in Fig. 4. The

TABLE 2 Currently Marketed Transdermal Systems[a]

Clonidine	Boehringer-Ingelheim
Estradiol	Ciba-Geigy
	Besins-Iscovesco
Etofenamate	Bayer
Fentanyl	Janssen
Isosorbide dinitrate	(a) Yamanouchi
	(b) Yamanouchi
Nicotine	Parke-Davis
	Ciba-Geigy
	Marion
	Lederle
Nitroglycerine	Several companies
	(a) Key Pharmaceuticals
	(b) Key Pharmaceuticals
	G. D. Searle
	Ciba-Geigy
	Wyeth
	Bolar, others
	Nippon Kayaku/Taiho
	3M Riker
Progesterone	Besins-Iscovesco
Scopolamine	Ciba-Geigy
	Myum Moon Pharm

[a]Reproduced in part from *Innovations in Drug Delivery, Impact on Pharmacotherapy,* A. P. Sam and J. G. Fokkens (eds.). The Anselmus Foundation, Houten, The Netherlands, 1995, p. 71.

drug is released from these devices through the skin and is taken up by the systemic circulation. Because the outer layer of the skin, the stratum corneum, is highly impermeable to most drugs, either very potent drugs must be used or the flux of the drug through the skin must be augmented by penetration enhancers. Transdermal systems currently approved by the FDA are shown in Table 2.

WATER PENETRATION CONTROLLED SYSTEMS

In this type of device, the rate of delivery is controlled by the penetration of water into the device. Two general types of devices are in use.

Osmotically Controlled Devices

The operation of an osmotic device can be seen schematically in Fig. 5 (Theeuwes and Yum, 1976). In this device, an osmotic agent is contained within a rigid housing and is separated from the agent by a movable partition. One wall of the rigid housing is a semipermeable membrane and when the

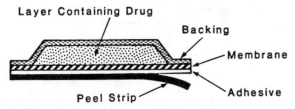

FIG. 4. Schematic of membrane-controlled transdermal device.

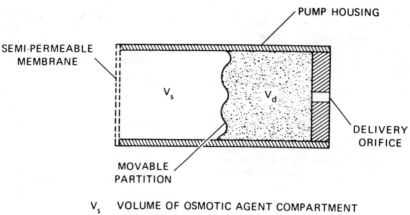

FIG. 5. Schematic representation of an osmotic pump. (Reprinted from F. Theeuwes and S. I. Yum, *Ann. Biomed. Eng.* **4**: 343–353, 1976.)

device is placed in an aqueous environment, water is osmotically driven across the membrane. The resultant increase in volume within the osmotic compartment exerts pressure on the movable partition, which then forces the agent out of the device through the delivery orifice.

The volume flux of water across the semipermeable membrane is given by:

$$\frac{dV}{dt} = \frac{A}{l} L_p [\sigma \, \Delta \pi - \Delta P] \qquad (7)$$

where dV/dt is the volume flux, $\Delta \pi$ and ΔP are, respectively, the osmotic and hydrostatic pressure differences across the semipermeable membrane, L_p is the membrane mechanical permeability coefficient, σ is the reflection coefficient, and A and l are, respectively, the membrane area and thickness.

The rate of delivery, dM/dt, of the agent is then given simply by

$$\frac{dM}{dt} = \frac{dV}{dt} C \qquad (8)$$

where C is the concentration of the agent in the solution that is pumped out of the orifice.

Swelling-Controlled Devices

In this type of delivery system, the agent is dispersed in a hydrophilic polymer that is glassy in the dehydrated state but swells when placed in an aqueous environment. Because diffusion of molecules in a glassy matrix is extremely slow, no release occurs while the polymer is in the glassy state. However, when such a material is placed in an aqueous environment, water will penetrate the matrix and as a consequence of swelling, the glass transition temperature of the polymer is lowered below the temperature of the environment and the drug diffuses from the polymer.

The process shown schematically in Fig. 6 is characterized by the movement of two fronts (Langer and Peppas, 1983). One front, the swelling interface, separates the glassy polymer from the swollen rubbery polymer and moves inward into the device. The other front, the polymer interface, moves outward and separates the swollen polymer from the pure dissolution medium. In systems where the glassy polymer is noncrystalline and linear, dissolution takes place. However, when the polymer is highly crystalline or cross-linked, no dissolution takes place. Although many swellable drug formulations are available, the term "swelling controlled" is only applicable to those formulations in which the release is actually controlled by the swelling phenomenon just described.

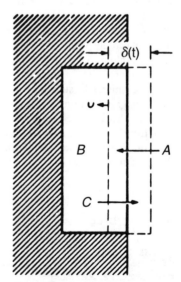

FIG. 6. Schematic representation of swelling-controlled release system. As penetrant A enters the glassy polymer B, bioactive agent C is released through the gel phase of thickness $\delta(t)$. (Reprinted from R. S. Langer and N. A. Peppas, *Rev. Macromol. Chem. Phys.* **C23**: 61–126, 1983.)

Applications

At present, two osmotically controlled drug release systems are commercially available. A number of swelling-controlled devices are under advanced stages of development and some have reached commercialization.

Two types of osmotic devices are in use. One device, known as Osmet, is a capsule approximately 2.5 cm long and 0.6 cm in diameter and is intended as an experimental device that can be implanted in the tissues of animals, where it delivers a chosen therapeutic agent at known, controlled rates. The agent is placed in an impermeable, flexible rubber reservoir that is surrounded by an osmotic agent, which in turn is surrounded and sealed within a rigid cellulose acetate membrane (Theeuwes and Yum, 1976).

In an aqueous environment, water is osmotically driven across the cellulose acetate membrane and the resultant pressure on the rubber reservoir forces the agent out of the orifice. The device is shown in Fig. 7. It is provided empty and can be filled with the desired therapeutic agent by the user. Because the driving force is osmotic transport of water across the cellulose acetate membrane, the rate of release of the agent from the device is independent of the surrounding environment.

A second type of device known as Oros is shown in Fig. 8 (Theeuwes, 1975). This device is intended for oral applications and is manufactured by compressing an osmotically active agent into a tablet, coating the tablet with a semipermeable membrane, and drilling a small hole through the coating with a laser. When the tablet is placed in an aqueous environment, water is driven across the semipermeable membrane and a

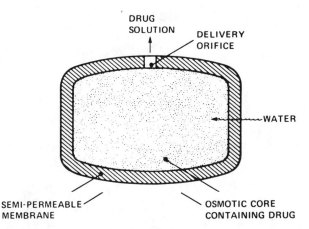

FIG. 8. Cross section of Oros. (Reprinted from F. Theeuwes, *J. Pharm. Sci.* **64:** 1987–1991, 1975.)

solution of the agent is pumped out of the orifice. A major advantage of this device is that a constant rate of release as it traverses the gastrointestinal tract is achieved. An over-the-counter osmotic device known as Acutrim used in appetite suppression is also available.

A swelling-controlled oral device commercially available as Geomatrix, is shown schematically in Fig. 9. In this device, a drug is dispersed in a swellable polymer such as hydroxypropyl methylcellulose, which is compressed into a tablet, and two sides are coated with a water-impermeable coating such as cellulose acetate propionate. This impermeable coating affects the swelling of the matrix and modifies diffusional release kinetics so that reasonably constant release kinetics are achieved.

CHEMICALLY CONTROLLED DEVICES

There is considerable confusion in the use of the terms "bioerodible" and "biodegradable," and no generally accepted definitions are availiable. In this chapter we only use the term "bioerodible." The prefix bio is used because the erosional process takes place in a biological environment and carries no implication that erosion is assisted by enzymes or other active species in the biological milieu.

Drug Release Mechanisms

Drug release from bioerodible polymers can occur by any one of three basic mechanisms shown schematically in Fig. 10

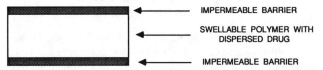

FIG. 9. Schematic representation of swelling-controlled oral system.

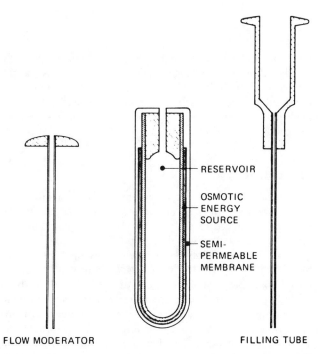

FIG. 7. Schematic representation of osmotic pump and components. (Reprinted from F. Theeuwes and S. I. Yum, *Ann. Biomed. Eng.* **4:** 343–353, 1976.)

MECHANISM A

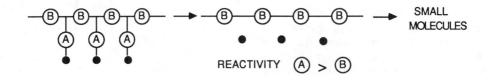

MECHANISM B

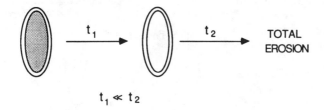

MECHANISM C

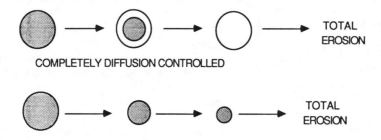

FIG. 10. Schematic representation of drug release mechanisms.

(Heller, 1985). In mechanism A, the active agent is covalently attached to the backbone of a biodegradable polymer and is released as its attachment to the polymer backbone cleaves by hydrolysis of bond A. Because it is not desirable to release the drug with polymer fragments still attached, the reactivity of bond A should be significantly higher than the reactivity of bond B. In mechanism B, the active agent is contained within a core and is surrounded by a bioerodible rate-controlling membrane. Release of the active agent is then controlled by its diffusion across the membrane. In mechanism C, the active agent is dispersed in a bioerodible polymer and its release is controlled by diffusion, by a combination of diffusion and erosion, or by pure erosion.

Drug Covalently Attached to Polymer Backbone

This delivery system has found two applications. In one, the polymer–drug adduct is water insoluble and is used as a subcutaneous or intramuscular implant. In the other application, the polymer–drug adduct is water soluble and is used in targeting applications. Such a polymer contains a covalently attached drug and targeting moieties so that when it is injected intravenously it collects at the target site, where the drug is released by cleavage of the labile bond (Kopecek, 1990). Both applications are still in the experimental stages and no commercial products are available.

Applications

An example of a depot-type system is poly(N^5-hydroxypropyl-L-glutamate) with the contraceptive steroid norethindrone covalently attached to the polymer by a carbonate link (Peterson *et al.*, 1980). Even though poly(N^5-hydroxypropyl-L-glutamate) is water soluble, attachment of the highly hydrophobic norethindrone yields a water-insoluble product. Then, as the hydrophobic steroid is released by hydrolysis of the carbonate link, polymer hydrophilicity increases and reaction rate in the hydrophilic region accelerates. As a consequence of this process, a hydrophilic front develops and moves through the solid, hydrophobic polymer. Because diffusion of the steroid through the hydrophilic layer is rapid relative to the movement of the hydrophilic front, the rate of drug release is determined by the rate of movement of this front and a fairly constant release can be achieved.

Drug Contained within a Biodegradable Core

This delivery system is identical to the reservoir system already discussed, except that the membrane surrounding the drug core is bioerodible. Therefore, such systems combine the advantage of long-term, zero-order drug release with bioerodibility.

Because constancy of drug release requires that the diffusion coefficient D of the agent in the membrane remain constant (Eq. 5), the bioerodible membrane must remain essentially unchanged during the delivery regime. Furthermore, the membrane must remain intact while there is still drug in the core to prevent its abrupt release. Thus, significant bioerosion cannot take place until drug delivery has been completed.

Application

The only system that uses this approach is a delivery device for contraceptive steroids. A device in the advanced stages of development is based on a poly(ε-caprolactone) capsule containing the contraceptive steroid, levonorgestrel. This device, known as Capronor, is designed to release levonorgestrel at constant rates for 1 year and to completely bioerode in 3 years (Pitt et al., 1980). Phase II clinical trials have recently been completed.

Drug Dispersed in a Bioerodible Matrix

A discussion of bioerodible matrix systems is conveniently divided into systems in which drug release is determined predominantly by diffusion, and systems in which drug release is determined predominantly by polymer erosion. When the polymer undergoes surface erosion, the rate of release is completely determined by erosion (Heller et al., 1978).

Drug Release Determined Predominantly by Diffusion

Drug release from polymers in which hydrolysis occurs at more or less uniform rates throughout the bulk of the polymer is determined predominantly by Fickian diffusion. When the rate of polymer hydrolysis is slow relative to drug depletion, $t^{1/2}$ kinetics identical to those observed with nondegradable systems are observed. When the rate of polymer hydrolysis is significant before drug depletion, then the $t^{1/2}$ kinetics are modified by the hydrolysis process.

The most extensively investigated bulk-eroding polymers are poly(lactic acid) and copolymers of lactic and glycolic acids. These polymers were originally developed as bioerodible sutures and because they degrade to the natural metabolites, lactic acid and glycolic acids, to this day they occupy a preemi-

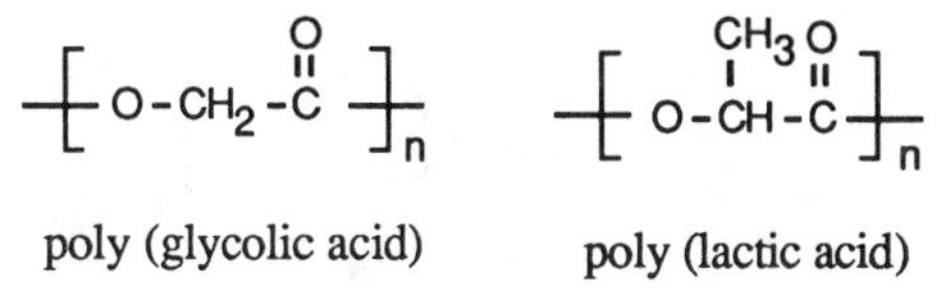

FIG. 11. Structure of polymer with uniform hydrolysis.

nent place among bioerodible drug delivery systems (Heller, 1984). Their structure is shown in Fig. 11

Drug Release Determined Predominantly by Erosion

Certain polymers can undergo a hydrolysis reaction at decreasing rates from the surface of a device inward and under special circumstances the reaction can be largely confined to the outer layers of a solid device. Two such polymers are poly(ortho esters) and polyanhydrides. Because the rates of hydrolysis of these polymers can be varied within very wide limits, considerable control over rate of drug release can be achieved.

Poly(Ortho Esters) Poly(ortho esters) are highly hydrophobic polymers that contain acid-sensitive links in the polymer backbone. At the physiological pH of 7.4, these links undergo a very slow rate of hydrolysis, but as the ambient pH is lowered, hydrolysis rates increase. Thus, the incorporation of small amounts of acidic excipients such as aliphatic dicarboxylic acids into such materials allows precise control over rates of erosion. With highly hydrophobic drugs, surface hydrolysis can take place because as water intrudes into the polymer, the acidic excipient ionizes and hydrolysis accelerates owing to the decreased pH in the surface layers. As a result of this process, an eroding front develops that moves into the interior of the device. However, when hydrophilic drugs are used, water is rapidly drawn into the polymer and bulk hydrolysis takes place. Three families of poly(ortho ester) have been developed and are shown in Fig. 12 (Heller, 1993).

Very long erosion times can be achieved by incorporating basic excipients such as $Mg(OH)_2$. In this case, and also with hydrophobic drugs, long-term surface erosion can take place because hydrolysis can only occur in the outer layers where the basic excipient has diffused out of the device and has been neutralized by the external buffer (Heller, 1985).

Polyanhydrides These materials were first prepared in 1909, and were subsequently investigated as potential textile fibers but found unsuitable owing to their hydrolytic instability. Although polyanhydrides based on poly[bis(p-carboxyphenoxy) alkane anhydrides] have significantly improved hydrolytic stability, they retain enough hydrolytic instability to prevent conmmercialization, despite their good fiber-forming properties.

The use of polyanhydrides as bioerodible matrices for the controlled release of therapeutic agents was first reported in 1983 (Rosen et al., 1983). Because aliphatic polyanhydrides

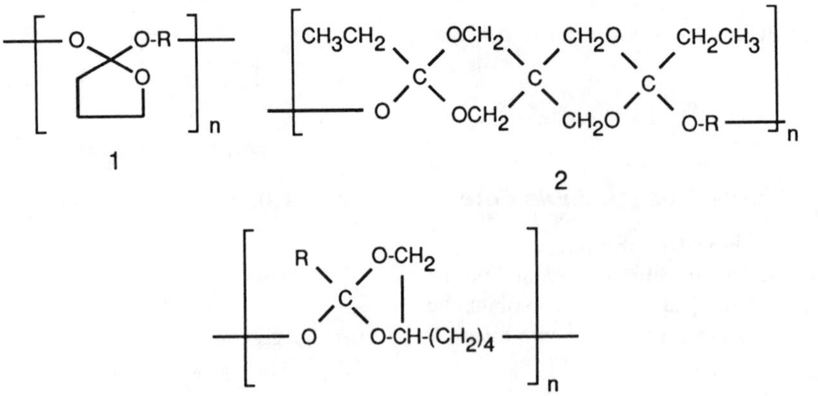

FIG. 12. Families of poly(ortho esters).

hydrolyze very rapidly while aromatic polyanhydrides hydrolyze very slowly, excellent control over hydrolysis rate can be achieved by using copolymers of aliphatic and aromatic polyanhydrides. In this way, erosion rates from days to years have been demonstrated (Leong *et al.*, 1985, 1986). The structure of a polymer based on bis(*p*-carboxyphenoxy) alkane and sebasic acid is shown in Fig. 13.

Applications of Bioerodible Polymers

Lactide–glycolide copolymer systems have been extensively investigated for the delivery of the contraceptive steroids norethindrone and levonorgestrel from injectable microspheres and for delivering synthetic analogs of luteinizing hormone-releasing hormone (LHRH). The contraceptive delivery system is in advanced stages of development and devices containing norethindrone have passed Phase II clinical trials and are in Phase III clinical trials. LHRH-releasing systems for control of prostate cancer are now commercially available.

Poly(ortho ester) system 2 has been used for delivering levonorgestrel, 5-fluorouracyl and naltrexone. It is patented by SRI International and rights are assigned to Merck, Sharpe, and Dohme, where it is being developed for a range of commercial applications.

Poly(ortho ester) 3 represents a unique polymer system that is a viscous, hydrophobic, pastelike material at room temperature even at fairly high molecular weights. The pastelike property allows incorporation of therapeutic agents under very mild conditions and it is under investigation for the continuous and pulsed release of proteins.

Polyanhydride has been patented by the Massachusetts In-

stitute of Technology and the rights assigned to Nova Pharmaceuticals, where it is being explored as a bioerodible implant for the release of BCNU [*N,N*-bis(2-chloroethyl)-*N*-nitrosourea] following brain cancer surgery. The polymer has been approved for use in terminally ill cancer patients, and clinical trial with excellent results are under way.

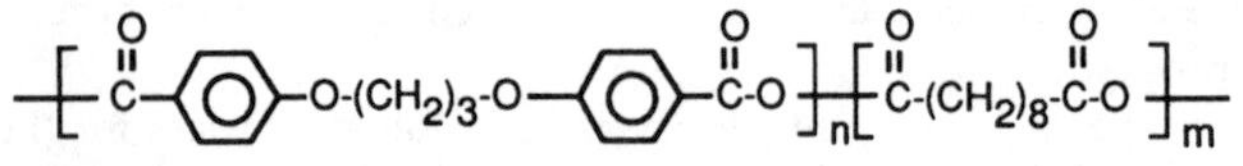

REGULATED SYSTEMS

During the past two decades, controlled-release administration of therapeutic agents from various types of delivery systems has become an important area of research, and significant advances in theories and methodologies have been made. However, even though devices that are capable of releasing therapeutic agents by well-defined kinetics are a significant improvement over conventional dosage forms, these devices do not yet represent the ultimate therapy because the agent is released without regard of the need of the recipient. Therefore, another very significant improvement can be realized if systems could be devised that are capable of adjusting drug output in response to a physiological need.

Regulated systems can be broadly grouped into externally regulated and self-regulated. In this classification, externally regulated devices can alter their drug output only in response to an external intervention while self-regulated devices can do so without external intervention (Heller, 1988).

Externally Regulated

Among the most advanced externally regulated devices are mechanical pumps that dispense drugs from a reservoir to the body by means of a catheter. Such pumps can be worn externally or can be implanted in a suitable body site. A major application is control of diabetes by delivering insulin in response to blood glucose levels. A number of such pumps, such as the CPI Lilly pump, have sophisticated control mechanisms and microprocessors that allow programmed insulin delivery (Brunetti *et al.*, 1991).

FIG. 13. Structure of polymer based on bis(*p*-carboxyphenoxy) alkane and sebasic acid.

Another means of externally regulating drug delivery is by means of magnetism (Hsieh *et al.*, 1989). In this procedure, small magnetic spheres are embedded within a polymer matrix that contains a dispersed therapeutic agent. When an oscillating magnetic field is applied to the polymer, the normal diffusional release is significantly and reversibly increased.

The rate of drug release can also be reversibly increased by the use of ultrasound (Kost *et al.*, 1981). This method has been applied to biodegradable polymers and it has been shown that the rate of bioerosion and drug release can be significantly enhanced. Unlike magnetically enhanced devices, ultrasound can be used with implants and represents a means by which enhanced insulin delivery could be realized.

Self-Regulated

Self-regulated devices are capable of altering drug output in response to an external change and can be classified into substrate-specific and environment-specific devices.

Substrate-Specific

Two different types of devices are under development. In one type, the substrate modulates drug release from the device, while in the other type the substrate triggers drug release from a passive device.

Modulated Devices The major driving force for developing such devices is a need to better control diabetes since it is now well recognized that replacement of insulin by periodic injections does not adequately control this disease. Thus, many approaches are under way in attempts to develop devices that can release insulin in response to blood glucose concentrations.

In one such approach, a sugar is covalently attached to insulin and the sugar–insulin conjugate is complexed with the plant-derived, carbohydrate-binding protein Concanavalin A (Con A). Because the sugar–insulin conjugate is displaced from Con A in proportion to the external glucose concentration, such a device can deliver insulin in response to blood glucose levels (Makino *et al.*, 1990).

Another approach utilizes pH changes resulting from the glucose oxidase conversion of glucose to gluconic acid and pH-sensitive polymers that can respond to that change. Such polymers can be membranes that reversibly alter their porosity by a protonation of tertiary amine functions in the polymer (Albin *et al.* 1985; Ishihara *et al.*, 1986) or by ionization of a grafted polyacrylic acid on a microporous substrate (Ito *et al.*, 1989; Iwata *et al.*, 1988), or they can be bioerodible polymers that alter their erosion rate in response to pH changes (Heller *et al.*, 1990).

Triggered Devices One important application is the development of a device capable of releasing the narcotic antagonist naltrexone in response to external morphine. Such a device would be useful in treating narcotic addiction by blocking the opiate-induced euphoria through a morphine-triggered release of the antagonist. Development of this device utilizes the reversible inactivation of an enzyme by hapten–antibody interaction

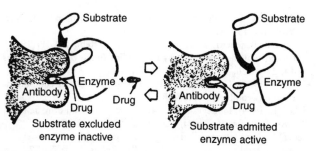

FIG. 14. Reversible enzyme inactivation by hapten–antibody interactions. In this particular case, the drug is morphine and the enzyme is lipase. (Reprinted from R. S. Schneider, P. Lindquist, E. T. Wong, K. E. Rubinstein, and E. F. Ullman, *Clin. Chem.* **19**: 821–825, 1973.)

(Schneider *et al.*, 1973) as shown in Fig. 14. In the device under development, naltrexone is contained in a core surrounded by an enzymatically degradable coating that prevents its release. In the absence of external morphine, the device is stable, but upon exposure to morphine, the reversibly inactivated enzyme is activated and degrades the protective coating, which results in naltrexone release. The actual device utilizes the enzyme lipase and a triglyceride protective coating (Roskos *et al.*, 1993).

Environment-Specific Devices

In an environment-specific approach, devices are constructed that can alter drug release in response to changes in temperature or pH. One example of polymers that can reversibly respond to temperature changes are hydrogels based on *N*-isopropylacrylamide (Yang *et al.*, 1990; Okano *et al.*, 1990). When this monomer is copolymerized with methylene bisacrylamide, the resulting cross-linked materials have been shown to demix with water and to shrink abruptly when heated just above 31°C. The collapse of the hydrogel occurs as a result of a phase transition of poly(*N*-isopropylacrylamide), which is soluble in water at its lower critical solution temperature (LCST) of 31°C but becomes insoluble when the temperature is increased just past this temperature. Such materials are useful for the delivery or removal of biological molecules triggered by temperature changes.

pH-Sensitive systems can be constructed from lightly cross-linked copolymers of *n*-alkyl methacrylates (hydrophobic monomer) and *N,N*-dimethylaminoethyl methacrylates (ionizable monomer) (Siegel, 1990). Such materials exhibit a strong dependence of swelling on external pH and can abruptly swell when the external pH is lowered. The pH at which the abrupt swelling occurs is a function of copolymer composition.

Bibliography

Albin, G., Horbett, T. A., and Ratner, B. D. (1985). Glucose sensitive membranes for controlled delivery of insulin: insulin transport studies. *J. Controlled Release* 2: 153–164.

Baker, R. W., and Lonsdale, H. K. (1974). Controlled release: mechanisms and rates. in *Controlled Release of Biologically Active*

Agents, A. C. Tanquary and R. E. Lacey, eds. Plenum Publ., New York, pp. 15–71.

Brunetti, P., Benedetti, M. M., Calabrese, G., and Reboldi, G. P. (1991). Closed loop delivery systems for insulin. *Int. J. of Artif. Organs* **14:** 216–226.

Heller, J. (1980). Controlled release of biologically active compounds from bioerodible polymers. *Biomaterials* **1:** 51–57.

Heller, J. (1984). Biodegradable polymers in controlled drug delivery. *CRC Crit. Rev. in Therap. Drug Carrier Syst.* **1:** 39–90.

Heller, J. (1985). Controlled drug release from poly(ortho esters): a surface eroding polymer. *J. Controlled Release* **2:** 167–177.

Heller, J. (1987). Bioerodible hydrogels. in Medicine and Pharmacy, N. A. Peppas, ed. CRC Press, Boca Raton, FL, Vol. III, pp. 137–149.

Heller, J. (1988). Chemically self-regulated drug delivery systems. *J. Controlled Release* **8:** 111–125.

Heller, J. (1993). Poly(ortho ester). *Advances in Polymer Science* **107:** 41–92.

Heller, J., Baker, R. W., Gale, R. M., and Rodin, J. O. (1978). Controlled drug release by polymer dissolution I. Partial esters of maleic anhydride copolymers. Properties and theory. *J. Appl. Polymer Sci.* **22:** 1991–2009.

Heller, J., Chang, A. C., Rodd, G., and Grodsky, G. M. (1990). Release of insulin from a pH-sensitive poly(ortho ester). *J. Controlled Release* **14:** 295–304.

Higuchi, T. (1961). Rates of release of medicaments from ointment bases containing drugs in suspension. *J. Pharm. Sci.* **50:** 874–875.

Hsieh, D. S., Langer, R., and Folkman, J. (1981). Magnetic modulation of release of macromolecules from polymers. *Proc. Natl. Acad. Sci. U.S.A.* **78:** 1863–1867.

Ishihara, KJ., and Matsui, K. (1986). Glucose-responsive insulin release from polymer capsule. *J. Polymer Sci., Polymer Lett. Ed.* **24:** 413–417.

Ito, Y., Casolaro, M., Kono, K., and Imanishi, Y. (1989). An insulin-releasing system that is responsive to glucose. *J. Controlled Release* **10:** 195–203.

Iwara, H., and Matsuda, T. (1988). Preparation and properties of novel environment-sensitive membranes prepared by graft polymerization onto a porous substrate. *J. Membrane Sci.* **38:** 185–199.

Kim, S. W., Petersen, R. V., and Feijen, J. (1980). Polymeric drug delivery systems. in *Drug Design,* A. Ariens, ed. Academic Press, New York, Vol. X, 193–250.

Kopecek, J. (1990). The potential of water-soluble polymeric carriers in targeted and site-specific drug delivery. *J. Controlled Release* **11:** 279–290.

Kost, J., Leong, K., and Langer, R. (1989). Ultrasound-enhanced polymer degradation and release of incorporated substances. *Proc. Natl. Acad. Sci. U.S.A.* **86:** 7663–7666.

Langer, R. S., and Peppas, N. A. (1983). Chemical and physical structure of polymers as carriers for controlled release of bioactive agents: a review *Rev. Macromol. Chem. Phys.* **C23:** 61–126.

Leong, K. W., Brott, B. C., and Langer, R. (1985). Bioerodible polyanhydrides as drug carrier matrices I: characterization, degradation and release characteristics. *J. Biomed. Mater. Res.* **19:** 941–955.

Leong, K. W., D'Amore, P. D., Marletta, M., and Langer, R. (1986). Bioerodible polyanhydrides as drug carrier matrices II: biocompatibility and chemical reactivity. *J. Biomed. Mater. Res.* **20:** 51–64.

Makino, K., Mack, E. J., Okano, T., and Kim, S. W. (1990). A microcapsule self-regulating delivery system for insulin. *J. Controlled Release* **12:** 235–239.

Okano, T., Bae, Y. H., Jacobs, H., and Kim, S. W. (1990). Thermally on-off switching polymers for drug permeation and release. *J. Controlled Release* **11:** 255–265.

Petersen, R. V., Anderson, R. G., Fang, S. M., Gregonis, D. E., Kim, S. W., Feijen, J., Anderson, J. M., and Mitra, S. (1980). Controlled release of progestins from poly(α-amino acid) carriers. in *Controlled Release of Bioactive Materials,* R. W. Baker, ed. Academic Press, New York, pp. 45–60.

Pitt, C. G., Marks, T. A., and Schindler, A. (1980). Biodegradable delivery systems based on aliphatic polyesters: applications to contraceptives and narcotic antagonists. in *Controlled Release of Bioactive Materials,* R. W. Baker, ed. Academic Press, New York, pp. 19–43.

Rosen, H. B., Chang, J., Wnek, G. E., Linhardt, R. J., and Langer, R. (1983). Bioerodible polyanhydrides for controlled drug delivery. *Biomaterials* **4:** 131–133.

Roskos, K. V., Tefft, J. A., and Heller, J. (1993). A morphine-triggered delivery system useful in the treatment of heroin addiction. *Clinical Materials* **13:** 109–119.

Schneider, R. S., Lidquist, P., Wong, E. T., Rubenstein, K. E., and Ullman, E. F. (1973). Homogeneous enzyme immunoassay for opiates in urine. *Clin. Chem.* **19:** 821–825.

Sigel, R. A. (1990). pH-sensitive gels: swelling equilibria, kinetics and applications for drug delivery. in *Pulsed and Self-Regulated Drug Delivery,* J. Kost, ed. CRC Press, Boca Raton, FL, pp. 129–157.

Theeuwes, F. (1975). Elementary osmotic pump. *J. Pharm. Sci.* **64:** 1987–1991.

Theeuwes, F., and Yum, S. I. (1976). Principles of the design and operation of generic osmotic pumps for the delivery of semi-solid or liquid drug formulations. *Ann. Biomed. Eng.* **4:** 343–353.

Yang, H. J., Cole, C. A., Monji, N., and Hoffman, A. S. (1990). Preparation of thermally phase separating copolymer, poly(N-isopropylacrylamide-co-N-acryloxysuccinimide) with a controlled number of active esters per polymer chain. *J. Polymer Sci., Part A., Polymer Chemistry* **28:** 219–226.

7.9 SUTURES

Dennis Goupil

Approximately 250 million sutures are used in the United States annually for a variety of surgical procedures ranging from routine skin lacerations to delicate organ transplants. Under the assumption that an equal number of sutures are used outside the United States, this total of nearly 500 million surgical devices represents the largest volume and most commonly used surgical device in the world.

A suture is a complicated medical product that must be designed and manufactured consistently to meet a range of physical and clinical demands. Its major functions are to bring and hold tissue together following separation by surgery or trauma. When one considers how sutures benefit the facial reconstruction of an accident victim, or the fatal consequences of suture breakage following a heart transplant, one begins to understand their complexity and function.

CATEGORIES AND CHARACTERISTICS

Sutures are broadly categorized according to the type of material (natural, synthetic) from which they are made, the

TABLE 1 Major Commercially Available Polymeric Sutures

Suture type	Generic chemical structure	Construction[a]	Sterilization method[b]	Representative commercial product (manufacturer)	Major clinical use
Natural materials					
Catgut	Protein	Tw	ETO/Rad.	Catgut (D+G, Ethicon)	Ob/Gyn, urology
Silk	Protein	B	ETO/Rad.	Silk (D+G)	Cardiovascular
				Surgical silk (D+G)	Vascular
Synthetic absorbable					
Poly(glycolic acid)	$[-OCH_2CO_2CH_2CO-]$	B	ETO	Dexon II (D+G)[c]	General, Ob/Gyn
Poly(glycolide-co-lactide)	$[-OCH_2CO_2CH_2CO-]_{90}$ $[OCH(CH_3)CO_2CH(CH_3)-CO-]_{10}$	B	ETO	Coated vicryl (Ethicon)[d] Polysorb (USSC)[e]	General, Ob/Gyn
Poly(p-dioxanone)	$[-O(CH_2)_2OCH_2CO-]$	M	ETO	PDS (Ethicon)[f]	General, Ob/Gyn
Poly(glycolide-co-trimethylene carbonate)	$[-OCH_2CO-]_{67}$ $[-OCH_2CH_2CH_2OCO-]_{33}$	M	ETO	Maxon (D+G)[g]	General, Ob/Gyn
Synthetic nonabsorbable					
Poly(butylene terephthalate)	$[-O(CH_2)_4OCOC_6H_4CO-]$	B, M	ETO/Rad.	Miraline (Braun)	Cardiovascular Orthopaedics
Poly(ethylene terephthalate)	$[-O(CH_2)_2OCOC_6H_4CO-]$	B, M	ETO/Rad.	Ti.Cron (D+G) Surgidac (USSC)	Cardiovascular Orthopaedics
Poly[p(tetramethylene ether) terephthalate-co-tetramethylene]	$[-(CH_2)_4OCOC_6H_4CO]_{84}$ $[-O(CH_2CH_2CH_2CH_2O-]n$ $COC_6H_4CO-]_{16}$	M	ETO/Rad.	Novafil (D+G)	Plastic/cuticular
Polypropylene	$[-CH_2CH(CH_3)-]$	M	ETO	Surgilene (D+G) Prolene (Ethicon) Deklene (Deknatel) Surgipro (USSC)	Cardiovascular Vascular
Nylon 66	$[-NH(CH_2)_6NHCO(CH_2)_4CO-]$	B, M	ETO/Rad.	Dermalon (D+G) Ethilon (Ethicon) Monosof (USSC)	Plastic/cuticular Ophthalmic

Note. Sources listed under Bibliography.
[a]Construction: Twisted (Tw); braid (B); monofilament (M).
[b]Sterilization method: Ethylene oxide (ETO); Gamma radiation (Rad.).
[c]Dexon II package claims, Davis+Geck.
[d]Vicryl package claims, Ethicon.
[e]Polysorb package claims, United States Surgical.
[f]PDS package claims, Ethicon.
[g]Maxon package claims, Davis+Geck.

permanence of the material (absorbable or nonabsorbable), and the construction process (braided, monofilament) used. As shown in Table 1, the most popular natural materials used for sutures are silk and catgut (animal intestine). A fair amount of art and effort is required in both cases to reduce the raw material to the finished product. The synthetic materials are exclusively polymeric, except for fine-sized stainless steel sutures. All sutures, regardless of material or construction, require special surgical needles for delivery through tissue.

Approximately half of today's sutures are nonabsorbable and remain indefinitely intact when placed in the body. Common engineering polymers like polypropylene, nylon, poly(ethylene terephthalate), and polyethylene are used as sutures. Copolymers of these materials have also been used clinically. Absorbable sutures were commercially introduced by Davis + Geck in 1970 with poly(glycolic acid) (PGA) sutures and were followed by copolymers of glycolide and lactide from Ethicon and U.S. Surgical. More recently, novel absorbable polymers of polydioxanone and poly(glycolide-co-trimethylene carbonate) have been developed for surgical use (see Chapter 2.5 for additional information on resorbable materials).

Regardless of whether a suture is made from a natural or a synthetic material, or if it is absorbable or permanent, it must meet the strength requirements necessary to close a wound under a given clinical circumstance. Almost all suture products will be efficacious for minor wounds or for

TABLE 2 Representative Mechanical Properties of Commercial Sutures

Suture type	St. pull (MPa)	Kt. pull (MPa)	Elongation to break (%)	Subjective flexibility
Natural materials				
Catgut	370	160	25	Stiff
Silk	470	265	21	Very supple
Synthetic absorbable				
Poly(glycolic acid)	840	480	22	Supple
Poly(glycolide-co-lactide)	740	350	22	Supple
Poly(p-dioxanone)	505	290	34	Mod. stiff
Poly(glycolide-co-trimethylene carbonate)	575	380	32	Mod. stiff
Synthetic nonabsorbable				
Poly(butylene terephthalate)	520	340	20	Supple
Poly(ethylene terephthalate	735	345	25	Supple
Poly[p(tetramethylene ether) terephthalate-co-tetramethylene terephthalate]	515	330	34	Supple
Polypropylene	435	300	43	Stiff
Nylon 66	585	315	41	Stiff
Steel	660	565	45	Rigid

Note. Sources listed under Bibliography.

normally healing wounds. Hence, a poly(glycolic acid) suture, which loses strength over a 28-day period, will be just as adequate as a permanent polypropylene suture. If a patient, however, suffers from a disease or conditions that retard healing (e.g., diabetic patients), a nonabsorbable or slower degrading suture may be more appropriate. Representative mechanical properties of some commercial sutures are listed in Table 2.

The construction of a surgical suture (i.e., braid or monofilament) is important to both the surgeon and the patient for objective and subjective reasons. In addition to out-of-package tensile strength and *in vivo* tensile strength, the surgeon considers a variety of other parameters before making a choice of sutures for the patient. As shown in Table 3, the parameters range from objective issues of knot security or the number of knots required to secure a suture, to the subjective issue of "feel" in the surgeons' hands. Braided sutures are generally more supple products compared with

TABLE 3 Suture Characteristics

Objective	Subjective
Tensile strength	Suppleness
Knot security	Ease of tying
Diameter	Ease on hands
Strength retention	
Flexibility	
Memory out of the package	
Tissue drag	
Infection potentiation (wicking)	

monofilaments and hence have an advantage in regard to out-of-package memory, ease of tying, and knot security if the same knot is used for both the braid and the monofilament. Monofilament sutures tend to be more wiry out of the package and can become tangled up with surgical instruments if the surgical team is not careful. The knot security issue is simply addressed by using different knots or more "throws" of a given knot to achieve security. The major advantage of a monofilament suture is its relatively low tissue drag compared with a braided suture. This low drag or friction between the tissue and the suture allows the surgeon to use different techniques in closing wounds (e.g., continuous or running closures). The low tissue drag is also less "abrasive" when the suture is being pulled through the tissue. This aspect is especially important for fragile cardiovascular, ophthalmic, and neurological tissue, where monofilament sutures are the products of choice.

PRODUCT DEVELOPMENT

It can take years to generate information to support clinical use of a new suture product, and usually 4–6 years for a completely new polymer suture. The steps required to develop a suture are outlined in Table 4. Starting with the concept or product design, several polymers are screened prior to choosing the optimum candidate. In the preclinical stage, extensive safety and animal efficacy testing is conducted to answer the questions of safety and effectiveness. This testing stage can take 2–3 years to complete. In addition to the basic requirement that the polymer must be able to be extruded and processed consistently to provide the fiber strength necessary to hold tissue together, the polymer must be able to withstand sterilization. Extensive toxicology testing is usually required, including acute and chronic toxicity, pyrogenicity, antigenicity, *in vitro* and *in vivo* infection potentiation, hemolytic potentiation, mutagenicity, and possibly carcinogenicity. In addition, suture strength and knot security are also evaluated in animal surgical procedures. If the suture is absorbable, the *in vivo* tensile strength over time and the absorption and metabolic fate of the polymer need to be established.

Following this testing, clinical evaluations are usually required to confirm the safety and efficacy of the product. The subjective issues of suture handling and knotting techniques need to be evaluated in the hands of many surgeons in scores of patients for each major clinical suture use. Again, this step can be lengthy, taking approximately 1–2 years to complete.

While the clinical studies are being conducted, manufacturing scale-up studies are completed and product and package stability data are assembled. The animal safety and efficacy data are combined with the clinical data and manufacturing documentation to prepare for registration or approval by a government agency (e.g., the U.S. Food and Drug Administration). The review process itself can be time consuming, often taking more than a year.

TRENDS

Several factors have influenced the suture industry in the past 5 years, including the changing regulatory environment in the United States and in Europe, the focus on cost containment in health care, the patent status of proprietary materials and processing methods, changing surgical practice toward minimally invasive techniques, and the advances in alternative wound-closure technologies (staplers, glues, etc.).

In the United States, most suture types have been reclassified from the stringent Class III (Pre-Market Approval) regulatory category to Class II (510.k) medical devices. This change will lead to more rapid commercialization of products and to the introduction of more generic sutures. The commercialization of sutures in Europe will become more difficult overall as the countries of the European Economic Community progress toward a more unified approach toward registration of medical devices. Previously, sutures could be commercialized in several countries without registration, but future use will necessitate that products meet the requirements of the CE process.

The majority of the proprietary technology that is germane to sutures is held by Ethicon, Davis + Geck, and more recently by U.S. Surgical. As the more traditional patents expire over the next 5–10 years, more manufacturers will enter an industry that is already highly competitive and cost constrained.

Sutures, which once constituted 100% of the wound closure market, are now receiving stiff competition from surgical staplers as these mechanical devices have become well accepted by surgeons. Mechanical devices have been especially useful in minimally invasive surgery and have allowed the surgeon to shift to laparoscopic surgery wherever possible. The removal of the gallbladder (cholecystectomy), which is the most frequently performed general surgical procedure, is now performed laparoscopically in approximately 90% of the cases in the United States. The promise of surgical glues and growth factors has been disappointing to date, but there are several research efforts ongoing that may find the appropriate chemistry to be both safe and efficacious.

Despite these changes, one major influence remains unchanged. By training, surgeons are conservative and must keep the well-being of their patient in mind at all times. As a result, despite commercial or technological changes that may emerge over the next decade, the surgeon will need a substantial body of clinical data before abandoning sutures. Therefore, the science and technology that support the development of sutures will remain viable for many years to come.

TABLE 4 Steps to Develop a Suture Product

Steps	Key activities
1. Concept	
Screening studies	Evaluation of polymers for extrusion ability and broad physical properties
2. Preclinical	
Manufacturing development	Extrusion optimization, including fiber annealing, braiding
Protocol physical testing	Tensile strength, knot pull strength, diameter, sterilization
Protocol animal efficacy testing	Knot security in various animal tissues, *in vivo* strength, handling
Protocol toxicology safety testing	Acute/chronic toxicity, pyrogenicity, antigenicity, *in vitro/in vivo* infection potentiation, hemolytic potentiation, mutagenicity, carcinogenicity.
3. Clinical	
Safety and efficacy confirmation	Clinical evaluation in representative patients (100–1000)
4. Registration	
Data summary and submission	Composition of preclinical and clinical data
Board of health approval	Agency review of documents and panel review (optional)
5. Manufacturing scaleup	Production qualification (a few lots) and validation (many lots/sizes)
6. Commercialization	Inventory stocking and promotional campaign

Bibliography

Brouwers, J. E., Oosting, H., de Haas, D., and Klopper, P. J. (1991). Dynamic loading of surgical knots. *Surg. Gynecol. Obstet.* **173**: 443–448.

Brown, R. P. (1992). Knotting technique and suture materials. *Br. J. Surg.* **79**: 399–400.

Casey, D. J., and Lewis, O. G. (1986). Absorbable and nonabsorbable sutures. in *Handbook of Biomaterials Evaluation: Scientific, Technical, and Clinical Testing of Implant Materials*, A. von Recum, ed. Macmillan Co., New York, pp. 86–94.

Chu, C. C. (1981). Mechanical properties of suture materials: An important characterization. *Ann. Surg.* **193**(3): 365–371.

Hermann, J. B. (1971). Tensile strength and knot security of surgical suture materials. *Am. Surg.* **37**(4): 209–217.

Holmlund, D. E. W. (1976). Physical properties of surgical materials: Stress-strain relationship, stress relaxation and irreversible elongation. *Ann. Surg.* **184**: 189–193.

Holmlund, D. E. W. (1974). Knot properties of surgical suture materials. *Eur. Surg. Res.* **6**: 65–71.

Moy, R. L., Lee, A., and Zalka, A. (1991). Commonly used suture materials in skin surgery. *AFP* **44**(6): 2123–2128.

Tera, H., and Aberg, C. (1976). Tensile strengths of twelve types of knots employed in surgery using different suture materials. *Acta Chir. Scand.* **142**: 1–7.

7.10 Burn Dressings

Jeffrey B. Kane, Ronald G. Tompkins, Martin L. Yarmush, and John F. Burke

Every year approximately 12,000 people die from severe burns and thermal injury. Most of these deaths are due to the catastrophic problems that ensue when the skin's integrity is disrupted. The major lethal problems are massive fluid losses and microbial invasion. Since the keratinocytes of the epidermis are primarily responsible for these activities, some investigators have focused on their replacement alone; however, as we will describe, replacement of the dermis or inner layer of skin is equally if not more important than the outer layer of epidermis.

Prompt replacement of the integrity of the skin is a cornerstone of therapy for these patients, but lack of available natural skin makes this an almost impossible task for those with large burns. The modern search for a suitable skin substitute has been under way since the 1940s, with steady progress as the principles of wound healing and the functions and properties of the skin became better understood. This chapter reviews those principles and examines their application as they relate to several burn dressings under development.

PRINCIPLES OF WOUND COVERAGE AND HEALING

To understand the requirements necessary to replace lost skin, we must first understand the general mechanism of wound healing. Burn injuries result in loss of skin structure to varying degrees. Usually burns are either first degree—the loss of the epidermal layer, second degree—the loss of the epidermal layer and a portion of the dermis, or third degree—the loss of tissue through the dermis, including the hair follicles and sweat glands, and extending into the hypodermis (subcutaneous) layers (Fig. 1). Occasionally, as often seen in electrical injuries, a deep, full-thickness wound is referred to as a fourth-degree burn, which is defined as extending downward through the subcutaneous tissues to involve tendon, bone, muscle, and other deep structures.

Second-degree burns can generally be classified as either superficial or deep. The interface between the epidermis and dermis is not linear, but consists of many papilla formed by the rete pegs of the epidermis. In superficial injury, enough of the deep epidermal or superficial dermal layers may remain to allow spontaneous healing of the wound by reepithelialization. Other sources of epidermal cells are the epidermal appendages, including the hair follicles and the sweat and sebaceous glands. These wounds can be treated by simple dressings without topical antibiotics and will heal spontaneously within 10 to 14 days. Deep second-degree burns have completely destroyed the epidermis and extend further into the dermis, with large amounts of necrotic tissue being present. Both the fluid and bacterial barriers are severely compromised, putting the patient at a much higher risk. These wounds, if allowed to heal on their own (because the dermis has been grossly distorted or destroyed), result in hypertrophic scarring, with nonoptimal cosmetic results. These types of wounds should be treated as if they were third-degree burns to allow for faster and better healing.

A third-degree burn is defined as having damaged the skin all of the way through the dermis and into the subcutaneous tissue. The wound is freely permeable to fluids, proteins, and bacteria. The constant proteinaceous exudate from these wounds, combined with an abundance of necrotic skin above, make an ideal media for bacterial growth. In these injuries, all the epidermal cells within the wound have been destroyed, including those in the epidermal appendages. Without skin replacement, epidermal cells must migrate from the wound edges; therefore, prompt débridement of the wound followed either by grafting or other methods of wound coverage is the treatment of choice. Small third-degree wounds (those of less than 2 cm) may be allowed to heal spontaneously by ingrowth from the wound edges, but this process requires at least 6 weeks, the wound may never heal completely, and always results in a generous scar.

When a patient suffers a third-degree burn, the necrotic layer of tissue that was once viable skin is referred to as the "eschar." This layer is removed surgically and any of the underlying dermis and hypodermis is then covered. If the freshly excised open wound is not immediately closed either with skin or skin replacement, then "granulation tissue" is formed by the local invasion of small blood vessels and fibroblasts from beneath the wound's surface. Granulating wounds become reddened with these new vascular structures and if no grafting is performed and the wound is simply allowed to spontaneously heal, the continued invasion of fibroblastic tissue will eventually form a hypertrophic

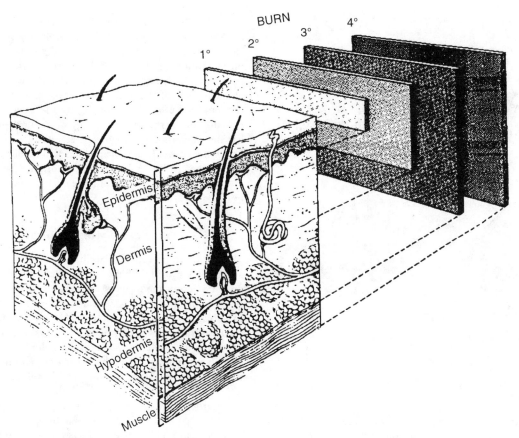

FIG. 1. Illustration of first, second, and third-degree burns. (From Tompkins and Burke, 1989.)

scar through the unorganized deposition of collagen. New epidermis seen around the edges of the wound is usually inadequate for reepithelialization in all but the smallest injuries. Myofibroblasts, specialized fibroblastic cells with the contractile properties of smooth muscle, invade the wound and begin to pull the edges inward and result in contractures and restriction in movement of the surrounding skin and its structures. This process can result in severe deformation of the surrounding features and is particularly troublesome in places such as the face, neck, and limbs. In some cases of massive wounds, even contraction of the wound will fail to bring the edges of the wound together, and the center of the wound remains open permanently.

Clearly the best coverage for the wound is natural skin taken from the individual himself (an autograft) to avoid specific immunological incompatibility. If the burn injury is anywhere from 35 to 50% of the total surface area of the body, it is frequently possible to transplant partial thickness skin grafts from other noninjured areas of the patient. These grafts are usually about 0.3 to 0.5 mm thick and include the epidermal layer and a thin portion of the underlying dermis. They are harvested from the donor site using a reciprocating blade such as the Padget electric dermatome or a hand-held Weck knife. The donor site epidermis regenerates in 2–3 weeks from the basilar epidermal elements left behind in the rete pegs. The graft is then placed on the freshly excised wounds over the

granulating wound beds and survives by simple diffusion of nutrients for the first 72 hr until neovascularization of the graft occurs. The graft must be placed on a wound bed free of dead tissue, infection, hemorrhage, or significant exudate or it will not survive. Wounds which have been freshly excised of all dead tissue offer the best short- and long-term results in all instances because granulation tissue prompts an inflammatory process that eventually results in exacerbation of scar formation.

Many times the graft is "meshed," which involves making many small linear incisions in the graft with a Tanner mesher so that it may be expanded in size (Fig. 2). By increasing the edge area of the epithelium, meshing allows the graft to cover from 1.5 to 9 times its initial area, although 1.5 to 3 expansions are most common. The mesh also allows fluids (tissue exudates and blood) to drain from the bed, which helps to increase the likelihood of graft survival. Unfortunately the mesh pattern is usually visible for extended periods of time after healing because the dermis does not properly regenerate within the interstices and therefore meshing makes the cosmetic results less desirable than with nonmeshed grafts.

For burns that encompass more than 50% of the body's surface area, obtaining enough autograft becomes difficult because the patients have a limited area available for donor sites. Using skin grafts from cadavers (allograft) provides some help,

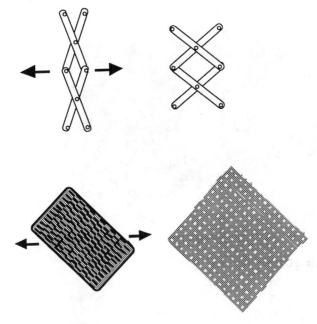

FIG. 2.   Illustration of stretched mesh graft.

but as with any tissue transplanted between individuals, allografts invoke an immune response from the patient and cause eventual rejection of the grafted skin. Allografts at least allow temporary wound closure and time for reepithelialization of a donor site, but all allografts must eventually be replaced by autograft from the regenerating donor sites. Allografts may be used fresh, stored in glycerol at 4°C, or preserved with glutaraldehyde. Although glutaraldehyde preservation eliminates the problems associated with sensitivity to foreign tissue, it kills all the cells of the allograft and eliminates many of their advantages, such as incorporation of the graft with concomitant regrowth and neovascularization from the wound bed. The shortage of skin donors, a problem common to all transplanted organs, severely limits the availability of allograft as a source of graft material. Porcine xenograft (i.e., a graft from a different species) have been used after preservation but problems with storage as well as limited biocompatibility restrict their usefulness. It is important to keep in mind that all natural grafts other than autografts must eventually be replaced with the patient's own skin.

DESIGN CRITERIA FOR WOUND COVERAGE

To replace the function of skin, it is important to specify properties that the skin substitute must possess.

Wetting the Wound Bed and Adherence

The skin replacement must wet the wound surface, conform to the wound surface in all areas, and adhere in all areas.

Without these properties, small air pockets exist where exudative fluid accumulates and bacteria proliferate. The graft must chemically adhere to the wound bed to avoid proliferation of the granulating layer and therefore control the subsequent contraction and deformity of the wound. Mechanical shearing or movement of the graft will destroy this interface, as will changes in the hydration of the wound area with edema or dehydration. If the surface energy of the graft–wound bed is less than that of the air–wound bed interface, then the wound bed will wet evenly and air pockets will be displaced (Yannas and Burke, 1980). A hydrophilic substance with the correct mechanical properties and surface energy would fulfill this requirement.

Porosity of the Graft

In order to prevent the swelling of the wound bed and the accumulation of fluid between the graft and its bed, the graft must be permeable to moisture. The exact permeability must be controlled, as too large a water flux of moisture will result in dehydration of the wound surface and disruption of the wound–graft interface, whereas too small a water flux will cause the accumulation of exudative fluid beneath the graft and lift it from the wound bed. The optimal rate of water flux through the graft is approximately 5 mg/cm²/hr (Yannas *et al.*, 1982b).

In order for a skin replacement to be successful, it must be incorporated into the soft tissues and replaced by the normal skin elements. This process allows the skin replacement to be vascularized and populated by normal healthy host fibroblasts. Although one could conceivably "seed" the dermal portion of the skin replacement with host fibroblasts, the proper pore structure is necessary to allow the fibroblasts to divide and migrate within the device. An impermeable graft without pores would not provide any means for such migration. Since most of the epithelial and mesenchymal cells found in the wound bed are on the order of 10 μm in diameter, pores consistent with that order of magnitude would allow free access into the graft material.

Dimensions of the Graft

Migration of cells from the wound edges appears to be a function of two parameters, the pore size and thickness of the graft (Yannas *et al.*, 1982b). If the mean pore size is under 10 μm in diameter, then cells (fibroblasts and epidermal cells) cannot migrate through the pores and the rate-limiting step becomes that of degradation of the matrix to allow cells to enter the material. On the other hand if the material has the appropriate porosity to allow migration of cells, but is too thick, then diffusion of nutrients from the wound bed to the upper area of the skin replacement becomes a limiting factor for cell migration.

The limiting distance that will maintain nutrients to the migrating cells is postulated by using the dimensionless expression as derived by Thiele, and modified by Wagner and Weisz:

$$S = \frac{rl^2}{Dc_0},$$

where r is the rate of utilization of critical nutrients by the cell [mole/cm³/sec], l is the distance that supports diffusional nutrition, D is the diffusivity of the nutrients in the hydrated membrane [cm²/sec], c_0 is the nutrient concentration at the wound bed [mole/cm³], and S is the dimensionless number.

The top portion of this ratio represents the reaction term for the utilization of the nutrient, whereas the bottom can be considered as the diffusive component. Thus when $S \gg 1$, the cells are consuming more than can be delivered by diffusion and similarly when $S \ll 1$, diffusion is greater than the amount consumed. Thus $S \cong 1$ corresponds to the distance when the two terms are balanced. Beyond this distance, other routes of delivery such as capillary ingrowth must occur in order to supply the cells with substrates and oxygen. If a graft material can induce neovascularization from the underlying wound bed, then these vessels can carry the appropriate nutrients to the cells convectively and extend the range of cell migration.

Another important consequence of graft size is the dependence of thickness on rigidity of the graft. Too thin a graft results in easy tearing and shearing, whereas if the graft is too thick, it becomes rigid and will not adhere to the wound evenly.

Strength of Graft

A graft must be supple enough to allow for some underlying movement of the wound but meanwhile it must maintain an even contact with its surface. The graft must also be durable and allow for easy handling without tearing. It must allow for suturing, and be durable enough to cover the wound for the entire healing period. As mentioned earlier, too thick a graft imparts rigidity and the graft will fail to adhere to the wound's surface. To design a membrane with appropriate modulus of elasticity and tear strength, the graft may be modeled as an elastic beam bonded to a rigid surface. An estimate of the normal strength perpendicular to the graft–wound bed interface is (Yannas and Burke, 1980):

$$\sigma_m = 0.45\alpha(\nu_2 - \nu_1)E,$$

where σ_m is the maximum normal stress, α is the coefficient of expansion for swelling the graft in water, ν_2, ν_1 is the initial and final volume fractions of water in the graft, and E is the modulus of elasticity.

It has been shown that the property which controls tear strength in rubber is the fracture energy. Two other important properties are the shear stress and peel strength. Shear stress is the force per unit area in a direction parallel to the wound bed that causes the graft to slide or buckle. Any disruption of the fragile wound surface breaks the newly formed blood vessels, retarding healing, and in the case of a skin graft, will lessen the likelihood of successful skin replacement. Peel strength can similarly be defined as the force per unit area needed to remove the graft when applied in a direction 90° to the wound's surface (Fig. 3).

Biodegradation Rate

If the skin replacement is nonbiodegradable, then it can fall into one of two categories. A device that is impervious to cell

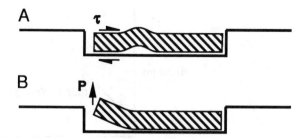

FIG. 3. Two possible forces on a graft: (A) shear stress and (B) peel force.

ingrowth from the wound bed could be used as a temporary cover for the wound. A number of simple polymer membranes have been formulated that take this approach. A second approach is a device that contains an open mesh or porelike structure that allows cellular integration. This would lead to attachment of the device to the wound bed but would necessitate eventual surgical removal of the device.

An alternative possibility would be if the dermal portion of the skin replacement in contact with the wound bed were biodegradable. Ingrowth of cells and vessels from below would allow firm initial attachment to the skin replacement and with time, the device would be broken down, leaving the newly formed tissue behind. In choosing a material it is important to have biodegradation proceed with minimal inflammatory response from the host, in order to provide controlled wound healing with minimal wound shrinkage, scarring, and contracture.

It is important to balance the degradation rate of the device with the time required for cellular regeneration. To balance these terms, the time constant for biodegradation (t_b) should be on the same order of magnitude as the time required for the wound to heal (t_h), approximately 25 days (Levenson *et al.*, 1965), or

$$\frac{t_b}{t_h} \cong 1.$$

Since t_h is constant for any patient, the design of the graft must be controlled for t_b. If the graft degrades too quickly, the wound is effectively left uncovered. On the other hand, if the graft remains intact for long periods of time after the wound is healed, then scar formation and incorporation of the device into the wound become problems.

It is also important to note that although cells will migrate through the device at a rate of 0.2 mm/day for fibroblasts, and 0.4 mm/day for epidermal cells, they must also migrate parallel to the wound, from the wound edges. Unfortunately this rate is only 0.1 to 1 mm per day, which would not be fast enough to cover a large wound within the time allotted by t_b. It might be necessary to provide another method of enhancing the lateral movement of epidermis (such as seeding the graft with epithelial cells or placing a thin epidermal graft).

Lack of Antigenicity

A major drawback to xenografts and allografts is the inherent antigenicity that they exhibit. Although some attempts have

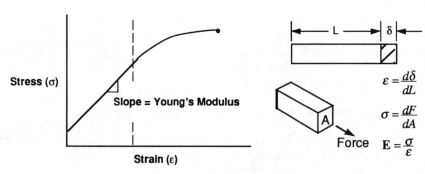

FIG. 4. Definition of Young's modulus.

been made to immunosuppress a burn patient in order to allow for the prolonged adherence of living, related donor skin, an ideal skin replacement material would have minimal potential for rejection or local inflammation. Some tolerance for foreign material is inherent in severely burned patients owing to the naturally diminished immune response resulting from their burn injuries, but further immunosuppression, particularly for adults with large burns, may leave them severely compromised and vulnerable to invasive microbial sepsis.

Macrostructure

It has been found that one of the functions of a wound cover is to provide a structural framework for the healing wound. If this structure is not present, fibroblastic tissues will lay down collagen in a random fashion, resulting in disordered dermis (scar). By providing scaffolding for the new vascular and mesenchymal tissue, an orderly and controlled restructuring of the dermis can take place without scar formation.

SELECTION OF MATERIALS

In the previous section several criteria important for the design of a skin substitute were discussed. Of the many biological materials that exist, only a small number have enough of these properties to be of interest.

The first group suitable for further examination includes macromolecular-based compounds. Polypeptide polymers and certain polysaccharide compounds have the advantage of being degraded by extracellular enzymes and yield nontoxic by-products. These materials are also structurally related to the natural macromolecules that populate the wound bed and may promote healing. In their hydrated form they also provide adequate wetting of the wound bed surface. It is important to compare each material's specific porosity, strength, and mechanism of bonding to the wound bed in evaluating these compounds. Various types of these compounds will be examined in more detail later.

Synthetic compounds form another important class. Common materials such as poly(dimethyl siloxane) are unfortunately neither hydrophilic nor biodegradable. Polyamides such as hexamethylene adipamid and saturated polyesters such as poly(ethylene terephthalate) are weakly hydrophilic and not

biodegradable. Poly 2-hydroxyethyl methacrylate is a relatively hydrophilic polymer but is also not biodegradable. Most of these compounds are suitable only for temporary superficial coverage of the clean wound under limited circumstances, but not as a means of permanent wound closure.

Other compounds that are biodegradable and wet the surface, such as poly n-butyl-α-cyanoacrylate, have toxic by-products when they are broken down. A good review of several biodegradable substances can be found in the work of Kronenthal (1975).

Collagen is a polypeptide compound, hydrophilic in nature, and subject to degradation by extracellular enzymes. The substance is very well studied, allowing for the control of many key physical parameters, including the small strain viscoelastic properties and the rate of degradation (which can be controlled either by cross-linking the collagen fibers or by incorporation of other materials in the suspension). Collagen also possesses the property of being a weak antigen, with minimal rejection potential.

The modulus of elasticity, or Young's modulus, is defined as the initial linear portion of the stress-strain diagram for a material. If the stress σ is defined as the force (F) applied divided by the area (A) over which that force is exerted, then the strain can be measured as the distance the material deforms (δ), divided by its original length (L). The Young's modulus for the material is then the ratio of the stress to the strain (Fig. 4).

By incorporating dilutants, the Young's modulus of collagen

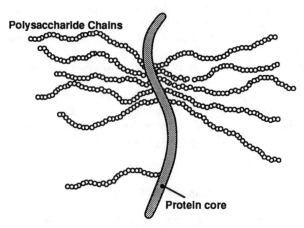

FIG. 5. Schematic diagram of chondroitin 6-sulfate.

can be varied over a range of 10^5. A problem noted with collagen was that by increasing the cross-linking (in order to add strength) the compound became stiff and brittle. This could be overcome by incorporating a second macromolecule, a glycosaminoglycan (GAG) into the membrane.

Glycoaminoglycans are found in significant quantity in the ground substance of extracellular matrix and may contribute up to 30% by dry weight of this intercellular region. GAGs consist of a family of macromolecules (molecular weight from 10^5 to 10^6) that share the characteristic of having multiple long polysaccharide chains covalently bound to a core protein by a glycosidic bond (Fig. 5). In contrast to glycoproteins, these compounds have a significant quantity of carbohydrate and exhibit behaviors closer to that of a polysaccharide rather than protein. Each of the various GAGs has its own unique repeating disaccharide which forms the long carbohydrate chains. In the case of chondroitin 6-sulfate, the sugar is N-acetyl galactosamine (Fig. 6). The molecules are highly charged and polyanionic in nature. They have multiple properties, including the ability to change the mechanical properties of their environment. Hyaluronic acid, which is found in the synovial fluid of joints, contributes to lubrication.

For the artificial skin or skin replacement developed by Yannas and Burke, the GAG, chondroitin 6-phosphate, was chosen because of its important properties: (1) low antigenicity; (2) nontoxic breakdown products; (3) ability to decrease the rate of collagen breakdown; and (4) ability to increase the strength of collagens while making them more elastic. Other GAGs investigated include heparin sulfate, chondroitin 4-sulfate, and dermatan sulfate, which share the first three properties in common with chondroitin 6-sulfate. A final advantage was noted by electron microscopy of collagen–GAG samples after coprecipitation, in that the pore structure of the resulting membrane was significantly more open than GAG–free collagen and one could control the pore size by adjusting the collagen–GAG mixture. Bonding of chondroitin 6-sulfate to the collagen is required because at neutral pH the GAG will dissociate from collagen and elute from the material in the absence of cross-linking.

SPECIFIC TYPES OF WOUND COVERAGE MATERIALS

Natural Graft Materials

Properties of natural autografts and some allografts include a decrease in the quantity of bacteria in the wetted surface of

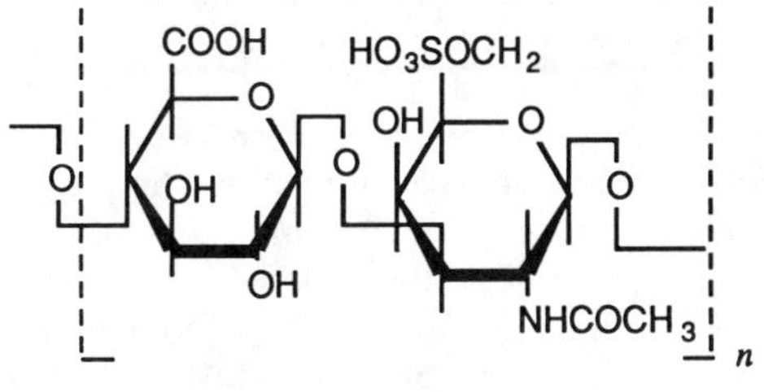

FIG. 6. Repeating disaccharide of chondroitin 6-sulfate.

the wound, a decrease in the loss of fluids and proteins, a limiting effect upon wound contracture when the dermal layer is included in the graft, a decrease in pain in second-degree burns, and an increase in the neovascularization of the wound area.

Major disadvantages of natural graft materials include a lack of autograft in severely burned patients, difficulty with storage of allograft material (typically 7–10 days when stored at 4°C and longer at −70°C), antigenicity, leading to the need for eventual replacement, and the potential to transmit diseases. Donors are scarce and must be carefully screened for transmissible diseases (including hepatitis and AIDS) and neoplasms before transplantation. Fresh-frozen and lyophilized grafts (rapidly freeze-drying the samples) prolong the shelf life of the tissues. However, if the samples are thicker than 0.38 mm, there is a propensity for the epidermal and dermal layers to separate in the lyophilized samples. Both techniques offer a lower graft survival than fresh graft material, and are less able to suppress microbial invasion.

Xenografts, especially porcine-derived commercial preparations, are used as a temporary wound closure material. As with the frozen and lyophilized allografts, porcine grafts at best only adhere to the wound bed with a fibrinous interaction but this can temporarily close a wound. The grafting success rate is lower than with fresh allograft and sterility of the xenografts must be ensured to protect the already immunocompromised burn patients. The porcine grafts must be immobilized for several days to allow bonding of the graft to the wound bed by growth of fibroblasts into the dermal layer of the graft. As with any skin replacement procedure, the wound bed must be debrided thoroughly and be free of significant bacterial contamination. Because porcine grafts will ultimately be rejected, they must be removed and replaced with autograft material.

Amnion has been tried as a temporary natural wound dressing, but again the source must be carefully screened for disease. Amnion is freely available from the delivery of babies, and is relatively inexpensive to prepare. The adherence of amnion to the wound is less successful than with dermal wound dressings and it is likely that a true incorporation of the material into the wound bed never occurs. The amnion must also be covered with a secondary occlusive dressing to prevent sloughing. It has been reported that the amnion increases the granulation of the wound bed, which makes subsequent skin graft survival more likely but ultimately also results in more scarring.

Synthetic Graft Materials

Single Laminar Grafts

In order to provide a simple method of closure, a number of simple, single-layer, polymer wound dressings have been developed. The Ivalon sponge was developed in 1962, and had the same problems as many of the simple coverage schemes. Bacteria present under the graft wound spread into deep structures when the sponge was used on third-degree burns. Also, the brittle nature of the material led to small pieces breaking off when the graft was removed, leading to foreign body reactions within the ultimately healed wound.

Spray-on materials, such as hydroxyvinylchloride-acetate and sebacic acid copolymer, produce the same problems of suppurative bacterial spread when placed over open third-degree wounds. If the film is nonpermeable to moisture, accumulation of serous exudate and suppurative material will result in uneven adhesion of the film to the wound. Other polymer materials have been tried over the years with similar results.

A spray-on silicone membrane, Hydron (Hydron Laboratories, Inc., New Brunswick, NJ), consists of poly(hydroxyethyl methacrylate) powder and liquid polyethylene glycol sprayed onto the wound surface, resulting in a thin film. The membrane is easily removed from the wound bed by contact with bed clothing, and infection can expand beneath its surface. These dressings would be best for treatment of donor sites from split thickness skin grafts and possibly clean superficial second-degree burns, but their high cost limits their use over simpler dressings.

A liquid gel made from agar and acrylamide copolymer has been introduced. This material is again best used as a temporary cover only because of its lack of adherence to the wound and its fragile nature. Some advantages of the gel are that it allows fluid to freely drain from the wound without accumulating and because the material is highly translucent, it allows easy inspection of the wound. As with Hydron, this should be used in clean wounds expected to heal rapidly, like donor sites and superficial second-degree burns.

Collagen (as discussed earlier) is an excellent material owing to many of its unique properties. A fairly pure form can be extracted in large amounts from many commercial sources. It is inherently low in antigenicity and exerts a hemostatic effect on the fragile and vascular wounds. Collagen can be formulated in many different ways from gels to films, depending on the properties desired and methods for application. Cross-linking the collagen fibers increases the tensile strength, but unfortunately it also tends to make the fibers stiff and brittle. The permeability of collagen materials may be controlled by their thickness, but mechanical properties must be considered if the material becomes too thick. Another property of collagen is that it naturally adheres to the wound initially because of its binding with fibrin. The particular form of collagen is important for success because various forms of collagen have been tried unsuccessfully over the years. Sponges of collagen dry slowly and behave as a serous crust or scab which prevents effective ingrowth of fibroblastic tissue from the bound bed; collagenases present both naturally in the wound bed and from wound bacteria result in the eventual shedding of these collagen sponges.

Multilaminar Grafts

In order to both control the ingrowth of fibroblasts and long-term adhesion to the wound, a large pore size must be maintained. On the other hand, the requirements of water permeability necessary to prevent the wound from drying out require a smaller pore diameter. The best approach to date has been to create bilaminar membranes. This approach makes designing the wound dressings simpler because differences in physical strength, adhesion, pliability, and other properties can be separated between the two layers of the device.

Biobrane is a synthetic bilaminate material developed by Woodroof Laboratories, Santa Ana, CA. This material consists of a finely knit nylon and hydrophilic type I porcine collagen which is covalently bound to the hydrophobic inert nylon. The outer layer of Biobrane is a silicone rubber which, through its porosity, controls the water permeability of the graft. The pore size of the outer layer is maintained small enough to provide a barrier for bacteria while being large enough to be permeable to topically applied antibiotics. If nylon mesh is placed against a clean wound (low bacterial count), then the Biobrane may adhere to the wound surface, and through its loose pore structure, permit ingrowth of cells from the wound bed below.

Biobrane is flexible enough to deform, maintaining adequate contact with the contours of the wound while still having the mechanical stability necessary for ease of handling and durability as a dressing. Biobrane is not biodegradable and serves at best only as a temporary closure material, lasting up to a month. Biobrane must eventually be removed either manually or by epithelial growth from below. Care must be taken not to prevent reepithelialization when Biobrane is used as a wound cover, as when the graft is placed over a finely meshed autograft. If Biobrane is left on for more than 6 days, it prevents epithelialization in the interstices of the graft (Lin et al., 1982). The problems of wounds with significant bacterial contamination have become apparent and development of Biobrane with the incorporation of antibiotics is under way. As with any inert skin substitute, Biobrane must eventually be removed and the wound then covered with autograft for those wounds without epidermal elements. Biobrane has been successfully used only in clean superficial second-degree burns and donor sites.

Levine et al. have developed a completely inert wound cover consisting of six layers of a nylon stocking fabric, covered with a 1-mm poly(tetrafluoroethylene) membrane with 0.1-μm pores. This dressing allowed fibroblasts to grow into the mesh while providing a semipermeable moisture and bacterial barrier. As there is no inherent inhibition of bacteria beneath the graft, the material was first soaked in an antimicrobial solution. This material is used more as a temporary wound dressing than as a skin replacement because these dressings must be changed every 48 to 72 hr to provide local debridement. Eventually they are replaced with a more definitive wound closure when it becomes available.

Yannas and Burke have developed a bilaminar replacement membrane consisting of an inner layer of collagen–chondroitin 6-sulfate, and an outer layer of silastic (Figs. 7 and 8). The collagen binds to the wound bed and is invaded by neovasculature and fibroblasts. The collagen chondroitin 6-sulfate very significantly inhibits or prevents wound contracture, and with time the collagen is rresorbed by collagenase and is completely replaced by remodeled dermis. Pore size of the collagen–GAG was found to be optimized at 50 μm, which supported active ingrowth of tissues from below. Smaller pores excluded fibroblasts and mesenchymal tissue, promoting a fibrouslike formation around the openings of the pores. The silastic layer is 0.1 mm thick and provides the bacterial barrier needed while maintaining the proper water flux through the membrane. In fact, the flux of 1 to 10 mg/cm^2/hr is similar to that of normal epidermal tissue. The silastic layer also provides the mechanical

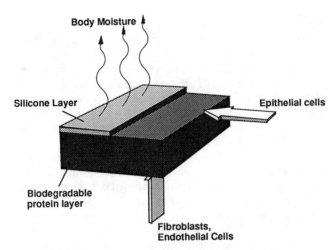

FIG. 7. Schematic diagram of a bilaminar artificial skin.

rigidity needed to suture the graft in place, preventing movement of the material during wound healing.

The GAG content of the collagen controls such things as the porosity and elasticity of the material, as explained in more detail in the section on materials. The collagen–chondroitin 6-sulfate adheres to the wound within minutes (in contrast to collagen sponge), and neovascularization can be observed within 3–5 days. The strong attachment of the collagen–GAG to the wound is evident by measuring the peel strength of the graft, which is 9 N/m at 24 hr and increases to 45 N/m by 10 days. Successful adherence of the artificial skin to the wound rresults in biological wound closure and its beneficial effects. Slow resorption of the collagen from native collagenase results in remodeling of the dermis. The three-dimensional structure furnished by the collagen–GAG material provides a scaffolding or suprastructure which results in controlled ordered dermal formation without the resulting scarring seen with simpler materials. When autografted epidermis (0.1 mm thick) becomes available from regenerating donor sites, the silastic layer can be removed from 3–4 weeks and the epidermal layer then can be applied directly to the developing neodermis with 95 to

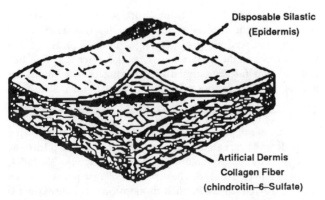

FIG. 8. Diagram of Burke and Yannas artificial skin (from Heimbach *et al.*, 1988).

100% successful grafting rates. If the graft wrinkles or fails to conform adequately for adherence to the wound, it must be trimmed to prevent seromas or hematomas from forming beneath it.

Histological cross sections of wounds closed with this artificial skin show complete replacement of bovine collagen at 7 weeks with remodeled human dermis. No evidence of hypertrophic scar formation has been noted, and the grafted areas are more supple than those areas simply covered with meshed autograft.

Yannas and Burke have extended their development by seeding dermal portions of a graft with autologous basal skin cells (stage II membranes). A small sample of cells is removed from the patient and the top layer of the epidermis is discarded. The basilar cells are then dissociated from one another with trypsin, and suspended in media. The basilar cells are introduced into the matrix, either by direct injection or by centrifugal force (this entire procedure can be accomplished in the guinea pig model in under 4 hr). The keratinocytes rapidly proliferate and form sheets of keratinized cells. Functional skin replacement has been achieved in guinea pigs in under 4 weeks (Yannas *et al.*, 1982a); preliminary studies in humans have been encouraging.

Because of the unique ability of this artificial skin to participate in the natural healing process and to allow remodeling of the dermis, infection rates are low and in fact are similar to those seen with autografted tissues. Cosmetic results are dramatic and result in a smooth, supple, homogeneous surface similar to that of normal skin. This artificial skin is the only long-term skin replacement with a large human experience. More than 100 patients have been treated with this artificial skin at Massachusetts General Hospital and the Boston Unit of the Shriner's Burns Institute. In addition, 106 patients have had this treatment in a U.S. Food and Drug Administration Phase II trial. Recent use of the artificial skin in 43 adults admitted to Massachusetts General Hospital showed a significant increase in survival, particularly in those patients over 40 years of age (64% survival of those receiving artificial skin, versus 22% of those receiving standard treatments only). This is particularly significant because those receiving the artificial skin generally had a larger percent of body surface area involved in their burn than those not treated (Tompkins *et al.*, 1989). This material received FDA approval recently.

The use of artificial skin allows for prompt dèbridement and closure of the largest of burn wounds, limiting wound contracture and providing permanent coverage with a readily available, easily stored skin replacement material. When seeded stage II material begins its routine clinical use, the advantages of cultured epidermis without the limiting delay of several weeks and the use of exogenous multiple mutagenic components will also be realized. A bilaminant artificial skin has also been developed by Boyce and Hansbrough (1987).

Culture Grafts

Grafts composed of epidermal cells grown from small samples taken from a patient have the theoretical advantage of significantly amplifying the donor sites available for epidermal cells. Studies have been conducted in which cells were grown on collagen film by seeding the film with basal cells. The sheets were grown by placing the film in the subcutaneous tissue or subcapsular region of the kidney. When a confluent layer of epithelial cells appeared, the sheets were applied to the wounds and an occlusive dressing was placed to immobilize the sheets. In those grafts that took, the epithelialization seemed complete after 1 week. Contraction was noted in the wounds and whether the cells were replaced by the animal's own cells was not established. Extensive experience in humans with long-term experience has not been reported with these methods.

Results in short-term human studies show histological sections with a mildly hyperkeratotic epidermal layer with no glandular structures. A mild chronic inflammatory response was noted in the underlying connective tissue, similar to that seen with scar formation. No stratum corneum or rete pegs were present, making the epidermal cells susceptible to trauma. Moist petroleum drressings were kept on the wounds for 4 weeks to prevent graft desiccation. A limitation to this technique is that the period required for cellular growth is 2–3 weeks; thus immediate closure of the burn wound must be provided by other means.

Green has extended the ability to generate cultured epidermal keratinocytes by coculturing them with lethally irradiated 3T3 cells (an immortal transformed mouse cell line) in the presence of epithelial growth factor (EGF), and cholera toxin—a nonlethal agent that increases intracellular cyclic AMP (cAMP) (Green et al., 1979). Using these techniques, the keratinocytes plated at a density of 5000 cells/cm^2, displaced the 3T3 cells, and grew to confluence in approximately 14 days. This technique improved yield and provided the ability to quickly culture cells, producing a monolayer of epidermal cells which could then be grafted over a healing wound site. When a modification of this technique (without the 3T3 coculture) was used on human patients, significant amplification of the patients' unburned epidermal areas could be achieved. Of course these grafted sites show no evidence of any stratum corneum because only the epidermal cell layer is being replaced, thus making them susceptible to desiccation.

Bell has developed a "living skin equivalent" consisting of a collagen fibrilar lattice seeded with autologous fibroblasts; epidermal cells can be cultured on its surface (Bell et al., 1981). This technology is being tested by Organogenesis, Inc. (Cambridge, MA). In rodent studies, rapidly neovascularization of the graft allowed the fibroblasts to proliferate. It was suggested that the graft prevented contracture of the wound edges by as much as 75 to 80%. In animal studies using small grafted areas, the dermal layer is remodeled after the graft is placed and over the next 10 weeks, the collagen layers thin to half the thickness of the surrounding dermis. Epithelial growth can be seen at the edges of the wound. There are no published human studies available to evaluate this approach.

More recently, another approach for an artificial dermal matrix has been taken in which human fibroblasts are grown on surgical mesh materials. The artificial dermal matrix is expected to provide a dermis for the interstices of meshed skin. In athymic mice studies (Hansbrough et al., 1992a), polyglycolic acid (PGA) and polyglactin-910 (PGL) mesh containing confluent, cultured human fibroblasts were applied to full-

thickness wounds. Expanded mesh, human, split-thickness skin grafts were placed over the artificial dermal matrix graft. During a 99-day period after graft placement, the PGA/PGL-fibroblast grafts were incorporated into the wound and epithelialization from the skin bridges proceeded rapidly across the surface of the PGA/PGL-fibroblast grafts. Basement membrane formation at the dermal–epidermal junction of the epithelialized interstices was seen and minimal inflammatory reaction to the PGA/PGL-fibroblast grafts was noted.

In a controlled clinical study with this approach, Hansbrough et al. (1992b) evaluated Dermagraft (Advanced Tissue Sciences, La Jolla, CA), which is composed of human neonatal fibroblasts grown on a PGA Vicryl mesh. The study tested the ability of Dermagraft to function as a dermal replacement material when placed beneath meshed, expanded, split-thickness skin grafts. Full-thickness burn wounds in 17 patients with burns (mean age, 31 years; range: 6–69 years; mean burn size, 23.8% total body surface area) were excised to subcutaneous fat (nine patients), to fascia (three patients), or to a combination of deep dermis and fat (five patients). The results showed that "take" of skin grafts on control sites was slightly better than take on the Dermagraft; however, the differences were not statistically significant. Mesh interstices epithelialized over the surface of the full-thickness wound in a fashion comparable to the surface of Dermagraft. No evidence of rejection of the allogeneic fibroblasts and minimal inflammatory reaction to the Vicryl fibers were seen. The Vicryl was hydrolyzed within the wound over 2–4 weeks although expulsion of the fibers was noted as the healing epithelium advanced to close the interstices. Further clinical trials with Dermagraft are in progress. Comparisons of currently available skin replacements are shown in Table 1.

CREATION OF A SYNTHETIC BILAMINAR MEMBRANE

The development of the Yannas and Burke artificial skin is briefly examined as an example of the technology needed to fabricate a skin replacement.

Extraction of Collagen

Collagen was obtained from several sources during the development of the artificial skin. The initial lots were extracted from rat tail tendon by suspending it in 0.05 M acetic acid solution as per the methods of Piez and Michaeli. The free collagen was further purified by centrifugation, with any remaining solid impurities being discarded with the pellet. NaCl (5% by wt.) was then added to salt precipitate the collagen; the collagen pellet was immediately redissolved in acetic acid. This cycle of salt precipitation, centrifugation, and redissolving the pellet was repeated two more times to further purify the collagen. The collagen solution was dialyzed against Na_2HPO_4 for 48 hr and centrifuged. The resulting collagen pellet was then lyophilized and stored. Purified chondroitin 6-sulfate extracted from shark cartilage was obtained commercially.

In order to bind the GAG into the collagen mixture, chon-

TABLE 1 Uses and Limitations of Wound Dressing Membranes and Skin Substitutes[a]

Skin replacements	Optimum use	Susceptibility to infection[b]	Submembrane fluid collection	Loss of integrity	Adherence
Hydron	Early coverage of small noncircumferential, superficial partial-thickness burns not involving joints	+++	+++	+++	Low on exudative wounds
Polymerized agar	Temporary dressing of clean wounds	+++	+	++++	Low
Fibrin film	Temporary dressing of partial-thickness burns	++	++	+++	Requires dressing
Collagen membranes alone	Temporary dressing of excised wounds	++	+++	++	Decreases with time
Biobrane	Dressing for donor site or excised wound	++	++		Less than allograft skin
(collagen-nylon)/ silicone rubber				+	
Silastic-collagen composite	Excised burn wounds (partial and full thickness)—permanent closure	+	++	+	Excellent
Nylon-microporous mesh composite	Wounds with granulation tissue temporary dressing	+ (when treated with 5% mafenide acetate)	+	0	Excellent on clean wound
Culture-grown epithelial sheets	Freshly excised wounds or those with granulation tissue	+++	+	+++	Good under occlusive dressing
Culture-grown composite	Excised wounds	+++	+	++	Good with sutures and an occlusive dressing

[a]Adapted from B. A. Pruitt and N. S. Levine, 1984. *Arch. Surg.* **119,** 312–322.

[b]0 indicates none; +, minimal; ++, moderate; +++, marked; ++++, severe.

droitin 6-sulfate was added slowly at a constant rate to a well-mixed acetic acid–collagen solution. The two substances coprecipitated at pH 3.2. The proportion of GAG incorporated, the temperature, and the acidity of the bath are important parameters in determining final coprecipitate membrane properties.

The pore size of the membrane was exquisitely sensitive to the method of drying. Dagalakis *et al.* (1980) showed that instantaneous freezing followed by slow sublimation at constant low temperatures was necessary to ensure the maximum mean pore size. Table 2 shows the comparative mean pore sizes and shrinkage ratios of the collagen–GAG membrane that were due to various method of drying.

Cross-linking of the collagen and GAG was performed by glutaraldehyde. The coherent membrane was removed from the glutaraldehyde and placed in distilled water to remove any

TABLE 2 Shrinkage Ratio and Pore Size of Membrane[a]

	Air dried, water	Air dried, ethanol	Critical point drying CO_2	Freeze drying of preformed membrane	Direct freeze drying
Shrinkage ratio, percent of untreated	41	48	75	94	97
Mean pore size, μm	<5	<5	15 ± 11	50 ± 46	100 ± 80

[a]Reprinted from N. Dagalakis *et al.* (1980). *J. Biomed. Mater. Res.* **14:** 511–528. Percent of original linear dimension, ±1%.

unreacted aldehydes. If the membrane was freeze dried, it was first necessary to partially cross link the collagen in a 105°C vacuum oven. After this step, it is possible to immerse the membrane in a glutaraldehyde solution without it collapsing; the remainder of the cross-linking is then carried out in the standard manner. The amount of cross-linking, which in turn affects the strength, elasticity, and pore size of the membrane, was controlled by adjusting several parameters. The time the membrane remained in the vacuum oven determined the initial retention of GAG by the membrane; the second step of cross-linking by glutaraldehyde was carried out under acidic conditions (pH 3.2) although they slow the cross-linking process because the GAG forms an unstable ionic complex with collagen at neutral pH. By adjusting the concentration of glutaraldehyde in the mixture, it was possible to tightly control the molecular weight between cross-links[1] (M_c) between a very loose network ($M_c = 30,000$) and a very tight one ($M_c = 10,000$). This processing resulted in a collagen–GAG complex with $8.2 \pm 0.8\%$ GAG by weight, a pore volume fraction of $96 \pm 2\%$, and a mean pore size of $50 \pm 20 \ \mu m$.

The silastic layer is applied as a liquid monomer and curing takes place at room temperature on the surface of the collagen after exposure to ambient moisture. This bilayer copolymer composite is then stored at room temperature either in 70% alcohol or after being freeze dried.

Stage II membranes may be created by seeding the collagen–GAG–silicone system with recently disaggregated epidermal cell suspensions. These epidermal cells are prepared by digesting a small piece of skin with trypsin in phosphate-buffered saline (pH 7.2) for 40 min at 37°C. The top of the epidermal layer is discarded and the bottom layer is placed in Eagle's minimum essential medium. Suspended basilar cells are separated from the tissue debris by filtration and centrifugation. The cells are then seeded into the artificial skin either by direct injection, or by placing a small piece of graft in a specially designed centrifuge in which centrifugal force is directed across the membrane thickness and drives the epidermal cells across one surface of the collagen–GAG membrane. A maximum of 4 hr are required to harvest and implant cells; thus no appreciable delay in wound closure has been encountered in either animal or limited human studies.

CONCLUSION

As our basic understanding of wound healing and skin properties continues to evolve, we can expect further progress toward early wound closure and replacement of damaged tissue. Advances in composite materials such as the Yannas and Burke artificial skin promise an "off-the-shelf" solution for seriously injured burn patients for which there is often not enough natural graft material available for immediate primary wound closure. As stage II membranes move from animal experiments to clinical application, we may see further advances, including initial and final closure without the need for a delayed

epidermal skin grafting. Further advances in our basic understanding of tissue–graft interactions will yield refinements and new approaches in the development of biomaterials suitable for incorporation in the healing dermis.

Bibliography

Bell, E., Ehrlich, H. P., Buttle, D. J. and Nakatsuji, T. (1981). Living tissue formed *in vitro* and accepted as skin-equivalent tissue of full thickness. *Science* 211: 1052–1054.

Boyce, S. T., and Hansbrough, J. F. (1987). Biological attachment, growth, and differentiation of cultured human epidermal keratinocytes on a graftable collagen and chondroitin-6-sulfate substrate. *Surgery* 103: 422–431.

Burke, J. F., and Bondoc, C. C. (1979). Burns: the management and evaluation of the thermally injured patient. in *Dermatology in General Medicine* T. B. Fitzpatrick and A. Z. Eisen, eds. McGraw-Hill, New York, pp. 931–936.

Dagalakis, N., Flink, J., Stasikelis, P., Burke, J. F. and Yannas, I. V. (1980). Design of an artificial skin. Part III. Control of pore structure. *J. Biomed Mater. Res.* 14: 511–528.

Green, H., Kehinde, O., and Thomas, J. (1979). Growth of cultured human epidermal cells into multiple epithelia suitable for grafting. *Proc. Natl. Acad. Sci. U.S.A.* 76: 5665–5668.

Hansbrough, J. F., Cooper, M. L., Cohen, R. *et al.* (1992a). Evaluation of a biodegradable matrix containing cultured human fibroblasts as a dermal replacement beneath meshed skin grafts on athymic mice. *Surgery* 111: 438–446.

Hansbrough, J. F., Dore, C., and Hansbrough, W. B. (1992b). Clinical trials of a living dermal tissue replacement placed beneath meshed, split-thickness skin grafts on excised burn wounds. *J. Burn Care Rehabil.* 13: 519–529.

Heimbach, D., Luterman, A., Burke, J., Cram, A., Herndon, D., Hunt, J., Jordan, M., McManus, W., Solem, L., Warden, G. and Zawacki, B. (1988). Artificial dermis for major burns, a multi-center randomized clinical trail. *Ann. Surg.* 208: 313–320.

Kroenthal, R. L. (1975). Biodegradable polymers, in *Medicine and Surgery.* Plenum Publ. New York.

Lin, S. D., Robb, E. C., and Nathan, P. (1982). A comparison of IP-758 and Biobrane in rats as temporary protective dressings on widely expanded meshed autografts. *J. Biomed. Mater. Res.* 3: 220–222.

Pruitt, B. A., and Levine, N. S. (1984). Characteristics and uses of biological dressings and skin substitutes. *Arch. Surg.* 119: 312–322.

Tompkins, R. G., and Burke, J. F. (1989). Burn wound. in *Current Surgical Therapy* B. C. Decker, pp. 695–702.

Tompkins, R. G., Hilton, J. F., Burke, J. F., Schoenfeld, D. A., Hegarty, M. T., Bondoc, C. C., Quinby, W. C., Behringer, G. E. and Ackroyd, F. W. (1989). Increased survival after massive thermal injuries in adults: Preliminary report using artificial skin. *Crit. Care Med.* 17: 734–740.

Yannas, I. V., and Burke, J. F. (1980). Design of an artificial skin. I. Basic design principles. *J. Biomed. Mater. Res.* 14: 65–81.

Yannas, I. V., Burke, J. F., Orgill, D. P. *et al.* (1982a). Wound tissue can utilize a polymeric template to synthesize a functional extension of skin. *Science* 215: 174–176.

Yannas, I. V., Burke, J. F., Warpehoski, M., Stasikelis, P., Skrabut, E. M., and Orgill, D. P. (1982b). Design principles and preliminary clinical performance of an artificial skin. in *Biomaterials: Interfacial Phenomena and Applications*, S. L. Cooper and N. A. Peppas, eds. pp. 476–481.

---

[1]An inverse measure of cross-link density.

7.11 BIOELECTRODES

Lois S. Robblee and James D. Sweeney

The term "bioelectrode" is broadly used to denote a class of devices which transmit information into or out of the body as an electrical signal. Those that transmit information out of the body generally comprise a category of electrodes called biosensors. Examples of these are sensors for oxygen, glucose, and urea. Chapter 7.12 provides a discussion of these devices. Bioelectrodes that transmit information into the body are found in electrical stimulation devices. Examples of these are the cardiac pacemaker, transcutaneous electronic nerve stimulators ("TENS" devices) for pain suppression, and "neural prostheses" such as auditory stimulation systems for the deaf, and phrenic nerve stimulators for artificial respiratory control. More complex experimental or possible future neural control devices include neuromuscular stimulation prostheses for restoration of hand, arm, or leg function in paralyzed individuals, visual prostheses for the blind, spinal cord stimulators for artificial bladder control, and low-level dc electric field systems for promoting regeneration of damaged nerves. In all of these devices, electrodes transmit current to appropriate areas of the body for direct control of or indirect influence over target cells. Hambrecht (1990) and Heetderks and Hambrecht (1988) provide overviews of the history and state-of-the-art of electrical stimulation devices.

ELECTRICAL STIMULATION OF EXCITABLE CELLS

Electrically excitable cells, such as those of the nervous system or heart, possess a potential difference across their membranes of approximately 60 to 90 mV, with the inside of the cell being negative with respect to the outside. Such cells are generally capable of transmitting information in the form of electrical signals (action potentials) along their lengths. Action potentials are self-propagating waves of depolarization. That is, during an action potential the cellular membrane transiently, for a period of less than 1 msec, changes polarity so that the inside of the cell becomes positive with respect to the outside.

Excitable cells can be stimulated artificially by electrodes that introduce a transient electric field of proper magnitude and distribution. In general, the field generated near a cathodically driven electrode can be used to depolarize most efficiently an adjacent excitable cell's axonal process above a "threshold" value at which an artificially generated action potential results. (Cathodes will tend to draw current outward through nearby cell membranes. In their passive state, such membranes can be modeled electrically as a parallel "cable" arrangment of resistors and capacitors. Outward current therefore elicits depolarizing resistive potential drops, and capacitive charging.) Depending upon the specific application and electrode properties, a single action potential in a nerve cell might be elicited using currents on the order of microamperes to milliamperes, introduced for periods of microseconds to milliseconds. Ranck (1975), Mortimer (1990), and Sweeney (1992) provide in-depth detail on excitable cell stimulation fundamentals.

It has been shown by a number of investigators that low-level dc electric fields can be used to alter nerve development or regeneration. "Stimulation" in this context refers to stimulation of axonal growth, not excitation. For example, neurites in tissue culture can be influenced to grow toward the cathode in a steady electric field (e.g., Jaffe and Poo, 1979). Researchers have also shown significant effects of low-level (cathodic) dc electric currents on regeneration in the lamprey (Roederer *et al.*, 1983) and guinea pig (Borgens *et al.*, 1987) spinal cord. This possibility of inducing central nervous system (CNS) neural regeneration following an injury is particularly exciting because significant CNS regeneration in humans does not normally occur. [See Borgens *et al.*. (1989) for an overview of the possible mechanisms involved in dc-enhanced nerve regeneration research (and related topics involving limb regeneration and wound healing).]

REQUIREMENTS FOR BIOELECTRODES IN ELECTRICAL STIMULATION DEVICES

The selection of materials for bioelectrodes in electrical stimulation devices is a challenging problem, with specifics that vary from one application to another. For example, intramuscular stimulation requires selection of an electrode material with excellent long-term strength and flexibility, yet this application does not require very high stimulation charge densities. Stimulation of the visual cortex through placement of electrodes on the cortical surface, however, requires little mechanical flexibility in an electrode material with relatively moderate charge injection needs. Intracortical stimulation electrodes must be fabricated on a very small scale and must be capable of injecting very high charge densities. In all applications, the electrode and any implanted conductors must be biocompatible both as passive implants and when passing current. In most neural control devices, the surface area of exposed metal of the stimulation electrode is relatively small, <1 cm^2, so that problems with passive biocompatibility of the metal are minor. However, the conductive wires or cables must be electrically isolated from the tissue with insulating layers of high dielectric materials. Since the volume and area of the dielectric materials in contact with the biological tissue are much greater than that of the active electrode surface, most of the passive biocompatibility issues relate to properties of the dielectrics. Thin layers of medical-grade silicone rubber, Teflon, polyimide, or epoxy are typically used for this purpose, although arrays of microelectrodes or sensors which are fabricated using lithography and thin film technology are typically insulated with thin layers of SiO$_2$ and/or Si$_3$N$_4$. The insulating materials must be carefully selected to provide pinhole-free coatings, good adhesion to the conductors, and biocompatibility with the tissue.

Each electrode of a neural prosthesis must be configured in dimensions that are compatible with its location in the target tissue; it must have long-term mechanical and electrical stability; and it must be capable of transferring the required coulombic charge without corroding the electrode or electrolyzing the tissue. The quantity of charge that can be delivered without these reactions occurring depends primarily on the surface area of the electrode–tissue interface. Therefore the size of an electrode will be determined primarily by the "charge injection capacity" of the electrode material.

Charge Injection Mechanisms

Charge transfer between a metal electrode and biological tissue requires a change in charge carriers from electrons in the metal to ions in the tissue fluid. This change in charge carriers occurs by two mechanisms. The first of these is a capacitive mechanism involving only the alignment of charged species at the electrode–tissue interface, i.e., the charging and discharging of the so-called "electrode double layer." This is the ideal mechanism for charge transfer across the interface because no chemical changes occur in either the tissue or the electrode. However, the amount of charge that can be transferred solely by capacitive charging is only about 20 μC per square centimeter of real electrode surface area for commonly used metal electrodes. "Real" electrode surface area is the actual surface area at the electrode–tissue interface, including areas created by pits, cracks, pores, etc. Calculations of "geometric" electrode surface areas are based on the simplifying assumption that the electrode is perfectly smooth. Real areas are often estimated from geometric areas using assumed roughness factors.

The second mechanism of charge transfer, i.e., faradaic charge transfer, involves the exchange of electrons across the electrode–tissue interface and therefore requires that some chemical species be oxidized or reduced. Metal electrodes used for electrical stimulation almost always inject charge by faradaic processes because the amount of charge required greatly exceeds that available from the capacitive mechanism alone. Brummer and Turner (1977) classified the faradaic processes available on platinum and other noble metals as "reversible" or "irreversible," depending upon whether new chemical species were generated in the tissue fluid. Reversible reactions are those that are, ideally, confined to the electrode surface such as monolayer oxide formation and reduction, and H-atom plating and oxidation reactions on platinum. These reactions can be quantitatively reversed by passing a current in the opposite direction, and they do not produce "new" chemical species in the bulk of the solution. However, they may produce potentially damaging transient shifts in H^+ or OH^- concentrations in the tissue fluid near the electrode surface.

Irreversible faradaic reactions include the electrolysis of water to produce oxygen or hydrogen, the oxidation of chloride ion, and corrosion or dissolution of the electrode to produce unstable surface films or soluble metal complexes in tissue fluids. These reactions are undesirable because they alter the chemical composition of the tissue fluid, produce toxic products, generate extremes of acidity or alkalinity, may evolve bubbles, or can destroy electrodes.

An important quantity in selecting a material for use as a stimulation electrode is its "reversible charge injection limit," i.e., the quantity of charge that can be injected or transferred using only capacitive charging and reversible faradaic processes. The charge injection limit of any material will depend upon which reversible processes are available during the time of the stimulus pulse, as well as the geometry of the stimulus waveform, the shape and surface morphology of the electrode, and the chemical composition of the surrounding medium. The relative importance of these factors in achieving reversible charge injection has been reviewed recently (Robblee and Rose, 1990).

Stimulation Waveforms

Before discussing the properties of individual metals typically used for stimulation electrodes, it is necessary to have some fundamental understanding about stimulation waveforms—the temporal patterns by which electric signals are presented to the nervous tissue. Examples of several waveforms are shown in Fig. 1. In general, for implanted electrode systems, regulated-current (as opposed to regulated-voltage) stimulators are preferred. As we have discussed, excitable cells can be stimulated artificially by introducing transiently appropriate electric fields. Such electric fields are directly proportional to the current density distribution created by an electrode. Regulated-voltage waveforms therefore suffer from the fact that impedance changes at the electrode–tissue interface, or within the tissues themselves, can produce often complex, nonreproducible fluctuations in current density. Stimulation waveforms for neural excitation are either pulsatile or sinusoidal. Pulsatile waveforms consist of a train of pulses which are usually of equal amplitude and duration. The charge, q, in a rectangular pulse is proportional to the pulse current, i, and the pulse duration, t, according to the relationship

$$q = it$$

The charge density, which is the charge per pulse divided by either the geometric surface area or the real surface area (if known) of the electrode, is the quantity most important in determining whether irreversible processes will be used in charge injection. The advantage of a pulsatile waveform is that the stimulus charge can be controlled precisely, and it is independent of the pulse repetition rate or frequency with which the stimulus is delivered. In contrast, the charge delivered in either half of a sinusoidal waveform is inversely proportional to frequency. Thus a current amplitude which gives an acceptable charge density at high frequency may lead to an excessively high charge density at low frequency.

The most favorable waveforms are those with no net dc current, and having charge densities below the level that will cause irreversible faradaic processes. Symmetric sinusoidal waveforms by their nature possess no net dc component. In a pulsatile waveform, the condition of no net dc is achieved by using a charge-balanced biphasic pulse pair which consists of consecutive pulses of equal charge but opposite polarity. As explained earlier, the physiologically preferred polarity is for the leading "stimulating" pulse of the biphasic pair to be cathodal. As will be seen, this is not always the preferred polarity from the standpoint of maximizing the quantity of charge that can be injected using only reversible processes.

Low-magnitude, constant-current regeneration enhancement systems generally pass a net current, virtually ensuring charge transfer by irreversible processes. Many researchers have therefore used so-called "wick" electrodes, as originally

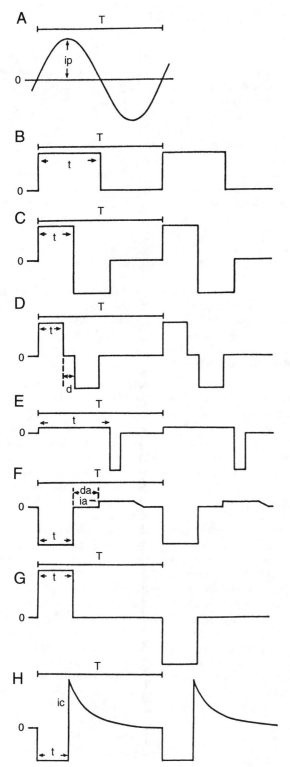

FIG. 1. Diagrammatic representation of current waveforms used in neural stimulation. (A) sinusoidal. (B) Square wave, pulsatile (may be unidirectional as shown, or bidirectional). (C) Charge-balanced, symmetrical biphasic (bidirectional). (D) Charge-balanced, biphasic with brief delay (d) between phases. (E) Charged-balanced, asymmetric biphasic. (F) "Anapol." (G) Charge-balanced, alternating bidirectional. (H) Monophasic, capacitively coupled. Note that T = time of one cycle, or period; t = pulse duration or pulse width; I_p = peak current of sinusoidal waveform; d_a = delay before anodization phase of Anapol waveform; i_a = anodization current following stimulation pulse of Anapol waveform; i_c = current due to capacitor discharge in monophasic, capacitively coupled waveform. [Reproduced, with permission, from Robblee and Rose (1990), Chapter 2, Fig. 2.5, p. 38, and legend, p. 39.]

developed by Borgens and colleagues (e.g., 1987) and subsequently by Kerns *et al.* (1987). These devices essentially isolate a metal electrode (usually Ag–AgCl) from the tissues of the body by passing current through a long insulated "wick" such as cotton thread. Over time, however, toxic silver by-products may be able to diffuse through the length of the wick into the body, or the wicks may become blocked.

Electrode Materials

Noble Metals

The noble metals—platinum, iridium, rhodium, gold, and palladium—are generally preferred for electrical stimulation because of their low chemical reactivity and high resistance to corrosion. However, even these metals undergo dissolution during both *in vitro* and *in vivo* stimulation. In the saline environment, the metals may dissolve as chloro-complexes such as $[PtCl_6]^{2-}$, $[PtCl_6]^{4-}$, and $[AuCl_4]^{1-}$, or they may form unstable surface films which will spall and leave metallic deposits in the tissue. The quantity of metal dissolution is related to the stimulation charge density, polarity of the waveform, and total coulombs delivered.

Platinum Platinum is used in many neural prostheses and cardiac stimulation applications, such as in cardiac pacemaker electrodes and neural stimulation devices for bladder control or respiration assistance. In these applications, the electrode is relatively large, approximately 0.1 cm² geom. or greater, so that even a high current in the range of 20–30 mA per pulse does not result in excessively high current densities or charge densities. Charge densities of 50–150 $\mu C/cm^2$ can be injected in 0.2-msec pulses with platinum electrodes without the onset of gas evolution or saline oxidation. Metal dissolution, however, is an unavoidable reaction and some platinum dissolution can occur at all charge densities. Platinum dissolution rates range from 100 to 1000 ng/C (nanograms per coulomb of aggregate charge injected in either anodic or cathodic pulses) in inorganic saline, and 0.1–1 ng/C in the presence of low concentrations of protein (McHardy *et al.*, 1980; Robblee *et al.*, 1980).

Pure platinum is a soft metal so that its alloys with iridium are used in applications such as intracortical microstimulation which require small-diameter electrodes strong enough to pierce the pial membrane of the brain surface without bending or breaking. Alloys containing up to 30% iridium have the same charge injection limits as pure platinum. However, because microstimulation electrodes must so small, even the very low currents on the order of 10–20 μA lead to charge densities far in excess of the charge injection limit of platinum and its alloys. For example, a 10-μA pulse of 0.2-msec duration is only 2 nC in charge. If this is injected with a microelectrode having a geometric surface area of 10^{-6} cm², the charge density is 2000 $\mu C/cm^2$ geom., > 10 times higher than the charge density that can be delivered with Pt or Pt–Ir alloy solely by reversible processes.

Iridium Iridium has much greater hardness than platinum and might be considered as a replacement for platinum–iridium alloy in applications requiring high mechanical strength. However, iridium metal has no advantage over pure platinum or platinum–iridium alloys in terms of charge injection properties. Its great advantage lies in its ability to store coulombic charge in the form of a multilayered surface oxide when it is appropriately "activated." Reversible valence changes within the surface oxide provide much higher charge injection levels than are possible with the bare metal (Robblee *et al.*, 1983).

Iridium Oxide   Iridium oxide is beginning to find use in a number of experimental neural prosthesis designs (Bak *et al.*, 1990). Iridium oxide layers can be formed on iridium metal or on films of sputtered iridium by electrochemical "activation," i.e., repetitive cycling of the iridium potential between 0.0 V and 1.5 V versus a reversible hydrogen electrode. Iridium oxide thus formed is one of a category of "valence change oxides." These oxides inject charge by highly reversible valence transitions between two stable oxides, e.g.,

$$Ir_2O_3 + H_2O \leftrightarrow IrO_2 + 2H^+ + 2e^-$$

While the multilayered oxide film may have as much as 100,000 $\mu C/cm^2$ of charge capacity available, nevertheless the total charge capacity cannot be utilized within the short time of a stimulus pulse because of ionic diffusion and conductivity limitations. Thus the measured charge injection limits of iridium oxide electrodes are very sensitive to stimulus waveform. When stimulated with cathodal or cathodal-first biphasic pulses, the charge injection limits with 0.2-msec pulses are ~1000 $\mu C/cm^2$ geom. However, if an anodic bias voltage is applied between stimulus pulses, then cathodic charge densities up to 3500 $\mu C/cm^2$ geom. can be injected with cathodal pulses utilizing only reversible charge injection processes (Agnew *et al.*, 1986; Kelliher and Rose, 1989). The higher charge injection is possible if all of the oxide is returned to its highest stable valence state between pulses. This can only be accomplished by applying a bias potential. Cathodal charge injection then occurs via reduction of some of this oxide to its lower valence state.

The iridium oxide film in its fully oxidized state has a lower interface impedance than the typical bare metal–saline interface (Glarum and Marshall, 1980). The low impedance of the oxide provides more efficient coupling between tissue and electrode, which is advantageous for recording as well as stimulation. The greater efficiency for charge flow may lower the compliance voltage required to drive the stimulus current, thereby prolonging the battery life of a stimulation device.

Non-noble Metals

Some non-noble metals are candidates for electrode applications requiring high mechanical strength and fatigue resistance such as is demanded of intramuscular electrodes. Stainless steel 316LVM has been the material of choice for these electrodes primarily because of its flexibility when fabricated in a coiled structure and its fatigue resistance. However, because its charge injection ability is quite limited, the surface area of the active electrode must be quite large to keep the stimulus charge densities low.

The stainless steels, as well as the nickel–cobalt alloy Elgiloy and the MP35N, rely on thin passive films for corrosion resis-

tance and must inject charge by faradaic processes involving reduction and oxidation of their passivating films. Corrosion will result when breakdown of the passive layer occurs. While loss of passivity and dissolution are certain to occur under conditions of large anodic polarization, the stainless steels and cobalt alloys are also susceptible to corrosion-related failure when used with cathodic-going pulses. The maximum charge density that can be accommodated by stainless steel is ~40 $\mu C/cm^2$ geom. and even at this level, corrosion-related failures occur (White and Gross, 1974).

Titanium, tantalum, zirconium, tungsten, and tungsten bronzes give strong surface reactions and changes in impedance under conditions of electrical stimulation. Anodic polarization of titanium and tantalum results in the formation of insulating oxide films and large increases in impedance. However, this property of titanium and tantalum is useful for their use as capacitor electrodes which inject charge without any faradaic reactions at the electrode–tissue interface. Silver and copper electrodes are generally not used because they produce tissue necrosis passively, even without active passage of electric current.

Capacitor Electrodes

Capacitor electrodes are considered the ideal type of stimulation electrode because the introduction of a dielectric at the electrode–solution interface allows charge flow completely by charging and discharging of the dielectric film. The oxide film that is produced on tantalum by anodic polarization withstands substantial voltage without significant dc leakage and permits charge flow without the risk of faradaic reactions. Anodized titanium has higher dielectric strength than anodized tantalum, but its dc leakage currents are too high for electrode applications. Both tantalum and titanium are used extensively as biological implant materials and have passive resistance to corrosion in a saline environment. Capacitor electrodes based on tantalum–tantalum pentoxide have a charge storage capacity of ~100–150 $\mu C/cm^2$ (Rose *et al.*, 1985). Their use is limited to neural prosthesis applications having electrodes about 0.05 cm^2 or larger in geometric area, such as peripheral nerve stimulators. It is not possible to obtain adequate charge storage capacity in electrodes smaller than 10^{-3} cm^2. An additional restriction on the use of tantalum capacitor electrodes is that they must always operate at a positive potential to prevent electronic conduction across the oxide. If cathodal stimulation is required, the tanalum capacitor electrode must be biased to an anodic potential and pulsed cathodally.

Bibliography

Agnew, W. F., Yuen, T. G. H., McCreery, D. B., and Bullara, L. A. (1986). Histopathologic evaluation of prolonged intracortical electrical stimulation. *Exper. Neurol.* **92**: 162–185.

Bak, M., Girvin, J. P., Hambrecht, F. T., Kufta, C. V., Loeb, G. E., and Schmidt, E. M. (1990). Visual sensations produced by intracortical microstimulation of the human occipital cortex. *Med. Biol. Eng. Comput.* **14**: 257–259.

Borgens, R. B., Blight, A. R., and McGinnis, M. E. (1987). Behaviorial recovery induced by applied electric fields after spinal cord hemisection in guinea pig. *Science* **238**: 366–369.

Borgens, R. B., Robinson, K. R., Yanable, J. W., Jr., McGinnis,

M. E., and McCaig, C. D. (1989). *Electric Fields in Vertebrate Repair: Natural and Applied Voltages in Vertebrate Regeneration and Healing.* Alan R. Liss, New York.

Brummer, S. B., and Turner, M. J. (1977). Electrochemical considerations for safe electrical stimulation of the nervous system with platinum electrodes. *IEEE Trans. Biomed. Eng.* BME-24: 59–63.

Glarum, S. H., and Marshall, J. H. (1980). The A-C response of iridium oxide films. *J. Electrochem. Soc.* 127: 1467–1474.

Hambrecht, F. T. (1990). The history of neural stimulation and its relevance to future neural prostheses. in *Neural Prostheses: Fundamental Studies,* W. F. Agnew and D. B. McCreery, eds. Prentice Hall, Englewood Cliffs, NJ, Ch. 1, pp. 1–23.

Heetderks, W. J., and Hambrecht, F. T. (1988). Applied neural control in the 1990s. *Proc. IEEE.* 76: 1115–1121.

Jaffe, L. F., and Poo, M.-M. (1979). Neurites grow faster toward the cathode than the anode in a steady field. *J. Exp. Zool.* 209: 115–127.

Kelliher, E. M., and Rose, T. L. (1989). Evaluation of charge injection properties of thin film redox materials for use as neural stimulation electrodes. *Mater. Res. Soc. Symp. Proc.* 110: 23–27.

Kerns, J. M., Pavkovic, I. M., Fakhouri, A. J., Wickersham, K. L., and Freeman, J. A. (1987). An experimental implant for applying a DC electrical field to peripheral nerve. *J. Neurosci. Meth.* 19: 217–223.

McHardy, J., Robblee, L. S., Marston, J. M., and Brummer, S. B. (1980). Electrical stimulation with Pt electrodes. IV. Factors influencing Pt dissolution in inorganic saline. *Biomaterials* 1: 129–134.

Mortimer, J. T. (1990). Electrical stimulation of nerve. in *Neural Prostheses: Fundamental Studies,* W. F. Agnew and D. B. McCreery, eds. Prentice Hall, Englewood Cliffs, NJ, Ch. 3, pp. 67–83.

Ranck, J. B., Jr. (1975). Which elements are excited in electrical stimulation of mammalian central nervous system: A review. *Brain Res.* 98: 417–440.

Robblee, L. S., and Rose, T. L. (1990). Electrochemical guidelines for selection of protocols and electrode materials for neural stimulation. in *Neural Prostheses: Fundamental Studies.* W. F. Agnew and D. B. McCreery, eds. Prentice Hall, Englewood Cliffs, NJ, Ch. 2, pp. 25–66.

Robblee, L. S., McHardy, J. Marston, J. M., and Brummer, S. B. (1980). Electrical stimulation with Pt. electrodes. V. The effects of protein on Pt dissolution. *Biomaterials* 1: 135–139.

Robblee, L. S., Lefko, J. L., and Brummer, S. B. (1983). An electrode suitable for reversible charge injection in saline solution. *J. Electrochem. Soc.* 130: 731–733.

Roederer, E., Goldberg, N. H., and Cohen, M. J. (1983). Modification of retrograde degeneration in transected spinal axons of the lamprey by applied DC current. *J. Neurosci.* 3: 153–160.

Rose, T. L., Kelliher, E. M., and Robblee, L. S. (1985). Assessment of capacitor electrodes for intracortical neural stimulation. *J. Neuroscience Meth.* 12: 181–193.

Sweeney, J. D. (1992). Skeletal muscle responses to electrical stimulation. in *Electrical Stimulation and Electropathology,* J. P. Reilly, ed., Cambridge Univ. Press, Ch. 8, pp. 285–327.

White, R. L., and Gross, T. J. (1974). An evaluation of the resistance to electrolysis of metals for use in biostimulation probes. *IEEE Trans. Biomed. Engin.* BME-21: 487–490.

7.12 BIOMEDICAL SENSORS AND BIOSENSORS

Paul Yager

A convergence of factors is now resulting in the rapid development of sensors for biomedical use. These factors include:

A. Increasing knowledge of the biochemical bases of normal and pathological biology,

B. Rapid advances of biotechnology and genetic engineering,

C. Steadily decreasing costs of sophisticated devices due to microprocessor technology,

D. Reduction in the size of sensors due to silicon microfabrication and fiber optics technology,

E. Rapidly advancing sensor technology in nonbiomedical fields and for *in vitro* use for clinical chemistry,

F. Advances in biomaterials for biocompatibility,

G. Pressure for reduction in costs of delivering medical care through more efficient treatment of patients and shorter hospital stays.

Advances in computer technology have reached the point that the control of devices and processes is often only limited by the ability to provide reliable information to the computer. Furthermore, the technologies developed by the microprocessor and fiber optics industries are now spawning a new generation of sensors. Areas benefitting from these new sensors include the automotive and aerospace industries, chemical and biochemical processing, and environmental monitoring. The introduction of such sensors to clinical practice is a slow process that has just begun because of the same regulatory factors that apply to any use of devices and materials *in vivo*. Furthermore, sensors are as subject to the problems of biocompatibility as are any other type of *in vivo* device, and initial optimism about the potential applicability of sensors to *in vivo* use has given way to a more sober appraisal of the potential of the field.

This chapter provides a brief overview of the application of sensors to clinical medicine, with an emphasis on chemical sensors and the emerging field of biosensors. Several excellent reviews may be found in the recent literature (Turner *et al.*, 1987; Collison and Meyerhoff, 1990; Rolfe, 1990).

PHYSICAL VS. CHEMICAL SENSORS

Sensors fall into two general categories: physical and chemical. Physical parameters of biomedical importance include pressure, volume, flow, electrical potential, and temperature, of which pressure, temperature, and flow are generally the most clinically significant, and lend themselves to the use of small *in vivo* sensors. Measurement of electrical potential is covered in Chapter 7.11. Chemical sensing generally involves the determination of the concentration of a chemical species in a volume of gas, liquid, or tissue. The species can vary in size from the H^+ ion to a live pathogen (Table 1). and when the species is complex, an interaction with another biological entity is required to recognize it. When such an entity is employed, the sensor is considered a biosensor. It is, in general, necessary to distinguish this chemical from a number of similar interferents, which can be technologically challenging, but this is an area in which biosensors excel. The clinical utility of monitoring a variety of marker compounds in the body has motivated great efforts to develop biosensors. The prime target is improvement upon current methods for determining glucose concentration for treatment of diabetes.

TABLE 1   Chemical Indicators of Health

Small, simple	pH (acidity)
↓	Electrolytes (ions)
↓	Blood gases, including general anesthetics
↓	Drugs and neurotransmitters
↓	Hormones
↓	Proteins (antibodies and enzymes)
↓	Viruses
↓	Bacteria
↓	Parasites
Large, complex	Tumors

INTERACTION OF THE SENSOR WITH ITS ENVIRONMENT

One helpful way to classify sensors is to consider the relationship between the sensor and the analyte, as shown schematically in Fig. 1. The more intimate the contact between the sensor and the analyte, the more complete is the information about the chemical species being measured. However, obtaining greater chemical information may involve some hazard to the physical condition of the system being studied. This is not a trivial problem when dealing with human subjects.

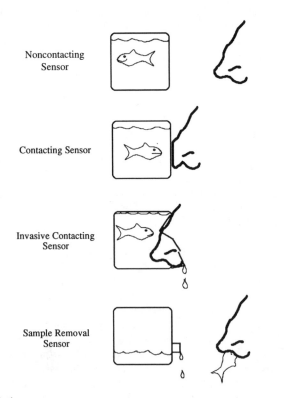

FIG. 1.   Different types of relationships may occur between a chemical sensor and the analyte to which it is sensitive.

Noncontacting sensors produce only a minimal perturbation of the sample to be monitored. In general, such measurements are limited to the use of electromagnetic radiation such as light, or sampling the gas or liquid phase near a sample. It may be even necessary to add a probe molecule to the sample to make the determination. Two examples are monitoring the temperature of a sample by its infrared emission intensity, and spectroscopic monitoring of pH dependence of the optical absorption of a pH-sensitive dye. No damage is done to the sample, but limited types of information are available, and chemical selectivity is difficult to achieve *in vivo*.

Contacting sensors may be either noninvasive or invasive. Direct physical contact with a sample allows a rich exchange of chemical information; thus much effort has been made to develop practical contacting sensors for biomedical purposes. A temperature probe can be in either category, but with the exception of removing or adding small quantities of heat, it does not change the environment of the sample. Few chemical sensors approach the nonperturbing nature of physical sensors. All invasive sensors damage the biological system to a certain extent, and physical damage invariably leads to at least localized chemical change. Tissue response can, in turn, lead to spurious sampling. Furthermore, interfacial phenomena and mass transport govern the function of sensors that require movement of chemical species into and out of the sensor. Restrictions on size of the invasive sensor allowable in the biological system can limit the types of measurements that can be made—even a 1-mm diameter pH electrode is of no use in measuring intracapillary pH values. Clearly, this is the most difficult type of sensor to perfect.

Sample removal sensors. Most contacting sensors are derived from chemical assays first developed as sample removal sensors. While it is certainly invasive and traumatic to remove blood or tissue from a live animal, removal of some fluids such as urine can be achieved without trauma. Once removed, a fluid can be pretreated to make it less likely to adversely affect the functioning of a sensor. For example, heparin can easily be added to blood to prevent clotting in an optical measurement cell. Toxic probe molecules can be added at will, and samples can be fractionated to remove interfering species. The sensor and associated equipment can be of any size, be at any temperature, and use as much time as necessary to make an accurate measurement. Furthermore, a sensor outside the body is much easier to calibrate. This approach to chemical measurement allows the greatest flexibility in sensor design and avoids many biocompatibility problems.

Consuming vs. Nonconsuming Sensors

There are at least two distinct ways in which a sensor can interact with its environment; these can be called consuming and nonconsuming (Fig. 2). A nonconsuming or equilibrium sensor is one that can give a stable reading with no net transport of matter or energy between the sensor and its environment. For example, while a thermometer is approaching equilibrium it takes up or releases heat, but when it has reached its ultimate temperature it no longer directly affects the tempeature of

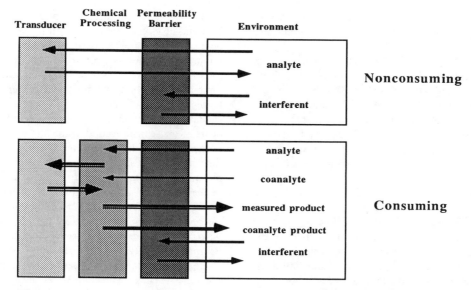

FIG. 2. A schematic diagram comparing consuming and nonconsuming sensors. The transducer is usually isolated from the environment by a permeability barrier to keep out interfering species, and in a consuming sensor, there is usually an intermediate layer in which active chemical processing occurs. In the consuming sensor there is a complex flux of analytes and products that makes it very sensitive to changes in permeability of the sensor–solution interface.

its environment. Some chemical sensors such as ion-selective electrodes and some antibody-based sensors work similarly, and have the advantage of being minimally perturbative of their environments. A nonconsuming sensor may become slower to respond after being coated by a biofilm, but may still provide the same ultimate reading.

The consuming or nonequilibrium sensors rely on constant unidirectional flux of energy or matter between the sensor and environment. The most common glucose sensor, for example, destroys glucose in the process of measuring it, thereby reducing its concentrations in the tissue in which it is measured, as well as reducing the pH and O_2 concentrations and generating toxic H_2O_2 as a byproduct. This chemical measurement cannot be made *in situ* without affecting the system in which the measurement is made, and while it may be tolerable in flowing blood, it may not be acceptable in tissue over the long term. The greater the size of the sensor, the greater its sensitivity, but the more seriously it perturbs its environment. Many of the most fully developed specific *in vitro* analytical techniques for the determination of biochemicals involve irreversible consumption of the analyte, so sensors based on such techniques are difficult to implement *in vivo*.

Site of Measurement

Sample removal sensors are the mainstay of the clinical chemistry laboratory. For example, the first commercially manufactured biosensor—an electrode for measuring glucose—was made by Yellow Springs Instruments for a clinical chemistry analyzer. The major problem with the use of such devices

is the time delay inherent in moving a sample to a distant location and waiting for delivery of the information derived from it, as well as the fact that useful instruments are often heavily utilized.

A major activity of modern bioanalytical chemistry is conversion of sample removal sensors into invasive contacting sensors. Not only can this approach improve care by reducing the aforementioned delays, but it can also allow new types of procedures, such as automated feedback control of delivery of drugs to patients. However, development of such an approach often requires inventing new ways of making the measurements themselves. Chemical measurement at bedside or *in vivo* is technologically more challenging than in a prepared chemical laboratory with highly trained technicians. Clinical personnel must attend to the critical needs of the patient and have little time for fiddling with temperamental instrumentation. The instrumentation must therefore be made nearly foolproof, rugged, safe, reliable, and, if possible, self-calibrating.

Duration of Use

Two major questions in the design of any sensor are how often and for how long it is expected to be used. There are several factors to be considered:

Length of time for which monitoring is required. Determination of blood glucose levels must be made for the entire lifetime of diagnosed diabetics, while intra-arterial blood pressure monitoring may only be needed during a few hours of surgery.

Frequency of measurements. Pregnancy testing may have

to occur only once a month, whereas monitoring of blood pCO_2 during an operation may have to be made several times a minute.

Reusability of the sensor. Some chemical sensors contain reagents that are consumed in a single measurement. Such sensors are usually called "dipstick sensors," such as are now found in pregnancy testing kits. High-affinity antibodies, for example, generally bind their antigens so tightly that they cannot be reused. On the other hand, most physical and chemical sensors are capable of measuring the concentration of their analyte on a continuous basis and are therefore inherently reusable.

Lifetime of the sensor. Chemical sensors all have limited lifetimes because of such unavoidable processes as oxidation, and while these may be extended through low-temperature storage, *in vivo* conditions are a threat to the activity of the most stable biochemical. Most sensors degrade with time, and the requirements for accuracy and precision usually limit their practical lifetime, particularly when recalibration is not possible. Mechanical properties can also limit lifetimes; while a thermocouple may have a shelf life of centuries, it can be broken on its first use by excessive flexing.

Appropriateness of repeated use. The need for sterility is the most important reason to avoid reuse of an otherwise resuable sensor. If it is not logistically possible or economically feasible to completely sterilize a used sensor, it will only be used on a single individual, and probably only once.

Biocompatibility. If the performance of a sensor is degraded by continuous contact with biological tissue (as discussed later), or if the risk to the health of the patient increases with the time in which a sensor is in place, the lifetime of the sensor may be much shorter *in vivo* than *in vitro*.

As a consequence of all of these factors in the design of an integrated sensing system, the probe—that part of the sensor that must be in contact with the tissue or blood—is often made disposable. Probes must therefore be as simple and as inexpensive to manufacture as possible, although it is often true that it is the sale of consumables such as probes that can be more profitable than the sale of the device itself.

Biocompatibility

The function of most chemical sensors is limited by the rate of diffusion through an unstirred layer of liquid at the interface, whether the alteration is rate control response or merely response time. However, interfacial flux in biological media can be altered further by processes unique to living systems. Biocompatibility is a serious problem for any material in contact with living systems, but is particularly so for chemical sensors. The properties of the interface are crucial to the ability of the sensor to make accurate reproducible readings.

IN VITRO USE

Biofilms often consist of bacteria embedded in a secreted matrix of complex polysaccharides. It is difficult to prevent biofilm formation on surfaces exposed to active media such as bacterial or eukaryotic cell cultures. The growth of a biofilm on a sensor degrades its performance in some way. In fact, in the most extreme case, the sensor can become sensitive only to conditions in the biofilm. While the microenvironment of these films may be beneficial to the bacteria, it is detrimental to the function of the sensor for several reasons. First, the chemistry of the film may damage structural or active components of the sensor. Second, if the film completely encloses the sensor, only the film's microenvironment is sensed, rather than the solution that surrounds it. If the living components of the film metabolize the analyte to be sensed, it may never reach the sensor. Even a "dead" film may exclude certain analytes by charge or size and thereby lower the concentration available for sensing at the surface below the film. If a sensor used *in vitro* is fouled, it can often be removed, cleaned, and restored to its original activity. For example, carbon electrodes containing immobilized enzymes can be restored by simply polishing away the fouled surface (Wang and Varughese, 1990). When sensors are used *in vitro* for monitoring the chemistry of body fluids, preprocessing can be used to reduce the accretion of biofilms that might impede the function of the sensor.

IN VIVO USE

Introduction of sensors into the body creates a complex set of problems. The sensor and the body act on each other in detrimental ways that must be minimized if not entirely avoided. Many of the problems (described later) worsen with time and may limit the utility of sensors to very short uses. In some cases it has been found that the problems are almost immediate, producing spurious results from the outset. Subcutaneous glucose needle electrodes have been found in give accurate results *in vitro* before *and after* producing erroneous values *in vivo*. The biological environment may simply make it impossible to perform accurate chemical measurements with certain types of sensors.

Effects of Sensors on the Body

The introduction to the body of a sensor is a traumatic event, although the degree of trauma depends on the site of placement. The gastrointestinal tract can clearly be less traumatically accessed than the pulmonary artery. Critically ill patients often must have catheters placed into their circulatory systems for monitoring blood pressure and administration of drugs, fluids, and food, so that no additional trauma is caused by including a small flexible sensor in that catheter. The size, shape, flexibility, and surface chemistry of the sensor probe are also important factors, although they are covered elsewhere in this volume. The outermost materials of the chemical sensor have at least one requirement not normally placed upon structural biomaterials: they must be permeable to the analyte in question. Thin films of polyurethanes are permeable to at least some analytes, and Nafion and porous Teflon work in other cases, but this issue is far from resolved. Some metals and graphite can be used directly as electrodes in the body.

When short-term implantation in tissue is possible, there is initial trauma at the site of insertion. Longer term implantation increases the risk of infection along the surface of the implant. When the site of implantation of the probe is the circulatory system, the thrombogenicity of the probe is of paramount importance. Surface chemistry, shape, and placement within the vessel have all been shown to be of great importance in reducing the risk of embolism. Also, sensors based on chemical reactions often contain or produce toxic substances during the course of their operation, so great care must be taken to ensure that these are either not released or are released at low enough levels to avoid significant risk to the patient.

Effects of Surface Fouling on Sensor Function

The response of a consuming sensor depends on the rate of flux of the analytes and products across the interface. Indeed, most consuming sensors achieve linearity of response by making the diffusion of analyte into the sensor the rate-limiting step in the chemical reaction on which the sensor relies. Any new surface film over the interface will slow the diffusion and reduce the sensor output signal, producing an apparent reduction in the concentration of the analyte. Only equilibrium sensors avoid this problem, and even they experience a reduction in their response time.

These aforementioned problems of inert biofilms pale in comparison to the problems that arise when the film contains living cells, which is usually the case *in vivo*. As mentioned in other chapters, enzymatic and nonenzymatic degradation of polymers can be greatly accelerated, resulting in mechanical and eventual electrical failure of a sensor. Large changes in local pH, pO_2, and pCO_2, and concentration of other chemically active species such as superoxide can either chemically alter analytes or damage the function of the sensor itself. If, as often happens when a foreign body is implanted in soft tissue, a complete capsule of polynuclear macrophages forms, the sensor within it may only be capable of sensing the microenvironment of activated macrophages, which is certainly not a normal sample of tissue.

Thrombus formation in blood is another common problem. While such schemes as the use of polyurethane coatings and covalent grafting of heparin to that surface that work for vascular grafts are also helpful for sensors, thrombus is a complex, active tissue that severely compromises the function of a sensor. In general, sensors in blood are used only for a few days at most and in the critical care setting, during which time the patient may be undergoing anticoagulant therapy that reduces the problem of thrombus formation. As yet there is no device for permanent implantation in the bloodstream, despite great efforts in this direction.

Because of the trauma involved with insertion and removal of *in vivo* sensors, it is not generally feasible to remove a sensor to calibrate it. Consequently, any calibration must be done *in situ*, and since this would require the delivery of a calibration solution to the sensor, it is not generally done. Successful *in vivo* sensors to date have been those that do not require recalibration during the lifetime of that sensor.

CLASSES OF SENSORS

New Technologies

As mentioned at the outset, the current rapid pace in development of biomedical sensors is fueled in part by a series of technological advances from other fields.

Fiber Optics

Advances in the field of optical communications have created a new technology for controlling light by using waveguides such as optical fibers. While most of the technology for communication uses near-infrared light and most chemical measurements are made with ultraviolet and visible light, the fiber optic sensor field is bridging the gap by developing fibers that work well in the visible and chemical techniques that employ near-infrared light. Optical fibers have an advantage over wires in that they do not conduct current, so sensors made from such fibers (optrodes) are intrinsically safer than electrochemical sensors or even thermistors in that they reduce the risk of electrical shock to the patient.

Microprocessor-Controlled Devices

The recent availability of powerful, small, and relatively inexpensive computers has permitted designers of sensors to include sophisticated analytical procedures as part of the normal function of the sensors. Many previously manual operations such as calibration can now be completely automated. Rather than building a custom analog circuit to perform a particular function, one can now assemble stock electronic parts and customize only the software for the application. This has also relaxed the once-stringent requirement for linearity of response for the sensor itself—nearly any form of response can be programmed into a lookup table kept in memory. Home treatment of the elderly and chronically ill has been aided by the acquisition of data from sensors in the home. The data can be sent by radio or phone lines to central locations for more sophisticated analysis or routing of emergency services. Improvements in telemetry have led to the development of sensors that have no wires penetrating the body at all, which avoids the route for infection provided by continuous penetration of the body by catheters and electrical leads.

Microsensors and Microfabrication

Because of progress in the computer industry, micromachining of silicon into complex three-dimensional shapes with dimensions of less than 1 μm is now commonplace. Devices can be electrical, such as electrodes, single transistors, and complex circuits; optical, such as photodiodes and optical waveguides; and mechanical, such as sensors, pumps, and microactuators. These diverse devices can be integrated into a single wafer, creating an entire "chemical laboratory on a chip." Furthermore, silicon microlithography is so successful in producing computers because it allows production of multiple copies of small and precisely manufactured devices, so the problems and costs inherent in manufacturing hand-made sensors can be avoided.

Physical Sensors

Temperature

The proper maintenance of a particular temperature, such as the 37°C of the human body core, is an indicator of health. Alteration of the temperature of the whole body or of a particular organ may be advantageous during certain medical procedures, such as surgery and preservation of organs, and this requires careful but minimally traumatic monitoring.

Thermometry began with development of devices based on calibrated changes in volume of liquids by Galileo in the 1600s and was perfected in the mid-eighteenth century. Such thermometers are inexpensive and can be quite accurate, but are generally fragile, bulky, slow to respond, and require reading by eye, so they have been largely replaced in the clinical setting. If determination of the surface temperature of the skin is adequate, there are inexpensive techniques available that rely on changes in the optical properties of a film of cholesteric liquid crystals, or expensive options such as the use of infrared radiometers. Modern methods of measuring the temperature of a bulk material such as blood or tissue are based on temperature-dependent electrical properties of matter. Devices include thermocouples, resistance temperature detector (RTD) sensors, thermistors, and silicon diodes, microcomputer-based applications of all of these. The sensor must be placed into tissue or blood, so to maintain accuracy there should be little transfer of heat into or out of the body along the leads to the thermometer. A small, self-contained telemetric temperature sensor with no external leads at all has been developed by Human Technologies, Inc. It is capable of continuous readings of core temperature for the duration of the device's residence in the gastrointestinal tract.

Thermocouples   The Seebeck effect is responsible for temperature-dependent potentials (or current, in a closed circuit) across the junction between two different metals. Responses (Seebeck coefficients) of 10 to 80 μV/°C are the range for commonly used thermocouple pairs. In potential measurement circuits, the size of the contact region is immaterial, so very small thermocouples can be made with response times as short as milliseconds. The response is not linear, and either lookup tables or high-order polynomials are needed to linearize the responses. Precision better than 0.1°C is not generally practical. The thermocouple is self-powered and thus introduces no heat to the system being measured, but a reference junction at a known temperature is required, usually within the housing of the digital voltmeter (DVM) sensor electronics (Fig. 3). Thermocouples are cheap and as reproducible as the chemistry of the metals used. Because they can be made extremely small, they continue to be the sensor of choice in some applications. The Cardiovascular Devices Inc. (CDI) Systems 1000 fiber optic sensor for pH, pCO_2, and pO_2 (see later discussion), ironically, uses a thermocouple for its required temperature reference at the tip of the optrode.

RTDs   The platinum RTD is based on the temperature dependence of the resistance of a metal. If care is taken to eliminate other sources of changes in resistance, chiefly me-

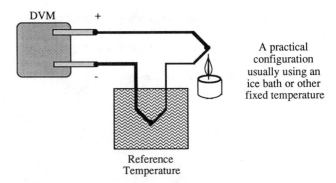

FIG. 3.   A schematic representation of the manner in which thermocouples are used to measure temperature.

chanical strain, it is possible to measure temperature quite accurately and reproducibly. When well treated, RTDs are the most stable temperature measurement devices. The best RTDs are still hand-made coils of Pt wire, but these tend to be very expensive and bulky. Recently, deposition of Pt film on ceramics has been developed as a smaller, cheaper alternative with faster response times and nearly the same stability. Nonlinearities require adjustment in software for accurate readings, and because some current must be supplied to make a measurement, there is joule heating of the device and the sample around it.

Thermistors   These are the most sensitive temperature measurement devices. They are generally made of semiconductive materials with negative temperature coefficients (decreasing R with increasing T). This effect is often quite large (several percent per °C), but also quite nonlinear. Very small devices can be fabricated with response times approaching those of thermocouples. Unfortunately, there are several drawbacks: the response of individual devices is highly dependent on processing conditions, the devices are fragile, and they self-heat. Nevertheless, because of their sensitivity they are the most common transducer for *in vivo* temperatures. For example, they are commonly used as the temperature sensors incorporated in the Swan–Ganz catheter used to determine cardiac output by thermodilution.

Optical Techniques   Optical techniques of temperature measurement have gained favor in recent years. This is partly because the use of optical fibers for measurements removes the necessity of using metal wires that both allow a path for potentially lethal shock and can perturb electromagnetic fields such as those used in magnetic resonance imaging and other techniques requiring the use of microwaves. The Model 3000 Fluoroptic Thermometer (Luxtron Corp.) is such a device; it is based on the temperature dependence of the lifetime of phosphorescent emission from an inorganic material (magnesium fluorogermanate) placed at up to four different locations along four 250-μm plastic optical fibers. This allows four nearly independent temperature measurements at spacings as close as 3 mm, which is useful for monitoring microwave-induced hyperthermia. Accuracy to ±0.2°C over a 40°C range is claimed. The expense of this type of device has initially limited its use to situations in which metal wires are not admissible, but any of several recent temperature measurement schemes may prove substantially cheaper. Particularly when integrated with other sensors in a single multichannel device, optical thermometry may prove more popular than the use of thermistors.

Pressure

The most common sensor placed into the circulatory system of hospitalized patients is a blood pressure monitor. Both static and pulsatile blood pressure are key signs for monitoring the state of patients, particularly those with impaired cardiac function or undergoing trauma such as surgery. While it is possible to measure the static blood pressure external to the body by using pressure cuffs and some acoustic or optical means of detection, such techniques are subject to a number of artifacts and do not work well in patients with impaired cardiac function. The site of measurement may be either an artery or vein, and the sensor may be implanted for short term use during an operation, or over a longer term in an intensive care unit setting. It is even possible to monitor intra-arterial pressure in ambulatory patients over long periods.

The pressure ranges generally seen in the circulatory system range from 0 to 130 mm Hg above the ambient 760 mm Hg. The most commonly available pressure transducers for accurate measurement in this range have been strain gauges, which are temperature-sensitive devices about an inch in diameter. The transducer itself is therefore placed outside the body, and pressure is transmitted to the transducer through a catheter. The transducer must be calibrated at least at turn on, and older models also required a two-point calibration against at least one calibrated pressure different from atmospheric and periodic rezeroing against ambient.

The mechanical properties of the catheter clearly can affect the accuracy of the waveform recorded by the transducer, as can the viscosity of the solution filling the catheter. Blood would normally clot in the stagnant interior of the catheter, degrading the sensor response and causing a risk of embolism, so it is necessary to flush the catheter periodically with heparinized saline to keep both the lumen of the catheter and the artery patent. This requires the presence of a fairly complex set of sterile tubing and valves attached to a saline reservoir. Because flushing the catheter perturbs the pressure, it is not possible to obtain truly continuous measurements.

Most transducers have been fragile and expensive, and a factor that contributes to the cost of their use is the requirement that all materials in contact with blood must be sterilized. There is generally a diaphragm that separates the blood and saline from the mechanical transducer. This diaphragm and associated dome-shaped housing has in the past been a permanent part of the apparatus that required cleaning and sterilization between uses, but recent advances in manufacturing inexpensive, mechanically reproducible diaphragms have allowed this part also to be made disposable.

Other pressures of clinical importance are intracranial pressure and intrauterine pressure. While in both these cases no direct invasion of the circulatory system is required, periodic flushing of all catheters is required. If, on the other hand, the

TABLE 2 Transducers Used in Chemical Sensors

Transducer	Mode of measurement
Ion-selective electrode, gas-selective electrode, FET	Potentiometry—determination of surface concentration of charged species
Oxygen electrode, electrochemical electrode	Amperometry—monitoring available concentration of electrochemically active species
Low impedance electrodes for monitoring conductance, impedance, admittance	Monitoring changes in bulk or surface electrical properties caused by altered molecular concentrations
Optical waveguides with detection of absorption, fluorescence, phosphorescence, chemiluminescence, surface plasmon resonance	Photometry
Thermistors, RTD, calorimeters	Monitoring temperature change induced by chemical reaction
Piezoelectric crystal SAW, BAW, etc., with chemically selective coating	Change in sound absorption or phase induced by binding to device

sensor can be placed directly in the cavity in which the pressure is to be measured, flushing may not be necessary. Only a thick, rigid, calcified surface coating (of a sort that would take weeks or longer to develop) would prevent an accurate monitoring of pressure by a sensor placed directly in the body. Advances in silicon processing have allowed the manufacture of extremely small pressure sensors based on thin diaphragms suspended above evacuated cavities. The position of the diaphragm depends on the instantaneous pressure differential across it, so these devices can be used as both microphones and pressure monitors. The pressure can be monitored electrically, based on changes of capacitance between the diaphragm and the apposing wall of the cavity, or optically by monitoring changes in the reflectivity of the resonant optical cavity between the diaphragm and wall. This latter approach is the basis for fiber optic sensors introduced by FiberOptic Sensor Technologies, Inc. and MetriCor Corporation.

Chemical Sensors

A variety of chemical sensors employing different transduction mechanisms are currently employed for *in vivo* and *in vitro* measurements of biological parameters. These are summarized in Table 2.

pH

The pH of blood and tissue are normally maintained within narrow ranges; even slight deviations from these values have great diagnostic value in critical care. Measurement of pH is

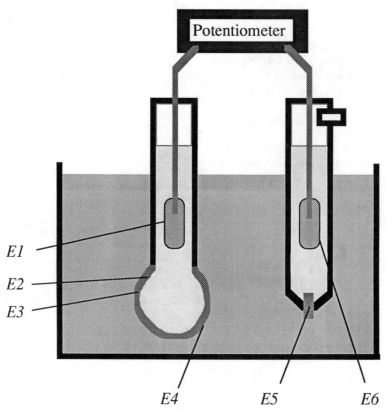

FIG. 4.   Schematic of a pH meter and the potentials generated at various interfaces. Independent potentials generated in a pH measurement system: E1, measuring internal AgCl electrode potential; E2, internal reference solution-glass potential; E3, glass asymmetry potential; E4, analyte-glass potential; E5, reference liquid junction potential; E6, reference internal AgCl electrode potential. Any contamination or fouling at any of those interfaces causes degradation and drift. E4 is the potential that varies with solution pH, but a key source of error is E5, the liquid junction potential generated at the point where the reference electrolyte must leak slowly from the reference electrode body.

made either by electrically monitoring the potential on the surface of pH-sensitive materials such as certain oxides or glasses (potentiometric sensor), or by optically monitoring the degree of protonation of a chemical indicator. The pH electrode has traditionally incorporated a thin glass membrane that encloses a reference solution (Eisenman, 1967) (Fig. 4), although of late some success in the use of pH-sensitive field effect transistors (pHFETs) has been reported and a solid-state pH electrode based on a pHFET is being marketed. Another new device is the light-addressable potentiometric sensor now produced by Molecular Devices, Inc., which has shown great promise for *in vitro* measurement for biosensing (Hafeman *et al.*, 1988). Electrical measurements require the use of a reference electrode that is often the source of problems and has a lifetime dependent on the volume of electrolyte that it contains. Consequently, miniaturization of pH electrodes for long-term *in vivo* use has been a difficult problem. Such sensors are nonconsuming, but surface fouling can influence the readings, and recalibration must be performed frequently.

Because of these drawbacks, there has been a recent shift toward optical indicator-based sensors using fiber optic detection (Saari and Seitz, 1982; Benaim *et al.*, 1986; Jordan *et al.*, 1987; Jones and Porter, 1988), and at least two such sensors are now in advanced clinical trials. In these sensors, a small amount of dye with a pH-dependent fluorescence spectrum is immobilized in a polymer or hydrogel at the end of an optical fiber. Exciting light is sent down the fiber and the fluorescence emitted is returned to a photodetector along the same fiber. The detector can discriminate between the reflected exciting light and the probe's emission because they are at different frequencies. Such a sensor requires no reference and can function until the dye leaches from the probe or is bleached by the exciting light. However, both electrical and optical measurements of pH are dependent on the temperature, so pH measurements are always performed in conjunction with a temperature measurement as close to the site of the pH sensor as possible. Also, no pH sensor is absolutely specific, so in an uncontrolled environment there is always the risk of interference. For example, the electrical sensors are subject to errors in the presence of biologically important metal ions, and fluorescence-based pH sensors are affected by fluorescence quenchers such as some inhalation general anesthetics.

Ions

Many simple ions such as K^+, Na^+, Cl^-, and Ca^{2+} are normally kept within a narrow range of concentrations, and the actual concentration must be monitored during critical care. Potentiometric sensors for ions (ion-selective electrodes or ISEs) operate similarly to pH electrodes; a membrane that is primarily semipermeable to one ionic species can be used to generate a voltage that obeys the Nernst (or more accurately the Nikolski) equation (Ammann, 1986):

$$E = \text{constant} + \frac{2.303\,RT}{zF} \log [a_i + k_{ij}(a_j)^{z/y}],$$

where E is the potential in response to an ion, i, of activity a_i, and charge z; k_{ij} is the selectivity coefficient; and j is any interfering ion of charge y and activity a_j.

Glasses exist that function as selective electrodes for many different monovalent and some divalent cations. Alternatively, a hydrophobic membrane can be made semipermeable if a hydrophobic molecule that selectively binds an ion (an ionophore) is dissolved in it. The selectivity of the membrane is determined by the structure of the ionophore. One can detect K^+, Mg^{2+}, Ca^{2+}, Cd^{2+}, Cu^{2+}, Ag^+, and NH_4^+ by using specific ionophores. Some ionophores are natural products, such as gramicidin, which is highly specific for K^+, whereas others such as crown ethers and cryptands are synthetic. S^{2-}, I^-, Br^-, Cl^-, CN^-, SCN^-, F^-, NO_3^-, ClO_4^-, and BF^- can be detected by using quaternary ammonium cationic surfactants as a lipid-soluble counterion. ISEs are generally sensitive in the 10^{-1} to $10^{-5}\ M$ range, but none is perfectly selective, so to unambiguously determine ionic concentrations it is necessary to use two or more ISEs with different selectivities. Also, ISEs require a reference electrode like that used in pH measurements. One can immobilize ionophore-containing membranes over planar potential-sensitive devices such as FETs, to create ion-sensitive FETs (ISFETs). As the potential does not depend on the area of the membrane, these work as well as larger bench-scale electrodes. An advantage of this approach is that a dense array of different ISFETs can be manufactured in a small area by using microfabrication techniques. Early probems with adhesion of the membranes to silicon have been largely solved by modifications of the design of the FETs themselves (Blackburn, 1987). The most typical membrane material used in ISEs is polyvinyl choride plasticized with dialkylsebacate or other hydrophobic chemicals. This membrane must be protected from fouling if an accurate measurement is to be made.

Electrochemically Active Molecules

If a chemical can be oxidized or reduced, there is a good chance that this process can occur at the surface of an electrode. Selectivity can be achieved because each compound has a unique potential below which it is not converted, so under favorable conditions a sweep of potential can allow identification and quantification of different species with a single elec-

TABLE 3 Electroactive Chemicals

Inorganic species
　Single electron transfers:
　Solvated metal ions such as Fe^{2+}/Fe^{3+}
　All M^0/M^{n+} pairs
　Many species undergo multielectron transfer reactions:
　The oxygen-water series $O_2/H_2O_2/H_2O/OH/H^+$
Organic species
　Most aromatics, generally nitrogen-containing aromatic heterocycles (reactions usually involve changes in number of atoms attached to molecule, and therefore are multistep, multielectron processes)
Metallo-organics
　Ferrocene/ferrocinium
Biochemical species
　Hemes, chlorophylls,
　quinones
　$NAD^+/NADH$　　　　(not affected by O_2)
　$NADP^+/NADPH$　　　(not affected by O_2)
　$FAD/FADH$
　$FMN/FMNH$

trode. This process is the basis for detection of a number of important biochemicals such as catecholamines (Table 3). Some species are determined directly at electrodes and others indirectly by interactions with mediator chemicals that are more easily detected as particular electrode surfaces.

Because detection involves conversion of one species to another, this is a consuming sensor, with all of the attendant problems. At least two electrodes are needed, and current must flow through the sample for a measurement to be made, although a precise reference electrode is not as necessary as in a potentiometric sensor. Near the electrode surface the concentration of either the oxidized or reduced species may differ greatly from the bulk concentration. This is partially because of depletion of the analyte near the surface (the diffusion layer), as well as attraction or repulsion of charged species from the charged electrode surface in the diffuse double layer (Fig. 5).

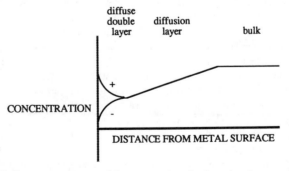

FIG. 5. A representation of the concentration of a charged analyte near the surface of an electrode (at left) at which it is being consumed. Note that there is a linear gradient of concentration in the unstirred diffusion layer and a further depletion or enrichment of the analyte when it is close enough to the electrode to sense the surface potential.

Since the current flow is the measured quantity and current is proportional to the number of molecules converted per unit time, the signal is controlled by mass transport, the electrostatics in the electric double layer, and specific chemical and physical interactions at the electrode surface. The great advantages of electrochemical detection are counterbalanced by its great sensitivity to surface fouling and any process that changes the resistance between the measuring electrodes. Also, there are interfering compounds present at high concentrations *in vivo* such as ascorbate that can swamp signals from more interesting but less concentrated analytes such as catecholamines. Tricks such as using selective membranes over the electrodes can solve some of these problems. For example, negatively charged Nafion allows passage of catecholamines but blocks access to the electrode of negatively charged ascorbate.

Blood Gases

Perhaps the most important physiological parameters after heart rate and blood pressure are the partial pressures of blood gases O_2 and CO_2 (pO_2 and pCO_2). It is also often useful to compare pressures of these gases in the arterial and venous circulation. The pulse oximeter, now manufactured by a number of vendors, allows noninvasive determination of the degree of saturation of hemoglobin in the arterial circulation, but does not give any information about the actual pO_2. It only works on the arterial circulation near the periphery and given no information about pCO_2. An invasive fiber optic probe developed by Abbott Critical Care Systems (the Oximetrix 3 SvO_2) allows measurement of oxygen saturation directly in veins by measuring the reflected light at three wavelengths (Schweiss, 1983). Since the affinity of hemoglobin for O_2 depends on the pH, which in turn depends on pCO_2, it is necessary in many cases to measure the actual pO_2, pCO_2, and pH simultaneously. Since all these sensors depend on temperature, a temperature probe is also required. The first successful fiber optic measurement of *iv vivo* pO_2 was reported by Peterson in 1984, and the principle employed has been used for most successful subsequent pO_2 sensors (Peterson *et al.*, 1984). The CDI System 1000 is a fiber optic sensor for pH, pCO_2, and pO_2, as shown in Fig. 6. It was the first complete blood gas sensor, and it combines the use of fiber optics with a smooth shape and size to avoid creating turbulence in the blood flow, and a covalent heparin coating to reduce thrombogenicity.

Two methods dominate for the measurement of pO_2. The first and most popular employs the amperometric Clark electrode, which consumes O_2 and generates H_2O_2 and OH^- as byproducts. The internal electrolyte in the sensor is separated from the external medium by a Teflon or silicone rubber membrane that readily passes O_2 but prevents both water and other electrochemically active species from passing. Severe fouling of the membrane can reduce the rate of O_2 diffusion and hence the response. The alternative optical approach relies on the efficient collisional quenching of most fluorophores by O_2. As the fluorescence intensity becomes inversely proportional to pO_2, the reduction in intensity can be used as the basis of a fiber optic sensor. Because this method relies entirely on the intensity of the fluorescent signal, it is very subject to drift and degradation from photobleaching and thus is not appropriate for long-term use.

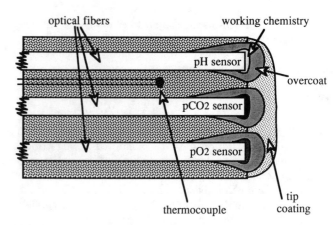

FIG. 6. A schematic drawing of probe of the Cardiovascular Devices System 1000 fiber optic blood gas sensor, based on three different combinations of selective membranes and fluorescent probes. Light enters and leaves from the left.

The same hydrophobic membranes that are permeable to O_2 are also permeable to CO_2, so they may be placed over pH electrodes or pH-sensitive optical probes containing bicarbonate buffer for selective determination of pCO_2 (Zhujun and Seitz, 1984; Gehrich, *et al.*, 1986).

Biosensors

Definition and Classification

The repertoire of chemicals that can be determined by the sensors mentioned here is relatively limited. To determine the presence or concentration of more complex biomolecules, viruses, bacteria, and parasites *in vivo*, it is necessary to borrow from nature (Fig. 7). Biosensors are sensors that use biological molecules, tissues, organisms, or principles. This definition is broad and by no means universally accepted, although it is more restrictive than the other common interpretation that would include all the sensors described in this chapter. The leading biological components of biosensors are summarized

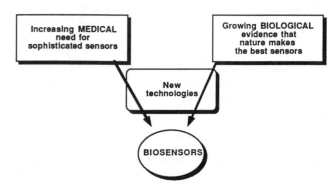

FIG. 7. The development of biosensors is driven by increased need for biochemical information in the medical community, the knowledge that nature senses these chemicals best, combined through emerging technologies to interface the biochemicals with physical transducers.

TABLE 4 Biological Components of Biosensors[a]

Binding	Catalysis
Antibodies	Enzymes
Nucleic acids	Organelles
Receptor proteins	Tissue slices
Small molecules	Whole organisms
Ionophores	

[a]The two categories are not mutually exclusive, because, for example, some enzymes may be employed for binding alone.

in Table 4. Enormous progress has been made in the development of biosensors in recent years, and this work has been recently and exhustively reviewed (Turner *et al.*., 1987). Most of the applications have been in the realm of analytical chemistry for use in chemical processing and fermentation, with the exception of the development of enzyme-based glucose sensors, on which we will focus.

Currently, biosensors are commercially available for glucose (used first in an automated clinical chemistry analyzer and based on glucose oxidase), lactate, alcohol, sucrose, galactose, uric acid, alpha amylase, choline, and L-lysine. All are amperometric sensors based on O_2 consumption or H_2O_2 production in conjunction with the turnover of an enzyme in the presence of substrate. A urea sensor is based on urease immobilized on a pH glass electrode (Turner, 1989). Most of these sensors are macroscopic and are employed in the controlled environment of a clnical chemistry analyzer, but the ExacTech device, manufactured by Baxter since 1987, is a complete glucose sensor containing disposable glucose oxidase-based electrodes, power supply, electronics, and readout in a housing the size of a ballpoint pen. One places a drop of blood on the disposable electrode and a few seconds later a fairly accurate reading of blood glucose is obtained. It is widely believed that much more frequent measurement of blood glucose with correspondingly frequent adjustments of the dose of insulin delivered could significantly improve the long-term prognosis for insulin-dependent diabetics. Increasing the frequency of the current sampling method (i.e, puncturing the finger for drops of blood) is not acceptable. Much progress has been made toward the goal of producing a glucose sensor that could be implanted for a period of time in the tissue or blood, but the problem is a formidable one that epitomizes the attempt to develop biosensors for *in vivo* use.

Background

The Utility of Biochemical Approaches to Sensing Some of the many advantages of using biochemicals for sensing are summarized in Table 5. The most important is that, despite the disadvantage of using chemically labile components in a sensor, they allow measurement of chemical species that cannot otherwise be sensed. Sensors have been fabricated that incorporate small biochemicals such as antibodies, enzymes and other proteins, ion channels, liposomes, whole bacteria and eukaryotic cells (both alive and dead), and even plant and animal tissue.

Immobilization One of the key engineering problems in biosensors is the immobilization of the biochemistry used to the transducing device. Approaches range from simply trapping an enzyme solution between a semipermeable membrane and a metal electrode, to covalently cross-linking several enzymes to a porous hydrogel coated on a pH electrode, to covalently cross-linking a complete monolayer of antibody to the surface of an optical fiber. The immobilization of a layer of material over the transducing device increases the response time, so for altered sensitivity and greatly enhanced selectivity, speed is often sacrificed. Monolayers do not contain much material, so to detect binding of so few molecules, it is generally necessary to employ some amplification scheme, such as attachment of an enzyme to an antibody that announces its presence by converting a subsequently added substrate to a large quantity of readily detected product. Such schemes add complexity and time to the detection. Immobilization also has unpredictable effects on the activity and stability of biochemicals.

TABLE 5 Use of Biochemicals for Chemical Detection

Advantages for binding
  "Uniquely" high selectivity
  Possibility of raising antibodies to nearly all antigens
  Antibodies and biotin-avidin system allow selective attachment of markers and reporters of binding
  High binding constants possible
  Several possible detection modalities
  Ion flux through gated channels can provide gain

Advantages for catalysis
  For every biochemical there is an enzyme that can be used to detect its presence
  High selectivity possible with some enzymes
  Several possible detection modalities
  Enzymatic cascades can provide gain
  Universality of redox coupling and pH effects permit common transduction schemes

Disadvantages of biosensors
  Biomolecules generally have poor thermal and chemical stability compared with inorganic materials
  The function of the biological component usually dictates that they must have narrow operating ranges in temperature, pH, ionic strength
  Susceptibility to enzymatic degradation is universal
  Need for bacteriostatic techniques in their fabrication
  Time-dependent degradation of performance is guaranteed with the use of proteins
  Production and purification can be difficult and costly
  Immobilization can reduce apparent activity of enzymes or kill them outright
  Most live organisms need care and feeding

Sensing Modalities

Potential-Based Sensors (pH and ISE) Some of the first biosensors employed enzyme-catalyzed reactions (such as those of penicillinase, urease, and even glucose oxidase) that affect pH. By putting a pH electrode into the solution containing the enzyme, it is possible to monitor the rate of enzymatic turnover. It is also possible to use pH to monitor the change in production of CO_2 by bacteria in the presence of substrates that they are capable of metabolizing (Simpson and Kobos, 1982). However, there is always a problem for *in vivo* use of pH-based sensors related to the fact that the external environment is capable of strong buffering of pH changes, and any change in pH in the immediate environment of the sensing surface is reduced toward the bulk pH by a degree that depends on the strength of that buffering.

Electrochemical Sensors Many enzymes perform oxidation and reduction reactions and can be coupled, if indirectly, to electrodes. The electrochemically active species in enzymes is generally a cofactor (Table 3) that, when bound, is not accessible to the electrode surface at which the electron transfer must take place for detection. In the case of the glucose oxidase reaction, the normal biological reaction is:

$$Glucose + O_2 + H_2O \Leftrightarrow Gluconic\ acid + H_2O_2.$$

The enzyme uses a flavin adenine dinucleotide (FAD) coenzyme to mediate the oxidation, and the resultant $FADH_2$ is directly oxidized by O_2 to return to FAD to prepare for the next catalytic reaction. Unlike nicotinamide adenine dinucleotide (NAD) and NAD phosphate (NADP), FAD is tightly bound to the enzyme, so normally only a small diffusible molecule like O_2 can gain access to it to alter its oxidation state. This means that under many circumstances, such as those present in tissue, the concentration of O_2 is rate limiting, so the sensor often measures not glucose but the rate at which O_2 is rate limiting, so the sensor often measures not glucose but the rate at which O_2 can arrive at the enzyme to reoxidize its cofactor. There are two electrochemical ways to couple the reaction to electrodes: monitoring depletion of O_2 by reducing what is left at an electrode, or monitoring buildup of H_2O_2 by oxidizing it to O_2 and protons. The latter approach is generally used to avoid direct effects of O_2 variation on the electrode, but this does not completely solve the problem. The electrode reaction for peroxide oxidation is as follows:

$$H_2O_2 \Rightarrow O_2 + 2H^+ + 2e^-.$$

The best solution to date to cast off the tyranny of the rate-limiting step of O_2 diffusion has been the use of electrochemical mediators that are at a higher concentration than O_2 and can therefore shuttle back and forth between the protein and the electrode faster than the enzyme is reduced, so that the arrival of the substrate such as glucose is always rate limiting. A typical chemical that works in this way is ferrocene, a sandwich of an iron cation between two cyclopentadienyl anions (Fig. 8). It exists in neutral and +1 oxidation states that are readily interconvertible at metal or carbon electrodes. A proprietary modified ferrocene is used in the aforementioned ExacTech

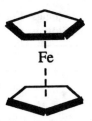

FIG. 8. The structure of the ferrocene-ferrocinium ion couple that allows one to overcome the dependence of the glucose oxidase reaction on pO_2. The two five-membered rings are cyclopentadienyl anions and the iron may be in either the Fe^{2+} or Fe^{3+} states, giving a total charge of 0 or +1.

carbon electrode-based glucose sensor. Other glucose oxidase-based electrodes have been employed on catheters for *in vivo* determination of blood glucose, with varying degrees of success (Gough *et al.*, 1986). Thrombus formation is generally a problem, as is the possible alteration in localized glucose levels in tissue traumatized by insertion of probes, no matter how small. It may well be that use of the techniques employed in keeping pressure catheters clear will also work with biosensors such as this.

Optical Waveguide Sensors Fiber optics can be used either as thin flexible pipes to transport light to and from a sensor at a remote site, or in a way that takes advantage of the unique properties of optical waveguides. The former mode still dominates, and the CDI blood gas sensor uses three fibers just to move photons to and from the small volumes of immobilized chemistry at the probe end. The Schultz fiber optic glucose sensors involve a more sophisticated use of the light path exiting the optical fiber combined with clever use of lectin biochemistry. In principle, these sensors allow continuous measurement of blood glucose. There are at least two features specific to waveguides that have been used for sensors for *in vitro* measurement that may soon find themselves ready for *in vivo* use as well. In one, the ability of light sent down two fibers to interfere with itself on return to the source allows sensitive measurement of changes in the length or phase velocity of the fibers. This, in turn can be altered by enzymatically induced changes in the temperature of the fiber or its cladding in the volume surrounding the fiber. Another approach is to use the light in the evanscent wave that exists in the region just outside the waveguide to probe a small volume adjacent to the surface. If binding species such as antibodies are immobilized on the surface, it is possible to selectively excite and collect fluorescence from the surface layer even in the presence of high concentrations of fluorophore or other absorbers in the bulk solution. This technique has allowed the use of antibody-based detection of analytes such as theophylline in whole blood in a sensor designed by the ORD Corporation. These sensors are primarily for single use, and one fiber is used for each measurement. Nonspecific adsorption to the fiber surface, which is a serious interference in such sensors, can be reduced by using surface passivating films of proteins such as bovine serum album.

Acoustic/Mechanical Sensors The binding of a material to a surface changes its mass, which can change either the object's resonant frequency or the velocity of vibrations propagated through it. This has allowed development of sensors called surface acoustic wave (SAW) or bulk acoustic wave (BAW) detectors that are based on oscillating crystals. Sensitive detection of analytes is relatively easy in the gas phase, and while there have been reports of selective detection of analytes using immobilized antibodies, there is still controversy as to how or if the technique works when the oscillating detector is in contact with liquid. It is, however, unlikely that this technique will prove applicable to *in vivo* use, where some nonspecific adsorption of protein is almost unavoidable.

Thermal and Phase Transition Sensors Chemical reactions can give up heat because they involve breaking and formation of chemical bonds, each of which has a characteristic enthalpy. There is also a strong effect of the heats of solution of the substrates and products, particularly charged species. Many enzymatic reactions release 25 to 100 kJ/mol, or 5 to 25 kcal/mol (Table 6). A 1-mM solution of substrate completely enzymatically converted to product with a 5 kcal/mol heat of reaction would increase in temperature by 0.005°C, which is readily measureable in laboratory conditions. Sensors based on this principle are in use as detectors in chromatography, and in principle could be applied to almost any enzymatic reaction. Some reactions have little or no heat production (e.g., ester hydrolysis, such as the acetylcholinesterase reaction) but can be observed using "tricks" such as coupling the reaction to the heat or protonation of a buffer such as Tris:

Acetylcholine $\Rightarrow H_3CCO_2H +$ choline $\quad \Delta H \approx 0$ kJ/mol
$H_3CCO_2H + Tris^- \Rightarrow H_3CCO_2^- + TrisH \quad \Delta H = -47$ kJ/mol.

Alternatively, a sequence of enzymes such as glucose oxidase followed by catalase can be used, which converts the hydrogen peroxide produced by the oxidase to O_2 and water in another exothermic reaction (Danielsson and Mosbach, 1987). However, the technical difficulties in making such measurements in the thermally noisy environment of the human body have so far prevented application of this principle to development of *in vivo* sensors.

An alternative thermal approach is to use the depression in phase transition temperatures of pure compounds by dissolving

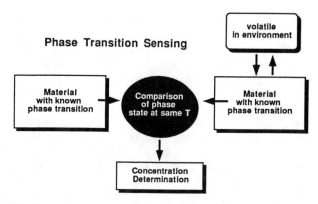

Phase Transition Sensing

FIG. 9. A schematic diagram of the process of phase transition sensing, which is an application of the well-known purity dependence of phase transition temperatures. Some diagnostic technique must be applied to allow a quantitative comparison of the phase states of two samples of the material, one of which is in equilibrium with small molecules in the environment, and another of which is at the same temperature, but chemically isolated.

in them dissimilar small molecules that prefer the fluid phase over the crystalline phase. If the temperature is known, the concentration of the small molecule can be determined by the extent of the freezing point depression (Fig. 9). This principle has been successfully used to detect general anesthetics and is now being used to develop a fluorescence-based fiber optic probe for *in vivo* use (Merlo, 1989; Merlo *et al.*, 1990).

Biomembrane-Based Sensors A very complex biological system that has been applied to the development of sensors is the biological membrane and the lipids and proteins that make it up. Numerous approaches have been made to use lipid bilayers to detect chemicals including at least two sensors for general anesthetics (Wolfbeis and Posch, 1985; Merlo, 1989; Merlo *et al.*, 1990). Membrane receptor proteins are responsible for transducing many important biological binding events and could be used to great advantage for monitoring such chemicals as hormones, neurotransmitters, and neuroactive drugs. Several schemes have been tried to this end, including immobilizing ligand-gated ion channel receptor proteins in fiber optic devices, measuring the binding of fluorescently labeled ligands, and reconstituting them into defined lipid monolayers on solid electrodes (Eldefrawi *et al.*, 1988) and across holes as bilayers (Ligler *et al.*, 1988). These electrical techniques promise the most sensitive detection, because a single channel opening can be monitored electrically, but they also involve some of the most difficult technical challenges, including stabilizing of the normally fragile lipid bilayer. An ancillary benefit of the use of biomembranes is that phospholipids have been reported to enhance the biocompatibility of biomaterials, so they may have a dual role in bilayer-based sensors.

THE FUTURE

In the near future, several factors can be expected to influence the development of sensors. Once it has been demon-

TABLE 6 Heat of Enzymatic Reactions

Enzyme	Substrate	$-\Delta H$, (kJ/mol)
Catalase	H_2O_2	100
Cholesterol oxidase	Cholesterol	53
Glucose oxidase	Glucose	80
Hexokinase	Glucose	28
Lactate dehydrogenase	Na-pyruvate	62
NADH-dehydrogenase	NADH	225
Penicillinase	Penicillin-G	67
Urease	Urea	61

strated that a given type of sensor has practical value, the process of making it cheaply and reproducibly becomes supremely important in determining whether it sees the marketplace. Automation of the fabrication of small *in vivo* probes will be a high priority in the next few years. There will probably be a great increase in chemical sensors that are manufactured from the start with silicon microfabrication in mind, rather than the current practice, which is generally scaling beaker chemistry down to the size of microchips. The problem of ensuring that a sensor is as biocompatible as possible, while maintaining its function, will continue to the be most pressing problem for *in vivo* use for some time. Advances must continue to be made in materials, probe shape and size, site of use, and manufacturing. There will continue to be a strong emphasis on development of noninvasive techniques that will avoid the difficulties of biocompatibility. It may well be that near-infrared spectroscopy and magnetic resonance imaging (MRI) techniques will be able to provide sufficient chemical information to diagnose some disease states. The biomolecules employed in biosensors have so far been restricted to natural enzymes and antibodies, but there is every reason to expect that as it becomes more common to tailor molecules for particular jobs, we will be able to improve on nature for transduction of chemical events. Regardless of the nature of the sensor, it is clear that the continuing reduction in the size and cost of computers will be reflected in increasing use of sensors to provide crucial input in "smart" devices, be they for the control of drug delivery or of prosthetic limbs.

SUMMARY

Physical and chemical sensors are already important in the diagnosis and treatment of the critically and chronically ill. New physical, chemical, biochemical, and biological sensing technologies are under development that could greatly augment our current *in vivo* capabilities. Biocompatibility remains the most important problem for all such sensors, particularly for biosensors and other chemical sensors in which transport of material is vital to function. The economic and human impact of improving the lot of diabetics is such that success of such a sensor will open the way for many other *in vivo* biosensors.

Acknowledgments

I would like to thank the faculty and students of the biomaterials group of the Center for Bioengineering for making me acquainted with the state of the art in solutions to the biocompatibility problem.

Bibliography

Ammann, D. (1986). *Ion Selective Microelectrodes: Principles, Design and Application*. Springer-Verlag, Berlin.

Benaim, N., Grattan, K. T. V. *et al.* (1986). Simple fibre optic pH sensor for use in liquid titrations. *Analyst* 111: 1095–1097.

Blackburn, G. F. (1987). Chemically sensitive field effect transistors. *Biosensors*. Oxford, Oxford Science Publ., pp. 481–530.

Collison, M. E., and Meyerhoff, M. E. (1990). Chemical sensors for bedside monitoring of critically ill patients. *Anal. Chem.* 62(7): 425A–437A.

Danielsson, B., and Mosbach, K. (1987). Theory and application of calorimetric sensors. *Biosensors*. Oxford, Oxford Science Publ., pp. 575–595.

Eisenman, G. (1967). *Glass electrodes for Hydrogen and Other Cations*. Marcel Dekker, New York.

Eldefrawi, M. E., Sherby, S. M. *et al.* (1988). Acetylcholine receptor-based biosensor. *Anal. Lett.* 21(9): 1665–1680.

Gehrich, J. L., Lubbers, D. W. *et al.* (1986). Optical fluorescence and its application to an intravascular blood gas monitoring system. *IEEE Trans. Biomed. Eng.* 33(2): 117–131.

Gough, D. A., Armour, J. C. *et al.* (1986). Short-term *in vivo* operation of a glucose sensor. *Trans. Am. Soc. Artif. Intern. Organs* 32: 148–150.

Hafeman, D. G., Parce, J. W. *et al.* (1988). Light-addressable potentiometric sensor for biochemical systems. *Science* 240: 1182–1185.

Jones, T. P., and Porter, M. D. (1988). Optical pH sensor based on the chemical modification of a porous polymer film. *Anal. Chem.* 60(5): 404.

Jordan, D. M., Walt, D. R. *et al.* (1987). Physiological pH fiber-optic chemical sensor based on energy transfer. *Anal. Chem.* 59(3): 437.

Ligler, F. S., Fare, T. L., *et al.* (1988). Fabrication of key components of a receptor-based biosensor. *Med. Instru.* 22(5): 247–256.

Merlo, S. (1989). "Development of a fluorescence-based fiber optic sensor for detection of general anesthetics." Master's Thesis, University of Washington, Seattle.

Merlo, S., Yager, P., *et al.* (1990). An optical method for detecting anesthetics and other lipid-soluble compounds. *Sensors and Actuators* A21–A23: 1150–1154.

Peterson, J. I., Fitzgerald, R. V. *et al.* (1985). Fiber-optic probe for *in vivo* measurement of oxygen partial pressure. *Anal. Chem.* 56: 62–67.

Rolfe, P. (1990). *In-vivo* chemical sensors for intensive-care monitoring. *Med. Biol. Eng. Comput.* 28(3): B34–B46.

Saari, L. A., and Seitz, W. R. (1982). pH sensor based on immobilized fluoresceinamine. *Anal. Chem.* 54(4): 821–823.

Schweiss, J. F. (ed.) (1983). *Continuous Measurement of Blood Oxygen Saturation in the High Risk Patient*. Abbot Critical Care Systems.

Simpson, D. L., and Kobos, R. K. (1982). Microbiological assay of tetracycline with a potentiometric CO_2 gas sensor. *Anal. Lett.* 15(B16): 1345–1359.

Turner, A. P. F. (1989). Current trends in biosensor research and development. *Sensors Actuators* 17: 433–450.

Turner, A. P. F., Karube, I. *et al.* (1987). *Biosensors: Fundamentals and Applications*. Oxford, Oxford University Press.

Wang, J., and Varughese, D. (1990). "Polishable and robust biological electrode surfaces." *Anal. Chem.* 62: 318–320.

Wolfbeis, O. S., and H. E. Posch (1985). Fiber optical fluorosensor for determination of halothane and/or oxygen. *Anal. Chem.* 57: 2556–2561.

Zhujun, Z., and Seitz, W. R. (1984). A carbon dioxide sensor based on fluorescence. *Anal. Chim. Acta* 160: 305–309.

8

Artificial Organs

Paul S. Malchesky, Kevin D. Murray, Don B. Olsen, and Frederick J. Schoen

8.1 INTRODUCTION
Frederick J. Schoen

Millions of people have benefited from prosthetic devices having structural, cosmetic, or simple passive mechanical functions. Artificial organs comprise complex medical devices that have active mechanical or biochemical functions such as heart, lung, kidney, liver, pancreas, or neurosensory organs. Artificial organs can be either surgically implanted or extracorporeal (in which blood is temporarily processed outside the patient's body). Although the range of devices that constitute artificial organs is at present limited in clinical use, considerable research and development has involved devices that have active mechanical, biologic, or mass exchange functions.

Considerable current research and development has the goal of developing implantable medical devices that use living cells (normal or genetically manipulated) together with extracellular components (either natural or synthetic). In this logical extension of biomaterials science, often called *tissue engineering,* the cells are either transplanted or induced in the recipient by implantation of the appropriate resorbable or permanent substrate (Galleti *et al.*, 1995; Langer, 1993). Representative examples of tissue engineering approaches include (1) repair of skeletal defects by cartilage regeneration using autologous chondrocyte transplantation (Paige and Vacanti, 1995), and (2) the biohybrid artificial organ, in which a device is in part synthetic, yet also has animal cells that elaborate specific substances; the cells are separated (and protected) from the recipient's blood by a semipermeable membrane (Colton, 1995). For example, an artificial pancreas using insulin-producing islet cells in this capacity has sustained diabetic dogs for more than 6 months (Sullivan, 1991). Anticipated generic problem areas include: (1) development of new cell lines and biomaterials; (2) evaluation of the optimal implant configuration; and (3) the reproducible manufacture of bioartificial devices and their preservation until use.

This section contains two chapters that describe some considerations related to devices having mechanical and chemical function, respectively, including implanted artificial organs, particularly cardiac assist devices and the total artificial heart, and extracorporeal systems, such as hemodialysis devices and extracorporeal oxygenators.

Bibliography

Colton, C. K. (1995). Implantable biohybrid artificial organs. *Cell Transplantation* 4: 415–436.
Galletti, P. M., Hellman, K. B., and Nerem, R. M. (1995). Tissue engineering: From basic science to products. *Tissue Eng.* 1: 147–161.
Langer, R., and Vacanti, P. (1993). Tissue engineering. *Science* 260: 920–925.
Paige, K. T., and Vacanti, C. A. (1995). Engineering new tissue: Formation of neo-cartilage. *Tissue Eng.* 1: 97–106.
Sullivan, S. J., Maki, T., Borland, K. M., Mahoney, M. D., Solomon, B. A., Muller, T. E., Monaco, A. P., and Chick, W. L. (1991). Biohybrid artificial pancreas: Long-term implantation studies in diabetic, pancreatomized dogs. *Science* 252: 718–721.

8.2 IMPLANTABLE PNEUMATIC ARTIFICIAL HEARTS
Kevin D. Murray and Don B. Olsen

CARDIAC SUPPORT AND REPLACEMENT

The first successful artificial heart was designed, fabricated, and eventually implanted into an animal by Kolff at the Cleveland Clinic in 1957 (Akutsu and Kolff, 1957). This initial success sparked a worldwide interest in the possibility of producing a successful replacement for the failing heart. The goal of this early laboratory effort was the development of a permanent replacement for the native heart. However, parallel with these experimental endeavors, routine clinical cardiac surgery was becoming a reality. The introduction of cardiopulmonary bypass (CPB) allowed successful repair of certain congenital cardiac defects and, much later, coronary artery bypass grafting, valve repair and replacement, orthotopic heart transplantation, and, most recently, artificial heart implants.

The advent of successful heart transplantation has provided cardiac replacement for nearly 2400 patients annually in the United States. However, this success falls far short of the minimum annual estimate of 30,000 Americans who need permanent replacement of their irreparable failed hearts. This mismatch of donor hearts and potential recipients provides an undisputed need for another type of long-term myocardial replacement. Congestive heart failure is the only class of heart disease that has continued to increase in incidence (O'Connell and Bristow, 1994). Total artificial hearts (TAHs) and ventricular assist devices (VADs) offer the greatest potential to fill this clinical need by providing permanent cardiac assist or replacement. Interestingly, successful, routine cardiac operative procedures have, on the one hand, limited the population requiring permanent replacement of the native heart, while simultaneously adding a new group of patients who need temporary cardiovascular support, either following an unsuccessful cardiac surgical procedure or to extend the patient's survival while awaiting a donor heart. An extensive review of the clinically used TAHs and VADs was published in 1993 by Rowles *et al.*

The first clinical use of a TAH was by Cooley in two patients who exhibited myocardial failure and were unweanable after cardiac surgical procedures (Cooley *et al.*, 1981). The goal of these initial TAH implantations was temporary use with eventual orthotopic heart transplantation. Both experiences demonstrated the feasibility of this clinical application of the TAH, despite poor results regarding the patient's ultimate survival.

This pneumatic artificial heart technology was developed in Dr. Kolff's laboratory at the University of Utah. There were numerous designs of pneumatic-powered TAHs. Most notable was the Kwan-Gett heart and the inherent increase in cardiac output in response to an increase in venous return (preload) (Kwan-Gett *et al.*, 1969). Great strides in available materials, design, methods of fabrication, implantation techniques, and noninvasive monitoring of the TAH occurred during the 1970s, culminating in FDA approval in 1981 for human implantation of the Jarvik-7 (J-7; 100-ml stroke volume) artificial heart.

The next clinical use of a TAH was a continuation of the FDA-approved trial of TAH implantation as a permanent replacement in patients with both end-stage heart failure and a contraindication to heart transplantation. The Utah team implanted a permanent TAH in patients beginning with Dr. Barney Clark in 1982 at the University of Utah (DeVries *et al.*, 1984). Similar to Cooley's experience, the procedure was successful in demonstrating that the TAH could replace the native heart with restoration of normal cardiovascular function. The University of Utah's TAH technology was subsequently licensed to Symbion, Inc., for worldwide marketing. The initial IDE was for the continuation of the IDE awarded to the University of Utah for nontransplant candidates only. A significant number of complications developed in all the recipients, which limited survival (ranging from 10 days to 619 days) and resulted in their deaths (Table 1) (Olsen, 1996). This initial clinical experience exposed problems with a pneumati-

cally powered TAH as a permanent implantable device, although recognized improvements in TAH design, materials, and postoperative patient care offered hope for eliminating these complications.

More recently, the TAH has been used, with increasing success, for temporary cardiac support in the role of a bridge to orthotopic heart transplantation. Despite this clinical success, the one limitation of the TAH, when used as a stopgap measure, is the need for removal of the native heart, necessitating either permanent support with the mechanical heart or the subsequent performance of transplantation. This shortcoming prompted the development, beginning in the 1970s, of VADs. The goal of VAD development was (as the name implies) the augmentation of ventricular performance rather than replacement of the native heart. A significant difference of the VAD system in comparison to the TAH was the potential to remove the VAD system if native myocardial function recovered adequately enough to sustain normal hemodynamics. The advantage of retaining the recipient's heart also contributed to the VAD's major limitation—fit. Since it was not replacing the native heart, there was no intrathoracic anatomic site available where it could be implanted and not interfere with adjacent organs and their function. Several approaches to this problem have included extracorporeal placement, intravascular placement, insertion between muscle layers of the abdominal wall, and overall downsizing of the VADs in an effort to minimize interference with adjacent organs when placed intracorporeally.

Implantable blood pumps, whether TAHs or VADs, have evolved significantly over the past 2 decades. Improvements in design, materials, fabrication, control, monitoring, and patient selection have all contributed to their clinical success. However, further refinements of currently available devices and the development of improved blood pump concepts are necessary in order to produce the perfect device. There are four questions that beg answers when discussing the future of implantable TAHs and VADs: (1) Are there patients who would benefit from implantable blood pumps? (2) Can implantable, reliable blood pumps restore normal cardiovascular hemodynamics? (3) Can highly reliable implantable blood pumps be designed which cause minimal (or no) complications for the recipient? And finally, (4) can society afford the cost (i.e., development and clinical use) of these devices? The first three questions can be answered "yes," but the answer to the fourth, and potentially the most challenging, is undecided at this time. One report strongly suggests that society can afford the technology when all the factors are considered (Poirier, 1991). It would be regrettable, but the future of the implantable permanent TAH and VAD may very well be determined by financial considerations rather than patient need or laboratory development.

TOTAL ARTIFICIAL HEART

The Natural Heart

The initial designs of total artificial hearts focused on mimicking the natural heart. The heart in adult humans weighs

TABLE 1   All Patients Receiving a Total Artificial Heart through 1985[a]

Date	Surgeon	Patient	Heart	Days on TAH	Transplant/survival
4/4/69	Cooley	HK	Liotta[b]	2	Yes/1 day
7/23/81	Cooley	WM	Akutsu[b]	2	Yes/8 days
12/2/84	DeVries	BC	Jarvik	111	No
11/25/84	DeVries	WS	Jarvik	619	No
2/17/85	DeVries	MH	Jarvik	477	No
3/6/85	Copeland	TC	Phoenix[b]	1	Yes/2 days
4/7/85	Semb	LS	Jarvik	227	No
4/14/85	DeVries	JB	Jarvik	10	No
8/29/85	Copeland	MD	Jarvik	9	Yes/3.1 years
10/18/85	Pierce	AM	Penn State	11	Yes/21 days
10/24/85	Griffiths	TG	Jarvik	4	Yes/10.5 years
12/18/85	Joyce	MI	Jarvik	45	Yes/194 days
16 years	7 surgeons	12	5 types	1518 days	5 no, 7 yes/14.2 years

[a]It should be noted that not one patient died from device failure.
[b]These devices were not approved by the Food and Drug Administration.

approximately 300–350 g and occupies approximately 300 cm³ of the thoracic cavity. Anatomically, there are four cardiac chambers: two atria and two ventricles are separated by valves and the heart is divided into right and left halves. The right ventricle pumps blood through the pulmonary valve into the low-resistance pulmonary circulation. The right ventricle is relatively thin walled, shares a common intraventricular septum with the left ventricle, and develops systolic pressures of approximately 17 to 30 mm Hg. The left ventricle ejects blood through the aortic valve into the higher resistance systemic circulation. The left ventricle is a thick-walled, conical muscle that produces systolic pressures of approximately 85 to 140 mm Hg. The flows (cardiac output) of the two ventricles are nearly equal, with the left ventricle pumping slightly more because of the left-to-left shunt of the bronchial circulation and perhaps other abnormal shunting. The atria provide a reservoir for ventricular filling, which function is further augmented by a synchronized contraction of the atrial muscle. The two atria are similar in anatomic appearance; however, the mean left atrial pressure exceeds the mean right atrial pressure by approximately 10 mm Hg. The right atrium receives venous blood from the superior and inferior venae cavae while the left atrium receives arterial (oxygenated) blood from the pulmonary veins (Guyton, 1981).

Electrical activity of the natural heart provides synchronized and efficient contraction of the atria and ventricles. The rate of electrical stimulation also plays a major role in the regulation of the volume of blood pumped by the heart. The volume of blood pumped in 1 min by the human heart is determined by the rate of ventricular contractions (i.e., heart rate) and the volume of blood pumped with each beat (i.e., stroke volume). The regulation of these two factors, heart rate and stroke volume, is a complex interaction of neurohumoral factors,

blood volume, and myocardial performance and physiologic demand for tissue perfusion. Under resting conditions, the human heart pumps approximately 2.5 liters/min/m² (body surface area) of blood, and with alteration of any of the flow-mediating factors listed above, the volume of blood ejected can instanteously change to several times its resting capacity. As a hydraulic pump, the ventricles require 3 W of energy and they operate at an efficiency of 10 to 20%.

Design

The TAH has taken many forms during its development. However, despite these external differences in shape and materials, all TAHs must fulfill specific criteria necessary for successful application in man. These areas of design are: (1) adequate volume of blood pumping necessary to meet the physiologic needs of the recipient, (2) proper anatomic alignment in relation to recipient structures that transport blood entering and exiting the TAH, (3) lack of interference with other organs and maintenance of the ability to approximate the chest wall structures, and (4) avoidance of any complications caused directly or indirectly by the TAH.

The TAH must provide a cardiac index (CI) of 2.5 liters/min/m² for recipients. This baseline blood flow requirement fulfills the physiologic demands of the recipient at rest. Additionally, the TAH must have the capability to increase its flow when physiologic demands necessitate such a change. A peak CI of 3.5 liters/min/m² or greater is a conservative minimum goal for a TAH that is designed for permanent implantation.

The initial Utah TAH provided a maximum stroke volume (SV) of 100 ml, a value that easily satisfied the flow demands for the vast majority of potential recipients. This 100-ml-SV device was successfully implanted in 70- to 100-kg animals, with no evidence of flow limitations. The transfer of this device to clinical use verified the adequacy of flow from the TAH to meet human physiologic needs. However, the external dimensions of this 100-ml-SV TAH was incompatible with the thoracic dimensions available in potential recipients who weighed less than 100 kg. This conflict of flow capabilities and external dimensions required a downsizing of the Utah TAH as it was made available to an increased number of patients.

The maximum SV was decreased to 70 ml, with proportionate changes in the Utah TAH's external dimensions. The change in SV, however, required an increased heart rate in order to provide the CI neeeded to satisfy the metabolic demands of the recipient, both at rest and during times of increased flow demands. Too severe a reduction in SV would undoubtedly provide an improved anatomic fit in patients with a smaller than normal chest, but the necessary increase in heart rate needed to maintain CI becomes excessive. When the heart rate (actually, the TAH pumping rate) of the TAH is increased excessively, two problems develop: inadequate time for filling and increased wear on the TAH and its various components. The TAH has passive filling from the recipient atria. There are no coordinated atrial contractions to augment venous filling during TAH ventricular diastole unless the TAH is driven from the electrical P wave of the ECG of the atrium (Iwaya et al., 1979). The use of an external vacuum system to collapse the TAH's blood pumping diaphragm can augment ventricular filling; however, it has a limited effect based upon the reservoir capacity of the atria, the size of the TAH and inflow valve, and the duration of diastole. If the TAH heart rate becomes excessive, a value that is variable for each recipient but is generally greater than 120 beats per minute, the TAH becomes ineffective for pumping adequate volumes of blood, either at rest or during moments of increased flow demands.

The 70-ml Utah TAH was found to provide satisfactory blood flow, both at rest and during increased physiologic demand, and also maintained external dimensions that permitted implantation in human recipients of more than 50 kg. This balance of size and function was critical for expanding the pool of potential TAH recipients.

The TAH must have unimpeded flow entering and exiting the artificial ventricles. The Utah TAH's ventricular design provides an unimpeded transition between the recipient's atria and great vessels. Severe angulation or other abrupt directional changes in flow cause several problems which include impaired ventricular filling, excessive restriction to outflow, hemolysis, and abnormal flow patterns which promote thrombus formation.

The Utah TAH has direct venous flow via polyurethane-covered atrial cuffs sewn to the native atria and connected to the TAH by a quick-connect snap-on system. The outflow pathways of the ventricles provide a smooth flow pattern. The Dacron graft connecting the TAH to the great vessels requires some angulation for the pulmonary artery to override the aorta. This configuration has not caused significant problems, particularly when the graft is contoured to the anatomic placement of the pulmonary artery and the outflow connector of the right ventricle.

In addition to this near-laminar flow alignment of the recipient–TAH interface, the ventricles must avoid compression of inflow and outflow vasculature. TAH ventricles that are too large or improperly designed can impede venous inflow. This mechanical interference of TAH diastolic filling results in inadequate CI for the recipient that is uncorrectable. Outflow restriction caused by TAH impingement of the great vessels necessitates increased force of ventricular ejection in an effort to maintain a satisfactory CI. The use of higher compressed-air pressures during systole imparts increased mechanical stress on the TAH components, as well as increasing blood velocity resulting in turbulence, hemolysis, and possibly thrombosis.

The pericardial space occupied by the recipient's native heart is ideally the dimension duplicated by the TAH. However, this simple concept must not be interpreted to mean the TAH must mimic the shape and function of the native heart. This design mistake was the failing for many early TAHs. More importantly, the TAH must meet the design characteristics previously described, in addition to unobtrusively occupying the mediastinum (i.e., pericardial space) based upon human fit trials (Kolff et al., 1984). The hemispherical shape of the Utah TAH's ventricles fits the mediastinal space in properly selected individuals. Excessive lateral dimensions impinge on the lungs, resulting in atelectasis, retention of secretions, and the increased risk of infectious pneumonias.

The Utah TAH was initially designated for permanent use as a heart substitute; however, more recently its success has been as a bridge to transplantation in patients with end-stage heart failure and the unavailability of a donor heart. Either clinical situation requires closure of the sternum and superficial soft tissues. Complete implantation of the TAH increases patient mobility and comfort and provides a natural barrier to infection. However, reapproximation of the sternum necessitates that the TAH fit the anterior–posterior dimensions of the mediastinum, a space that is uncompromising because of the bony limitations of the spine and anterior chest wall. An excessively large TAH or inappropriately designed TAH ventricular configuration can cause vascular compromise (as noted previously) or impinge on other organs such as the esophagus, trachea, lungs, or diaphragm.

These considerations were paramount in the downsizing of the Utah TAH to a 70-ml-SV ventricle. This size reduction permitted proper alignment of the TAH in smaller recipients without impinging upon surrounding structures. The TAH design and the materials used to fabricate the device must minimize the rate of complications, while simultaneously providing durability to device function. Significant complications include thrombosis formation, component failure, infection, and hemorrhage.

The formation of thrombus within the TAH has two prominent effects: (1) dislodgment with embolization (particularly to the central nervous system) and (2) interference with TAH function (i.e., obstruction of blood flow or TAH and valve malfunction). Thrombus within a TAH is primarily formed by a complex interaction of blood coagulability, blood flow pattern abnormalities, and blood contact with artificial surfaces.

The use of unlimited anticoagulation agents to paralyze the

recipient's coagulation cascade is an unreasonable approach to this problem. The TAH recipient receives either heparin or coumadin in doses that prolong the time for clot formation, while simultaneously avoiding the complication of clinically significant hemorrhage. In addition to interfering with the protein-mediated clotting mechanisms, drugs are administered which inhibit platelet adherence, aggregation, and release of thrombus-promoting chemicals. This use of anticoagulation and antiplatelet drugs is unable to independently prevent TAH thrombus; however, they are a necessary component of TAH patient care. Without these drugs, there would be uncontrolled thrombosis and failure of the device, resulting in the death of the recipient.

The blood-contacting surfaces of the Utah TAH are fabricated from a segmented polyurethane. This material, when made into an ultrasmooth, blood-contacting surface, is resistant to thrombus formation. Microscopic defects in the polyurethane surfaces can allow the development and attachment of thrombus, the deposition of calcium, or both. This was an early problem for the Utah TAH; however, modification of the fabrication process has significantly minimized this problem. The polyurethane is hand-poured over molds in order to create the TAH ventricular housing and blood-contacting surface which includes the multiple-layered diaphragm. A clean room environment for fabrication avoids contamination of the polyurethane by extraneous material. Proper ratios of liquid polyurethane and solvent allow curing of the blood-contacting polyurethane without surface pitting caused by bubble formation.

Description of the Total Artificial Heart

The original J-7 TAH had two pneumatically powered ventricles, each with a 100-ml stroke volume. The ventricular components included an injection-molded (Isoplast-301) circular base; a housing made by solution casting multiple layers of Biomer, a segmented polyurethane; and a flexing, four-layer, blood-contacting diaphragm also made of Biomer. Two Medtronic Hall (Medtronic, Inc., Minneapolis, MN) tilting-disk valves per ventricle (one inflow and one outflow) were mounted in Isoplast-301 rings. These valve-holding rings formed part of the quick-connector snap-on system linking the TAH to the atria and great vessels. Air tubes (2 m long) connected each ventricle percutaneously, via infection-resistant skin buttons to the external drive system. The UtahDrive console was a patented device that regulated the delivery of compressed air to the TAH as well as allowing adjustment of the heart rate, percentage of time in systole for each cardiac cycle, and diastolic vacuum. The cardiac output monitor and diagnostic unit displayed and recorded each ventricle's stroke volume by measuring the diastolic air exhaust (Nielsen *et al.*, 1983).

Symbion developed and manufactured a scaled down J-7 TAH (100-ml stroke volume) with a 70-ml stroke volume (J-7-70). The smaller J-7-70 TAH was created in response to the need for a TAH that fit the thoracic dimensions of both averge-sized and petite patients without compromising venous return or interfering with adjacent organs. The J-7-70 became the TAH of choice based upon its clinical success restoring

cardiovascular homeostasis, combined with its uncompromising fit in patients with widely differing body sizes.

The TAH anticoagulation regimen varied, but most commonly consisted either of heparin or of coumadin combined with an antiplatelet agent(s), usually aspirin and dipyridamole. Infections in TAH recipients were dramatically decreased by careful screening (and exclusion) of candidates for established infections and the prophylactic use of broad-spectrum antibiotics. Perhaps contributing the most to improving the Symbion TAH success was the trend toward more thoughtful patient selection. The early clinical experience included a significant number of temporary TAH recipients with preoperative, irreversible organ dysfunction, multiple comorbid factors (in addition to their end-stage heart disease), and, occasionally, near-moribund conditions. These mortally ill patients were not salvageable despite a successful TAH implantation and postoperative function. An elimination of these heroic (and all too often futile) TAH implantations dramatically improved the rate of TAH patients successfully transplanted with eventual recovery and long-term survival.

There are limitations associated with the TAH. The four initial J-7 TAH implants in the United States were performed by DeVries, who used the device as a permanent replacement of the patient's heart. Despite survivals as long as 620 days, all patients developed serious (and eventually fatal) complications. Postoperative hemorrhage, infection, and thromboembolic events were the most prominent sources of morbidity and mortality.

The TAH as a temporary cardiac replacement had the potential for similar complications, as had the permanent application of the artificial heart. Early use of the Symbion TAH as a bridge to transplant was plagued by numerous infections, as reported by the University of Pittsburgh group and others. However, better patient selection produced a reduction in the incidence of infectious complications. Similarly, improved anticoagulation regimens all but eliminated thromboembolic events resulting in permanent neurologic deficits after 1988 in Symbion's bridge-to-transplant patients. Increasing experience with surgical implantation of the TAH also decreased the incidence of serious perioperative hemorrhagic complications.

The original J-7 TAH design permitted implantation only in patients with large thoracic dimensions. Improper matching of the J-7 TAH to patient size resulted in numerous problems including venous inflow obstruction and interference with pulmonary function. The introduction of the downsized J-7-70 artificial heart reduced these problems while simultaneously increasing the pool of potential TAH candidates based on size. Only one patient died (0.4%) as a direct result of a mechanical failure of a Symbion TAH or from its external drive and monitoring system. All other hardware problems were successfully corrected by various methods, including reoperation, with no clinical morbidity directly linked to these malfunctions.

Function

The method of TAH blood pumping must meet the criteria discussed in the previous section dealing with design. Efforts to precisely duplicate the native heart have been universally unsuccessful when considering device function and design.

The Utah TAH is powered by compressed air delivered individually to the right and left ventricles. The compressed air is transmitted through Tygon tubing (i.e., drive lines) that originates at the external heart driver (see below), traverses the skin and soft tissues into the inferior thorax region, and connects with the base of the TAH ventricle.

The right and left ventricles have identical pumping mechanisms. The hemispherical ventricle has an internal polyurethane diaphragm that separates the compressed air from the recipient's blood. The systolic phase of TAH function involves the delivery of compressed air with expansion of the internal pumping diaphragm. The diaphragm extension increases intraventricular pressure which causes closure of the mechanical inflow valve and the opening of the mechanical outflow valve. These unidirectional valves direct the outflow of blood to either the aorta (left ventricle) or the pulmonary artery (right ventricle).

The pressure of the compressed air is externally regulated and separated for the right and the left ventricle. Sufficient pressure is applied to the pumping diaphragm to eject all end-diastolic blood and fully extend the diaphragm. The forces that determine this regulated compressed air pressure include the volume of end-diastolic blood, the compliance of the pumping diaphragm, the size (and functional status) of the mechanical outflow valve, the alignment of the blood pathway of the TAH ventricle and the recipient's great vessels, and the vascular resistance of the circulatory system. Alteration of any or all of these parameters can cause significant changes in the delivery pressures of the compressed air needed for complete systolic diaphragm excursion.

Proper function of the TAH necessitates complete emptying of the ventricles at end systole. If this goal is not met, then the diastolic volume capacity of the TAH is suboptimal because of residual systolic blood. This reduction in diastolic filling can decrease the efficiency of the TAH, resulting in increased venous pressures and eventually, if uncorrected, the symptoms of heart failure.

At end systole, with the diaphragm fully extended, the external heart driver allows evacuation of the compressed air to the atmosphere. The external release of the compressed air causes diaphragm collapse and a dramatic decrease in intraventricular pressure. This intra-TAH pressure drop results in closure of the mechanical outflow valve and opening of the mechanical inflow valve. The native atria, now filled with blood, provide an inflow pressure which exceeds the TAH intraventricular pressure, resulting in inflow valve opening and TAH ventricular filling. The atrial pressure alone is sufficient to permit significant TAH diastolic filling; however, as previously noted, this is not as efficient as native heart diastolic function. The limitations of diastolic TAH filling are decreased atrial size, lack of coordinated atrial contraction, the size and function of the mechanical inflow valve, the compliance of the TAH pumping diaphragm, and the rate of compressed air evacuation from the TAH. Similar to systolic alterations, any changes in these parameters can dramatically effect the diastolic (filling) capabilities of the TAH.

This cycle of pulsed air delivery with diaphragm expansion

(systole) and compressed air exhaust with diaphragm collapse (diastole) is repeated at a rate that satisfies the flow demands of the recipient. The external heart driver provides independent compressed air regulation of the right and left TAH ventricles. The TAH heart rate is adjusted to provide complete ejection of the blood volume and partial filling of the maximum TAH diastolic volume. The percentage of time that each TAH cycle (systole plus diastole) spends in systole is also adjustable (about 33%). The inherent restrictions to diastolic TAH filling, as previously noted, require as much time as possible to be devoted to this phase of each cardiac cycle. A decrease in the percentage of time spent in systole can often be overcome by the use of higher systolic compressed air pressures. Also, the diastolic phase filling of the TAH can be augmented by the addition of external vacuum. This negative pressure is applied during diastole to the compressed air exhaust system of the TAH ventricles. The vacuum-assisted evacuation of the ventricular compressed air collapses the pumping diaphragm more rapidly, causing a steeper decline in intraventricular pressure. This augmented decline in early diastolic ventricular pressure provides a greater difference in atrial ventricular pressures, resulting in more rapid early diastolic filling.

The selection of settings for the external heart driver controls is guided by the need to provide complete systolic ejection and partial filling (approximately 70 to 80%) of the maximum TAH ventricular diastolic volume. Adherence to these goals of systolic and diastolic TAH performance both provides steady-state blood pumping to meet metabolic needs and allows automatic adjustment of TAH blood flow (within a broad limit) in response to increased venous return, for instance, during exercise.

The complete ejection of all diastolic blood provides a forward flow of all blood that is presented to the TAH during diastole. Also, the complete ejection of this diastolic volume permits the TAH ventricles to have available their maximum filling capacity for the subsequent diastolic period. Any residual ventricular end systolic volume would reduce the filling capacity for the TAH and potentially result in increasing atrial volumes (and pressures) that, if left unchecked, could result in signs and symptoms of venous congestion (heart failure). The complete ejection of all diastolic blood is achieved by adjustment of individual ventricular compressed-air pressure and the percentage of time spent in systole. As stated previously, the percentage systole adjustment must be guided by the need to preserve as much time for diastole as possible. The pressure of the compressed air can provide full ejection and limited percentage systole. However, excessive pressure provides stress to the polyurethane diaphragm and could cause damage or failure.

The adjustment of TAH function to partial filling of the maximum ventricular diastolic volume achieves both satisfaction of steady-state flow demands and the ability for autoregulation of the TAH to increase venous flow. Following the complete ejection of end-diastolic volume, the TAH has a potential for a maximum of 70 ml to enter the device during the next diastolic period. The determination of the ventricular filling volume (assuming complete systolic ejection) is controlled primarily by the heart driver's adjustment of TAH heart rates. Although diastolic vacuum can augment diastolic filling, it does not increase the diastolic volume available for pumping by the TAH. For this reason, the use of vacuum will be avoided for this discussion of partial diastolic filling of the TAH ventricles.

Under steady-state conditions there is a relatively constant demand for forward flow from the TAH, with a similar finding for venous return from the recipient's systemic and pulmonary vascular beds. The TAH heart rate can be increased to a value where all available diastolic venous blood equals the maximum diastolic volume of the TAH ventricle (i.e., 70 ml). This balance of venous inflow and TAH filling provides TAH output that meets the baseline metabolic needs of the recipient. However, any sustained increases in venous return cannot be accommodated by the TAH ventricles, since they are operating at their maximum pumping capacity for that particular TAH heart rate. If left unchecked, this scenario results in increasing venous pressures and the pathologic consequences of this problem.

The two methods of correction require an active input from an outside observer: (1) the TAH heart rate can be increased, which allows an increased capacity for pumping blood (HR × SV = CO), or (2) the increased vascular volume can be decreased or redistributed, either by drugs or by mechanical means. This need for potentially frequent external intervention in TAH control is impractical and possibly dangerous. This problem can be overcome by proper adjustment of the baseline TAH heart rate. Instead of selecting a TAH heart rate that balances venous return and TAH ejection volume, the heart rate is increased until the volume equals 70 to 80% of the maximum TAH ventricular volume. This partial filling of the maximum TAH ventricular capacity, when the recipient is at rest (or more accurately at steady state, whether it be physical rest or during activity) provides a diastolic reserve volume (20 to 30%). Any increases in venous return can be self-adjusted by the TAH using this ventricular filling reserve. This autoadjustment, based upon the volume of venous return, permits the TAH recipient to change position, exercise, and perform other routine daily activities without the need for external adjustments of the TAH heart driver or the administration of drugs. If the increased venous return is sustained or beyond the diastolic TAH filling reserve, then the TAH heart rate must be increased or other measures taken to decrease intravascular volume. Although this self-regulation of the Utah TAH is limited and without sophisticated physiologic sensing, it has proven to be reliable and successful in maintaining satisfactory blood flow without external adjustment of TAH controls.

Monitoring

The pneumatic TAH requires reliable, accurate, and easy to interpret external monitoring. The proper adjustment of external TAH control parameters necessary to achieve full systolic ejection and partial diastolic filling requires such a monitoring system. Additionally, TAH monitoring should quantitate device performance and determine mechanical malfunction or failures. Ideally, this system of evaluating TAH performance should be noninvasive. The Utah TAH

system has two monitoring methods that meet these criteria. The two systems are complementary, one being only qualitative while the other supplies both quantitative as well as qualitative data.

The delivery of compressed air to the TAH via the external heart driver produces a characteristic pressure waveform. This pressure waveform provides no quantitative formation about the TAH pumping characteristics. However, it does have characteristics that indicate particular aspects of individual right and left ventricular performance. This information allows proper adjustment of external control parameters necessary to meet the partial fill/full eject mode of operation.

An external pressure transducer independently monitors the right and the left compressed-air pressure waveform. The onset of TAH systole produces a rapid rate of rise in pressure as compressed air is delivered to the ventricle. When the pressure exceeds the forces resisting the movement of blood (diaphragm compliance, blood mass, outflow valve impedance, and vascular resistance), the outflow valve opens and blood is ejected from the TAH ventricle. This initiation of blood flow changes the pressure waveform to a slow rate of pressure rise for the majority of the systolic period. As systole continues, if the total volume of end-diastolic blood is ejected, the pressure waveform achieves a peak value (or plateau, if the duration is sufficiently long), which indicates full emptying of the ventricle with the stretching of the diaphragm. This maximum (and sudden) pressure spike usually appears near the end of systole and indicates to the observer that complete emptying of the TAH ventricle has been achieved. If this surge of maximum pressure is not observed in the drive-pressure waveform, then full emptying of the ventricle has not occurred. However, the quantity of residual end-systolic blood is impossible to determine from this monitoring system.

The onset of TAH diastole results in evacuation of the intraventricular compressed air to the atmosphere. The drive-line pressure waveform shows a rapid rate of descent of the pressure during the early diastolic period. As diastolic filling of the TAH ventricle proceeds, the incoming blood displaces the small residual volume of intraventricular compressed air. This process is reflected by a gradual decline in the air drive-line pressure; over the majority of the diastolic period the ventricle continues to fill and the compressed-air pressure waveform has an abrupt decline to its baseline value. This late diastolic, rapid decline in air pressure indicates full filling of the pneumatic ventricle. Since the goal of TAH control is the achievement of partial ventricular filling, the finding of full filling on the drive-line pressure waveform requires an adjustment in TAH heart rate. However, despite this adjustment in heart rate, abolishing complete ventricular filling, the residual end-diastolic volume now available for TAH self-regulation is unknown.

In addition to determining full filling and full ejection of the TAH, the drive-line pressure waveform can be used to diagnose specific abnormalities of ventricular function and compressed-air delivery. The diagnostic capabilities of the drive-line compressed-air waveform, albeit limited, can allow corrective action before lethal consequences develop for the TAH recipient.

Initially, quantitative measurement of the Utah TAH's function required implanted flow probes or other invasive devices that were often unreliable. A novel approach to determine TAH blood flow employed a simple but accurate concept. When the TAH ventricles fill with blood during diastole, there is an equal displacement of air from the ventricular chamber. Quantitating this exhausted air volume provides an indirect measure of the stroke volume for the next systolic cycle. An inexpensive system was developed to monitor and record the volume and flow rate of the air during diastole (Nielsen *et al.*, 1983). This was called a cardiac monitor and diagnostic unit (COMDU) and was used to monitor the first clinical artificial heart patient (Hastings *et al.*, 1984). Correction factors for the loss of blood volume through the inflow valve further refined this indirect calculation of TAH stroke volume. A volume curve of exhausted compressed air in relation to the duration of diastole is plotted on a personal computer. This flow curve (in liters/second) provides both a quantitative measure of TAH stroke volume as well as a large amount of quantitative information regarding TAH performance. The COMDU has been valuable in making early diagnoses of device failures (or impending failures) and other patient-related complications (Taenaka *et al.*, 1985; Mays *et al.*, 1988).

Clinical Problems

Infection remains one of the most feared complications of TAH application in man, as it is for other implanted devices (Didisheim, *et al.*, 1989). The combination of the intrathoracic TAH with its polyurethane blood-contacting surface, mechanical valves, and multiple sites for bacterial adhesion and proliferation, combined with the anticipated performance of an orthotopic heart transplantation with its requirement for radical immunosuppression, makes infection of the TAH a life-threatening event. Methods to prevent infection have been successful in minimizing this complication. Prior to TAH implantation, a careful survey for any established infections in the recipient is mandatory. Pneumonia, intravenous catheter contamination, urosepsis, or other sites of infection must be eradicated before the TAH is considered for use as a bridge to transplantation and then must be carefully monitored postimplantation.

During and immediately after TAH implantation, the judicious use of prophylactic antibiotics is necessary. Daily surveillance of surgical wounds, intravenous catheter sites, temperature elevations, or symptoms of infection is vital to successful early detection and treatment of infections. Periodic monitoring of the white blood cell count is also a vital component of this monitoring. Even when infections are detected, prompt antibiotic treatment combined with other measures (i.e., removal of iv catheters, etc.) has permitted the vast majority of these patients to proceed to heart transplantation without significant posttransplant problems related to these infectious events.

The exit site of the transcutaneous compressed-air drive lines have been the subject of intense scientific investigation. There has been a vast array of devices designed to protect these large tubes from serving as a nidus and conduit for infection. There was a scientific consensus that when used in man, these

TAH drive lines would permit the migration of bacteria from the skin to the mediastinum. There were animal recipients of the TAH that appeared to confirm this theory. The design of devices to protect the cutaneous entrance site of the drive lines focused on elimination of tissue injury from tension stresses and the rhythmic pulsation of these compressed-air drive lines, as well as the promotion of a secure barrier to cutaneous bacterial migration to deeper tissues. Years of animal work was moderately successful in realizing these goals. Interestingly, when the TAH was used in man, this potential site for life-threatening infection did not develop as a frequent clinical problem. Most recently the devices (termed skin buttons), designed to protect the Utah TAH drive lines from causing tissue injury and promote tissue ingrowth, have been abandoned in human applications of the TAH. The elimination of the skin buttons has resulted in no rise in the incidence of cutaneous infections or the development of subsequent mediastinitis.

There have been scientific reports that TAH recipients have impaired immune systems. The sophisticated evaluations have documented abnormalities in the function of both the cellular and the humoral components of the immune system. Interpretation of this information and its application to TAH human recipients is complex, with overlap of immunologic defects caused by the patient's preoperative heart failure, malnutrition, and medical therapy. Additionally, the known immunosuppressive effects associated with CPB also contribute to this problem. Whether the materials used in the blood-contacting portions of the TAH, or other factors related to the TAH, are directly immunonsuppressive requires further investigation. This scientific reformation is critical if interventions are to be successful in reversing the potentially disastrous suppression of the immune system.

Postoperative hemorrhage remains a significant problem in TAH recipients. Multiple factors interact in the development of significant hemorrhage. Many of the TAH recipients have been on oral anticoagulants and antiplatelet medications, preoperatively, as therapy for their severely depressed native myocardial function and often accompanying coronary artery disease. Partial reversal of the anticoagulant's direct effect is possible. However, the primary therapy is the replacement of the protein clotting factors and platelets from random donors. Although successful in restoring coagulation capabilities of the patient, the transfusions provide a source of potential infectious contamination as well as immune reactions (antibody titer) to the transfused foreign proteins.

The surgical technique of TAH implantation provides long suture lines where native tissue is approximated to woven Dacron grafts and polyurethane-covered Dacron (Levinson and Copeland, 1987). Although the magnitude of suture lines is no greater than an orthotopic heart transplantation, there is a significant difference in the technique to achieve a blood-tight anastomosis. The TAH's Dacron cuffs, which interface with the native atria and great vessels, are relatively rigid (particularly the atrial cuffs). The anastomosis must be constructed with closely spaced sutures that avoid crimping or long gaps. Unlike suture lines that approximate native tissue to native tissue, the TAH suture lines are unforgiving, and any crimps or gaps will leak, causing significant hemorrhage. Because of this situation, the four TAH suture lines (two atrial and two great vessel) are tested under physiologic pressures for leaks before the two artificial ventricles are snapped into place, since the two ventricles restrict direct visualization of some of the suture lines.

The use of CPB requires total anticoagulation using the drug heparin. Heparin paralyzes the protein cascade that is necessary for the development of thrombus. However, in addition to the protein-blocking effects of heparin, the contact of blood with the foreign surfaces of the CPB circuit and oxygenator and the use of nonpulsatile flow, various drugs, and hypothermia combine to cause a myriad of effects that interfere with clot generation. Strategies to minimize these effects have included the pre-CPB infusion of aprotinin, a proteinase inhibitor of plasmin, thrombin, and other components of normal coagulation and platelet function. Other drugs have also been administered in an effort to minimize the CPB effects on postoperative bleeding. The success of these interventions has been variable and incomplete. Despite the magnitude of the problem, the vast majority of patients will not have significant hemorrhage if attention is directed to the basic surgical principles of meticulous hemostasis at sites of cardiovascular anastomosis and planes of dissection, combined with thoughtful replacement of coagulation factors and platelets post-CPB. The use of plasmacologic intervention should be viewed as an adjunct to these measures and not as a primary therapy.

Tears of the internal TAH diaphragm have occurred early on in the laboratory. However, because of its four-layer design, these disruptions of individual layers have been soft failures, allowing for successful replacement of the effected ventricle. Critical to this preterminal replacement of a malfunctioning ventricle is the ability to diagnose the diaphragm problem before irreparable harm comes to the recipient. The COMDU has proven itself as a device that can accurately diagnose diaphragm tears before there is any recipient morbidity or mortality.

Mechanical valves are TAH components that have a potential for failure. The first human recipient of the Utah TAH had a failure of the left ventricular inflow valve. The accurate COMDU diagnosis of this potentially hard failure provided immediate diagnosis and prompt management with follow-up surgical intervention which required left ventricular replacement. Investigation of this inflow valve failure revealed two distinct but interrelated problems. The mechanical valve used in the inflow position was a 27-mm Bjork Shiley mitral valve. The metal struts, which hold the tilting carbon disk in place, were welded to the metal valve ring. These weld points were found to be susceptible to failure with the applied stress generated by closure of the valve disk during ventricular systole. This fracture potential was combined with the TAH pneumatic driver generating superphysiologic intraventricular pressure changes during systole. Measurement of the TAH's ventricular dP/dT during the ejection phase indicated a value in excess of 7500 mm Hg/sec pressure change. These two problems, one an inherent characteristic of the mechanical valve and one a feature of TAH ventricular function, were complementary to the development of mechanical failure of the inflow valve. Both areas of concern were addressed for corrective changes.

The Bjork Shiley valve was replaced with a Medtronic Hall tilting disk valve. The major engineering advantage of this new valve was the machining of the metal valve ring and diskholding struts from one block of metal. This manufacturing technique avoids the inherent weak areas of weld points found in the Bjork Shiley valve, thereby avoiding the risk of strut breakage and disk escape. Simultaneously, the external heart driver, which regulates the pressure of compressed air delivered to the TAH ventricle, was modified to decrease the dP/dT of the TAH ventricle to less than 5000 mm Hg/sec. The combination of these corrective actions has eliminated the problem of mechanical valve failure in the Utah TAH.

Hemolysis, the development of a pathologic rate of red blood cell destruction, is a concern for all blood pumps. The use of four mechanical valves in the Utah TAH caused an inevitable amount of hemolysis. However, the rate of red blood cell destruction is a level that is tolerable to the recipient, avoiding the need for transfusions and also avoiding the overloading of the recipient's reticuloendothelial system by the

TABLE 2 Patient Data on Utah-Developed Total Artificial Hearts

	Symbion 1985–1992		CardioWest 1993–1996	
Number of implants	198		69	(4 on the device)
Average age (years)	42	(13–64)	44	(16–63)
Average duration	24	(1–670)	31	(1–186)
Female/male	28/170		5/64	
Total implant days	4742	(13 years)	2002	(5.5 years)
Number transplanted	143	(72)	47	(72)
Number discharged	85	(43% IMP, 59% TX)[a]	43	(66% IMP, 92% TX)[a]
Number alive post-TX	Unknown		42	(98%)
Accumulation post-TX	Unknown		59.5 years	

[a]IMP, TAH implants; TX, cardiac transplantation.

TABLE 3 Total Artificial Heart Implants by Year

Heart	Number and year of total artificial heart implants															Total
	1969	1981	1982	1984	1985	1986	1987	1988	1989	1990	1991	1992	1993	1994	1995	
Liotta	1															1
Akutsu		1														1
Jarvik[a]			1	1	6	37	45	61	22	14	17	5	19[b]	18	26	272
Penn State					1	1	1	1	1							4
Phoenix					1											1
Berlin						1	6		1							8
Unger						4										4
Brno								3	2							5
Total	1	1	1	1	8	43	52	65	25	14	17	5	19	18	26	296

[a]The first five patients received the total artificial heart as a permanently implanted device.

[b]CardioWest manufactured and sold the Utah-developed total artificial heart from 1993 to the present.

products of red blood cell destruction. The object of TAH design and function is to avoid any additional hemolysis that would be additive to the valve hemolysis. Design features of the Utah TAH that minimize hemolysis include smooth, unobstructed transitions from native atria to the ventricle and from the TAH to the great vessels; minimal intraventricular turbulence; avoidance of diaphragm–ventricular housing contact, thereby avoiding red blood cell crushing; and minimizing changes in blood flow velocity and turbulence (i.e., maximizing orifice size at the inflow and outflow portions of the ventricle). All of these factors have been successfully incorporated into the design and fabrication of the Utah TAH.

CLINICAL USE OF THE ARTIFICIAL HEART

The early designs of artificial hearts attempted to mimic all anatomic features of the natural heart. This concept was found to be flawed for a variety of reasons. Replacement of the native atria was unnecessary, retention of the recipient's atria provided the compliant reservoir capacity needed to fill the ventricles and, more importantly, it simplified the surgical implantation. The external configuration of the artificial ventricles did not need to be "heart shaped." The shape was dictated by stroke volume, fit, intracavitary flow patterns, and engineering considerations. The artificial ventricles did not require contraction of the entire ventricular wall during systole, and there was no need to design individual right and left ventricles.

The University of Utah TAH technology was subsequently licensed to Symbion, Inc., for worldwide marketing. The Symbion TAH was used as a permanent cardiac replacement in four additional nontransplant candidates (the criteria specified by the FDA-approved IDE), DeVries implanted four

TAHs in the United States (1982–1985), and Semb implanted one TAH in Sweden (1985). Survival in this limited experience ranged from 10 to 620 days. Table 1 lists the clinical TAH implants over the first 16 years (Olsen, 1996). The FDA approved a new IDE in 1985 for use of the Symbion TAH as a bridge to cardiac transplantation based on extensive experience in calves (Olsen *et al.*, 1981). Subsequently, the Symbion TAH was used temporarily in 198 patients (1985 through 1992) as a bridge to transplantation. A total of 203 Symbion TAHs were implanted with five patients receiving 2 devices (Table 2). The 2nd TAHs were all used as replacements for failed transplants, 4 of them for acute rejection and 1 for chronic rejection episodes.

On January 12, 1990, the FDA withdrew its approval of the IDE for both the permanent and the temporary use of the Symbion TAH and VAD. The FDA cited manufacturing, monitoring, and reporting deficiencies as the basis for suspension of Symbion's IDE. However, the FDA permitted the Symbion-sponsored U.S. TAH and VAD centers to implant previously purchased devices (which continued until December of 1992). Symbion initially struggled to correct its deficiencies and seek reinstatement of their FDA-approved IDE in order and to resume their clinical U.S. TAH and VAD programs. However, after careful review, Symbion's board of directors recommended to its shareholders on March 15, 1991, the disposal of all operating assets and liquidation of the company. The company went out of business in July 1991, less than 10 years from its inception.

With the closure of Symbion, Dr. Copeland and Dr. Olsen started a new company, CardioWest, which licensed the technologies from the University of Utah and took over supporting the clinical centers until January 1993 when all implanted devices were manufactured and sold by CardioWest. The Utah-developed TAH was named and referred to as the C-70. The bridge to cardiac transplantation with the Utah-developed TAH is compared and presented in Table

TABLE 4 Total Artificial Heart Bridging to Cardiac Transplantation

Device	Years	Number of patients	TX	Number (% IMP)	Discharged	Number (% IMP)	(% TX)
Liotta	1969	1	1	(100)	0	(0)	(0)
Akutsu	1981	1	1	(100)	0	(0)	(0)
Phoenix	1985	1	1	(100)	0	(0)	(0)
Penn State	1985–1988	3	1	(33)	0	(0)	(0)
Jarvik	1985–1992	198	143	(72)	85	(43)	(59)
Berlin	1985–1989	8	2	(25)	1	(25)	(50)
Vienna	1985–1989	4	3	(75)	2	(50)	(67)
Brno	1988–1989	6	0	(0)	0	(0)	(0)
CWT	1993–1996	69[a]	47	(72)	43	(66)	(92)
9 devices	27 years	291	199	(69)	131	(46)	(66)

[a]Four patients were supported on the device at the time of writing.

2. Some criticized the early use of VADs and TAHs in the bridge to cardiac transplantation patient on the basis that the success would be far lower and thus would be a waste of the very limited donor organs. The more recent data on the CardioWest TAH would disprove that feat and this is supported by the results with both the TCI Heartmate and the Baxter/Novacor VADs.

This chapter describes the Utah TAH experience; however, there have been other pneumatic TAHs developed and used clinically as of this writing. Table 3 is a listing of all the known human implantations by year (Olsen, 1996). Table 4 presents the best available data on the number of patients implanted with the nine different types of hearts and the numbers transplanted and those that were discharged alive from the hospital.

Bibliography

Akutsu, T., and Kolff, W. J. (1957). Pneumatic substitutes for valves and hearts. *Trans. Am. Soc. Artif. Intern. Organs* 4: 230–235.

Cooley, D. A., Akutsu, T., Norman, J. C., *et al.* (1981). Total artificial heart in two staged cardiac transplantation. *Cardiovasc. Dis. Bull. Texas Heart Inst.* 8: 305–319.

DeVries, W. C., Anderson, J. L., Joyce, L. D., Anderson, F. L., Hammond, E. H., Jarvik, R. K., and Kolff, W. J. (1984). Clinical use of the total artificial heart. *N. Engl. J. Med.* 310: 273–278.

Didisheim, P., Olsen, D. B., Farrar, D. J., Portner, P. M., Griffith, B. P., Pennington, D. G., Joist, J.H., Schoen, F. J., Gristina, A. G., and Anderson, J. M. (1989). Infections and thromboembolism with implantable cardiovascular devices. *ASAIO Trans.* 35(1): 54–70.

Guyton, A. G., ed. (1981). *Textbook of Medical Physiology*, 6th Edition. Saunders, Philadelphia.

Hastings, W. L., Joyce, L. D., DeVries, W. C., Olsen, D. B., Nielsen, S. D., and Kolff, W. J. (1984). Drive system management in the clinical implantation of the JARVIK-7 total artificial heart. *Prog. Artif. Organs*—1983 1: 217–221.

Iwaya, F., Fukumasu, H., Olsen, D. B., Nielsen, S. D., Lawson, J.,

White, S., Mochizuki, T., and Kolff, W. J. (1979). Studies of the remnant atria of the total artificial heart: P-wave response to the surgery, treadmill exercise and drugs. *J. Artif. Organs* 3(Suppl): 324–328.

Kolff, J., Deeb, G. M., Cavarocchi, C., Riebman, J. B., Olsen, D. B., and Robbins, P. S. (1984). The artificial heart in human subjects. *J. Thorac. Cardiovasc. Surg.* 87: 825–831.

Kwan-Gett, C. S., Wu, Y., Collan, R., Jacobsen, S., and Kolff, W. J. (1969). Total replacement artificial heart and driving system with inherent regulation of cardiac output. *ASAIO Trans.* 15: 245–250.

Levinson, M. M., and Copeland, J. G. (1987). Technical aspects of total artificial heart implantation for temporary applications. *J. Cardiac. Surg.* 2: 3–19.

Mays, J. B., Williams, M. A., Barker, L. E., Pfeifer, M. A., Kammerling, J. M., Jung, S., and DeVries, W. C. (1988). Clinical management of total artificial heart drive systems. *JAMA* 259(6): 881–885.

Nielsen, S. D., Willshaw, P., Nanas, J., and Olsen, D. B. (1983). Noninvasive cardiac monitoring and diagnostics for pneumatic pumping ventricles. *Trans. Am. Soc. Artif. Intern. Organs* 29: 589–592.

O'Connell, J. B., and Bristow, M. R. (1994). Economic impact of heart failure in the United States: Time for a different approach. *J. Heart Lung Transplant.* 13: S107.

Olsen, D. B. (1996). Artificial heart registry maintained by the Institute for Biomedical Engineering, University of Utah, Salt Lake City.

Olsen, D. B., DeVries, W. C., Oyer, P. E., Reitz, B. A., Murashita, J., Kolff, W. J., Daitoh, N., Jarvik, R. K., and Gaykowski, R. (1981). Artificial heart implantation, later cardiac transplantation in the calf. *Trans. Am. Soc. Artif. Intern. Organs* 27: 132–136.

Poirier, V. L. (1991). Can our society afford mechanical hearts? *Trans. Am. Soc. Artif. Intern. Organs* 37: 540–544.

Rowles, J. R., Mortimer, B. J., and Olsen, D. B. (1993). Ventricular assist and total artificial heart devices for clinical use in 1993. *ASAIO J.* 39(4): 840–855.

Taenaka, Y., Olsen, D. B., Nielsen, S. D., Dew, P. A., Holmberg, D. L., and Chiang, B. Y., (1985). Diagnosis of mechanical failures of total artificial hearts. *Trans. Am. Soc. Artif. Intern. Organs* 31: 79–83.

8.3 EXTRACORPOREAL ARTIFICIAL ORGANS
Paul S. Malchesky

Historically, the term "extracorporeal artificial organs" has been reserved for life support techniques requiring the on-line processing of blood outside the patient's body. The substitution, support, or replacement of organ functions is performed when the need is only temporary or intermittent support may be sufficient. The category of extracorporeal artificial organs does not include various other techniques which may justifiably be considered as such, such as infusion pumps or dermal patches for drug delivery, artificial hearts used extracorporeally, eyeglasses and contact lens for vision, and orthotic devices and manipulators operated by neural signals to control motion.

This chapter focuses on artificial organ technologies which perform mass transfer operations to support failing or impaired organ systems. The discussion begins with the oldest and most widely employed kidney substitute, hemodialysis, and outlines other renal assist systems such as hemofiltration for the treatment of chronic renal failure and fluid overload and peritoneal dialysis. The blood treatment process of hemoperfusion, and apheresis technologies which include plasma exchange, plasma treatment, and cytapheresis—used to treat metabolic and immunologic diseases—are also discussed. In addition, blood–gas exchangers, as required for heart-lung bypass procedures, and bioartificial devices that employ living tissue in an extracorporeal circuit are addressed. Significant concerns and associated technological considerations regarding these technologies, including blood access, anticoagulation, the effects of the extracorporeal circulation, including blood cell and humoral changes, and the biomodulation effects of the procedure and materials of blood contacts are also briefly discussed.

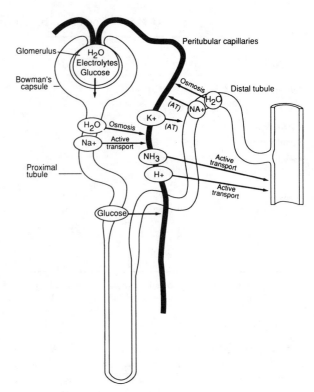

FIG. 1. Schematic representation of the nephron.

on dialysis. About 200,000 people in the United States and 600,000 worldwide are on dialysis.

Dialysis

Dialysis is the process of separating substances in solution by means of their unequal diffusion through a semipermeable membrane. The essentials for dialysis are (1) a solution containing the substance to be removed (blood); (2) a semipermeable membrane permeable to the substances to be removed and impermeable to substances to be retained (synthetic membrane in hemodialysis or the peritoneal membrane in peritoneal dialysis), and (3) the solution to which the permeable substances are to be transferred (dialysate).

The device containing the semipermeable membrane is called a dialyzer. Hemodialysis is performed on the majority (approximately 85%) of the patients while peritoneal dialysis is used for the rest.

Hemodialysis

Hemodialysis is the dialysis of blood. Approximately 170,000 patients in the United States and about 525,000 worldwide undergo hemodialysis typically three times a week, totaling about 80 million treatments per year. The patient's blood is typically anticoagulated with heparin throughout the extracorporeal treatment. Blood access is usually from a permanent fistula made in the patient's forearm. The blood is drawn from

KIDNEY ASSIST

The kidney's function is to maintain the chemical and water balance of the body by removing waste materials from the blood. The kidneys do this by sophisticated mechanisms of filtration and active and passive transport that take place within the nephron, the single major functioning unit in the kidney, of which there are about 1 million per kidney (Fig. 1). In the glomerulus, the blood entry portion of the nephron, a blood filtration process occurs in which solutes up to 60,000 Da are filtered. As this filtrate passes through the nephron's tubule system, its composition is adjusted to the exact chemical requirements of the individual's body to produce urine. In addition to these functions, the kidneys support various other physiological processes by performing secretory functions that aid red blood cell production, bone metabolism and blood-pressure control. Renal failure occurs when the kidneys are damaged to the extent that they can no longer function to detoxify the body. The failure may be acute or chronic. In chronic renal failure or in the absence of a successful transplant, the patient must be maintained

A

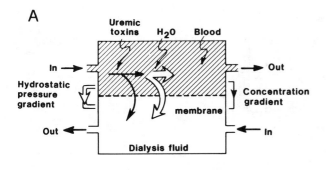

B

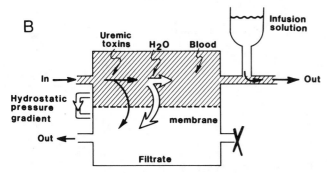

FIG. 2. Comparative operation principles in (A) hemodialysis and (B) hemofiltration with postfiltration addition of fluid.

the body at a flow rate of 50 to 400 ml/min, depending on patient conditions and procedural requirements, and then pumped with a roller pump through the extracorporeal circuit which includes the dialyzer. The transport process of water, electrolytes, and simple metabolites between the blood and the dialysate takes place in the dialyzer (Fig. 2A). The dialysate is a water solution of electrolytes of about the same concentration as normal plasma.

The membranes used are primarily cellulosic but various synthetic membranes, including polyacrylonitrile, poly(methylmethacrylate), ethylvinyl alcohol, and polysulfone are also being used. The membranes permit the low-molecular-weight solutes (typically less than about 5000 Da) such as urea, creatinine, uric acid, electrolytes, and water to pass freely, but prevent the passage of high-molecular-weight proteins and blood cellular elements. However, some membranes utilized will pass appreciable amounts of solutes of a greater molecular weight (below about 20,000 Da) and with high rates of water transport. Such membranes are called high-flux membranes.

The stimulus for the introduction of synthetic membranes has primarily come from the need to improve biocompatibility, although recent modifications of cellulosics have also addressed this need. The term "biocompatibility" is loosely defined and generally refers to the response and effects of blood contact with the material. However, the influence of the flow, the properties of the blood, the choice of anticoagulant and the dialyzer design and its method of sterilization may also be important. Major indices of biocompatibility studied include variations in blood coagulation, blood cell changes, variations

in platelet and leukocyte activation, and complement system activation. Different membranes elicit different responses. The degree of the response is related also to the surface area of blood contact.

Most dialyzers in use are of the hollow-fiber design, employing hollow-fiber membranes. Also available are parallel-plate designs employing sheet membrane films. Many variations of these basic designs are commercially available. Generally, blood and dialysate flow in opposite directions (countercurrent).

Water is removed in hemodialysis primarily by ultrafiltration. Convective flux of water and associated solutes takes place under a hydraulic pressure gradient between the blood and the dialysate compartments. The functions of the dialysate delivery system are to prepare and deliver dialysate of the required chemical makeup for use in the hemodialyzer. Monitoring and control equipment is included as part of the system to ensure that the dialysate composition is correct and ready for use by the hemodialyzer and that the procedure is carried out in a safe manner.

Methods of preparing and delivering dialysate are either batch or continuous. In a continuous dialysate-supply system, concentrate and processed tap water (filtered to remove particulate, ions, and organic matter) are continuously mixed during the course of the dialysis and delivered to the dialyzer. This type of system eliminates the space required for mixing an entire batch of dialysate. However, to be effective, the system must be closely controlled, since any malfunction can result in an improperly mixed dialysate. Dialysate can be recirculated in the dialyzer, although fresh dialysate is most effective because it produces a high concentration gradient with the blood and therefore a high removal rate for solutes. In a single-pass system, the flow of dialysate is usually kept low (about 500 ml/min) to limit the amount required.

Dialysate delivery systems also include monitoring, control, and safety equipment. These systems range from the use of simple components to automated systems capable of operating without an attendant. Equipment includes flow meters, temperature and pressure monitors, dialysate-conductivity probes, and display meters. Control equipment includes thermostat-controlled heaters and mixing valves for regulating dialysate temperatures and composition, valves for regulating flow rates and patient fluid loss, and adjustable high and low limits on various safety and monitoring devices. Water conditioning and treatment equipment is usually available separately from the dialysate delivery system. Safety equipment includes devices designed to indicate or correct any factor in dialysis which exceeds the established limits for safe operation. This equipment includes audio and/or visual alarms and failsafe shutdown sequences that would be employed during the course of dialysis. Present-day equipment allows one nurse to oversee several dialysis stations at one time.

Hemodialysis is generally performed for periods of from 3–5 hr, three times a week. Hemodialysis may be performed in the hospital, a dialysis center, or at home. Equipment requirements vary, depending upon where and by whom the dialysis is performed. Figure 3 schematically shows the circuitry with on-line dialysate preparation, and Table 1 outlines the most common factors monitored during hemodialysis with reference

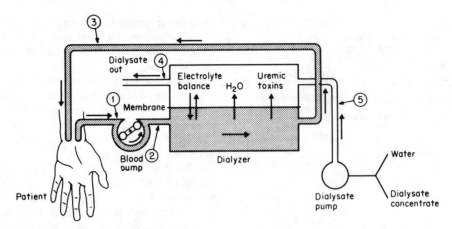

FIG. 3.   Schematic of hemodialysis circuitry with on-line dialysate preparation. Numbered locations refer to instrumentation listed in Table 1.

TABLE 1   Factors Most Commonly Monitored during Hemodialysis

Factor	Equipment position (see Fig. 1)	Operation	Remarks
Extracorporeal blood pressure	1, 2, and/ or 3	Measures pressure in drip chambers by mechanical or electronic manometer; abnormal pressure indicates any one of several malfunctions (increased line resistance, clotting, blood leak) and has high and low alarm limits	Installation in location 3 is considered mandatory and provides most meaningful information with respect to changes in flow (such as due to clots) or bloodline leak
Blood-leak detector	4	Photoelectric pickups in effluent dialysate line detect optical transmission changes due to presence of blood	Detection threshold is adjustable with alarm circuit to shut off blood pump or bypass the dialysate flow
Dialysate pressure	5 and/or 4	Measures pressure of dialysate inlet and/or outlet by mechanical or electronic manometer; usually has high and low alarm limit; abnormal operation can result in membrane rupture or improper ultrafiltration	Transmembrane pressure may be displayed with possible control of blood-side pressure
Dialysate temperature	5	Thermostatic measurement generally used to control electric heaters; dial thermometer or thermocouple gauge readout; out-of-range operation can result in patient discomfort or fatal blood damage	In central dialysate delivery systems, the dialysate is centrally heated with trimming at individual stations
Dialysate flow rate	5	Measures and displays flow in a rotameter; unless extremely low, cannot result in undue harm to patient	Dialysate is normally used at a rate of 500 ml/min
Dialysate concentration	4 and 5	Flow differential is used to determine patient fluid loss and ultrafiltration rate	Ultrafiltration control is particularly important with the use of high-flux dialyzers
	5	Measures and displays electric conductivity of dialysate; improper concentration can result in blood cell and central nervous system damage	Continuous concentration measurement is a necessity in delivery systems using continuous proportioning of dialysate
Air trap and air bubble detector	3	Collects air prior to blood return to patient; if air is detected, clamp is activated to stop flow to patient	Mandatory

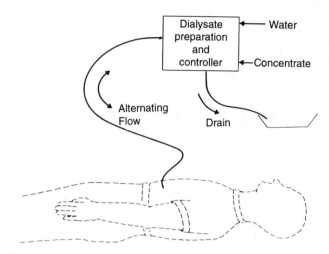

FIG. 4. Schematic of intermittent peritoneal dialysis with on-site dialysate preparation.

to the site of monitoring shown in Fig. 3. Portable or wearable systems have also been developed.

Peritoneal Dialysis

Peritoneal dialysis is carried out in the peritoneal cavity of the patient. The peritoneum is a thin membrane lining the abdominal cavity and covering the abdominal organs. It forms a closed sac. Through a cannula placed through the skin or a catheter permanently implanted, dialysate solution (about 2 liters in an adult) is infused, allowed to dwell for a designated time period, and drained. This process is repeated according to the needs of the patient. This semipermeable membrane permits transfer of solutes from the blood to the dialysate.

The efficiency of the process is strongly dependent upon the blood flow through the peritoneal membrane; the permeability of the peritoneal membrane; and the dialysate conditions of flow, volume, temperature, and net concentration gradient. Peritoneal dialysis may be performed intermittently or continuously.

The low volume of dialysate, the low degree of agitation of the dialysate in the peritoneal cavity, and blood flow dynamics in the peritoneum contribute to a lower efficiency for small solute transfer than hemodialysis. Thus, peritoneal dialysis requires longer treatment times than does hemodialysis.

In intermittent peritoneal dialysis, a single cycle for dialysate solution infusion, dwell, and drainage is generally accomplished in less than 30 min and the procedure continued for 8 to 12 hr per treatment (Fig. 4). Dialysate is available in commercially prepared bags or bottles or can be made on site, as in hemodialysis, from dialysate concentrate and water. Additional precautions must be taken with its preparation in comparison with the dialysate used in hemodialysis because of the sterility requirement for infusion of the dialyzing solution into the body. Monitoring and control equipment

for the preparation and delivery of the dialyzing fluid is generally automated and capable of operating without an attendant: it includes automatic timers, pumps, electrically operated valves, thermostat-controlled heaters, conductivity meter, and alarms.

In the more popular form of peritoneal dialysis referred to as continuous ambulatory peritoneal dialysis (CAPD), the infused dialysate is allowed to reside for only a few hours. Generally, five to six exchanges are made per day. The difference between this technique and that of intermittent peritoneal dialysis and hemodialysis is that body chemistries are more stable and not fluctuating between the extremes of the pre- and postdialysis periods.

In peritoneal dialysis, the hydraulic pressure difference between the blood and dialysate is insufficient to cause any appreciable water removal so osmotic forces are utilized. The necessary osmotic gradient between the blood and dialysate is achieved by utilizing high concentrations of dextrose in the dialysate. Typically, solutions containing 1.5 to 4.5% dextrose are used.

An advantage of peritoneal dialysis over hemodialysis is that direct blood contact with foreign surfaces is not required, eliminating the need for anticoagulation. A disadvantage includes the increased risk of peritonitis. Approximately 15% of the U.S. patients with chronic renal failure are supported by peritoneal dialysis.

Hemofiltration

Hemofiltration refers to the removal of fluid from whole blood. Solute and water are removed strictly by convective flux. The standard hemodialysis membrane, however, is a poor reproduction of the filtration properties of the nephron's glomerulus. The standard hemodialysis membrane can separate from the blood solutes that are typically less than about 5,000 Da. In an effort to make the process more like that of the natural glomerulus and to duplicate the process in the natural kidney, membranes of higher permeability (complete passage of solutes less than about 50,000 Da and more typically less than 20,000 Da) have been developed. In addition to their applicability for dialysis under controlled ultrafiltration conditions (referred to as high flux dialysis), such membranes have also found use as ultrafilters of fluid from blood. Membranes used for hemofiltration therapy are made of polysulfone, polyacrylonitrile, poly(methyl methacrylate), poly(ether sulfone), polyamide, and the cellulosics.

Hemofiltration was originally designed for treating chronic renal failure by removing a substantial fraction of fluid and reconstituting blood by adding fluid either before or after filtration (Fig. 2B). Owing to the high physiological sensitivity to changes in circulatory volume, very sensitive monitoring and control equipment for fluid withdrawal and infusion must be used to maintain the fluid flow balance accurately, as required for net removal of fluid from the patients. Only a few centers in the world are investigating this technique for treating chronic renal failure.

Modes of hemofiltration applications include continuous

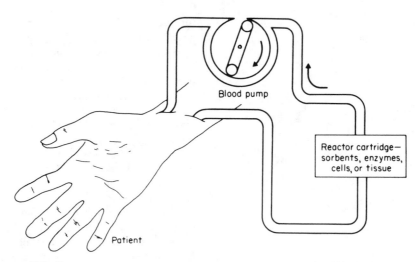

FIG. 5. Circuitry for sorbents or reactor (enzymatic, cellular, or tissue) hemoperfusion.

arteriovenous hemofiltration (CAVH), continuous venovenous hemofiltration (CVVH), or slow continuous ultrafiltration (SCUF). In these extracorporeal procedures, the patient's blood, under the driving force of arterial pressure (CAVH), or by an extracorporeal pump (CVVH), passes through a low resistance device (typically of hollow fiber design) and then is returned to the patient. Under force of the hydraulic resistance and aided by applied vacuum pressure if necessary on the filtrate side, fluid (ultrafiltrate) is removed from the blood. CAVH, CVVH, or SCUF is used to treat acute renal failure and fluid overload, particularly in critically ill patients who are in renal failure and require fluids, as in the form of parenteral nutrition. To enhance the solute transfer, the circuit may be modified to include the pumping of dialysate. This technique is referred to as hemodiafiltration. A further application of hemofiltration is the on-line removal of excess fluid during extracorporeal heart-lung bypass.

Hemoperfusion

Hemoperfusion refers to the direct perfusion of blood. In general, it refers to the perfusion of whole blood over a sorbent or reactor and therefore is different in this respect from hemodialysis (Fig. 5). In principle, the sorbent or reactor may consist of a wide variety of agents such as activated charcoal, nonionic or ionic resins, immunosorbents, enzymes, cells, or tissue. The purpose of biologically active agents may be to remove specific toxins from the blood or to carry out specific biochemical reactions. For practical reasons, such as biocompatibility concerns, hemoperfusion has been limited to the use of a small number of sorbents such as activated charcoal or resins in the treatment of drug intoxication or hepatic support.

The sorption of solute from the blood is based primarily upon its chemical affinity for the sorbents and less on its molecular size. The surface area for sorption may be as high as hundreds of square meters per gram of sorbent.

The use of hemoperfusion is limited primarily by biocom-

patibility concerns, including particulate release from the sorbent or reactor system. Sorbent and reactor systems are also applied to plasma perfusion (see the next section).

APHERESIS

Apheresis[1] refers to a procedure of separating and removing one or more of the various components of blood. This procedure can be used either therapeutically to remove cells or solutes which are considered harmful, or preparatively to obtain plasma or blood cells from normal donors. The procedure is referred to as plasmapheresis when plasma is separated and removed from the whole blood. If the plasma is discarded and replaced with donor plasma or a protein solution, the process is referred to as plasma exchange. If the separated plasma is processed and returned to the patient, the process is referred to as plasma treatment. Cytapheresis refers to the removal of cells and includes various types: leukocytapheresis, which involves the separation and removal of white blood cells; lymphocytapheresis, in which the lymphocytes are removed; erythrocytapheresis, in which the red blood cells are removed; and thrombocytapheresis, in which the platelets are removed.

Plasma Exchange

Plasma is routinely separated from whole blood in the medical laboratory and clinics. Millions of liters of plasma are collected annually in the United States for transfusion and the production of plasma products. Therapeutic plasma replacement or on-line plasma treatment has been applied in a number

---

[1]Definitions taken from Report of the Nomenclature Committee in *Plasma Separation and Plasma Fractionation*, M. J. Lysaght and H. J. Gurland, eds., Karger, Basel, pp. 331–334.

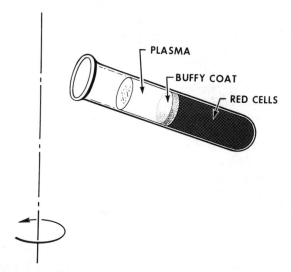

FIG. 6. Plasma separation in a test tube placed in a centrifugal field. The centrifugal separation of plasma from whole blood results in generally three distinct layers: the plasma, the buffy coat consisting of platelets and white blood cells, and the packed red cell layer.

of disease states, including renal, hematological, neurological, autoimmune, rheumatological, and hepatic. Plasma exchange is now a recognized therapy in selected disease states and is being investigated in many others. Traditionally, plasma separation has been carried out by centrifugal techniques. However, within the past decade, membrane devices have been used increasingly in therapeutic and bulk plasma collection. In centrifugal separation, the difference in density between blood cells and plasma is exploited. In membrane separation, the plasma is separated from the blood cells by the application of a low (typically less than 50 mm Hg) transmembrane pressure gradient across a microporous membrane permeable to all plasma components and impermeable to blood cells.

A number of membrane materials in the flat-film and hollow-fiber configuration have been utilized, including polyethylene, polypropylene, cellulose diacetate, poly(methyl methacrylate), polysulfone, poly(vinyl chloride), poly(vinyl alcohol), and polycarbonate. The membranes are of the tortuous path type and differ significantly in their physical properties such as porosity, mean pore size and its distribution, and pore number. Differences in the chemical nature of the membranes affect the amount and types of proteins that are deposited on the membrane surface. Such differences in protein adsorption can affect the filtration and biocompatibility properties of the membranes. Extensive filtration studies with bloods of various species and varying compositions show that membranes of different polymer types and varying microstructures have different plasma separation rates. Plasma separation efficiency for all membrane or module designs correlates with blood shear rate. Biocompatibility considerations similar to those used in assessing hemodialysis apply.

In therapeutic plasmapheresis, the separated plasma is discarded and replaced with an oncotic plasma substitute such as albumin solution or fresh-frozen plasma. Plasma exchange

has several limitations, including the reliance on plasma products, reactions or contaminations (particularly viral) caused by plasma product infusion, and loss of essential plasma components.

Centrifugal Plasma Separation

The centrifugal separation of plasma from blood is schematically depicted in Figs. 6 and 7. As the blood volume is rotated, the contents of the fluid exert a centrifugal force outward to the walls. Since the density of the blood particulate components, the cells, is higher than that of the plasma medium, sedimentation of the cells occurs outwardly through the plasma. The centrifugal acceleration exceeds that of gravity by a factor of a thousand.

For particle settling in the Stokes's law range (Reynolds number is less than 1 for most blood centrifugal processes) the terminal velocity, V_t, at a radial distance, r, from the center of rotation is:

$$V_t = \frac{\omega^2 r D_p^2 (\rho_p - \rho)}{18\mu},$$

where ω is the angular velocity, D_p is the particle diameter, ρ_p the particle density, ρ the liquid density, and μ the liquid viscosity. This basic formula serves as the theoretical foundation for investigations of the phenomena of the centrifugal separation of particulate matter from the suspending fluid. Corrections to this equation for the nonspherical nature of blood cells and the effects of cell concentration have been proposed by various authors and serve as a basis of continued investigations of this process.

Blood is a complex mixture consisting of plasma, plasma

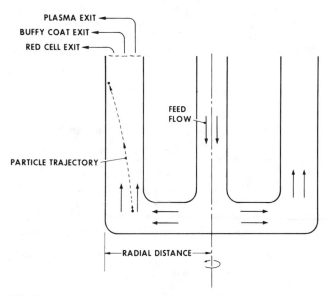

FIG. 7. Continuous plasma separation in a rotating bowl. The blood components are continuously or discontinuously ported off. Several commercial systems are available for clinical use.

TABLE 2   Comparison of Cell Properties

	Normal concentration in blood	Cell diameter (μm)	Mean density (g/ml)	Sedimentation coefficient (S $\times$ 10^7)
Red blood cell	4.2–6.2 $\times$ 10^{12}/liter	8	1.098 (1.095–1.101)	12.0
White blood cell	4.0–11.0 $\times$ 10^9/liter			
Lymphocyte	1.5–3.5 $\times$ 10^9/liter	7–18	1.072 (1.055–1.095)	1.2
Granulocyte	2.5–8.0 $\times$ 10^9/liter	10–15	1.082	
Monocyte	0.2–0.8 $\times$ 10^9/liter	12–20	1.062	
Platelet	150–400 $\times$ 10^9/liter	2–4	1.058	0.032
Plasma	—	—	1.027 (1.024–1.030)	

solutes in a broad range of sizes, and cells of varying densities and sizes (Table 2). When blood is allowed to sit in a tube or is centrifuged, the higher density red blood cells will settle to the bottom of the tube first, followed by the white blood cells and platelets (which are of intermediate density between the red cells and the plasma). Based on the sedimentation coefficients, the order of separation is red cell > white cell > platelet. For centrifugal speeds which are nondestructive to the cells, no appreciable separation of the individual plasma constituents occurs (proteins, amino acids, electrolytes). The high sedimentation rate of red cells is related to their ability to aggregate, forming rouleaux. With rouleaux formation, the effective cell diameter increases, augmenting the settling rate, (see equation).

Table 3 lists the operational parameters and blood properties that affect the sedimentation rate of blood cells and therefore that dictate the plasma separation rate. From an operational point of view, the force produced during centrifugation is proportional to the radial distance of blood in the centrifugal field and the square of the angular velocity, or revolutionary rate, of the centrifugal apparatus. Whole blood flow rates in and component flow rates out, which will determine cell residence times (and therefore particle radial travel), establish the degree of separation between the various cell groups in a given centrifugal design.

In clinical operation, visual inspection or optical sensors are employed to select component pumping rates. For specific collection purposes, two or more separation steps may be employed. For example, the buffy coat is drawn from the whole blood and subjected to a second centrifugal separation or another process which concentrates specific white blood cell types.

Owing to the varying densities and sedimentation velocities of the blood cells in routine operation of clinical centrifuges, a distinct separation of the cell types from plasma is not achieved. Thus in routine plasma collection, the plasma contains cells, most typically platelets.

Membrane Plasma Separation

Membrane separation of plasma is schematically shown in Fig. 8. Under low hydraulic transmembrane pressures, the

TABLE 3   Plasma Separation Considerations in Centrifugation

Operational and system-related parameters
  Whole blood and component flow rates
  Dimensions of centrifugal apparatus
  Angular velocity
  Radius of cell–plasma interface

Blood properties
  Cell diameter
  Cell concentration
  Specific weight of cell
  Specific weight of blood
  Viscosity of blood and plasma

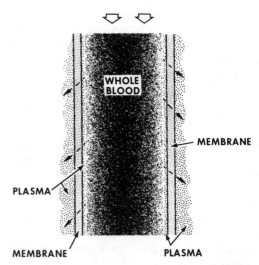

FIG. 8.   Plasma separation from blood by a membrane. Based upon the membrane pore structure and operating conditions, the plasma is separated free of cells. Membranes of differing polymer types have been developed for clinical applications.

TABLE 4 Plasma Separation Considerations in Membrane Systems

Operational and module-related parameters
    Blood flow rate (shear rate)
    Plasma flow rate (filtration velocity)
    Transmembrane pressure
    Membrane structure
        (pore number and pore size)
    Filtration area
    Number of fluid paths
    Blood channel dimensions

Blood properties
    Cell properties
    Cell concentration
    Macromolecule concentration
    Viscosity of blood and plasma

plasma is separated. With the fine pore size membranes employed, the cells do not pass; thus the separated plasma is devoid of any cells. Plasma solute removal, as assessed by its sieving coefficient (ratio of the concentration in the separated plasma to the concentration in the plasma entering the separator), has been shown to be near complete for all macromolecular weight solutes (albumin, immunoglobulins, lipids) of clinical interest. It has been shown that exceeding the safe limits of blood and plasma flow rates and transmembrane pressure leads to deterioration of plasma flux and solute sieving, and may cause hemolysis. In comparison with other membrane filtration processes such as reverse osmosis, dialysis, and ultrafiltration, the separation fluid flux (ml/min-m^{-2}) and solute sieving rates are high for membrane plasma separation. This is due to the membrane styles employed. Such microporous membranes have relatively high porosities and high mean membrane pore diameters. Studies have also shown that the cellular and macromolecule concentrations of the blood will to a significant extent dictate operational limits of the membrane plasma separation process. Thus in the control of the process, membrane and module properties, blood properties, and operating conditions should be considered (Table 4).

Conventional filtration theories have been applied to the use of membranes to separate plasma from blood with only limited success. With increasing amounts of data collected on modules and membranes of varying properties, the theories have had to be modified to improve experimental correlations.

Operation of a hollow-fiber membrane plasma separation system is shown in Fig. 9. This schematic may be used to depict the control necessary for operation of membrane systems. For a given membrane or module design, and operation with a blood of given specification (cell and solute concentrations), it has been determined from studies on multiple types of membrane units that a unique relationship exists among the plasma flow and the transmembrane pressure for a given membrane type. Exceeding specified plasma flows causes the transmembrane pressure to rise unstably and the plasma flux to decline (Fig. 10). The plasma flux–transmembrane pressure relationship has been found to be highly dependent on module design and membrane properties, blood flow and membrane properties, blood flow conditions, and blood properties. It is recognized that membrane polymer type and module sterilization method influence the degree of complement activation, coagulation, and solute–material interactions (such as adsorption) that occur at the membrane–plasma interface and that can influence the net pore size of the membrane during use. However, there is no evidence to suggest that the membrane chemical composition directly influences filtration.

Plasma separation is controlled through the blood and plasma flow rates (Fig. 9). The maximum value of the transmembrane pressure $(P_i - P_F)$ is an indicator of the stability of the plasma separation process, and the feedback control for the blood and plasma pumps. The value of $(P_i - P_F)$ is generally a more sensitive indicator than the mean transmembrane pressure $(P_i + P_o)/2 - P_F)$ because P_o may be monitored at a point downstream that will cause a significant difference between the mean and maximum values. As shown in Fig. 10, when the maximum allowable plasma flow rate is exceeded, the transmembrane pressure will rise and not stabilize. This has been shown to be the case for all membrane modules evaluated.

Increases in transmembrane pressure with increases in plasma flow in the stable transmembrane pressure region have been shown to depend on the membrane microstructures (pore size and pore number). An unstable operation is recognizable when the maximum transmembrane pressure does not stabilize with time for a given constant plasma flow. The plasma separation rate can be controlled manually or automatically through the transmembrane pressure change, or by preselecting a maximum operating transmembrane pressure.

The pore properties of the membranes employed for plasma separation are too small to pass even the smallest of the blood cells, the platelet, at operating conditions for which the plasma separation rate is stable. A plasma product devoid of particulate is deemed more desirable when on-line treatment of the plasma is carried out since the particulate may be interfering with the on-line treatment processes. Furthermore, the separation of cellular elements with plasma may result in biological systemic reactions. On-line treatment systems being employed at present include membrane filtration, sorption, and processing by enzyme and biological reactors.

Plasma Treatment

The rationale for therapeutically applying plasmapheresis is that the removal from plasma of macromolecules accumulated in a disease will correct the abnormality. For plasmapheresis techniques to be effective, the macromolecules must be accessible through the circulation and exhibit favorable kinetics with respect to their rates of production and intercompartmental transfer (tissue, or interstitial, to plasma). In theory, plasma exchange can remove all solutes and achieve a postexchange solute concentration that is dependent on the initial solute concentration, the patient's plasma volume, and rate of plasma exchange as described by the equation:

$$C_f = C_i \exp(-Q_p t / V),$$

where C_f is the postexchange solute concentration, C_i is the

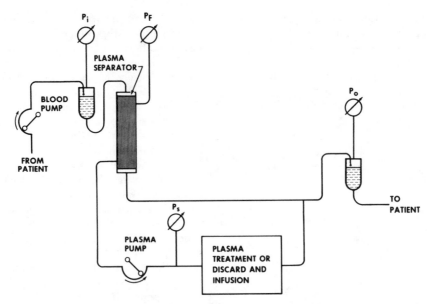

FIG. 9.   Scheme of operation with a hollow-fiber membrane plasma separator.

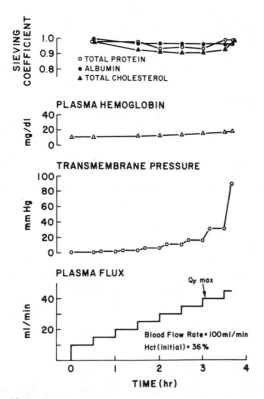

FIG. 10.   *In vitro* test with bovine blood of a polyvinyl alcohol hollow-fiber plasma separator. Note that sieving for macromolecules is close to unity and that plasma flow below a given value (in this case, 40 ml/min) shows stability with respect to the transmembrane pressure. Unstable plasma separation is readily noted by the rising transmembrane pressure when the plasma flow is increased above 40 ml/min.

initial plasma solute concentration, Q_p is the mean plasma exchange flow rate for the treatment time period t, or $Q_p t$ is the volume exchange per treatment period, and V is the patient's plasma volume (one-compartment model).

Plasma exchange is carried out when the specific factors in the plasma to be removed are not known or a specific technique for their selective removal is not available, or a group of solutes are to be removed and exchange is the simplest technique to be employed. However, plasma exchange has certain limitations. Since a large volume of plasma is discarded, many plasma constituents which are important to maintain homeostasis are discarded. This necessitates reliance on plasma products to replace the discarded plasma. Typically, albumin is used to replace the discarded plasma but in some cases whole plasma or other plasma fractions such as immunoglobulins are required. In 1987, 12 million liters of plasma were collected in the United States, yet in recent years there has been a critical shortage of albumin, thus placing limits on its use. In addition to the limited supply of plasma product, the products are a potential source of contamination.

Schemes of plasma treatments which would remove specific plasma factors, minimize total volume and mass of plasma solutes removed, and eliminate or minimize the need for replacement protein solution could be particularly beneficial and more cost effective. A number of systems for on-line treatment have been developed. These are generally classified into their broad categories: membrane plasma fractionation, plasma sorption, and other physicochemical methods.

Membrane Plasma Filtration

Membrane filtration is the technique by which plasma is filtered through single or multiple membranes following its separation from the blood. The principle of macromolecule solute removal from plasma by membrane filtration is based on solute size differences. Molecules which are smaller than the pores of the membrane pass through, whereas molecules larger than the size of the pores are retained by the membrane. The membranes used for plasma filtration are of the microporous types, with permeability and pore sizes typically much larger than those of the standard hemodialysis membranes, and smaller than those of the plasma separation membranes. Most membranes available have a broad range of pore sizes and therefore produce plasma solute separation efficiencies that are a function of the membrane media and the conditions of operation. Separation efficiency may be defined in terms of sieving coefficient (SC) or rejection coefficient (R) where $R = 1 - SC$. Sieving coefficient is calculated as the ratio of the concentration of the solute in filtrate to that in the incoming plasma. A sieving coefficient of 1 means the solute passes completely; a sieving coefficient of 0 means the solute is completely rejected and therefore retained by the membrane. Most macromolecular weight plasma solutes have sieving coefficients between these extremes.

On-line membrane plasma fractionation techniques can be classified according to the temperature range of their separation, namely: (1) cascade (double) filtration (filtration at ambient temperature); (2) cryofiltration (filtration below physiological temperature), and (3) thermofiltration (filtration at or above physiological temperature). These techniques have been developed and used in clinical plasma therapy. The temperature and membrane chosen for various procedures are dependent on the targeted macromolecules to be removed. For instance, cryofiltration is used in treating disease states associated with cold-aggregating serum solutes. Thermofiltration has been used to selectively remove low-density lipoproteins (LDL) in hyperlipidemic patients.

Sorption Plasma Fractionation

Sorption methods have been developed and used in plasma therapy to selectively remove pathological solutes to eliminate the need for substitution fluids. As in membrane plasma fractionation, the procedure involves the separation of plasma followed by perfusion through a sorbent column. Following treatment, the plasma is mixed with the concentrated cell fraction from the plasma separator and returned to the patient. Over the past two decades the development and use of sorbents in hemoperfusion or on-line plasma perfusion has been applied in treating drug overdose, uremia, liver insufficiency, autoimmune disorders, and familial hypercholesterolemia. Table 5 outlines the sorbents evaluated. Only a limited number of sorbents have reached clinical use.

Other Physicochemical Methods

Plasma fractionation may also be carried out by other methods. For example, on-line precipitation of LDL or the globulin fraction of plasma has been studied. Convective electrophoresis has also been investigated for protein separation.

TABLE 5 Sorbents[a]

Ligand or material for adsorption	Agent sorbed
Polylysin methylated albumin	T4 phage DNA
Anion-exchange resin Polyanion	bilirubin[b]
Dextran sulfate	LDL[b]
Heparin: heparin agarose	LDL
Tryptophan IM-TR	anti-acetycholine receptor Ab,[b] IC, RF[b]
Phenylalanine IM-PH	anti-MBP Ab,[b] IC, RF
Modified PVA gel I-02	RF, IC, anti-DNA Ab anti-RNP Ab anti-SM Ab
Oligosaccharide	anti-blood type AB
Charcoal sorbent	bilirubin, creatinine, urea, potassium
DNA	anti-DNA Ab[b]
Ag Blood-type Ag	anti-blood type Ab[b]
Insulin	anti-insulin Ab
Factor VIII	anti-factor VIII Ab[b]
Factor IX	anti-factor FIX Ab
Anti-LDL Ab	LDL[b]
Ab Anti-α Feto Ab	α-fetoprotein
Anti-HBs Ab	HBs
Anti-IgE Ab	IgE
Clq	IC
Protein A	IC, IgG,[b] C_1

[a]PVA, polyvinyl alcohol; RF, rheumatoid factor; IC, immune complexes.

[b]Clinical application stage.

Cytapheresis

Cytapheresis is most commonly performed by centrifugation (see the section on centrifugal plasma separation). The most common application of cytapheresis has been in the treatment of leukemia by leukocytapheresis. Lymphocytapheresis has been investigated in a number of autoimmune disorders, including rheumatoid arthritis and renal allograft rejection. Erythrocytapheresis has been employed in sickle cell anemia and severe parasitemia such as malaria. Thrombocytapheresis has been used to treat patients whose platelet concentrations are greater than 1×10^{12}/liter.

In addition to the centrifugal technique, filters have been developed specifically for the removal of leukocytes and lymphocytes. It has been shown that granulocytes adhere to various kinds of fibers such as nylon or cotton. Acrylic and polyester fibers have been found to bind granulocytes and lymphocytes. In such procedures, the blood is generally perfused on-line through a device containing the fibers. Several liters of blood may be processed over a couple of hours. The adhesion of cells to the fiber materials is related to the diameter of the fiber. Fiber devices are also used in blood collections to reduce certain cell populations. Specific adsorbent materials and magnetic separation techniques are also being developed to more specifically separate and collect cells.

A unique white blood cell treatment scheme referred to as photopheresis is now clinically available. In photopheresis, patients are orally administered the drug methoxsalen. The drug enters the nucleus of the white blood cells and weakly binds to the DNA. Blood is drawn, centrifugation is carried out, and the separated plasma and the lymphocytes are combined and exposed to ultraviolet A light. The photoactivated drug is locked across the DNA helix and blocks its replication. After the irradiation, this fraction is combined with the remainder of the blood and reinfused. The technique is used to treat skin manifestations of cutaneous T-cell lymphoma.

LUNG SUBSTITUTES AND ASSIST

The lungs are the body's interface with the air. They provide the means of getting oxygen into the blood and carbon dioxide out. The lungs perform their transport processes via a membrane that separates the air from the blood. As the blood cells deform to pass through the intricate and very fine capillary network, oxygen and carbon dioxide transport occur through the membrane. Failure of the lungs or their inability to carry out these functions requires artificial respiratory support in the form of mechanical assistance or substitution.

A ventilator provides mechanical assistance in breathing. It can be of two types, volume or pressure controlled. The cycling of the ventilator may be controlled and in most cases ventilators are set to operate continuously and independently of the patient's inspiratory efforts.

Artificial lungs are of two general types; direct blood contacting and membrane. Direct blood contacting devices may be of two types: bubble or film. In bubble-type oxygenators, oxygen gas is bubbled through the blood. The large surface area of the bubbles and their intimate contact with blood promote high gas-transport rates. Since the blood cells can be damaged by the mixing action, this type is restricted to short-term use for routine cardiac surgery. In film-type oxygenators, the gas contacts the blood, which is spread or distributed in films. These devices generally incorporate a mechanical mixing device so that the blood film is continually renewed. Since such complex devices are not made in disposable form, they are no longer employed clinically.

In membrane oxygenators, the blood and gas are separated by a membrane. The membrane may be of two types, diffusion (as silicone rubber) or microporous (as polypropylene), and may be in the form of film or hollow fibers. Combination-type membranes (thinly coated microporous membrane) are also available. Such membranes are designed to reduce water vapor transfer and improve biocompatibility. In hollow-fiber devices, the blood flow may be on the inside or outside of the fiber. In membrane oxygenators, the gas transport is similar to that in the natural lungs where a thin tissue layer separates the blood and gas. Damage to blood in membrane oxygenators is generally considered to be less than in direct-contacting devices. This allows them to be used for extended periods of time, even up to several weeks as in extracorporeal membrane oxygenation (ECMO) used in lung disorders, such as acute respiratory distress syndrome (ARDS), carbon dioxide retention, and in particular, ARDS caused by immature lung development in the newborn. Such extended uses are rare at present.

More than 400,000 open heart surgery procedures are performed each year worldwide and these procedures require a blood–gas exchanger or so-called artificial lung or oxygenator. Membrane oxygenators are being used more frequently for routine cardiopulmonary bypass. More than 80% of the cases use a membrane oxygenator. In open-heart surgery, the oxygenator is used in conjunction with a pump ("heart-lung machine") as shown in Fig. 11. Membrane oxygenators are also used in organ preservation to provided oxygen and remove carbon dioxide from the perfusate circulated through the organ.

Most recent has been the investigation of membrane oxygenators as intravascular devices. When placed in the vascular system, such devices can provide partial support.

BIOARTIFICIAL DEVICES

Because of the limitations of mechanical and passive mass transport devices, there is a growing interest in the development of hybrid artificial organ technologies. Hybrid artificial organs combine some form of biological material and nonbiological material to make a substitution or assist an organ system. The biological system may consist of enzymes or cells which carry out the biological and chemical functions absent or destroyed in the diseased organ system. The nonbiological material is used to encapsulate or enable the utilization of the biological material. Such applications typically involve the use of membranes. Two areas of investigation involve hepatic (liver) and pancreatic assist.

The liver is a complex organ, often referred to as the body's chemical factory. Its many functions can be classified as secretion of bile; metabolism of proteins, fats, and carbohydrates; detoxification of drugs and metabolites; and storage of vital substances such as iron and vitamins. Under certain clinical conditions the liver may regenerate itself in a short span of time, thereby making assist systems practical. However, substitution for the required metabolic support is not possible with present knowledge. The culturing of hepatic cells or the use of liver cells or tissue in membrane modules that permit extracorporeal perfusion but prevent direct blood to hepatic cell contact is being experimentally investigated. The membrane may provide a means for containing the cells; ensure the transport of oxygen, carbon dioxide, nutrients, and cell products; and provide a barrier against immunologic linteractions between the cells or tissue and the patient. Because of this latter concern, membranes with a molecular weight below about 70,000 Da and typically below 50,000 Da are employed.

A major function of the pancreas is the secretion of insulin by its beta cells in the islet tissue in response to glucose increases in the blood. The lack of such control leads to diabetes.

In addition to mechanical and chemical modes of insulin delivery, efforts are directed at transplanting islet cells. Because

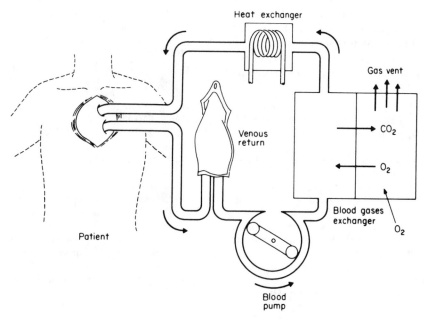

FIG. 11. Circuitry for heart-lung bypass.

of the immunologic-related problems discussed earlier, macro-encapsulation (as in semipermeable hollow-fiber membranes) or microencapsulation of islets within thin semipermeable films is being investigated. Such systems may be implanted or used outside the body.

IMPORTANT AREAS OF CONCERN IN DEVELOPMENT OF EXTRACORPOREAL ARTIFICIAL ORGANS

Significant progress has been made and applications not considered possible have become reality; however, the field of extracorporeal artificial organs is still considered to be in its infancy. For example, chronic hemodialysis has only been possible since 1960. At present, all extracorporeal applications require the use of anticoagulants. While efforts are being expanded to develop materials and coatings that will not require anticoagulation, few practical applications are possible at present. By design, extracorporeal artificial organs are invasive; blood access in many cases limits carrying out the procedures safely and effectively.

Extracorporeal artificial organs require the use of varied types of materials in contact with whole blood or plasma cells. Blood–materials interactions, through generally of only intermittent and short duration, can greatly affect various physiological systems, including the complement and coagulation cascades and the humoral and cellular immunological systems. While investigations point to changes and disturbances in such systems, the short- and long-term impacts are not known. The effects of a procedure may extend beyond the treatment times.

Since the systems are applied in a variety of disease states,

the effects are also clouded by the variables related to the clinical situation (concomitant drug therapy, stage of the disease, type of disease, and frequency and duration of the treatment). An extracorporeal artificial organ should not be viewed in engineering analysis as a black box but must be viewed as actively interacting with the physiological system. Therapeutic applications tend not to be tailored to the needs of the patients. Hemodialysis is commonly prescribed without quantitation of the procedure. The question of adequacy is still debated. The prescription requirements for other extracorporeal therapies are even more vague and there are no markers or guidelines to assess adequacy and efficacy. Even so, dialysis for renal assist and some of the other technologies discussed are viewed as mature developments, and generally taken for granted.

Transplantation benefits only a small percentage of those patients with chronic renal failure. Dialysis continues to support the large majority of patients, yet research on this technique or on improved technologies is poorly funded. Optimization of the devices used for extracorporeal therapies and processes is rarely achieved. These problems point to the need for a better understanding of the machine–human interface.

Regulatory requirements for new devices have considerably increased the time from concept to clinical application through marketing. Reimbursement issues on new technologies also greatly affect the research and development cycles. Concerns about animal research and the relationship of animal testing to the clinical situation also affect the development cycle.

The move to the development of hybrid organs emphasizes the lack of knowledge on organ system design as well as the lack of technology to substitute for the organ functions, and the need to make technologies more user friendly. Studies on organ substitution have been made possible by progress in such technologies as biomaterials and separation science. Developments in such fields will advance the science of and applications of extracorporeal artificial organs. This field is challenging. With a better understanding of the functions of the body, it holds promise for providing organ substitutes not only in end organ failure or as a bridge to transplantation, but also prophylactically.

Bibliography

Buckwald, H. (1987). Insulin replacement: bionic or natural. *Trans. Am. Soc. Artif. Intern. Organs* **33**:806.

Galletti, P. M., and Brecher, G. A. (1962). *Heart-Lung Bypass: Principles and Techniques of Extracorporeal Circulation.* Grune & Stratton, New York.

Gurland, H. J., Dau, P. C., Lysaght, M. J., Nosé Y., Pussey, C. D., and Siafaca, K. (1983). Clinical plasmapheresis: who needs it? *Trans. Am. Soc. Artif. Intern. Organs* **29**:774.

Henderson, L. W. (1980). Historical overview and principles of hemofiltration. *Dialysis Transpl.* **9**:220.

HEW, (1977). Evaluation of Hemodialyzers, and Dialysis Membranes. DHEW Publication No. (NIH) 77-1294, U.S. Department of Health, Education and Welfare.

Hori, H. (1986). Artificial liver: present and future. *Artif. Organs* **10**:211.

Nolph, K. D., Ghods, A. J., Brown, P. *et al.* (1977). Factors affecting peritoneal dialysis efficiency. *Dialysis Transpl.* **6**:52.

Nosé, Y. (1969). *The Artificial Kidney.* Mosby, St. Louis.

Nosé, Y. (1973). *The Oxygenator.* Mosby, St. Louis.

Maher, J. F. (1980). Pharmacology of peritoneal dialysis and permeability of the membrane. *Dialysis Transpl.* **9**:197.

Malchesky, P. S. (1986). Immunomodulation: bioengineering aspects. *Artif. Organs* **10**:128–134.

Malchesky, P. S. and Nosé, Y. (1987). Control in plasmapheresis. *In Control Aspects of Biomedical Engineering.* M. Nalecz, ed. Int. Fed. Automatic Control, Pergamon Press, Oxford, pp. 111–122.

Matsushita, M. and Nosé, Y. (1986). Artificial liver. *Artif. Organs* **10**:378.

Michaels, A. S. (1966). Operating parameters and performance criteria for hemodialyzers and other membrane separation devices. *Trans. Am. Soc. Artif. Intern. Organs* **12**:387.

Mito, M. (1986). Hepatic assist: present and future. *Artif. Organs* **10**:378.

Practical Aspects of Biomaterials

Implants and Devices

James M. Anderson, Brian Bevacqua, A. Norman Cranin, Linda M. Graham, Allan S. Hoffman,
Michael Klein, John B. Kowalski, Robert F. Morrissey, Stephen A. Obstbaum, Buddy D. Ratner,
Frederick J. Schoen, Aram Sirakian, and Diana Whittlesey

9.1 INTRODUCTION

Frederick J. Schoen

This section contains chapters that describe biomaterials used in implanted medical devices. The areas chosen illustrate the wide range of technical and medical considerations and include the most frequent types of medical devices surgically implanted today, including cardiovascular, dental, ophthalmological, and orthopedic prostheses. Clinical and device-specific biomaterial issues are integrated in order to produce a "consumer's" perspective on the use of implanted biomaterials in patients. As you will see, each area has its own unique problems and potential solutions.

9.2 STERILIZATION OF IMPLANTS

John B. Kowalski and Robert F. Morrissey

Materials implanted into the body of a human or an animal must be sterile to avoid subsequent infection that can lead to serious illness or death (see Chapter 4.7). In this chapter, we discuss the meaning of the term "sterile," the various sterilization methods, including their advantages and disadvantages, and the validation of sterilization processes from the point of view of sterility as well as the integrity of the physical and biological properties of the implant (Dempsey and Thirucote, 1989).

STERILITY AS A CONCEPT

"Sterile" is defined as the absence of all living organisms. This especially includes the realm of microorganisms, such as bacteria, yeasts, molds, and viruses. The presence of even one bacterium on an implant renders it nonsterile. Sterility should not be confused with cleanliness. A shiny stainless steel surface may easily be nonsterile (contaminated with numerous unseen microorganisms), while a rusty nail will be sterile after exposure to an appropriate sterilization method. Even fecal material can be rendered sterile.

How then is sterility measured or proved? For relatively small numbers of implants (assuming the implant is not too large to test in its entirety), sterility can be determined by immersing the item into liquid microbiological culture media. If it is sterile, no microbial growth will be observed; if it is nonsterile, the culture medium will become turbid as a result of microbial proliferation (Fig. 1). Testing small numbers of samples, however, does not give very meaningful information about the sterility of a large batch of implants that have been subjected to an industrial-scale sterilization process. Sterilization validation studies are used to determine what is called the sterility assurance level (SAL).

The SAL is the probability that a given implant will remain nonsterile following exposure to a given sterilization process. The generally accepted minimum SAL for implants is 10^{-6} or a probability of no more than one in one million that the implant will remain nonsterile.

The determination of a SAL begins with the enumeration of the bioburden, the number of viable microorganisms on the implant just prior to sterilization (Morrissey, 1981). Bioburden is usually determined on 10–30 samples and involves washing, shaking, or sonicating the microorganisms off the implant into a sterile recovery fluid such as a saline solution. By using conventional microbiological techniques, the number of microorganisms in the recovery fluid can be determined.

Once the bioburden is known, fractional-run sterilization studies can be performed to determine the microbial rate of kill or process lethality. In a fractional sterilization run, implant samples (in packages) are exposed to a fraction of the desired

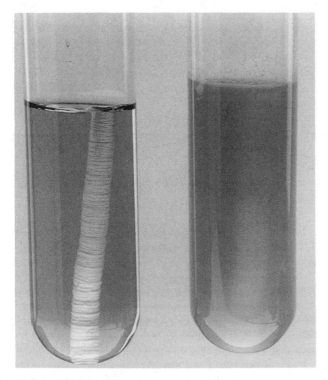

FIG. 1. Sterility tests of an experimental vascular graft illustrating negative (sterile, left) and positive (nonsterile, right) results.

sterilization cycle or dose. For example, if the proposed process exposure time is 2 hr, the fractional runs may have exposure times of 30, 40, and 50 min. Samples from these runs are tested for sterility and the results plotted to determine the exposure time required to achieve a 10^{-6} SAL. The results from such a study are shown in Fig. 2. In this example, the average bioburden per sample was 240. After the 30-, 40-, and 50-min fractional cycles, there were, respectively, 28/50, 7/50, and 1/50 samples that were still nonsterile. The calculated time to achieve a 10^{-6} SAL for this hypothetical bioburden and sterilization process is approximately 100 min, which is within the proposed 120-min exposure time. Note that when there is less than one surviving organism per unit, there is obviously not 0.01 of an organism on each unit, for example, but a probability of 1 in 100 that the unit is nonsterile.

OVERVIEW OF STERILIZATION METHODS

The first sterilization method to be used for implants was autoclaving, which involves exposure to saturated steam under pressure. Owing to the relatively high temperature of the process (121°C), most nonmetallic implants and packaging materials cannot be sterilized by this method. This limitation led to the development and use of ethylene oxide (EtO) gas and ionizing

radiation (gamma rays, accelerated electrons) to sterilize medical products (Association for the Advancement of Medical Instrumentation, 1988; Block, 1983).

Rationale for Choosing a Method

The choice of a sterilization method involves two key issues: implant and packaging compatibility, and the ability to attain the required SAL.

IMPLANT AND PACKAGING COMPATIBILITY

The first concern when choosing a sterilization method is the documentation that the method is compatible with the implant itself as well as the packaging required to maintain its sterile integrity prior to delivery to the operative site. The integrity of the implant and the packaging system must be demonstrated immediately after sterilization and also after aging studies (often at elevated temperatures) to prove the absence of delayed deleterious effects. Such delayed effects are most commonly encountered with radiation sterilization. During the implant and packaging compatibility studies, "worst case" processing conditions must be used to ensure that the implant and packaging are tested following the most rigorous conditions to which they may be exposed. For example, if the sterilization process specification allows a temperature range of 52° to 57°C, the tests must be conducted on samples exposed to a 57°C process. Also, the effect of multiple sterilization exposures must be considered in the event that an

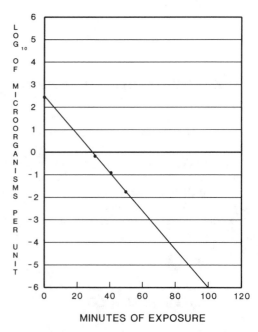

FIG. 2. Microbial kill curve based on data from a series of fractional sterilization runs. Time to achieve a 10^{-6} SAL is approximately 100 min.

implant may be exposed to a second sterilization process as a result of a sterilizer malfunction where the initial sterilization cycle was determined to be invalid.

When sterilizing with EtO gas, the amount of residual EtO and its byproducts in the implant and associated packaging must be determined. If required, an aeration regimen must be developed to ensure that the residuals are below safe limits and personnel are not exposed to airborne EtO.

SAL

Sterilization validation studies must also be performed to document the SAL of the proposed process. By using the techniques of bioburden determination and fractional sterilization studies described earlier, the ability of the proposed process to consistently deliver at least a 10^{-6} SAL must be conclusively documented. For these studies to be valid, the implants must be produced under actual manufacturing conditions and be exposed to the process in their final packaging configuration. Also, the fractional sterilization studies must represent the least lethal conditions (52°C in example above) allowed by the relevant process specification.

STEAM STERILIZATION

The first sterilization method applied to biomedical implants was steam sterilization or autoclaving.

Process and Mechanism of Action

With this method, sterilization is achieved by exposing the implant to saturated steam at 121°C. The use of steam at this temperature requires a pressure-rated sterilization chamber; a typical industrial steam sterilizer in shown in Fig. 3. The design of the implant must ensure that all surfaces are contacted by the steam, and the packaging must allow steam to penetrate freely. A typical steam sterilization process lasts 15 to 30 min *after all surfaces of the implant reach a temperature of at least 121°C.*

Steam sterilization kills microorganisms by destroying metabolic and structural components essential to their replication. The coagulation of essential enzymes and the disruption of protein and lipid complexes are the main lethal events.

Applications—Advantages and Disadvantages

Currently, the main use of this method for implants occurs in hospitals with the intraoperative steam sterilization of metallic devices. It is the method of choice for the sterilization of metallic surgical instruments and heat-resistant surgical supplies (linen drapes, dressings). A specialized form of steam sterilization is also used for many intravenous solutions.

The advantages of steam sterilization are efficacy, speed, process simplicity, and lack of toxic residues. The high temperature and pressure limit the range of implant and packaging compatibility.

EtO STERILIZATION

EtO sterilization has been exploited as a low-temperature process that is compatible with a wide range of implant and packaging materials.

Process and Mechanism of Action

Below its boiling point of 11°C, EtO is a clear, colorless liquid. It is toxic and a suspected human carcinogen. Contact with the skin and eyes and inhalation of the vapors should be avoided. EtO is used in the pure form or mixed with N_2, CO_2, or a non-ozone-depleting chlorofluorocarbon (CFC)-like compound (HCFC$_1$HFC. Pure EtO and mixtures without a proven inexting compound are flammable and potentially explosive. Owing to the negative effects of CFC compounds (Freon) on the earth's ozone layer, alternative inerting compounds have been developed (see later discussion).

In commercial use, implants contained in gas-permeable packaging are loaded into a sterilization vessel. A vacuum that is a compatible with the implant and the packaging is drawn to remove air, and the gas mixture is injected to a final EtO concentration of 600–1200 mg/liter. To be effective, a somewhat moist environment is required within the sterilizer; therefore, the relative humidity is maintained between approximately 40 and 90%. The sterilizer is maintained at the desired temperature (typically 30–50°C) for a sufficient time to achieve a minimum 10^{-6} SAL. The chamber is reevacuated to remove the EtO, and "air washes" are performed to reduce the EtO levels to below acceptable limits. Often, further aeration outside of the sterilization chamber (in some instances, at elevated temperatures) is required to effectively remove the residual EtO.

For EtO sterilization, a vessel similar to that shown in Fig. 3 is employed. It is somewaht more complex than its steam counterpart to allow for evacuation, moisture, and gas addition and control, and air washes. A typical EtO sterilization cycle can range from 2 to 48 hr in duration, depending on the time required for implant aeration. The lethal effect of EtO on microorganisms is mainly due to alkylation of amine groups on the nucleic acid.

Applications

EtO is used to sterilize a wide range of medical implants, including surgical sutures, intraocular lenses, ligament and tendon repair devices, absorbable and nonabsorbable meshes, neurosurgery devices, absorbable bone repair devices, heart valves, and vascular grafts.

EtO Residuals Issues

Because they are toxic and potentially carcinogenic, residual EtO and its by-products, ethylene chlorohydrin and ethylene glycol, are of concern when they are used with implants and their packaging (Glaser, 1979). Also, release of EtO into the environment is a concern owing to the potential for exposing

FIG. 3. An example of an industrial-scale steam sterilizer.

personnel. Depending upon the size of the implant and the site of implantation, permissible residual EtO levels range from 5 to 250 ppm (Table 1) (Food and Drug Administration, 1978). Current Occupational Health and Safety (OSHA) regulations dictate that a worker may not be exposed to more than 1 ppm of EtO during an 8-hr time-weighted average work day.

Advantages and Disadvantages

The advantages of EtO are its efficacy (even at low temperatures), high penetration ability, and compatibility with a wide range of materials. The main disadvantage centers on EtO residuals with respect to both the implant and release into the environment.

The current concern over the impact of CFC compounds on the earth's ozone layer will force facilities using the 12%/88% EtO/Freon mixture to switch to pure EtO, a non-CFC gas mixture, or one of the alternative inerting compounds that have been developed.

TABLE 1 Proposed EtO Residue Limits on Medical Devices

Device type	Maximum ppm
Implant:	
Small (<10 g)	250
Medium (10–100 g)	100
Large (>100 g)	25
Intrauterine devices	5
Intraocular lenses	25
Devices contacting:	
Blood (*ex vivo*)	25
Mucosa	250
Skin	250
Surgical scrub sponges	25

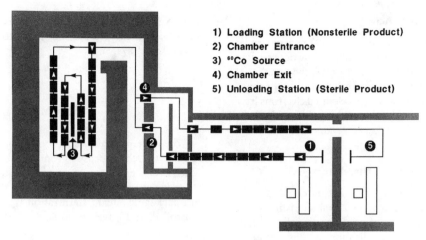

1) **Loading Station (Nonsterile Product)**
2) **Chamber Entrance**
3) **⁶⁰Co Source**
4) **Chamber Exit**
5) **Unloading Station (Sterile Product)**

FIG. 4.   A schematic top view of a typical industrial ⁶⁰Co irradiator.

RADIATION STERILIZATION

This method of sterilization utilizes ionizing radiation that involves either gamma rays from a cobalt-60 isotope source or machine-generated accelerated electrons. In this method, if a sufficient radiation dose is delivered to all surfaces of the implant, the implant can be rendered sterile.

⁶⁰Co Sterilization

Of the radiation sterilization methods, exposure to ⁶⁰Co gamma rays is by far the most popular and widespread method. Gamma rays have a high penetrating ability, and the doses required to achieve sterilization are readily delivered and measured.

Process and Mechanism of Action

A schematic top view of a typical industrial ⁶⁰Co irradiator is shown in Fig. 4. The ⁶⁰Co isotope is contained in sealed stainless steel "pencils" (1×45 cm) held within a metal source rack in a planar array. When the irradiator is not in use, the source rack is lowered into a deep (~20–25 ft) water-filled pool. At this depth, the radiation cannot reach the surface, and it is therefore safe for humans to enter the radiation cell. The outside walls and ceiling of the cell are constructed of thick, reinforced concrete for radiation shielding.

In use, materials to be sterilized are moved around the exposed source rack by a conveyor system to ensure that the desired dose is uniformly delivered. Radiation measuring devices called dosimeters are placed with the materials to be sterilized to document that the minimum dose required for sterilization was delivered and that the maximum dose for product integrity was not exceeded. The maximum dose divided by the minimum dose is referred to as the "overdose ratio." Irradiators are designed and product loading patterns are configured to minimize this ratio. The most commonly validated dose used to sterilize medical products is 25 kGy.

The radioactive decay of ⁶⁰Co (5.3 year half-life) results in the formation of ⁶⁰Ni, the ejection of an electron, and the release of gamma rays. The gamma rays cause ionization of key cellular components, especially the nucleic acid, which results in the death of microorganisms (Hutchinson, 1961). The ejected electron does not have sufficient energy to penetrate the wall of the pencil and therefore does not participate in the sterilization process.

Applications—Advantages and Disadvantages

⁶⁰Co radiation sterilization is widely used for medical products, such as surgical sutures and drapes, metallic bone implants, knee and hip prostheses, syringes, and neurosurgery devices. A wide range of materials are compatible with radiation sterilization, including polyethylene, polyesters, polystyrene, polysulfones, and polycarbonate. The fluoropolymer, polytetrafluoroethylene (PTFE), is not compatible with this sterilization method because of its extreme radiation sensitivity.

⁶⁰Co radiation sterilization approaches being the ideal sterilization method. It is a simple process that is rapid and effective, and it is readily controlled through straightforward dosimetry methods. The main disadvantages are the very high capital costs associated with establishing an in-house sterilization operation and the incompatibility of some materials with this method. Another disadvantage is the continual decay of the isotope (even when the irradiator is idle), which results in longer processing times and the periodic need for additional isotope.

Electron Beam Sterilization

Medical implants may also be sterilized with machine-generated accelerated electrons. With this method, radioactive isotopes are not involved because the electron beam is generated by an accelerator. The accelerator is also located within a concrete containment room but, in this case, when the accelerator is turned off, no radiation is present and a water-filled pool is unnecessary.

Process and Mechanism of Action

With this method, articles to be sterilized are passed under the electron beam long enough to accumulate the desired dose (again, usually 25 kGy). As with gamma rays, the lethality of this process is related to ionization of key cellular components. In contrast to gamma rays, however, accelerated electrons have considerably less penetrating ability, making this method unsuitable for thick or densely packaged products.

Applications—Advantages and Disadvantages

Electron beam sterilization has the same potential range of applications and material compatibility characteristics as the ^{60}Co process. However, because of its penetration distance, its use has been lmited. A unique application for this method is the on-line sterilization of small, thin products immediately following primary packaging.

OTHER STERILIZATION PROCESSES

Traditional Methods

Owing to the extreme temperatures involved (>140°C), dry heat sterilization is rarely if ever used for medical implants. Occasionally, implants are sterilized in hospitals by immersion in an aqueous glutaraldehyde solution. This procedure is used only in special circumstances where the implant is sensitive to heat and the aeration time after EtO sterilization is not acceptable. Achieving an acceptable SAL with this method requires meticulous attention to detail and relatively long immersion times.

New Technologies

Several new technologies are emerging that have potential utility for the sterilization of medical implants. These include gaseous chlorine dioxide (Kowalski *et al.*, 1988), low-temperature gas plasma, gaseous ozone, vapor-phase hydrogen peroxide, and machine-generated X-rays (Cleland, 1993; Morrissey and Phillips, 1993). The first four methods are being examined as potential alternatives to EtO. Machine-generated X-rays have the advantage of a nonisotopic source and penetrating power similar to gamma rays.

CHALLENGE TO THE BIOMATERIALS SPECIALIST

The challenge to the future biomaterials specialist will be to develop materials and implants that have the desired mechanical and biological properties and that can be sterilized by cost-effective methods having minimum potential effects upon the patient, manufacturing personnel, and environment. Reliance upon EtO will diminish in favor of radiation steriliza-

tion. New technologies that pose less risk to people and the environment should be exploited whenever possible (Jorkasky, 1987).

SUMMARY

Whenever possible, radiation sterilization should be used because it is reliable, easy to control, and has no toxic residues. When EtO is used, care must be taken to ensure that residuals are reduced to acceptable levels. For thermostable materials and packaging, steam sterilization is effective, rapid, readily controlled, and leaves no toxic residues.

Bibliography

Association for the Advancement of Medical Instrumentation. (1988). National Standards and Recommended Practices for Sterilization; 2nd ed. Association for the Advancement of Medical Instrumentation, Arlington, VA.

Block, S. S. (1983). *Disinfection, Sterilization, and Preservation*. Lea & Febiger, Philadelphia.

Cleland, M. R., O'Neill, M. T., and Thompson, C. C. (1993). Sterilization with accelerated electrons. in *Sterilization Technology* R. F. Morrissey and G.B. Phillips, eds. Van Nostrand Reinhold, New York.

Dempsey, D. J., and Thirucote, R. R. (1989). Sterilization of medical devices: a review. *J. Biomat. Appl.* **3**: 454–523.

Food and Drug Administration. (1978). EO, ECH, and EG, proposed maximum residue limits and maximum levels of exposure. *Federal Register* **43**(122): June 23.

Glaser, Z. R. (1979). Ethylene oxide: toxicology review and field study results of hospital use. *J. Environ. Pathol. Toxic.* **2**: 173–208.

Hutchinson, F. (1961). Molecular basis for action of ionizing radiation injury. *Science* **134**: 533.

Jorkasky, J. F. (1987). Medical product sterilization changes and challenges. *Med. Device Diagn. Ind.* **9**: 32–37.

Kowalski, J. B., Hollis, R. A., and Roman, C. A. (1988). Sterilization of overwrapped foil suture packages with gaseous chlorine dioxide. *Dev. Ind. Microbiol.* **29**: 239–245.

Morrissey, R. F. (1981). Bioburden: a rational approach. in *Sterilization of Medical Products, Vol II* E. R. L. Gaughran and R. F. Morrissey, eds. Multiscience Publications Limited, Montreal.

Morrissey, R. F., and Phillips, G. B., eds. (1993). *Sterilization Technology: A Practical Guide for Manufacturers and Users of Health Care Products*. Van Nostrand Reinhold, New York.

9.3 CARDIOVASCULAR IMPLANTATION
Linda M. Graham, Diana Whittlesey, and Brian Bevacqua

VASCULAR GRAFTS

Clinical Overview

The modern era of vascular surgery was ushered in by the development of prosthetic grafts in the 1950s, making aortic

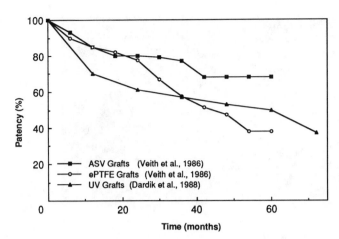

FIG. 1. Graph demonstrating patency of femoropopliteal grafts of various materials: autogenous saphenous vein (ASV), expanded poly(tetrafluoroethylene) (ePTFE), and umbilical vein (UV). Note that the patency of grafts of all materials drops immediately after implantation, reflecting poor outflow, thrombogenicity of the graft material, and technical errors. Unlike saphenous vein grafts, prosthetic grafts demonstrate a somewhat constant rate of decline in patency, which is probably a reflection of the progression of distal atherosclerotic disease and the development of anastomotic intimal hyperplasia.

replacement feasible. Voorhees (1952) successfully replaced the aorta of a dog with a tube fashioned from a sheet of Vinyon "N" cloth. Prosthetic grafts for clinical use followed and were fabricated of woven and knitted Dacron and Teflon. Despite their thrombogenic surface and lack of compliance, Dacron grafts, still the most commonly used aortic prosthesis, function well in a high flow setting, with cumulative patency rates for aortofemoral grafts of close to 90% at 5 years and 80% at 10 years (Lindenauer *et al.*, 1984, Sladen *et al.*, 1986). These grafts are durable, become incorporated in fibrous tissue to varying degrees depending on porosity and texture, and seldom become infected.

As the techniques of vascular surgery have improved, reconstructions have been carried to smaller arteries in the distal lower extremity. Autogenous saphenous vein (ASV) is the graft material of choice for revascularization of lower extremities with a 5-year patency of approximately 70 to 85% for femoropopliteal bypasses (Tilanus *et al.*, 1985; Veith *et al.*, 1986; Taylor *et al.*, 1987). When ASV is unavailable or of inadequate size or length, a prosthetic graft is often used. However, the inadequacy of synthetic grafts becomes apparent in the reconstruction of medium- and small-caliber arteries for cardiac and lower extremity revascularization. The only prosthetic graft materials with tolerable patency in reconstructions extending below the knee are expanded poly(tetrafluoroethylene) (ePTFE) and glutaraldehyde-stabilized human umbilical vein (HUV), both of which have significantly lower patency rates than ASV. The 5-year primary patency of ePTFE femoropopliteal and HUV grafts is approximately 40 to 50% (Tilanus *et al.*, 1985; Veith *et al.* 1986; Dardik *et al.*, 1988) (Fig. 1). The difference in patency between ASV and other graft materials is even more marked when the bypass is carried to tibial arteries. The patency of femorotibial bypasses with vein is approximately 85% at 5 years (Taylor *et al.*, 1987) compared with 12% for ePTFE

at 4 years (Veith *et al.*, 1986), and 30% for HUV at 5 years (Dardik *et al.*, 1988). HUV may have some advantage over ePTFE in low-flow situations, but has fallen into disfavor due to aneurysmal degeneration in up to 36% of grafts by 5 years (Dardik *et al.*, 1988).

Basic Precepts

The reasons for graft failure, apart from technical errors at the time of implantation, include the inherent thrombogenicity of the prosthetic surface, proliferation of tissue at the junction of the host artery and graft (anastomotic intimal hyperplasia), and progression of atherosclerosis in the native artery downstream from the graft. After a graft is implanted in the arterial system and blood flow is restored, proteins adsorb on the graft surface, the coagulation and complement systems become activated, and blood cells and platelets adhere to the graft surface with release of platelet granule contents. These blood–materials interactions are described in detail in Chapter 4.5. With time, the flow surface becomes less reactive, with decreased activation of blood elements. However, regardless of the length of implantation, prosthetic grafts in humans never develop a complete endothelial cell surface. This lack of complete endothelialization may be responsible for the increased thrombogenicity compared with normal blood vessels, as demonstrated by uptake of indium-labeled platelets even years after graft implantation (Stratton and Ritchie, 1986). If blood flow decreases below a critical level, thrombosis of the graft will ensue.

Anastomotic intimal hyperplasia can cause graft failure months to years after implantation. It occurs at all anastomoses with all types of graft materials, but it is more pronounced at the downstream anastomosis of a prosthetic graft (Fig. 2). The exact cause of this hyperplastic tissue response is not known. Contributing factors may include hemodynamic disturbances at the anastomosis, activation of platelets and other blood elements by the luminal surface of the graft, leading to release of growth factors, or mechanical stress due to a compliance mismatch between the graft and host artery. Although the role of compliance mismatch as a sole cause of anastomotic intimal hyperplasia has not been established, it appears to be a contributing factor (Abbott *et al.*, 1987; Okuhn *et al.*, 1989).

Another important cause of later graft failure is progression of atherosclerosis in the downstream vasculature. This accounts for over 50% of ePTFE femoropopliteal graft failures in which a cause can be identified (O'Donnell *et al.*, 1984). Atherosclerotic disease seems to progress more rapidly distal to a prosthetic graft than a vein graft, but this is difficult to quantitate. One possible explanation is the continued activation of blood elements by the inner surface of the graft. Platelet release products include growth factors for smooth muscle cells (SMC), and SMC proliferation is an essential part of atherosclerotic plaque formation.

Characteristics of the Ideal Graft

The characteristics of an ideal prosthetic graft include acceptable mechanical properties such as ease of handling and

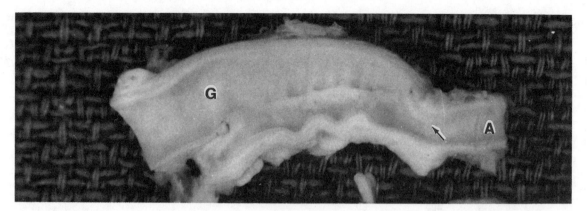

FIG. 2. Photograph of an anastomosis between a Dacron graft (G) and an iliac artery (A) removed from a dog 16 weeks after implantation. Development of anastomotic intimal hyperplasia is evident at the toe of the anastomosis (arrow).

suture placement, good suture retention, flexibility without kinking across joints, biocompatibility with tissue and blood elements, durability of the graft with its connective tissue ingrowth, and resistance to the formation of aneurysms. In addition, prosthetic grafts must be economical to manufacture, of uniform quality, available in a variety of sizes, easily stored, and sterilizable and resterilizable. Compliance should match that of the native vessel, and this compliance match should be maintained after tissue ingrowth. The graft should be capable of being "healed" or incorporated by the body to decrease the risk of infection as well as later formation of hematomas or seromas.

The extent of tissue growth through the interstices of the graft is determined by its porosity. However, optimal porosity of the graft wall is not clearly defined and must be a balance between porosity sufficiently low to provide structural integrity and limit extravasation of blood at implantation, and sufficiently high to permit incorporation. Porosity can be decreased temporarily by either preclotting the graft with the patient's blood prior to implantation or by applying an albumin or collagen coating on the graft when it is manufactured.

The flow surface of the graft must have low thrombogenicity and must be nonreactive so complement and the cellular elements of the blood are not activated. In addition, the flow surface must be stable and prohibit embolization from the graft. Ideally, the luminal surface should become covered with endothelium that is actively antithrombogenic.

Future Directions

The field of vascular surgery is rapidly expanding as a consequence of technical advances and the need to respond to an aging population. Although autogenous vein grafts fulfill most requirements for an ideal graft for small-caliber applications, suitable vein is not available in 10 to 20% of cases requiring reconstruction of medium- and small-caliber arteries. The need for a better synthetic graft is apparent.

It is now recognized that a useful prosthetic graft must not only be a durable conduit capable of being sutured into a blood vessel, but must also mimic the complex characteristics of an artery. The goal of prosthetic graft development is to construct a vascular substitute that has acceptable patency in small-vessel, low-flow, and low-pressure applications, including coronary artery bypass procedures, lower extremity revascularizations, and venous reconstructions.

One approach is to modify the surfaces of currently available grafts to minimize activation of blood elements. This has been accomplished by physicochemical means such as changing the surface charge of the graft and thereby altering protein adsorption and cellular deposition. Chemical bonding of pharmacologic agents, such as heparin, has been shown to improve graft patency in animal models. Establishing a functional biologic surface of endothelial cells on grafts minimizes activation of blood elements and promotes patency in experimental studies.

Another approach is to use polymers with unique properties such as elastomeric materials to produce a compliant graft or biodegradable materials to produce a graft that would be absorbed slowly as ingrowth of host tissue occurs. Testing is under way in these areas. However, success in animal models must be viewed with caution since application of such data to humans is unreliable because of the differences in healing processes between humans and experimental animals. Furthermore, the real value of a new graft or biomaterial cannot be ascertained without long-term implant data extending up to 5 years.

CARDIAC VALVES

Characteristics of the Ideal Valve

From a clinical perspective, the ideal cardiac valve should be nonthrombogenic, hemodynamically competent, and non-obstructive; cause minimal damage to blood elements; be durable, completely reliable, and easy to insert; and not annoy the patient with noise. Although replacement of intracardiac valves has been common for well over two and a half decades and has prolonged thousands of lives, the perfect valve substitute has yet to be developed. Currently, valve-related complications are the primary cause of death (over 50%) for patients followed on a long-term basis after heart valve replacement (Knott *et al.*, 1988). The following factors must be taken into account in designing and developing a heart valve.

Low thrombogenicity is highly important since the first arterial branches downstream from left-sided heart valves are the coronary arteries, followed by the arch vessels that supply the carotid arteries. Bits of debris and thrombus dislodged from a valve can cause a massive and fatal heart attack or serious stroke. Furthermore, acute thrombosis of the valve itself results in cessation of forward cardiac output. Even with long-term anticoagulation, some valves have exhibited excessive rates of thrombi and emboli and are no longer in use.

Hemodynamic adequacy is also of paramount importance. Replacement of a diseased native valve with a hemodynamically obstructive and/or incompetent prosthetic valve will not completely relieve cardiac strain and will further weaken cardiac function. For example, turbulence around the valve and differential rates of flow through the orifice can cause shear stress to blood elements and result in significant anemia and destruction of platelets. If a valve causes unrelenting destruction of blood cells, it must be removed and replaced with a different type of valve, at significant risk to the patient.

Lack of long-term durability has been a major problem in the design of cardiac valves and has resulted in the withdrawal of many valves from clinical use. Wear and irregularities of valve leaflets, discs, and balls can cause incompetence or stenosis, and fractures of valve struts can cause complete failure after several years of use. Durability is also crucial because replacement of a malfunctioning prosthetic valve carries a significant risk of death. Valves that need to be replaced for a third time can have an operative mortality of up to 33% for aortic valves, 50% for mitral valves, and 65% for multiple valves (Lytle *et al.*, 1986). This increased risk is due in part to (1) the presence of adhesions, making dissection planes difficult and causing more intra- and postoperative blood loss; (2) the difficulty in removing a prosthetic valve and leaving enough cardiac tissue to sew in another valve; and (3) the further myocardial damage secondary to a malfunctioning prosthesis. Thus, there is strong impetus for developing improved durable biomaterials for use in heart valves.

Ease of insertion and patient satisfaction are also important criteria. The sewing ring to which the valve is attached must be flexible and wide enough to allow sutures to be placed, but still narrow enough to allow the largest valve size possible for the patient's annulus. Because of the precise engineering of the disc leaflet and annulus, the sewing ring also has to be placed so that the cut sutures do not get caught between the prosthetic annulus and the prosthetic leaflet, causing the valve to jam. Finally, the amount of noise of the functioning valve is also important. Excessive valve noise can be extremely irritating to patients and their families, especially those patients who are in atrial fibrillation and have an irregular heart beat.

Surgical Implantation

From a surgical standpoint, initial removal and replacement of a valve is fairly straightforward. This is done with the patient on cardiopulmonary bypass and the heart cooled and arrested. If the aortic valve is to be removed, the aorta is opened above the coronary arteries. The diseased valve is excised, leaving a 2 to 3 mm circumference of native annulus. The resulting orifice is then sized with valve sizers corresponding to the particular type of valve to be used. Mattress sutures are then placed through the sewing ring, and the valve is seated into the native annulus by using a specific valve holder. Occasionally, the native valve annulus can be very calcified, making a hard, unyielding sewing ring difficult to seat without leaving a small perivalvular orifice that will leak and cause hemolysis and/or significant valve insufficiency. Prior to aortic closure, the valve is examined to ensure that it is not obstructing the orifices of the coronary arteries, which are usually 1 cm or less from the annulus.

The mitral valve is approached through the left atrium. Exposure of the mitral valve is somewhat more difficult owing to its position and tethering of the chordae tendinae, the fibrous cords inserting into the left ventricle and papillary muscles which keep the valve leaflets in place. The native valve is removed and sutures placed as in the aortic valve replacement. A potential problem unique to the mitral valve is that a valve with large struts or a caged ball device may protrude into the ventricular cavity. This can obstruct blood flow through the outflow track of the left ventricle to the aortic valve. In a small left ventricle, the struts can also impinge on the ventricular muscle and cause either bending of the struts or damage to the ventricular wall.

Currently Available Valves

All current cardiac valves lack some of the criteria for an ideal prosthesis. Therefore, specific patient needs must be matched to the attributes of a particular valve. There are essentially two classes of valve substitutes available at present: the bioprosthetic gluteraldehyde-preserved valve and the mechanical valve (Fig. 3). Both types are in common use with fair to good long-term use data.

Bioprosthetic gluteraldehyde-preserved valves are less thrombogenic, are easy to place in the valve annulus, cause minimal hemolysis, and make no abnormal noise in the patient. Unfortunately, these valves significantly obstruct blood flow, especially in the smaller sizes. A 19-mm aortic porcine prosthesis can have a resting gradient of 20 torr, rising to 54 torr with exercise (Yoganathan *et al.*, 1984), which obviously can place a significant strain on the ventricle. Another serious drawback is their lack of long-term durability, especially in younger patients.

The actuarial valve survival rate is estimated as 51% ± 15% after 12 to 14 years (Magilligan *et al.*, 1985). Patients under 20 years of age have a 50 to 80% failure rate after 6 years, and patients over 60 have a failure rate of 8% at 8 years. There is also some evidence to suggest that the failure rate increases after 10 years in position. The mode of failure is usually subacute, with partial cuspal tear or calcification and

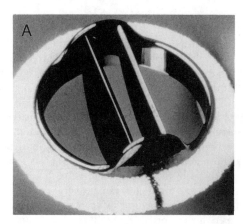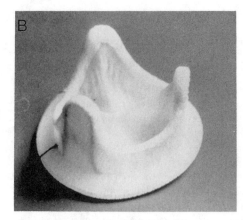

FIG. 3.   Photograph of a mechanical heart valve and a porcine bioprosthetic heart valve.

stenosis. This may be caused by processing of the tissue component of the leaflets, resulting in deformation of the normal collagen. There is usually sufficient time before acute failure to electively replace these valves. Currently, surgeons use the bioprosthetic valves for patients who have significant contraindication to long-term anticoagulation (e.g., a history of cerebral vascular or gastrointestinal bleeds, geographical or social constraints on anticoagulation monitoring, or a desire for pregnancy). They are also used in older patients whose natural life expectancy is more limited and who have a higher long-term valve survival rate. Factors against use of the bioprosthesis include a small annulus and young age. Advances in design are being directed toward methods of preserving prosthetic tissue that result in greater tissue longevity.

The mechanical valves currently available (caged ball, tilting disc, and bileaflet) are durable, but have a somewhat higher rate of acute failure than bioprosthetic valves. Their major drawback is an increased incidence of thromboembolism, requiring the indefinite use of an anticoagulant. As a consequence, these valves are also associated with significant hemorrhagic complications. In fact, 75% of all mechanical valve-related complications have been reported to occur from thromboembolism and hemorrhage (Hammond *et al.*, 1980). The thrombotic problems associated with mechanical valves are believed to be due to a combination of flow stasis around hinge points and the sewing ring, and a lack of endothelialization of the devices. The rates of thromboembolism are relatively consistent, from 0.7 to 3.1% per patient year. This works out to a cumulative risk of around 30% at 20 years after implantation. By definition, this includes transient ischemic attacks (TIAs) as well as peripheral emboli and strokes. The least thrombogenic material that is sufficiently sturdy for valve use currently is pyrolytic carbon.

The tilting disc and bileaflet valves are so-called low-profile valves, meaning that the valve itself is less than 1 cm in total depth, and hence does not impinge on the ventricular outflow tract in the mitral position, or on the aortotomy closure or coronary artery orifices in the aortic position. Thus, they are especially suitable for patients with a small left ventricle or aortic root. The bileaflet valves have the best forward hemody-

namic profile of the prosthetic devices, especially in the smaller sizes, although they have a slightly increased insufficiency rate, especially at slow heart rates. Thus, patients who are young, have no history of hemorrhagic problems, do not wish to become pregnant, and have small valve annuli or outflow tracts, are best suited to mechanical valves. Advances in mechanical valve design must include the development of reliable, nonthrombogenic materials and refinements of the flow dynamics to minimize turbulence and areas of flow stasis.

CATHETERS

Clinical Overview

Intravenous catheters are among the most widely implanted biomaterial devices in use today. It has been estimated that more than 25% of all hospitalized patients receive intravenous infusions (Maki *et al.*, 1973). In addition, there is a rapidly growing number of nonhospitalized patients whose long-term therapy (e.g., antimicrobial, anticancer chemotherapy) requires chronic (weeks to months) indwelling vascular catheters.

Characteristics of the Ideal Catheter

From the clinician's perspective, the ideal vascular catheter should elicit minimal inflammatory response even with long-term use, inhibit infection, be flexible, and be easy to place and secure. In addition, the usual responses of the vascular system to a foreign body, such as clot formation, white cell migration with cellular destruction, and vessel wall inflammation and damage, should be minimized or avoided. For example, some catheters are impregnated with heparin to attenuate blood clotting. Steel (scalp vein) needles are another example of a partially successful solution to these reactions. For a number of reasons (some still not explained), steel needles have been associated with a lower rate of microbial colonization and bloodstream infection. However, the drawbacks of steel needles include the lack of large needle sizes and their tendency

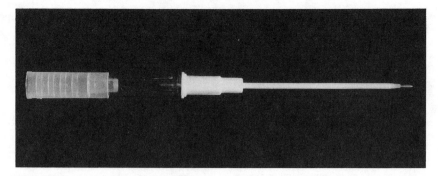

FIG. 4. Photograph of a commonly utilized type of catheter with a bevel-tipped needle and overriding catheter.

to become extravascular more easily than plastic catheters, thus requiring frequent changing of catheters and vascular sites.

Because the skin is one of the body's most effective defenses against bacterial entry, the design and construction of the catheter should aid in combatting the migration of bacteria through the percutaneous insertion site. Numerous studies have shown that intravenous catheters are rapidly colonized (beginning with insertion but increasing markedly after 48 hr of continuous use) and are a major source of bloodstream infections. This has led to the recommendation that catheters and catheter sites be changed every 48 hr, increasing patient discomfort and hospital costs. One possible solution is to attach to the catheter a silver-impregnated collagen cuff that is placed subcutaneously. These cuffs apparently act both as a physical barrier (the collagen inducing tissue growth that seals the subcutaneous tract) and a chemical barrier (the silver acting as a broad-spectrum antimicrobial). Silver-impregnated collagen cuffs reduced bloodstream infections from 13.8% (with catheters alone) to 0% in one study (Flowers *et al.*, 1989).

The ideal vascular catheter also must be flexible enough to allow vein and patient movement without becoming extravascular and damaging both the vessel and the surrounding structures. Furthermore, the catheter must maintain this flexibility indefinitely. Catheters that are initially supple may become brittle over time, resulting in vascular wall damage or shearing of the tip followed by catheter embolization. In addition, catheters must be available in a variety of sizes and lengths in order to accommodate the wide range of vessels to be cannulated.

Currently, the most common type of vascular cannulation involves using a needle with a beveled tip and an overriding catheter (Fig. 4). The catheter and needle length are coordinated so that only the needle bevel is introduced into the vessel before the catheter is also inserted. In this case, the ideal catheter should slide easily over the needle into the vessel without requiring extreme effort. The catheter hub should be designed with an area that allows both suturing and taping when attaching the catheter to skin. Finally, catheter hubs should be color coded using an industrywide standard so that the clinician can visually determine the size of the indwelling catheter.

Because intravenous catheters are relatively inexpensive and ubiquitously used, their significance is often overlooked in clinical situations. However, their importance is underscored by the disastrous results that can occur when they fail (loss of intravenous access, catheter emboli, etc.). Furthermore, it is not overstating the issue to say that without readily available intravenous access, modern medical practice would be impossible.

Bibliography

Abbott, W. M., Megerman, J., Hasson, J. E., L'Italien, G., and Warnock, D. F. (1987). Effect of compliance mismatch on vascular graft patency. *J. Vasc. Surg.* 5: 376–382.

Benson, R. W., Payne, D. D. and DeWeese, J. A. (1975). Evaluation of prosthetic grafts of different porosity for arterial reconstruction. *Am. J. Surg.* 129: 665–669.

Dardik, H., Miller, N., Dardik, A., Ibrahim, I. M., Sussman, B., Berry, S. M., Wolodiger, F., Kahn, M., and Dardik, I. (1988). A decade of experience with the glutaraldehyde-tanned human umbilical cord vein graft for revascularization of the lower limb. *J. Vasc. Surg.* 7: 336–346.

Echave, V., Koornick, A. R., Haimov, M., and Jacobson, J. H., III (1979). Intimal hyperplasia as a complication of the use of the polytetrafluoroethylene graft for femoral-popliteal bypass. *Surgery* 86: 791–798.

Flowers, R. H., Schwenzer, K. J., Kopel, R. F., Fisch, M. J., Tucker, S. I., Farr, B. M. (1989). Efficacy of an attachable subcutaneous cuff for the prevention of intravascular catheter-related infection. *J.A.M.A.* 261: 878–883.

Hammond, G. L., Laks, H., Geha, A. S. (1980). Development of aortic valve prostheses, *Conn. Med.* 44: 348–352.

Hobson, R. W. II, Lynch, T. G., Jamil, Z., Karanfilian, R. G., Lee, B. C., Padberg, F. T. Jr., and Long, J. B. (1985). Results of revascularization and amputation in severe lower extremity ischemia: A five-year experience. *J. Vasc. Surg.* 2: 174–185.

Knott, E., Reul, H., Knoch, M., Steinseifer, U., Rau, G. (1988). In vitro comparison of aortic heart valve prostheses Part 1: mechanical valves. *J. Thorac. Cardiovasc. Surg.* 96: 952–961.

Lindenauer, S. M., Stanley, J. C. Zelenock, G. B., Cronewett, J. L., Whitehouse, W. M. Jr., and Erlandson, E. E. (1984). Aorto-iliac reconstruction with Dacron double velour. *J. Cardiovasc. Surg.* 25: 36–42.

LoGerfo, F. W., Soncrant, T., Teel, T., and Dewey, F. C. Jr. (1979). Boundary layer separation in models of side-to-end arterial anastomoses. *Arch Surg.* 114: 1369–1373.

Lytle, B. W., Cosgrove, D. M., Taylor, P. C., Gill, L. L., Goornastic, M., Golding, L. R., Stewart, R. W., Loop, F. D. (1986). Reoperations for valve surgery: Perioperative mortality and determinants of risk for 1,000 patients, 1958–1984. *Ann. Thorac. Surg.* 42: 632–643.

Magilligan, D. J., Lewis, J. W. Jr., Tilley, B., Peterson, E. (1985). The porcine bioprosthetic valve: twelve years later. *J. Thorac. Cardiovasc. Surg.* 89: 499–507.

Maki, D. M., Goldman, D. A., Rhame, F. S. (1973). Infection control in intravenous therapy. *Ann. Int. Med.* 79: 867–887.

Malone, J. M., Moore, W. S., and Goldstone, J. (1975). The natural history of bilateral aortofemoral bypass grafts for ischemia of the lower extremities. *Arch. Surg.* 110: 1300–1306.

O'Donnell, T. F. Jr., Mackey, W., McCullough, J. L. Jr., Maxwell, S. L. Jr., Farber, S. P., Deterling, R. A., and Callow, A. D. (1984). Correlation of operative findings with angiographic and noninvasive hemodynamic factors associated with failure of polytetrafluoroethylene grafts. *J. Vasc. Surg.* 1: 136–148.

Okuhn, S. P., Connelly, D. P., Calakos, N., Ferrell, L., Man-Xiang, P., and Goldstone, J. (1989). Does compliance mismatch alone cause neointimal hyperlasia? *J. Vasc. Surg.* 9: 35–45.

Sladen, J. G., Gilmour, J. L. and Wong, R. W. (1986). Cumulative patency and actual palliation in patients with claudication after aortofemoral bypass. Prospective long-term follow-up of 100 patients. *Am. J. Surg.* 152: 190–195.

Stratton, J. R., and Ritchie, J. L. (1986). Reduction of indium-III platelet deposition on Dacron vascular grafts in humans by aspirin plus dipyridamole. *Circulation* 73: 325–330.

Taylor, L. M. Jr., Edwards, J. M., Phinney, E. S., and Porter, J. M. (1987). Reversed vein bypass to infrapopliteal arteries. Modern results are superior to or equivalent to *in-situ* bypass for patency and for vein utilization. *Ann. Surg.* 205: 90–97.

Tilanus, H. W., Obertop, H., and Van Urk, H. (1985). Saphenous vein or PTFE for femoropopliteal bypass. A prospective randomized trial. *Ann. Surg.* 202: 780–782.

Veith, F. J., Gupta, S. K., Ascer, E., White-Flores, S., Samson, R. H., Scher, L. A., Towne, J. B., Bernhard, V. M., Bonier, P., Flinn, W. R., Astelford, P., Yao, J. S. T., and Bergan, J. J. (1986). Six-year prospective multicenter randomized comparison of autologous saphenous vein and expanded polytetrafluoroethylene grafts in infrainguinal arterial reconstructions. *J. Vasc. Surg.* 3: 104–114.

Voorhees, A. B. Jr., Jareszki, A. III, and Blakemore, A. H. (1952). The use of tubes constructed from vinyon "N" cloth in bridging arterial defects. A preliminary report. *Ann. Surg.* 135: 332–336.

Yoganathan, A. P., Chaux, A., Gray, R. (1984). Bileaflet, tilting disc and porcine aortic valve substitutes: *In vitro* hydrodynamic characteristics. *J. Am. Col. Cardiol.* 3: 313–320.

9.4 DENTAL IMPLANTATION

A. Norman Cranin, Aram Sirakian, and Michael Klein

The most frequent surgical procedure performed upon humans is the dental extraction. Without question, the most frequently employed prosthesis is the complete or partial denture (Fig. 1), These are available as both fixed and removable devices. Most often they are supplied in the latter form owing to their relative ease of production as well as economy. There are numerous instances where, due to a paucity of natural teeth, a lack of posterior teeth (required as abutments), or in the presence of anatomic aberration resulting from traumatic, congenital, or metabolic causes, the most desirable type of prosthesis, the fixed or nonremovable type, cannot be used. In such circumstances, surgically implanted abutments may be placed to anchor prosthetic denture superstructures. Depending on the areas requiring such devices and the purposes which they must serve, dental implants are available in a variety of materials and an abundant number of designs (Cranin *et al.*, 1987; Worthington, 1988).

HISTORY

The Egyptians were known to have secured teeth to the jawbone with gold wire, and through the centuries, surgeons have sought techniques for implanting tooth substitutes, with little reported success. In the days of the American Revolution, seamen were attacked in darkened waterfront areas and their teeth extracted, to be implanted into the mouths of the wealthy and more fortunate gentry. Those biologic specimens were lost quickly owing to the immune responses of the recipients (Guerini, 1969; Cranin, 1970; Driskell, 1987).

In the early twentieth century, Greenfield reported on the successful implantation of circular gold and platinum cribs as artificial dental roots, to which porcelain teeth were attached by a slotted coupling device. Later, the Strock brothers, using an Austenal alloy cast in the shape of self-tapping wood screws (after the work of Venable and Stuck), implanted over a dozen cast alloy screws as free-standing dental implants. One was reported to have survived for over 20 years (Strock, 1939; Strock and Strock, 1949). The most consistent dental implant reported in the first half of this century, and which initiated the current era of enthusiasm, was the subperiosteal device. It was introduced by Gershkoff and Goldberg (Gershkoff and Goldberg, 1949). Enthusiasts flourished, and reports of long-term success came from all parts of the world. These castings, made from nonferrous, cobalt–chromium-based alloys, were employed at first only for completely edentulous mandibles, and in so doing, satisfied a significant and popular need. The individual who required single or unilateral implants, however, was not served well by adaptations of these early subperiosteal implants (which, for the first few years were not fabricated from direct bone impressions, but rather, by using a "guesswork" X-ray or intraoral replica template measuring system) (Weinberg, 1950). In 1951, Berman introduced the direct bone impression, which created opportunities for more accurate, longer lasting, and more predictable subperiosteal implants (Berman, 1951) (Fig. 2).

In 1947, Formaggini discovered a technique for placing the first of the modern-day, endosteal root-form implants (Formaggini, 1947). His method was improved by Rafael Chérchève, whose system was heralded universally (Chérchève, 1956) (Fig. 3). Those root forms, however, failed to serve all patients who were in need because their large dimensions could not be accommodated by the commonly found posterior alveolar ridge, which was thin because of atrophy after tooth loss.

Roberts in 1967 made a laminar-shaped stainless steel device that he called a blade implant. It was placed in a narrow osteotomy that all but the thinnest ridges could accept and showed considerable promise (Roberts and Roberts, 1970). Linkow developed this device into a globally acceptable system by creating a wide variety of sizes and shapes designed to fit

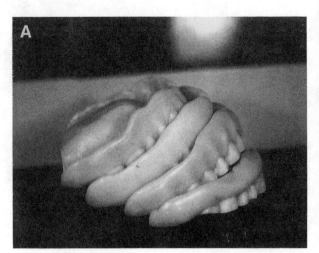

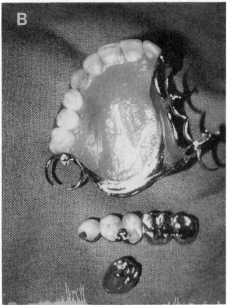

FIG. 1. (A) Complete, removable dentures are the prosthetic replacement used by most edentulous people. (B) Removable partial dentures, fixed partial dentures (bridges), and crowns are widely accepted replacements for missing teeth.

almost any clinical requirement (Linkow, 1968). The 1980s was the era of popularity for the blade endosteal implant (Fig. 4).

The concept of osseointegration was developed and the term coined by Professor Per-Ingvar Bränemark at the University of Göteberg, Sweden. Osseointegration has been defined as a direct structural and functional connection between ordered, living bone and the surface of a load-carrying implant (Bränemark, 1983). The initial concept stemmed from vital microscopic studies of bone and other tissues, which involved the use of an implanted titanium chamber containing an optical system for transillumination. The investigators observed that

the screw-shaped titanium chambers became completely incorporated within the bone tissue, which actually grew into the irregularities of the titanium surface (Bränemark, 1969). These findings led to research exploring the possibilities of artificial root replacement. The experimental studies that followed involved extraction of teeth in dogs and their replacement by osseointegrated screw-shaped implants. Fixed prostheses were connected after an initial healing period of 3 to 4 months during which they bore no load (Fig. 5). Radiologic and histologic studies showed that integration could be maintained for 10 years in dogs, without inflammatory reactions. Further, the

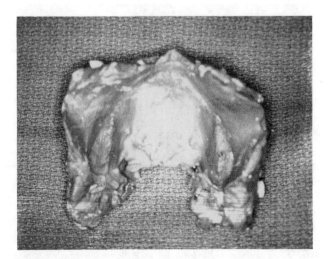

FIG. 2. This direct bone impression of a maxilla using a polysulfide material shows the accurate anatomic reproduction produced.

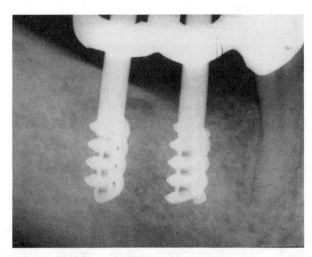

FIG. 3. These several spiral alloy Chérchève implants are being used to help support a fixed (nonremovable) bridge in conjunction with the natural teeth.

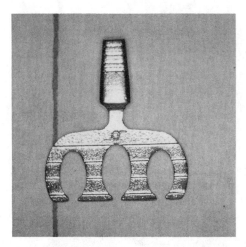

FIG. 4. The one-piece blade implant made of commercially pure titanium is a laminar appliance designed to fit into jawbone ridges too thin to accommodate the three-dimensional root-form types.

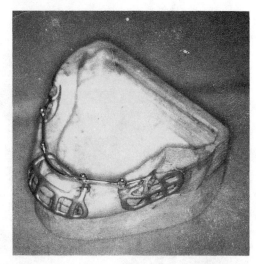

FIG. 6. CAD/CAM-generated models of the mandible eliminate bone impression surgical procedures for subperiosteal implants. Seen here is the mandibular tripodal subperiosteal implant that fits on the bone beneath the gum. Only the post and bars protrude to serve as retainers for dental prostheses.

load-bearing capacity of the individual implants was shown to be extremely high. Based on these findings, the Foundation for Osseointegration and the Bränemark Implant System were established in 1962. Basic research continued in the years that followed and the first edentulous patient was treated in 1965 (Bränemark , 1983).

Clinical data on osseointegrated implants were presented to North America at a scientific meeting held at Toronto in 1982 (Zarb, 1983). Success rates of 90–100%, initially presented by the Swedish group, were confirmed by longitudinal follow-up studies from over 50 osseointegrated implant centers around the world (Adell, 1983; Laney *et al.*, 1986; Albrektsson *et al.*, 1988).

Today, osseointegrated implants are used to treat both partially edentulous and completely edentulous ridges. Severely atrophied ridges, however, require several surgical procedures involving bone augmentation, to prepare the host site for acceptance of endosteal devices. A staunch group of implantologists

continues to utilize the subperiosteal implant for those patients who are considered untreatable by endosteal root-form or blade-type devices.

The main disadvantage of the subperiosteal implant continued to be the necessity of performing a two-stage operation—the first, a direct bone impression in order to make the implant casting, and the second, 24 hr to 6 weeks later, in order to insert the casting surgically. Using computer-based design and machining technology (CAD/CAM), Truitt and James, in 1982, introduced a method by which computerized axial tomography (CT) scans of facial skeletons could be employed to generate relatively accurate models of the jawbones. These models then were used to develop subperiosteal implant

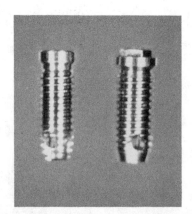

FIG. 5. The Nobelpharma implant is completely buried under the gum with its superior border level with the crest of bone. These submergible implants are left buried for a healing period of 3–6 months.

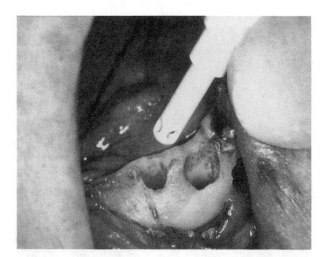

FIG. 7. Exact size and shape holes are made to accommodate implants placed within the bone. This hydroxylapatite (Integral) coated press-fit implant is stabilized initially by a precise frictional grip.

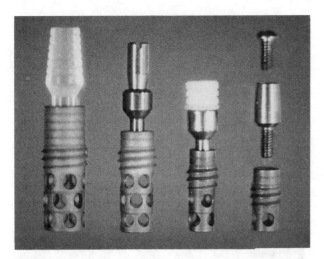

FIG. 8. These root-form (core-vent) implants have various attachments that are chosen according to the final design of the crowns or dentures.

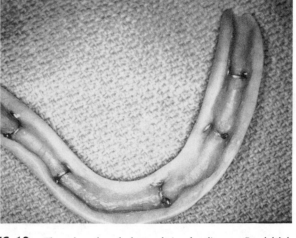

FIG. 10. The undersurface of a denture designed to clip onto a Brookdale bar.

castings (Truitt *et al.*, 1986). Such castings have served successfully and have eliminated the need for first-stage, bone-impression surgery (Fig. 6). The 1980s also witnessed a wide range of rootform, two-stage endosteal implants following the work of Bränemark, using numerous adaptations of his original threaded device. These have incorporated materials and designs that include press-fit (nonthreaded) configurations, coatings of alumina, hydroxylapatite, and titanium oxide plasma spray (Lemons, 1977; Niznick, 1982; Kay *et al.*, 1986; Kirch, 1986) (Figs. 7 and 8). Self-tapping implants have also broadened the field to a level that now offers the clinician a large and often confusing spectrum of implant varieties, the benefits of some remain unclear, and the selection of which is dependent upon a number of poorly defined criteria and prejudices (English, 1987).

CURRENTLY USED DENTAL IMPLANT MODALITIES

Subperiosteal

Subperiosteal dental devices may be used for partially and completely edentulous jaws and are the implants of choice for those regions that contain insufficient bone to accommodate endosteal implants of either the blade- or root-form varieties (Schnitman, 1987). They consist of a mesh-type infrastructure cast of a surgical-grade cobalt–chromium–molybdenum alloy to which are attached from four to seven permucosal cervices. Atop these cervical protrusions may be prosthetic abutments to serve as retainers for fixed bridge protheses or retentive bars, which connect the abutments together into a single structure. These are used to gain the retention of overdentures (Yurkstas, 1967; Dalise, 1979; Cranin *et al.*, 1978) (Figs. 9 and 10). Subperios-

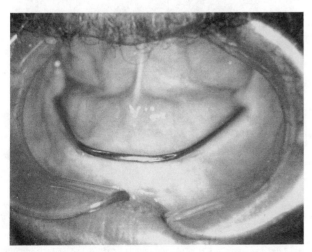

FIG. 9. The Brookdale bar is attached to an underlying subperiosteal infrastructure (see Fig. 6) by four to seven permucosal posts. A denture prosthesis is designed with clips for stabilization.

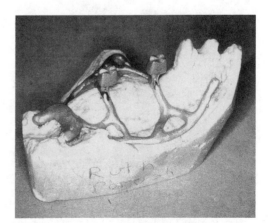

FIG. 11. The maxillary unilateral pterygohamular subperiosteal implant is used to provide posterior abutments for nonremovable bridges in the maxilla. It is designed to fit against the palatal zygomatic and pterygoid bones, which are reliable sources of support.

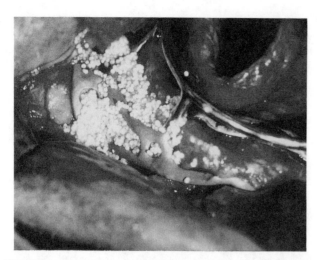

FIG. 12.   Minor discrepancies that occur between the subperiosteal implant fabricated on a CAD/CAM model and the jaw it rests on are grafted with synthetic bone replacement materials (e.g., hard tissue replacement, hydroxylapatite). These osteoconductive materials serve as a reliable physiological graft.

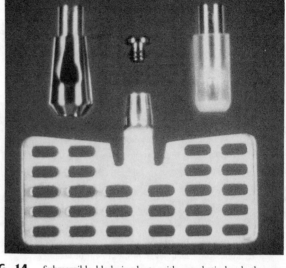

FIG. 14.   Submergible blade implants with prosthetic heads that may be attached to the implant at a date after the surgery allow for an unloaded healing period.

teal implants are cast to models of maxillae or mandibles made either by direct bone impressions using polysulfide or poly(vinyl siloxane) elastomeric impression materials or by CAD/CAM-generated models. They are designed to rest on the cortical bone, and are entrapped and fixed by a reattachment of periosteal fibers through the numerous interstices incorporated into their infrastructural designs. When successful, they are purely incidental to local tissue physiology (James, 1983).

Relatively high levels of success have been attributed to these subperiosteal designs, particularly the newer mandibular tripodal configurations as well as the pterygohamular maxillary designs (Linkow, 1986; Cranin *et al.*, 1985) (Fig. 11). When

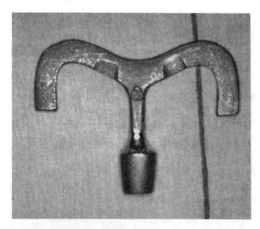

FIG. 13.   One-piece blade implants such as the Cranin anchor design have been used successfully as distal abutments or pier abutments for fixed (nonremovable) bridges. The entire structure, with the exception of the protruding dental abutment, is placed within the confines of the jawbone. The abutment protrudes to serve as a prosthetic retainer.

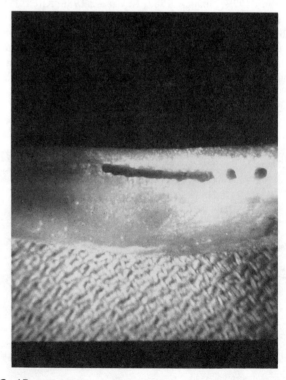

FIG. 15.   The receptor site for blade implants (such as the one shown in Figs. 13 and 14) is made freehand with a saline-cooled high-speed bur (250,000 rpm). First a dotted line following the length and position of the implant is made. Then the dots are connected to form the groove receptor site.

TABLE 1 Examples of Frequently Used Endosteal Implants

Name	Material	Description	Stages	Surface	Primary retention	Length (mm)	Diameter (mm)
Bränemark (Nobelpharma)	Titanium, commercially pure (CP)	Threaded screw	2	Machined	Threading (bone tapping, self tapping)	7, 10, 13, 15, 18, 20 7, 10, 13, 15, 18	3.75 4.0
Core-vent	Titanium alloy ($T_{14}VA_{10}A_1$)	Threaded hollow basket	2	Sand blasted	Threading, core (self-tapping)	8, 10.5, 13, 15	4.3, 5.3, 6.3
TPS screw	Titanium (CP)	Threaded screw	1	Plasma spray	Threading	8, 11, 14, 17, 20	3.5, 4.0
IMZ (Interpore)	Titanium (CP)	Bullet shaped	2	Plasma spray, optional hydroxylapatite	Press fit	8, 10, 13, 15 8, 11, 13, 15	3.3 4.0
Integral (Calcitek)	Titanium (CP)	Bullet shaped	2	Hydroxylapatite	Press fit	8, 10, 13, 15	3.25, 4.0
Kyocera (Bioceram)	Single crystal sapphire,	Threaded screw	1	Smooth, porous	Threading	E: 19.5, 22, 25 S: 19.5, 23 20.5, 24	3.3, 4.2 3.0, 4.0 4.2, 4.8
Tubingen	Aluminum oxide polycrystal porous	Stepped	1	Aluminum oxide stippled	Press fit	21, modifiable	4.0, 5.0 6.0, 7.0
Stryker	Titanium alloy ($TiAl_6Va_4$)	Finned	2	Machined, optional hydroxylapatite	Press fit	8, 11, 14	3.5, 4.0 5.0
Bausch and Lomb (Steri-Oss)	Titanium (CP)	Threaded screw, bullet shaped	2	Machined, optional hydroxylapatite	Threaded screw, Press fit	12, 16, 20 8, 10, 12, 14, 16 12, 16, 21	3.5 3.8 4.0

CAD/CAM-generated castings are used, slight inaccuracies are sometimes experienced which may be treated successfully with particulate bone substitute materials such as hydroxylapatite that are used as fillers (Fig. 12).

Endosteal

Blade Type

The blade- or plate-type designs are available in one-piece (with attached abutment) or submergible (with separate to-be-attached abutment) configurations (Figs. 13 and 14). They are fabricated in a great variety of designs of relatively pure surgical-grade titanium or titanium–aluminum–vanadium alloy. With the aid of clear plastic templates, these may be employed in an almost limitless number of host sites. In addition, several manufacturers supply the surgeon with one- and two-piece titanium blade-form blanks. These may be fashioned and cut (unfortunately, with stainless steel snippers) to almost any shape or size that the potential surgical site demands. If the implantologist cannot find an implant that will satisfy a need within the broad spectrum of manufactured designs, or believes in the anchor ("shoulderless") philosophy, a custom casting may be made by using distortion-free dental and CT scan radiographs to incorporate

dimensional accuracy (Cranin, 1980; Baumhammers, 1972). Chrome–cobalt alloy is the material most often used for such castings.

There is a question as to whether true osseointegration (a direct bony interface) occurs when intraosseous slot-osteotomies are created by using high-speed fissure bur operative techniques, or if two-stage submergible blades fare better than the one-piece implants. Nonetheless, there are a reasonable number of host sites, as demonstrated by CT scan analyses, that do not have the width to accommodate root-form implants and that, on a basis of time, and reports from the literature, support blade-form endosteal implants with success for periods exceeding 5 and even 10 years. James and others have reported that the mechanism of support is fibro-osseous integration. They indicate that a sling of connective tissue fibers suspends the implant from its bony crypt and that upon a stimulus of tension placed on these fibers caused by mastication, an environment of osteogenesis and continuous bone anabolism is created (James, 1988, Weiss, 1986). These implants, which require a reasonable level of surgical skill to insert, demand the use of freehand, high-speed, saline-cooled capabilities. (Fig. 15). They may be used as unilateral posterior abutments for fixed bridge prostheses or in a wide variety of prosthetic functions, such as supports for full coping bar/overdentures or even as single tooth supports.

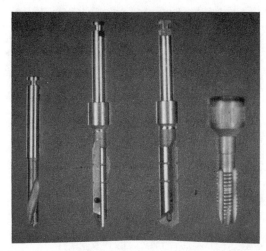

FIG. 16. A variety of instruments are used to create precise holes within the bone for the root-form implants. Twist drills, bi- and tri-spade drills (two cutting edges and three cutting edges, respectively), as well as bone taps to thread the bone all help to make an exacting atraumatic procedure. The two drills in the center, which can be rotated at speeds as low as 6 rpm, have hollow cores that allow passage of a cooling irrigant.

Root-Form

Long-term clinical experience has proven osseointegration to be a reliable modality for treating edentulism, making root-form implants the most popular and compelling implant design. Currently more than 100 companies are manufacturing implants and implant-related products. Most of the root-form implants available today follow the basic principles of design for successful osseointegration which were outlined by the Swedish group at Göteborg (Albrektson, 1983) (Table 1).

The first factor is the materials used in manufacturing such implants. The Nobelpharma implant is made from commercially pure titanium. Upon exposure to air or water, it quickly forms an oxide layer of 3–5 nm at room temperature. This oxide layer makes titanium extremely corrosion resistant, particularly in the biologic environment. The alloyed form of titanium (Ti6Al4V) is also widely used to manufacture these devices. Titanium alloy has been shown to be as biocompatible as commercially pure titanium and to have superior mechanical properties (Williams, 1977).

The second factor is the design of the implant. It must be designed to encourage initial stability in the osteotomy. It has been shown that micromotion or displacement of the implant relative to its host site during initial healing can destroy the network of connective tissue which serves as the scaffold for bone development, leading to repair instead of regeneration (Brunski, 1979). In the Nobelpharma (Bränemark) system, initial stability is achieved by the helical shape of the implant which is inserted in a threaded osteotomy. Other designs include cylindrical or press-fit configurations, spiral shapes, finned cylinders, and hollow cylinders. Cylindrical implants may be coated with either titanium

plasma spray or hydroxylapatite. These surface treatments allow for precision of fit of the implants in the osteotomies, preventing micromotion during the initial healing period. Hydroxylapatite, due to its chemical similarity to osseous tissues, forms a biochemical bond with bone, independent of any mechanical interlocking mechanism. This concept has been described by the term "biointegration" (Kay, 1987; Meffert, 1987; Niznick, 1982; Kirch, 1986).

The third factor involves the use of a surgical technique which prevents excessive heat generation during preparation of the host site. It has been shown that the heating of bone to a temperature of 47°C or above for 30 sec causes vascular damage and alters its vitality (Ericsson et al., 1983). To minimize mechanical and thermal trauma, the surgical protocol involves the use of specially designed spiral, helical, or spade drills of graduated lengths and diameters (Fig. 16). All bone preparations are carried out at a maximum rotational speed of 2000 rpm and under profuse irrigation with saline.

Finally and perhaps the most important factor for achieving a direct bone-to-implant interface is the requirement of maintaining newly placed implants within bone spared from mechanical forces or loads. It has been shown that for successful osseointegration the individual implant should not be loaded for a period of at least 3 months in the mandible and 6 months in the maxilla. The Bränemark system and most other systems achieve this by having two-piece designs. As such, these submerged implant systems require two-stage surgical procedures. At the first-stage surgery, the implants (submergible portion) are installed into the bone and the overlying mucosa is closed, protecting the implants from mechanical loads and oral microflora. After a 3- to 6-month period of healing, the second surgical stage of abutment connection is carried out. The bone remodeling process continues after the implant is loaded with dental prostheses for an additional period of 3 to 6 months. There are a few nonsubmerged implant systems available (ITI, Kyocera). These are placed in a single surgical procedure and immediately are permitted to protrude into the oral cavity. The critical requirement of unloaded healing is still observed, however, by carefully protecting the implants from stress during the healing period. Clinical longitudinal studies have demonstrated successful osseointegration using such one-stage implantation techniques (Buser et al., 1990) (Fig. 17).

There are a wide variety of other endosteal implant designs that employ jaw bone anchorage to stabilize loose teeth (endodontic implants), to serve as prosthetic abutments (transosteal implants), and to anchor long-span bridges to the bone (C-M pins).

PACKAGING AND PREPARATION

Of the three basic implant types—blades, root forms, and subperiosteals—the former two are prefabricated and the latter is custom cast. Blades are supplied in titanium or titanium alloy directly from the manufacturer either in sterile packages

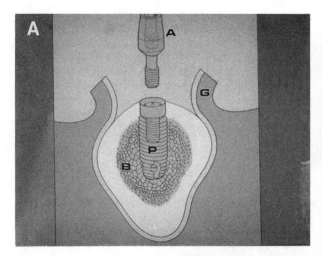

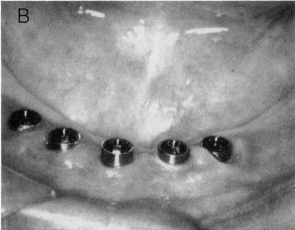

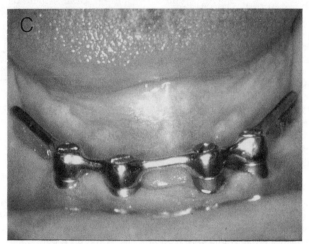

FIG. 17. (A) An abutment is connected to the implant after a 3–6-month healing period. The implant (P) rests between the gum (G) while the bone (B) grows into and around the metal structure, granting the firm anchorage known as osseointegration. This cross section shows the gum being opened surgically for placement of the abutment (A). (B) The abutment attaches to the implant, which is buried within the bone. The abutment passes through the gum into the oral cavity. (C) A final anchoring prosthesis such as this bar with four copings is screwed into the implant abutments, which have internally threaded holes. This permits the practitioner to remove this prosthesis for evaluation, hygiene, and repair.

or in plastic envelopes. Handling them is difficult, because even if they are prepared properly by the maker or user, the benefits of presurgical treatment such as radiofrequency glow discharge, sterility, and freedom from surface lipids or contaminants (e.g., talc from rubber gloves) will be lost owing to the frequent and sometimes aggressive manipulations necessitated during implantation procedures.

In most instances root forms are successfully placed on their first insertion, so innovative sterile packaging in small glass or plastic vials is the current method of presentation. These implants need not be touched by the surgeon's fingers, and are transferred from package to host site either by the vial covers into which they are inserted and which serve as handles, or in the case of Bränemark implants, in a more complex series of steps, using only titanium-tipped forceps, mounting platforms, and fixture mounts (Fig. 18).

Because subperiosteal implants are individually cast of cobalt–chromium–molybdenum alloy, they must be treated individually by the laboratory and the surgical team. They require passivation with dilute acids, defatting using acetone or other organic solvents, ultrasonic cleaning, and, finally, sterilization by autoclaving at 270°F for 20 min (Baier *et al.*, 1986). During surgical insertion, no other metals (stainless steel retractors, suction tips, or seating tappers) should be permitted to touch their surfaces. Baier and co-workers contend that such implants may achieve a more compatible surface condition with the use of radiofrequency glow discharge, which creates an environment of maximum wettability.

The implant surgeon, the manufacturer, and the biomaterials scientist, in addition to their basic skills, must acquire a thorough understanding of metallurgy, biocompatibility, and the benefits of proper cleanliness and sterilization so that an

FIG. 18. Root-form implants are sterile packaged within glass or plastic vials ready for use. Many companies have the cover of the vials attached to the implants for easy handling and to prevent metallic contamination from transport instruments.

appropriate host-site environment will be created to ensure the long-term success of implants.

Bibliography

Adell, R. (1983). Clinical results of osseointegrated implants supporting fixed prosthesis in endentulous jaws. *J. Prosthet. Dent.* **50:** 251.

Adell, R., Lekholm, U., Rockler, B., and Bränemark, P. I. (1981). A 15-year study of osseointegrated implants in the treatment of the edentulous jaw. *Int. J.Oral Surg.* **10:** 387.

Albrektsson, T. (1983). Direct bone anchorage of dental implants. *J. Prosthet. Dent.* **50:** 255–261.

Albrektsson, T., Dahl, E., and Endom, L., (1988). Osseointegrated oral implants. A Swedish multicenter study of 8,139 consecutively inserted Nobelpharma implants. *J. Periodontol.* **59:** 287.

Albrektsson, T., Jansson, T., and Lekholm, U. (1986). Osseointegrated dental implants. *Dent. Clin. North Am.* **30:** 151.

Baier, R. E., Natiella, J. R., Meyer, A. E., and Carter, J. M. (1986). Importance of implant surface preparation for biomaterials with different intrinsic properties in *Tissue Integration in Oral and Maxillo-Facial Reconstruction,* D. van Series 29, Excerpta Medica, Amsterdam, pp. 13–40.

Baumhammers, A. (1972). Custom modifications and specifications for bladevent implant designs to increase their biologic compatibility. *Oral Implantol.* **2:** 276.

Berman, N. (1951). An implant technique for full lower denture. *Dent. Digest* **57:** 438.

Bränemark, P. I. (1983). Osseointegration and its experimental background. *J. Prosthet. Dent.* **50:** 399.

Bränemark, P. I., Breine, U., Adell, R., Hansson, B. O., Lindstorm, J., and Ohlsson, A. (1969). Intra-osseous anchorage of dental prostheses. I. Experimental studies. *Scand. J. Plast. Reconstr. Surg.* **16:** 17.

Bränemark, P. I., Hanssen, B. O., Adell, R., Brien, U., Lindstrom, J., Hallen, O., and Ohman, A. (1977). Osseointegrated implants in the treatment of the edentulous jaw. *Scand. J. Plast. Reconstr. Surg. Suppl. 16.*

Brunsky, J. B., Moccia, A. F., Pollack, S. R., Korostoff, E., and Trachtenberg, D. I. (1969). The influence of functional use of endosseous implants on the tissue-implant interface: Histological aspects. *J. Dent. Res.* **58:** 1953.

Buser, D., Weber, H. P., and Lang, N. P. (1990). Tissue integration of non-submerged implants: 1-year results of a longitudinal study with ITI hollow-screw and hollow-cylinder implants. *Clin. Oral. Implantol. Res.* **1:** 33.

Chérchève, R. (1956). Nouveaux aperçus sur le probleme des implants dentaires chez l'edente comple. Implants et tuteurs. *Rev. Fr. Odont. Stomatol.* **July.**

Cranin, A. N. (1970). *Oral Implantol.* Thomas, Springfield, IL.

Cranin, A. N. (1980). The anchor endosteal implant. *Dent. Clin. North Am.* **24:** 505.

Cranin, A. N., Schnitman, P., and Rabkin, M. (1978). The Brookdale bar subperiosteal implant. *Trans Soc. Biomater.* **2:** 331.

Cranin, A. N., Satler, N., and Shpuntoff, R. (1985). The unilateral pterygohamular subperiosteal implant: Evolution of a technique. *J. Am. Dent. Assoc.* **110:** 496.

Cranin, A. N., Gelbman, J., and Dibling, J. (1987). Evolution of dental implants in the twentieth century. *Alpha Omegan* **80:** 24–31.

Dalise, D. (1979). The micro-ring for full subperiosteal implant and prosthesis construction. *J. Prosthet. Dent.* **42:** 211.

Doundoulakis, J. H. (1987). Surface analysis of titanium after sterilization: Role in implant tissue interface and bioadhesion. *J. Prosthet. Dent.* **58:** 471.

Driskell, T. D. (1987). History of implants. *CDIA* **15:** 10.

English, C. (1988). Cylindrical implants. *J. Calif. Dent. Assoc.* **1:** 18.

Formaggini, M. (1947). Prostesti dentaria a mezzo di infibulazione directa endoalveolare. *Rev. Ital. Stomatol.* **March.**

Gershkoff, A., and Goldberg, N. I. (1949). The implant lower denture. *Dent. Digest* **55:** 490.

Guerini, V. A. (1969). *History of Dentistry.* Milford House, New York.

James, R. A. (1983). Subperiosteal implant designs based on periimplant tissue behavior. *N. Y. J. Dent.* **53:** 407.

James, R. A. (1988). Connective tissue dental implant interface. *J. Oral Implantol.* **13:** 607.

Kay, J. L., Logan, G., and Liu, S. T. (1986). The structure and properties of HA coatings on metal in *Trans. 12th Annual Meeting Society of Biomaterials.*

Kirch, A. (1986). Plasma-sprayed titanium IMZ implants. *J. Oral Implantol.* **12:** 494.

Laney, W. R., Tolman, D. E., Keller, E. E., Desjardins, R. P., Van Roekel, N. B., and Branemark, P. I. (1986). Dental implants: Tissue integrated prostheses utilizing the osseointegration concept. *Mayo Clin. Proc.* **61:** 91.

Lemons, J. E. (1977). Surface conditions for surgical implants and biocompatibility. *J. Oral Implantol.* **3:** 362.

Linkow, L. I. (1968). Prefabricated endosseous implant prostheses. *Dent. Concepts* **11:** 3.

Linkow, L. I. (1986). Tripodial subperiosteal implants. *J. Oral Implantol.* **12:** 228.

Niznick, G. A. (1982). The Core-vent implant system. *Oral Implantol.* **10:**

Roberts, H. D., and Roberts, R. A. (1970). The ramus endosseous implant. *J. S. Calif. Dent. Assoc.* **38:** 571.

Schnitman, P. A. (1987). Diagnosis, treatment planning, and the sequencing of treatment for implant reconstructive procedures. *Alpha Omegan* **80:** 32.

Strock, A. E. (1939). Experimental work on direct implantation in the alveolus. *Am. J. Orthol. Oral Surg.* **25:** 5.

Strock, A. E., and Strock, M. S. (1949). Further studies on inert metal implantation for dental replacement. *Alpha Omegan.*

Truitt, H. P., James, R. A., and Lozado, J. (1986). Noninvasive technique for mandibular subperiosteal implants: A preliminary report. *J. Prosthet. Dent.* **55**: 494.

Weinberg, D. D. (1950). Subperiosteal implantation of Vitallium (Cobalt-chromium alloy), artificial abutment. *J. Am. Dent. Assoc.* **40**: 549.

Weiss, C. M. (1986). Tissue integration of dental endosseous implants: Description and comparative analysis of fibro-osseous integration and osseous integration systems. *J. Oral. Implantol.* **12**: 169.

Willaims, D. F. (1977). Titanium as a metal for implantation. Part 1. Physical properties. *J. Med. Eng. Technol.* **6**: 195–198, 202–203.

Worthington, P. (1988). Current implant usage. in *Proc. NIH Consensus Development Conference on Dental Implants,* June 13–15.

Yurkstas, A. A. (1967). The current status of implant prosthodontics. *Newsl. AAID* **16**:

Zarb, G. A. (1983). Introduction to osseointegration in clinical dentistry. *J. Prosthet. Dent.* **49**: 824.

Zarb, G. A., and Symington, J. M. (1983). Osseointegrated dental implants: Preliminary report on a replication study. *J. Prosthet. Dent.* **50**: 271.

9.5 OPHTHALMIC IMPLANTATION

Stephen A. Obstbaum

It has been suggested that 38% of the fibers transmitting information to and from the human brain are within the optic nerves—a sum of 2.5 million axons (Bruesh and Arey, 1942). While the magnitude of involvement of vision in the function of the brain is staggering, it emphasizes vision's essential role as the primary sensory input to the human organism.

The crystalline lens is a component of the mechanism that focuses light rays on the neurosensory retina. Cataract formation, an aging process involving the lens, disturbs this function and alters normal visual physiology. In this chapter, the structure of the eye, its component parts, and in particular the lens are examined and discussed to understand the process of vision. Current techniques of cataract surgery, emerging trends, and the role of the intraocular lens (IOL) as a replacement prosthesis for the natural lens are discussed.

ANATOMIC CONSIDERATIONS

To assess the design and fabrication of an artificial lens, anatomic relationships, physiologic requirements, and biocompatability criteria should all be satisfied. The structure of the eye is depicted schematically in Fig. 1 of chapter 7.6.

The outer coat of the eye, the sclera, is composed of a tough collection of collagen fibers that are arranged irregularly, and therefore the sclera assumes an opaque, white color. The extraocular muscles, which are responsible for ocular motility, are attached to it, and a canal at its posterior pole permits external passage of the optic nerve. This coating protects the delicate internal ocular structures.

The anterior aspect of the eye is the cornea. This is a transparent structure composed mainly of collagen fibrils arranged

in an orderly fashion that permits the transmission of light energy. The cornea serves as a lens, since it refracts or bends light rays to focus them on the retina. The anterior corneal curvature provides a refractive power of 48.8 diopters, while the posterior corneal surface has a refractive power of -5.8 diopters. Thus, the cornea has an overall refractive power of about 43 diopters or 70% of the total refractive power of the eye.

The innermost layer of the cornea, the endothelium, is composed of a single layer of hexagonal cells. These cells do not replicate and over a lifetime decrease in number as a result of attrition. Cataract and IOL surgery can also reduce the endothelial cell population. However, modern cataract/IOL surgery techniques and surgical adjuncts, such as viscous sodium hyaluronate solutions to protect the cornea, have reduced the incidence of endothelial cell loss as a consequence of surgery. The function of the endothelium is to maintain corneal clarity through its barrier and pumping functions. When these functions are compromised, the cornea decompensates and becomes hazy (Waring *et al.*, 1975).

The boundaries of the anterior chamber are defined by the posterior aspect of the cornea and the anterior surface of the iris. This area is filled with a fluid, the aqueous humor, which is secreted by the ciliary body. Aqueous fluid passes from its posterior origin behind the iris, through the pupil, into the anterior chamber, and exits through the trabecular meshwork, a complex of lacunae that lies in the peripheral part of the external eye structure at the transitional zone between the scleral and corneal tissues. This area, the anterior chamber angle, is used as a site of implantation for one type of IOL.

The iris is the anteriormost extension of the uveal tract. This structure contains muscles that dilate and constrict the pupil, thereby influencing the depth of field and focus, much like the diaphragm of a camera. During cataract surgery, the pupil is dilated by pharmacologic means to gain access to the cataractous lens behind it.

The ciliary body is a part of the uveal tract, and the iris originates from its anterior surface. Its structure serves several functions. Its epithelial cells secrete the aqueous humor, and its tripartite muscular structure influences the shape of the lens by altering tension on the suspensory ligaments attached to the capsule surrounding the lens. In its posterior aspect, the ciliary body becomes continuous with the choroid.

The choroidal layer is the posterior portion of the uveal tract that extends from the end of the ciliary body to the optic nerve. It is interposed between the outer wall of the eye, the sclera, and the retina. The choroid is composed of blood vessels that nourish the outer layer of the retina and serves as a conduit for transmitting arteries and nerves to the anterior portion of the eye.

The retina is the innermost layer of the eye. It is the sensory structure that converts light energy to photochemical energy, which then transmits electrical stimuli through the visual pathway in the brain to the occipital cortex, where vision is recognized. The focusing mechanisms of the eye—the cornea, lens, and fluid media—work in concert to provide a clear image to the retina.

The vitreous is a mesenchymal tissue composed of 99% water. The molecular structure of the vitreous is composed of hyaluronic acid and a collagen framework. The interstices of this framework are filled with water. The vitreous transmits more than 90% of visible light to the retina. The high water content also allows easy passage of nutrients through the vitreous from the ciliary body to the retina.

From the foregoing descriptions, the crystalline lens can be considered to be suspended within a fluid-filled vessel. It is bathed by aqueous humor anteriorly and by the vitreous body posteriorly. The suspensory ligaments, the zonules, are acellular fibers that originate at the ciliary body and insert onto the anterior and posterior capsule of the lens. The tension of these fibers, as a result of ciliary muscle relaxation or contraction, influences the shape of the crystalline lens. When the ciliary muscle contracts, the zonular fibers are under lessened tension and permit the elastic capsule surrounding the lens to relax. The lens becomes more spherical and produces greater refractive power. This process, known as accommodation, is responsible for the additional lens power that permits near objects to be clearly imaged on the retina (Fincham, 1937). As we age, the lens hardens and is less deformable, and the effect of accommodation is, therefore, diminished. This condition, called presbyopia, is the normal recession of the near point that occurs during our mid-40s. Eyeglasses are prescribed to compensate for the loss of accommodation.

The structure of the natural lens can be simplistically divided into three components: (1) an outer elastic capsular bag that invests the entire lens; (2) the cortex, the outer zone of lens substance; and (3) the nucleus, the hardened centralmost core of the lens. Opacification of any of these structures is termed a "cataract." When opacification affects visual function enough to disrupt an individual's daily activities and nonsurgical means of improvement have been exhausted, surgical cataract removal becomes a viable option.

The lens does not have a homogeneous structure. Beneath the anterior capsule is an epithelial layer that extends to the equatorial zone of the lens. The epithelium at the equatorial lens bow continues to produce new lens fibers throughout life. Addition of these fibers to the peripheral portion of the lens increases its size. The older fibers in the central portion of the lens become more densely compacted and hardened. This condition, termed "nuclear sclerosis," occurs commonly with age, and produces greater optical heterogeneity of the lens. The continued sclerosing of the nucleus often results in brunescence, the development of a yellowish-red to red-brown color of the nucleus.

Another type of cataract formation involving the cortex of the lens is the development of small, round water droplets and water clefts. Frequently, these alterations develop in the peripheral portion of the lens that is masked by the iris. It is only when the pupil is pharmacologically dilated (e.g., during an ophthalmic examination) that these cataractous changes are observed. A particular type of cataract formation that affects the posterior cortical area is the posterior subcapular cataract. This opacification, when situated axially along the path of light from the external environment to the retina, is one of the most visually disabling, producing significant glare and reducing visual acuity in direct light. As a general rule, the type of surgical treatment of a cataract is similar, regardless of the type of cataract formation.

CATARACT SURGERY METHODS

Cataract surgery has passed through several evolutionary steps during this century. These changes reflect the development of new technologies to facilitate cataract removal. During the past few decades, two major forms of cataract surgery have been practiced, as described in the following section.

Intracapsular Extraction

Intracapsular cataract extraction is the method of removing the entire lens, including the capsular bag. This method separates the lens from its zonular attachments and from attachments to the anterior portion of the vitreous. This latter attachment, called "Weigert's ligament," generally wanes as we age, but is quite strong in childhood and young adulthood. When the intracapsular technique was performed on younger age groups, disruption of the vitreous, termed "vitreous loss," frequently occurred. This is a complication of cataract surgery that has long-term effects, even when it is adequately managed intraoperatively. The instruments used to perform intracapsular surgery were initially forceps, which were used to grasp the anterior capsule and remove the lens by either tumbling it from behind the pupil or sliding it from the eye. In each instance, the mechanical action of removing the lens caused traction on and release of the zonules that held the lens in position. In some cases, a specific enzyme that digested the zonules was instilled within the eye before the cataract was removed.

In the late 1960s, another method to remove the lens intracapsularly was introduced. This technique employed cryosurgery that, in essence, produced an adhesion between the cryoprobe and a portion of the lens substance. Once this adhesion had formed, the cataractous lens was slid from the eye.

The advantage of this intracapsular technique was that it removed the entire lens without leaving any remnants. There was clear passage of light from the anterior corneal surface to the retina, although the light was not fully focused because the refractive power of the lens was removed along with the cataract. The disadvantages of this method were (1) the greater potential for vitreous disruption (vitreous loss); (2) the higher incidence of retinal detachment; (3) the higher incidence of fluid accumulations in the central region of the retina, the macula (aphakic cystoid macular edema); and (4) the greater instability of the internal ocular milieu. Because of these complications, other types of cataract surgery were adopted.

Extracapsular Extraction

Extracapsular cataract surgery obviated many of the concerns posed by the intracapsular techniques. This method has several variants but, in principle, maintains a portion of the surrounding capsular bag. In this procedure, the anterior capsule (i.e., the portion of the lens capsule facing the pupil) is opened. This anterior capsulotomy then permits the contents of the lens capsule (i.e., the nucleus and cortex) to be removed. When the nucleus is totally removed as a single structure, it is called nucleus expression, and the technique is termed planned extracapsular cataract extraction. Another method called phacoemulsification removes the nucleus by fragmenting it with an ultrasonic needle. In recent years, phacoemulsification has become the favored method for cataract removal.

Advantages of the extracapsular technique are related to maintenance of an intact posterior capsule. An eye with an intact capsule has greater internal stability and lower incidence of retinal detachment and cystoid macular edema. The major consequence of the procedure, however, is that the posterior capsule can become hazy or opacified in the late postoperative period. Epithelial cells in the equatorial zones of the lens that are not removed during cataract surgery proliferate, undergo metaplasia, and migrate along the capsule to reside in the axial zone of the eye. This reduces vision, and an additional procedure is required to create an opening in the capsule.

Another consideration in retaining the posterior capsule and, if possible, even greater portions of the capsular bag, is related to the implantation of IOLs. It was this aspect of visual rehabilitation that served as the major impetus to encourage ophthalmologists to adopt the extracapsular technique.

VISUAL REHABILITATION OF THE APHAKIC EYE

The lens is a major component of the light-focusing apparatus of the eye. When the lens becomes cataractous, it reduces the amount of light effectively reaching the retina and decreases the quality of the visual image. The refracting power of the lens, however, remains an inherent property, although its function is eroded by the visual distortion produced by lens changes. Removing the cataract surgically creates an unimpeded passage of light to the retina, but these light rays are influenced basically by the cornea and are still somewhat defocused.

To achieve a clear retinal image, the natural lens must be assisted or replaced by another focusing device. Eyeglasses, contact lenses, and IOLs have all been used to rehabilitate the aphakic (lensless) eye. Each of these devices is capable of focusing light to create a clear retinal image. However, there are particular optical and functional considerations that favor the use of the IOL as the prime method for aphakic rehabilitation.

The cataract eyeglass is a convex, high-plus lens that also magnifies an image about 25%. This means that, if a cataract extraction is performed on one eye and the patient's other eye is unaffected, there will be a differential in the image size between the two eyes that exceeds the compensatory ability of the brain. The patient will typically complain of an inability to fuse the images between the two eyes, will have significant discomfort, and will often abandon the use of the eyeglasses.

These high-plus lenses also produce optical aberrations that are disconcerting. A spherical aberration is induced, so that straight lines assume a curved orientation. This has been termed the "pincushion effect." For example, sides of a doorway tend to bow inward rather than maintain their straight vertical orientation. The visual field is also restricted, mainly because of the appearance of a ring scotoma, a ring-shaped blurred area surrounding the clear central visual field. A patient's vision becomes distorted unless he or she is looking through the central portion of the lens. Patients often complain that images jump in and out of the field of view. It is evident that, for the patient with aphakia affecting both eyes, the quality of vision is compromised unless these glasses are worn constantly.

Another alternative for aphakic rehabilitation is the use of contact lenses. Optically, this method has several advantages over those of aphakic or cataract eyeglasses. Since the lens is applied to the tear film of the cornea, it is closer to the nodal point of the eye and produces magnification of only about 8%. The image disparity between the two eyes of a person who has had an operation in only one eye is much better tolerated. The aberrations of spatial orientation and visual field restrictions observed with cataract glasses do not occur. The basic concern with the use of a contact lens involves elderly patients who have not worn one previously and must now learn to manipulate and maintain it. This can cause problems for younger patients as well. Problems inherent to the eye of an older individual, such as dry eyes, blepharitis (an infection involving the lids), and corneal degenerative conditions, also limit the use of contact lenses. For example, if an elderly patient has arthritis and his or her manual dexterity is impaired to some degree, contact lenses can be difficult to deal with. Finally, visual function is significantly limited when the lens is not worn.

Although cataract glasses and contact lenses are optically competent methods to focus light on the retina, the practicality of using either of them is less than desirable. The IOL has emerged as a solution to many of the failings of these other alternatives. Implanting a lens within the eye, in effect, mimics the role of the human crystalline lens. Optical problems related to removal of the cataract of one eye, while the unoperated eye still sees, are resolved. Since the IOL is placed near the nodal point of the eye, there are no problems related to image disparity. The IOL will focus light on the retina when the eyelid is open, and the patient does not have to manipulate it because it resides securely within the eye. IOL implantation is at present the accepted method for rehabilitating the aphakic eye.

DEVELOPMENT AND TYPES OF IOLs

Figure 1 illustrates some of the more common types of IOLs and identifies their components. The first IOL was implanted in 1949 and was termed the "Ridley posterior chamber lens." It was made of poly(methyl methacrylate) (PMMA) and designed to resemble the characteristics of the human lens. It had two dissimilar radii of curvature, was biconvex in its shape, and weighed about 112 mg in air, making it an extremely heavy lens. To benefit from the support of an intact posterior capsule, the IOL was implanted after an extracapsular cataract extraction (Ridley, 1951). Although this original Ridley posterior chamber lens restored visual function, problems related to the weight and size of the implant, as well as its displacement into the vitreous body, limited its usefulness. The fact that the IOL produced reasonable results when complications did not ensue stimulated ophthalmologists to design other types of IOLs.

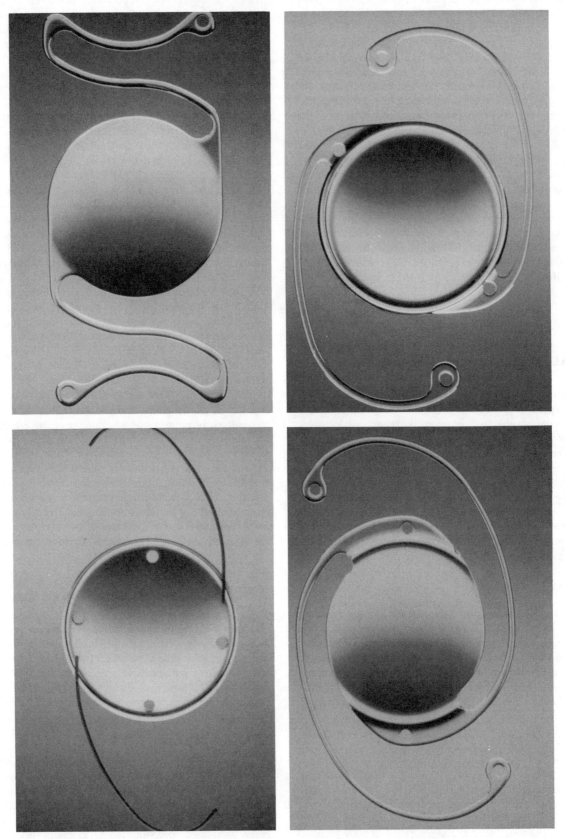

FIG. 1. Some common types of IOLs.

Typically, the Ridley lens was placed behind the iris and in front of the intact posterior capsule. This area, known as the posterior chamber, offers several optical and mechanical advantages. Despite the fact that early posterior chamber IOL implantation was abandoned shortly after its inception, it currently is the most popular type of IOL implantation. Other sites selected for IOL implantation were the anterior chamber and the pupillary space. The anterior chamber IOL relies on support of the peripheral iris structure and the collagen fibers of the scleral tissue in the anterior chamber angle. The first anterior chamber lenses were very heavy and generally had three or four thick footplates that rested in the peripheral anterior chamber (Strampelli, 1961; Choyce, 1958). Contact between the thick PMMA footplates and the cornea caused corneal decompensation because the endothelial cells of the peripheral cornea were damaged, thereby compromising their barrier and pump functions. As much a problem with the designs of these implants was the problem of properly sizing the diameter of the IOL to fit the dimensions of a particular eye (Moses, 1984). If the anterior chamber IOL was too long, it produced erosion into the iris and ciliary body, as well as corneal damage. If the IOL was too short, it would fail to be supported properly, causing a propelling effect in the anterior chamber, often damaging the anterior surface of the iris and the corneal endothelium.

A number of anterior chamber IOLs were developed later, each with features designed to obviate the problems of their predecessors. However, even these newer anterior chamber IOLs produced the type of tissue erosion, hemorrhage, inflammation, and corneal decompensation seen with earlier anterior chamber IOL designs (Ellingson, 1978). These lenses used finely looped haptics that were associated with micromovement in the anterior chamber angle, breakdown of the blood–ocular barrier, adhesions around the haptic, iris chafing, and inflammatory symptoms. Although the lens optics were made of PMMA, the edges were not adequately polished. The haptics were fabricated from the kind of polypropylene used for suture material, or extruded PMMA (Apple *et al.*, 1987). Problems with these anterior chamber lenses resulted in similar clinical complications and prompted their withdrawal from the market.

The latest anterior chamber lens designs incorporate one-piece PMMA technology, open haptics to impart greater flexibility, vaulting to lessen the likelihood of contact between the posterior surfaces of the IOL and the iris, discrete points of contact between the ends of the haptics and angle structures, and excellent polishing and finishing of the surfaces and edges of the anterior chamber lenses. Attention to modification in lens design and fabrication has made the present generation of anterior chamber IOLs acceptable.

Following early dissatisfaction with the Ridley posterior chamber IOL and anterior chamber IOLs, several surgeons used IOLs that were implanted in the pupil. These pupil- or iris-supported IOLs could be implanted after intracapsular or extracapsular surgery, did not rely on the anterior chamber angle for support, and were readily centered in the visual axis. Among the earliest designs were the Epstein collar button and Maltese cross, which was later modified as the Copeland lens (Epstein, 1959). However, the more popular style of these

lenses—the Binkhorst, Worst, and Fyodorov IOLs—had PMMA optics and haptics made variously of nylon 6, nylon 66, and polypropylene (Binkhorst, 1959; Fyodorov, 1977; Worst, 1977). Evidence that nylon sutures degenerated within the eye contributed to the demise of this material (Kronenthal, 1977). Later, polyimide was used as a haptic material to support a glass optic pupil-supported IOL (Barasch and Poler (1979). For short periods, some manufacturers used platinum–iridium or titanium haptics. However, these metal-loop lenses were heavy and caused lacerations of the iris, anterior chamber hemorrhages, and corneal decompensation (Shepard, 1977).

Pupil-supported lenses popularized the concept of IOL implantation. Although they were more difficult to implant than anterior chamber IOLs, there were no problems with sizing of these lenses, thereby reducing the need to maintain a large inventory of IOLs. With the passage of time, however, late corneal decompensation occurred, often necessitating corneal transplantation and IOL removal. In addition, problems related to erosion of the iris, especially at the pupil border, developed. At times, this condition mimicked that observed with anterior chamber IOLs. Perhaps the most significant concern was the reliance on the pupil for support that not infrequently caused IOL dislocation, either partial or total, when the pupil dilated (Obstbaum, 1984). Thus, despite an acceptance of this generic-style IOL, the pupil-iris-supported lens was supplanted once the posterior chamber IOL wasl introduced.

Pearce modified the Ridley tripodal anterior chamber IOL for use in the posterior chamber. This was a solid, one-piece PMMA IOL (Pearce, 1976). Once again, as with the Ridley posterior chamber IOL, the presence of a posterior capsule was essential for implanting this lens. The two lower haptics were placed within the capsular bag, while the superior haptic was sutured through the posterior surface of the iris. This lens enjoyed good success in the hands of relatively few surgeons but, more important, served as a reawakening to the benefits of posterior chamber implantation. Shortly after this, Shearing, modifying a Barraquer anterior chamber lens, announced a new era of IOL implantation (Shearing, 1979). This lens had a PMMA optic and two J-shaped polypropylene haptics designed for placement behind the iris in front of the posterior capsule. This concept of IOL implantation achieved such success that further modifications of the initial design followed and resulted in manufacturing techniques to create one-piece PMMA IOLs with flexible haptics.

COMPLICATIONS OF POSTERIOR CHAMBER IOLs

In recent years, more than one million posterior chamber IOLs were implanted annually in the United States. These IOLs have been significantly modified from the style of the original Shearing lens. The haptics of these lenses are no longer the conventional J-shape but have evolved to a C-shaped, softer configuration. Optic sizes have also been modified to accomodate newer techniques for cataract removal, and drilled positioning holes in the optic have all but disappeared. The finish of these one-piece PMMA lenses is excellent. Yet, despite the technological advances in IOL manufacture, clinical complications still exist, as discussed in the next section. Although the magnitude of these problems is relatively small, our goal is for a higher degree of excellence in this procedure, and therefore we should strive to overcome these problems. While several of these complications are related to the surgical procedure, others arise from fabrication of the IOL and its existence within a biologic environment.

Dislocations or Malpositions of IOLs

This complication (Obstbaum, 1989) is significant because recent surveys suggest that in about 50% of cases in which a posterior chamber IOL must be removed postoperatively, improper positioning of the IOL plays a significant role. There are several types of posterior chamber IOL malpositions. In some instances, the problem can be managed by purely medical means, while in other cases surgical intervention, either to reposition the implant or remove it, must be undertaken.

The most common cause of posterior chamber IOL malpositioning is asymmetric placement of the lens haptics. This generally occurs when one haptic is placed within the capsular bag, while the other is placed outside the bag, generally in the region behind the iris. When fibrosis of the capsule occurs postoperatively, it exerts a force on the haptic within the bag, while the haptic outside the bag does not have an equal and opposing force exerted upon it. The lens implant will shift in the direction of the lesser force, frequently causing only a portion of the optic to be within the pupil. Other consequences of this type of malposition may result in an optical disturbance because the lens optic is decentered with respect to the pupil.

Another type of dislocation is associated with surgical trauma that results in disruption of the zonular support of the capsular bag. In this complication, zonular weakness or absence fails to support the IOL, which eventually dislocates into the vitreous so that only the superior aspect of the IOL optic is observed in the pupil. This has been termed the "sunset syndrome" and represents a type of dislocation that usually requires surgical intervention. Other situations where the IOL may dislocate inferiorly or posteriorly occur with the intraoperative rupture of the posterior capsule, which may not be recognized.

These are the major causes of malpositioning of posterior chamber lenses. There are other lesser causes that are often of more cosmetic concern than physiologic importance. The majority of these are managed by either careful observation or medical therapy. In the past few years newer techniques to ensure IOL placement have reduced the incidence of these occurrences.

Iris Chafing

This complication has occurred with every type of IOL and reflects an undesirable interaction between plastic and ocular tissue. The iris- or pupil-supported IOLs caused erosion of the pupil sphincter as the haptic rubbed against the pupillary margin. This clinical finding was often totally asymptomatic. In other situations, the edge of the haptic chafed the anterior iris surface, resulting in disturbances of the blood–ocular barrier,

bleeding into the anterior chamber, and inflammation (Liepmann, 1982; Nicholson, 1982).

The iris chafing that developed with anterior chamber IOLs was termed the "UGH syndrome" because of the components of uveitis (inflammation), glaucoma (high intraocular pressure), and hyphema (bleeding into the anterior chamber) (Ellingson, 1978). The greatest incidence of this complication occurred when a roughened edge of an injection-molded anterior chamber IOL chafed the anterior surface of the iris as the pupil normally dilated and contracted in response to physiologic stimuli (Keates and Ehrlich, 1978). The edge defect corresponded to the site of introduction of the plastic into the mold. The condition was remedied by polishing the edges of these one-piece PMMA anterior chamber IOLs. However, several years later, when multipiece anterior chamber IOLs with fine tubular haptics were introduced, the problem reemerged. Subsequently, these IOLs were removed from the market.

Another form of iris chafing—loss of pigment from the posterior or back surface of the iris—was observed with some of the iris-supported IOLs because the haptics rested behind, as well as in front of, the iris. Only rarely did these changes occur progressively enough to cause pathologic damage to the eye (Ballin and Weiss, 1982). More significant pigmentary dispersion occurred with posterior chamber IOLs, especially those implanted in the so-called ciliary sulcus, the area outside the capsular bag. These cases were manifest as pigment lost from the posterior iris, either over the ends of the polypropylene haptics or in the midperipheral iris from the edge of the optic. These findings, which were exaggerated with posterior chamber lenses that were planar in design, reflected contact between iris tissue and the PMMA IOL. In some instances, these conditions were asymptomatic but demonstrated iris transillumination defects (i.e., areas where light reflected from the retina was visible through the iris defects) (Johnson et al., 1984). Other cases showed components of the UGH syndrome, only in these instances with posterior iris chafing and, in a smaller number of cases, a condition known as "pigmentary glaucoma" developed (Masket, 1986; Samples and Van Buskirk, 1985).

Posterior iris chafing remains a significant concern in IOL implantation. Current implantation techniques favor capsular bag IOL implantation that sequesters the lens within capsular tissue to reduce the area of potential contact between the IOL and the iris. However, even under this circumstance, a portion of the PMMA will come into contact with the iris. There is a need to modify the PMMA surface or to develop newer biomaterials that can lessen this complication.

Inflammation

Ocular surgical procedures produce inflammation (Obstbaum, 1984). This is an acknowledged short-term complication of surgery and is related to disruption of the blood–ocular barrier. Inflammatory mediators are released at the time of surgery. Their actions are generally transitory and can be ameliorated by steroidal and nonsteroidal anti-inflammatory agents. After the early postoperative period of surgically induced inflammation, some eyes continue to demonstrate low-grade inflammatory signs. Often, there is deposition of cellular material on the surface of the implanted IOL, and adhesions can develop between the iris and the IOL. The clinical importance of late inflammation is that it may lead to corneal and retinal complications (Obstbaum and Galin, 1979). Moreover, signs of inflammation indicate compromised biocompatability and disturbed physiology. If their is a mechanical basis for this altered relationship, such as previously described, correction of this condition by mechanical or surgical means is indicated. In some cases, the cause is not readily identified, and IOL removal might be required.

Inflammation reflects an organism's response to noxious stimuli in its environment. Although an IOL is composed of plastic material that is presumed to be inert, it is a foreign body that evokes an inflammatory response. The extent of this reaction will vary, depending on mechanical factors related to the surgery and implantation, activation of inflammatory mediators, and protective mechanisms that influence the biologic behavior of the implanted lens (Tuberville *et al.*, 1982). Our concerns center on design and fabrication of IOLs that reduce mechanical trauma and irritation, surgical techniques that minimize iatrogenic trauma, ways to render IOLs more biocompatible, and methods to sequester the IOL within the eye.

Capsular Opacification

Extracapsular cataract extraction retains an intact posterior capsule, a portion of the anterior capsule, and the lens equator. As previously described, there is a region of epithelial cells that proliferate and migrate along the posterior capsule. When this occurs to a significant extent, the visual axis is obscured and results in reduced visual acuity (Obstbaum, 1988). Since some degree of capsular opacity frequently develops, this event is considered to be a consequence rather than a complication of extracapsular surgery.

Several ways to reduce the incidence of capsular opacification have achieved moderate success, but at best, they only retard the progression of opacification. Complete inhibition of cellular metaplasia and migration, however, has not yet been accomplished, although promising work using monoclonal antibodies directed against lens epithelial cells is being investigated. At present, it appears that placing PMMA firmly against the posterior capsule slows the rate of posterior capsule opacification related to epithelial proliferation but not that caused by fibrosis (Lindstrom and Harris, 1980; Downing, 1986; Sterling and Wood, 1986; Sellman and Lindstrom, 1988; Hansen *et al.*, 1988). IOL designs that accomplish this type of apposition are those with biconvex and reverse optics. There is further evidence that angulated haptics that push the posterior surface of the optic more tightly against the posterior capsule enhance this effect (Downing, 1986). It has also been suggested that capsular bag implantation of the IOL retards capsular opacification, perhaps by contributing to the barrier effect.

A benefit of fibrosis in the peripheral region of the capsular bag is that it keeps the IOL in position. Although fibrosis is associated with capsular opacification, an undesirable effect, it also helps retain the stability of the IOL. If a method evolves that completely removes the epithelial cells, the incidence of

capsular opacification will decrease, but so will the desired stabilizing effect. We will then require lens designs that maintain the IOL in its proper position independent of capsular fibrosis.

RECENT ADVANCES

One-Piece PMMA IOLs

There is a growing trend to use one-piece PMMA IOLs rather than three-piece designs, which use optics made from PMMA and haptics made from either polypropylene or extruded PMMA fibers. The haptic materials in three-piece designs, although widely used, behave differently from those of one-piece lenses. For example, surface cracking occurs on polypropylene haptic fibers, especially when placed against uveal tissue, which may represent biodegradation by hydrolysis (Drews, 1983). Also, these haptics display a tendency to permanently deform owing to the force of the tissues within the eye, shortening the overall diameter of the IOL (Drews and Kreiner, 1987). This finding is clinically relevant in the surgery for a malpositioned IOL. In addition, because the haptics in three-piece designs are inserted into drill hole sites in the optic, areas are created that are potentially rough and irregular and require greater attention to quality control and finishing. On the other hand, one-piece IOL fabrication offers the advantages of precisely made IOLs, excellent surfaces, highly polished lenses, and haptics that resist permanent deformity (Hansen *et al.*, 1988).

Viscoelastic Agents

Viscoelastic agents are a class of materials that were initially used in anterior segment surgery to protect the corneal endothelium when pupil-supported IOL implantation and intracapsular surgery were performed. These agents are used to occupy space, coat and protect tissue, move tissues within the eye, and prepare and expand the capsular bag for precise IOL placement. With the development of extracapsular surgery and posterior chamber IOL implantation, viscoelastic materials have became useful tools to facilitate delicate IOL surgical maneuvers (Obstbaum, 1987). New preparations of these agents are being formulated in response to emerging demands.

Surface Modification

Surface modification of IOLs offers an effective means to alter biological interactions with a particular material and to produce a device that approaches biological inertness and compatibility. Studies have demonstrated that the interactions of corneal tissue and PMMA relative to surface adhesiveness and endothelial damage are considerably lessened by surface modification (Yalon *et al.*, 1983). Other studies have shown that surface-modified PMMA IOLs produce less iris tissue damage and pigment dispersion, (Hofmeister, 1988; Burstein, 1988). Clinical studies to confirm the findings of laboratory investigations are continuing.

Small Incision Surgery

Extracapsular cataract surgery, either by planned extracapsular techniques or by phacoemulsification, achieves satisfactory visual results. Over the past 7 years there has been a progressive trend toward using phacoemulsification for cataract removal, although planned extracapsular procedures are still performed by many ophthalmologists. Phacoemulsification permits cataract removal through an incision that is approximately 2/3 smaller than with conventional extracapsular surgery. The benefits of small incision surgery include mechanical stability of the incision, rapid return to normal activities, and rapid visual functional rehabilitation. With advances in the techniques of phacoemulsification and the technological developments of phacoemulsification equipment, the results of the surgery have favored this method of cataract surgery. Phacoemulsification has also hastened the development of IOLs that enhance the benefits of the smaller incision. These include narrow profile PMMA IOLs and foldable lenses of silicone and acrylic materials.

Small Incision IOLs

One-piece PMMA IOLs with smaller optics and shorter tip-to-tip diameters, designed for capsular bag implantation, have gained popularity over the past several years. Implantation of these IOLs requires minimal widening of the phacoemulsification incision and maintains wound stability, thereby hastening visual rehabilitation. The most frequently implanted IOLs have variations of haptic configurations associated with optic sizes of 5.0 or 5.5 mm.

Another trend is for implantation of foldable IOLs. These offer the advantage of being inserted through an unenlarged phacoemulsification incision. The IOLs currently implanted are the silicone lenses of either plate haptic design or with PMMA haptics and acrylic lenses with PMMA haptics. Hydrogels have also been used but several surgical complications associated with the mechanical properties of this lens have prompted modifications in lens design. The hydrophilic nature of this lens material makes it biologically desirable. The popularity of foldable lenses has increased recently because of renewed interest in phacoemulsification and the use of a particular anterior capsulotomy technique, the continuous curvilinear capsulotomy. This method retains the configuration of the capsular bag which sequesters the IOL from contacting ocular structures as well as insuring centration of the IOL. Some of the previously experienced problems related to soft, foldable IOLs, such as pigmentary dispersion and malposition, have been obviated by this capsulotomy technique.

SUMMARY

Cataract surgery with IOL implantation is the most frequently performed ophthalmic procedure. The success of this operation has occurred as a consequence of the rehabilitative benefits afforded by the IOL. We believe that over time this operation will become even safer and more effective, and the IOLs of the future will more closely match the original physiology of the eye. New designs, surface modifications, and newer materials will all be seen. The challenges provided by these tasks should stimulate exciting new advances in the field of biomaterials.

Bibliography

Apple, D. J., Brems, R. N., Park, R. B., Norman, D. K. V., Hansen, S. O., Tetz, M. R., Richards, S. C., and Letchinger, S. D. (1987). Anterior chamber lenses. Part I: Complications and pathology and a review of design. *J. Cataract Refract. Surg.* **13**: 157–174.

Ballin, N., and Weiss, D. M. (1982). Pigment dispersion and intraocular pressure elevation in pseudophakia. *Ann. Ophthalmol.* **14**: 627–630.

Barasch, K. R., Poler, S. (1979). A glass intra-ocular lens. *Amer. J. Ophthalmol.* **88**: 556–559.

Binkhorst, C. D. (1959). Iris-supported artificial pseudophakia: A new development in intra-ocular artificial lens surgery (iris-clip lens). *Trans. Ophthalmol. Soc. U.K.* **79**: 569–584.

Bruesh, S. R., and Arey, L. B. (1942). The number of myelinated and unmyelinated fibers in the optic nerve of vertebrates. *J. Comp. Neurol.* **77**: 631–665.

Burstein, N. C., Ding, M., and Pratt, M. V. (1988). Intraocular lens material evaluation by iris abrasion *in vitro*: A scanning electron microscope study. *J. Cataract Refract. Surg.* **14**: 520–525.

Choyce, D. P. (1958). Correction of uni-ocular aphakia by means of anterior chamber acrylic implants. *Trans. Ophthalmol. Soc. U.K.* **78**: 459–470.

Downing, J. E. (1986). Long-term decision rate after placing posterior chamber lenses with the convex surface posterior. *J. Cataract Refract. Surg.* **12**: 651–654.

Drews, R. C. (1983). Polypropylene in the human eye. *Am. Intraocular Implant Soc. J.* **9**: 137–142.

Drews, R. C., and Kreiner, C. (1987). Comparative study of the elasticity and memory of intraocular lens loops. *J. Cataract Refract. Surg.* **13**: 525–530.

Ellingson, F. T. (1978). The uveitis-glaucoma-hyphema syndrome associated with the Mark VIII anterior chamber lens implant. *Am. Intra-ocular Implant Soc. J.* **4**: 50–53.

Epstein, E. (1959). Modified Ridley lenses. *Br. J. Ophthalmol.* **43**: 29–33.

Fincham, E. (1937). The mechanism of accommodation. *Br. J. Ophthalmol.* **8**: 1–9.

Fyodorov, S. N. (1977). Long-term results of 2,000 operations of implantation of Fyodorov intraocular lenses performed in the Soviet Union. *Am. Intrac-ocular Implant Soc. J.* **3**: 101.

Hansen, S. O., Solomon, K. D., McKnight, G. T., Wilbrandt, T. H., Gwin, T. D., O'Morchoe, D. J. C., Tetz, M. R., and Apple, D. J. (1988). Posterior capsular opacification and intraocular lens decentration. Part I: Comparison of various posterior chamber lens designs implanted in the rabbit model. *J. Cataract Refract. Surg.* **14**: 605–614.

Hofmeister, F. N., Yalon, M., Iida, S., Goldberg, E. P. (1988). *In vitro* evaluation of iris chafe protection afforded by hydrophylic surface modification of PMMA intraocular lenses. *J. Cataract Refract. Surg.* **14**,: 514–519.

Johnson, S. H., Kratz, R. P., Olson, P. F. (1984). Iris transillumination defect and microhyphema syndrome. *Am. Intra-ocular Implant Soc. J.* **10**: 425–428.

Keates, R. H., and Ehrlich, D. R. (1978). "Lenses of chance," complications of anterior chamber implants. *Ophthalmology* 85: 408–414.

Kronenthal, R. L. (1977). Intraocular degradation of nonabsorbable suture. *Am. Intra-ocular Implant Soc. J.* 3: 222–228.

Liepmann, M. E. (1982). Intermittent visual white-out: A new intraocular lens complication. *Ophthalmology* 89: 109–112.

Lindstrom, R. L., and Harris, W. S. (1980). Management of posterior capsule following posterior chamber lens implantation. *Am. Intraocular Implant. Soc. J.* 6: 255–256, 258.

Masket, S. (1986). Pseudophakic posterior iris chafing syndrome. *J. Cataract Refract. Surg.* 12: 252–256.

Moses, L. (1984). Complications of rigid anterior chamber implants. *Ophthalmology* 91: 819–825.

Nicholson, D. H. (1982). Occult iris erosion: A treatable cause of recurrent hyphema in iris-supported intraocular lenses. *Ophthalmol* 84: 113–120.

Obstbaum, S. A. (1984). Complications of intraocular lenses. Membranes, dislocations, inflammation and management of the posterior capsule. in *Cataract Surgery: Current Options and Problems*, J. M. Engelstein, ed. Grune & Stratton, Orlando, FL, pp. 509–533.

Obstbaum, S. A. (1987). Viscosurgery as a soft surgical tool. *Clin. Ocul.* 8: 477–479.

Obstbaum, S. A. (1988). The posterior capsule. *Implants in Ophthalmology* 2: 110–116.

Obstbaum, S. A., and Galin, M. A. (1979). Cystoid macular edema and ocular inflammation. The corneoretinal inflammatory syndrome. *Trans. Ophthalmol. Soc. U.K.* 99: 187–191.

Obstbaum, S. A., and To, K. (1989). Posterior chamber intraocular lens dislocations and malpositions. *Aus. N. Zealand J. Ophthalmol.* 17: 265–271.

Pearce, J. L. (1976). A new light-weight sutured posterior chamber lens implant. *Trans. Ophthalmol. Soc. U.K.* 96: 6–10.

Ridley, H. (1951). Intra-ocular acrylic lenses. *Trans. Ophthalmol. Soc. U.K.* 71: 617–621.

Samples, J. R., Van Buskirk, E. M. (1985). Pigmentary glaucoma associated with posterior chamber intraocular lenses. *Am. J. Ophthalmol.* 100: 385–388.

Sellman, T. R. and Lindstrom, R. L. (1988). Effect of plano-convex posterior chamber lens on capsular opacification from Elschnig pearl formation. *J. Cataract Refract. Surg.* 14: 68–72.

Shearing, S. P. (1979). Mechanism of fixation of the Shearing posterior chamber intra-ocular lens. *Contact Intraocul. Lens Med. J.* 5: 74–77.

Shepard, D. D. (1977). The dangers of metal looped intraocular lenses. *Am. Intraocular Implant Soc. J.* 3: 42.

Sterling, S., Wood, T. O. (1986). Effect of intraocular lens convexity on posterior capsule opacification. *J. Cataract Refract. Surg.* 12: 655–657.

Strampelli, B. (1961). Anterior chamber lenses: Present technique. *Arch. Ophthalmol.* 66: 12–17.

Tuberville, A. W., Galin, M. A., Perez, H. D., Banda, D., Ong, R., and Goldstein, I. M. (1982). Complement activation by nylon- and polypropylene-looped prosthetic intraocular lenses. *Invest. Ophthalmol. Vis. Sci.* 22: 727–733.

Waring, G. O., III, Bourne, W. M., Edelhauser, H. F., Kenyon, K. R. (1975). The corneal endothelium: Normal and pathologic structure and function. *Ophthalmology* 89: 531–590.

Worst, J. G. F. (1977). Complications, complicating factors and adverse conditions in lens implantation surgery: Prevention and management of surgical problems and technical complications. *Am. Intra-ocular Implant Soc. J.* 3: 20–27.

Yalon, M., Sheets, J. W., Reich, S., and Goldberg, E. P. (1983). Quantitative aspects of endothelium damage due to intraocular contacts: Effect of hydrophylic polymer graft coatings. *Int. Cong. Ophthalmol.* 24: 273–276.

9.6 IMPLANT AND DEVICE FAILURE

Allan S. Hoffman

The purpose of this short chapter is to summarize briefly the different causes which could lead to the failure of a device such as an extracorporeal device, or the failure of an implant. This chapter is not meant to cover biologic responses to foreign materials, which are discussed adequately in other chapters (e.g., see Chapters 3–5).

Most often when one sees reports in the lay literature of the mechanical or biological failure of a medical device or an implant, the materials are implicated as being the major cause of the problem. It is true that in some cases the material is indeed the major cause of the failure, since there is a wide range of different material "breakdown" mechanisms which can cause the implant or device to fail. Table 1 summarizes some of these mechanisms. However, the main objective here is to discuss the potential contributory causes of failures which are due to factors other than *material* breakdown.

MATERIALS

Table 2 lists a number of these "other" factors which can lead to the kinds of failures that we are considering. First on the list is the **materials** themselves. It is clear that if the wrong material is selected for a particular application, a device or implant can fail simply because the material components do not have the requisite physical or chemical properties. For example, the earliest silicone balls in the Starr–Edwards ball-in-cage heart valves were so highly cross-linked that when they absorbed lipids, they became brittle and fractured during their functioning, and this led to blockage of blood vessels down-

TABLE 1 Some Mechanisms of Biomaterial "Breakdown"

Mechanical
  Creep
  Wear
  Stress cracking
  Fracture

Physicochemical
  Adsorption of biomolecules, as proteins (leading to irreversible fouling)
  Absorption of H_2O or lipids (leading to softening or hardening)
  Desorption of low materials with low molecular weight, as drug or plasticizer (leading to weakening or embrittlement)
  Dissolution

Biochemical/Chemical Reactions
  Hydrolysis of amide, ester bonds
  Oxidation, reduction
  Mineral deposition
  Excessive fibrous deposition

Electrochemical
  Corrosion

stream from the heart. This problem stimulated development of a better curing protocol for the silicone rubber materials, which are still the material of choice for ball-in-cage heart valves.

TESTING OF MATERIALS

Another potential cause of failure could be that the *in vitro* or *in vivo* material evaluation tests which are carried out do not adequately test for the actual end-use conditions. In addition, certain tests may have been neglected or overlooked since the problems encountered later were not anticipated. Furthermore, there may be physiologic responses for which appropriate tests are not available and this could also eventually lead to failure because of a lack of important information in screening and selecting the materials. One example is the occurrence of stroke with artificial heart patients, due to emboli thrown off by the artificial heart itself. It is difficult to design a meaningful test for this event, whether *in vitro* or *in vivo* in an animal model. Another example is the (unanticipated) potential for a significant immunogenic response, such as has been implicated in some patients with a silicone gel breast implant.

DESIGN

The next issue which arises is the **design of the device** itself. There are a multitude of designs for heart valves, each one supposedly having a specific advantage over the others. Clearly, the design of a heart valve is extremely important for the successful functioning of such an implant. In another example, in the early days of vascular grafts, parachute materials were sewn together longitudinally and implanted. The sewing threads along the length of the implant acted as sites for thrombus formation and embolization. Clearly that early design was inadequate. Another example is design of the femoral stem in

TABLE 2 Potential Contributory Causes of Implant or Device Failure

Materials
In vitro and *in vivo* tests on materials
Design
Fabrication
Sterilization method
In vitro and *in vivo* tests on device or implant
Manufacturing, packaging, shipping, and storage
Clinical handling
Surgical procedure
Patient
"Abnormal" biological responses

the hip joint, which has evolved from clinical experience with fractured stems in human implants to computer-aided design and manufacturing techniques.

FABRICATION

Even if the materials are perfect for the application, and the tests are appropriate and the design is adequate, **the fabrication process** can introduce defects or contaminants which can lead to failure. A famous example in this case is the weld in the Bjork–Shiley stent on the tilting disk carbon heart valve. The weld on the metal stent failed in some cases because of defects introduced during the weld fabrication process.

STERILIZATION METHOD

Even if all of these factors have been adequately controlled and are appropriate to the end use, **the method of sterilization** could introduce problems from incomplete sterilization or damage of the material during sterilization. There are a number of serious limitations that certain materials place on sterilization conditions. For example, the intraocular lens cannot be irradiated since it introduces coloration in the lens, nor can it be heat sterilized because the shape (and optics) of the rigid poly(methyl methacrylate) (PMMA) lens would change above the glass transition temperature of the PMMA. Ethylene oxide also cannot be used in a number of cases because of solubility within and/or reactivity with the biomaterial. These limitations are sometimes so severe that the sterilization protocol may not adequately sterilize the device. Incomplete sterilization, of course, can lead to infection and failure.

TESTING OF IMPLANT

After the device has been designed and fabricated and sterilized, a number of biologic tests are normally used to gain more information for eventual regulatory approval and introduction of the device into the clinic. In this case there are a number of *in vitro* and *in vivo* tests which are possible candidates **for testing the finished device or implant**. Once again, as in the tests on the materials themselves, the *in vitro* and *in vivo* tests on the final device or implant (or in some cases, a prototype model) could be poorly chosen, or key tests could be overlooked, or the animal model used may not be appropriate. In the case of the animal model, there have been continuing arguments over the past 20 years among biomaterials scientists as to whether the dog is an appropriate animal model for blood interactions, especially since dog platelets are so much more adherent to foreign surfaces than are human platelets. Nevertheless, the dog is still the major test animal of choice, mainly because of availability and cost. Sometimes an inappropriate animal model may be used, or animal studies may be limited due to animal welfare issues.

PACKAGING AND SHIPPING

Then, even if all of the processes described have been adequate and appropriate, the device must be **manufactured, packaged, stored, shipped, and stored once again.** As in the fabrication of the test devices, contamination or degradation can be introduced during these steps. Transdermal drug delivery patches have to be carefully protected so the drug does not leak out of the device. In all medical devices and implants, packaging and storage are very important, and contribute to a limited shelf life for many biomedical products. Packaging and storage are also critical issues for the condom since most condoms are made from natural rubber, which is sensitive to oxygen and can degrade in the package during storage on the shelf.

CLINICAL HANDLING AND SURGICAL PROCEDURE

The next critical issue, specifically for implants or devices used clinically, is handling by surgical assistants and **how the device is brought into contact with the patient's organs or tissues** (e.g., insertion, connection, or implantation). This is a critical step, because even if the implant is made of the appropriate materials, which have been properly designed, fabricated, sterilized, tested, manufactured, packaged, shipped and stored, the "moment of truth" will be the moment the package is opened and the device is handled and eventually implanted in or contacted with the patient's organs or tissues. Thus handling in the clinic up until the implantation and/or the connection of the device to a patient (e.g., for insertion of a catheter or for extracorporeal blood treatment) can have a significant impact on the eventual success or failure of that implant or device. The surgical procedure is similarly critical because of the possibility of introduction of bacteria during contact with the patient, and eventual infection.

THE PATIENT

Another cause of failure is related to **the patient.** The patient can abuse or misuse the implant, or can exhibit an unexpected "abnormal" physiologic response. One example of such abuse could be a hip implant patient who physically exercises excessively before complete healing, which could cause a loosening of the implant. An example of misuse could be rough handling of a condom which causes a rip or pinhole to form. In terms of an "abnormal" physiologic response, one must remember that, similar to drug responses, there will be a dose-response type of curve for every implant or device. For example, a silicone breast implant may affect a very small fraction of the population in an unexpected and undesirable way, relative to the great majority of the women receiving the implant.

ABNORMAL RESPONSES

The previous example leads to the last item in Table 2, for situations when there may be **"abnormal" biological responses** due to events which were not observed previously (and which probably could not be tested for, in most cases). This cause of implant or device failure is related to an "abnormal" or "skewed" physiologic response in a large population of patients. Every device will have a certain "failure rate" based on normal population statistics, even if each implanted or inserted device is always exactly the same as every other device. Nevertheless, such events can be very unfortunate for the rare patient who encounters such unexpected responses. Furthermore, the legal consequences can be serious for any one (or all) of those who were involved in each step listed in Table 2, from the materials manufacturer to the clinical user. Nevertheless, such "abnormal" events can lead to useful changes in one or more of the steps listed in Table 2, which can be useful (perhaps even life-extending) for the majority of "normal-responding" patients. Finally, it should be recognized that the "experimental animal" in this last cause of failure is the human. This entire ethical-legal issue is beyond the scope of this chapter, but it cannot be ignored by the biomaterials scientist or engineer.

In summary, this short chapter is meant to indicate the wide range and diversity of things that can contribute to the failure of a medical device or implant. It is important to realize that **the material is not always the cause of failure,** but the materials scientist or engineer can play a major role in ensuring the success of a device or implant.

9.7 CORRELATION OF MATERIAL SURFACE PROPERTIES WITH BIOLOGICAL RESPONSES

Buddy D. Ratner

Physical or chemical measurements that can reliably predict the *in vivo* biocompatibility of biomaterials have not yet been identified. It would be ideal, when considering a new material for medical applications, to use a spectroscopic technique to measure the composition, and, from that physical measurement, to predict how well the material will work in a particular application. Animal experiments are expensive, of questionable value for predicting performance in humans and raise ethical issues (see Chapter 5.5). Human clinical trials are very expensive and also raise ethical issues. Can we predict or prescreen *in vivo* or *in vitro* performance from measurements of surface and other properties? This chapter examines this question and offers suggestions for future exploration.

BIOCOMPATIBILITY AND PERFORMANCE OF MEDICAL DEVICES

A reexamination of definitions is appropriate at this point. The "biocompatibility" of a medical device can be defined

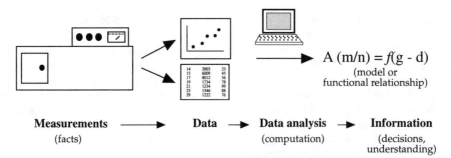

$$A (m/n) = f(g - d)$$
(model or
functional relationship)

Measurements ⟶ **Data** ⟶ **Data analysis** ⟶ **Information**
(facts)   (computation)   (decisions,
                                                        understanding)

FIG. 1. The relationship between data and information.

in terms of the success of that device in fullfilling its intended function. Thus, the blood filtration module of a hemodialysis system might be defined by its ability to appropriately fractionate soluble blood components, its robustness over its intended lifetime, and its nondamaging interaction with the patient's blood. Alternatively, we can define a "blood compatibility" for the membrane, a "blood compatibility" for the silicone rubber header, a "blood compatibility" for the tubing, and a "blood and soft tissue compatibility" for the percutaneous connection between the apparatus and the patient's blood stream. Similarly, for a hip joint prosthesis, we can discuss the fatigue resistance of the device, its corrosion resistance, the distribution of stresses transferred to the bone by the device, the solid angle of mobility provided and the overall success of the device in restoring a patient to an ambulatory state. The performance of a hip joint prosthesis might also be couched in terms of the tissue reaction to the bone cement, to an uncemented titanium prosthesis stem, and to the acetabular cup. In the first case, the performance of a whole system (device) is assessed, and in the second case, the biological reaction to specific components of the device (biomaterials) is examined. The difference between the consideration of the whole device and the materials that comprise it is a critically important point. In certain contexts, only the performance of the complete device is defined as biocompatibility. This definition can be inferred from the U.S. Food and Drug Administration policy that only complete devices, and never materials, receive "approval." The performance of the individual materials is referred to sometimes as "biocompatibility" and sometimes as "bioreaction." This is a book on biomaterials, so it is appropriate to focus on the materials. "Bioreaction" is a much simpler word than "biocompatible," and this term will largely be used here. Bioreaction is discussed here and defined by example.

Toxicology assays, often inappropriately referred to as "biocompatibility assessment," were presented in Chapter 5.2. These assays deal with measurement of substances that leach from materials, most of which will induce some cell or tissue reaction. Such assays are discussed here because the published correlations are clear, interpretable, and measurable. If we concentrate on bioreactions to implanted materials that do not intentionally leach substances (i.e., most biomaterials), the surface properties immediately assume a high profile as the prime candidate to control bioreaction. However, hardness, porosity, shape, and specific implant site are also important.

These issues are addressed in working through important concepts in correlating material properties with bioreaction.

DATA, INFORMATION, AND STATISTICS

Laboratory experiments lead to the acquisition of data from the testing of the properties (biological and physical) of implant materials. Frequently we are presented with columns of numbers or a plot. Staring at these numbers or plots is often unhelpful in understanding the system under study. What we really desire is not data, but useful information, about our system. This idea is illustrated in Fig. 1. Correlation is one way to reduce data so that it yields information.

Correlation

Correlation is a tool in statistical methodology that is useful for analyzing data so as to appreciate its variation and its significance. A general introduction to statistics is not presented here, but every student of biomaterials science should be versed in these mathematical tools. A few useful, general books on statistics as applied to scientific problems are Bevington (1969), Mosteller and Tukey (1977), and Anderson and Sclove (1986).

A correlation is a dependence (i.e., a relationship) between two or more variables. It does not necessarily imply cause and effect. For example, most people walking with open umbrellas will have wet feet. However, in this situation, it is obvious that umbrellas do not cause wet feet. We have a high correlation, but it misses the controlling factor (causative factor) in this example—the rain. Thus, we can propose that A (the umbrella) causes B (wet feet), B causes A, they cause each other, or they are both caused by C (the rain). We cannot prove causation, but it can be strongly suggested. Often, where correlations are observed, the causative factor is obscured, and so we have data but little useful information. Where relationships are established between a dependent variable and an independent (or explanatory) variable, this is referred to as regression analysis.

It may be more productive to look at this problem in terms of calibration and prediction. Calibration in a practical (e.g., analytical) sense has been defined by Martens and Naes (1989) as the use of empirical data and prior knowledge for determin-

TABLE 1 Some Bioreactions[a]

Protein adsorption	Macrophage adhesion
Protein retention	Phagocytosis
Lipid absorption	Macrophage release
Bacterial adhesion	Neutrophil attachment
Platelet adhesion	Biodegradation
Hemolysis	Angiogenesis
Platelet activation	Cell spreading
Expression of new genes	Fibrous encapsulation

[a]These reactions are commonly observed with implant materials. However, their relationship to "biocompatibility" is not direct.

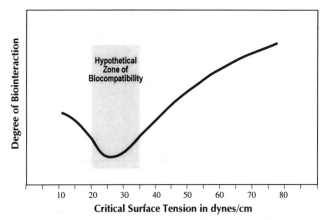

FIG. 2. A hypothesis for a minimum biointeraction for surfaces with critical surface tensions around 22 dynes/cm.

ing how to predict unknown quantitative information, Y, from available measurements, X, via some mathematical transfer function. This could be as simple as plotting Y versus X and using a least-squares fit to deal with modest levels of random noise, or as complex as a multivariate calibration model to accommodate noisy data, interferents, multiple causes, nonlinearities, and outliers. Multivariate methods are elaborated upon toward the end of this chapter.

Aspects of the Bioreaction to Biomaterials

Bioreaction, a process related to, but more general than, "biocompatibility," can have many manifestations. Some of these are listed in Table 1. A bioreaction is most simply defined as response observed upon the interaction of a material with a biological sysem or system containing biomolecules. Can simple measured physical properties of materials be correlated with bioreactions? There are many examples where this is indeed the case. Table 2 lists some of the surface physical measurements for materials that one might hypothesize as influencing bioreaction or biocompatibility.

TABLE 2 Physical Parameters of Biomaterial Surfaces That Might Correlate with Bioreactivity

Wettability	Subsurface features
Hydrophilic/ hydrophobic ratio	Distribution of functional (receptor sites)
Polar/dispersive character	Modulus
Surface chemistry	Hydrogel (swelling) character
Specific functional groups	
Surface electrical properties	Mobility
Roughness/porosity	Adventitious contamination
Domains of chemistry	Trace quantities of groups

THE CASE FOR CORRELATION—A BRIEF REVIEW OF THE LITERATURE

An early and influential paper demonstrating that physical measurements might be correlated with observed reactions to biomaterials concerned data extractable from biomaterials (Homsy, 1970). Many materials were examined in this study. Each was extracted in a pseudoextracellular fluid. The extract was examined by infrared (IR) absorbance spectroscopy of hydrocarbon bands that are indicative of organic compounds. A strong, positive correlation was observed between the strength of the IR absorbances and the reaction of the materials with a primary tissue culture of newborn mouse heart cells. This paper was important in scientifically justifying *in vitro* cell culture analysis for screening the toxicology of biomaterials, and for reinforcing the principle that biomaterials should not unintentionally leach substances.

In the early 1970s, Robert Baier and colleagues offered an intriguing hypothesis concerning surface properties and bioreaction that continues to stimulate new experiments to this day (Baier, 1972). This hypothesis is based upon interfacial energetics of surfaces as measured by contact angles, and suggests that materials with critical surface tensions (see Chapter 1.3) of approximately 22 dynes/cm will exhibit minimum bioreactivity (Fig. 2). Support for this hypothesis has been generated in a number of experiments spanning many different types of biointeractions (Baier *et al.*, 1985; Dexter, 1979). However, in a larger number of cases, this minimum has not been observed, raising questions about the generality of this concept (Neumann, 1979; Yasuda *et al.*, 1978; Mohandas *et al.*, 1974; Lyman, 1970; Chang *et al.*, 1977).

Some of the clearest biointeraction correlations have been observed in simple, nonproteinaceous media. Linear trends of cell (mammalian and bacterial) adhesion versus various measures of surface energy have been noted (Chang *et al.*, 1977; Neumann *et al.*, 1979; Yasuda *et al.*, 1978; Mohandas *et al.*, 1974). For example, Chang and co-workers (1977) found that the adhesion of washed pig platelets to solid substrates increased with increasing water contact angle, a parameter that generally correlates well with solid surface tension. It is interesting that these simple linear trends often vanish or diminish if

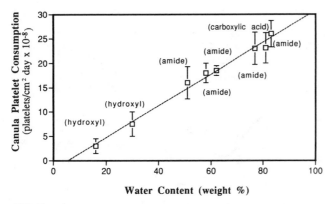

FIG. 3. A correlation between baboon platelet consumption as measured in an *ex vivo* shunt system and hydrogel water content (From Hanson *et al.*, 1980, used with permission.)

protein is present in the attachment medium (Neumann *et al.*, 1979; Chang *et al.*, 1977; van der Valk *et al.*, 1983). More complex surface energetic parameters have also been explored to correlate bioreaction to surface properties (Kaelble and Moacanin, 1977).

Correlations between material properties and long-term events upon implantation are less frequently seen in the literature. However, some important examples have been published. The baboon A-V shunt model of arterial thrombosis has yielded a number of intriguing correlations (Harker and Hanson, 1979). Using an *ex vivo* femoral-femoral shunt, this model measures a first-order rate constant of platelet destruction induced by the shunt material (the units are platelets destroyed/cm²·day). The values for this surface reactivity parameter are independent of flow rate (after the flow rate is sufficiently high to ensure kinetically limited reaction), length of time that the reaction is observed, blood platelet count, and surface area of the material in contact with the blood. In one experiment, a series of hydrogels grafted to the luminal surfaces of 0.25-cm i.d. tubes was studied (Hanson *et al.*, 1980). The platelet consumption (see Chapter 5.4) was found to increase in a simple, linear fashion with the equilibrium water content of the hydrogels. This correlation, illustrated in Fig. 3, is particularly intriguing because the hydrogel materials studied were amide-, carboxylic acid-, and hydroxyl-based. The only clear, correlating parameter was equilibrium water content. In another study, the platelet consumption of a series of polyurethanes was observed to decrease in a linear fashion as the fraction of the polyurethane C1s ESCA surface spectrum that was indicative of hydrocarbon moieties increased (Hanson *et al.*, 1982).

A material parameter that lends itself to correlation is roughness or surface texture. Roughness is readily measured using a scanning electron microscope, a profilometer, or an atomic force microscope (see Chapter 1.3). Relationships between roughness and blood hemolysis (Wielogorski *et al.*, 1976) or thrombogenicity (Hecker and Scandrett, 1985) have been reported. Textures and roughness are also extremely important to the fixation of materials into hard tissue and to the nature of the foreign body response observed (Brauker *et al.*, 1992; Thomas and Cook, 1985; Schmidt and von Recum, 1991).

A complication in the use of roughness parameters is a differentiation between porosity and roughness, and also an appreciation of the difference between the average feature amplitude (often called R_a) and the nature of the roughness (e.g., are rolling hills and jagged rocks of the same height also of the same roughness? See Fig. 4).

Clinical results correlated with material properties are rare, in part because materials used in clinical studies are generally not fully characterized. However, a few such studies have been published. For example, the wettability of rigid gas-permeable contact lenses was correlated with the discomfort of subjects and a predictive trend was noted (Bourassa and Benjamin, 1989). Catheters that are used in humans were evaluated in a test system closely simulating clinical application (Wilner *et al.*, 1978). The catheters were classified into three groups related to their probable success, but clear relationships with surface properties were not discerned. The complications in performing control studies, the difficulties in assembling a sufficiently large experimental population, and the complexity of the materials (devices) and the human biology make clinical correlation a difficult problem.

ISSUES COMPLICATING SIMPLE CORRELATION

It should be clear by now that, although many correlations have been noted, there is considerable contradictory evidence about what the correlating factors are and the nature of the correlations. Also, in many systems, no correlations have been noted.

The most widely used correlating factor has been surface energetics, possibly because contact angles can be readily measured in any laboratory. Surface energetic parameters all relate back to the second law of thermodynamics and it is well established that the interactions of simple colloidal particles can be modeled using thermodynamic and electrostatic arguments. If living cells are treated as simple colloid particles with fixed mass, charge density, polar forces, and hydrophobic interactions, thermodynamic (energetic) modeling may be appropriate (Gerson and Scheer, 1980; Fletcher and Pringle, 1985). However, living cells most often cannot be viewed as "hard, charged spheres." Living cells can change their surface characteristics in response to surfaces and other stimuli. Also, specific (e.g., receptor) interactions do not lend themselves to this simple thermodynamic modeling. For example, two surfaces, with similar immobilized peptides (and hence essentially the same surface energy) may interact very differently with cells, if one of the peptides represents a minimal recognized sequence for the cell-surface receptor. This was observed with fibroblast cell attachment where an immobilized peptide containing an RGD unit (arginine-glycine-aspartic acid) and a closely related immobilized peptide containing an RGE segment (where the E indicates glutamic acid) were compared (Massia and Hubbell, 1990). The RGD peptide was highly active in inducing cell spreading while the RGE peptide was not. Finally, the nature of the correlations may be multivariate rather than univariate. This concept is discussed in the next section.

MULTIVARIATE CORRELATION

We are trained throughout our science education to appreciate cause-and-effect correlations. For example, if the temperature of a solution is increased, the reaction rate of two reactants in the solution will increase in some relatively simple manner. Unfortunately, many systems, particularly the multicomponent systems so often found in biomaterials science, have competing reactions that are dependent upon each other (e.g., the product of one reaction may influence the rate of another reaction). Thus, we do not see a simple relationship, but rather, many things changing simultaneously. Our eye cannot discriminate the key trend(s) in this "stew" of changing numeric values. Multivariate statistics is a class of statistical methods that looks for trends, patterns, and relationships among many variables. Also, where contemporary analytical instrumentation produces large amounts of complex (e.g., spectral) data, multivariate statistics can assist in examining the data for similarities, differences, and trends. Where large amounts of data overload our ability to appreciate relationships by "eyeballing" them, multivariate methods thrive on large amounts of data and, in fact, become more accurate and useful. This class of statistical methods has come into its own only with the introduction of powerful computers since the methods are computation-intensive. Many general introductions to multivariate statistics are available (Martens and Naes, 1989; Sharaf et al., 1986; Brereton, 1990; Massart et al., 1988). Multivariate statistics applied to problems involving chemistry is often referred to as "chemometrics."

An important general principle in multivariate analysis is dimensionality reduction. A plot of x versus y requires us to think in two dimension. A 3-D plot of x, y, and z can still be easily visualized. Where we have w, x, y, and z as the axes, we lose the ability to absorb the information in graphical form and identify trends in the data. However, if we take our 3-D example, we can visualize a projection (shadow) of the 3-D data cluster in two dimensions. We have reduced the dimensionality from 3 to 2. Similarly, a 4-D data set can be projected (by a computer) into a 3-D space. Thus we have a data representation we can visualize in order to look for trends. This dimensionality reduction is readily performed by computers using standard linear algebraic methods. The number of dimensions that the computer can work with is, for all practical purposes, unlimited.

There are many multivariate statistical algorithms useful for analyzing data. They are sometimes divided into classification methods (also called cluster analysis methods) and factor analysis methods (Mellinger, 1987). Classification methods find similarities in groups of data points and arrange them accordingly. Factor analysis methods take data and transform them into new "factors" that are linear combinations of the original data. In this way the dimensionality of the problem is reduced. Factor analysis methods that are useful for multivariate correlation with data sets such as are acquired in biomaterials research include principal component analysis (PCA) (Wold et al., 1987) and partial-least squares (PLS) regression (Geladi and Kowalski, 1986). Two important points about these methods are that they do not require a hard model (rarely do we have such a quantitative model) and that they make use of all the data (i.e., we do not have to choose which data we want to put into the correlation model). Examples of the application of these methods to biomaterials research are few (Perez-Luna et al., 1994; Wojciechowski and Brash, 1993), but, as the power of these tools is recognized, these methods will become standard data analysis tools. This is because they make efficient use of all data; they thrive on large amounts of data produced by modern instruments; they are objective in that we do not have to choose which data to use; and they reduce the influence of noise and irrelevant variables, thereby effectively increasing the signal-to-noise ratio.

Multivariate statistical methods can be a great boon to data analysis, but they will not solve all our problems in biomaterials science. They should be considered as powerful hypothesis generators. The correlations and trends noted using such analysis represent a new view of the significance of data that we

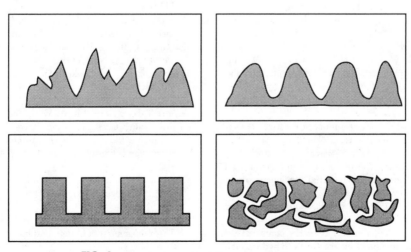

FIG. 4. Four surfaces with similar average roughnesses (R_a).

could not appreciate staring at spectra or tables. New hypotheses about the importance of materials variables can now be formulated, and then must be tested. Multivariate statistical methods also provide powerful tools for experimental design, but these are not discussed here.

CONCLUSIONS

Unfortunately, biomaterials science is not yet at the state where we can assemble a handbook of design data about the relationships between surface structures and biological reactions. We do not yet understand the controlling variables of the biology or the materials science sufficiently well to generalize our observations. What this chapter does is suggest that such relationships probably exist, and point out that there are powerful mathematical methods that have the potential to help generalize data into correlations and trends useful in the design of biomaterials and medical devices.

Bibliography

Anderson, T. W., and Sclove, S. L. (1986). *The Statistical Analysis of Data*. Scientific Press, Palo Alto, CA.

Baier, R. E. (1972). The role of surface energy in thrombogenesis. *Bull. New York Acad. Med.* 48(2): 257–272.

Baier, R. E., De Palma, V. A., Goupil, D. W., and Cogen, E. (1985). Human platelet spreading on substrata of known surface chemistry. *J. Biomed. Mater. Res.* 19: 1157–1167.

Bevington, P. R. (1969). *Data Reduction and Error Analysis for the Physical Sciences* McGraw-Hill, New York.

Bourassa, S. and Benjamin, W. J. (1989). Clinical findings correlated with contact angles on rigid gas permeable contact lens surfaces *in vivo. J. Am. Optom. Assn.* 60(8): 584–590.

Brauker, J., Martinson, L., Young, S., and Johnson, R. C. (1992). Neovascularization at a membrane-tissue interface is dependent on microarchitecture. *Trans. Fourth World Biomaterials Cong.* Berlin, April 24–28, p. 685.

Brereton, R. G. (1990). *Chemometrics*. Ellis Horwood Publishers, New York.

Chang, S. K., Hum, O. S., Moscarello, M. A., Neumann, A. W., Zingg, W., Leutheusser, M. J., and Ruegsegger, B. (1977). Platelet adhesion to solid surfaces. The effect of plasma proteins and substrate wettability. *Med. Progr. Technol.* 5: 57–66.

Chang, S. K., Neumann, A. W., Moscarello, M. A., Zingg, W., and Hum, O. S. (1977). Substrate wettability and *in vitro* platelet adhesion. *Abstr. of the 51st Colloid and Surface Science Symposium,* Grand Island, N.Y., June 19–22. American Chemical Society, Washington, D.C.

Dexter, S. C. (1979). Influence of substratum critical surface tension on bacterial adhesion—*in situ* studies. *J. Coll. Interf. Sci.* 70(2): 346–354.

Fletcher, M., and Pringle, J. H. (1985). The effect of surface free energy and medium surface tension on bacterial attachment to solid surfaces. *J. Coll. Interf. Sci.* 104(1): 5–14.

Geladi, P., and Kowalski, B. R. (1986). Partial least-squares regression: a tutorial. *Anal. Chim. Acta* 185: 1–17.

Gerson, D. F., and Scheer, D. (1980). Cell surface energy, contact angles and phase partition III. Adhesion of bacterial cells to hydrophobic surfaces. *Biochim. Biophys. Acta* 602: 506–510.

Hanson, S. R., Harker, L. A., Ratner, B. D., and Hoffman, A. S. (1980). *In vivo* evaluation of artificial surfaces with a nonhuman primate model of arterial thrombosis. *J. Lab. Clin. Med.* 95(2): 289–304.

Hanson, S. R., Harker, L. A., Ratner, B. D., and Hoffman, A. S. (1982). Evaluation of artificial surfaces using baboon arteriovenous shunt model. in *Biomaterials 1980, Advances in Biomaterials,* G. D. Winter, D. F. Gibbons, and H. Plenk, Jr., eds. Wiley, Chichester, England, Vol. 3, 519–530.

Harker, L. A., and Hanson, S. R. (1979). Experimental arterial thromboembolism in baboons. Mechanism, quantitation, and pharmacologic prevention. *J. Clin. Invest.* 64: 559–569.

Hecker, J. F., and Scandrett, L. A. (1985). Roughness and thrombogenicity of the outer surfaces of intravascular catheters. *J. Biomed. Mater. Res.* 19: 381–395.

Homsy, C. A. (1970). Bio-compatibility in selection of materials for implantation. *J. Biomed. Mater. Res.* 4: 341–356.

Kaelble, D. H. and Moacanin, J. (1977). A surface energy analysis of bioadhesion. *Polymer* 18: 475–482.

Lyman, D. J., Klein, K. G., Brash, J. L., and Fritzinger, B. K. (1970). The interaction of platelets with polymer surfaces. *Thrombos, Diathes, Haemorrh.* 23: 120–128.

Martens, H., and Naes, T. (1989). *Multivariate Calibration*. Wiley, New York.

Massart, D. L., Vandeginste, B. G. M., Deming, S. N., Michotte, Y., and Kaufman, L. (1988). *Chemometrics: A Textbook*. Elsevier, Amsterdam.

Massia, S. P., and Hubbell, J. A. (1990). Covalent surface immobilization of Arg-Gly-Asp- and Tyr-Ile-Gly-Ser-Arg-containing peptides to obtain well-defined cell-adhesive substrates. *Anal. Biochem.* 187: 292–301.

Mellinger, M. (1987). Multivariate data analysis: its methods. *Chemometrics and Intelligent Laboratory Systems* 2: 29–36.

Mohandas, N., Hochmuth, R. M., and Spaeth, E. E. (1974): Adhesion of red cell to foreign surfaces in the presence of flow. *J. Biomed. Mater. Res.* 8: 119–136.

Mosteller, F., and Tukey, J. W. (1977). *Data Analysis and Regression*. Addison Wesley, Reading, MA.

Neumann, A. W., Moscarello, M. A., Zingg, W., Hum, O. S., and Chang, S. K. (1979). Platelet adhesion from human blood to bare and protein-coated polymer surfaces. *J. Polym. Sci. Polym. Symp.* 66: 391–398.

Perez-Luna, V. H., Horbett, T. A., and Ratner, B. D. (1994). Developing correlations between fibrinogen adsorption and surface properties using multivariate statistics. *J. Biomed. Mater. Res.* 28: 1111–1126.

Schmidt, J. A., and von Recum, A. F. (1991). Texturing of polymer surfaces at the cellular level. *Biomaterials* 12: 385–389.

Sharaf, M. A., Illman, D. L., and Kowalski, B. R. (eds.) (1986). *Chemometrics* Wiley, New York.

Thomas, K. A., and Cook, S. D. (1985). An evaluation of variables influencing implant fixation by direct bone apposition. *J. Biomed. Mater. Res.* 19: 875–901.

van der Valk, p., van Pelt, A. W. J., Busscher, H. J., de Jong, H. P., Wildevuur, C. R. H., and Arends, J. (1983). Interaction of fibroblasts and polymer surfaces: relationship between surface free energy and fibroblast spreading. *J. Biomed. Mater. Res.* 17: 807–817.

Wielogorski, J. W., Davy, T. I., and Regan, R. J. (1976). The influence of surface rugosity on haemolysis occurring in tubing. *Biomed. Eng.* 11: 91–94.

Wilner, G. D., Casarella, W. J., Baier, R., and Fenoglio, C. M. (1978). Thrombogenicity of angiographic catheters. *Circ. Res.* 43(3): 424–428.

Wojciechowski, P. W., and Brash, J. L. (1993). Fibrinogen and albumin adsorption from human blood plasma and from buffer onto chemically functionalized silica substrates. *Coll. Surf. Biointerfaces* **1**: 107–117.

Wold, S., Esbensen, K., and Geladi, P. (1987). Principal component analysis. *Chemometrics Intelligent Laboratory Systems* **2**: 37–52.

Yasuda, H., Yamanashi, B. S., and Devito, D. P. (1978). The rate of adhesion of melanoma cells onto nonionic polymer surfaces. *J. Biomed. Mater. Res.* **12**: 701–706.

9.8 IMPLANT RETRIEVAL AND EVALUATION

James M. Anderson

Implant retrieval and evaluation offers the opportunity to investigate and study the intended use of biomaterials. Implant retrieval and evaluation programs, in general, are designed to determine the efficacy and safety or biocompatibility of biomaterials, prostheses, and medical devices. Appropriate programs include assessment of safety and/or biocompatibility. The goal of safety testing is to determine if a biomaterial presents potential harm to the patient; it evaluates the interaction of the biomaterial with the *in vivo* environment and determines the effect of the host on the implant. Biocompatibility assessment is the determination of the ability of a biomaterial to perform with an appropriate host response in a specific application. In the *in vivo* environment, biocompatibility assessment is considered to be a measure of the degree (magnitude) and extent (duration) of adverse alteration(s) in homeostatic mechanism(s). The terms "safety assessment" and "biocompatibility assessment" may or may not be considered to be synonomous.

In this chapter, implant retrieval and evaluation are considered from the perspective of human clinical use. Therefore, this chapter draws upon many important perspectives presented in chapters dealing with materials science and engineering, biology, biochemistry, and medicine, host reactions and their evaluation, the testing of biomaterials, the degradation of materials in the biological environment, the application of materials in medicine and dentistry, the surgical perspective of implantation, and the correlation of material surface properties with biological responses of biomaterials. While this chapter focuses on clinical implant retrieval and evaluation, many of the goals and perspectives presented are important to preclinical evaluation programs. In this chapter, implants are considered to be

TABLE 1 General Goals of Implant Retrieval and Evaluation

Determine modes and mechanisms of implant failure or success

Determine dynamics and temporal variations of tissue–materials and blood–materials interactions

Develop design criteria for future implants

Determine adequacy and appropriateness of animal models

TABLE 2 Important Components of an Implant Retrieval and Evaluation Program

Appropriate accessioning, cataloging, and identification

Review of patient's medical history and radiography

Gross examination and photography

Analytical protocols and techniques for assessing host and implant responses

Development of correlative and cause-and-effect relationships between material, design, mechanical, manufacturing, clinical, and biological variables

Statistical and multivariant analyses

composed of one or more biomaterials that are part of a specific design for a specific application. For these purposes, implants are considered to be biomaterials, prostheses, medical devices, or artificial organs.

GOALS

The general goals of such programs are presented in Table 1. While many implant failures can be characterized as implant- or material-dependent or clinically or biologically dependent, many modes and mechanisms of failure are interdependent, that is, they are dependent on both implant and biological factors. To appropriately appreciate the dynamics and temporal variations of tissue–materials and blood–materials interactions of implants, a fundamental understanding of these interactions is important. An implant retrieval and evaluation program should provide for the appreciation of materials, design, and biological factors and result in design criteria for future development. Finally, implant retrieval and evaluation programs should offer the opportunity to determine the adequacy and appropriateness of animal models used in preclinical testing of implants and biomaterials. Thus, the strengths and weaknesses as well as the advantages and disadvantages of animal models can be evaluated.

These programs may also serve as a teaching or research resource. They may be used to educate patients, their families, physicians, residents, students, engineers, and materials scientists, as well as the general public. As a research resource, an implant retrieval and evaluation program can be used to develop and test hypotheses and to improve protocols and techniques.

COMPONENTS OF AN IMPLANT RETRIEVAL AND EVALUATION PROGRAM

An implant retrieval and evaluation program is an interdisciplinary effort by scientists with expertise in materials science, materials engineering, biomechanics, biology, pathology, microbiology, radiology, medicine and surgery. Table 2 lists the

TABLE 3 Retrieved Implants, Division of Surgical Pathology, Institute of Pathology, University Hospitals of Cleveland

Category	1979–1995	1994	1993	1992	1991	1990	1989	1988–1984	1983–1979
Hip	1030	62	72	102	130	114	79	311	119
Knee	476	29	32	45	71	65	37	108	69
Catheter	836	48	54	95	73	110	95	199	131
Graft	566	42	42	57	35	46	51	166	99
IUD	154	7	11	5	10	11	18	60	27
General	686	42	42	40	47	49	65	284	89
Breast	334	47	67	61	23	30	19	42	11
Finger	87	3	4	5	6	1	8	34	24
Orthopedic	2174	189	179	180	140	143	154	683	416
Heart valve	214	7	7	13	14	10	9	96	56
Total	6557	476	510	603	549	579	535	1983	1041

important components of such a program. Given these components, the importance of an interdisciplinary or team perspective is apparent.

The first component is the appropriate accessioning, cataloging, and identification of retrieved implants. Patient anonymity is required in any implant retrieval and evaluation program and this can be achieved through appropriate accessioning and cataloging. In our program at University Hospitals of Cleveland/Case Western Reserve University, this is carried out through the Surgical Pathology Division of the Institute of Pathology (see Table 3). (The Autopsy Division of the Institute of Pathology has a similar program for implant retrieval and evaluation.)

This program uses the general accessioning and cataloging scheme for surgical pathology specimens. Gross and microscopic diagnoses in a standardized format are provided for patients' charts, the appropriate clinicians and surgeons, and for the database for Implant Retrieval and Evaluation Program. The Department of Surgery provides specimens to the Division of Surgical Pathology, with the patient's name and hospital identification number, clinical diagnoses, and notes on the patient and/or implant history.

As can be seen from Table 3, the number of types of implants retrieved is large and covers the orthopedic, cardiovascular, gynecological, and soft tissue areas. Second, the number of retrieved implants of each type has increased over the past decade. Several factors are responsible for this. They include increased recognition by surgeons of the importance of implant retrieval and evaluation as well as the increased numbers of implants.

An in-depth evaluation of retrieved implants requires a review of the patient's medical history and radiographs, where pertinent. Table 4 provides a partial list of conditions that may influence the failure or success of orthopedic and cardiovascular implants. The identification of acute and chronic problems presented by the patient will provide guidance on how the evaluation of a specific implant should be carried out. For

example, the clinical diagnosis and identification of infection is most helpful in determining techniques to be used in the evaluation. Gross examination and photography play an important role and must be carried out before specific techniques are used to evaluate implants.

The next step in the program is the development of a strategy that will allow an optimum amount of information to be obtained using appropriate analytical protocols and techniques to assess host and implant responses. This strategy is directed toward developing correlative and cause-and-effect relationships among material, design, mechanical, manufacturing, clinical, and biological variables. Finally, analytical protocols and techniques should produce quantitative information that can be analyzed statistically. These analyses may also include clinical information.

TABLE 4 Patient Conditions Influencing Implant Failure or Success

Orthopedic	Cardiovascular
Polyarthritis syndromes	Atherosclerosis
Connective tissue disorders	Diabetes
Osteoarthritis	Infection
Trauma	Hypertension
Infection	Ventricular hypertrophy
Metabolic disease	
Endocrine disease	
Tumor	
Primary joint disease	
Osteonecrosis	

TABLE 5 Techniques for Implant Evaluation

Implant	Tissue
Atomic absorption spectro-photometry	Atomic absorption spectrophotometry
ATR-FT-IR	Autoradiography
Burst strength	Biochemical analysis
Compliance studies	Cell culture
Contact angle measurements	Chemical analysis
Digestability	Enzyme histochemistry
EDAX	Gel electrophoresis
ESCA	Histology
Extractability	Immunocytochemistry
Fatigue studies	Immunofluorescence
Fracture analysis	Immunohistochemistry
FT-IR	Immunoperoxidase
Gel permeation chromatography	In situ hybridization
Glass transition temperature	Microbiologic cultures
Hardness studies	Morphometry
Light microscopy	Radiographic analysis
Macrophotography	Scanning electron microscopy
Melt temperature	Tissue culture
Metallographic examination	Transmission electron microscopy
Particulate analysis	Ultrastructure studies
Polarized light microscopy	
Porosity analysis	
Scanning electron microscopy	
Shrink temperature	
SIMS	
Stereomicroscopy	
Stress analysis	
Tensile studies	
Transmission electron microscopy	
Topography analysis	

TABLE 6 Complications of Dental Implants

Adverse foreign body reaction	Loosening
Biocorrosion	Metal allergy
Electrochemical galvanic coupling	Pain
Fatigue	Particulate formation
Fixation failure	Sinus formation
Fracture	Surface fracture
Infection	Torsion
Interface separation	Wear
Loss of mechanical force transfer	

nondestructive and destructive testing procedures. Only after appropriate accessioning, cataloguing and identification, and complete review of the patient's medical history and radiography can the analytical protocols and techniques for implant evaluation be specified.

It should be noted that the techniques for implant evaluation are most commonly destructive techniques, that is, the implant or portions of the implant must be destroyed or altered to obtain the desired information on the properties of the implant or material. The availability of the implant and tissue specimens will dictate the choice of technique.

Dental Implants

The complications of dental implants that lead to failure are presented in Table 6. These complications may be related to the mechanical–biomechanical aspects of force transfer, or the chemical–biochemical aspects of elements transferred across biomaterial and tissue interfaces, or both. The complex synergism which exists between tissue and biomaterial responses presents a significant challenge in the identification of the failure mechanisms of dental implants. Lemons and others have provided appropriate perspectives to be taken in the evaluation of dental implants.

ASSESSMENT OF HOST AND IMPLANT RESPONSES

Simple evaluation of the implant with no attention to the tissue will produce an incomplete evaluation and no understanding of the host response. Table 5 provides a partial list of techniques for evaluating implants and tissues. The table is incomplete because each type of implant and the materials of which it is made will dictate further techniques to be used in evaluating its success.

In general, analytical protocols and techniques for assessing host and implant responses can be divided into two categories:

TABLE 7 Complications of Orthopedic Implants

Bone resorption	Loosening
Corrosion	Mechanical mismatch
Fatigue	Metal allergy
Fibrosis	Pain
Fixation failure	Particulate formation
Fracture	Plastic deformation
Incomplete osseous integration	Stress riser development
Infection	Stress-shielding
Interface separation	Surface wear

TABLE 8 Complications of Cardiovascular Devices

Heart valve prostheses	Vascular grafts	Cardiac devices
Paravalvular leak	Thrombosis	Thrombosis
Infective endocarditis (PVE)	Embolism	Embolism
Thrombosis	Infection	Endocarditis
Embolism	Perigraft erosion	Extraluminal infection
Extrinsic dysfunction	Perigraft seroma	Component fracture
Mechanical valves	False aneurysm	Hemolysis
Cloth wear	Anastomotic hyperplasia	Calcification
Poppet wear	Disintegration/degradation	
Cage wear		
Incomplete valve closure		
Tissue valves		
Cusp degeneration		
Cusp tearing		
Cusp calcification		

Orthopedic Implants

Wright *et al.* (1982; 1988) and Cook *et al.* (1985) have provided significant information on the analysis of retrieved orthopedic implants. Table 7 presents a list of complications commonly found with retrieved orthopedic implants; of particular interest are metal allergy and tissue ingrowth (Merrit and Brown, 1981; Cook *et al.*, 1988).

Cardiovascular Implants

Cardiovascular implants commonly involve both blood and soft tissue interactions with materials. The complications most commonly found with these implants, i.e., heart valve prostheses, vascular grafts, and cardiac devices, are listed in Table 8. Schoen and others have provided useful information on the perspectives, approaches, and techniques for evaluating cardiovascular implants.

TABLE 9 Implant Retrieval and Evaluation:
Clinical Benefits

Provides prosthesis selection criteria

Enhances patient–prosthesis matching

Facilitates patient management

Provides recognition of device complications

Reduces device complications by progressive prosthesis development

Identifies subclinical patient–prosthesis interactions

Elucidates patient–prosthesis interactions

Determines tissue–material interaction mechanisms

CONCLUSIONS

While the focus of this chapter has been on retrieval and evaluation of implants in humans, the perspectives, approaches, techniques, and methods may also apply to evaluation of new material, preclinical testing for biocompatibility and premarket clinical evaluation. Each type of *in vivo* or clinical setting has its unique implant–host interactions and therefore, requires the development of a unique strategy for retrieval and evaluation.

Implant retrieval and evaluation must take bias into account. The findings from individual groups or hospitals on specific implants may not necessarily represent the norm regarding the implant–host interactions for that implant. In particular, the surgeon's experience, technique, and preference for a specific type of implant must be considered. In addition, the clinical failure of an implant versus failure of the implant itself must also be considered.

Implant retrieval and evaluation should be viewed as a continuous process that provides numerous clinical benefits. Table 9 provides a partial list of such benefits and highlights the interdisciplinary nature of this process.

Bibliography

Anderson, J. M. (1986). Procedures in the retrieval and evaluation of vascular grafts. in *Vascular Graft Update: Safety and Performance*, H. E. Kambic, A. Kantrowitz, and P. Sung, eds. ASTM STP 898 American Society for Testing and Materials, Philadelphia, pp. 156–165.

Anderson, J. M. (1993). Cardiovascular device retrieval and evaluation. *Cardiovasc. Pathol.* 2(3)(Suppl.): 199S–208S.

Cook, S. D., Renz, E. A., Barrack, R. L., Thomas, K. A., Harding, A. F., Haddad, R. J., and Milicic, M. (1985). Clinical and metallurgical analysis of retrieved internal fixation devices. *Clin. Orthop.* **184**: 236–247.

Cook, S. D., Barrack, R. L., Thomas, K. A., and Haddad, Jr., R. J. (1988). Quantitative analysis of tissue growth into human porous total hip components. *J. Arthroplasty* **3**: 249–262.

Fraker, A. C., and Griffin, C. D. (eds.). (1985). *Corrosion and Degradation of Implant Materials: Second Symposium*. ASTM Special Technical Publication 859, Philadelphia, PA.

Lemons, J. E. (1988). Dental implant retrieval analyses. *J. Dent. Educ.* **52**: 748–756.

Merritt, K., and Brown, S. A. (1981). Metal sensitivity reactions to orthopaedic implants. *Int. J. Dermatol.* **20**: 89–94.

Ratner, B. D. (ed.). (1988). *Surface Characterization of Biomaterials.* Elsevier, Amsterdam.

Schoen, F. (1989). *Interventional and Surgical Cardiovascular Pathology.* Saunders, Philadelphia, PA.

Schoen, F. J., Anderson, J. M., Didisheim, P., Dobbins, J. J., Gristina, A. G., Harasaki, H., and Simmons, R. L. (1990). Ventricular assist device (VAD) pathology analyses: guidelines for clinical study. *J. Appl. Biomat.* **1**: 49–56.

Weinstein, A., Gibbons, D., Brown, S., and Ruff, W. (eds.). (1981). *Implant Retrieval: Material and Biological Analysis.* U.S. Department of Commerce, National Bureau of Standards, Rockville, MD.

Wright, T. M., Hood, R. W., and Burstein, A. H. (1982). Analysis of material failures. *Orthop. Clin. N. Am.* **13**: 33–44.

Wright, T. M., Rimnac, C. M., Faris, P. M., and Bansal, M. (1988). Analysis of surface damage in retrieved carbon fiber-reinforced and plain polyethylene tibial components from posterior stabilized total knee replacements. *J. Bone Jt. Surg.* **70**-A: 1312–1319.

USPHS (1985). U.S. Department of Health and Human Services, Public Health Service, National Institutes of Health. Guidelines for Blood-Material Interactions. NIH Publication No. 85-12185, National Institutes of Health, Bethesda, MD.

10

New Products and Standards

STANLEY A. BROWN, JACK E. LEMONS, AND NANCY B. MATEO

10.1 INTRODUCTION

Jack E. Lemons

A very dynamic situation continues to exist related to the development and introduction of new surgical implant devices. This, in part, relates to product and business competition. Significant amounts of proprietary research and development is conducted annually throughout the industrial sector, often first seen in product-related literature or within regulatory submissions that become public domain.

Surgical implant standards within the United States are based on consensus organizations with the American Society for Testing and Materials Committee F-4 (ASTM F-4) representing most implant devices. These standards are published annually in Volume 13.10 from the ASTM. Those interested in biomaterials and surgical implant devices should refer to the various national and international sources of product literature and standards as a key source of information. The section components of Chapter 10 will provide related references and sources for obtaining the industrial and standard related literature.

10.2 VOLUNTARY CONSENSUS STANDARDS

Stanley A. Brown

WHAT ARE STANDARDS?

Consensus standards are documents which have been developed by committees to represent a consensus opinion on test methods, materials, devices, or procedures. Most standards organizations review their documents every 5 years to ensure that they are up to date. The mechanisms by which they are developed are described in subsequent sections.

A *test method standard* describes the test specimen to be used, the conditions under which it is to be tested, how many

are to be tested, and how the data are to be analyzed. Many specimens are validated by "round robin testing," meaning that several laboratories have followed the method and their results agree to a specified degree of precision. Once a test method has been standardized, it can be used in any other laboratory; the details are sufficient to ensure that different facilities will obtain similar results for the same samples. Stating that a test was "conducted in accordance with . . .", ensures that the results can be duplicated. Some representative test methods are listed in Table 1.

A *material standard* describes the chemical and physical properties of the material. Any test method standards cited are to be used to ensure that a significant sample meets the requirements of the standard. Some representative material standards are listed in Table 2.

For implant materials, there is also a requirement that the materials meet general biocompatibility test criteria. There are two formats for the biocompatibility language in the material standards of the American Society for Testing and Materials (ASTM). For materials that can be well characterized by chemical and physical tests, and have demonstrated a well-characterized biological response, reference to the published biological testing data is sufficient. For materials which are not well characterized, for example, epoxy resins, biological test methods are cited, and each particular formulation must be tested independently.

A *device standard* describes the device and its laboratory-based performance. General design aspects, dimensions, and dimensional tolerances are given using schematic drawings. The materials to be used are described by reference to materials standards. Methods for testing the device are also cited. Since test methods only describe how to do a test, it is in the device standards that performance is addressed. For example, the fatigue life requirements or biocompatibility requirements of the device and its materials would be stated in a device standard. Some representative device standards are listed in Table 3.

A *procedure standard* describes how to do something which would not be considered a test. Examples include standards for surface preparation and standardized procedures for sterilizing implants. Table 4 lists some typical procedure standards.

Biomaterials Science

TABLE 1 Some Representative ASTM Standard Test Methods

A. Mechanical testing standards
   ASTM D412   Test methods for rubber properties in tension
   ASTM D638M   Test method for tensile properties of plastics (metric)
   ASTM D695M   Test method for compressive properties of rigid plastics (metric)
   ASTM D790M   Test methods for flexural properties of unreinforced and reinforced plastics and electrical insulating materials (metric)

B. Metallographic methods
   ASTM E3   Preparation of metallographic specimens
   ASTM E7   Terminology relating to metallography
   ASTM E45   Determining the inclusion content of steel
   ASTM E112   Determining the average grain size

C. Corrosion testing
   ASTM G3   Conventions applicable to electrochemical measurements in corrosion testing
   ASTM G5   Reference test method for making potentiostatic and potentiodynamic anodic polarization measurements
   ASTM G59   Conducting potentiodynamic polarization resistance measurements
   ASTM F746   Pitting and crevice corrosion of surgical alloys
   ASTM F897   Fretting corrosion of osteosynthesis plates and screws

D. Polymer testing
   D 2238   Test methods form absorbance of polyethylene due to methyl groups at 1378 cm^{-14}
   D 3124   Test method for vinylidene unsaturation in polyethylene by infrared spectrophotometry

WHO USES STANDARDS?

The term "voluntary standards" implies that the documents are not mandatory; anyone can use them. Standards are used by manufacturers, users, test laboratories, and, in some instances, college professors and their students. One's use or compliance with a standard is voluntary. Using them is often to everyone's advantage.

Manufacturers often use standards as guidelines in making

TABLE 2 Some Typical ASTM Materials Standards (4)

ASTM F75   Cast cobalt–chromium–molybdenum alloy for surgical implant applications

ASTM F139 stainless steel sheet and strip for surgical implants (special Quality)

ASTM F451   Acrylic bone cements

ASTM F603   High-purity dense aluminium oxide for surgical implant applications

ASTM F604   Silicone elastomers used in medical applications

ASTM F641   Implantable epoxy electronic encapsulants

TABLE 3 Some Representative AAMI and ASTM Device Standards

AAMI CVP3   Cardiac valve prostheses

AAMI VP20   Vascular graft prostheses

AAMI RD17   Hemodialyzer blood tubing

AAMI ST8   Hospital steam sterilizers

E667   Clinical thermometers (maximum self-registering, mercury-in-glass)

ASTM F367M   Holes and slots with spherical contour for metric cortical bone screws (metric)

ASTM F703   Implantable breast prostheses

ASTM F623   Foley catheters

and testing their materials and devices. Manufacturers also cite standards in their sales literature as a concise way of describing their product. Stating that a device is made from cast cobalt–chromium–molybdenum alloy in accordance with ASTM F75 tells the user precisely what the material is. For example, after purchasing a piece of plastic pipe at the hardware store labeled with "ASTM D1784," one could go to ASTM Volume 8.02 and find that this is a specification for rigid poly(vinyl chloride) compounds. If you have "DIN" stamped on the bottom of your ski boots, you know they conform to the standards of the Deutsches Institut für Normung, and the ski shop will have standards for adjusting your bindings.

As an example of why one would use device standards, consider screws for fixing bone fractures. There are device standards for bone screws, plates, taps, and screwdrivers. One can purchase a screw and a screwdriver, and be confident that the components will fit as intended. A surgeon about to remove a plate implanted at another hospital can evaluate radiographs and see that the device has 4.5-mm bone screws of a specific design. Knowing this, a standard 4.5-mm screwdriver can be used to remove the screws.

TABLE 4 Some Representative AAMI (1) and ASTM (4) Procedure Standards

AAMI ROH-1986   Reuse of hemodialyzers

AAMI ST19   Biological indicators for saturated steam sterilization processes in health care facilities

AAMI ST21   Biological indicators for ethylene oxide sterilization processes in health care facilities

ASTM F86   Surface preparation and marking of metallic surgical implants

ASTM F561   Analysis of retrieved metallic orthopaedic implants

ASTM F565   Care and handling of orthopaedic implants and instruments

ASTM F983   Permanent marking of orthopaedic implant components

Standardized test methods should simplify life. For example, the author used to teach two laboratory sessions as part of an undergraduate biomedical engineering course. One was on mechanical testing, the other on metallography and implant analysis. In the mechanical testing laboratory, several ASTM standard test methods for mechanical testing, such as D790m, "Flexural properties of plastics and electrical insulating materials (metric)," were used. This method describes the samples, test apparatus, test speeds, and equations used to calculate the results. The students were requested to follow the test directions. In writing the methods section of their reports, all they had to write is that "the test was done according to D790m."

WHO WRITES STANDARDS?

In the United States, voluntary consensus standards are developed by a number of organizations. In the medical electronics, sterilization, vascular prostheses, and cardiac valve areas, most standards are developed by committees within the Association for the Advancement of Medical Instrumentation (AAMI). In the implant materials and implants area, most standards are set by ASTM Committee F-4 on medical and surgical materials and devices. These documents may then be reviewed and accepted by the American National Standards Institute (ANSI). ANSI is the official U.S. organization which interacts with other national organizations in developing international standards within the International Standards Organization (ISO).

Dental material standards are written by the American Dental Association (ADA). Similar committees exist in other countries: the Canadian Standards Association (CSA), the British Standards Institute (BSI), the Association Francaise de Normalisation (AFNOR) in France, and the Deutsches Institut für Normung (DIN) in Germany, which is a voluntary organization.

Performance Standards within ASTMF.4

Committee F-4 on medical implants includes MD users of devices. These groups have been leading the development of voluntary consensus standards on the clinical performance of certain types of implant systems. These groups which operate within the arthroplasty section provide a specialized extension of normal ASTM standards.

THE ASTM SYSTEM

The following discussion uses the development of ASTM standards as an example of the process in general. The term "voluntary standard" also implies that those developing the standard are volunteers. The ASTM committees are composed of "producers" (manufacturers and suppliers), "users" (physicians, dentists, etc.), and "general interest" representatives (professors, consumer advocates, etc.). The society regulations ensure that the voting on each standard is not dominated by any one faction. Consensus means that documents must be approved by a majority vote at each level of the committee. All negative votes must be discussed. Overriding a negative requires a two-thirds vote by each committee level.

There are more than 160 committees within ASTM. The letter designation indicates the type of product: "A" committees develop standards on steels and test methods for steels; Bs are for nonferrous alloys; Ds are plastics; and Fs are for consumer products.

Committee F4 was organized in 1971 by orthopedics manufacturers and orthopedic surgeons in response to a need to standardize the materials and devices being used. Surgeons perceived a need for standards on materials and such things as screws, drills and screwdrivers, rods and plates, so they could be confident that their instruments would fit any manufacturer's device. The scope of F4 has been expanded over the years to include cardiovascular, neurosurgical, maxillofacial, plastic and reconstructive surgical devices, and surgical instruments.

Within the F4 main committee, there are several divisions: I on resources; II on orthopedic devices; IV on medical/surgical devices; V on administration. Within each division there are subcommittees. The biomaterials and biocompatibility standards are under the jurisdiction of the subcommittee of the resources division. Subcommittee F04.11 is on polymeric materials, F04.12 is on metallic materials, F04.13 is on ceramics, F04.16 is on biocompatibility, etc. Within each subcommittee there are task forces (TFs) which initiate the documents. The TF documents are then voted on at each level, starting at the subcommittee and main committee. At each level, the negatives must be addressed, as described previously.

BIOCOMPATIBILITY STANDARDS

There is a wide range of tests that may be used to determine the biological response to materials. Short-term uses require only short-term tests. Long-term uses require tests applicable to the particular device and tissue type. Since not all tests are necessary for all applications, national and international standards organizations have developed matrix documents which indicate what methods are appropriate for specific applications. These documents can be used as guidelines in preparing a submission to the U.S. Food and Drug Administration (FDA) for approval of a new material or device. Similar matrix documents have been standardized by the CSA, BSI, and ISO. Test method documents have also been developed by the National Institutes of Health (NIH), the U.S. Pharmacopeia (USP), and other national organizations such as the Health Industry Manufacturers Association (HIMA). Guidelines for dental materials have also been developed by the ADA and ISO.

Much of the standards activity is now associated with the International Standards Organization (ISO) with biological evaluation of medical devices under the consideration of TC 194 and presented in the developing documents of ISO 10993. There are various parts to this document. Part 1 is definitions and the guidance on selection of evaluation test categories that should be done. The other parts of 10993 give more discussion and detail on the selection of individual tests that should be done for a particular biological interaction or biological effect (e.g., contact with blood, systemic toxicity, genotoxicity). In general, details of test methods are not given in the ISO documents and reference is made to other documents such as ASTM and USP standards for procedures and methodology.

In the following section we review some of the steps taken to establish the biocompatibility of a new material for a specific application, in this case a long-term orthopedic implant. We use ASTM standard F748 "Practice for selecting generic biological test methods for materials and devices" as a guideline. The standard test methods described are those used within ASTM.

In Vitro Tests

F619. Practice for extraction of medical plastics. A method for extraction of medical plastics in liquids that simulate body fluids. The extraction vehicle is then used for chemical or biological tests. Extraction fluids include: saline, vegetable oil (sesame or cottonseed), and water.

F813. Practice for direct contact cell culture evaluation of materials for medical devices. A cell culture test using ATCC L929 mouse connective tissue cells.

This method or this type of cell culture method can be used as the first stage of biological testing. It is also used for quality control in a production setting. There are other ASTM standard cell culture methods, and others not standardized by ASTM which could also be used.

F756. Assessment of hemolytic properties of materials. An *in vitro* test to evaluate the hemolytic properties of materials intended for use in contact with blood. Procedure A is static, procedure B is done under dynamic conditions.

Short-Term in Vivo Testing

F719. Testing biomaterials in rabbits for primary skin irritation. A procedure to assess the irritancy of a biomaterial in contact with intact or abraded skin. This test would be indicated for surgical glove material, or skin dressings.

F720. Practice for testing guinea pigs for contact allergens: guinea pig maximization test. A two-stage induction procedure employing Freund's complete adjuvant and sodium lauryl sulfate, followed two weeks later by a challenge with the extract material. Ten animals per test material.

F749. Practice for evaluating material extracts by intracutaneous injection in the rabbit. Extraction vehicles (as per F619) of saline and vegetable oil are injected intracutaneously and the skin reaction graded for erythema, edema and necrosis. Two rabbits per extraction vehicle.

F750. Practice for evaluating material extracts by systemic injection in the mouse. Intravenous injection of saline extracts and intraperitoneal injection of oil extracts. Animals are observed for evidence of toxicity. Five mice per extract and five mice per extract vehicle controls.

F563. Practice for short-term screening of implant materials. This method provides for several implant types and sites for short term screening *in vivo*.

This method is essentially the first stage of animal testing of solid pieces of the implant material.

There are additional *in vitro* tests which have not yet been standardized by ASTM:

Thrombogenicity. Tests for the propensity for materials to cause blood coagulation have not been standardized. Guidelines for such tests have been developed by the NIH Heart Lung and Blood Institute.

Mutagenicity. There are a number of *in vitro* tests to determine if chemicals cause cell mutations. While not specifically developed for implants, guidelines do exist as part of the OECD guidelines for testing of chemicals, and within ASTM, e.g., E 1262, guide for the performance of the chinese hamster ovary cell/hypoxanthine guanine phosphoribosyl transferase gene mutation assay; E1280, guide for performing the mouse lymphoma assay for mammalian cell mutagenicity; E 1397, practices for the *in vitro* rat hepatocyte DNA repair assay; and E 1397, practices for the *in vivo* rat hepatocyte DNA repair assay.

Pyrogenicity. A pyrogen is a chemical that causes fever. The USP rabbit test is a standard *in vivo* test. One can also test for bacterial endotoxins, which are pyrogenic, using the *Limulus* amebocyte lysate (LAL) test. The oxygen-carrying cell (amebocyte) of the horseshoe crab, *Limulus polyphemus,* lyses when exposed to endotoxin.

Long-Term Testing in Vivo

There are two aspects to the long-term testing issue. One is the response of tissue to the material; the other is the response of the material (degradation) to implantation.

F981. Practice for assessment of compatibility of biomaterials (nonporous) for surgical implants with respect to effects of materials on muscle and bone. Long term implantation of test materials in the muscle and bone of rats, rabbits and dogs. Two species are recommended. For rabbit muscle implants: the standard calls for 4 rabbits per sacrifice period, with one control and two test materials placed in the paravertebral muscles on each side of the spine. For bone implants in rabbits: the standard calls for 3 implants per femur.

A general necropsy is performed at the time of sacrifice. Muscle and bone implant sites are removed at sacrifice and the implants left *in situ* until the tissue has been fixed in formalin. Implants may be removed prior to embedding and sectioning.

The ASTM has not established any long-term standardized tests for devices. However, for a device intended for a particular application, it is essential to conduct a functional device test. For a fracture fixation plate, it could be proposed to use plates to fix femoral osteotomies in dogs. This study would consider the effects of the implant on the tissues, as well as the effect of implantation on the properties of the device, i.e., material degradation.

The methodology for long-term carcinogenicity testing of implants also has not yet been standardized by the ASTM, although F1439 (standard guide for performance of lifetime bioassay for the tumorigenic potential of implant materials) does provide guidelines for test selection. This is normally a life survival and tumor production test, typically in rats.

SUMMARY

Standard test methods allow results to be reproduced or verified by other researchers. In the biomaterials field, standards can be used for physical, mechanical, chemical, and biological testing of materials. By using materials that conform to standards, a manufacturer can tell the user what is in the material and what to expect from it in terms of properties.

Bibliography

AAMI Resource Catalogue, Association for the Advancement of Medical Instrumentation, Arlington, VA 1989.

ADA Document 41 for recommended standard practices for biological evaluation of dental materials, Chicago, 1979.

AFNOR S90-700 Choice of tests enabling assessment of biocompatibility of medical materials and devices, Paris.

ASTM Annual book of standards, West Conshohocken, PA (published annually).

BSI BS5736, Evaluation of medical devices for biological hazards. British Standards Institution.

CSA CAN3-Z310.6-M84, Tests for biocompatibility. Canadian Standards Organization, Ontario.

DIN 13930, Biological prufungen von dentalwerkstoffen, DIN, Berlin, 1983.

HIMA Guidelines for the preclinical safety evaluation of materials used in medical devices, Health Industry Manufacturers Association, Washington, DC 1985.

ISO DIS 7405, Preclinical evaluation of biocompatibility of medical devices used in dentistry-test methods. International Organization for Standardization, Geneva.

ISO 10993, Implants for surgery, Biocompatibility selection of biological test methods for materials and devices, International Organization for Standardization, Geneva.

NIH HLBI Guidelines on characterization of biomaterials. 1984.

OECD Guidelines for testing of chemicals: 451, carcinogenicity studies, and 453, combined chronic toxicity/carcinogenicity studies.

OECD Organization for economic cooperation and development, Washington, DC.

USP, The U.S. Pharmacopeia XXII, United States Pharmacopeial Convention, Rockville, MD, 1990.

10.3 PRODUCT DEVELOPMENT AND REGULATION

Nancy B. Mateo

Preparing a new medical device for the market requires more than good research and development. Medical devices are regulated in the United States by the Food and Drug Administration (FDA) to ensure that they are safe and effective and, where applicable, that their clinical utility has been sufficiently demonstrated. Because the success of a product often depends on the timeliness of its introduction into the market, regulatory requirements must be considered early in the product's development to avoid delays in the approval process later on.

The approach to FDA approval differs from device to device, depending on the classification of the specific device and its uniqueness. Developing a product hand-in-hand with the regulatory and marketing functions of a company helps to ensure that the appropriate approach is taken, that necessary nonclinical testing is performed, and that critical clinical questions are addressed.

Each country has its own regulations concerning the marketing of medical devices. Devices manufactured in the United States must meet FDA requirements and the requirements of the country to which the device is exported. In this chapter, only U.S. regulatory issues are discussed.

HISTORICAL OVERVIEW OF MEDICAL DEVICE REGULATION

Federal Food, Drug, and Cosmetic Act of 1938

Although the FDA was authorized by the Federal Food, Drug, and Cosmetic Act of 1938 to regulate medical devices, the Act did not require, as it did for drugs, any type of premarket approval for devices. The focus of the FDA's device authority was to ensure proper labeling and to remove fraudulent devices from the marketplace.

After World War II, medical device technology rapidly progressed in the wake of war and postwar advances in metallurgy, ceramics, plastics, and engineering design. As a result of these advances, medical devices such as heart valves, pacemakers, renal catheters, and replacement joints were developed and marketed. However, the lack of premarketing regulations allowed these types of complex devices to be implanted without any standardized guidelines or testing, and problems associated with device design and function were not always identified until after implantation. During the 1960s, the court system began to define certain devices as "drugs" in order to bring them under the FDA drug-approval process (Parr and Barcome, 1989).

1976 Medical Device Amendments

In 1976, the Medical Device Amendments to the 1938 Act were passed, which offered a clear definition of a medical device (see Table 1) and gave the FDA the authority to regulate devices during their development, testing, production, distribution, and use. Under these amendments, approximately 1700 types of medical devices, ranging from bandages to hemodialysis machines, require regulation and approval by the FDA (Parr and Barcome, 1989). The amendments created three regulatory classes (Table 2) based on the extent of control necessary to ensure that devices are safe and effective: Class I, which is the least regulated class; Class II; and Class III, which is the most regulated. The type of regulation required for each class was specified, and existing devices were assigned classifications. The purpose of these classes was to create a system for the FDA to evaluate and categorize devices intended for human use, and for manufacturers and importers to identify the controls that apply to their devices.

TABLE 1 What Is a Medical Device?

As defined in the 1976 Medical Device Amendments, a medical device is:

. . . any instrument, apparatus, implement, machine, contrivance, implant, *in vitro* reagent, or other similar or related article, including any component, part, or accessory

• which is recognized in the official National Formulary, or the U.S. Pharmacopeia, or any supplement to them

• which is intended for use in the diagnosis of disease or other conditions, or in the cure, mitigation, treatment, or prevention of disease, in man or other animals; or intended to affect the structure or any function of the body of man or other animals

• which does not achieve any of its principal intended purposes through chemical action within or on the body of man or other animals and which is not dependent upon being metabolized for the achievement of any of its principal intended purposes.

Safe Medical Devices Act of 1990

The Safe Medical Devices Act (SMDA) of 1990 significantly expanded the FDA's authority to regulate medical devices in the premarketing and postmarketing stages. Under the SMDA, the FDA has greater enforcement powers, controls on certain types of devices are increased, and premarket approval and notification submissions are significantly affected (Holstein, 1991, 1992; Kahan *et al.*, 1991). Some of the requirements for each of the three classes created under the 1976 Medical Device Amendments were modified by the SMDA.

DOMESTIC DEVICE APPROVAL

The requirements for approval and the extent of regulatory control are determined by the classification of the device. The following discussion briefly outlines the three FDA classifications and the approval processes involved. Table 2 lists examples of FDA-classified medical devices.

TABLE 2 Examples of Class I, II, and III Devices

Class I	Class II	Class III
Root canal post	Oxygen mask	Intraocular lens
Dental floss	Blood pressure cuff	Heart valve
Enema kit	Powered wheelchair	Infant radiant warmer
Tongue depressor	Skull clamp	Ventricular bypass device
Surgeon's glove	Obstetric ultrasonic imager	Automated blood cell separator

Device Classification

Class I

Class I is the least stringent classification. Devices in this category are subject only to general controls to ensure safety and effectiveness (21 *CFR* 860.3).[1] General controls, the basic requirements that apply to all medical devices, include regulations regarding adulteration, misbranding, banned devices, and restricted devices. General controls also require manufacturers to follow good manufacturing practices and to maintain records and reports to show that their devices are not adulterated or misbranded. Also, manufacturers must list their commercially distributed devices and annually register their facilities.

Class II

Although the 1976 Amendments required Class II devices to meet performance standards that ensure their safety and effectiveness (21 *CFR* 860.3), until the SMDA was passed, no real regulatory distinction existed between Classes I and II. In fact, when the 1976 amendments were passed, no standards for Class II devices were created and the only regulations that applied to these devices were the general controls. The SMDA introduced special controls for Class II devices, including postmarket surveillance, patient registries, and the development of guidelines and other "appropriate actions" that the FDA deemed necessary to ensure safety and effectiveness. In addition, since the SMDA removed the hindering safeguards for establishing performance standards, the FDA can now follow simple rule-making procedures to create and amend standards.

Class III

Class III devices are those devices for which there is not enough information to show that general controls and performance standards would ensure their safety and effectiveness (21 *CFR* 860.3). In general, they are defined as devices that are implanted in the body, are life-sustaining or life-supporting, or present unreasonable risk of injury or illness. Class III devices cannot be marketed without premarket approval from the FDA.

Panel Reviews

Panels of experts serve as advisory committees to the FDA in recommending device classifications. To determine whether a device is safe and effective, the panel reviewers consider who the users are, the conditions of use, the benefit-to-risk ratio, and the reliability of the device. After receiving the panel's recommendation, the FDA decides on the device classification. Depending on the classification granted, the device manufacturer can then follow the appropriate regulatory path to marketing a safe and effective product.

Approval Processes

Regardless of the type of medical device, a manufacturer must submit a premarket notification to the FDA at least 90

---

[1]*CFR* is the abbreviation for the *Code of Federal Regulations*.

days prior to commercially distributing a new or substantially modified device (21 *CFR* 807, Subpart E). This type of notification is usually referred to as a "510(k)" submission because it is mandated in Section 510(k) of the 1976 amendments. The premarket notification system enables the FDA to identify the proper classification and to determine whether the device is substantially equivalent in terms of safety and effectiveness to another device that is or has been marketed.

According to the 1976 amendments, "substantial equivalence" refers to equivalence to a device already on the market in the United States prior to the 1976 amendments, to a device marketed after that date that has been deemed substantially equivalent, or to a device that has been reclassified as Class I or II. However, the SMDA allows the 510(k) submission to be based on an equivalence determination of a legally marketed device. This change removes the need to find a pre-1976 predicate device. If the FDA determines that a device is substantially equivalent, then it can be marketed. If it is nonequivalent, the device is automatically classified as a Class III device. To change a classification, the manufacturer can submit a second 510(k) with new data, file a reclassification petition (to change the device class to I or II), or submit a premarket approval (PMA) application.

510(k) Submissions

A 510(k) submission must include a description of the device and an explanation of the similarities and differences between the new device and comparable devices already on the market (21 CFR 807, Subpart E). According to the SMDA, companies claiming substantial equivalence to Class III preamendment devices must provide the FDA with a summary of and citation to all adverse safety and effectiveness data. Also, submission of clinical data concerning both the predicate device and the new device is becoming more common for 510(k) submissions.

To learn about the specific FDA requirements for a particular type of device, a manufacturer can obtain copies of 510(k) submissions for competitive products and check with the FDA for any guidance documents that may be available.

PMA Applications

Until the passage of the SMDA, an approved PMA application was essentially a private license granted to the manufacturer for marketing a particular medical device. However, according to the SMDA, information contained in a PMA application that concerns the safety and effectiveness of the device should be made public 1 year after the approval of four devices of a kind. These data are used by the FDA to establish performance standards and to approve other devices.

A PMA application must contain full reports of investigations, both nonclinical and clinical, that show whether the device is safe and effecitve (21 *CFR* 814). In addition, the components of the device and the principle of operation must be described in detail. Manufacturing and quality control procedures, proposed labeling, and actual samples of the device (or the location of where one can be examined) are also required. PMA applications undergo an extensive review process: a filing review, an in-depth scientific and regulatory review, a panel review by the appropriate advisory committee, and a

final deliberation by the FDA. After the filing review, which is an administrative and limited scientific review, the FDA officially files the application if it is considered complete. A 180-day review period then begins, during which the FDA approves or disapproves the application.

Investigational Device Exemptions

PMA applications and some 510(k) notifications for new devices require submission of clinical data that establish device safety, effectiveness, and clinical utility. To conduct clinical studies in the United States with a nonapproved device, a manufacturer must obtain an investigational device exemption (IDE) for that device. For some diagnostic devices and custom devices, IDEs are not required (21 *CFR* 814).

Before an IDE can be obtained, a device must first be categorized as presenting either a nonsignificant or significant risk. A significant risk device is defined as a device that presents the potential for serious risk to a human subject and that falls into one of three categories: (1) is an implant; (2) is used in supporting or sustaining life; or (3) is important in preventing the impairment of health or in diagnosing, curing, or treating disease. Nonsignificant risk devices are defined as not usually presenting a serious risk to human subjects.

If the device is classified as a nonsignificant risk, an IDE application is not required. However, for a significant risk device, an application must contain all manufacturing and quality control procedures, complete reports of nonclinical and prior clinical studies, a full investigational plan for the clinical study, and lists of the investigational review boards (committees designated by a university or an institution to review biomedical research with human subjects) involved in the study. Within 30 days of the receipt of the application, the FDA approves, with conditions, or disapproves the application.

DEVICE TESTING

Testing requirements differ from device to device. Both nonclinical and clinical testing schemes must be well planned to answer questions about safety and effectiveness.

Nonclinical Testing

The nonclinical studies required for a PMA are described in 21 *CFR* 814 as "including microbiological, toxicological, immunological, biocompatibility, stress, wear, shelf life, and other laboratory or animal tests as appropriate." Although there is no legal requirement to conduct a particular profile of tests, the FDA does expect adherence to their guidance document, *Tripartite Agreement for Biocompatibility Testing of Devices*. This generic document and other FDA guidance documents specific to particular types of devices suggest the types of physical, chemical, and biological testing to be done. However, it is left to the manufacturer to design a profile of tests suited to the specific device and its intended use.

A number of organizations participate in defining standards and testing methods. Among them are the International Standards Organization (ISO), the Health Industry Manufacturers Association (HIMA), the American Association of Medical Instrumentation (AAMI), the American Society for Testing and Materials (ASTM), and the U.S. Pharmacopoeia (USP). As a guideline developed to combine and harmonize testing requirements of the European community, *ISO Standard 10993-1: Biological Evaluation of Medical Devices* gives suggestions for developing a testing plan for different categories of devices. Other publications also describe testing profiles and methods (Henry, 1985; Williams, 1981; Von Recum, 1986).

Clinical Testing

Clinical protocols must be designed to generate data that demonstrate the safety and effectiveness of the device. However, protocols need to be flexible because safety concerns can change during the course of a clinical study. Meetings between the manufacturer and the FDA prior to an IDE submission are useful for discussing the study design; the types of data to be gathered; the statistical methods to be used; and what the study should accomplish in terms of illustrating safety, effectiveness, and clinical utility.

In determining how to test their devices, manufacturers must balance the indications for use against study and market sizes. In other words, although the market size might be large for a particular indication, that indication might not be the first one to be studied for approval. Selecting skilled investigators who will gather data properly and in a timely manner is a critical aspect of setting up a clinical study. Also, the manufacturer must monitor the study carefully to follow the progress of the data collection and analysis.

SUMMARY

The manufacturer is responsible for demonstrating the safety, effectiveness, and clinical utility of a medical device. If these aspects are considered carefully during the developmental stages, the regulatory issues involved in getting that device to market are greatly simplified.

Bibliography

Code of Federal Regulations (CFR), Title 21, Parts 800 to 1299. Revised as of April 1, 1991.

Henry, T. J. (ed.) (1985). *Guidelines for Preclinical Safety Evaluation of Materials Used in Medical Devices,* HIMA Report 85-1. Health Industry Manufacturers Association, Washington, D.C.

Holstein, H. M. (1991). The Safe Medical Devices Act of 1990. *Regulatory Affairs* 3: 91.

Holstein, H. M. (1992). Practical considerations in filing product applications. *Medical Device & Diagnostic Industry* 14(2): 45.

ISO Standard 10993-1: Biological Evaluation of Medical Devices (1992). International Organization for Standardization, Geneva.

Kahan, J. S., Holstein, H. M., and Munsey, R. (1991). The implications of the Safe Medical Devices Act of 1990. *Medical Device & Diagnostic Industry* 13(2): 44.

Parr, R., and Barcome, A. (1989). *Regulatory Requirements for Medical Devices: A Workshop Manual,* HHS Publication FDA 89-4165, FDA, Rockville, MD.

Tripartite Agreement for Biocompatibility Testing of Devices. FDA guidance document.

Von Recum, A. F. (1986). *Handbook of Biomaterial Evaluations.* Macmillan, New York.

Williams, D. F. (1981). *Techniques in Biocompatibility Testing.* CRC Press, Boca Raton, FL.

11

Perspectives and Possibilities in Biomaterials Science

BUDDY D. RATNER

The field of biomaterials is a young one, with perhaps 40 years of formal history. We have come from an earlier era in which research and development in biomaterials was driven by surgeon-visionary-entrepreneurs (the "surgeon hero"), to the 1990s where these development activities have largely been transferred to university and industrial laboratories. We are entering an era of exciting discoveries in molecular biology, cell biology, and materials science. The old-style, empirical research driven by the urgency to address immediate patient needs (i.e., construct a device and get it into clinical use as quickly as possible) is now, more and more, being replaced by systematic, hypothesis-driven investigation by teams of engineers, basic scientists, and physicians. This approach leads to better science and better engineering, and ultimately, better medicine and patient satisfaction. Such an approach also offers the possibility to incorporate new, scientific ideas into the development of biomaterials.

This brief epilogue offers those interested in the future of biomaterials a list of new ideas that have the potential to revolutionize the field. This chapter is intended to stimulate vision, dreaming, and planning. The ideas here are largely predicated upon (1) molecular engineering (engineering from the molecules up rather than the bulk properties down), and (2) the rigorous application of principles derived from physics, chemistry, materials science, and biology. Many of these basic ideas have been expanded upon elsewhere (Drexler, 1992; Lehn, 1988; McGee, 1991; Prime and Whitesides, 1991; Ratner, 1993; Tirrell *et al.*, 1991; Ulman, 1991). Here are some areas of contemporary exploration (in alphabetical order) that have the potential to replace empirical biomaterials science with precision design and engineering.

Assembly The spontaneous, facile fabrication of molecularly self-assembled perfect organic and inorganic surfaces is now a reality. The order provided by these systems can assist us in engineering recognition surfaces and making templates for assembling other systems. Organized assemblies of molecules and molecular templates are so widely used in nature that it is surprising we have not exploited them in our synthetic biomaterials.

Biocompatibility When will we finally understand, at a

molecular level, the phenomenon called biocompatibility? There is mounting evidence that it is closely associated with cellular activation driven both by geometry and chemistry, and also is closely linked to chemical communication between cells.

Biomimetics Can we mimic the superb combinations of strength, toughness, and flexibility found in sea urchin spines or spider silk? Can the phase changes, lubricity, and water retentiveness of mucus be applied to the design of biomaterials? These questions suggest new approaches to designing biomaterials by observing how nature does its work.

Chemometrics The application of multivariate statistics to problems in biomaterials science can break us free from the trap of univariate correlation. Chemometrics permits us to design better experiments, make efficient use of the large amounts of data generated by contemporary analytical instrumentation, and to generate new hypotheses that were not obvious from a simple "eyeballing" of a data set.

Composite materials Composite materials combine the desired mechanical properties of each of two phases into one materials system. New records for mechanical performance are now being set using nanocomposites.

Computational chemistry In the past, our ideas for new biomaterials came from implantation in animals, observation, and optimization of responses. Now we have the possibility of designing materials on a computer to perform specific functions or to resemble natural macromolecules. By planning materials on the computer, synthesizing them, and then studying their interactions with biological systems, we can efficiently explore hypotheses about the mechanisms of bioreactions and use this information to continue the optimization on the computer screen.

Cytokines The chemical messengers between cells, also called growth factors or interleukins, will help us to understand healing, growth and cell function. Can we use them (via controlled release from synthetic materials or cells) to engineer a new generation of biomaterials that performs exactly as designed?

Drug delivery This field is an offshoot from biomaterials that is expanding by leaps and bounds. Important ideas that are central to drug release are stabilization of the biological

Biomaterials Science
Copyright © 1996 by Academic Press, Inc.
All rights of reproduction in any form reserved.

function of the molecules being released, degradable polymers, swellable polymers, intelligent materials and, of course, biocompatibility.

Extracellular matrix components This former sleepy backwater of biology is now a hot area of research as new molecules and functions are discovered daily for the glue that holds the cells together. In healing responses that are engineered into biomaterials, the formation of a new extracellular matrix must be considered. Also, can we learn from the composite structure of the extracellular matrix some new molecular architecture tricks that can be used to design improved biomaterials?

Fullerenes A whole new field of chemistry has appeared in a five-year period based on the unique reactivity and structure of this C_{60} form of carbon and related shell, tube, and sheet forms. Biomaterials scientists have long used carbon in implant materials. What can we do with these structures?

Genetic engineering Site-specific mutagenesis in *Escherichia coli* can make unique proteins to order. One can imagine designing a new protein variant on a computer, expressing it in bacteria, isolating it, and self-assembling it into a new class of designed materials.

Heparin This polysaccharide molecule has functions and pharmacologic actions far more diverse than once imagined. Synthetic heparinoid structures, natural heparin fractions, and related polysaccharides will provide researchers with a varied palette from which to choose new components (structural and bioactive) to be used in biomaterials.

Human genome The effort to sequence the human genome can lead to a direct understanding of disease, aging, individual variation, healing mechanisms, and human potential. The impact of this knowledge on the design of biomaterials is unclear, but it will certainly be profound. In the short term, those working on gene sequencing exploit the principles governing the interaction of biological materials with synthetic materials. For this reason, biomaterial scientists are appropriate as team players in the exciting gene sequencing project.

Hybrid artificial organs Composite systems composed of living cells or tissues and synthetic biomaterials can provide a degree of functionality for our materials that we could not imagine without the living component, namely responsive biochemical factories called cells. Liver, pancreas, skin, and nerve prostheses are all possible with hybrid systems. The challenges here include stabilization of the phenotype of the cell, cell viability, inhibition of fouling, metabolite delivery rate, and biocompatibility.

Intelligent materials A word frequently used in biomaterials science to describe materials that interact minimally with the body has been "passive." More recently, we have come to realize that there are no passive materials, and furthermore, that materials can be designed to be active. This activity can be manifested in surface processes, or it can relate to changes in the bulk of the materials. Systems that rapidly shrink, swell, or change their optical or electrical properties in response to their environment have been developed. Such systems have potential for new drug delivery devices, bioseparations, bioreactors, and artificial muscles.

Interfaces The majority of bioreactions occur at interfaces, yet the formal study of interfaces is just developing in biomaterials. Interfaces can concentrate components and trigger new reactions that would not occur in homogeneous solution phases. New, more powerful methods to characterize the order and composition of minute amounts of material at interfaces are continually being developed.

Josephson tunnel junction These structures on a chip, and other developments in microelectronics, will lead to steady increases in the speed of computers. Increases in computer power are important to molecular design, pattern recognition, and improved instrumentation.

K_m The Michaelis–Menten constant assists us in understanding the kinetics and specificity of enzyme reactions. There are many examples of immobilized enzymes used in biomaterial devices. Still, the high reactivity and specificity of enzymes encourage further development. Can we also use catalytic antibodies?

Life The phenomenon called life might be thought of as a modestly clever chemical system that, among other things, provides biomechanistic inspiration for bioengineers. We must always look to living systems to understand how, over evolutionary history, nature has optimized functionality, efficiency, and performance. Also, improving the quality of life through improved therapies is a goal of biomaterials designers.

Microscopy The field of microscopy is undergoing a revolution. Optical techniques in the visible range such as laser confocal and near field are significantly improving depth of field and resolution. Fluorescence microscope methods give us quantitation and specificity. Scanning electron microscopy (SEM) methods (low voltage and environmental) permit hydrated systems to be studied and minimize electron damage. Surface analytical imaging (TOF-SIMS and scanning Auger microprobe) will generate compositional maps with resolutions finer than 0.1 μm. The scanning probe microscopies (scanning tunneling microscope (STM), atomic force microscope (AFM), and other techniques) offer unprecedented resolution of surface conductivity, friction, topography, and electrochemistry. AFM images of living cells show remarkable detail, such as subsurface features.

Mobility Surface molecular mobility has long been postulated to play a role in biointeraction. Analytical methods including SIMS, electron paramagnetic resonance, nuclear magnetic resonance, fluorescence measurements, and contact angles permit us to directly measure or infer information about surface mobility and correlate it with biological response.

Nanofabrication Proposed by Richard Feynman in 1959 and just now becoming an exploitable reality, nanofabrication offers the possibility of constructing systems and devices on a scale of sizes between cells and proteins. Nanomachines that are the size of red blood cells might ream out occluded arterioles or operate nanovalves to control peptide delivery. Patterning of surfaces on the submicron scale (a technology that is, at this time, approachable) might control protein adsorption and cell function.

Orientation Most of the biomaterials now synthesized are random and amorphous at the molecular level. Yet, biology elegantly exploits molecular order and orientation. Self-assembly and electrical fields might be used to create ordered surfaces that mimic biological systems.

Polymer synthesis Synthetic polymer chemists now have techniques to make novel polymers with greater control than ever before. Methods such as metathesis polymerization, carbocationic polymerization, anionic polymerization, and group transfer polymerization can produce polymers with narrow molecular weight distributions and few side reactions. Chemistries mimicking aspects of natural proteins [e.g., pseudopoly (amino acids)] and polynucleotides are being explored. Peptidic polymers synthesized by genetic manipulation of bacteria are composed of only one molecular weight and can be engineered for desired mechanical and surface properties. Template polymerization and dentridic polymerization provide new options for controlling structure. This combination of versatility and control makes polymer synthesis an important tool in the design and development of new biomaterials.

Polymerase chain reaction (PCR) PCR permits the analysis of trace levels of nucleotides, and also offers the possibility of scaling up the synthesis of genetic material. PCR will have application in measuring cell up- and downregulation via mRNA, and in manufacturing antisense DNA that might be useful in diagnostics and biosensors.

Porosity and texture Evidence is accumulating that architecture (porosity and texture) at the micron scale can dramatically influence the nature of cellular reactions to materials. What are the rules governing this process, and how can we use them to make improved prostheses? How can we best fabricate controlled pore structures and surface nanotextures?

Quantum tunneling This technique provides us with a startling new perspective on the molecular world via the scanning tunneling microscope. Polycrystalline gold with large facets of (111) crystals can be an excellent substrate for observing adsorption and assembly at the molecular scale. The STM can also be used to manipulate individual molecules and thin films to create interesting nanostructures.

Receptors and recognition Biomaterials as we know them today are largely nonspecific in their interactions with biological systems. This is in contrast to nature, which performs most of its function by a high specificity, receptor-recognition route. Receptor biology has revolutionized our mechanistic understanding of cellular processes. The specificity of receptor sites on cell surfaces (often, they are proteins called integrins), and the close relationship of those sites with the internal machinery of the cell, offers a route to control of cell function. Also, the research on disintegrins, proteins that strongly interact with the integrin molecules, has provided insights into cell control. We can use disintegrins, minimal peptide sequences derived from disintegrins, or synthetic models of the binding complement to integrins, to direct cell processes. In the future, such cell-recognized molecules (synthetic or natural) will be built onto our biomaterials, or released from them, to cleanly and precisely control cell function and hence biocompatibility.

Sensors Biosensors are a growing, vital technology that very much takes advantages of biomaterials ideas, recognition concepts, and new technologies. There are myriad opportunities for *in vitro* and *in vivo* diagnostic and monitoring biosensors. Solutions to fouling problems and biocompatibility issues are critical for a robustly designed *in vivo* biosensor.

Sol-gel The sol-gel process is a synthetically flexible route to new ceramics. Such materials can be processed at low temperatures and have excellent purity. These are clearly desirable characteristics for biomaterials systems.

Titanium Are the biointeractions of titanium unique to all metals? What is the nature of the oxide layer that may drive these interactions? What is the biological significance of the peroxide-induced surface gel that forms on titanium?

PolyUrethanes This older class of polymers is still the primary contender for the artificial heart, if we can engineer blood compatibility and biostability. Polyurethanes have superb mechanical properties and offer many options for synthesizing new structures.

Vitronectin and the other molecules in the class called adhesins (fibronectin, laminin, etc.) continue to amaze researchers as their ubiquity, functionality, and potency are explored. We will certainly use these molecules, and minimal peptide sequences from them, in the design of new biomaterial surfaces. Also, learning to control the adsorption of these molecules, and their conformations once adsorbed, remains a frontier worthy of exploration.

Water Since 98 mole% of living systems is water, we do ourselves a disservice by considering water merely a solvent instead of engineering with it.

Wear New insights are developing on wear from computer simulations and studies with the atomic force microscope and surface force apparatus. Understanding and reducing wear may be critical for improving the longevity of hip and knee prostheses.

X-rays National synchrotron sources provide X-rays that are tunable, bright, and polarized. These new sources offer the possibility of probing matter for orientation, composition, and phase separation. New characterization methods, such as those being developed with the availability of these photon sources, enhance our ability to understand materials and to engineer improved materials.

Yeast Recombinant gene technology using yeasts permits the large-scale synthesis of proteins that are glycosylated. Thus, many human proteins and variants of those proteins are now accessible, and potentially manufacturable in large quantities. Consider, for example, the engineering possibilities for large quantities of proteins such as fibrinogen that are free of viral contamination.

Zeolites Zeolites can be thought of as inorganic molecular recognition systems. Just as efforts are under way to develop organic-based synthetic recognition systems, why can we not exploit this specificity in our bioceramics?

This list is hardly complete. There are many other new ideas, paradigms, and algorithms that can be equally influential. The important point is that biomaterial science can freely use many ideas—we passed beyond single academic disciplines long ago, and we now concentrate on solving problems relating to human health and the interactions of biological systems with materials.

Perhaps the final frontier in biomaterials science will be in the areas of ethics. Ethical questions are frequently raised and stimulate us to think about such issues as control and shaping of our own bodies; life and death; relationships among patient, doctor, and attorney; relationships with federal regulatory agencies; relationships between academic scientists and entrepreneurs; the cost of health care; and the use of animals in research. The ubiquity of these ethical issues in our continuing effort highlights a special strength and excitement in the field of biomaterials science: we have a direct impact on people.

The editors and authors of this volume hope that you become as excited about biomaterials—its intellectual challenges, humanitarian aspects and rewards—as we are.

Bibliography

Drexler, K. E. (1992). *Nanosystems*. Wiley, New York.

Lehn, J. M. (1988). Supramolecular chemistry—scope and perspectives: molecules, supermolecules, and molecular devices (Nobel lecture). *Angew. Chem. Int. Ed. Engl.* 27(1): 89–112.

McGee, H. A., Jr. (1991). *Molecular Engineering*. McGraw Hill, New York.

Prime, K. L., and Whitesides, G. M. (1991). Self-assembled organic monolayers: model systems for studying adsorption of proteins at surfaces. *Science* 252: 1164–1167.

Ratner, B. D. (1993). New ideas in biomaterials science—a path to engineered biomaterials. *J. Biomed. Mater. Res.* 27: 837–850.

Tirrell, D. A., Fournier, M. J., and Mason, T. L. (1991). Genetic engineering of polymeric materials. *MRS Bull.* 16(7): 23–28.

Ulman, A. (ed.) (1991). *An Introduction to Ultrathin Organic Films*. Academic Press, Boston.

Properties of Biological Fluids

Steven M. Slack

The physicochemical properties of several human biological fluids have been measured by numerous investigators and the results are shown in Table A1. Where possible, ranges of values are given, but the reader should realize that significant variations are possible, particularly in states of disease. Values for cerebrospinal fluid (CSF) refer to the lumbar region and values for synovial fluid refer to the knee joint, unless otherwise specified.

Next to water, proteins comprise the bulk of most biological fluids, and the concentrations of various protein fractions in several human biological fluids are shown in Table A2. The concentrations of specific proteins are listed where possible.

Otherwise, the amount of protein migrating with a given mobility (α, β, γ), as determined from electrophoretic studies, is given. Table A3 shows the concentrations of the major electrolytes in several human biological fluids, and Table A4 lists concentrations of various organic compounds in the same fluids. The lymph is that obtained from the thoracic duct. Finally, Table A5 lists the major proteins in plasma and gives their concentration, molecular weight, isoelectric point, sedimentation constant, diffusivity, extinction coefficient at 280 nm, electrophoretic mobility, and carbohydrate content. Additional information regarding the composition of human biological fluids can be found in the references.

TABLE A1 Physicochemical Properties of Several Biological Fluids

Property	Whole blood	Plasma (serum)	Cerebrospinal fluid	Synovial fluid	Saliva	Tear fluid
Freezing-point depression[a]	0.557–0.577	0.512–0.568	0.540–0.603	—	0.07–0.34	0.551
Osmolality[b]	—	281–297	306	—	—	—
pH[c]	7.39–7.46	7.35–7.43	7.35–7.70	7.29–7.45	5.6–7.6	7.3–7.7
Refractive index[d]	16.2–18.5	1.3485–1.3513	1.3345–1.3351	—	—	1.3361–1.3379
Relative viscosity[e]	2.18–3.59	1.18–1.59	1.020–1.027	>300	—	—
Specific conductivity[f]	—	0.0117–0.0123	0.0119	—	—	—
Specific gravity[g]	1.052–1.060	1.024–1.027	1.006–1.008	1.008–1.015	1.002–1.012	—
Specific heat[h]	0.92	0.94	—	—	—	—
Surface tension[i]	55.5–61.2	56.2	60.0–63.0	—	15.2–20.6	—

[a]Units are °C.
[b]Units are mosm/kg H_2O. Calculated from freezing-point depression.
[c]pH measured from arterial blood and plasma, and from cisternal portion of CSF.
[d]Measured at 20°C.
[e]Measured *in vitro* at 37°C for whole blood, plasma, and synovial fluid, and at 38°C for CSF. The viscosity of serum is slightly less than plasma owing to the absence of fibrinogen.
[f]Units are S/cm. Measured at 25°C for plasma, 18°C for CSF.
[g]Relative to water at 20°C.
[h]Units are cal/g °C.
[i]Units are dyne/cm. Measured at 20°C.

Biomaterials Science
Copyright © 1996 by Academic Press, Inc.
All rights of reproduction in any form reserved.

TABLE A2 Protein Concentrations (mg/100 ml) in Several Biological Fluids

Protein	Plasma (serum)	Cerebrospinal fluid	Synovial fluid	Saliva	Tear fluid	Lymph
Total	6500–8000	20–40	500–1800	156–630	136–592	2910–7330
Albumin	4000–4800	11–22	315–1130	—	20–100	1500–2670
Amylase		—	—	38	—	—
α_1-Globulins	380–870	1.1–2.2	35–130	—	—	260
α_2-Globulins	570–940	1.3–2.8	35–130	—	16–70	250
β-Globulins	730–1380	2.4–4.8	45–160	—	37–160	400
γ-Globulins	590–1450	2.2–4.4	70–250	—	42–185	780
Transferrin	—	—	48–77	—	—	—
Ceruloplasmin	—	—	4.3–6.2	—	—	—
α_2-Macroglobulin	—	—	7.3–12.5	—	—	—
IgA	—	—	—	19	—	—
IgG	—	—	—	1.4	—	—
IgM	—	—	2.3–8.6	0.2	—	—
Fibrinogen	200–400	—	—	—	—	—
Lysozyme	—	—	—	<150	18–80	—

TABLE A3 Concentrations of Major Electrolytes (mEq/l) in Various Biological Fluids

Electrolyte	Whole blood	Plasma (serum)	Cerebrospinal fluid	Synovial fluid	Saliva	Tear fluid	Lymph
Bicarbonate	19–23	24–30	21.3–25.9	—	3.5–10.7	26	—
Calcium	4.8	4.0–5.5	2.0–2.60	2.3–4.7	2.3–5.5	—	3.4–5.6
Chloride	77–86	100–110	100–129	87–138	15.1–31.6	118–138	87–103
Magnesium	3.0–3.8	1.6–2.2	0.45–4.0	—	0.16–1.06	—	—
Phosphate	0.76–1.1	1.6–2.7	—	Same as serum	—	—	2.0–3.6
Potassium	40–60	4.0–5.6	2.06–3.86	3.5–4.5	14–41	7.7–22.1	3.9–5.6
Sodium	79–91	130–155	129–153	133–139	5.2–24.4	126–166	118–132
Sulfate	0.1–0.2	0.7–1.5	—	Same as serum	—	—	—

TABLE A4 Concentrations of Various Organic Compounds (mg/100 ml) in Several Biological Fluids

Species	Whole blood	Plasma (serum)	Cerebrospinal fluid	Synovial fluid	Saliva	Tear fluid	Lymph
Amino acids	38–53	35–65	1.0–1.5	—	—	7.6	—
Bilirubin	0.2–1.4	0.2–1.4	<0.1	—	—	—	0.8
Cholesterol	115–225	120–200	0.16–0.77	5–14	2.5–50	—	34–106
Creatine	2.9–4.9	2.5–3.0	0.46–1.9	—	—	—	—
Creatinine	1–2	0.6–1.2	0.6–1.4	—	0.5–2	—	0.8–8.9
Fat, neutral	85–235	25–260	0–0.9	—	—	—	—
Fatty acids	250–390	150–500	—	—	—	—	—
Glucose	80–100	60–130	50–80	—	10–30	10	140
Hyaluronic acid	—	—	—	250–365	—	—	—
Lipids, total	445–610	285–675	0.77–1.7	—	—	—	—
Nonprotein N	25–50	19–30	11–20	22–43	7–35	—	13.4–139
Phospholipid	225–285	150–250	0.2–0.8	13–15	—	—	—
Urea	20–40	20–30	13.8–36.4	—	14–75	20–30	—
Uric acid	0.6–4.9	2.0–6.0	0.5–2.6	7–8	0.5–2.9	—	1.7–10.8
Water	81–86 g	93–95 g	94–96 g	97–99 g	99.4 g	98.2 g	81–86 g

TABLE A5   Physicochemical Properties of the Major Plasma Proteins

Species	Plasma concentration (mg/ml)	Molecular weight (daltons)	pI	S^a	D^b	E_{280}^c	V_{20}^d	Mobility	CH_2O^e
Prealbumin	10–40	54,980	4.7	4.2	—	13.2	—	α	0
Albumin	35–45	66,500	4.9	4.6	6.1	5.8	0.733	α	0
α_1-Seromucoid	0.5–1.5	44,000	2.7	3.1	5.3	8.9	0.675	α_1	41.4
α_1-Antitrypsin	2.0–4.0	54,000	4.0	3.5	5.2	5.3	0.646	α_1	12.2
α_2-Macroglobulin	1.5–4.5	725,000	5.4	19.6	2.4	8.1	0.735	α_2	8.4
α_2-Haptoglobin									
Type 1.1	1.0–2.2	100,000	4.1	4.4	4.7	12.0	0.766	α_2	19.3
Type 2.1	1.6–3.0	200,000	4.1	4.3–6.5	—	12.2	—		—
Type 2.2	1.2–2.6	400,000	—	7.5	—	—	—		—
α_2-Ceruloplasmin	0.15–.60	160,000	4.4	7.08	3.76	14.9	0.713	α_2	8
Transferrin	2.0–3.2	76,500	5.9	5.5	5.0	11.2	0.758	β_1	5.9
Lipoproteins				(S_f)					
LDL ($\rho < 1.019$)	1.5–2.3	5–20×10^6	—	>12	5.4	—	—	β	—
LDL ($\rho = 1.019$–1.063)	2.8–4.4	3,200,000	—	0–12	—	—	—	α_2	—
HDL$_2$ ($\rho = 1.093$)	.37–1.17	435,000	—	4–8	—	—	—	α_1	—
HDL$_3$ ($\rho = 1.149$)	2.17–2.70	195,000	—	2–4	—	—	—	α_1	—
IgA (monomer)	1.4–4.2	162,000	—	7	3.4	13.4	0.725	γ	7.5
IgG	6–17	150,000	6.3–7.3	6.5–7.0	4.0	13.8	0.739	γ	2.9
IgM	0.5–1.9	950,000	—	18–20	2.6	13.3	0.724	γ	12
C1q	.1–.25	400,000	—	11.1	—	6.82		γ_2	8.5
C3	1.5–1.7	180,000	6.1–6.8	9.55	4.5	—	0.736	β_1	1.5
C4	0.2–0.5	206,000	—	10.1	—	—	—	β_1	6.9
Fibrinogen	2.0–4.0	340,000	5.5	7.6	1.97	13.6	0.723	β_2	2.5

[a]Sedimentation constant in water at 20°C, expressed in Svedberg units, i.e., 10^{-13} cm s^{-1} dyne^{-1}. For the lipoproteins, the flotation constant (S_f) in a medium of relative density 1.063 at 26°C is listed (Svedberg units).

[b]Diffusion coefficient in water at 20°C, expressed in 10^{-7} cm^2/s.

[c]Extinction coefficient for light of wavelength 280 nm traveling 1 cm through a 10 mg/ml protein solution.

[d]Partial specific volume of the protein at 20°C, expressed as ml g^{-1}.

[e]Carbohydrate content of the protein, expressed as the percentage by mass.

Bibliography

Bing, D. H. (ed.) (1978). *The Chemistry and Physiology of the Human Plasma Proteins,* Pergamon Press, Boston, MA.

Ditmerr, D. S. (ed.) (1961). Federation of American Societies for Experimental Biology, Bethesda, MD, 1961.

Geigy Scientific Tables, (1970). K. Diem and C. Lentner, eds. Ciba-Geigy, Basle, Switzerland, 1970.

Fullard, R. J. (1988). Identification of proteins in small tear volumes with and without size exclusion HPLC fractionation. *Curr. Eye Res.* 7: 163–179.

Hermens, W. T., Willems, G. M., and Visser, M. P. (1982). *Quantifica-tion of Circulating Proteins: Theory and Applications Based on Analysis of Plasma Protein Levels.* Martinus Nijhoff, The Hague.

Iwata, S. (1973). Chemical composition of the aqueous phase. *Int. Ophthalmol. Clin.* 13: 29–46.

Jenkins, G. N. (1978). *The Physiology and Biochemistry of the Mouth.* Blackwell Scientific Publ., Oxford.

Kugelmass, I. N. (1959). *Biochemistry of Blood in Health and Disease.* Charles C. Thomas, Springfield, IL.

Schultze, H. E. and Heremans, J. F. (1966). *Nature and Metabolism of Extracellular Proteins.* Elsevier, Amsterdam.

Sokoloff, L. (ed.) (1978). *The Joints and Synovial Fluid.* Academic Press, New York.

INDEX